DATE DUE

NATO ADVANCED STUDY INSTITUTES SERIE

*Proceedings of the Advanced Study Institute Programme, which aims
at the dissemination of advanced knowledge and
the formation of contacts among scientists from different countries*

The series is published by an international board of publishers in conjunction
with NATO Scientific Affairs Division

A	Life Sciences	Plenum Publishing Corporation
B	Physics	London and New York
C	Mathematical and Physical Sciences	D. Reidel Publishing Company Dordrecht, Boston and London
D	Behavioural and Social Sciences	Sijthoff & Noordhoff International Publishers
E	Applied Sciences	Alphen aan den Rijn and Germantown U.S.A.

Series C – Mathematical and Physical Sciences

Volume 74 – Identification of Seismic Sources – Earthquake or Underground Explos

Identification of Seismic Sources – Earthquake or Underground Explosion

Proceedings of the NATO Advanced Study Institute
held at Voksenåsen, Oslo, Norway, September 8-18, 1980

edited by

EYSTEIN S. HUSEBYE

and

SVEIN MYKKELTVEIT

NTNF/NORSAR, Kjeller, Norway

D. Reidel Publishing Company

Dordrecht : Holland / Boston : U.S.A. / London : England

Published in cooperation with NATO Scientific Affairs Division

Library of Congress Cataloging in Publication Data

NATO Advanced Study Institute (1980: Oslo, Norway)
    Identification of seismic sources—earthquake or underground
explosion

    (NATO advanced study institutes series. Series C, Mathematical
and physical sciences ; v. 74)
    Includes index.
    1. Earthquakes—Congresses.   2.   Explosions.—Congresses.
3.   Seismometry—Congresses.   I.   Husebye, Eystein Sverre, 1937–
II.   Mykkeltveit, Svein, 1950–   .   III.   Title.   IV.   Series.
QE531.N37   1980   551.2′2   81-10709
ISBN 90–277–1320–0   AACR2

Published by D. Reidel Publishing Company
P.O. Box 17, 3300 AA Dordrecht, Holland

Sold and distributed in the U.S.A. and Canada
by Kluwer Boston Inc.,
190 Old Derby Street, Hingham, MA 02043, U.S.A.

In all other countries, sold and distributed
by Kluwer Academic Publishers Group,
P.O. Box 322, 3300 AH Dordrecht, Holland

D. Reidel Publishing Company is a member of the Kluwer Group

TABLE OF CONTENTS

* Key Lecturer

SEISMIC WAVE ANALYSIS

SCATTERING AND EARTH HETEROGENEITIES

SEISMIC DATA CENTERS

APPENDIX

IDENTIFICATION OF SEISMIC SOURCES – EARTHQUAKES OR UNDERGROUND EXPLOSIONS

Sponsored by and organized on behalf of the

SCIENTIFIC AFFAIRS DIVISION
NORTH ATLANTIC TREATY ORGANIZATION

Co-sponsored by:

The Norwegian Seismic Array (NORSAR)

U.S. Department of the Air Force, Air Force
Office of Scientific Research (AFSC)

U.S. Department of Energy/Lawrence Livermore National Laboratory

Norwegian Office of Foreign Affairs, Disarmament Committee

The Royal Norwegian Council for Scientific
and Industrial Research (NTNF)

Scientific Directors:

E. S. Husebye
NTNF/NORSAR
Norway

H. C. Rodean
Lawrence Livermore
Laboratory, USA

W. Best
AFOSR/NP
Bolling AFB, USA

Organizing Committee:

S. Mykkeltveit
NTNF/NORSAR
Norway

L. Tronrud
NTNF/NORSAR
Norway

G. Young
AFOSR
Bolling AFB, USA

M. Shore
VELA Seismological Center
USA

PREFACE

The subject of this NATO Advanced Study Institute was seismic
monitoring under a nuclear test ban - an application of scientific knowledge and modern technology for a political purpose. The
international political objective of a comprehensive nuclear
test ban provided in turn the motivation for our technical and
scientific discussions. In order to obtain a historical perspective on the progress of the work towards a comprehensive test-ban
treaty (CTB), it is necessary to go back to 1958, when a conference of scientific experts in Geneva made the first steps toward
an international seismic monitoring system. However, agreement
on actual capabilities of a monitoring system for verifying
compliance with such a treaty was not achieved, and thus the
conference did not lead to immediate political results.

After the Partial Test Ban Treaty of 1963, which banned
nuclear explosions in the atmosphere, outer space and under the
seas, renewed interest in the seismological verification of a
CTB took place. A number of countries initiated large-scale
research efforts toward detecting and identifying underground
nuclear explosions, and it was in this context that the large-
aperture seismic arrays NORSAR and LASA were established. This
type of development resulted in excellent seismic data in digital
form and was thus of great improtance to the seismological community. These efforts have signficantly improved the technical
possibilities of verifying a potential nuclear test-ban treaty,
although the most efficient source discriminants still reflect
observational experience and to a lesser extent a true in-depth
physical understanding of the various seismic signal-generating
processes.

The main theme of these proceedings is the identification
of seismic sources. In this context we consider different aspects
of technical methods that can be used to answer on a routine
basis the question: Was a recorded seismic event an earthquake
or an underground nuclear explosion?

Now, to answer this question confidently and correctly,
we must have a proper understanding of the physical processes
associated with both earthquakes and explosions. Therefore, an
extensive review is given of the present state of the art with
respect to these phenomena with due regard to the compiling of
observational data and to the theoretical developments - the
hallmarks of the seismological science.

E. S. Husebye and S. Mykkeltveit (eds.), Identification of Seismic Sources - Earthquake or Underground
Explosion, xi–xii.

The ultimate experience for any seismologist would be to have a direct, unobstructed view of the earthquake or explosion process, so the digital/analog seismometer recordings available represent a relatively poor substitute, though the only one available. Also, complexities of the earth strongly affect the seismic signal, thus presenting us with a blurred and distorted observational image of the source. To improve this image we have to remove complicating wave-propagation and recording effects, and consequently these topics are subject to extensive presentations in the proceedings. In particular, recent approaches to wave propagation modeling and also the use of observational data in constructing structural images of the earth's interior are reviewed.

These proceedings also comprise extensive contributions on the present state of the art of signal processing, seismic source identification, seismometer developments and finally seismological data center operations. These items are partly considered in the context of seismic monitoring of a possible nuclear test ban. The recent trends in this area of seismology are indeed fascinating, with the traditional, faithful seismogram anlayst being replaced by specialists screening waveform data from a network of seismograph stations, and with a battery of computers to perform any conceivable kind of seismogram analysis. The hugh data bases likely to be collected at the seismological centers of the future would undoubtedly be an excellent research tool for dedicated seismologists.

The planning of a conference such as this one calls for a considerable amount of effort, both technical and administrative, from a number of people. We would like to express our sincere thanks to all who have assisted in making this conference possible. Particular thanks go to K. Frydenlund, Foreign Minister of Norway and G.V. Bulin, DARPA, USA, who gave the opening addresses of the Institute. Also, we are most indebted to our fellow scientific directors H.C. Rodean and W. Best and organizing committee members L.B. Tronrud, G. Young and M. Shore without whose interest, work efforts and advice these conference proceedings would not have been possible.

Kjeller, 15 March 1981

Eystein S. Husebye   Svein Mykkeltveit

SEISMIC SOURCE IDENTIFICATION: A REVIEW OF PAST AND PRESENT RESEARCH EFFORTS

A. Douglas

Ministry of Defence, Blacknest, Brimpton,
Near Reading, U.K.

ABSTRACT

The first work on the problem of discriminating between
the signals from underground explosions and those from earthquakes was done using the first arriving short period (SP) P
signals recorded at distances out to about 2000 kms (regional
distances). In 1961-1962 it was demonstrated that unless
stations can be established within 500 kms of a seismic source
the best distance for detecting SP P waves is the range 3000
to 10,000 kms. At these teleseismic distances the P signals
from explosions were observed to be simpler than at regional
distances suggesting that it might be easier to distinguish
between earthquakes and explosions at long rather than short
range. From 1962-1977 most work on discrimination used where
possible teleseismic observations, in particular much effort
was put into the search for discrimination methods that use
only SP P signals. The most important of these methods are
described in this paper; they are: location and depth of source,
complexity of the signal, polarity of first motion and spectral
methods. Since 1977 there has been renewed interest in regional
discriminants: this work is described by Blandford in another
paper presented to this Institute.

The main advance in discrimination since 1962 has been
the development of the so-called m_b:M_s criterion based on the
observations that for a given bodywave magnitude, m_b (estimated
from the amplitude of the SP P waves) explosions have a much
smaller surface wave magnitude, M_s (estimated from the amplitude
of the long period Rayleigh waves) than do earthquakes. The

1

*E. S. Husebye and S. Mykkeltveit (eds.), Identification of Seismic Sources - Earthquake or Underground
Explosion, 1–48.*
Copyright © 1981 by D. Reidel Publishing Company.

Rayleigh waves from explosions are about 10 times smaller than
those of an earthquake of the same m_b. Thus to identify small
magnitude sources on the $m_b:M_s$ criterion requires the detection
of weak Rayleigh waves and for this the nearer the station to
the source the better.

The main conclusions that can be drawn from research on
discrimination over the past 20 years are listed below.
(i) There is no SP P wave discriminant that can identify the
large majority of earthquakes and explosions and it is unlikely
that such will be found; consequently a purely teleseismic
system for detecting and identifying underground explosions is
not possible for acceptable thresholds of identification.
(ii) The most important discriminants are the obvious ones of
location and depth together with the $m_b:M_s$ criterion.
(iii)A few earthquakes each year cannot be identified on the
$m_b:M_s$ criterion as currently defined but by making use of
higher mode observations most of these can be identified as
earthquakes; some of these earthquakes can also be identified
because the P waves can be shown to be compatible with a double-
couple source at a depth of greater than 5 kms.
(iv) There remain a few earthquakes that can only be identified
at present by the first motion method.

These conclusions assume that no attempt will be made to
deliberately evade a treaty and carry out an explosion in
secret. Two methods of possible evasion are discussed in the
paper, these are: (1) attempts to mimic an earthquake using
explosions; and (2) attempts to hide an explosion in the coda
of an earthquake. Several methods appear to be available to
recognise multiple explosions. The only way to be confident
that no tests are being carried out by hiding explosion signals
in the coda of earthquakes is to have stations as close as
possible to sites for such tests.

The main remaining work on discrimination is to find a
physical explanation for those criteria for which there is no
well-founded explanation. Some possible explanations of the
$m_b:M_s$ and complexity criteria are given in this paper.

INTRODUCTION

In July and August 1958 a Committee of Experts met in
Geneva to consider the design of an inspection system for
nuclear tests. They concluded that 180 control posts - 10 of
them shipbourne - would be required to monitor tests in all
environments. Underground explosions, they concluded, could
only be detected at long range by their seismic effects, so
the equipment of the control posts would include seismographs.
As seismic means were the only ones available for the detection

of underground explosions this led immediately to the problem
of distinguishing between the signals from underground explosions
and those of earthquakes. The problem has proved remarkably
difficult to solve and it is the work on this over the past
20 years that I review in this paper.

The experts who met in Geneva in 1958 placed great emphasis
on one particular method of recognising earthquakes, the so-
called "first motion" criterion which requires the observation
of the direction of ground motion at the onset of the P signal.
To observe first motion unambiguously requires large signal to
noise ratios and the only data available to the Experts - from
one underground explosion Rainier (1.7 kt) fired at the Nevada
Test Site (NTS) - showed that the amplitudes of P_n, the first
arriving wave, at least in the distance range 0-1000 kms fell
off rapidly; the decay rate being roughly proportional to Δ^{-3}
where Δ is distance. It thus seemed obvious that if first
motion was to be observed satisfactorily then stations should
be placed as close as possible to the source, say within 700 km
of the epicentre. The Hardtack II series of underground
explosions (of which Logan (5 kt) and Blanca (19 kt) had the
largest yields) carried out in October 1958 seemed to confirm
the rapid decay of P_n to 1000 kms. In addition the results
seemed to show that the seismic magnitude for an explosion of
given yield was less than had been predicted from the Rainier
results. If this were true then the results had important
consequences.

It is important to note that the methods of identification available in those early days only allowed some earthquakes
to be identified: there was no criterion for identifying explosions. Thus for example if negative first motions were
observed for a particular source this was taken to identify a
source as an earthquake - if only positive first motions were
observed then the source could be either an earthquake with an
orientation such that the observing stations lay only in the
positive quadrants or an explosion. Thus it appeared that
there would always be a residue of sources that could not be
identified as earthquake or explosion. These sources could
only be identified it was thought by on-site inspection.
Suppose now we say we wish to inspect all unidentified sources
with magnitudes greater than some threshold m_b^T, where m_b^T is the
magnitude given by the smallest yield one wishes to identify
with certainty. Then we can estimate the number of earthquakes
in a given region that will have magnitudes m_b^T or greater and
given an estimate of the percentage of earthquakes that would
be identified on the available criteria come to an estimate of
the number of unidentified earthquakes that would be detected
annually. If now we find that the estimates of the number of
earthquakes has been based on an erroneous m_b-Yield relationship

and that m_b^T is lower than was assumed, then the number of
unidentified sources and hence the requirements of on-site
inspection is increased. It was arguments about the magnitude-
yield relationship and the numbers of earthquakes that would
go unidentified each year in the USSR that occupied the experts
following the Hardtack II series and disagreements on these
subjects that led to the abandonment of these discussions.

To make any progress towards a treaty banning underground
tests it seemed obvious in 1959–1961 that methods of improving
signal to noise ratio (S/N) were required and work began both
in the USA and UK on the development of such methods. Most
of the effort went into development and evaluation of arrays.
The Experts had in fact recommended that arrays be installed.
These so-called Geneva arrays were to have 10 vertical component
seismometers for detecting short period signals, distributed
over an aperture of about 3 kms; because of the small aperture,
signals with high apparent surface speeds would be passed with
little attenuation even on simple summing whereas, it was
hoped, noise would be suppressed. Such arrays give little
velocity information. In the UK a different approach was
used: experiments were carried out using arrays with apertures
of about 9 kms which is about the wavelength of P_n at 1 Hz
and attempts were made to correct for the travel time of the
signal across the array to produce what is now referred to as
the delay and sum output.

With the co-operation of the USA the UK established such
an array at Pole Mountain in Wyoming. The site was chosen to
be about 1000 kms from the site of the Gnome explosion
(10 December 1961); it was also fortuitously about 1000 kms
from the NTS. The reason for choosing this distance was that
it appeared to be the distance at which the P signal was
weakest and if arrays could be used to extract signals from
noise at this distance then, this being the worst case, arrays
could presumably do better still at shorter distances. Work
on signals from the Gnome explosion confirmed that arrays could
be used to improve signal to noise ratios at this critical
distance. The work also showed that it was possible to
identify other crustal phases by their apparent speeds and
improve estimates of focal depth thus aiding in identification
(Thirlaway (1)). The most important results as it turned out
however, came from other observations. In February 1962 the
USSR fired an explosion at E Kazakh and this was clearly
observed at Pole Mountain a distance of about 89°. Later that
year – in May 1962 – France detonated an explosion in the
Sahara and again clear signals were recorded at Pole Mountain
at a distance of 88.7°. By the time Pole Mountain was closed
in 1963 a total of 6 explosions had been recorded with epicentres at teleseismic distances from the array. The array had

meanwhile been enlarged to an aperture of 18 km to allow
velocity filtering techniques to be applied to teleseismic
sources. These recordings focussed attention on the fact that
the amplitude of the SP P waves observed at distances of around
3000-10,000 kms is about the same as at 500 kms and that from
around 500 kms to 1000 kms there is a shadow zone (at least
for sources in the western US) so that unless stations can be
established within 500 kms of an explosion site the best site
for a station for P-wave detection is at distances of 3000-
10,000 kms.

    A further advantage of using arrays revealed by the Pole
Mountain recordings was that delay and sum processing of the
array data suppressed arrivals generated by scattering in the
vicinity of the array. With these scattered signals reduced,
the delay and sum output for explosion signals proved to be
simple in agreement with what might be expected intuitively.
(SP P signals from underground explosions had been recorded at
long range before the Pole Mountain data became available -
Logan and Blanca for example had been recorded in the USSR
at distances of 10,000 kms - but the S/N was so low that the
form of the signal was not clear). The observation that
explosions recorded at long range were usually simple whereas
it was observed earthquakes often gave complex signals, marked
the beginning of the idea of using complexity of SP P signals
as a method of recognising earthquakes. The value of complexity
for discrimination is discussed in full later. A detailed
account of these early developments in array seismology is
given by Birtill and Whiteway (2).

    Following from these observations and their implications
the UK switched its effort in 1962 from 1st Zone and 2nd Zone
recording (0-10° and 10°-20° respectively) to Teleseismic or
Third Zone recording (strictly 20°-180°) and particularly to
recording in what has been called the source window that is
30°-90°. The US also began to shift some of its effort to
teleseismic distances in 1962 and much of the work in the US,
UK and other countries until a few years ago has concentrated
on detection and identification at these distances. The
interest in teleseismic recording led to the establishment of
several seismometer arrays around the world the largest of
which, the Large Aperture Seismic Array (LASA) in Montana and
the Norwegian Seismic Array (NORSAR) had at their maximum
extent apertures of $\approx$ 200 kms.

    The advantages of teleseismic recording, if reliable
identification criteria can be established, are obvious:
(i) explosions can be monitored from outside the country
conducting the tests;
(ii) a much smaller number of recording stations is needed
than the 180 suggested by the experts;

(iii) a much wider range of possible sites are available for
stations and from these the best sites can be selected; the
best sites being those where losses due to scattering and
anelastic attenuation in the crust and upper mantle are low,
where the S/N for a given magnitude is high and for array
sites where the signals are coherent over the aperture of the
array.

For the 15 years 1962–1977 most of the work on identification has been done using as much as possible, teleseismic
observations. In practice of course a full teleseismic system
is probably not possible because the most stable criterion
for identifying earthquakes and explosions is the m_b:M_s
criterion (for details see later) and this requires that the
surface waves from explosions be observed. The surface waves
generated by an explosion are about 10 times less in amplitude
than those from an earthquake of the same m_b so to identify
small magnitude sources on the m_b:M_s criterion requires the
detection of weak surface waves. As surface wave decay is
roughly proportional to $\log \Delta$ out to at least $150°$ then the
nearer the station to the source the better.

Recently there has been a resurgence of interest in
detection and identification at regional distances that is the
1st and 2nd Zones. Much of this effort appears to have focussed
on the use, not of P_n, but on the other arrivals in the seismogram particularly the SP surface wave train L_g, which often
has an amplitude many times larger than P_n. Thus the problem
of small signals, that made work in the early days difficult,
is avoided. This new look at regional discrimination is covered
by Dr Blandford in a later paper.

SOME PROBLEMS OF EXPERIMENTAL DESIGN

The procedure for using seismic methods to monitor compliance with a treaty banning underground nuclear tests is
usually divided into three parts: (1) detection of signals;
(2) location of the source of the signals; and (3) identification of the source as either earthquake or explosion. Ideally
one would hope that the sources of all signals detected could
be located and identified but in practice there will always be
detected signals that are so weak they cannot be identified;
there will always be a gap between the detection threshold and
identification threshold.

The main requirements of a good identification criterion
are that it:
(i) give complete separation between earthquakes and explosions;

(ii) be understood theoretically so that it is possible to
extrapolate from available observations to lower magnitudes and
to all parts of the earth;
(iii)be difficult to evade by using explosions or a series of
explosions to mimic earthquake signals;
(iv) be easy to apply so that a lot of computer time is not
required to identify each source.

Most of the evidence shows that earthquakes are the result
of slip on a fault plane whereas explosions can be modelled at
least roughly as point compressional sources. In general then
it would appear that there should not be much difficulty in
finding criteria to distinguish between the two types of source.
There are however several practical problems that must be kept
in mind; the most important ones are the following:
(i) Much of the earth's surface is covered by sea and the
establishment of permanent stations on the sea bed is difficult
so that it is not easy to achieve the optimum distribution of
stations for identification.
(ii) Because of the well known peak in the seismic noise
spectrum recording of seismic signals has usually to be confined
to two narrow frequency bands one centred on 1 Hz (SP) and one
on 0.05 Hz (LP) yet to distinguish between signals from different source types requires in general as wide a band of
frequencies as possible.
(iii)The explosion data available come from a small number of
test sites mostly in aseismic zones so that there is no guarantee
that methods that work at these sites will work elsewhere.
(iv) SP S waves which might provide the clearest way of distinguishing earthquake from explosion are rapidly attenuated
with distance and so cannot be detected at long range.

I now look at some of the most important discrimination
criteria that have been investigated over the past 20 years.

CLASSICAL DISCRIMINANTS USING SP P WAVES ONLY

At long range the most easily recorded seismic signal from
an underground explosion is the vertical component of the SP
P wave. As a consequence of this most array stations that
have been installed to detect weak signals have measured only
the vertical component of ground motion. This in turn has
led to the search for methods of identifying explosions and
earthquakes using only SP P waves; some of the main methods
studied are described below.

Location and depth

If the epicentre of a seismic disturbance lies over the
sea then according to Evernden (3) the source is either

an earthquake below the sea bed or an explosion in the water.
Evernden (3) asserts that an explosion in the sea can be
identified from the hydroacoustic signal generated, so that
if there is a high probability that a given epicentre lies
over ocean and there is no hydroacoustic signal the source is
identified as an earthquake. Evernden (3) shows that in this
way the large majority of shallow earthquakes can be identified
as such. The criterion Evernden (3) uses for identification
is that there is a high probability that the epicentre is
more than 25 kms from land. A distance of 25 kms was chosen
to allow for the possibility of azimuthal variations in travel
time around the epicentre which would result in a bias in the
estimated location.

The most reliable method of determining the epicentre of
a seismic disturbance is the well known Geiger's method (see
for example Bullen (4)). The observations used are the arrival
time of P waves at four or more stations. The procedure is to
start with rough estimates of origin time, epicentre (and depth
if required) and compute the predicted arrival time of P waves
at the observing stations, using standard tables of travel time
against distance and depth. A non-linear least square procedure
is then used to compute corrections to these trial estimates
in such a way that the sum of the squares of the differences
between the observed and predicted arrival times is minimised.
Provided that travel times are a function of distance only,
this method gives reliable estimates of the epicentre in the
sense that the $\alpha\%$ confidence limits will on average cover the
true epicentre α times in every hundred. However there is
some evidence (see for example Herrin and Taggart (5)) that
travel times vary with azimuth in some regions of the earth
and this could introduce a systematic bias into the epicentre
so that the confidence limits may never cover the true epicentre;
the maximum value of this bias appears to be about 25 kms –
hence the definition of the identification criterion on location
adopted by Evernden (3).

If it can be shown that there is a high probability that
a given seismic source occurred at a depth greater than a few
(say 5) kilometres then it can be concluded that the source is
most likely an earthquake. Thus if a reliable estimate of
source depth can be obtained it is obviously a powerful method
of identifying earthquakes. Much thought has therefore gone
into devising reliable methods of depth estimation. Geiger's
method does in fact allow source depth to be estimated but
these estimates often have large uncertainties so that only
foci with estimated depths of greater than say 50-60 km can be
positively identified as earthquakes.

The most reliable method of estimating depth is to use
the time difference between direct P and the surface reflections (pP and sP) assuming the surface reflections can be
identified. The classical way of identifying pP and sP is to
show that the time between P and the possible surface reflection varies with distance as predicted by travel time tables.
However for shallow sources the variation in P-pP and P-sP
times with distance is so small – particularly at teleseismic
distances – that this method of identification can rarely be
used.

The main danger in accepting a second arrival as a surface
reflection without some other supporting evidence is that the
second arrival may simply be a second explosion. One way to
show that this is unlikely is to demonstrate that the second
arrival is of opposite polarity to direct P. Determining the
polarity of a second arrival from an SP seismogram may not be
easy but Douglas et al (6) have shown that by deconvolving the
SP seismogram to take out some of the effects of the seismograph
and anelastic attenuation on the path from source to receiver
it is sometimes possible to observe the polarity of the main
arrivals more clearly. Examples of this deconvolution technique are shown in Figure 1. Figure 1a is an SP P seismogram
with two prominent arrivals which might be P and pP. Both
arrivals appear to have the same (positive) polarity and
Figure 1b, the deconvolved seismogram, shows more clearly
that this is true. Thus there is no way of deciding whether
the second arrival from these seismograms is pP or a second
explosion. In fact the two arrivals are from two chemical
explosions in the USSR – the MEDEO explosions – fired 3.6s
apart (see Marshall (7) for details). Figure 1c again shows
an SP P seismogram with two clear arrivals – the polarity of
the first arrival is clearly positive, the polarity of the
second may be negative but it is not easy to draw a definite
conclusion. Figure 1d shows the deconvolved seismogram: here
it is clearly demonstrated that the second arrival has negative
polarity and hence cannot be a second explosion and so is most
likely pP.

Few shallow earthquakes show surface reflections and fewer
still show reflections as clearly as 1c and d. However attempts
have been made to reduce the uncertainty in depth estimates
obtained by Geiger's method by calibrating a given region of
the earth using the earthquakes that do show identifiable pP
and sP phases.

The simplest way of doing this is to restrain the depth
of focus of an earthquake with a depth obtained from the P-pP
and P-sP times and determine the epicentre and origin time of
the earthquake using Geiger's method. The differences between

the observed times and the times computed from the estimated
epicentre, origin time and restrained depth are then used as
corrections to the observed P arrival times of nearby earthquakes. These corrections are thus assumed to be a measure of
differences between the true travel times and the travel time
tables used and hence departures in the earth structure from
the ideal model in which wave speeds are functions only of
depth. The method is usually called the "master event" method
and has been developed and much used by Evernden (8). The
method can be systematised by extending Geiger's method to
allow corrections for departures in the travel times from
standard tables to be estimated along with the other unknowns;
this is an extension of the method of joint epicentre determination proposed by Douglas (9) and has been used by Dewey (10)
for studying the distribution of earthquakes with depth but
it does not appear to have been extensively used in the identification of shallow earthquakes.

Evernden (8) claims that depth estimation using the
"master event" method is valuable for identifying shallow
earthquakes. However the method can fail, for when corrections
derived from Aleutian earthquakes are used to estimate the
depth of the explosion LONGSHOT, Evernden (8) finds that the
uncertainty in the depth estimate is not reduced. As Blandford
(11) points out the "master event" method works well except
when it is most needed.

The complexity criterion

One of the most striking features of the SP P signals
from explosions as recorded at teleseismic distances is their
simplicity; most signals showing one or two cycles of relatively
large amplitude followed by a tail or coda of much lower
amplitude. The simplicity of explosion P waves is shown most
clearly on the delay and sum output of arrays, this is because
at some stations much of the coda consists of scattered arrivals
generated in the vicinity of the recording station. Combining
the individual channels to form the delay and sum output tends
to cancel out these scattered arrivals and results in a simpler
signal than is seen on single channels. To illustrate how
scattering in the vicinity of the receiver can add to the
complexity of single seismometer recordings consider Figure 2
taken from Key (12); the seismograms shown are from an earthquake in Kamchatka recorded at the Eskdalemuir array (Scotland).
Figure 2a, b and c are single channel recordings and two of
these traces, (b) and (c), show a prominent arrival about
5 seconds after onset which might be interpreted as a surface
reflection, pP or sP. However when the array is phased for
signals from Kamchatka (azimuth 12°, speed 19 km s^{-1}) this
second arrival is suppressed and the P signal is shown to be

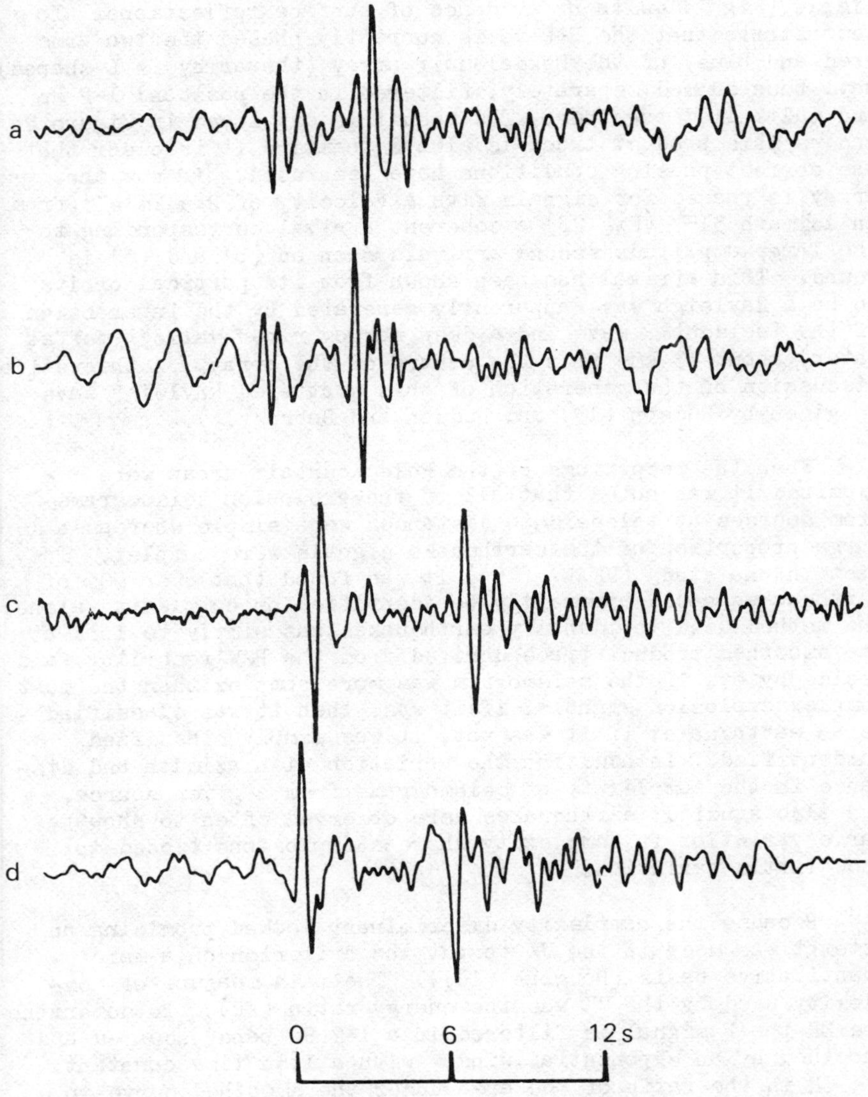

Fig.1. P seismograms of the MEDEO explosions and an earthquake
(1968 July 1; origin time 19h 14 min 54.7s) recorded at
Yellowknife, Canada. Both the earthquake and the explosions
took place near Alma Ata, USSR. a, SP signal recorded from
the MEDEO explosions; b, deconvolved seismogram of the MEDEO
explosions; c, SP signal recorded from the Alma Ata earthquake;
d, deconvolved seismogram of the Alma Ata earthquake.

simple (Fig 2d) with no evidence of surface reflections. To
demonstrate that the P-wave is correctly phased the two arms
(red and blue) of the Eskdalemuir array (the array is L-shaped)
have been summed separately, filtered in the passband 1-2 Hz
and multiplied together. The result is displayed in Figure 2e
and as this product trace is always positive it is clear that
the correct phasing conditions have been used. If now the
array is phased for signals with a velocity of 2.5 km s^{-1} from
an azimuth 315° (Fig 2f) a coherent arrival corresponding to
the large amplitude second arrivals seen on (b) and (c) is
found. This arrival has been shown from its partical orbits
to be a Rayleigh wave apparently generated by the interaction
of the incident P wave and a deep narrow river valley (Moffat
Water) about 13 kms from the centre of the array. A theoretical
discussion of the generation of this scattered Rayleigh wave
is given by Hudson (13) and Hudson and Boore (14).

When the recordings of the Pole Mountain array were
examined it was noted that all of the explosion seismograms
from sources at teleseismic distances were simple whereas a
large proportion of the earthquake signals were complex. In
fact in one study (UKAEA (15)) it was found that over 90% of
earthquakes could apparently be identified on complexity alone.
The method used to identify earthquakes was simply to look at
the smoothed product trace derived from the PMW recordings and
decide by eye if the seismogram was more complex than the most
complex explosion signals: if it was, then it was classified
as an earthquake; if it was not, it was simply classified
unidentified. In addition the variation with azimuth and distance in the complexity of seismograms from a given source,
was also studied: earthquakes were observed often to show
large variation in complexity whereas explosions tended to
show little variation.

Because the complexity discriminant looked promising an
attempt was made in the UK to put the criterion on a more
quantitative basis (Douglas (16)). The main measure of complexity used by the UK was the energy ratio (ER). To determine
the ER the P signal is filtered in a 1-2 Hz band, squared and
smoothed by an exponential window with a 1.5s time constant.
The ER is the ratio of the area under the smoothed curve in
the first 5s of the signal to the area under the curve from
5-35s making allowances for the effects of noise. For simple
records this ratio is roughly unity or greater: for complex
records the energy ratio is less than 1.0; ER is thus inversely
proportional to complexity.

Until 1964 no UK type array had ever recorded a complex
explosion signal but in September of that year the array at
Yellowknife (YKA) Canada recorded a complex P seismogram from

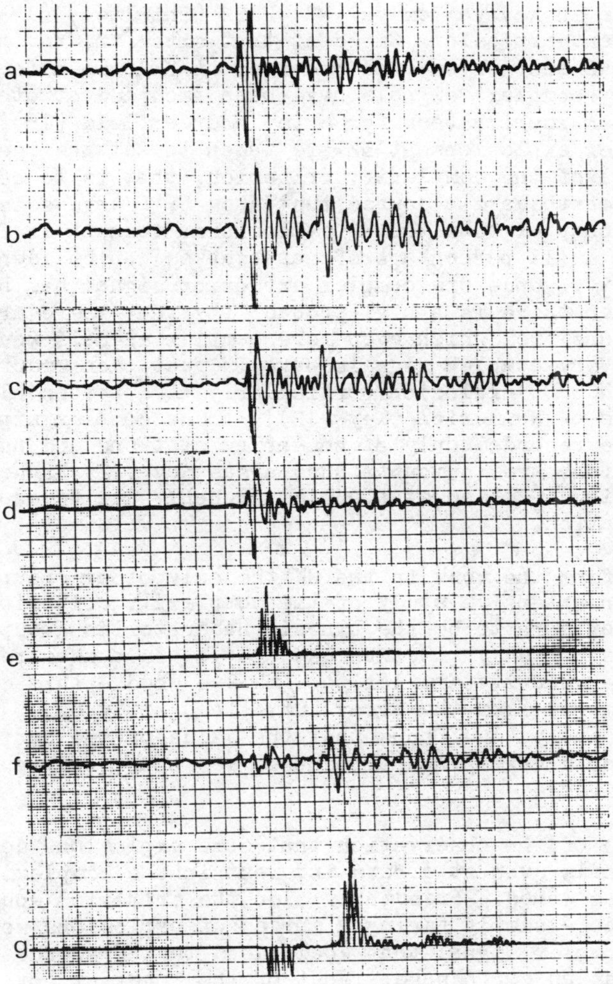

Fig.2. Example of a Rayleigh wave generated by P-wave scattering near the Eskdalemuir array (Scotland). The P wave is from an earthquake in Kamchatka. a, b and c, single channels; d, delay and sum output of the array phased for signals from Kamchatka (azimuth 12° speed 19 km s^{-1}); e, correlogram derived by forming the delay and sum output (phased for signals from Kamchatka) for each of the two arms of the array separately, filtering these in the pass band 1-2 Hz and multiplying the filtered outputs together; f, delay and sum output for the array phased for the scattered Rayleigh wave which has a velocity of 2.5 km s^{-1} from an azimuth of 315°; g, correlogram as in e except that array is phased for Rayleigh wave.

an explosion in Novaya Zemlya (NZ) USSR (Thirlaway (17)).
All subsequent explosions fired at that site (the northernmost
of the two NZ test sites) and recorded at YKA are also complex
and similar (complex) signals have been recorded at other
stations in N America (see for example Greenfield (18)). If
these complex explosions at NZ are taken to set the identification level of the complexity criterion, that is to be classified as an earthquake a source must show SP P wave seismograms more complex than the NZ explosions, then it is found
that only a small percentage of earthquakes can be identified
on complexity. Once one complex explosion signal has been
recorded then there is a high probability that there are other
source regions from which even more complex signals would be
recorded. Very complex signals have in fact been recorded
from the LONGSHOT explosion particularly at stations along the
western side of N America (Key (12)); these complex signals
however were recorded only at non-array stations and as many
of the stations are in mountainous areas Key (12) argues that
the complexity of these stations is probably due to scattering
near the station.

     Apart from the work in the UK the only other group who
have made an extensive study of the complexity criterion using
array data appears to be the Lincoln Laboratories group who
used LASA data (Kelly (19), Lacoss (20)). As a measure of
complexity a quantity similar to 1/ER was used - this quantity
being determined from the array sum. In a study of 48 shallow
earthquakes Kelly (19) found that the value of the complexity
criterion for recognising earthquakes was poor, only about 20%
being identified.

     Because of the observation that some explosions do give
complex signals even at arrays interest in the complexity
criterion has waned. Presumably also the criterion would be
easy to evade - simply firing a series of explosions would
produce a complex signal that would lead to the source being
identified as an earthquake. Most of the interest now centres
on the origins of the complexity and this is discussed below.

First motion

     On the first motion criterion a source is identified as
an earthquake if a negative (rarefactional) first motion is
observed at some specified number of stations. It is well
known however that first motion particularly on SP P seismograms is difficult to read unambiguously so that in practice
the first motion method is difficult to use except with seismograms with large S/N. Work over the past 15 years has shown
that first motion observations can be read more reliably from
LP than SP seismograms but, as the LP detection threshold is

higher than that of the SP, LP seismograms are not available
for many small magnitude sources.

From time to time apparently genuine negative first
motions are observed from explosions. The SP P seismogram
recorded at a station 714 kms from the underground explosion
Logan, for example, appears to show such a negative motion.
Also Nowroozi (21) published a fault plane solution for an
explosion in the USSR (using observations taken from LP
records) which shows both positive and negative polarity observations. It is difficult to know if any of the apparent
negative first motions seen from explosions are truly negative
or if the true first motion is positive but has been missed
by the analyst. Enuscu et al (22) have assumed that where
negative first motions have been reported for explosions
these are correct and they have tried to explain the observed
distributions of polarities (Enuscu et al (22) take these
data from Bulletins) using an explosion source that is not a
point compressional source but three unequal dipoles without
moment each dipole being combined with a simple force. If
it can be clearly demonstrated that the true first motion of
some explosion P seismograms is negative then the whole idea
of first motion would have to be abandoned at least as presently
used. As it is there are a few earthquakes which without the
first motion criterion would go unidentified (see
below).

Spectral methods

The P waves of many of the explosions in the USSR and
particularly E Kazakh have higher predominant frequencies than
most shallow earthquakes. Consequently many attempts have
been made to establish a method of identifying earthquakes and
explosions based on differences in the spectrum of their P waves.
Kelly (19) for example has attempted to use a spectral ratio
(SR) for discrimination defined by:

$$SR = \int_{h_1}^{h_2} A(f)df \left/ \int_{l_1}^{l_2} A(f)df\right. ;$$

$A(f)$ is the spectral amplitude at frequency f, h_1 and h_2 are
the integration limits of a high frequency band and l_1 and l_2
the limits of a low frequency band. Typical bands used by
Kelly (19) are l_1 = 0.35 Hz, l_2 = 0.85 Hz, h_1 = 1.45 Hz and
h_2 = 1.75 Hz.

An alternative discriminant is the third moment of frequency (TMF) defined by Wiechert (23):

$$TMF = \left\{ \frac{\int_0^5 f^3 \, A(f)df}{\int_0^5 A(f)df} \right\}^{\frac{1}{3}}$$

where f is expressed in Hz. The more high frequency energy in a signal the larger TMF. Evernden (24) and Evernden and Kohler (25) have defined some very complex discriminants that are constructed using the energy in a number of pass bands; these discriminants seem to be designed to achieve the same result as the TMF that is to give the greatest weight to the highest frequency.

Another approach to discrimination using the P-wave spectrum is suggested by Sandvin and Tjøstheim (26) who show that P signals can be described by models of the autoregressive type. The autoregression coefficients summarise the main features of the signal spectrum and if there are differences in the spectra of earthquakes and explosions these coefficients should form a convenient basis for constructing spectral discriminants; Sandvin and Tjøstheim (26) show how such discriminants might be constructed.

The problem with all discriminants based on the P-wave spectrum is that they seem to be regionally dependent. For some populations of Eurasian earthquakes and explosions complete identification of the two source types has been achieved using spectral discriminants (Kelly (19); Evernden and Kohler (25)). Discrimination for earthquakes and explosions in southwest US however is not as good (Blandford (11)). Spectral discriminants thus probably have little value in the routine identification of earthquakes and explosions, however they may be of value as discussed below in recognising attempts to simulate an earthquake using multiple explosions.

$m_b : M_s$

The main advance in discrimination since the 1958-1962 discussions in Geneva has been the development of the so-called $m_b : M_s$ criterion: on this criterion both earthquakes and explosions can be identified. The discriminant is based on the observation that for a given m_b, explosions have a much smaller M_s than do earthquakes: as m_b is measured from the SP P waves and M_s from the LP Rayleigh waves then the criterion is

essentially a measure of the relative excitation of P and
Rayleigh waves. The first solid evidence to show that an m_b:M_s
criterion might allow earthquakes and explosions to be identified appears to be that published by Brune, Espinosa and
Oliver (27) although suggestions that discrimination in this
way might be possible had been made earlier (see Bolt (28)
for further details on the origins of the m_b:M_s criterion).
The method used by Brune et al (27) was not strictly m_b:M_s
as it has subsequently been developed. Like later workers
Brune et al (27) do use m_b (measured on a Wood–Anderson seismograph) for the P wave amplitudes but for the surface waves
they used AR defined as the area of the envelope of the surface
wave recorded on a three component LP system – the areas of
the envelopes of the three components were simply computed and
summed. Plotting AR (normalised to a fixed distance) against
m_b showed that many of the earthquakes were well separated
from the explosions; there was however not complete separation
of the two source types. The use of AR rather than M_s has
three main advantages which are listed below.
(i) Use of AR allowed surface waves recorded at distances of
less than $20°$ to be used for discrimination. The M_s scale as
defined by Gutenberg (29) requires that the amplitude of the
surface wave with period around 20s be measured but for this
period to be visible the surface waves must be well dispersed
which is only possible if the signal is observed at $\Delta \geqslant 20°$.
(ii) By making use of all three components of ground motion
in the determination of AR it was possible to take account of
all types of surface wave.
(iii)By using the envelope rather than a single amplitude
measurement it was hoped that the observations would be more
clearly related to the total energy of the seismic surface
waves. In fact Press, Dewart and Gilman (30) show that using
simple amplitude measurements from surface wave trains (at
short range: 250-650 kms) as a measure of the surface wave
excitation did not result in the separation of earthquakes and
explosions when plotted against m_b. However by using, like
Brune et al (27), a measure of the energy in the whole wave
train – they used the energy density in the period range
7-30s – Press et al (30) confirmed that the relative excitation
of P and surface waves looked to be a promising way of identifying both earthquakes and explosions.

The main disadvantage of using the AR is that it is much
more difficult to compute than a magnitude M_s. Also for AR
to be determined as defined by Brune et al (27) requires the
horizontal component of ground motion but horizontal component
seismographs tend to be noisy so that it is difficult to obtain
reliable readings from them.

Because of the difficulties in measuring AR compared to
making a magnitude determination, most workers since Brune
et al (27) have used the surface wave magnitude M_s as a measure
of the surface wave energy; the difficulties of determining M_s
at short range being avoided either by using only observations
for distances $\Delta \geqslant 20°$ or more recently by redefining M_s to
allow its computation from observations at any distance. Discrimination between earthquakes and explosions on the relative
excitation of P waves and surface waves has thus become known
as the $m_b:M_s$ criterion.

The breakthrough to the wide acceptance of the $m_b:M_s$
criterion and the realisation that it is possible to achieve
almost complete separation of earthquakes and explosions
appears to have come when the data from the LONGSHOT explosion
were analysed (Marshall, Carpenter, Douglas and Young (31);
Liebermann, King, Brune and Pomeroy (32)). Three factors
appear to have helped with the widespread acceptance of the
$m_b:M_s$ criterion at around this time:
(i) the establishment of well calibrated standardised networks
of SP and LP seismographs (such as the World Wide Standard
Station and the Canadian networks) from which reliable amplitude observations could be obtained;
(ii) the availability of data from an increasing number of
explosions particularly with yields of several 10's of kilotons
which allowed surface waves to be well recorded at long range;
(iii) the refinement of the magnitude scale for the USA
(Evernden (33a)) which allowed more reliable estimates of m_b
to be obtained at regional distances.

Numerous papers have been published over the years confirming the value of the $m_b:M_s$ criterion for discriminating
between earthquakes and explosions (see for example Liebermann
and Pomeroy (33,34); SIPRI (35); Molnar, Savino, Sykes,
Liebermann, Hade and Pomeroy (36); Evernden (3,8,37); Evernden,
Best, Pomeroy, McEvilly, Savino and Sykes (38); Marshall and
Basham (39)). The main developments in the $m_b:M_s$ criterion in
recent years has been in refinements to the measurement of
magnitude particularly M_s where the definition has been
generalised to allow the maximum amplitude observed at any
distance and over any path to be used (Marshall and Basham (39))
thus lowering the threshold at which the $m_b:M_s$ criterion can
be applied.

A typical plot of m_b against M_s for a sample of earthquakes
and explosions is shown in Figure 3. Note that there is no
overlap between the explosion and earthquake population
although some studies do appear to show that the two populations
are tending to converge at low magnitude ($< m_b$ 4.5). More
recent work however, using observations close in to the Nevada

Test Site, shows that the populations do not in fact converge
at least down to $m_b \approx 3.5$ (Peppin and McEvilly (40)).

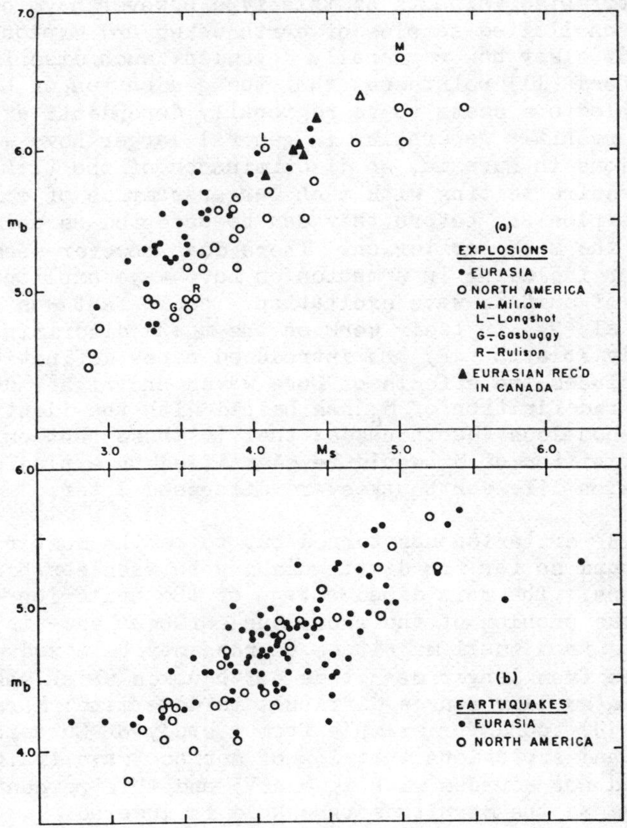

Fig.3. m_b:M_s plot of: a, explosions; b, earthquakes.

Tjøstheim (41) has pointed out that m_b and M_s "were not
originally intended for discrimination purposes and it is not
clear how much of the discriminating information contained in
the raw data they represent or in what sense, if any, they can
be considered to be optimal". Tjøstheim (41) has therefore
investigated some other discriminants that basically measure
the relative excitation of P and surface waves but in which
m_b and M_s are supplemented by features derived from fitting
models of the autoregressive type to the observed wave trains.
Such methods can be looked on as discrimination by pattern
recognition. Tjøstheim (41) finds that in general better
separation can be obtained between Eurasian earthquakes and

explosions using such discriminants particularly when the
surface wave features are derived from the Love waves - what
Tjøstheim (41) refers to as the L1 (Love wave): L2 (Love wave)
discriminant. Discriminants of this type however have only
been tested on limited samples of earthquakes and explosions
and it is not clear how regionally dependent such discriminants
are. Blandford (11) points out that the generation of Love
waves by explosions seems to be regionally dependent: explosions
at NTS and Armchitka generating in general larger Love waves
than explosions in Eurasia, so discriminants of the L1:L2
type will require testing with much larger samples of earthquakes and explosions before they can be accepted as being as
reliable as the $m_b:M_s$ criterion. There does however seem to
be a case for including information on Love wave amplitudes in
any measure of surface wave excitation - as in fact was done
by Brune et al (27) in their work on the $m_b:AR$ discriminant.
More recently Forsyth (42) has introduced a new definition of
M_s which includes the effects of Love waves and higher mode
waves: this redefinition of M_s has helped with the identification of "anomalous" earthquakes: that is those that on the
standard definition of M_s would be classified as explosions -
these explosion-like earthquakes are discussed later.

The $m_b:M_s$ criterion has turned out to be the most robust
criterion found so far for distinguishing between earthquakes
and explosions. The main disadvantage of the criterion -
apart from the problem of the anomalous earthquakes - is that
surface waves from small magnitude sources may be mixed with
surface waves from larger magnitude earthquakes which makes
the M_s of the smaller source difficult to determine: Marshall
and Basham (39) found for example from a study of Eurasian
earthquakes and explosions that 14% of surface wave trains
were obscured for sources with $m_b \geq 4.75$ and this percentage
will increase as the magnitude threshold is lowered.

ANOMALOUS EARTHQUAKES

Although the vast majority of earthquakes can be identified
by the $m_b:M_s$ criterion there appear to be some that generate
only weak surface waves and which are usually referred to as
anomalous earthquakes. Brune et al (27) in fact noted that
there were some earthquakes in the population they studied that
would be classified on the $m_b:AR$ criterion as explosions.
Other well documented examples of these anomalous earthquakes
or earthquakes that look like explosions have been discussed
by Landers (43). The most comprehensive list of anomalous
earthquakes is that given in a working paper tabled by the USA
at the Conference of the Committee of Disarmament (CCD(44)).
Most of the anomalous earthquakes in this list occurred in

south-central USSR and Tibet. Some of these earthquakes have
body wave seismograms that show one or two definite arrivals
only: Figure 4 shows the three P wave seismograms from such an
earthquake that occurred in E Kazakhstan (Figure 1c shows another);
Figure 4 is taken from Douglas, Marshall, Young and Hudson (45).
Douglas et al (45) demonstrate that the E Kazakh source was
in fact an earthquake because the second arrivals shown on
the YKA and WRA seismograms are of opposite polarity to P −
this is most clearly seen on the deconvolved seismograms,
Figures 4a and 4d − and so the second arrival is a surface
reflection probably pP. The source depth must therefore have
been about 25 km so the source was an earthquake and not one
or more explosions. Douglas et al (45) also attempted to
model the SP seismograms as shown in Figures 4c, f and i.
The model seismograms predict rather closely the relative
amplitudes of the main arrivals shown by the seismograms and
demonstrate that the second arrival on the observed GBA seismogram which occurs 10s after P is probably sP not pP and that
the absence of pP at GBA (expected at 7.5s after P) can be
accounted for because for the source orientation used pP
leaves the source close to a node in the P radiation pattern.
The agreement between observed and model seismograms is further
evidence that the E Kazakh source was an earthquake. This is
one example; other earthquakes that radiate signals of the
type shown in Figure 4a, d and g should also be rather easy
to identify.

     Tatham, Forsyth and Sykes (46) have investigated some
anomalous earthquakes that occurred in eastern Tibet and which
are listed in CCD (44). Many of these earthquakes show no
clear depth phases and so cannot be identified on depth.
Tatham et al (46) find that many of these sources have been
incorrectly identified as anomalous either because the m_b
measured was inaccurate or because, they claim, the surface
wave magnitude did not adequately represent the LP character
of the source. The M_s used in computing the list given in
CCD (44) was based on the amplitude of the 20s period fundamental mode Rayleigh wave and Tatham et al (46) argue that
because of the particular depth, orientation of the double
couple mechanism and possibly great thickness of the crust
under the Tibetan plateau the fundamental mode was poorly
excited. The thick crust however leads to a more efficient
generation of Love waves and higher mode Rayleigh waves and
by making use of M_s as defined by Forsyth (42) to include
the effects of Love waves and higher mode waves it is possible
to get good discrimination on m_b:M_s for most of the sources
on the CCD (44) list. One sequence however remains unidentified on m_b:M_s and Tatham et al (46) have to resort to the use
of first motion measured from LP seismograms to demonstrate
that these sources are in fact earthquakes. So for this

particular sequence one has to fall back on the discriminant
first proposed over 20 years ago and which cannot be regarded
as Tatham et al (46) acknowledge as a satisfactory way of
identifying earthquakes.

DISCRIMINATION USING S WAVES

     Numerous other methods have been suggested for discrimi-
nating between earthquakes and explosions, the most successful
appear to be those that make use of S-wave observations.
Evernden (3,8,37,47) has emphasised the value of S observations
at distances out to 10° in improving depth estimates. Given
a P and an S phase that have followed the same path - Evernden (8)
uses short period P_g and S_g - then the origin time of a seismic
disturbance can be estimated from the S-P time given the ratio
of the P and S wave speeds. Geiger's method is then used to
estimate epicentre and depth of focus from the P-arrival times;
the origin time being restrained to that estimated from the
S-P time.

     Other discriminants that make use of S observations are
listed below.
(a) Ratio of long period S to Rayleigh amplitude: this ratio
is larger for earthquakes than explosions (von Seggern (48);
Blandford and Clark (49)).
(b) Ratio of short period S to short period P amplitude:
this ratio is larger for earthquakes than explosions (von
Seggern (48); Blandford and Clark (49)).
(c) $m_b : m_b^S$ where m_b^S is the magnitude derived from the long
period S-waves: $m_b - m_b^S$ is larger for explosions than earthquakes
(Evernden (37)).

     That explosions and earthquakes can be separated using
the discriminants listed above is readily understood from
simple models of the earthquake and explosion source: on such
models, earthquakes are more efficient generators of S waves
than are explosions. The disadvantage of these methods is
that the detection threshold for long period S is high at
teleseismic distances and short period S is attenuated so
rapidly with distance that it is rarely seen beyond 10° from
the epicentre. Blandford (11) however has pointed out that
discriminants (a) and (b) may be useful in recognising attempts
to simulate an earthquake using multiple explosions (see later)
where some of the explosions in the group have yields large
enough to generate detectable S waves. Evernden (37) finds
that given observations within about 10° of the epicentre the
$m_b : m_b^S$ discriminant can make a significant contribution to the
identification of earthquakes down to around m_b 4.5.

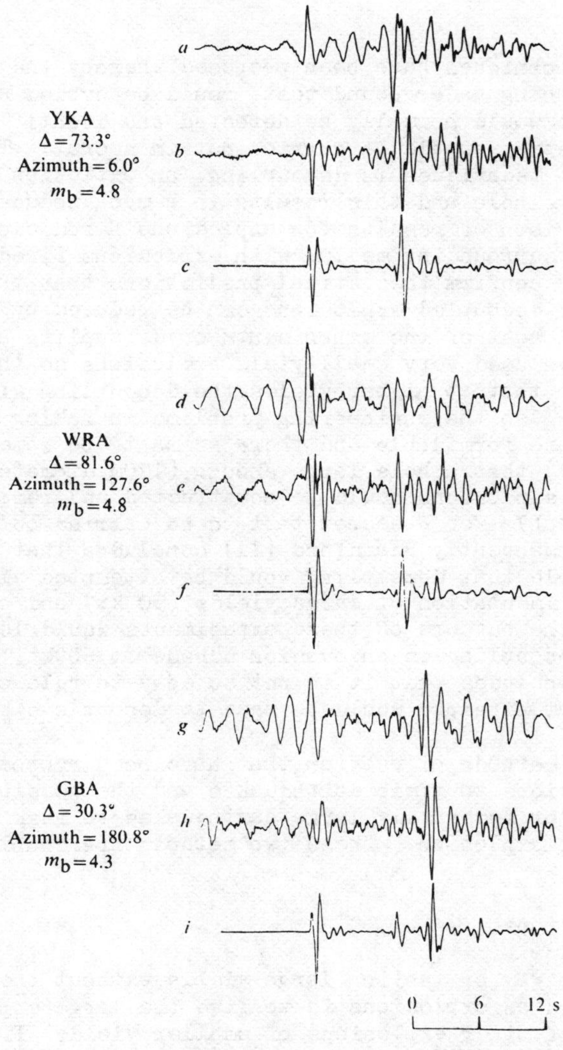

Fig.4. Seismograms for the East Kazakhstan event of May 1, 1969
(origin time 04h 00 min 08.7s). YKA, Yellowknife: a, deconvolved
trace: b, observed trace: c, theoretical trace. WRA, Warramunga:
d, deconvolved trace, e, observed trace: f, theoretical trace.
GBA, Gauribidanur, g, deconvolved trace: h, observed trace:
i, theoretical trace. Theoretical seismograms were computed
for a dip-slip double couple source, with fault plane dipping
at 50° towards YKA. (Reprinted by permission from Nature,
Vol. 248, No. 5451, pp 743-745 Copyright (C) 1974 Macmillan
Journals Limited).

EVASION

Several techniques have been proposed whereby the provisions
of a treaty banning underground tests could be evaded and
explosions that would normally be detected and identified by
a monitoring network could be carried out in secret. The
oldest of these techniques is decoupling: an explosion is
fired in a large hole and this results in a much reduced seismic
signal. Comparison of results for explosions fired closely
coupled to the surrounding medium with explosions fired in a
cavity seems to confirm theoretical predictions that the amplitude at 1 Hz of decoupled explosions can be reduced by factors
of almost 100. Most of the experiments on decoupling as far
as is known have used very small yield explosions so that there
is no guarantee that at higher yields the decoupling will be
as effective. Also the engineering problems in making a large
stable cavity are formidable and there seems to be some doubt
(Blandford (11)) that a hole large enough (170m diameter) to
decouple 50 kt explosions could be constructed and remain
sufficiently stable for a secret test to be carried out successfully. Consequently Blandford (11) concludes that "it
seems implausible that decoupling would be attempted without
extensive experimentation at large yields (50 kt) and on balance
it seems that the outcome of these experiments would lead to a
rejection of decoupling as an evasion scheme at 50 kt." Blandford
(11) however concludes that it is not so easy to rule out the
possibility that attempts would be made to decouple 5 kt.

Two other methods of evasion that have been proposed are
multiple explosions to mimic earthquakes and the possibility
of using large or moderately large earthquakes to mask the
signals from an explosion. These two methods are considered
below.

Multiple explosions

A possible way of testing large yields without the tests
being identified as explosions is to fire the large explosions
in a sequence of other explosions of smaller yield. If the
first explosions in the series have small yields the m_b would
be small because the magnitude is conventionally computed from
the amplitude in the first 3-4 seconds of the SP signal. If
several large explosions are fired late in the sequence within
a few seconds of each other the Rayleigh waves will sum in-
phase thus increasing M_s so on the m_b:M_s criterion the sequence
of explosions would be made earthquake like. The SP P signal
would also tend to be complex and the onset of the signal
would appear emergent and so would appear to be earthquake
like. Kolar and Pruvost (50) have made a study of this method
of evasion and conclude it could be successful for yields up

to 100 kt. Marshall and Hurley (51) however point out that
on broad band seismograms the P signals from such multiple
explosions would be seen to have the characteristic period of
explosions, that is 1-2s: earthquakes on the other hand when
recorded on broad band instruments have dominant periods of
5-10s. The reason that the characteristic period of the P
seismogram is the same for multiple explosions as a single
explosion is that each explosion contributes both P and pP
and the interference of these two signals cancels out the low
frequency energy in explosions (Molnar (52), Helmberger and
Harkrider (53) and Blandford (11)).

A multiple explosion is also likely to be revealed on the
m_b:M_s criterion if m_b is measured on the largest amplitude in
the P seismogram rather than on the arrivals in the first
few seconds. This however raises the possibility that if m_b
were measured on all earthquakes in this same way some earthquakes would be moved into the explosion population. Blandford
(11) however claims that this change of definition of m_b moves
very few earthquakes if any into the explosion population.
So here is a second possible method of detecting multiple
explosions. Further the ratio of long period S to Rayleigh
wave amplitude and short period S to short period P amplitude
should be no different for multiple explosions than for single
explosions so the discriminants based on these ratios might be
useful in identifying multiple explosions (Blandford (11)).

To be successful the P signals from the multiple explosions
must appear emergent. Blandford (11) however points out that
the number of places where emergent earthquakes are observed
may be extremely limited and so the evaders will have to choose
the epicentral region for these explosions with care if they
are to avoid raising any suspicions that clandestine tests have
been carried out. Even with careful selection of the firing
site however the evaders will not be able to predict the exact
form of the signal that will be seen at every station so there
remains the possibility that a seismologist will notice the
signal from the explosions because it is not characteristic of
the source region where it occurred.

Masking an explosion with an earthquake

In this evasion technique – usually referred to as the
Hide-In-Earthquake technique – the evader waits for a moderate
earthquake near the test site or a large earthquake farther
away and then fires the explosion, hoping that the signal from
the test will be hidden in the coda of the earthquake.
Blandford (11) suggests two ways that such attempts at evasion
might be discovered. One is to try and use processing techniques
to extract possible explosion signals from the earthquake coda.

The other is to obtain seismograms from stations where the
amplitude of the earthquake signal is a minimum because the
station lies at a distance that is close to a minimum in the
amplitude-distance curve for P waves: explosions fired at
distances from the station that lie close to a maximum in the
amplitude-distance curve may then stand out above the earthquake coda. For example seismograms from stations in the core
shadow zone for the earthquake could be examined to look for
signals from explosions with epicentres at shorter distances
from the station than the earthquake. Alternatively at stations
at the PKP focus (142°-155°) from the explosion the signal will
be enhanced relative to the earthquake provided that the
earthquake epicentre is not at the same distance from the
station as the explosion. Blandford (11) states that this
method of searching for explosion signals in the coda of
earthquakes is worth applying even when the earthquake and
explosion epicentre are only separated by 2 or 3 degrees.

Evernden (54-56) has made a detailed study of the Hide-
in-Earthquake technique and concludes that there are real
possibilities for evasion in this way and that only by having
stations close to possible test sites can one be confident
that evasion is not being practised.

THE THEORETICAL BASIS OF THE m_b:M_s AND COMPLEXITY CRITERIA

Of the discriminants discussed above three have no well
established theoretical basis: complexity, spectral ratios
and m_b:M_s. Little progress appears to have been made towards
a detailed understanding of spectral ratios, particularly why
some samples of earthquakes and explosions can be separated
using spectral ratios and some cannot. Blandford (11) suggests
that it may simply be that where the spectral ratio discriminant
works the earthquakes are in a region of higher absorption
than the explosions. In addition interference between P & pP
may act to enhance or reduce the ability of a spectral ratio
to identify the source as an explosion. For interference
between P & pP produces a series of minima in the amplitude
spectrum of the signal centred at frequencies nf_o, n=o, 1,
2 ...∞ where $f_o = \tau^{-1}$ and τ is the time delay between P & pP.
The position of these minima in relation to the high and low
frequency bands used to define the spectral ratios will strongly
influence the actual value computed for the ratio: spectral
ratios can thus be changed by changing the source depth.
Further understanding of spectral ratios must await a better
knowledge of the earthquake and explosion source.

Some progress has been made towards understanding the $m_b:M_s$ criterion and this is summarised below. I then consider possible explanations of complexity over which there has been much debate.

Some explanations of the $m_b:M_s$ criterion

Brune et al (27) suggest that earthquakes generate larger amplitude surface waves than explosions of the same m_b because earthquakes have large source dimensions: interference between seismic signals radiated from different points on a fault would then reduce the shorter periods relative to the longer; consequently the SP body waves on which m_b is measured would be reduced relative to the longer period surface wave amplitudes from which M_s is derived. Wyss and Brune (57) from a study of earthquakes in the California-Nevada region conclude that most earthquakes - I will refer to these as normal earthquakes - in the magnitude range $3 < M_L < 5$ have source dimensions very much greater than for NTS explosions and they thus regard the explanation of the $m_b:M_s$ criterion put forward by Brune et al (27) as proved. Wyss and Brune (57) note however that in their study there are some earthquakes - which they term high stress drop earthquakes - that generate only weak surface waves from which they deduce that these have much smaller source dimensions than normal earthquakes.

Press et al (30) see no evidence that the source dimensions of small magnitude earthquakes and explosions are greatly different and they conclude that source dimensions cannot account for the separation of earthquakes and explosions on the $m_b:M_s$ criterion. The source dimensions estimated by Press (58) for normal earthquakes are about the same as those estimated by Wyss and Brune (57) for high stress drop earthquakes.

Other suggestions put forward to explain the $m_b:M_s$ discriminant include the following.
(i) Differences in the shape of the source spectrum. An earthquake is assumed to be a step of displacement so that the spectrum of the radiated P wave is flat, whereas explosions are assumed to be roughly impulses of displacement (outward motion followed by a return to zero or near zero displacement) thus the radiated spectrum tends to zero at low frequency (reducing M_s) and is peaked at high frequencies (enhancing m_b).
(ii) Differences in the seismic rise time. An earthquake is assumed to be due to slip on a fault plane with the slip taking place over several seconds. The displacement close to an explosion on the other hand takes place in a very short time. On this model the signal radiated by an earthquake thus contains less high frequency energy relative to the low frequencies than does an explosion signal.

Peppin (59) argues that neither differences in source
size, source spectrum or seismic rise time explains $m_b:M_s$.
Peppin (59) reaches this conclusion after studying the P-wave
spectra of NTS explosions and earthquakes at near and very near
distances. The study showed that there are no differences between the spectra of the P waves radiated by explosions and
those radiated by shallow earthquakes. Peppin (59) concludes
that the suggestion put forward by Douglas, Hudson and
Kembhavi (60) amongst others may explain the $m_b:M_s$ criterion.
Douglas et al (60) point out that the source mechanism of
earthquakes and explosions differ markedly and this might
account for the differences in the relative excitation of LP
Rayleigh waves and SP body waves by the two types of source.
To test this Douglas et al (60) evaluated $\chi^R(\omega_R)/\chi_o^P(\omega_P)$
where $\chi^R(\omega_R)$ is the ratio at frequency ω_R of the Rayleigh-
wave amplitude of a point double-couple source to that of a
point dilatational source and $\chi_o^P(\omega_P)$ is the ratio at frequency ω_P of the P-wave amplitudes (neglecting the contribution from the surface reflections); both sources being assumed
to be in a uniform half-space. Three orientations of a double-
couple source were examined: vertical dip-slip; vertical
strike-slip and dip-slip on a plane dipping at an angle of 45°;
Burridge, Lapwood and Knopoff (61) show that the response to a
double-couple source of any orientation may be represented as
a linear combination of these three basic source types.
Figure 5 shows $\chi^R(\omega_R)/\chi_o^P(\omega_P)$ for a vertical dip-slip fault
plotted against $\omega_R h/C_R$ where h is the depth of the earthquake
and C_R is the Rayleigh wave speed. It is assumed that the
material is a Poisson solid and that the P waves have a take-

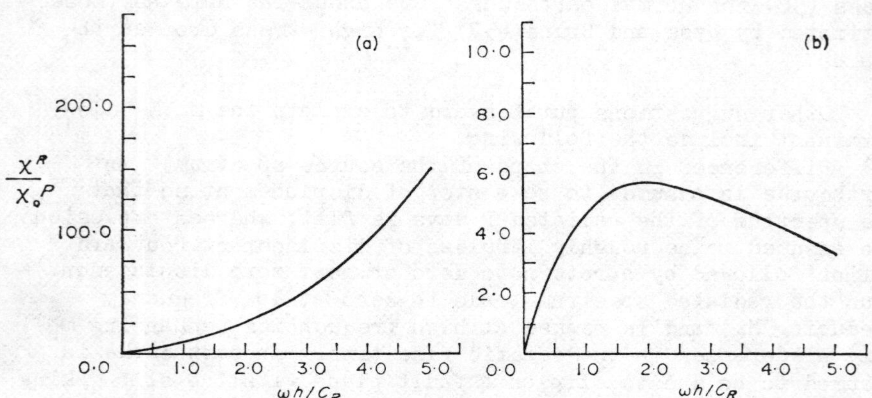

Fig.5. χ^R/χ_o^P for a vertical dip slip fault with $\theta = 10^\circ$ and
(a) h=h', (b) h=10h'.

off angle Θ , into the half space of 10° to the vertical. Two cases are shown one for which $h=h'$ the other $h=10h'$; h' is the depth of the explosion. It can be seen that the ratio is well above unity over most of the range of $\omega_R h/C_R$; this is partly due to the fact that the direction $\Theta=0^{\circ}$ is a node for P-waves for the vertical dip-slip source and that $\Theta=10^{\circ}$ is relatively close to the node. Note that the deeper the earthquake the less the amplitude of the Rayleigh waves whereas the body wave amplitude is not much affected so χ^R/χ_0^P decreases in general as h increases.

Figure 6 shows χ^R/χ_0^P against $\omega_R h/C_R$ for a vertical strike-slip fault in a Poisson solid and with $\Theta=10^{\circ}$. Again the ratio is well above unity except around $\omega_R h/C_R=1.2$ where this ratio goes to zero; this is because at this point the Rayleigh wave amplitude is zero. Figure 7 shows χ^R/χ_0^P against $\omega_R h/C_R$ for the 45° dip-slip fault. For this case the ratio is a function of azimuth ϕ where ϕ is measured from the strike of the fault. The direction $\Theta=10^{\circ}$ now lies close to the maximum ($\Theta=0^{\circ}$) in the P wave radiation for the double-couple source. The result of this is that unless the double-couple source is at a similar depth to the explosion the ratio χ^R/χ_0^P is less than unity except at small $\omega_R h/C_R$.

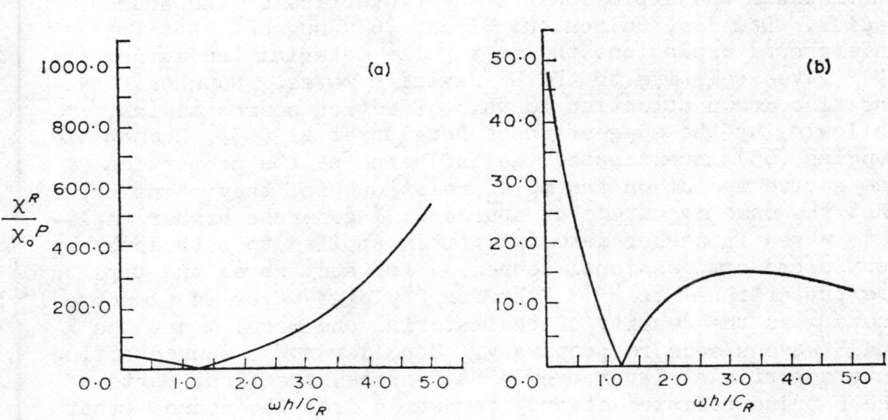

Fig.6. χ^R/χ_0^P for a vertical strike slip fault with $\Theta=10^{\circ}$ and (a) $h=h'$, (b) $h=10h'$.

Usually the explosion will be shallow enough for the
effect of the surface reflection to influence the body-wave
amplitude whereas most earthquakes will have depths of a few
kilometres or more so that the effect of the surface reflections
can be neglected provided the P amplitude is measured near the
onset of the record. Depending on the depth of the explosion,
P and pP may interfere constructively or destructively at
frequency ω_P, or they may be so widely separated that the
amplitude of P is unchanged, so that the ratio χ^R/χ^P which
includes the effects of the surface reflections may be greater
than, less than or equal to χ^R/χ^P_o. If the earthquake is so
shallow that P, pP and sP interfere, Douglas et al (60) conclude that χ^R/χ^P will tend to be larger than χ^R/χ^P_o.

The main conclusion of the work of Douglas et al (60) is
that in general double-couple sources generate more Rayleigh
waves than do compressional sources. Gilbert (62) in a short
elegant paper confirms this general conclusion.

From the work of Douglas et al (60) it is clear that other
things being equal the most difficult earthquakes to identify
on the m_b:M_s criterion will be those that result from dip-slip
on a plane dipping at 45° and that the difficulty will increase
as the earthquake depth increases. The presence of frequencies
in the Rayleigh wave spectrum where the amplitude goes to zero
may also result in a low value for M_s thus making an earthquake appear explosion-like.

Another possible influence on the m_b:M_s relationship for
earthquakes and explosions is the properties of the source
medium. Douglas, Hudson and Blamey (63) suggest that for
underground explosions the softer the material the larger the
SP P waves relative to the LP Rayleigh waves. Bouchon (64)
has also drawn attention to this effect of source medium.
Following up the suggestion of Douglas et al (63), Hudson and
Douglas (65) investigated the influence of the properties of
the source medium on the m_b:M_s relationship; they point out
that the same magnitude of source will generate higher amplitude waves in weaker materials (this applies to both double-
couple and compressional sources): for body waves the amplitudes
are proportional to $(\rho \alpha^3)^{-1}$ for Rayleigh waves to $(\rho \alpha^2 \beta^{\frac{5}{3}})^{-1}$;
where ρ is the density of the material and α and β are the P
and S wave speeds respectively. Consider now a source acting
in a superficial layer over a half-space, the first parts of
the P pulse radiated steeply downwards from the source (that
is the short period part of the signal) will be the same as if
the source were set in an unbounded layer of the source material
except in so far as it is modified by the coefficients of
transmission through the layers and by geometrical spreading.
The LP Rayleigh waves on the other hand interact with the

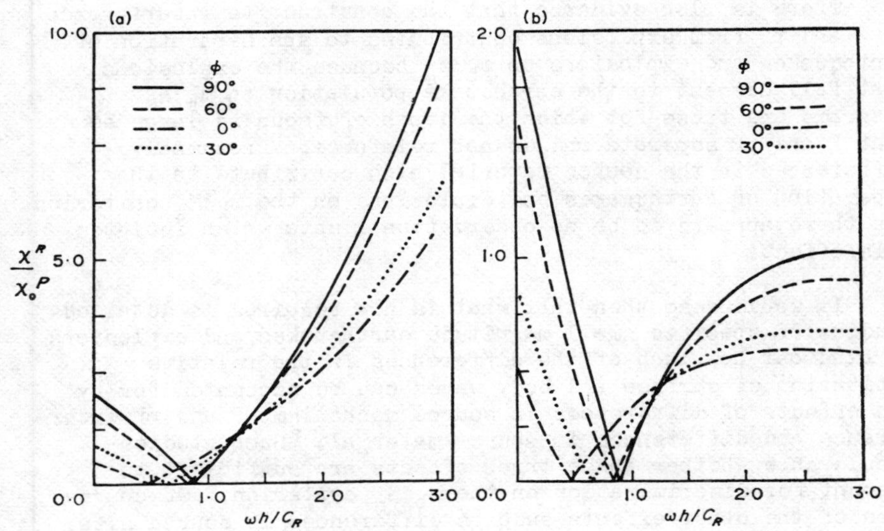

Fig. 7. X^R/X_0^P for a dip slip at $45°$ with $\theta = 10°$, $\phi = 0$, $30°$, $60°$, $90°$, and (a) $h = h'$, (b) $h = 10h'$.

layer and the half space and so the existence of the superficial layer will barely change the LP Rayleigh waves from
what would be generated in the halfspace without the layer.
The effect of the superficial layer of soft material is thus
to enhance m_b relative to M_s. Now most explosions are fired
in superficial layers of soft material overlying hard, whereas
most earthquakes probably occur in hard material, consequently
m_b will be increased relative to M_s for the explosion but not
for the earthquake. From the work of Douglas, Hudson and
Blamey (63) it appears that $m_b - M_s$ for a source (explosion) in
alluvium ($\rho = 1.8$ g/cm^3; $\alpha = 1.7$ km s^{-1}) is 0.8 magnitude units
greater than for the same source in granite ($\rho = 2.7$ g/cm^3;
$\alpha = 4.8$ km s^{-1}).

There are thus several possible factors tending to separate
earthquakes and explosions on the $m_b : M_s$ criterion. The most
important factor at least at low magnitudes appear to be the
difference in source mechanism, the main observational evidence
in favour of this explanation is that it is possible to demonstrate that most other explanations are incorrect. If source
mechanism is the main explanation of the $m_b : M_s$ criterion then
we would expect anomalous earthquakes to be of the $45°$ dip-slip
type with depths greater than say 10 times that of explosions and
there is evidence that some of the anomalous earthquakes are of
this type.

There is also evidence that the constructive interference
of P and pP from explosions contributes to the separation of
earthquakes and explosions on m_b:M_s because the explosions
that fall closest to the earthquake population on m_b:M_s
diagrams are those for which the depth of focus is large so
that P and pP separate and do not reinforce. Presumably
differences in the source material also contribute to the
separation of earthquakes and explosions on the m_b:M_s criterion
but there appears to be no observational data which isolates
this effect.

It would seem then that what is now required is detailed
studies of specific small magnitude earthquakes and explosions
to find out how much of the differences in the relative
excitation of surface and body waves can be accounted for by
the effects of differences in source mechanism, P and pP interference and differences in source material. Such studies
should show whether these three effects are sufficient to
account for discrimination on the m_b:M_s criterion. Of course
some of the other effects such as differences in source size,
source spectra and source time history may also contribute to
differences in m_b-M_s for specific earthquakes and explosions.
The m_b:M_s lines for earthquakes do diverge towards large magnitude and presumably this is an effect of the increase in
the source dimensions of the earthquake - as the source size
increases, M_s for earthquakes increase faster than m_b.

Finally Ward and Toksoz (66) have shown that m_b-M_s for
NTS explosions is less than for those in E Kazakh. The
reason for this appears to be twofold: (1) NTS explosions are
often accompanied by tectonic release - effectively an earthquake is triggered by the explosion - which enhances M_s but
leaves m_b unchanged; and (2) E Kazakh is underlain by an upper
mantle with lower anelastic attenuation than NTS thus m_b is
reduced at NTS compared to E Kazakh. Effects such as these
must obviously be taken into account when trying to interpret
m_b:M_s relationships.

Some explanations of complexity

When complexity was first studied it seemed that what had
to be explained was the difference in complexity between earthquake and explosion P signals. And as complex earthquake
signals were recorded on paths that lay close to paths over
which simple explosion signals had been observed it was
naturally assumed that complexity was related to the differences
in the source mechanism of earthquakes and explosions - the
effects of the path were assumed to be negligible (Thirlaway (1,
17)). Several lines of evidence can be adduced to support this
view that P signals are transmitted to teleseismic distances
with little distortion; some of the evidence is listed below.

(i) Explosion P signals are observed to be simple and for a
given explosion, signals at stations widely spread over the
earth are similar (Jansson and Husebye, (67)); explosions in
fact produce SP P seismograms at teleseismic distances with a
characteristic shape: that of a W.
(ii) Earthquake P signals are observed to be coherent over
large areas. Husebye and Jansson (68) for example point out
that over the length of Sweden "rather strong correlations
exist between signals recorded at different stations" - this
is so even though not all the stations used in the study had
seismographs with the same response; some of the sources were
at distances of less than 30° from the nearest station and
much of the study was carried out with hand digitised data.
Jansson and Husebye (67) extended the work and concluded that
even over the Fennoscandian seismograph network for which the
greatest separation of stations is 1800 kms the coherence of
the signals between stations is large enough to make array
processing worthwhile. Jansson and Husebye (67) point out that
this is true despite the fact that stations of the network
lie on structures that range from the sedimentary rocks of
Denmark to the homogeneous crust of the Baltic Shield.
(iii)Attempts to compute the P waveforms for explosions using
simple source and earth models produce waveforms that show
good agreement with the observed simple explosion seismograms;
this was first demonstrated by Carpenter (Carpenter and
Thirlaway (69); Carpenter (70-71)) using very simple models
and has been confirmed by later more detailed calculations.
Explosion seismograms modelled in this way are simple because
the explosion source used generates no S waves and being
shallow the P to pP time is short: the main reverberations are
generated by very shallow boundaries so they follow closely
on P. The marked variations in complexity with azimuth and
distance shown by some earthquake seismograms has also been
simulated without the need to assume most of the complexity is
due to path. In these model studies the complexity of the
earthquake seismogram arises mainly from the reverberations in
the source layering of the P and S waves radiated upwards from
the source: simple seismograms are recorded at those stations
for which the direct P amplitude radiated towards the station
is much larger than the amplitude radiated along pP and sP
paths.

The observation of complex P wave seismograms from explosions however shows that the earth is not everywhere as
simple as suggested by the evidence listed above and there
seems to be general agreement that these complex seismograms
must be the result of scattering in some way. However to come
up with a satisfactory model of the distribution and form of
the scattering region has proved to be extremely difficult, for
not only is it necessary to explain the few complex seismograms,

but any model must also be able to explain why simple seismograms are also observed at some stations. To illustrate the
problem consider the P seismograms from the LONGSHOT explosion
recorded at two stations in western Canada (SI-BC Smithers,
British Columbia and PG-BC; Prince George, British Columbia).
These stations record very complex P seismograms (Figure 8)
but this is at first sight not surprising as the stations are
in an orogenic belt and in a mountainous region. Scattering
either at the free surface or in complex structures beneath
the station might thus account for the complexity. Note
however that PcP is a much simpler signal than P (PcP has in
fact the characteristic shape expected of P) and so if there
are large regions of intense scattering in the vicinity of
these two stations one has to ask why only P is affected and
PcP is left untouched.

There are some further features of these seismograms
that are worth noting: the average body wave magnitude of the
LONGSHOT explosion is 5.8 whereas the magnitude at SI-BC and
PG-BC are 4.95 and 4.75 respectively. Douglas et al (72,73)
have shown that it is a characteristic of most complex records
that the observed magnitude is below average and that for
simple records the observed magnitude tends to be above average.
This suggests that perhaps in seeking an explanation of the
complexity of explosion signals one should not be concerned
with the question "why is the coda so large relative to the
first arrival?" But rather "why is the absolute amplitude of
the first arrival so small?" At YKA (which is at a similar
distance (36.4°) from the LONGSHOT explosion as SI-BC (31.8°)
and PG-BC (34.5°)), the signal from the LONGSHOT explosion is
simple and has an amplitude of about 200 nm, 4 times the amplitude of PcP seen at SI-BC and twice the amplitude of PcP
at PG-BC. It is clear from Figure 8 that had the direct first

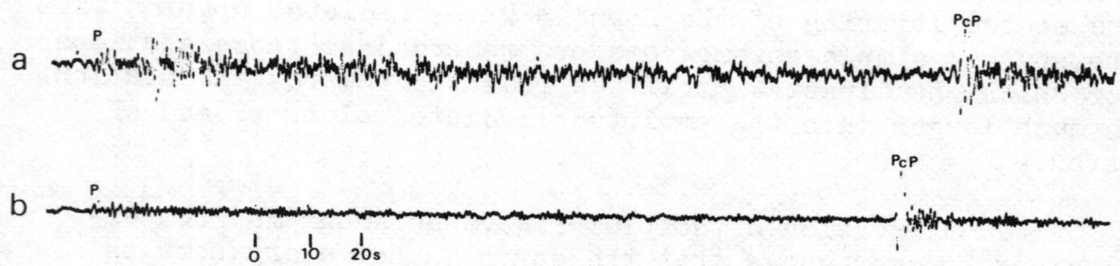

Fig.8. Short period P-wave seismograms from the LONGSHOT
explosion recorded at: a, Smithers, British Columbia (SI-BC);
b, Prince George, British Columbia (PG-BC). Note that PcP has
a much larger amplitude on both these seismograms than P.

arriving P signal been the same amplitude at SI-BC and PG-BC
as at YKA and the coda amplitude unchanged, the P seismograms
at the two British Columbia stations would also have been
classed as simple.

Studies of complex explosion seismograms such as those
shown in Figure 8 has led Douglas et al (72,73) to conclude
that in general complex explosion seismograms can be accounted
for by assuming that direct P has been reduced by attenuation
in passing through a low Q region which is missed by the
arrivals in the coda. On this explanation of complexity the
magnitude of the explosion source computed from complex records
should be less than that computed from a simple record of the
same explosion, the first arrivals of the complex records
should show a lower proportion of high frequency energy than
those of simple records and the coda of complex records should
contain a higher proportion of high frequency energy than the
first arrival. Douglas et al (72,73) show complex explosion
seismograms do have these properties; the low magnitude of the
complex LONGSHOT seismogram at SI-BC and PG-BC has already
been noted above.

A similar mechanism to that of Douglas et al (72,73) is
proposed by Davies and Julian (74) who suggest that complex
seismograms are recorded when P is reduced by defocussing and
so the other arrivals that are not similarly reduced may appear
large relative to P. Davies and Julian (74) have attempted
to explain in this way the variation in complexity of the
seismograms recorded from the LONGSHOT explosion. This mechanism may also explain the variation in the complexity of P
signals across the NORSAR array. Frazier (75) shows that for
the same explosion some NORSAR subarrays record simple seismograms whereas others record very complex seismograms and that
for the subarrays that record the simple signals the first
arrival is almost an order of magnitude larger than at the
subarrays that record complex signals: in fact the coda of
the simple and complex seismograms are of about equal amplitude: only the amplitude of the first arrival differs between
the two types of seismograms. The variation in complexity at
the NORSAR could thus apparently be explained by assuming that
the structure in the upper mantle beneath the NORSAR produce
a partial shadow zone at the free surface (Haddon and Husebye
(76) have recently determined possible structures that will
account for such a shadow zone) and that the coda is generated
by scattering into the shadow zone by scatterers both in the
crust and upper mantle and at the free surface. So it is the
weakness of the first arrival as seen on the complex records
rather than the amplitude of the scattered waves that accounts
for the complexity.

At present there is no widely accepted explanation of
complexity. There is much evidence of scattering within the
earth and at the free surface but there is also much apparently
contradictory evidence that much of the mantle is relatively
uniform. More detailed work is required on inverting the
observed distribution of complex and simple seismograms for say
LONGSHOT, to try and find a realistic distribution of scatterers
that can account for the observed variation in complexity. It
may be that the observations at the NORSAR can help here:
Haddon and Husebye (76) have studied the case of plane waves
incident on the lithosphere beneath NORSAR and have deduced
that a lens-like scatterer will account for the variation in
arrival times and amplitudes at the surface. Perhaps now the
reciprocal case should be considered and computations made of
the distribution of high and low magnitude regions on the
surface of the earth at teleseismic distances that would be
produced by sources at various points of the NORSAR array.
Such computations would show if observed distributions of
complex and simple seismograms at teleseismic distances can
be accounted for by the presence in the source region of a
lens-like scatterer such as that which appears to be present
in the lithosphere beneath NORSAR.

If the apparent inverse correlation of complexity and
station magnitude is accepted then this suggests an explanation of why when complexity was first proposed there appeared
to be no exceptions to the rule that explosion seismograms are
simple: complex explosions will in general have a lower S/N
than simple signals and so fewer complex signals will be
observed with adequate S/N for the complexity to be clearly
seen. Note also that at low yields near the detection level
of the most sensitive stations only simple signals will usually
have adequate signal to noise to be detected.

Although complexity is no longer of use for identifying
earthquakes it is important that it be explained. For only by
understanding how the earth distorts the SP P signal to produce
complex records can an assessment be made of how this will
limit our ability to identify the source of a signal as earthquake or explosion. If for example there are sites from which
the P signals radiated to all distant stations would be complex
and of low amplitude given the yield of the explosion, then
explosions of a few kilotons fired at such sites might be difficult to detect or, if detected, identify: m_b would be reduced
relative to M_s and so the explosion would appear earthquake-
like on the $m_b:M_s$ criterion. Knowledge of the size and distribution of scattering regions would also show whether there
are regions of the earth's surface over which signals from distant sources are sufficiently coherent for a large aperture
array to be operated efficiently.

DISCUSSION

From the foregoing it is clear that under any treaty the
bulk of the detected earthquakes would most likely be identified on location, depth of focus and m_b:M_s where M_s is the
normal surface wave magnitude computed from the fundamental
mode Rayleigh waves. The discrimination line for the m_b:M_s
criterion would have to be drawn conservatively to make sure
that multiple shots were not identified as definite earthquakes.
The residue after applying these three criteria could then be
studied in detail. Use could be made of the higher mode surface waves and Love waves as suggested by Tatham et al (46)
or of the pattern recognition techniques proposed by Tjøstheim
(41) to try and decide if any of these are truly explosions.
An alternative approach - which has been called the deterministic
method - and the one I personally would prefer, is to look at
all the available data and try and answer the question "Is
the data more compatible with an explosion source than an
earthquake source?" An example of this approach is that used
by Douglas et al (45) to study anomalous earthquakes: there
the main arrivals following P on the SP seismograms at three
stations, were shown to have amplitudes and arrival times
relative to P that are consistent with the source being a
double couple at a depth of about 25 kms. Douglas et al (45)
found an orientation that was compatible with observed seismograms by informed trial and error. A systematic way of determining the source orientation from such seismograms -
assuming the seismograms are from an earthquake - is given by
Pearce (77-79) who suggests that the amplitude of P and possible
surface reflections pP and sP (with the polarities of each
phase if this can be read unambiguously) be specified with
error bands and that a search then be made through the possible
orientations of the double-couple source to find what orientations, if any, are compatible with these relative amplitudes
for all seismograms. If orientations can be found this tends
to confirm that the phases have been correctly identified and
so a reliable estimate of depth can be obtained. With depth
and orientation it should then be possible to compute P and
the surface wave seismograms (including higher modes and Love
waves) using information on crustal structure in the source
region. If the computed seismograms are shown to be consistent
with the observed then this will be conclusive evidence that
the source is in fact an earthquake: if not then it would be
necessary to see if the seismograms were more easily explained
assuming the source was one or more explosions.

Note that it is not necessary for the paths to all stations
to be so simple that surface reflections stand out clearly. A
selection can be made of the simplest seismograms and the remainder ignored on the assumption that the added complexity

is due to path effects. For example consider the application
of the method to the identification of the LONGSHOT explosion:
all the complex records from LONGSHOT can be ignored and only
the simple ones used. If we make the assumption that LONGSHOT
is in fact an earthquake at a depth of say more than 5 kms
then we can put a range of possible values on pP and sP from
the amplitude of the coda of these simple records. Using the
LONGSHOT seismograms from the four arrays, Eskdalemuir (EKA),
Scotland, YKA, WRA and GBA it is found that there is no possible
orientation of double-couple compatible with these four seismograms. The conclusion is then that the source must be very
shallow so that P and the surface reflections are contained in
the first second or so. With this information on depth one
could then go on to predict the surface waves from an explosion
and various orientations of earthquake source to see which are
most consistent with the observed waves. With this kind of
deterministic method I would hope that it would not be necessary
to rely solely on the first motion method for the identification
of anomalous earthquakes.

To carry out the identification of the large majority of
earthquakes, signals recorded on narrow band SP and LP seismographs are sufficient; such seismographs allow weak signals to
be detected. It is essential however to have broad band signals
available to help in the identification of attempts at evasion
using the multiple explosion technique. Further if higher mode
surface waves are to be used for identification then broad
band instruments are also required to record these modes, as
Landers (43) has pointed out.

It is possible to estimate the detection and identification thresholds of any given network: the network of stations
is specified along with such parameters as noise levels,
amplitude-distance curves, S/N required for a detection and so
on and estimates made say, of the smallest magnitude source
that can be detected by some specified number of stations from
various points on the earth. The ability of station networks
to locate sources and to detect Rayleigh waves can also be
estimated. Examples of maps showing the detection thresholds
for P and Rayleigh waves for a variety of networks are shown
in CCD (80). The results of these studies show that, for
example, for a network of 50 stations with high quality instrumentation there are places in the northern hemisphere where
the threshold magnitude is m_b 4.4; where the threshold magnitude is defined as the magnitude of a source that has 0.9
probability of being detected at 4 stations with a S/N of 3
from distances of greater than 2200 km; the threshold magnitude
for some places in the southern hemisphere on the other hand
is greater than m_b 5.0. For surface waves the threshold
magnitude for parts of the northern hemisphere is as low as

M_s 3.0 (equivalent to about 10 kt) when the threshold magnitude
is now defined as the magnitude that has 0.9 probability of
being detected by 2 stations with S/N of 2. For the southern
hemisphere the surface wave threshold magnitude is greater
than M_s 3.8 (equivalent to about 60 kt) over some parts of the
Pacific: the yield estimates have been made assuming the
relationship of M_s = log Y + 2 (Marshall et al (81)).

For established networks it is possible to determine
directly their ability to detect and identify earthquakes and
explosions. Most discrimination studies use only small samples
from the total population of seismic disturbances that take
place each year and so are of little value in estimating the
detection and identification thresholds of existing networks.
Only Evernden (37) appears to have made a study of the ability
of the current network of stations to identify earthquakes:
he studied all the earthquakes reported in the ISC bulletins
for the period 1 January – 10 June 1972. The data used, which
includes observations at epicentral distances of < 10°, are
those routinely available in bulletins, together with seismograms from the WWSS network and the high gain narrowband LP
network. The identification criteria used are depth of focus
(using where possible the S-P time to reduce the uncertainty in
the depth estimates) $m_b:m_b^s$ and variants of the $m_b:M_s$ criterion.
Evernden (37) finds that out of 977 earthquakes with $m_b \geq 4.5$,
22 had depths less than 50 kms and surface wave trains that
are mixed with those from other earthquakes and so could not
be identified on the criteria used. Of the remainder only 2
could not be identified as earthquakes. Mixing of surface
waves is thus not a great problem in this study as Evernden (37)
points out, this is because there are almost always stations
adjacent to each of the earthquakes studied where the surface
waves of one earthquake predominates over those from the other.
The 22 earthquakes that had mixed surface waves were aftershocks where the waves were mixed with those of the main shock.

CONCLUSIONS AND RECOMMENDATIONS FOR FURTHER WORK

The main conclusions that can be drawn from research on
discrimination over the past 20 years are listed below.
(i) There are no short period P-wave discriminants currently
available that can identify the large majority of earthquakes
and explosions and there seems to be little prospect that such
a discriminant will be found; consequently a purely teleseismic
system for detecting and identifying underground explosions is
not possible for acceptable thresholds of identification.
(ii) The most important criteria for discriminating between
earthquakes and explosions are the obvious ones of location and
depth together with the $m_b:M_s$ criterion.

(iii)There seems to be a few earthquakes each year that are
difficult to identify on the m_b:M_s criterion if M_s is measured
on fundamental mode Rayleigh waves but there is some evidence
that by making use of higher modes it is possible to identify
most of these earthquakes; some of these earthquakes can also
be identified because the P-waves can be shown to be compatible
with double-couple sources at depths greater than 5 kms.
(iv) There remain a few earthquakes that can only be identified
at present by the first motion method.
(v) Recordings from narrow band SP and LP seismographs are
sufficient for the detection, location and identification of
most earthquakes and explosions but to identify anomalous
earthquakes using higher modes or attempts to mimic an earthquake using multiple explosions wider band recordings are
required.
(vi) The only way to be confident that no tests are being
carried out by hiding explosion signals in the coda of an
earthquake is to have stations close to possible sites for
such tests.

Much work remains to be done on discrimination. In particular work is required to demonstrate conclusively why the
m_b:M_s criterion works so well for most earthquakes and perhaps
more significantly why a few "anomalous" earthquakes are
classified as explosions on this criterion. Such understanding
should show what are the optimal methods of measuring the
relative excitation of P and surface waves and why at present
even with the most refined methods of estimating m_b and M_s,
m_b:M_s diagrams show such a large scatter in the observations.
If deterministic methods are to be used to aid identification
of suspicious sources then we need to develop modelling techniques to allow for departures from the simple source and
earth models currently used.

Most discrimination studies carried out so far have had
to make use of magnetic tape recordings from the output of
narrow band seismographs and visual recordings from wider band
(WWSSN LP) seismographs. Research on seismological methods of
discrimination - as on other aspects of seismology - would be
made easier given digital recordings from which a wide range
of frequencies could be recovered. Frequency filtering could
then be applied to the recordings to obtain the best wide band
estimate of signal shape. Given array recordings differences
in the spatial properties of signals and noise could be used
to extract broad band signals.

At this ASI we will hear of the latest developments on,
amongst other things, regional discriminants, magnitude estimation, seismic source theory, wave propagation and wide band
digital recording - these developments will strongly influence
research on discrimination in the future.

ACKNOWLEDGEMENTS

I thank the Editors of the Geophysical Journal of the
Royal Astronomical Society for permission to reproduce
Figures 2, 3, 5, 6 and 7 and the Editor of Nature for per-
mission to reproduce Figure 4.

REFERENCES

1   Thirlaway, H.I.S., 1963. "Earthquake or Explosion?"
    New Scient, 18, pp 311-315.

2   Birtill, J.W., and Whiteway, F.E., 1965. "The application
    of phased arrays to the analysis of seismic body waves."
    Phil Trans Roy Soc, Series A, 258, pp 421-493.

3   Evernden, J.F., 1975. "Further studies on seismic dis-
    crimination." Bull Seis Soc Am, 65, pp 359-391.

4   Bullen, K.E., 1963. "An Introduction to the theory of
    seismology." 3rd Edition, Cambridge University Press.

5   Herrin, E., and Taggart, J., 1968. "Source bias in
    epicentre determinations." Bull Seis Soc Am, 58,
    pp 1791-1796.

6   Douglas, A., Hudson, J.A., Marshall, P.D., and Young, J.B.,
    1974. "Earthquakes that look like Explosions." Geophys
    J R astr Soc, 36, pp 227-233.

7   Marshall, P.D., 1970. "Some Seismic Results of the MEDEO
    Explosions in the Alma Ata Region of the USSR." AWRE
    Report No. 0-33/70. HMSO, London.

8   Evernden, J.F., 1969. "Identification of earthquakes and
    explosions by use of teleseismic data." J Geophys Res,
    75, pp 3828-3856.

9   Douglas, A., 1967. "Joint Epicentre Determination."
    Nature, 215, pp 47-48.

10 Dewey, J.W., 1972. "Seismicity and Tectonics of Western
    Venezuela." Bull Seis Soc Am, 62, pp 1711-1751.

11 Blandford, R.R., 1977. "Discrimination between Earth-
    quakes and Underground Explosions." Ann Rev Earth Planet
    Sci, 5, pp 111-122.

12   Key, F.A., 1968. "Some Observations and Analyses of
      Signal Generated Noise." Geophys J R astr Soc, 15,
      pp 377-392.

13   Hudson, J.A., 1967. "Scattered surface waves from a sur-
      face obstacle." Geophys J R astr Soc, 13, pp 441-458.

14   Hudson, J.A., and Boore, D.M., 1980. 'Comments on
      "Scattered surface waves from a surface obstacle" by
      J.A. Hudson.' Geophys J R astr Soc, 60, pp 123-127.

15   UKAEA, 1965. "The detection and recognition of underground
      explosions." United Kingdom Atomic Energy Authority,
      HMSO, London.

16   Douglas, A., 1967. "P-signal complexity and source radia-
      tion patterns," in VESIAC Report 7885-1-X. University of
      Michigan.

17   Thirlaway, H.I.S., 1966. "Interpreting array records:
      explosion and earthquake P-wave trains which have traversed
      the deep mantle." Proc Roy Soc A, 290, pp 385-395.

18   Greenfield, R.J., 1971. "Short Period P-wave generation
      by Rayleigh wave scattering at Novaya Zemlya." J Geophys
      Res, 76, pp 7988-8002.

19   Kelly, E.J., 1968. "A study of two short period discri-
      minants." Massachusetts Institute of Technology, Lincoln
      Laboratory, Technical Note 1968-8.

20   Lacoss, R.T., 1969. "A large population LASA discrimina-
      tion experiment." Massachusetts Institute of Technology,
      Lincoln Laboratory, Technical Note 1969-24.

21   Nowroozi, A.A., 1972. "Focal Mechanisms of Earthquakes
      in Persia, Turkey, West Pakistan and Afghanistan and plate
      tectonics of the Middle East." Bull Seis Soc Am, 62,
      pp 823-850.

22   Enuscu, D., Georgescu, A., Jianu, D., Zamara, I., 1973.
      "Theoretical models for the process of underground explosions.
      Contributions to the problem of the separation of large
      explosions from earthquakes." Bull Seis Soc Am, 63,
      pp 765-785.

23   Wiechert, D.H., 1971. "Short period spectral discriminant
      for earthquake and explosion differentiation." Z Geophys,
      37, pp 147-152.

24   Evernden, J.F., 1977. "Spectral characteristics of the
      P codas of Eurasian earthquakes and explosions." Bull
      Seis Soc Am, 67, pp 1153-1171.

25   Evernden, J.F., and Kohler, W.M., 1979. "Further study
      of spectral composition of P codas of earthquakes and
      explosions." Bull Seis Soc Am, 69, pp 483-511.

26   Sandvin, O., and Tjøstheim, D., 1978. "Multivariate
      Autoregressive Representation of Seismic P-wave Signals
      with Application to Short-Period Discrimination." Bull
      Seis Soc Am, 68, pp 735-756.

27   Brune, J., Espinosa, A., and Oliver, J., 1963. "Relative
      excitation of surface waves by earthquakes and underground
      explosions in the California-Nevada region." J Geophys
      Res, 68, pp 3501-3513.

28   Bolt, B.A., 1976. "Nuclear Explosions and Earthquakes:
      The Parted Veil." W.H. Freeman and Co., San Francisco,
      p309.

29   Gutenberg, B., 1945. "Amplitudes of surface waves and
      magnitudes of shallow earthquakes." Bull Seis Soc Am,
      35, pp 3-12.

30   Press, F., Dewart, G., and Gilman, R., 1963. "A study of
      techniques for identifying earthquakes." J Geophys Res.,
      68, pp 2909-2928.

31   Marshall, P.D., Carpenter, E.W., Douglas, A., and Young, J.B.
      1966. "Some Seismic Results of the LONGSHOT Explosion."
      AWRE Report No. 0-67/66. HMSO, London.

32   Liebermann, R.C., King, C.Y., Brune, J.N., and Pomeroy, P.W.
      1966. "Excitation of surface waves by the underground
      nuclear explosion LONGSHOT." J Geophys Res, 71, pp 4333-
      4339.

33   Liebermann, R.C., and Pomeroy, P.W., 1967. "Excitation
      of surface waves by events in Southern Algeria." Science,
      156, pp 1098-1100.

33a   Evernden, J.F., 1967. "Magnitude determination at regional
      and near-regional distances in the United States." Bull
      Seis Soc Am, 57, pp 591-639.

34   Liebermann, R.C., and Pomeroy, P.W., 1969. "Relative
      excitation of surface waves by earthquakes and underground
      explosions." J Geophys Res, 74, pp 1575-1590.

35   SIPRI, 1968. "Seismic methods of monitoring underground
     explosions." International Institute for Peace and
     Conflict Research, Stockholm.

36   Molnar, P., Savino, J., Sykes, L.R., Liebermann, R.C.,
     Hade, G., and Pomeroy, P.W., 1969. "Small Earthquakes
     and Explosions in western North America recorded by new
     high gain, long period seismographs." Nature 224,
     pp 1268-1273.

37   Evernden, J.F., 1977. "Adequacy of routinely available
     data for identifying earthquakes of $m_b > 4.5$." Bull
     Seis Soc Am, 67, pp 1099-1151.

38   Evernden, J.F., Best, W.J., Pomeroy, P.W., McEvilly, T.V.,
     Savino, J.M., and Sykes, L.R., 1971. "Discrimination be-
     tween small-magnitude earthquakes and explosions."
     J Geophys Res, 76, pp 8042-8055.

39   Marshall, P.D., and Basham, P.W., 1972. "Discrimination
     between Earthquakes and Underground Explosions Employing
     an Improved M_s scale." Geophys J R astr Soc, 28,
     pp 431-458.

40   Peppin, W.A., and McEvilly, T.V., 1974. "Discrimination
     among small-magnitude events on Nevada Test Site."
     Geophys J R astr Soc, 37, pp 227-243.

41   Tjøstheim, D. 1978. "Improved seismic discrimination
     using pattern recognition." Phys Earth Planet Int, 16,
     pp 85-108.

42   Forsyth, D.W., 1976. "Higher-Mode Rayleigh Waves as an
     Aid to Seismic Discrimination." Bull Seis Soc Am, 66,
     pp 827-841.

43   Landers, T., 1972. "Some Interesting Central Asian events
     on the M_s:m_b diagram." Geophys J R astr Soc, 31,
     pp 329-339.

44   CCD 1972 United States of America. "A review of current
     progress and problems in seismic verification." Conference
     of the Committee on Disarmament, CCD/358.

45   Douglas, A., Marshall, P.D., Young, J.B., and Hudson, J.A.,
     1974. "Seismic Source in East Kazakh." Nature, 248,
     pp 743-745.

46   Tatham, R.H., Forsyth, D.W., Sykes, L.R., 1976. "The
     occurrence of anomalous seismic events in eastern Tibet."
     Geophys J R astr Soc, 45, pp 451-482.

47   Evernden, J.F., 1976. 'Additional data on "anomalous"
      events listed in "Further studies on seismic discrimination."
      Bull Seis Soc Am, 66, pp 349-352.

48   von Seggern, D.H., 1972. "Seismic shear waves as a dis-
      criminant between earthquakes and underground nuclear
      explosions." SDL Report 295, Teledyne Geotech, Alexandria,
      Virginia.

49   Blandford, R.R., and Clark, D., 1974. "Detection of
      Long Period S from Earthquakes and Explosions at LASA
      and LRSM Stations with application to Positive and Negative
      Discrimination of Earthquakes and Underground Explosions."
      Report SDAC-TR-74-15, Teledyne Geotech, Alexandria,
      Virginia.

50   Kolar, O.C., and Pruvost, N.L., 1975. "Earthquake Simu-
      lation by Nuclear Explosions." Nature, 253, pp 242-245.

51   Marshall, P.D., and Hurley, R.W., 1976. "Recognising
      Simulated Earthquakes." Nature, 259, pp 378-380.

52   Molnar, P., 1971. "P-wave spectra from underground
      nuclear explosions." Geophys J R astr Soc, 23, pp 273-287.

53   Helmberger, D.V., and Harkrider, D.G., 1972. "Seismic
      source descriptions of underground explosions and a depth
      discriminate." Geophys J R astr Soc, 31, pp 45-66.

54   Evernden, J.F., 1976a. "Study of seismological evasion,
      Part I. General discussion of various evasion schemes."
      Bull Seis Soc Am, 66, pp 245-280.

55   Evernden, J.F., 1976b. "Study of seismological evasion,
      Part II. Evaluation of evasion possibilities using normal
      microseismic noise." Bull Seis Soc Am, 66, pp 281-324.

56   Evernden, J.F., 1976c. "Study of Seismological evasion,
      Part III. Evaluation of evasion possibilities using
      codas of large earthquakes." Bull Seis Soc Am, 66,
      pp 549-594.

57   Wyss, M., and Brune, J.N., 1968. "Seismic Moment, Stress
      and Source Dimensions for Earthquakes in the California-
      Nevada Region." J Geophys Res, 73, pp 4681-4694.

58   Press, F., 1967. "Dimensions of the source region for
     small shallow earthquakes." Proceedings of the Vesiac
     Conference on the Current Status and Future Prognosis for
     Understanding the Source Mechanism of Shallow Seismic
     Events in the 3-5 Magnitude Range. VESIAC Report
     7885-1-X, Willow Run Laboratories, University of Michigan.

59   Peppin, W.A., 1976. "P-wave spectra of Nevada Test Site
     Events at Near and Very Near Distances: Implications for
     a Near-Regional Body Wave-Surface Wave Discriminant."
     Bull Seis Soc Am, 66, pp 803-825.

60   Douglas, A., Hudson, J.A., and Kembhavi, V.K., 1971. "The
     Relative Excitation of Seismic Surface and Body Waves by
     Point Sources." Geophys J R astr Soc, 23, pp 451-460.

61   Burridge, R., Lapwood, E.R., and Knopoff, L., 1964.
     "First motions from seismic sources near a free surface."
     Bull Seis Soc Am, 54, pp 1889-1913.

62   Gilbert, F., 1973. "The Relative Efficiency of Earthquakes
     and Explosions in Exciting Surface Waves and Body Waves."
     Geophys J R astr Soc, 33, pp 487-488.

63   Douglas, A., Hudson, J.A., and Blamey, C., 1972. "A
     Quantitative Evaluation of Seismic Signals at Teleseismic
     Distances - III. Computed P and Rayleigh Wave Seismograms."
     Geophys J R astr Soc, 28, pp 385-410.

64   Bouchon, M., 1976. "Teleseismic body-wave radiation
     from a seismic source in a layered medium." Geophys J R
     astr Soc, 47, pp 515-530.

65   Hudson, J.A., and Douglas, A., 1975. "On the Amplitudes
     of Seismic Waves." Geophys J R astr Soc, 42, pp 1039-1044.

66   Ward, R.W., and Toksoz, M.N., 1971. "Causes of regional
     variation of magnitudes." Bull Seis Soc Am, 61, pp 649-670.

67   Jansson, B., and Husebye, E.S., 1968. "Application of
     Array Data Processing Techniques to a Network of Ordinary
     Seismograph Stations." PAGEOPH, 69, pp 80-99.

68   Husebye, E.S., and Jansson, B., 1964. "Applications of
     Array Data Processing Techniques to the Swedish Seismo-
     graph Stations." PAGEOPH, 63, pp 82-104.

69   Carpenter, E.W., and Thirlaway, H.I.S., 1966. "Seismic
     signal anomalies, travel times, amplitudes and pulse shapes."
     Proceedings of the Vesiac Special Study conference on
     Seismic Signal Anomalies, Travel Times, Amplitudes and
     Pulse Shapes. VESIAC Report 4410-99-X, Willow Run
     Laboratories, University of Michigan.

70   Carpenter, E.W., 1966. "A quantitative evaluation of
     teleseismic explosion records." Proc R Soc Lond, A, $\underline{290}$,
     pp 396-407.

71   Carpenter, E.W., 1967. "Teleseismic signals calculated
     for underground, underwater, and atmospheric explosions."
     Geophys, $\underline{32}$, pp 17-32.

72   Douglas, A., Marshall, P.D., and Corbishley, D.J., 1971.
     "Absorption and the complexity of P signals." Nature,
     $\underline{223}$, pp 50-51.

73   Douglas, A., Marshall, P.D., Gibbs, P.G., Young, J.B.,
     and Blamey, C., 1973. "P-signal Complexity Re-examined."
     Geophys J R astr Soc, $\underline{33}$, pp 195-221.

74   Davies, D., and Julian, B.R., 1972. "A study of short
     period P-wave signals from LONGSHOT." Geophys J R astr
     Soc, $\underline{29}$, pp 185-202.

75   Frazier, C.W., 1972. "Short-period amplitude and waveform
     studies at NORSAR," in Seismic Discrimination Semi-Annual
     Technical Summary. Lincoln Laboratory, MIT (31 December
     1972).

76   Haddon, R. and Husebye, E.S., 1978. "Joint Interpretation
     of P-wave time and amplitude anomalies in terms of litho-
     spheric heterogeneities." Geophys J R astr Soc, $\underline{55}$,
     pp 19-43.

77   Pearce, R.G., 1977. "Fault plane solutions using relative
     amplitudes of P & pP". Geophys J R astr Soc, $\underline{50}$, pp 381-394.

78   Pearce, R.G., 1979. "Earthquake Focal Mechanisms from
     Relative Amplitudes of P, pP and sP: Method and Computer
     Program." AWRE Report No. O-41/79. HMSO London.

79   Pearce, R.G., 1980. "Fault Plane solutions using relative
     amplitudes of P and surface reflections: further studies."
     Geophys J R astr Soc, $\underline{60}$, pp 459-487.

80   CCD 1978 "International Co-operative measures to detect and
     to identify seismic events." Conference of the Committee
     on Disarmament, CCD/558.

81   Marshall, P.D., Douglas, A., and Hudson, J.A., 1971.
      "Surface Waves from Underground Explosions." Nature,
      234, pp 8-9.

THE NATURE OF THE EARTHQUAKE SOURCE

Leon Knopoff

Institute of Geophysics and Planetary Physics
University of California, Los Angeles

Abstract.

The dynamics of crack formation, extension, and suturing is
reviewed. Finite strength of materials regulates the
rate of extension of cracks and allows for termination of the extension of cracks. Strength of materials is described mathematically by singular cohesive forces.

I. Introduction

Of the various proposals for methods of discrimination between underground explosions and earthquakes, the most promising
has been that which uses m_b-M_S differences. In principle, the
differences between these two measures of magnitude are correlated
with the frequency spectrum of the seismic signal radiated from
the two kinds of sources. We may suppose that the greater seismic
radiation of earthquake sources at long periods compared with
short, is due to the extended dimensions of the earthquake source;
the earthquake source is probably the better "antenna" for seismic
radiation because of its two-dimensionality.

Perhaps the proposal that has given the greatest insight into
the determination of the dimensions of the earthquake source has
been that which converts the d.c. level and the corner frequency
of the far-field seismic spectrum into three d.c. properties of
the source (1). The three properties are the fault length,
static   stress drop and seismic moment of the earthquake. Unless they are correlated, one cannot derive three quantities
uniquely from two pieces of data without applying an additional

E. S. Husebye and S. Mykkeltveit (eds.), Identification of Seismic Sources - Earthquake or Underground
Explosion, 49–69.

constraint to restrict the problem further. The constraint that has been introduced involves an assumption regarding the nature of the fracture process: specifically, it is assumed that the velocity of rupture is that of S-waves and that the high-frequency part of the radiation spectrum derives from the onset and not from the later history of the fracture process.

It can be argued that the three d.c. properties and the nature of the radiation spectrum are both derivative from the characterization of the fracture process. We may imagine that failure of materials near or on an earthquake fault is a consequence of the response of these materials to an application of large stresses. The ensuing rupture leads to radiation of seismic waves; the spectrum of the radiation can be determined. We can suppose, hypothetically of course, that an imaginary, leisurely inspection of the site before and after the earthquake will provide enough information to obtain the three d.c. quantities. Thus the inversion from spectrum to d.c. properties involves the presumption of a relationship between two sets of derivative quantities coupled, as remarked, through an assumption regarding the source.

I propose to discuss the fracture process with a view toward assessing the quality of the assumption used in the inversion of spectral to d.c. properties. Whether the rupture model used by Brune and co-workers is inappropriate and some alternative may have to be suggested, or conversely that it is quite appropriate, the basic idea to do this kind of inversion I have already characterized as insightful. I shall not discuss the details of the data analysis that leads to the determination of the spectra, a process involving corrections for attenuation, geometrical broadening and frequency response of the instruments.

II. Critical Shear Stress

Knowledge of the properties of matter during fracture is crucial to the understanding of the nature of the seismic focus. The prestressed material in the vicinity of the incipient focus has a greater stored potential energy than the same material has after the earthquake. Significant amounts of energy can be stored in the shallow earth through elastic or gravitational processes. We ignore the latter, since the release of gravitational energy implies large scale vertical motions and/or significant changes in density; in shallow earthquakes this term is small in comparison with the elastic term. The release of elastically stored prestress-energy in earthquakes was described in 1910 by Reid (2) as his 'elastic rebound' model of earthquake occurrence. A rapid release of elastically stored deformational energy, that has been accumulated slowly over long periods of time, requires that the physical properties of materials must change in a time scale that

is small compared with the time scale of the duration of an earthquake and the change of properties must be a natural process consequent to the achievement of the critical state of prestress.
Qualitatively, the physical state must change from one which allows
for the storage of significant amounts of deformational energy
without much motion, to a lower energy state such that the decrease of energy takes place through relative motions that occur
in a short interval of time. If the transition to the new state
is slowed by viscous creep, the dissipation of energy in
viscous processes at the source may be too large to permit significant high frequency radiation; nevertheless, this possibility
exists as a cause of "silent earthquakes".

The most plausible assumption that we can make is that the
elastic properties of materials are instantly reduced and, as a
consequence, that the motion ensues. A process of fracture in
which the reduction of elasticity takes place instantaneously is
unreasonable on physical grounds, but certain mathematical simplifications of the details of the rupture process are a convenient
consequence. If we imagine that fracture takes place under shear
deformations on a fault plane that will be described mathematically
as an infinitesimally thick surface, we suppose that the strength
of the interatomic bonds across this surface change abruptly. Such
a model was proposed in a one-dimensional case by Burridge and
Knopoff (3). In that paper, the interatomic bonds across the fault
surface were modeled as springs that could store prestress energy
elastically. When the energy stored in these bonding springs exceeded a critical threshold, these springs were presumed to break;
the spring constants of these bonding springs then dropped to zero.
The rigidity across the fault plane in the pre-critical state was
replaced, after the transition, by a frictional force such as
Coulomb friction or by a more general viscous force which is a
function of the relative velocity of slip. The presence of a
frictional force implies that influences are present that tend to
slow down the relative motions during rupture and that these re-
tarding influences may depend on large displacements and their
rates of change. The one-dimensionality of the system was an
effort to simplify the treatment of the complex features of the
elasticity of the system at locations removed from the fault zone.

A simple model of uniform elasticity up to the breaking point
is inappropriate. One finds in the literature sketches of the
radial forces between pairs of atoms and these appear schematically
as in Figure 1a. The repulsive force between atoms becomes infinitely large for large compression; the attractive force vanishes
for large extension. We can sketch the shear force on an atom as
a function of lateral displacement from its equilibrium position
as in Figure 1b. At the equilibrium position, the force on a
particle is zero. It is also zero when two particles on opposite
sides of the fault plane are at infinite separation. When the

shear force between adjacent atoms on opposite sides of the fault
plane is small, the force-distance relationship is approximately
linear and Hooke's Law of ordinary elasticity applies. If the magnitude of the applied force exceeds the extreme value B, the two
particles move into a regime of relative displacement where the
applied force exceeds the force of attraction of the cross-fault
bonds   that would tend to restore the particles to their equilibrium separation. In this circumstance, the particle accelerates
and its displacement increases even more. The process now becomes
catastrophic and at large distance the particle separation accelerates under the influence of a force equal to the difference between the critical force B and the ever decreasing interatomic
forces. At large separation, the particles are free of one another
and accelerate under the influence of the force B. A method of
slowing down the accelerated motions is by collisions, which we
model as viscous drag or dynamical friction on the torn part of
the fault plane, or by seismic radiation. In the Burridge-Knopoff
model, the force-distance model for the cross-fault coupling
springs was linear up to the separation x_c; for $x > x_c$ the restoring
force was assumed to be zero. This model is sketched in Figure 1c.

The freedom of a particle to move laterally depends on the
disposition of other particles in its neighborhood. If the nearest
lattice site in the direction of motion is occupied, the unstable
motion of one particle may cause a dynamical increase of force on
its neighbor that may ultimately cause the cross-fault bond
strength of the neighbor to be exceeded. A cascade of adjoining
particles that are triggered into free or almost free motion, un-
fettered or lightly fettered by cross-fault spring restoring forces
that wane as the shear displacement increases, simulates the

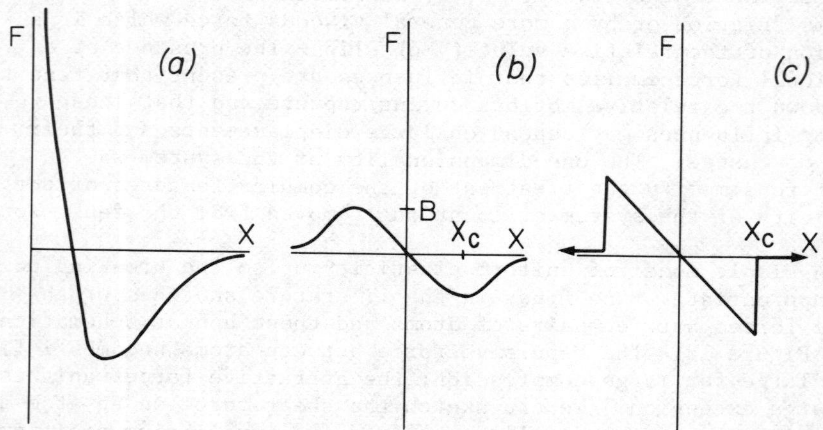

Fig. 1. Interatomic forces across a fault plane: a) tension,
     b) shear, c) idealized version of b).

growth of an extending crack. On the other hand if a vacancy
exists at a neighboring site, the force-distance relation for a
particle is much altered and the history of the particle-vacancy
pair depends on the acceleration of the particle: for small
accelerations an interchange with the vacancy is possible; macro-
scopic manifestation of this latter process is one of creep. For
larger accelerations, cascade of motion beyond the vacancy is
possible.

III. Cohesion

Under the above model, the fracture will continue to extend
as long as the dynamically accumulated force on each particle in
its turn, exceeds its own critical bonding strength B. Conversely,
the fracture will cease its extension if the total dynamical force
on any one particle, that is the sum of the pre-existing static
force plus the additional force due to the motion of the already-
ruptured particles, is less than the critical breaking strength
for the cross-fault bond for that particle. In the one-dimensional
model of nearest neighbor elastic interactions, the dynamical force
is accumulated in the single spring that adjoins the first particle
that has an unbroken bond; this defines the edge of the crack. In
the continuum limit, this force becomes a delta function multiplied
by a certain coefficient; if the coefficient of the delta function
describing the dynamically accumulated force exceeds the coefficient
of the delta function of the breaking strength of the bonds, rupture
continues, and the converse. In the case of two- or three-
dimensional fractures, the tearing edge is a curve in space, rather
than a point. If the radius of curvature of the curve in space
describing the edge is large, the singular stress near the edge of
the crack varies as $r^{-\frac{1}{2}}$ in the continuum limit, where r is the
distance from the edge; in this case, whether the crack continues
or ceases to extend depends on the comparison between the coefficients of two singular functions; both functions vary as $r^{-\frac{1}{2}}$: one
describes the dynamically accumulated stress and the other describes
the bond strength. The critical bond strength is called the cohesion. For the purposes of the later discussion, we shall modify
the definition of the cohesion, or the cohesive forces, to describe
the difference between the critical bond strength B and the prestress.

What has this idealized model of rupture in a perfect or almost perfect crystal to do with fractures on real materials? The
answer lies in our appreciation that functions such as $\delta(r)$ or
$r^{-\frac{1}{2}}$ are the consequences of certain important mathematical simplifications and have no real significance. To paraphrase Spinoza,
"nature abhors an infinity". The infinite values of the above
singular functions cannot arise in any real case arising in nature.

These are the consequences of the convenience, for mathematicians, of formulating problems involving a sharp edge of crack in either one-or two-dimensions. In addition, the result that the continuum limit of the edge stress is a delta function in the one-dimensional case is a consequence of the assumption that the force of inter- action has a nearest-neighbor character; for crystalline solids, nearest-neighbor forces are patently impossible, since this cannot account for the long-range order, extending to distances of many orders of magnitude of lattice dimensions.

The passage to the continuum limit should take the longer range character of the forces into account. Qualitatively we may assume that this implies that the edge of the crack is no longer perfectly sharp but instead that there must be some taper between the totally cracked and the totally uncracked states (4). The ac- tual shape of the taper for given interatomic forces is unsolved. We can however parameterize it by arguing that the function of the taper is to relieve the mathematical singularity at the edge of the crack as we take the continuum limit. In one dimension, this relief probably takes the form of suppression of the delta function character of the cohesive force and deforms it into a bump of finite size such as gaussian or the function $(a^2+x^2)^{-1}$; the specific form of this function depends on the shape of the taper which is un- known, as already remarked. In two dimensions, the best that can be argued is that for most mathematical purposes it suffices to assume that the stresses vary as $r^{\frac{1}{2}}$ near the edge and as $r^{-\frac{1}{2}}$ at longer range. The cross-over between the two terms depends on the shape of the taper. The range over which the stresses can be taken to vary as $r^{-\frac{1}{2}}$ is less than the dimensions of the crack. At dis- tances large compared with the dimensions of the crack, the crack looks like a point source and the stresses in this range vary as $r^{-\frac{1}{2}}$. The coefficients that convert these functions into stresses depend mostly on the sizes of the strains just behind the crack tip; these in turn are largely functions of the stress drop and the dimensions of the crack, as well as of the rupture history. In the case of short range stresses, where the stresses are expressed as $Kr^{-\frac{1}{2}}$, the quantity K is called the stress-intensity factor. We note that as the scale of the taper in the edge becomes smaller, the magnitude of the cohesive stress barrier at the edge soon over- whelms the prestress so that the distinction introduced earlier between the two definitions of cohesion, the one as the measure of the strength and the other as the difference between the strength and the prestresses, loses its significance; both measures are re- duced to a comparison of the coefficients of singular functions.

The presence of a taper in the shape of the edge has several consequences: it moderates the infinity in the singularity of the stress beyond the edge, it moderates the sharpness of the edge and introduces a dimension of distance into the edge condition, which is the order of the width of the taper or of the width of the

force-distance law of Fig. 1. Rigorously, we should compare the
peak values of the stress generated by the dynamics of the crack
with the peak of the stress-distance law but, as remarked, for
mathematical reasons we prefer to compare the coefficients of certain singular functions.

With this discussion in mind, it is now apparent that the cohesion at the point of initiation of a crack is zero, since the
prestress is exactly equal to the strength of materials at that
point. This implies that if the cohesion is the coefficient of a
singular function, and the cohesion is zero at the origin, it cannot be an intrinsic property of material, but must be measured by
the difference between the strength of materials and the prestress.
this difference being scaled by some function of the characteristic
dimension of the taper.

What has this idealized model of rupture to do with fractures
on polycrystalline materials or with fractures on real earthquake
faults? On the atomic scale the distance scale for the edge taper
or the width of the stress-distance curve of Fig. 1 must be of the
order of the atomic lattice constant. The range of a singularity
that varies as $r^{-\frac{1}{2}}$ cannot be significant for fractures in a perfect
crystal, if the fractures are as large as a millimeter, since this
is many orders of magnitude larger than atomic spacings. Does some
process exist in polycrystalline matter or earthquake faults that
simulates cohesion in ideal crystals? A qualitative indication of
the physics of fracture of geologically significant polycrystalline matter is given in the experiments of Griggs (5) on alabaster
(Figure 2). The axial strain under constant axial compressional
load stress as a function of time shows three distinct regimes of
creep. In the first, dislocations present in the material prior to
the experiment are swept out of the system or to grain boundaries;
the process later proceeds at a reduced rate when fewer dislocations remain in the solid. The regime of secondary creep is one of
almost constant viscosity and is due to a constant flux of dislocations through the sample, developed at the load-sample contacts or
at grain boundaries. The final or tertiary stage, is an accelerated
creep leading to ultimate fracture. Stress concentrations at crystal junctions are great enough to cause abrasion of the corners:
this acts as a source of dislocations that propagate along the interfaces radiating from the junctions. Linkage of these degrada-
tions of crystal boundaries results in ever increasing displacements that lead to ultimate catastrophic failure.

Faulting in materials such as alabaster takes place in narrow,
but not infinitesimal, shear zones. We shall rely on a model of
linkage of nearby, but initially decoupled cracks to describe the
process of generation of ever larger cracks. On the scale of earthquake faults, the geometrical detail that is found has
considerable complexity. Fault zones are not infinitesimal in

width. The projections of faults on the surface of the earth show
changes in directions, bifurcations, and echeloning. We may assume
the three-dimensional geometry is correspondingly complex. Recent
statistical studies of the distributions of earthquake loci show
that the simple planar model of an earthquake is inappropriate (6).
Certainly the distributions of earthquakes both in space and in
time show self-similarity on all scales (7,8). If self-similarity
exists even down to the microscopic level, then large scale phenomena such as felt earthquakes on faults are only scaled manifesta-
tions of the same process as fracture of a polycrystalline material
We can imagine that the irregularities and complexities of the
shear zone in a laboratory specimen are generated by the same
stochastic process that generates irregularities on large earthquake fault zones.

The role of grain junction points in the rupture process in
crystals or the role of echeloning or bifurcations or more complex
geometrical arrangements of cracks, is to provide barriers to the
rupture process. When the regional stress exceeds the strength of
the barrier, a fracture initiates. The fact that it involves a
gradual reduction in strength through the tertiary creep process

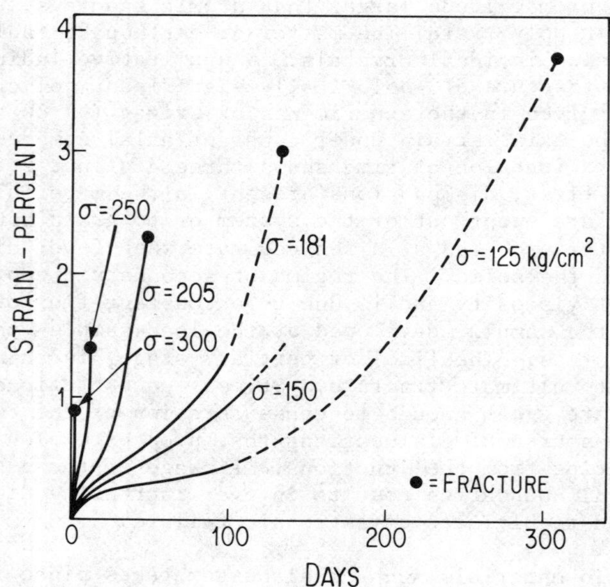

Fig. 2. Creep of alabaster in aqueous environment (after Griggs).
        Fracture occurs sooner and at smaller strain for larger
        load stress.

i.e., a weakening of the barrier, is of consequence to those who
would model the catastrophic part of the fracture. It is also of
importance for earthquake prediction, since it is this episode of
the pre-fracture history that is most likely to be accompanied by
changes in other physical and geophysical properties that may
presage an earthquake. Thus we can imagine that the triggering
of an event generates the following chronology: the stresses
generated in the rupture are added to the pre-stress and raise
the general level of stress near the advancing edge of a crack.
When the crack encounters one of the obstacles, such as a bifurca-
tion, an echelon discontinuity, etc., the rupture can break through
the high-strength barrier, if the total dynamical stress exceeds
the strength of the barrier. If not, the crack stops. The analogy
with fracture on a crystal at the atomic level is quite good,
since we now model the cohesion in terms of the stress required to
trigger motion on the continuation of an echelon structure, for
example, and a distance scale determined by the interval between
the two nearest edges of the echelon structure.

In fresh materials there is some debate whether a finite
strength exists. In metals, in which dislocations can move about
with relative ease, if finite strength is present, it seems to be
rather low. In the case of earthquake faults, where strength is
likely to be dominated by geometrical effects, we can certainly
take it as being non-zero; the magnitudes of the strengths and
cohesions are measured by the gaps that must be bridged or the
angles that must be turned, as well as by particle sizes and
orientations.

It will not have escaped notice that our models of stress are
scalar ones; the three-dimensionality of the true rupture surface
demands that we ensure that the full tensor character of the problems be elucidated. On the scale of grain sizes this is not possible in detail. We seek some mathematical simplifications. These
mathematical conveniences are hazardous and must be considered
cautiously.

Most mathematical models that have led to accessible solutions
have been for either one- or two-dimensional cracks. Fracture in
three dimensions may involve growth of a crack on one part of its
perimeter while on another part, growth may have terminated and
healing begun due to the encounter with an obstacle of high cohesive strength. Before embarking on a description of the properties
of these solved problems. I digress to discuss an alternative
fracture criterion to that of the condition of critical stress
described to this point.

IV.   Griffith Fracture Criterion

The criterion that a critical state of stress must exist for
a fracture to begin or continue (9) has not gone unchallenged.
An alternative criterion for fracture of a prestressed material is
attributed to Griffith (10) and involves an assessment of the prestress energy density available at the crack tip and the energy
density required for continuation of the fracture. If and only
if the first exceeds the second, the crack continues to extend.
Growth of the crack implies an energy flux from the region of prestress outside the crack into the crack through the crack tip.
This implies that the critical energy density for extension is a
characteristic quantity at any point along the fault. For static
cracks it can be shown that the two criteria are equivalent. For
dynamical one-dimensional cracks, it can be shown that the two
criteria are internally inconsistent (11,12) and lead analytically
to a statement that the critical time rate of energy density flux
into the crack is equal to the product of the critical cohesion
(shear stress per unit length) and one-half the particle velocity
immediately behind the crack tip. Since the particle velocity
behind the crack tip is dependent on the previous dynamical history
of faulting it follows that the two criteria cannot both be properties of the function of the crack tip only. In several cases of
two-dimensional shear cracks, Das and Aki (13) have found numerically that the two criteria lead to different crack histories.

I have considered the following numerical model, which I
think is relevant to this problem. I have considered numerically
the fracture of a discrete chain of masses and springs, in the
spirit of the model of Burridge and Knopoff (3), such that the
uniformly prestressed chain will fracture unilaterally, and with
a strength of the bonds holding the particles to the substrate
increasing linearly with distance from the origin. The one-
dimensional continuum limit to this problem has an exact solution
(11). What is found is that the discrete problem involves an oscillation of the particles about the continuum solution (Fig. 3).
The amplitude of the oscillation near the extending edge of the
crack is proportional to the value of the particle velocity in
the continuum solution; the oscillation decreases with distance
from the edge of the crack. The mean particle velocity is the
same as the continuum solution. It is evident in this case that
the available energy from the prestressed region in front of the
advancing crack edge goes partly into changing the kinetic energy
of the chain in a manner consistent with the continuum solution
and partly into kinetic energy of vibration of the lattice around
the continuum solution. In a Boltzmann sense, we have excited a
wave of elevated temperature that travels with the crack tip. In
this case the temperature in this wave grows with the advance of
the crack because of the extension into a region of ever increasing cohesion. Thus, in this example the energy budget flowing

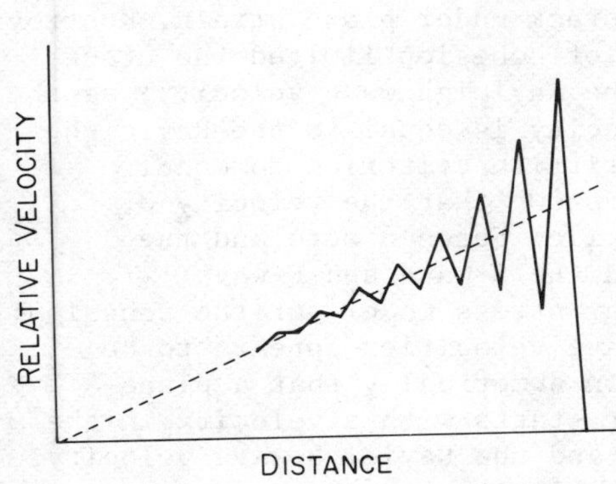

Fig. 3. Numerical particle velocity profile for a rupture
event in a discrete lattice
(solid) in comparison with the
corresponding problem for a continuum (dashed)

through the crack tip goes to
provide thermal energy, a term
which is not considered in establishing the Griffith criterion. Of course, we must now
ask what happens to this thermal
wave in the continuum limit.
Does the thermal wave disappear
in this limit, or is there still
a component of the energy that
goes into a singular thermal
pulse that travels with the
crack tip, hitherto ignored in
the application to the dynamical
case of a criterion derived from
consideration of static cracks?
Is there also a singular particle
velocity at the crack tip in the
continuum limit, which has
hitherto been ignored in the
problems using critical shear
stress conditions? No definitive answer to these questions
is available at this time. I conclude that the establishment of a
definitive fracture criterion for dynamical crack propagation is
still unsolved.

It is important to emphasize that either of the two criteria
for edge conditions are descriptive of the phenomenon of cohesion
and provide a method for stopping the extension of a crack. If
either the critical shear stress or the critical energy density
is too high, the crack will stop. Physically, the energy from the
prestressed region is used to break the bonds in the tapered region
of the crack tip, where the properties of materials are such as
to inhibit the futher propagation of cracks.

V. Dynamical Extension of Cracks

I now turn to a partial summary of generalizations that can
be made to date, derived from both analytical and numerical
solutions to problems of dynamical crack extension.

In the absence of cohesive forces at the crack tips, i.e., if
the prestress is exactly equal to the yield strength of the fault
contact everywhere along the length of the fault, or if negligible
energy is required to break the cross-fault bonds in the vicinity
of the crack tip, the crack extends unhindered by cohesion. The
stress drop behind the crack tip accelerates the particles
astride the ruptured segment. In this case of zero cohesion, the
rate of extension of a one-dimensional crack is exactly that of
sound waves on the elastic chain, and it is that of S-waves in

the case of a two-dimensional anti-plane shear crack. In the
case of a two-dimensional shear crack under plane strain, Kostrov
(14) indicated that the presence of cohesion limited the crack
velocities to values less than the Rayleigh wave velocity; as the
cohesion vanishes, the crack velocity is equal to the Rayleigh
wave velocity. Kostrov used a Griffith criterion to model
the cohesion. Burridge (15) has shown that the velocity of
extension of plane-strain cracks lies between zero and the
Rayleigh wave velocity or between the S-wave and P-wave
velocities, using a critical shear stress model for the cohesion;
the gap between Rayleigh and S-wave velocities appears to be
forbidden. Andrews (16) has shown numerically that a plane-
strain two-dimensional crack that starts with a velocity in the
lower region, i.e., between zero and the Rayleigh wave velocity,
accelerates to the Rayleigh wave velocity and then may jump into
the upper interval of rupture velocity. Burridge, et al. (17)
have used a critical shear stress criterion for the cohesion and
have shown that cracks initially in the rupture velocity range
between $2^{\frac{1}{2}}V_s$ and V_p or between 0 and $V_{Rayleigh}$ are stable in the
sense that their rupture velocities will increase with increasing
ratio of stress drop to cohesion: if the rupture velocities
are in the range between V_s and $2^{\frac{1}{2}}V_s$ they will slow down (speed
up) with increasing (decreasing) stress ratio and move into the
lower (upper) stable region.

VI.   Velocity of Extension

     I have remarked that the mechanism for stopping the further
extension of a crack is due to the encounter between the crack
and an "unbreakable barrier". There is another mechanism for
stopping the extension. If the crack moves into a region of
negative stress drop, wherein the dynamical friction exceeds the
prestress, the crack edge starts to decelerate and can ultimately
come to rest. An example of the history of an antiplane crack of
this type, with sharp edges, is given by Burridge and Halliday
(18). It is noteworthy that the static stresses at the edges of
the crack, after termination of all particle motions, have no
singularities; the stress beyond the edge, after the crack has
healed, varies as $r^{\frac{1}{2}}$, a result consistent with the assumption
that the cohesive force is everywhere zero. The absence of
singular stresses at the edges in the case of cracks stopping
solely under the influence of negative stress drops argues
against it, and in favor of cohesive forces as a mechanism for
causing aftershocks (19). Thus the presence of aftershocks is
perhaps the most powerful argument in favor of cohesion as an
edge condition for extension of a crack. The physical significance
of the absence of cohesion is that the prestress is equal to the
strength over the entire range of extension of the crack. Conversely, the presence of cohesion implies that the strength is

greater than the prestress and that this difference must be supplied by the stresses at the edge of the crack in order that the
crack continue to extend.

The influence of cohesion on the rate of extension of cracks
is to reduce the velocity from the limiting values indicated above.
It can be shown that if the spatial gradient of the cohesion is
large enough, a shear crack will not initiate even though the
prestress be at the critical value for rupture (11).

The mathematics of shear cracks with and without cohesive
forces has been restricted to consideration of problems in which
a jump discontinuity of the tangential component of particle displacement occurs across a plane surface of zero thickness. These
cracks are presumed to end abruptly on a moving curve in the plane
of the fault. Beyond the edge of the crack, the jump in the displacement across the fault plane is zero. Everywhere except at
the plane of the crack itself, the equations of linear elasticity
are assumed to hold. These conditions of a system of linear
partial differential equations, with boundaries that move macro-
scopically, give rise to problems that are non-linear· the proto-
type of non-linear problems arising from linear systems with moving
boundaries is the Stefan problem, originally proposed in the preceding century to discuss the conjunct thermal and mechanical
equilibration of what is usually an isochemical, two-phase system
in disequilibrium. The greatest progress in the solution of problems of the Stefan type other than direct numerical integration,
has come from the use of Green's functions to construct convergent
iterations to the integral equations. Green's function methods
have proved to be powerful and valuable techniques in obtaining
analytical solutions to the problems of growth of cracks as well.

Suppose that we have an earthquake fault that is prestressed
exactly to the local yield strength B everywhere along its length.
In this case the cohesive forces are zero everywhere since the
dynamics of rupture need generate no additional stresses (or
energy) to overcome the barrier of finite strength. The discussion that has been given in connection with Figure 1 implies that
the shear stress that opposes the motion cannot drop instantly
from the yield strength to the dynamical friction. Andrews (16)
has proposed that a condition of gradual reduction of resistive
stress be imposed in an effort to make the frictional model more
realistic. To model this, let us make the stress drop f on the
ruptured part of the fault a function of the relative displacement
$x-x_c$ in consonance with the right hand part of Figure 1b, for
the region $x>x_c$. Let us assume $f(0)=0$. Then such a crack will
never initiate at a point (19). The problem is similar to the
unstable equilibrium of a pendulum pointing exactly vertically
upwards. It cannot move; if displaced infinitesimally, it
falls in the usual way. In our case, the crack will grow from a

point if $f(0)$ differs from zero by an infinitesimal quantity. Wher
the pendulum in our analog begins to move with infinitesimal velocity for an infinitesimal initial displacement, the crack begins
to move with the sonic (or S-wave) velocity for an infinitesimal
initial stress drop. This latter result is in consequence of our
statement that in the absence of cohesion, cracks extend at sonic
velocities. However, in this case, the displacements behind the
crack edges will be infinitesimal.

The effect of non-zero cohesion is to reduce the rate of extension of shear cracks. In the cases of one-dimensional and anti-
plane two-dimensional shear cracks, the situation is clear. Knopof
et al. (11) have shown that such cracks extend with subsonic velocities that are regulated by the cohesion; the cracks may be accelerated or decelerated by extension into a region where the cohesion
decreases or increases; some special cases can be found in which
the cracks extend with constant suubsonic velocity. The crack
will not initiate if the gradient of the cohesion is too large.
Once started, these cracks will stop if the gradient of the cohesion is too large or if a sufficiently strong barrier in the
cohesion is encountered.

With considerably greater mathematical difficulty, Knopoff
and Chatterjee (21) have made a calculation of the fracture history
of two-dimensional anti-plane shear cracks with cohesion at the
crack tips, and have been able to show that the rate of advance
of these cracks is also less than or equal to the S-wave velocity,
and that these cracks can also be limited in their growth if the
cohesion becomes too large. As in the one-dimensional cases
these cracks can accelerate, decelerate or move with constant sub-
shear wave velocity.

In the case of two-dimensional plane-strain shear cracks,
the demonstration is considerably more difficult. Burridge et
al. (17) have given an analysis of the properties of the extension
of semi-infinite cracks under the influence of cohesive forces;
I have described some of the kinematic properties of the edge
above; semi-infinite cracks can be imagined as prestressed semi-
infinite sawcuts that are maintained in some deformed state up to
t=0, at which time they are released and allowed to extend.
Since they are initially in some non-equilibrium state, these
cracks begin their advance with some initial velocity and have a
subsequent history that depends on the equilibration of the
stresses and/or displacements along the entire extend of the pre-
existing crack surfaces. Chatterjee and Knopoff (22) have considered two-dimensional plane-strain shear cracks initiating at a
point under an influence of cohesive stresses that vary with distance as $r^{\frac{1}{2}}$. We find, as with the cases of one-dimensional

cracks and two-dimensional anti-plane shear cracks, that plane-strain two-dimensional cracks can accelerate, decelerate or move at constant crack velocities in the regime of crack velocities less than the Rayleigh wave velocity. For small cohesive forces, the cracks speed up and for large cohesive forces the cracks slow down; these results are consistent with both the results quoted by Andrews (16) and Burridge et al. (17). For cracks moving with constant velocity and initiating at a point, we have been unable to produce cracks that will move faster than the Rayleigh wave velocity, a result not inconsistent with that cited by Kostrov (14). Because of our inability to generate cracks with super shear wave speeds that initiate spontaneously, we are not able to comment at this time on the properties of cracks in the higher crack velocity regime.

We speculate (at this time in the absence of a detailed investigation) that our inability to produce plane-strain shear cracks that initiate at a point, and move with a constant super-shear velocity is due to back-edge radiation effects. In the presence of cohesion, finite two-dimensional cracks advance into a region that is not only in a prestressed state due to the static field present prior to the rupture but also the edge advances into a region dynamically stressed by the elastic wave radiation mainly from the advancing edge, as well as from the opposite edge of the crack; in fact radiation is developed by all points between the two edges. The coupling between the two edges lends additional complexity to the already non-linear mathematical character of the problems. The complexity can be moderated by introducing an approximation for the dynamical state of stress beyond the opposite edge of the crack, which is that this stress varies as $\eta^{-\frac{1}{2}}$ near the opposite edge and vanishes on the elastic wave front beyond the opposite edge, where η is the characteristic coordinate for the edge, $\eta = (x \pm V_s t)/2^{\frac{1}{2}}$. The introduction of characteristic coordinates into the problems of dynamics has been made by Kostrov (14) and has been of noteworthy value in assisting in the solution of the non-linear integral equations.

In general, the back-edge radiation in the case of anti-plane shear cracks serves to reduce the velocity of extension of such cracks even more than such velocities are already reduced by cohesion. These effects are most pronounced when the velocity of extension of the "forward edge" of the crack is slow. The effects, though small, are not negligible (21) and we speculate that they may be sufficient to prevent the Rayleigh wave barrier for rupture velocities from being penetrated, although the barrier is indeed penetrable for semi-infinite cracks. i.e. cracks with no back edge radiation.

VII. Termination of Extension and Cessation of Motion

The encounter of a growing crack with an unbreakable barrier
initiates the process of suturing of the walls of the crack. We
must not forget that the unbreakable character of a barrier depends
on the momentum developed by a crack: in general, a small crack
will be unable to break through a large barrier, while a larger
crack may have enough stresses concentrated at its tip to break
through the barrier. After the encounter with the barrier, in
general a wave of suturing moves away from the barrier, progressively freezing the walls of the crack in the position of relative
displacement at the moment of arrival of the "freezing wave". The
condition for freezing that most mathematicians use is the condition that the particle velocity be zero. Unfortunately this is
not a necessary physical condition for freezing; it merely provides for the prevention of overshoot, that is the reversal of the
direction of relative particle motions on the crack faces. There
is no reason why overshoot cannot occur, except that in many cases
we may expect that the accompanying reversal of the direction of
dynamical friction will impede the reversal of motion; overshoot
is least likely to occur if the dynamical friction is a substantial
fraction of the prestress. We note that the freezing condition,
as stated above, is a kinematical condition, while the rupture
criteria we have described are dynamical conditions. A lower
bound for the freezing wave velocity is the sonic velocity in one
dimension and the S-wave velocity in two-dimensional antiplane
cases. The upper bound is an infinite velocity of freezing, i.e.,
freezing takes place simultaneously over the entire extent of the
fault.

Freezing is triggered by a stress wave that travels from the
obstacle back into the already ruptured fault segment. The velocity of the stress wave is at least the S-wave velocity. The
freezing wave velocity may be greater than the S-wave velocity if
the particles were already about to come to rest, due to a negative stress drop; the arrival of a freezing event merely hastens
the process.

If a crack encounters a change in the gradient of the cohesion
or a step change in the cohesion, it may tear through these obstacles. Nevertheless a stress wave will be returned into the interior of the crack, with the S-wave velocity, which may initiate
freezing at an interior point. In one- or two-dimensions, interior
freezing will cause fission of the fault to occur and the two
fragments will then contract separately.

In figure 4a, the locus of points that are in relative motion
for a unilateral fault event is shown. The fault tears subsoni-
cally along the line OA. At coordinate $x = L$ it encounters an
unbreakable obstacle and freezing is initiated which progresses

along the sonic line AT. At time t_o all motion ceases. The
source-time function for the earthquake is presumed to have lasted
t_o.

In Figure 4b, a locus of rupture is sketched for a bilateral
fault event that tears sonically initially. Again as "breakable"
cohesion barriers are encountered at L and L' the velocity of rupture is decreased. The maximum extent of rupture is AB; these
points are either the sites of unbreakable barriers or the loci
where the fault has decelerated to the extent that it has ex-
hausted its stress drop fuel to propel it further. During the
encounter with the obstacles at L and L', stress waves propagate
into the interior and collide at time t_1 which initiates a process
of fission. The two fragments then contract separately with cessation of motion of each taking place at times t_2 and t_3.

The abrupt cessation of extension as at A in Figure 4a generates   a strong stopping phase. The termination of motion at
point T generates an extinction phase. The seismic signal at a
distant point has a duration greater than t_o in two-dimensional

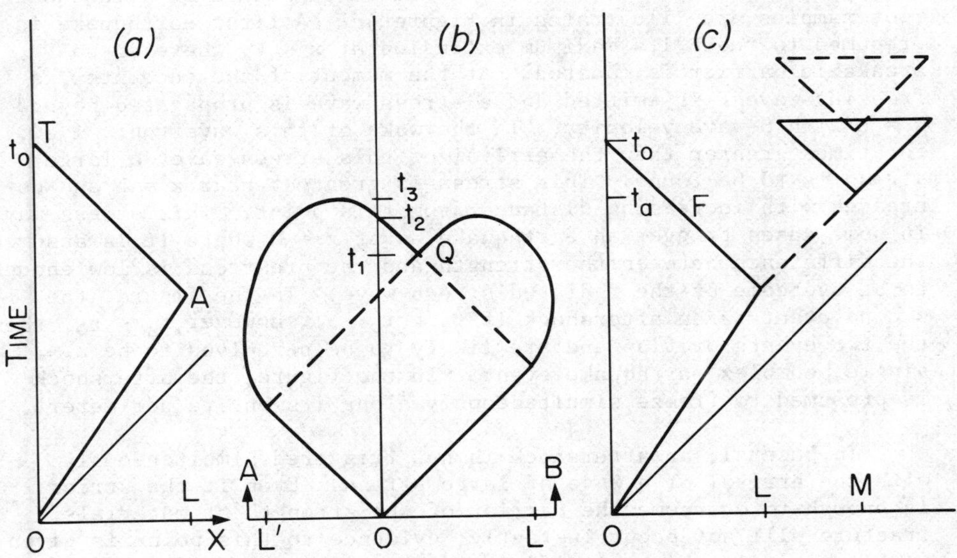

Fig. 4. Rupture loci: a) unilateral fracture, b) bilateral fracture with fission source freezing initiating at Q, c) aftershock
triggered at M by stress wave from L or delayed (dashed)
due to creep or stress corrosion.

cases and is in fact infinite because of the infinite tail of the
two-dimensional Green's function. The high frequency spectra of
the seismic signals in one- and two-dimensions are dominated by
the stopping phases, because it is at the loci of abrupt termination of extension that the maximum particle velocity is likely to
be found; at the origin of motion the motion is likely to be small,
partly because the size of the fault "antenna" is small near the
origin. Stopping phases are likely to be considerably less important features of three-dimensional faulting events because the
maximum outward extension of the fault is not likely to occur
simultaneously over the entire perimeter (23).

As already remarked, a high-cohesion site such as $x = L$ in
Figure 4a will be found to be the point of singularity of a large
static stress concentration. For $x \gtrsim L$ a large stress concentration is to be found for times $\infty > t > t_L$. For times $\infty > t > t_o$
the stress is low in the interval $0 < x < L$.

VIII. Aftershocks

High cohesion barriers provide large static stresses beyond
the edges of a fractured segment, as at L in Figure 4a. These
large residual stresses can be invoked as the cause of aftershocks.
Two examples are illustrated in Figure 4c. A first earthquake is
presumed to reach its maximum extension at $x = L$, where an unbreakable barrier is located. At the moment of the encounter, a
freezing wave F is emitted and a stress wave is propagated beyond
$x = L$ with S-wave velocity. In the wake of this wavefront, i.e.,
for times greater than the arrival of this stress wave; a large
stress is to be found. This stress is greatest near $x = L$ and decreases with increasing distance from this point. This stress may
in some cases trigger an earthquake as at $x = M$ where it is assumed
the difference between the strength and the prestress is low enough
to be overcome by the radiated stress wave. In the figure, the
second event is an aftershock if $t_a > t_o$. If however $t_a < t_o$, then
the two events overlap and are likely to be perceived to be a
single, complex earthquake event. In the figure, the aftershock
is presumed to freeze simultaneously along its entire perimeter.

In general, an aftershock is not triggered simultaneously
with the arrival of a wave of large stress. Even if the stress
is enough to overcome the barrier of the strength of materials,
fracture will not occur instantly; evidence to this point is given
in Figure 2, in which significant delays of rupture occur following the application of what is presumed to be a super-yield stress.

It may be a restatement of the same phenomenon to indicate
the possibility that some form of erosion of the strength of the
materials can take place after the application of a sustained

sub-critical stress. We suspect the strength decreases largely
because of the agency of intergranular water.

IX.   Concluding Comments

     It is my belief that the complexity of the fracture process
that has been described above, leads one inescapably to the conclusion that a priori assumptions regarding the nature of the
fracture history -- including initiation, extension, and con-
traction, are probably going to be inappropriate if the purpose
is to describe such things as source-time functions and their
spectra. Among the assumptions that have been made are the following: that cracks may be assumed to fracture with the S-wave
velocity, that they tear with some other constant velocity, that
the freezing process is not too important, that the process of
rupture is a smooth one, etc. The problems of fracture dynamics
are extremely complex, partly because of the non-linearity of the
process, and partly because of the richness of the variety of the
physical processes that are likely to influence the rupture history
and hence the seismic signal. My personal prejudices are that the
source must be treated as a complex, multiple event in which the
complexity arises due to the variation of physical properties
along the fracture surface, both at the advancing edge and behind
it. The treatment of three-dimensional sources is important. To
date we do not have the capability to do more than handle these
sources numerically, and because of the non-linearity, it is difficult to draw generalizations from the numerical solutions. I
further believe that the processes of cohesion and the nature of
the strength of materials are strong influences on the onset,
extension and termination of fracture, features which adds considerable complexity to the analysis. There are rich problems of
investigation for students willing to take the risk of venturing
into a difficult area.

     With regard to the problem cited at the start of this paper,
we can assert that it is indeed possible to derive the seismic
moment from the spectral properties. The other two source parameters still elude us.

References

1. Brune, J.N.:   1970, J. Geophys. Research 75, pp. 4997-5009.

2. Reid, H.F.:   1910, in California State Earthquake Commission
   Report, The California Earthquake of April 18, 1906, 2,
   Carnegie Institution of Washingon.

3. Burridge, R. and Knopoff, L.:   1967, Bull. Seismol. Soc.
   Amer. 57, pp. 341-371.

4. Griggs, D.:   1940, Bull. Geol. Soc. Amer. 51, pp. 1001-1022.

5. Barenblatt, G.I.:   1962, Adv. Appl. Mech. 7, pp. 55-129.

6. Kagan, Y.Y.:   in press, Geophys. J. Roy. Astron. Soc.

7. Kagan, Y.Y. and Knopoff, L.:   1980, Geophys. J. Roy. Astron.
   Soc. 62, pp. 303-320.

8. Kagan, Y.Y. and Knopoff, L.:   in press, J. Geophys. Research.

9. Irwin, G.R.:   1957, J. Appl. Mech. 24, pp. 361-364.

10. Griffith, A.A.: 1920, Phil. Trans. Roy. Soc. A221, pp. 163-198.

11. Knopoff, L., Mouton, J.O. and Burridge, R.:   1973, Geophys.
    J. Roy. Astron. Soc. 35, pp. 169-184

12. Burridge, R. and Keller, J.B.:   1978, SIAM Review, 20,
    pp. 31-61.

13. Das, S. and Aki, K.:   1977, J. Geophys. Research, 82,
    pp. 5658-5670.

14. Kostrov, B.V.:   1966, J. Appl. Math. Mech. (English
    Translation) 30, pp. 1241-1248.

15. Burridge, R.:   1973, Geophys. J. Roy. Astron. Soc. 35,
    pp. 439-455.

16. Andrews, D.J.: 1976, J. Geophys. Research 81, pp. 5679-5687.

17. Burridge, R., Conn, G. and Freund, L.B.:   1979, J. Geophys.
    Research 85, pp. 2210-2222.

18. Burridge, R. and Halliday, G.S.:   1971, Geophys. J. Roy.
    Astron. Soc. 25, pp. 261-283.

19. Knopoff, L.:   1972, in Flow and Fracture of Rocks,
    Geophysical Monograph 16, American Geophysical Union,
    pp. 259-263.

20. Landoni, J.A. and Knopoff, L.:   1981, Geophys. J. Roy.
    Astron. Soc. 64, pp. 151-161.

21. Knopoff, L. and Chatterjee, A.K.:   in press, Geophys. J.
    Roy. Astron. Soc.

22. Chatterjee, A.K. and Knopoff, L.:   in preparation.

23. Blandford, R.:   Personal Communication.

DYNAMICS OF SEISMIC SOURCES

Raul Madariaga

Institut de Physique du Globe de Paris and
Departements de Sciences de la Terre, University of
Paris, France

ABSTRACT

The term dynamic source models usually refers to seismic source
models based on fracture mechanics. We study in this paper the
relation between the source rupture history and the far-field
radiation on the basis of a few simple fracture models. A plane,
two-dimensional crack model is used to show the main features of
crack models and of the body waves radiated at the initial and
final moments of rupture. Two-dimensional models are not very
satisfactory for the radiation of seismic waves because the
finiteness of the fault is one of the main factors controlling
the farfield signals. We study then a circular crack model
and its radiation; this is the model most seismologists use to
invert source parameters. Finally, we review some recent propo-
sitions to explain earthquake complexity and multiple events.
Seismic moments, stress drops and corner frequencies may be
strongly affected by source complexity.

1.   INTRODUCTION

It is well established that shallow earthquakes are due to
brittle fracture along more or less planar fault surfaces.
Theoretical models of the rupture process have evolved considerably in the last twenty years, from the early point couples
(or double couples) up to sophisticated numerical simulations of
fault slip in three space dimensions. The purpose of these notes
is to review a subset of the seismic source models, the so-called
dynamic or crack models based on linear fracture mechanics. The
earliest work was on simple two-dimensional or circular shear
cracks for which analytical solutions could be found (1,2,3,4).

71

*E. S. Husebye and S. Mykkeltveit (eds.), Identification of Seismic Sources - Earthquake or Underground
Explosion, 71–96.*
Copyright © 1981 by D. Reidel Publishing Company.

These solutions were very useful to understand the essential
features of dynamic fracture: the role of stress concentrations,
energy absorption at the rupture front, 'correct' slip functions,
etc... However, they neglected the finiteness of the earthquake
sources. A finite crack is a multiple diffraction problem for
which no analytical solution techniques exist. Recently, a
number of authors have proposed numerical methods to study shear
fracture (5,6,7,8), with these techniques it should in principle
be possible to study quite complicated source geometries and
stress release histories. Yet a number of important basic problems
remain on the proper way of modelling the spontaneous growth of
rupture (9,10). Even if these problems were solved, a basic
question remains. The rupture zone being inaccessible to direct
observation, all information about the rupture process has to
be deduced from seismic radiation in the near or far field. Yet,
seismic waves lose a significant part of their information in the
trajectory to the seismic station because of multiple reflection,
seismic attenuation and instrument response. For instance, when
observed at long period WNSSN stations, the body waves radiated
by two very different fault models – a circular shear crack and
a rectangular dislocation – are practically indistinguishable.
This lack of resolution of far–field observations explains why
the simple dislocation models are so successful even if we know
that they contain a number of unphysical features. Given this
essential non–uniqueness of the inverse source problem, how
can fracture models help to understand the physical processes
at the seismic source Knopoff (11) reviews the work done on
fracture models and the mechanics of rupture growth. We shall
concentrate on the radiation of seismic waves and the relation
between these and the rupture processes at the source.

2.   SEISMIC RADIATION

Radiation from a seismic source is a classical problem in elasto-
dynamics. The most recent developments include the study of
generalized distributed seismic sources (12), which we shall
review briefly and in an overly simplified fashion.

     Let $\underline{f}(\underline{r},t)$ be a general distribution of body forces in the
earth, where $\underline{r}$ and t are position and time, respectively. Seismic radiation may be easily calculated from a representation
theorem. The displacement $\underline{u}$ is given by:

$$u_i(\underline{r},t) = \int_{-\infty}^{t} dt \int_V G_{ij}(\underline{r},t|\underline{r}_o,t_o)\, f_j(\underline{r}_o,t_o)\, dV \qquad (1)$$

where V is the volume of the earth, G_{ij} is the elastodynamic
Green tensor (13) for the earth, which may include all near

and far-field waves. The problem now is to find the body force
distribution equivalent to a realistic seismic source. Let
$\mathcal{M}(\underline{r},t)$ be a seismic moment tensor distribution corresponding
to a stress glut as defined by (12), or to inelastic stress in
the more traditional nomenclature. The body force distribution
equivalent to this moment tensor field is:

$$\underline{f} = -\nabla\mathcal{M} \tag{2}$$

a relation that expresses the equilibrium of forces. Inserting
this definition of $\underline{f}$ in the representation theorem and integrating
by parts assuming that $\mathcal{M}$ is different from zero only for $t > 0$
and in a finite region we find:

$$u_i(\underline{r},t) = \int\limits_{-\infty}^{t} dt \int\limits_{V} G_{ij,k}(\underline{r},t|\underline{r}_o,t_o)\,\mathcal{M}_{jk}(\underline{r}_o,t_o)\,dV \tag{3}$$

where the comma indicates differentiation with respect to x_k.
From the symmetry of $\mathcal{M}_{jk}$ we see that the radiation from a
point moment tensor source is equivalent to a set of dipoles.
A seismic fault may be described as a simple layer distribution
of moment tensors of the form:

$$\mathcal{M}_{jk}(\underline{r},t) = \mu(\Delta u_j n_k + \Delta u_k n_j)\,\delta(D) \tag{4}$$

where $\delta(D)$ is a surface Dirac delta function, $\Delta\underline{u}$ is the slip
at the fault and n is the normal to it at the point $\underline{r}_o$.

Inserting this in the representation theorem we obtain:

$$u_i(\underline{r},t) = \int\limits_{S} \mu\,\Delta u_j *(G_{ij,k} + G_{ik,j})\,n_k\,dS \tag{5}$$

where the time integration has been replaced by the convolution
indicated by *. Using the true Green tensors for the earth is
a formidable problem so that we usually take only a part of it,
either surface waves or body waves. In the near field, where the
different waves are not well separated, the complete Green tensor
has to be used and only numerical solutions are possible. The
difficulty of performing these integrations and the interference
of source information with the structure has prevented seismologists from obtaining simple relations between near-field records
and source processes. Most of the information comes at present
from the far field, and we shall direct our efforts to its
study.

In the far field the Green function takes a much simpler
form. First, P and S waves are uncoupled so that they may be

approximation in which the source appears to radiate only plane
waves.

In a homogeneous medium the far-field waves for plane fault
may be written in the form:

$$u(\underline{r},t) = \frac{1}{4\pi\rho c^3} \, \mathcal{R} \, \frac{1}{|\underline{r}|} \, \Omega \, (\hat{R}, t - \frac{|\underline{r}|}{c}) \tag{6}$$

where ρ is density, c is either P or S wave velocity according
to the type of wave under consideration, $\mathcal{R}$ is the radiation
pattern and Ω is the source time function containing all the
information about the source slip (13). This expression may
be generalized to waves obeying simple geometrical optics, re-
placing the geometrical decay term $|\underline{r}|^{-1}$ by the calculated
one. $\mathcal{R}$ is the radiation pattern which depends on the take-off
angle of the ray at the source and its azimuth. In actual source
studies the far-field waves are also affected by attenuation,
instrument response and reflections. Of the latter, the most
important ones are the near-source surface reflected waves
pP, sS, sP and pS. In the following we shall assume that we have
corrected for these propagation effects and we shall concentrate
on the far-field pulse Ω.

$$\Omega(\hat{R},t) = \mu \int_S \Delta\dot{u}_i(\underline{r}, t + \frac{\underline{r}\cdot\hat{R}}{c})dS \tag{7}$$

where $\hat{R}$ is a unit vector in the direction of the observer.
Most properties of the radiation in the far field may be deduced
from this equation. $\Delta\underline{\dot{u}}$ is the slip velocity at the source. For
three-dimensional sources $\Delta\underline{\dot{u}}$ is usually a vector with two components, but it is frequently assumed by seismologists that
$\Delta\underline{\dot{u}}$ is perfectly parallel to the principal shear traction on the
fault plane. In fact, in most numerical calculations of shear
faulting in three dimensions, small components of slip perpendicular to the shear traction are usually found. These
small transverse components of slip are present even in the
simple circular shear crack. They are however small and in the
following we shall assume that they do not significantly affect
far-field radiation, so that we shall simply write $\Delta\dot{u}$ for $\Delta\dot{u}_x$.
The far-field pulse Ω as defined in eq. (7) is an integral of
the slip velocity at the fault, therefore from a single observation of a full seismogram one cannot invert the slip velocity
at the fault. Kostrov (14) has shown that this is not possible
even if we know the seismogram pulses radiated in all directions.

Let us introduce now the far-field radiated spectrum by
means of the Fourier transform

$$\hat{\Omega}(\hat{R},\omega) = \mu \int \Delta\dot{u}(\underline{r},\omega)\, e^{\,i\omega/c\ \hat{R}\cdot\underline{r}}\, dS \qquad\qquad (8)$$

where ω is the circular frequency. Some simple, but very useful properties of the seismic spectrum are the following:

1)   for $\omega \to 0$ the spectrum is independent of $\hat{R}$ and:

$$\hat{\Omega}(\hat{R},0) \to \mu \int \Delta\dot{u}(\underline{r},\omega=0)\, dS \qquad\qquad (9)$$

$$\to M_o$$

Since

$$\Delta\dot{u}(\underline{r},\omega=0) = \Delta u_i(\underline{r})$$

That is, the low frequency limit of the spectrum is the static seismic moment M_o, actually in our case the $\mathcal{M}_{zx}$ component of the static seismic moment (16).

2)   If $\Delta\dot{u}_x$ does not change sign (17) on the fault, then $\hat{\Omega}(\hat{R},\omega)$ takes its maximum at $\omega = 0$ and:

$$M_o \geqslant \hat{\Omega}(\hat{R},\omega) \qquad\qquad \text{for all } \omega \qquad\qquad (10)$$

3)   At high frequencies the behaviour of Ω is controlled by the singularities of $\Delta\dot{u}(\underline{r},t)$. For complex rupture histories a number of high frequency asymptotes of the form $\omega^{-\gamma}$ should superpose yielding a complicated envelope.

4)   The corner frequency is the intersection of this asymptote with the low frequency level. Brune proposed, on the basis of a simple source model, that the corner frequency is inversely related to the dimensions of the source (18, 19); this has been used practically to estimate source dimensions. The problem is that these simple theories are used to interpret complex source data and the relation between corner frequencies and signal duration in that case may be more complicated.

The high frequency properties of the spectrum are in fact properties of the discontinuities of the seismic signals; when a pulse has a number of these discontinuities, the high frequency behaviour is controlled by a complex interplay between the asymptotes associated with each one of them. For a number of single, unipolar far-field pulses that include triangles, rectangles, half circles, Brune's far-field pulse, etc., we have found that the corner frequency is very approximately given by

$$f_o = \frac{1}{\pi \langle T \rangle}$$

(11)

where $\langle T \rangle$ is the half-width of the pulse (see Figure 1).

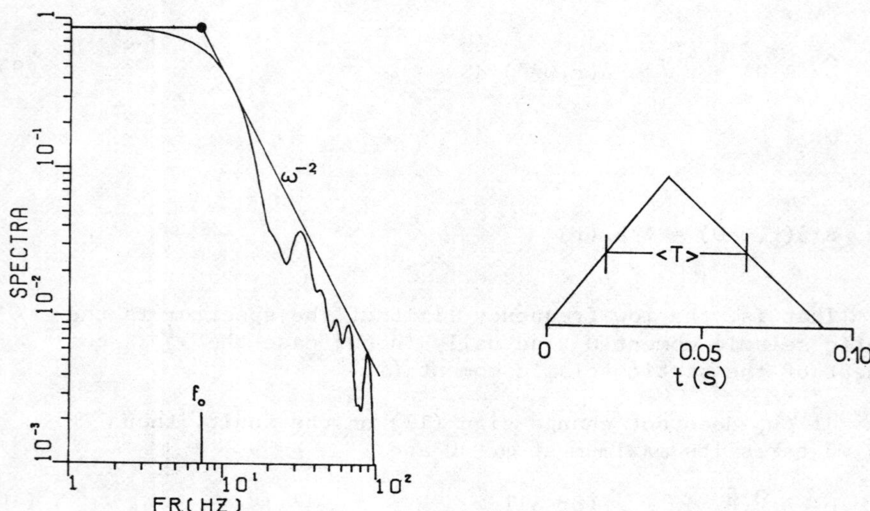

Figure 1. A simple triangular impulse and its spectrum. The corner frequency f_o is related to the half-height pulse width as in eq. (11).

3.   CRACK AND DISLOCATION MODELS

Radiation, as seen in the previous section, is proportional to slip velocity on the fault. The simplest models of the source process assumed that slip had a simple form; it was the same at every point on the fault and its time dependence was a step function with a finite rise time. In these models the rupture process itself was also specified kinematically in the form of a fixed geometry and rupture velocity. Thus no fracture mechanical information was incorporated in these models except that the result of rupture was the slip on the fault. At long periods, where details of the slip function are not important, radiation is essentially controlled by the seismic moment and the corner frequencies. Since the dislocation models do also have corner frequencies associated with the source dimensions, they are basically undistinguishable from crack models at

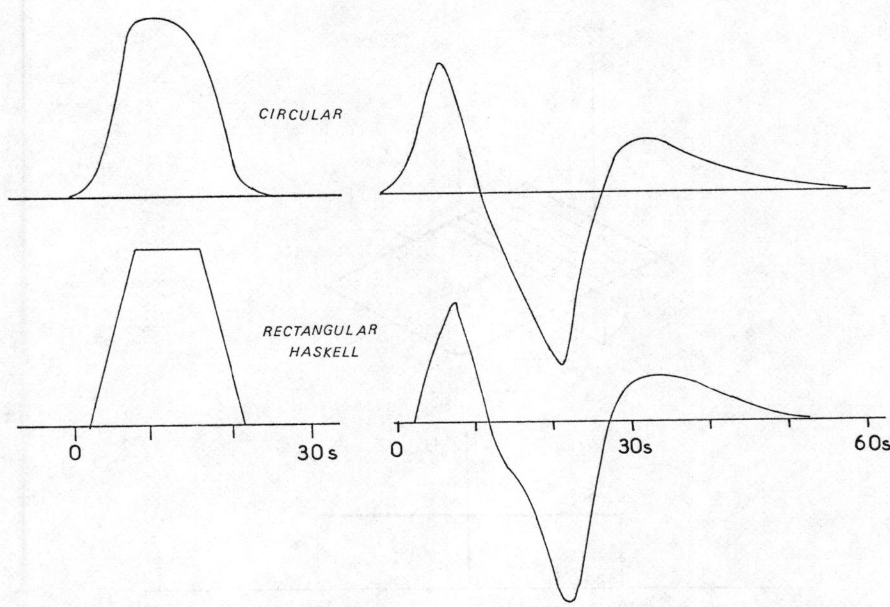

Figure 2. Comparison of synthetic far-field seismograms for
a trapezoidal and a circular crack pulse. The instrument is
a long period WWSSN instrument.

periods of the order of the corner frequencies and longer. As
an example in Figure 2 we compare the synthetic long period
seismograms obtained from a Háskell dislocation model and a
circular crack model, for a large M = 7.5 earthquake of
northern Chile. There is obviously no way to decide between
these two models from this type of observation.

The dislocation models are associated with very strong
singularities which are physically unacceptable. Stresses near
the edges of the dislocation behave like r^{-1} where r is the
distance from the edge. This is shown in Figure 3 for the case
of Háskell's rectangular dislocation. What is even more serious
is that the vertical displacement u_z associated with the dislocation has a logarithmic singularity at the edge, that is,
the fault edge moves vertically to infinity! The solution to this
paradox is well known in dislocation theory: the elastic solutions are not valid inside a core region around the edges of
the dislocations. In order to calculate correct solutions we
have to eliminate these singularities by considering the rupture
process itself. This is precisely what fracture mechanics does.
The essential result of crack models is that slip near the edges
has to have an $r^{\frac{1}{2}}$ behaviour and the stress concentrations have

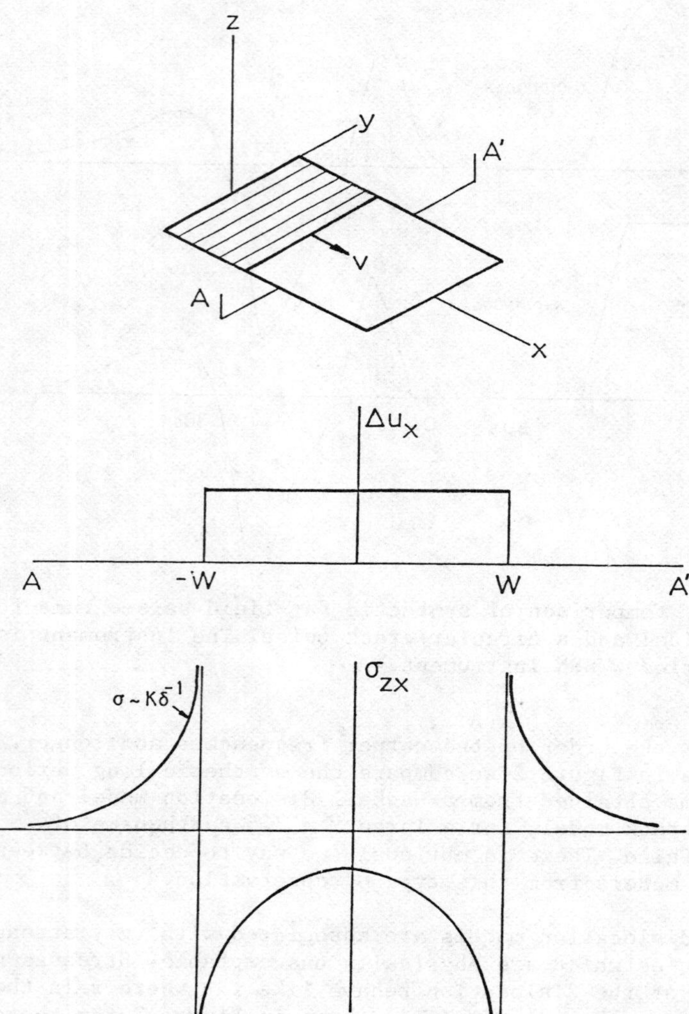

Figure 3. Haskell's rectangular dislocation model. At the top is shown the geometry of the rupture. At the center is the slip on the AA' cross-section. At the bottom is the stress change on the same section.

and $r^{-\frac{1}{2}}$ type of singularity. At an even finer scale these
singularities do also have to be smoothed out taking into account the cohesive forces acting near the crack tip (11, 20, 21).
The $r^{\frac{1}{2}}$ type of smoothing of the slip on the crack differs from
the constant slip assumed in dislocation models at scales of the
order of the dimensions of the fault. It is clear then that
conclusions about high frequency radiation drawn from dislocation
models are at least suspect if not wrong. We conclude that crack
models are essential to an understanding of high frequency radiation from seismic sources.

What is a crack model then? The fundamental feature of a
crack model is that stress drop inside the crack, that is, the
difference between the initial traction before the earthquake
and the final traction is finite everywhere <u>inside</u> the crack.
The main features of the stress, slip velocity and slip fields
near the crack tip are determined from these simple requirements
(14,22). For a plane shear crack advancing at an instantaneous
rupture velocity v the stress singularity ahead of the crack tip
may be written:

$$\sigma_{xz}(x,t) = A(v)\ K^{\dagger}\ (\ell(t),t)(x-\ell(t))^{-\frac{1}{2}}\qquad x > \ell(t) \qquad\qquad (12)$$

where $\ell(t)$ is the instantaneous position of the crack tip, $A(v)$
is a function that depends only on the instantaneous rupture
velocity $v = \dot{\ell}(t)$ and $K^{\dagger}$ is a part of the stress intensity
factor that depends on the stress drop and dimensions of the
crack. For a semi-infinite crack $K^{\dagger}$ is equal to the stress intensity of a static crack with the same stress drop and the
tip at $\ell(t)$. $A(v)$ is a very complicated function but very roughly
it behaves like $1-v/c_R$ where c_R is Rayleigh's wave velocity.
Stress intensity decreases steadily with rupture velocity. This
is the source of instability of brittle cracks: as the load (the
stress intensity $K^{\dagger}$) grows, the cracks accelerate to the limit
velocity c_R. It is remarkable that the stress concentration is
independent of the instantaneous acceleration, this is frequently interpreted as a lack of inertia of the crack tip. Subject to abrupt changes in fracture resistance, the crack changes
its velocity instantaneously radiating strong high frequency
waves (23).

Associated with the stress singularity there is a slip velocity singularity behind the rupture front which is of the form:

$$\Delta\dot{u}_x(x,t) = 2B(v)\ \frac{K^{\dagger}}{\mu}\ v\left[\ell(t)-x\right]^{-\frac{1}{2}}\qquad x < \ell(t) \qquad\qquad (13)$$

where $B(v)$ is a complicated function given in (23) but which
may be approximated by:

$$B(v) = \frac{1}{1.8(1-\alpha^2/\beta^2)^{\frac{1}{2}}} \frac{1}{(1+v/c_R)^{\frac{1}{2}}} \tag{14}$$

where α, β and c_R are P,S and Rayleigh wave velocity, respectively. $B(v)$ varies by a factor of 1.4 within the range $0 < v < c_R$, that is, even at the Rayleigh wave velocity there is a velocity concentration behind the crack tip. When, as proposed above, the rupture velocity changes rapidly in response to changes in rupture resistance, the slip velocity intensity also changes abruptly due to its almost linear proportionality to the rupture velocity v. These jumps in slip velocity intensity produce high frequency waves because, as we have seen above, seismic radiation is proportional to slip velocity and the slip velocity concentration is the strongest singularity of the slip velocity field at the source. The role of crack features in controlling seismic radiation appears clearly: radiation depends on slip velocity, the dominating slip velocity feature is the singularity at the rupture front, changes in this singularity control radiation. At low frequencies the singularities are integrated and the waves see essentially the slip on the fault, thus the crack and dislocation models are not very different. At high frequencies the waves see only the singularities and therefore they control high frequency radiation. In dislocation models the singularities are stronger and radiation of high frequency waves is exaggerated.

4.   A PLANE CRACK MODEL

In order to study the relation between crack models and seismic radiation, we shall consider first a rather simple crack problem that admits a complete analytical solution. Let a semi-infinite static plane shear crack extend along the negative x-axis. The stress inside the crack has been relaxed to the frictional level and this will be taken as the reference for measuring stresses. Since we shall only be interested in the initial moments of fracture, we shall not concern ourselves with the other crack tip nor with the stress field far from the crack tip. A similar model, but under simpler antiplane stress, was studied by Eshelby (24). In front of the crack tip, along the positive x-axis the σ_{xy} component of the stress field has a stress concentration of the form

$$\sigma_{xy} = K_o \, x^{-\frac{1}{2}} \tag{15}$$

remembering again that this stress is measured from the frictional level. We assume that the fault is in unstable equilibrium, the stress concentration is just enough to set the crack tip in rapid motion. The stress intensity factor K_o is

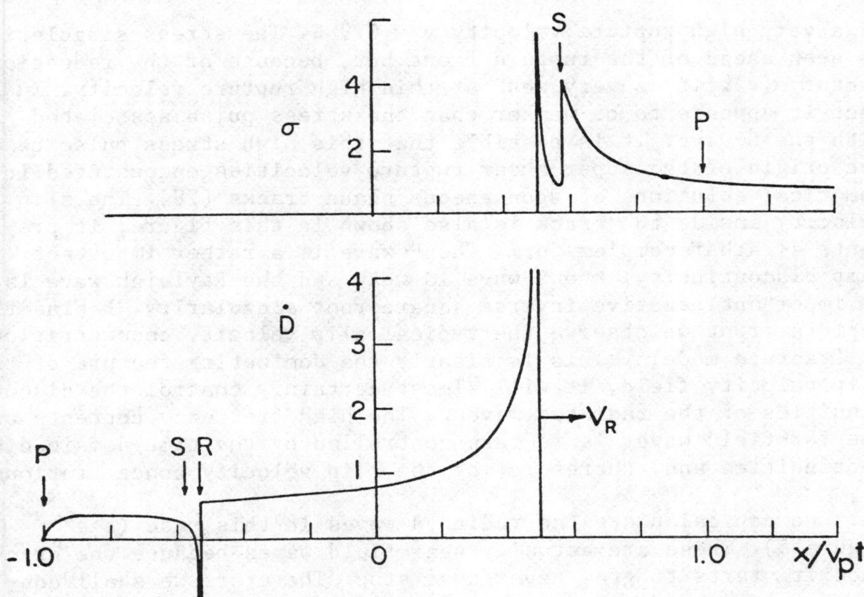

Figure 4. Stress (top) and slip velocity (bottom) for a plane
crack running at constant rupture velocity V_R. Crack starts
suddenly from the origin at time t = 0. Slip velocity and stress
are on arbitrary units (from (24)).

equal to the critical stress intensity for mode II loading
(Irwin criterion). Subject to a small perturbation the crack
will start growing with a velocity that depends on the details
of the rupture process at the tip. In order to calculate a correct rupture history we have to abandon the large-scale approach
and look at the cohesive forces near the crack tip. The initial
stages of rupture may be calculated if these cohesive forces are
specified in the form of a slip weakening function (25). Rice
calculated a series of such models during the initial quasi-
static growth of rupture. In order to calculate the actual
rupture history into the full dynamic range, a nonlinear boundary
value problem incorporating the slip weakening model in the
boundary conditions would have to be solved. Ida (4) has treated
some of these problems in the antiplane mode; for plane cracks
there is probably no other method than fully numerical calculations. To simplify we shall study here the case when the crack
starts to grow abruptly with a fixed rupture velocity v. This
problem admits an analytical solution (23,27) that may be found
using the method developed by Kostrov (26). Full details of the
solution may be found in (23). Figure 4 shows the stress field
and the velocities calculated on the crack plane for the case

of a very high rupture velocity $v = 0.9 \, \beta$. The stress singularity
is seen ahead of the rupture front but, because of the reduction
factor $A(v)$, it is very weak at this high rupture velocity. In
fact it appears to be weaker than the stress pulse associated
with the S-wave. It is possible that this high stress pulse be at
the origin of the super-shear rupture velocities encountered in
numerical solutions of spontaneous plane cracks (28). The slip
velocity inside the crack is also shown in this figure; it presents a rather complex form. The P wave is a rather important
ramp discontinuity; the S wave is weak and the Rayleigh wave is
an important negative inverse square-root singularity. Behind the
rupture front we observe the typical slip velocity concentration
of fracture models. This is clearly the dominating feature of the
slip velocity field, it will almost certainly control the discontinuities of the radiated waves. The high frequency contents of
the far-field waves is in turn controlled by the time-domain discontinuities and, therefore, by the slip velocity concentrations.

We can calculate the radiated waves in this case (see
Figure 5). These are actually near-field waves because the crack
once it starts to grow never does stop. Therefore we shall concentrate on the discontinuities which will be preserved when
the finite nature of the crack will become evident. For most
angles of radiation we observe step discontinuities in radiated
velocity. At shallow angles with respect to the crack surface
($\psi > 150°$) the S waves become complicated by the presence of
a logarithmic pulse. This is due to the SP waves refracted by
the crack surface (see Figure 6). This is also a near-field
wave that will in fact be multiply diffracted by the crack tips
in the case of a finite crack. The far-field radiation presents
step discontinuities in particle velocity at both the P and S
wave fronts. In terms of far-field displacement they correspond
to ramp discontinuities and they are therefore associated with
an ω^{-2} spectral decay at high frequencies. It is not difficult
to show (23,24) that the form of the discontinuities due to a
sudden arrest of a crack is the same as that of starting phases
seen in Figure 5, but with an opposite sign. Thus we find a
simple expression for the waves radiated by a sudden arrest or
start of plane rupture front:

$$\dot{u}(r,\psi) = \frac{K_o}{\mu} \, v \, \frac{1}{\sqrt{r}} \, \frac{R(\psi)}{1 - v/c \, \cos\psi} \, F(\psi) \, H(t - r/c) \qquad (16)$$

where r, ψ are polar coordinates centered around the point on the
crack plane where the jump in rupture velocity took place. $F(\psi)$
is a complicated function of the radiation angle which was
found in (23) for both P and S waves. It represents a modification of the radiation patterns.

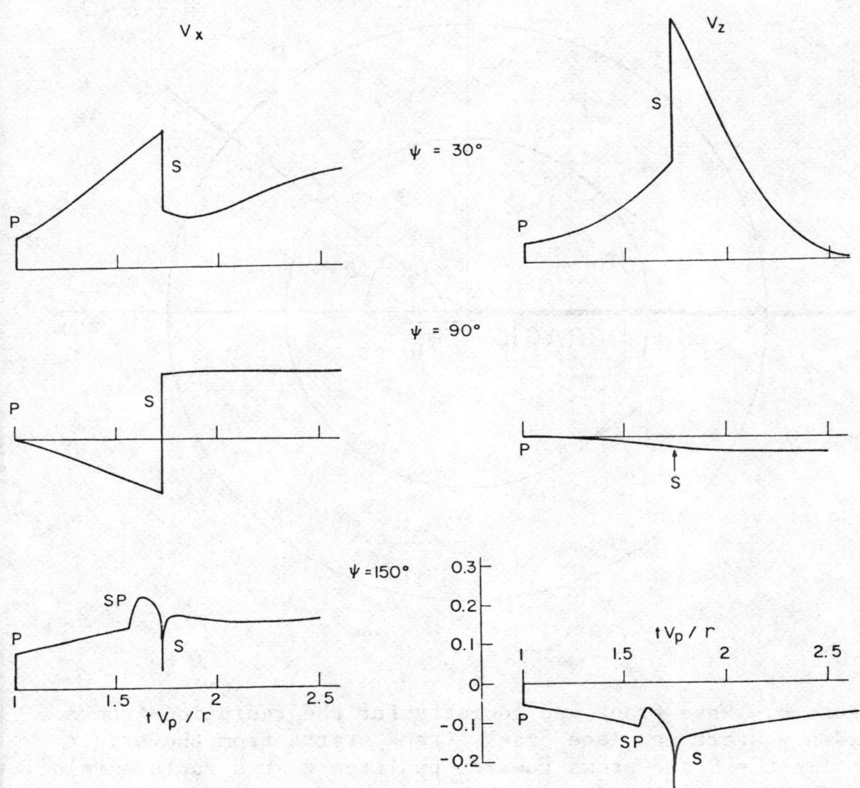

Figure 5. Waves radiated by the plane crack of the previous
figure. Jumps at the P and S wave fronts are due to the sudden
start of rupture. ψ is the azimuth with respect to the fault
plane as indicated in Figure 6. At ψ = 150° logarithmic pulses
are seen at the S wave front and the SP diffracted wave is also
seen. Time is in units of r/v_p. r is the distance of the observer
from the tip of the crack at t = 0 (see Figure 6). Particle
velocity is in arbitrary but homogeneous unit for all seismograms.

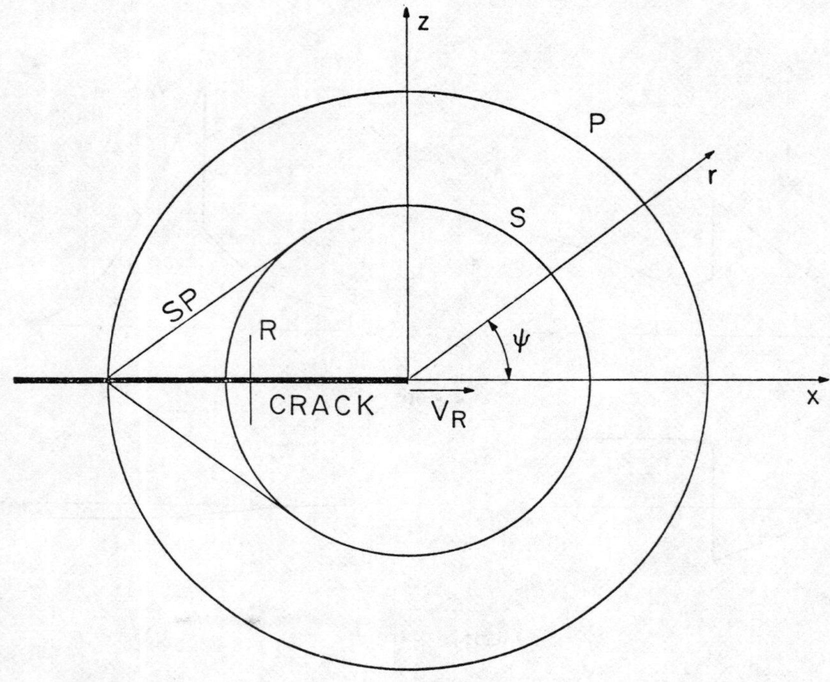

Figure 6. Wave front and geometry for the radiation from a
suddenly starting plane crack. Crack starts from the origin
at time t = 0 and grows towards positive x at a rupture velocity
V_R. The principal radiated phases are shown in this figure.

$$R(\psi) = \sin 2\psi \qquad \text{for P waves}$$
$$R(\psi) = \cos 2\psi \qquad \text{for S waves} \qquad (17)$$

The denominator, $(1 - v/c \cos\psi)$ is the usual directivity where c
is either the P or the S wave velocity.

A rapid change in velocity produced, for instance, by a
rapid change in rupture strength, may be modelled by a superposition of the discontinuities due to a sudden stopping and
starting of the rupture front. This simple result is a consequence of a very special property of the elastodynamic fields
near the crack tip: they do not depend on the acceleration
of the rupture front but only on the instantaneous rupture

velocity. Thus, in two-dimensional crack models, the dominating
high frequency radiation will come from these abrupt changes in
rupture velocity which are in turn due to rapid changes in fracture strength. These results are valid only for wave lengths
longer than the dimensions of the cohesive zone near the crack
tip. If the cohesive zone has a length shorter than a hundred
meters, as is usually assumed (25), then our theory applies
to the radiation of waves longer than this dimension; i.e.,
for frequencies over 50 Hz for P waves and 35 Hz for S waves.
If, as we shall discuss in section 6, earthquakes are highly
heterogeneous processes and the fault has barriers or asperities, then the high frequency spectrum should be controlled
by these ω^{-2} asymptotes.

Earthquakes are three-dimensional ruptures and the theory
for antiplane and plane cracks has to be adapted with great
care. In (23) we showed that this may be done, under certain
restrictions, by means of geometrical diffraction theory. But
this is possible only if a finite segment of the rupture front
changes velocity simultaneously. This is the case of the circular crack model we shall review in the next section. In a
general situation, we expect that crack arrest will initiate
from a point and propagate around the rupture front. If the
propagation of this stopping front is supersonic, radiation
shall be adequately approximated by the step functions obtained
for the plane cracks. But if the stopping front propagates sub-
sonically around the crack periphery, the shape of the crack
will change during the radiation, the stopping phases will
interfere and the discontinuities will be smoothed out. It is
possible then for rather smooth ruptures to radiate high frequency waves weaker than ω^{-2}.

5.   THE CIRCULAR CRACK MODEL

A circular shear crack is the simplest geometry of a fault that
may be directly compared to real seismic sources in the earth.
The static solution for the slip on the fault for a uniform
stress drop is given by (29):

$$\Delta u_x(r) = \frac{24}{7\pi} \frac{\Delta\sigma_{zx}}{\mu} (a^2 - r^2)^{\frac{1}{2}} \tag{18}$$

where a is the radius of the crack and $\Delta\sigma_{zx}$ its stress drop.
From Δu_x we can find the average dislocation and the seismic
moment

$$M_o = \frac{16}{7} \Delta\sigma_{zx} a^3 \tag{19}$$

an expression which is very often used to calculate stress drop from estimates of the seismic moment M_o and of the source radius a.

Stress concentrations on the fault plane, just outside the fault's edge are given by

$$\Delta\sigma_{xz}(r) = (K_{II}\cos\phi + K_{III}\sin\phi)\,(r-a)^{-\frac{1}{2}} \qquad r > a \tag{20}$$

where ϕ is azimuth on the fault measured from the x-axis, and

$$\begin{aligned} K_{II} &= 0.515\ \Delta\sigma_{xz}a^{\frac{1}{2}} \\ K_{III} &= 0.385\ \Delta\sigma_{xz}a^{\frac{1}{2}} \end{aligned} \tag{21}$$

where the coefficients were calculated assuming a Poisson ratio of 1/4.

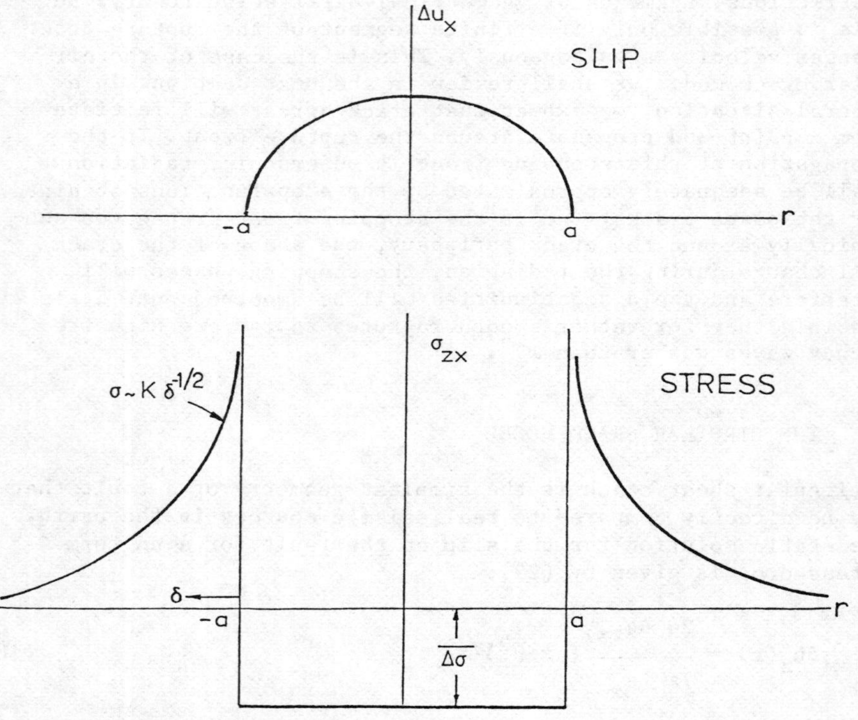

Figure 7. Slip and stress from a circular crack model of radius a. Stress drop inside the crack is uniform. Stress concentrations are seen outside the crack.

A dynamic version of this model was approximated by Brune (18) in his well known formula relating corner frequency to source radius a:

$$f_o = 0.37 \, \beta/a \tag{22}$$

in fact this is essentially a dimensional expression for the scaling of f_o and the factor 0.37 was obtained by a rough estimate of the far-field pulse width.

Let us consider now dynamic source models; the simplest model admitting an analytical solution is a self-similar circular shear crack. This is a shear crack that starts from a point and then grows symmetrically with a constant rupture velocity without ever stopping. The slip is driven by a drop in the σ_{xz} component of the pre-existing stress tensor. The solution to this problem is very simple and may be completely calculated by Cagniard's method (30). The slip in this special crack model is everywhere parallel to the x-axis ($\Delta u_y = 0$) and is given by

$$\Delta u(r,t) = \frac{\Delta\sigma_{xz}}{\mu} \, C(v) \, (v^2t^2 - r^2)^{\frac{1}{2}} \qquad\qquad r < vt \tag{23}$$

where $C = C(v)$ is an almost constant function of the rupture velocity v, $C \simeq 1$ for the whole subsonic rupture velocity range. The shape of the slip function is of the same form as that of the static solution given by eq. (18). Since this crack never stops, we cannot directly compare its radiation with actual far-field observations in the earth. It may however approximate the initial behaviour of the far-field pulses radiated by finite cracks. By straightforward application of eq. (7)

$$\Omega(t,\theta) = \Delta\sigma_{xz}v^3C \, \frac{2\pi}{\left[1-v/c \, \sin^2\theta\right]^2} \, t^2 H(t) \tag{24}$$

where θ is the angle of radiation of a ray with respect to the fault normal. Thus, the far-field pulses behave initially like t^2. The frequency behaviour of these pulses is then of the ω^{-3} type.

The circular self-similar shear crack is very useful since it permits verification of many of the properties of crack models reviewed in the previous sections. Let δ be a distance along the fault measured from the instantaneous position of the rupture front (vt), i.e.,

$$r = vt + \delta \tag{25}$$

Around the edge of the crack there are stress concentrations

$$\sigma_{xz}(r) = (K_{II}\cos\phi + K_{III}\sin\phi)\ \delta^{-\frac{1}{2}} \qquad \delta > 0 \qquad (26)$$

where ϕ is azimuth on the fault plane measured from the x-axis. K_{II} and K_{III} are stress intensity factors given by

$$K_{II} = \Delta\sigma_{xz}\ \sqrt{vt}\ k_2(v)$$
$$K_{III} = \Delta\sigma_{xz}\ \sqrt{vt}\ k_3(v) \qquad\qquad (27)$$

for the inplane (mode II) and antiplane (mode III) stress intensities. The velocity dependent functions

$$k_2(v) = \sqrt{2}\ \frac{\beta^2}{v^2}\ \{(\frac{v^2}{2\beta^2} -1)^2 - \sqrt{(1-v^2/\alpha^2)(1-v^2/\beta^2)}\}\ \cdot$$
$$\cdot\ (1 - v^2/\beta^2)$$
$$\qquad\qquad (28)$$
$$k_3(v) = \frac{C}{2\sqrt{2}}\ \sqrt{1-v^2/\beta^2}$$

tend to the static values $k_2 = 0.515$ and $k_3 = 0.385$ as the velocity tends to zero. These are the correct coefficients of the stress intensity factor of the static circular shear crack of radius a = vt, as may be verified comparing with eq. (21). At high rupture velocities $k_2(v)$ and $k_3(v)$ decrease monotonically and go to zero at the Rayleigh wave velocity and shear velocity, respectively. These are the terminal velocities for ideal shear cracks, where the detailed distribution of cohesive forces is replaced by a critical stress intensity factor. Since the k_i factors vary rapidly only at very high rupture velocities, cracks with increasing load accelerate very rapidly to rupture velocities close to the terminal ones. For finite cohesive forces it has been numerically found that v may become larger than the shear velocity (8,28).

The slip inside the crack was given in eq. (23); the slip velocity concentration just inside the crack edge is:

$$\Delta u_x(r) = \frac{\Delta\sigma_{xz}}{\mu}\ v\ \sqrt{2vt}\ C\ |\delta|^{-\frac{1}{2}} \qquad \delta < 0 \qquad (29)$$

thus, the velocity concentration does not depend on ϕ, the azimuthal position around the crack edge.

As shown by eq. (26) around the crack edges there is a mixed

mode of rupture with a variable stress intensity. An obvious
consequence of the mixed mode of rupture is that the apparent
simplicity of the solutions should disappear as soon as a rupture
criterion would be introduced. The crack will immediately deform
into a more or less elliptical shape. For identical resistance
to plane and antiplane rupture it will grow preferentially in the
x-direction, i.e., it will elongate in the plane stress direction
since $K_{II} > K_{III}$ in eq. (26). This has been verified in three-
dimensional numerical modelling of shear cracks (10).

The next step in approximating actual seismic sources is
to assume that the self-similar crack stops suddenly once it
reaches a certain final radius a. This problem was numerically
solved in (6) and shall not go into details here because a complete review of its consequences has been recently published
(31,13). An example of the slip function for this model is
shown in Figure 8 as a function of time for several positions
along the crack radius. The role of the P waves radiated by the
stopping of rupture is very important. Healing, defined as slip
arrest when the slip velocity becomes zero, proceeds with a
velocity that is probably related to the Rayleigh waves on the
crack.

Radiation from this crack model may be calculated by numerical evaluation of the Fourier Transforms in eq. (8). The results
are the well-known pulses shown in Figure 9, where the arrival
times of the stopping phases, the wave front discontinuities associated with rupture arrest, are shown by the arrows. It is
these stopping phases, with their associated particle velocity
jumps that control the high frequency contents of the far-field
pulses. To stress this point in Figure 10 we show the particle
velocity far-field pulses; the stopping phases, indicated by
arrows, stand out very clearly although in this numerical solution the jumps are smoothed out, specially for the last stopping
phases. Theoretical values for these discontinuities were calculated in (23) and (31).

Finally, let us note that the jumps in particle velocity
correspond to a slope discontinuity in displacement. Slope discontinuities or ramps correspond, in turn, to ω^{-2} high frequency asymptotes. This is clearly the dominating high frequency
behaviour in the spectra shown in Figure 9. We have theoretically
calculated the spectral behaviour associated with the stopping
phases and this is shown by lines marked theoretical in Figure 9.
We see that the theoretical high frequency slope is a very good
estimate of the slope obtained numerically. The holes in the
spectrum are due to destructive interference of the stopping
phases and they are a distinctive feature of roughly symmetrical
pulses. Definition of the corner frequency is difficult, the
S waves show the clear presence of an intermediate frequency

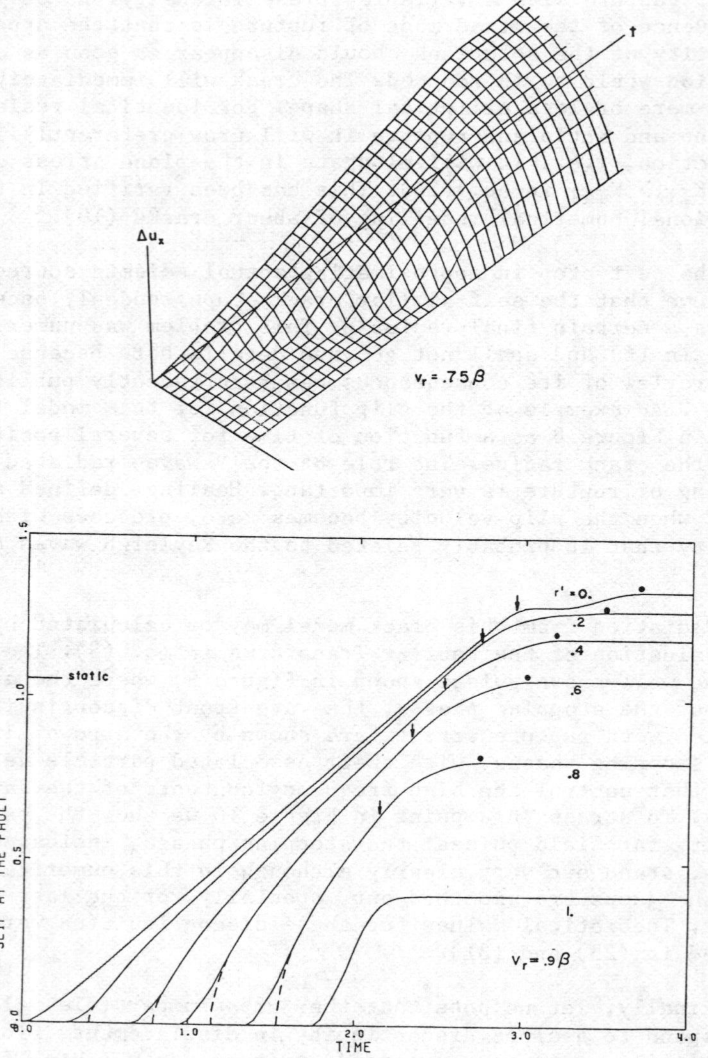

Figure 8.   Slip as a function of relative radius r' = r/a
on a circular fault of radius a. Crack grows from the origin
at constant rupture velocity V_r until it stops. At top, three-
dimensional plot of slip vs time and radius for $V_r = 0.75\beta$.
Bottom, slip as a function of radius; time is in units of a/β,
slip is units of $1/3\ \Delta\sigma/\mu\ a$.

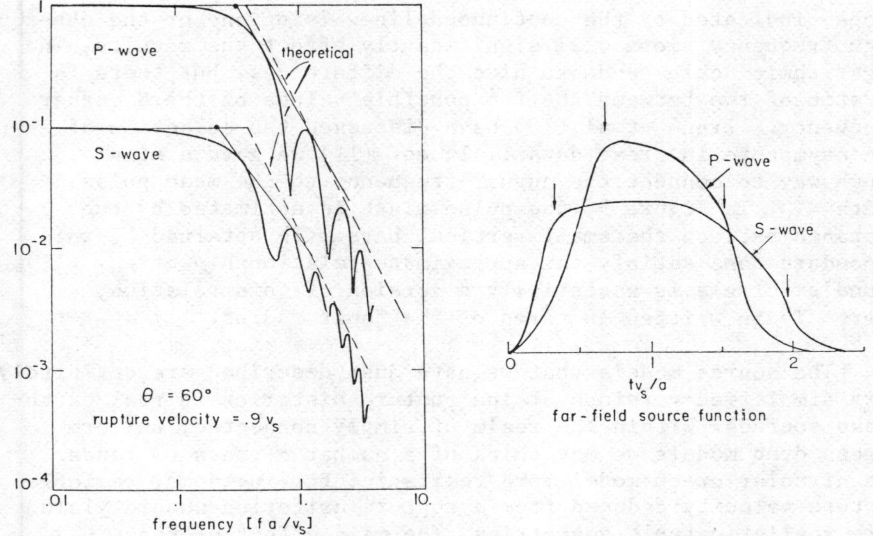

Figure 9.   Far-field pulses and spectra for a circular crack.
Amplitude units are arbitrary. θ is the angle of radiation between
the ray and the normal to the fault. Arrows indicate the theoretical position of the stopping phases.

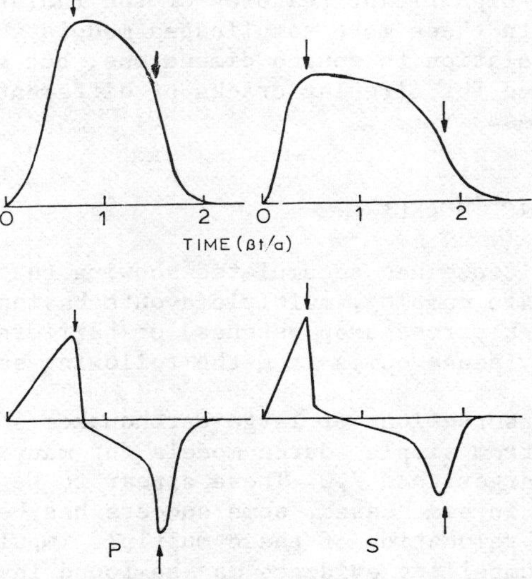

Figure 10. . Far-field displacement (top) and particle velocity
(bottom) pulses radiated by the circular fault model. Arrows indicate stopping phases as in Figure 9. Radiation angle θ = 60º.
Amplitudes are in arbitrary units.

slope, indicated by the continuous line. Using one or the other
high frequency slope will significantly affect the results. The
logarithmic scale tends to hide the differences, but there is
a ratio of two between the two possible values of the S corner
frequency. Brune et al (19) have discussed the selection of
the asymptote in great detail. In eq. (11) we gave a simple,
rough way to connect the corner frequency to the mean pulse
width $\langle T \rangle$. In Figure 9, the pulse width is estimated by the
distance between the small vertical bars. $\langle T \rangle$ obtained by this
procedure does satisfy the approximate relationship of eq. (11).
Brune's formula is essentially a version of this relation,
where $\langle T \rangle$ is written in terms of the fault radius.

The source models that we have just described are definitely
very simplified versions of the rupture histories in real earthquake sources. Within the realm of simply connected, uniform
stress drop models we may think of a number of ways to render
the circular crack model more realistic; for instance, variable
rupture velocity deduced from a rupture criterion should yield
more realistic fault geometries. The main effect of these
variable velocities is to elongate and add some character to
the initial part of the far-field pulses. As we have seen, this
part of the signal plays a minor role in controlling the high
frequency radiation. As long as the fault stops more or less
abruptly and simultaneously around the edge, stopping phases
will be the most significant features of the radiated signals.
Source duration in these more complicated models should have
a more complex relation to source dimensions, but still within
the range obtained for circular cracks of different constant
rupture velocities.

6.   COMPLEX SOURCE MODELS

A substantial evidence has accumulated showing that most natural
seismic sources are complex, multiple events having asperities
(concentrated high stress drop patches) or barriers (unbreakable
patches). This evidence comes from the following sources:

a)   Far-field observations of large earthquakes show significant
      departures from simple source models for many events of
      magnitude larger than 7.0. These appear to be multiple
      events and, in some cases, some success has been reported
      in relative relocation of these multiple impulses (32,33).
      Even more compelling evidence may be found in short period
      records of events with moment near 10^{25} to 10^{26} dynes-cm;
      within this range both short and long period WWSSN records
      are on-scale and a direct comparison of apparently simple
      long period records with complex short period ones may be
      made (34).

b)   Direct evidence from fracture in mines shows that faults
     are usually very complex with side steps and unbroken but
     highly deformed ligaments (35). Similar evidence of discon-
     tinuous slip in actual earthquake faults has been reviewed
     by Aki (36).

c)   A theoretical argument by Andrews (37) shows that if sources
     were simple, tectonic stress would drop to the frictional
     level during rupture and no aftershocks would occur. Tec-
     tonic loading of the fault would not create the small-scale
     heterogeneities necessary for aftershock generation and some
     other mechanism would have to be found for aftershocks. Al-
     though the arguments in (37) are rather complex, it is easy
     to see that variations of strength, expressed for instance
     in terms of variable critical stress intensity, can create
     segments of the fault that remain unbroken or that do not
     completely reach the instability limit. Subsequent rupture
     of these segments would generate aftershocks.

We can get some rough idea of the effects of complexity on
the estimation of source parameters by means of simple static
models. Complexity will affect significantly the usual seismic
moment – stress drop relation given by eq. (19). A very general
relation between seismic moment and stress drop was obtained in
(39) by means of Betti's reciprocity theorem. The implications
of these results are clearer if we consider a simple case that
retains all the physical properties of the more general equations. Consider again a plane fault perpendicular to the z axis
whose external shape is a circle. The stress drop inside the
fault may be highly heterogeneous due to the presence of the
asperities or barriers. If we assume that the σ_{zx} component
of stress is the only one that changes during the rupture,
then the seismic moment tensor component $M_{zx} = M_0$ is given by:

$$M_0 = \frac{24}{7\pi} \int_o^a \int_o^{2\pi} \Delta\sigma(r,\phi)(a^2-r^2)\, r\,dr\,d\phi \qquad (30)$$

where a is the radius of the overall fault surface. For a uniform
stress drop, eq. (30) reduces to the expression eq. (19), which
is used by the majority of seismologists to determine stress
drop. In the case of heterogeneous stress drop, eq. (30) shows
that the stress estimated by eq. (19) is in fact an average with
an ellipsoidal weighting function. Unless the stress drop is
very high near the borders of the fault, the average calculated
in this way should not be too different from the true arithmetic
mean of the stress drop on the fault. Consider the asperity model
(38), i.e., a model in which stress drop is very heterogeneous,
with patches where stress drop is high (asperities), and patches
which have already broken during previous events and have zero

stress drop. In this case the stress drop estimated by eq.
(19) will not be very different from the true average stress
drop on the fault.

Another possible model for earthquake complexity is the
barrier model of Das and Aki (8,9). In this model the rupture
leaves unbroken patches on the fault plane. In these high-
strength ligaments between the subfaults in the stress is highly
concentrated and the stress drop is negative (stress rises).
Since the stress rises may be very high, the estimated stress
may severely underestimate the stress drop on each individual
subfault. In (39) we showed that the estimated and the true stress
drop are in the ratio ρ/a, where ρ is radius of an individual subfault and a is the overall radius of the complex fault.

Let us consider now the effect of heterogeneity on the estimation of corner frequencies. For circular faults where heterogeneity is randomly distributed on the fault surface, the pulses
radiated by each segment of the fault (subfault or asperity)
will arrive in a random sequence to the seismic station and will
superpose yielding a complex signal. Its corner frequency will
probably satisfy the relation in eq. (11), and the total pulse
length will be a measure of the overall fault dimensions. If
the fault is elongated, like a long strike-slip fault, pulses
radiated from different patches of the fault will arrive separately at the recording stations. This is the case of the so-called
multiple events reported in the literature (32,33). Far-field
radiation and spectra for a few examples of this kind of model
have been calculated (9,39). It is easy to see that in this
case the corner frequency is not inversely proportional to
overall pulse width. In Figure 11 a complex event is modelled
as a series of two triangular pulses separated by a time T.
Each triangular pulse is of the form shown in Figure 1. Thus
for T = 0 the spectra are the same as for the single impulse.
As T grows the corner frequency decreases slightly for T =
0.03 but for T = 0.06 and T = 0.09 it is almost the same as
that for T = 0. This shows that the corner frequency reflects
not the total duration of the signal, but the width of each of
the triangles. This result is well known in time series analysis,
the spectra are dominated by the simplest common pulse. The
multiplicity shows up in the spectral holes that move to lower
frequency as T increases.

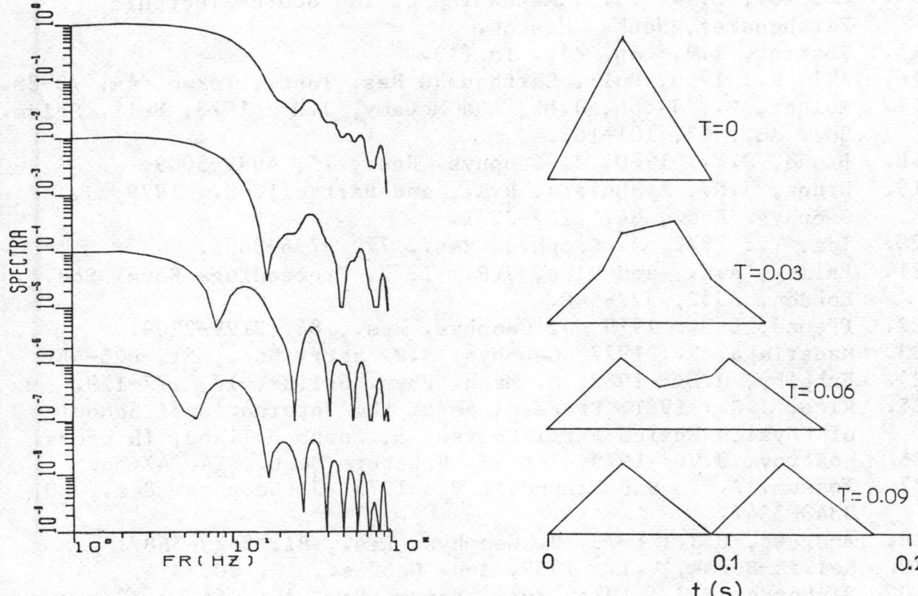

Figure 11. Spectra for a series of multiple pulses formed by the superposition of two triangular functions of the type shown in Figure 1. T indicates the time interval between the two successive pulses in seconds. Amplitudes are in arbitrary units

REFERENCES

1.   Kostrov, B.V.:1964, J.Appl. Math. Mech., 28, 1077-1087.
2.   Burridge, R., and Willis, J.R.: 1969, J.R. Proc. Camb. Phil. Soc., 66, 443-468.
3.   Kostrov, B.V.: 1966, J. Appl. Math. Mech., 30, 1241-1248.
4.   Ida, Y.: 1972, J. Geophys. Res., 77, 3796-3805.
5.   Andrews, D.J.: 1975, Bull. Seism. Soc. Am., 65, 163-182.
6.   Madariaga, R.: 1976, Bull. Seism. Soc. Am., 66, 639-666.
7.   Archuleta, R.J., and Frazier, J.A.: 1978, Bull. Seism. Soc. Am., 68, 573-598.
8.   Das, S., and Aki, K.: 1977, J. Geophys. Res., 82, 5658-5670.
9.   Das, S., and Aki, K.: 1977, Geophys. J.R. astr. Soc., 50, 643-648.
10. Virieux, J., and Madariaga, R.: 1980, Subm. to Bull. Seism. Soc. Am.
11. Knopoff, L.: 1981, Nature of the earthquake source, This volume.
12. Backus, G., and Mulcahy, M.: 1976, Geophys. J.R. astr. Soc. 46, 341-362; 47, 301-329.
13. Aki, K., and Richards, P.G.: 1980, Quantitative Seismology, W.H. Freeman, San Francisco.

14. Kostrov, B.V.: 1975, Mechanics of the Source Tectonic
     Earthquakes, Nauka, Moscow.
15. Kostrov, B.V.: op. cit. in (1).
16. Aki, K.: 1966, Bull. Earthquake Res. Inst., Tokyo, 44, 73–88.
17. Molnar, P., Jacob, J.H., and McCamy, J.H.: 1973, Bull. Seism.
     Soc. Am., 63, 101–103.
18. Brune, J.N.: 1970, J. Geophys. Res., 75, 4999–5009.
19. Brune, J.N., Archuleta, R.J., and Hartzell, S.: 1979, J.
     Geophys. Res., 84, 2262–2272.
20. Ida, Y.: 1972, J. Geophys. Res., 77, 3796–3805.
21. Palmer, A.C., and Rice, J.R.: 1973, Proceedings Royal Soc.
     London, A332, 527–548.
22. Freund, L.B.: 1979, J. Geophys. Res., 84, 2199–2209.
23. Madariaga, R.: 1977, Geophys. J.R. astr. Soc., 51, 625–651.
24. Eshelby, J.D.: 1969, J. Mech. Phys. Solids, 17, 177–179.
25. Rice, J.R.: 1981, Proceedings of the International School
     of Physics Enrico Fermi Course 78, North Holland, in press.
26. Kostrov, B.V.: 1975, Int. J. Fracture Mech., 14, 47–56.
27. Fossum, A.F., and Freund, L.B.: 1975, J. Geophys. Res., 80,
     3343–3347.
28. Andrews, D.J.: 1976, J. Geophys. Res., 81, 5679–5687.
29. Keilis-Borok, V.I.: 1959, Ann. Geofis., 12, 205–214.
30. Richards, P.J.: 1976, Bull. Seism. Soc. Am., 66, 1–32.
31. Madariaga, R.: 1979, SIAM-AMS Proceedings, 12, Am. Math.
     Soc., Providence, RI.
32. Rial, J.A.: 1978, J. Geophys. Res., 83, 5405–5414.
33. Kanamori, H., and Stewart, G.S.: 1978, J. Geophys. Res., 83,
     3427–3434.
34. Lyon-Caen, H.: 1980, Thesis, Universite de Paris VII, Paris.
35. McGarr, A., Spottiswoode, S.M., Gay, N.C., and Ortlepp,
     W.D.: 1979, J. Geophys. Res., 84, 2251–2261.
36. Aki, K., 1979: J. Geophys. Res., 84, 6140–6148.
37. Andrews, D.J.: 1978, J. Geophys. Res., 83, 2259–2264.
38. Rudnicki, J.W., and Kanamori, H.: 1981, J. Geophys. Res.,
     in press.
39. Madariaga, R.: 1979, J. Geophys. Res., 84, 2243–2250.

INELASTIC PROCESSES IN SEISMIC WAVE
GENERATION BY UNDERGROUND EXPLOSIONS

Howard C. Rodean

University of California
Lawrence Livermore National Laboratory

ABSTRACT

There are similarities and differences between chemical and nuclear explosions underground. Most of the differences are in the early stages of the explosions. The later stages are similar with respect to seismic wave generation. Three sources of seismic waves from explosions are coincident in space and time, or nearly so: the explosion itself, explosion-induced tectonic strain release, and (probably) spall-closure following explosion-produced spall. Cavity collapse and explosion-induced aftershocks are two sources of delayed seismic signals. Theories, computer calculations, and measurements of spherical stress waves from explosions are described and compared, with emphasis on the transition from inelastic to *almost*-elastic relations between stress and strain. Two aspects of nonspherical explosion geometry are considered: tectonic strain release and surface spall. Tectonic strain release affects the generation of surface waves; spall closure may also. The forward problem in seismology can, in principle, be solved by calculations beginning with explosive detonation and ending with the synthetic seismogram. The inverse problem can also, in principle, be solved by inverting observed seismic data to obtain an "equivalent elastic source," but the solution cannot extend backward in space and time into the nonlinear inelastic processes of the explosion. The reduced-displacement potential is a common solution (the "equivalent elastic source") of the forward and inverse problems, assuming a spherical source. Measured reduced-displacement potentials are compared with potentials calculated as solutions of the direct and inverse problems; there are significant differences between the results of the two types of calculations and between calculations and measurements. The simple

E. S. Husebye and S. Mykkeltveit (eds.), Identification of Seismic Sources - Earthquake or Underground Explosion, 97–189.

spherical model of an explosion is not sufficient to account for observations of explosions over wide ranges of depth and yield. The explosion environment can have a large effect on explosion detection and yield estimation. The best sets of seismic observations for use in developing discrimination techniques are for high-magnitude high-yield explosions; the identification problem is most difficult for low-magnitude low-yield explosions. Most of the presently available explosion data (time, medium, depth, yield, etc.) are for explosions in a few media at the Nevada Test Site; some key questions concerning magnitude vs yield and m_b vs M_S relations can be answered only by data for explosions in other media at other locations.

1. INTRODUCTION

This Advanced Study Institute is concerned with the means for answering the question, "Do the characteristics of a set of observed seismic signals indicate that the source of those signals was an earthquake or an underground explosion?" The subject of this lecture is the inelastic processes that are involved in the generation of seismic waves by underground explosions. The emphasis is on how these inelastic processes determine (or may determine) seismic signal characteristics that are pertinent to seismic identification and two other important aspects of seismic monitoring: event detection and (if the event is identified as an explosion) yield estimation.

The parenthetical phrase, "or may determine," indicates that our knowledge of these inelastic processes is imperfect and incomplete. In this lecture, we will try to distinguish between proven knowledge and informed speculation, and to establish what is known about this subject.

2. PHENOMENA IN CHEMICAL AND NUCLEAR EXPLOSIONS

In the context of this Institute, we may define an underground explosion as a sudden release of concentrated energy, the response of the surrounding rock to this energy, and the consequent radiation of seismic waves. Two sources of concentrated energy are chemical explosives and nuclear explosives.

As illustrated in Fig. 1, there are analogies between the different phenomena in chemical and nuclear explosions at early times, but the mechanical effects at later times are identical --except as may be determined by differences of scale. The nuclear explosive is analogous to the primary chemical explosive or initiator; and the vaporization of the rock in the nuclear explosion is analogous to the detonation of the secondary or high

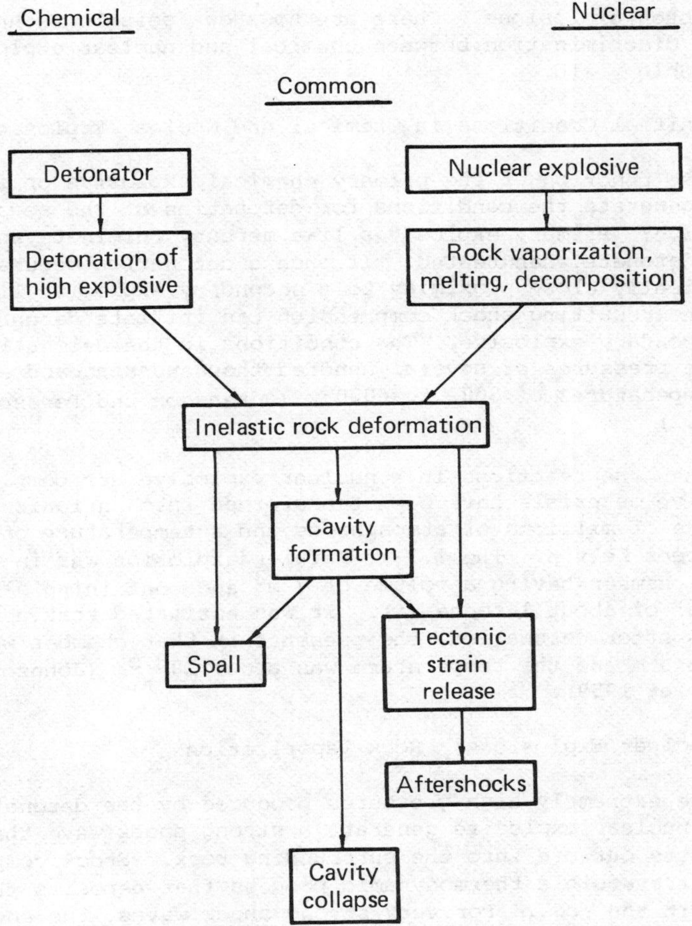

FIG. 1. Processes in underground chemical and nuclear explosions.

explosive. The subsequent inelastic rock deformation and seismic wave generation are fundamentally identical in both types of explosions.

There are generally large differences of scale between chemical and nuclear explosions. There have been only a few chemical explosions with yields of more than a kiloton [1 kiloton (kt) = 4.18 terajoules (TJ)], but there have been several nuclear explosions with yields of a megaton or more (Springer and Kinnaman 1971, 1975). The problems of seismic detection, identification, and yield estimation are comparatively simple for megaton-class explosions, but are challenging and often difficult for subkiloton to kiloton explosions. There are no known seismic techniques for remote discrimination between chemical and nuclear explosions of comparable yield.

2.1 Initial Conditions in Chemical and Nuclear Explosions

The function of the primary chemical explosive or initiator is to generate the conditions for detonation of the secondary explosive. Primary explosives like mercury fulminate and lead azide detonate when heated. If such a detonation occurs in sufficiently close proximity to a secondary explosive like TNT or RDX, the resulting shock compression can initiate detonation of the secondary explosive. The conditions in the detonation wave in RDX are pressures of several hundred thousand standard atmospheres and temperatures of 5000 to $6000^{\circ}K$ (Johansson and Persson 1970, Chs. 1,2).

When the reactions in a nuclear explosive are completed, the explosive materials have been transformed into an ionized gas at a pressure of millions of atmospheres and a temperature of millions of degrees Kelvin. The 1.7-kt Rainier explosion was in an underground chamber having a volume of 7 m^3 and containing a mass of material of about 1 tonne (t). It was estimated that a few microseconds after detonation, the pressure in that chamber was about 7 million atm and the temperature was about 10^6 $^{\circ}K$ (Johnson, Higgins, and Violet 1959).

2.2 Nuclear Explosions: Rock Vaporization

The extremely high pressures produced by the detonation of a buried nuclear explosive generate a strong shock wave that propagates outward into the surrounding rock. Shock compression is an irreversible thermodynamic process that deposits thermal energy in the rock. For very strong shock waves, the energy deposition is sufficient to vaporize the rock. For a nonporous silicate rock like granite, the shock pressure required for vaporization is estimated to be about 2 million atm; for porous silicate rocks like tuff or alluvium, the required shock pressure is estimated to be about one-third to one-half that value (Butkovich 1967).

A shock wave is attenuated as it propagates outward from the center of an explosion, and the thermal energy deposition decreases accordingly. There is a range in which the energy deposition is insufficient for vaporization but sufficient to melt the rock. Some rocks, such as dolomite, do not melt but decompose and sublime instead.

The vaporized rock, which is initially at a pressure of a million or more atmospheres, expands and does mechanical work on the surrounding rock until an equilibrium is established between

the cavity pressure and the resisting stresses in the surrounding
rock. This expansion process is described further in Section 2.7.

2.3 Chemical Explosions: Explosive Detonation

The detonation of a buried charge of secondary explosive is
analogous in one sense to the vaporization of rock in a nuclear
explosion: the result of both processes is a gas that expands and
does mechanical work on the surrounding rock. There are also
differences between the two processes. The rock vapor is initially
at much higher pressures and temperatures than the detonation
products of the chemical explosive. Rock vaporization absorbs
explosive energy, but chemical explosive detonation releases
energy. The energy release in underground chemical explosions is
not sufficiently intense to vaporize any surrounding rock. But
special chemical explosive assemblies can be designed to vaporize
metals. See Zel'dovich and Raizer [1967, Ch. XI (4) 21-22] and
Ahrens and Urtiew (1971).

2.4 Inelastic Rock Deformation

In underground explosions, a strong shock wave is propagated
outward into solid rock from the cavity containing either explosive
detonation products (in the case of a chemical explosion) or rock
vapor and melt or decomposition products (in the case of a nuclear
explosion). In either case, the initial shock wave pressure is
generally on the order of several hundred thousand atmospheres
(except, for example, when the explosive is emplaced in a large
cavity for seismic-decoupling purposes; this special condition is
discussed in Section 6.1). The shock pressure is attenuated as
the shock wave propagates outward; partly because of the geometry
of a diverging wave, but also because of the irreversible thermodynamic processes in the shock wave. At very high pressures,
solid-solid phase changes may occur because of thermodynamic
nonequilibrium during shock loading and unloading. An example is
the transition of quartz into coesite or stishovite. At
sufficiently high pressures, the shear stresses are negligible and
the rock deforms as if it were a fluid. At somewhat lower
pressures, the shear stresses are significant and the rock is
plastically deformed. At still lower pressures, the rock may
fracture in shear or tension. Porous rocks will be crushed and
permanently compacted at sufficiently high pressures. If the
pores are partially- or fully-saturated with a liquid (usually
water), the crushing and compaction are reduced and other mechanical properties are modified. Finally, there is some ill-defined
boundary called the "elastic radius" beyond which nonlinear
inelastic deformation does not occur, but linear anelastic
processes continue to attenuate the "elastic" waves from the
explosion. Inelastic deformation and the "elastic radius" are
discussed further in Section 3.4.

2.5 Tectonic Strain Release

Early in the history of underground nuclear explosions, seismological and geological evidence began to accumulate that some underground nuclear explosions released preexisting tectonic strain. Most of the evidence was in the form of Love- and Rayleigh-wave radiation patterns. Two principal hypotheses have been advanced to explain the observed tectonic strain releases. One is release of the preexisting tectonic strain around the inelastic zone surrounding the explosion (the tectonic strain field responds to the creation of a weak fractured zone around the explosion). The other hypothesis is triggering of a dislocation along a nearby fault plane. The first hypothesis appears to be satisfactory in many cases, but the triggering hypothesis is favored in cases with relatively large amounts of tectonic strain release. Tectonic strain release is discussed further in Section 4.1.

2.6 Spall

The stress waves that propagate upward from an underground explosion are reflected at the free surface because of the impedance mismatch between the surface layer and the atmosphere. Other reflections may occur below the surface if the geological structure is layered, with different densities and elastic moduli in adjacent layers.

The principal outgoing wave from an explosion is a compressional wave; it is reflected from the free surface as a rarefaction wave. Rocks generally have low tensile strength; consequently, the downgoing reflected wave may cause the rock to fail in tension at some point below the surface. This failure occurs while the mass of rock above the region of tensile failure has a net upward momentum. The spalled layer enters a ballistic trajectory that eventually ends in "slapdown" when the gap opened by the spalling process is closed. Some evidence suggests that spall "slapdown" is a significant seismic wave source under some conditions. Spall is discussed further in Section 4.2.

An interesting related phenomenon, observed in connection with several nuclear tests in the Yucca Flat testing area of the Nevada Test Site, is the occurrence of multiple reflections of compressional and rarefaction waves between the free surface and the underlying rocks of Paleozoic age beneath Yucca Flat. There is a considerable impedance mismatch between these Paleozoic rocks and the overlaying aluminum and tuff. Displacement time histories from several velocity gauges from one nuclear test are shown in Fig. 2, and the Fourier spectrum of the signal from one of these gauges is given in Fig. 3. It was found that the wave period is

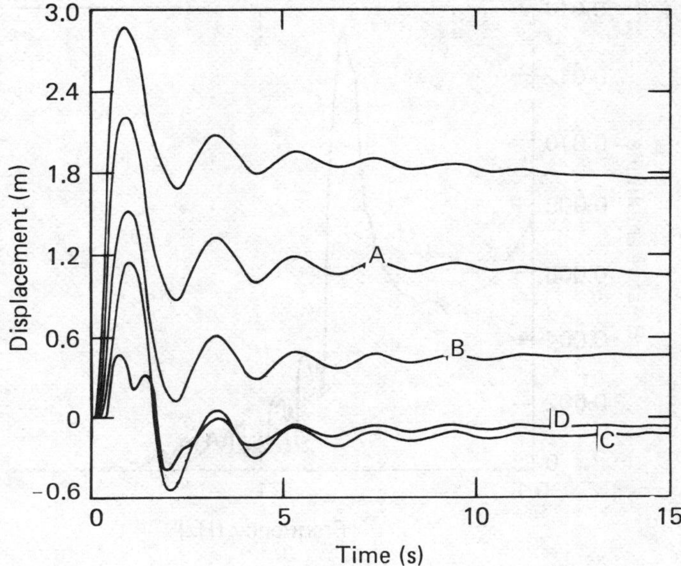

FIG. 2. Vertical compressional-wave resonance
in Yucca Flat at the Nevada Test Site. Integrated velocity-gauge signals from Starwort:
(top) gauge 22 UV, (A) gauge 23 UV, (B) gauge
24 UV, (C) gauge 25 UV, and (D) gauge 51 UV
(from Wheeler, Preston, and Frerking, 1976).

equal to four compressional wave transit times between the free
surface and the Paleozoic rock surface, and this resonance period
is independent of explosion yield and depth. The velocity gauge
data were examined for a total of eight tests; this resonance was
found in seven of the eight cases (Wheeler, Preston, and Frerking
1976).

This resonance phenomenon may be related to Rayleigh wave
generation by explosions. In a theoretical study, Hudson and
Douglas (1975) noted that when a sharp impedance contrast exists
in a plane-layered model of the crust, the Rayleigh wave group
velocity minimum in the fundamental mode occurs close to a period
equal to four times the travel time of P-waves from the surface to
the interface.

2.7 Cavity Formation and Residual Stresses

The cavity gases, either rock vaporized in a nuclear
explosion or detonation products of a chemical explosion, are
initially at a pressure of several hundred thousand atmospheres or
more. This pressure is applied to the surrounding rock in the

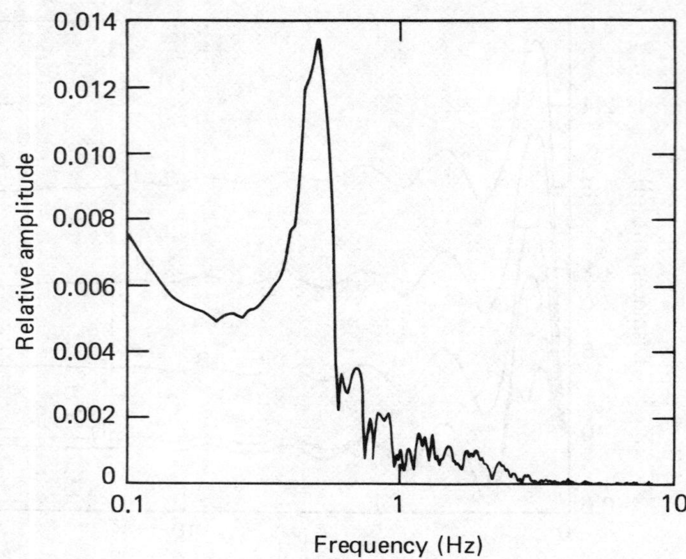

FIG. 3. Vertical compressional-wave resonance
in Yucca Flat at the Nevada Test Site. Fourier
spectrum of the signal from velocity gauge 23
UV (Fig. 2) in Starwort: time interval is 4
to 30 s of signal record (from Wheeler,
Preston, and Frerking, 1976).

case of a "tamped" explosion (the special case of an explosion in
a large preexisting cavity is discussed in Section 6.1). The
effects of this pressure are not only the generation of a shock
wave that inelastically deforms the rock (discussed in Section
2.4), but also the outward displacement of the surrounding rock as
the cavity expands. The outward propagation of the principal
inelastic stress wave and the cavity growth are intimately related
at early times when rock is being vaporized. When rock vaporization no longer occurs, the stress wave "breaks away" from the
cavity boundary, and cavity expansion becomes independent of the
principal outgoing stress wave--until this wave returns as a
rarefaction wave reflected from the free surface. Except for
perturbations by the surface reflection, the later phases of
cavity dynamics are functions of the thermodynamic properties of
the cavity gases, the mechanical properties of the surrounding
rock, and the overburden pressure. The final dynamic cavity
pressure of a contained explosion may range from one-third to
twice the initial overburden pressure, depending upon the mechanical properties of the surrounding rock. The creation of this
cavity results in a significant change in the stress pattern in
the surrounding rock. These stress changes are probably related

to the aftershocks that are sometimes observed following an underground explosion.

The phenomena of rock vaporization, inelastic rock deformation, and cavity expansion are schematically illustrated in Fig. 4.

2.8 Cavity Collapse and Aftershocks

After the explosion, any coincident tectonic strain release, and spall have occurred, two explosion-related phenomena can generate seismic signals at later times: cavity collapse and aftershocks.

In general, the cavities produced by contained explosions are not stable. The only known exceptions are cavities in salt formations; e.g., those produced by the 3-kt Gnome (Rawson 1963) and 5.3-kt Salmon (Rawson, Taylor, and Springer 1967) explosions in the U.S., and by 1.1- and 25-kt explosions in the USSR (Kedrovskii 1970).

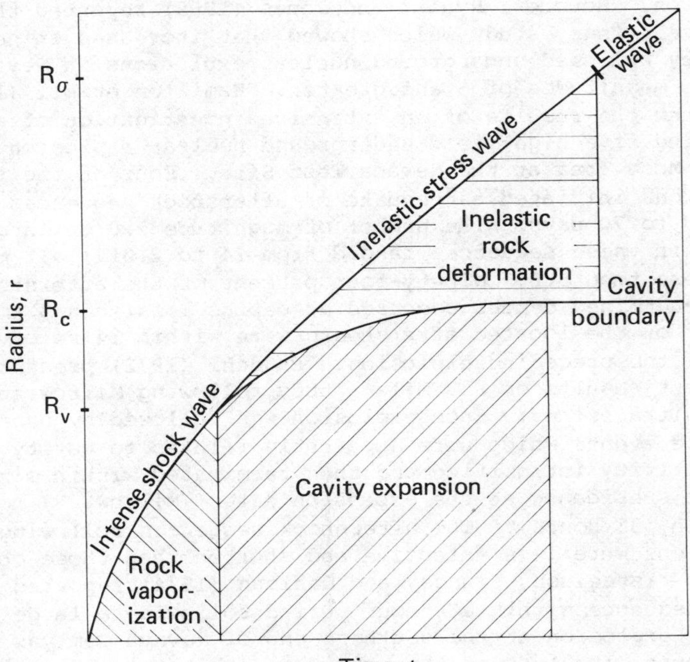

FIG. 4. Radius-vs-time relation for an underground nuclear explosion with regard to rock vaporization, cavity expansion, and inelastic rock deformation (from Rodean 1971a).

Cavity collapse forms a "chimney" which may or may not extend to the surface, forming a subsidence crater (Boardman, Rabb, and McArthur 1964). Springer and Kinnaman (1971, 1975) list surface collapse intervals that have been observed following US nuclear explosions; the intervals range from minutes to hours--and even years in a few exceptional cases. Smith (1963) compared P, SV, and Rayleigh waves from explosions and subsequent cavity collapses at the Nevada Test Site. He noted that the amplitudes of these phases from collapses are significantly smaller than from the corresponding explosions. He also noted that the phases of Rayleigh waves from collapses are reversed relative to those from the corresponding explosions. These observations have been confirmed by others, including McEvilly and Peppin (1972) who also found that the Rayleigh to P_n amplitude relations for collapses are similar to those for earthquakes and explosion aftershocks, and different from those for explosions.

It was noted in the preceding section that the formation of a cavity by an underground explosion changes the stress distribution in the surrounding rock. Brune and Pomeroy (1963) found that small earthquakes were triggered near the site of the Hardhat explosion. Boucher, Ryall, and Jones (1969) reported the results of a preliminary study which showed that increases in seismic activity followed underground nuclear explosions in Nevada with seismic magnitudes of 5 and greater. Hamilton et al. (1972) presented the results of an extensive investigation of aftershocks following five high-yield underground nuclear explosions in the Pahute Mesa area at the Nevada Test Site. Four of the five explosions initiated earthquake or aftershock sequences lasting from 10 to 70 days. The number of magnitude 2.0 or larger aftershocks in these sequences ranged from 24 to 2,012; all magnitudes were less than 5.0. Ninety-four percent of the aftershocks with well-determined depths occurred at depths less than 5 km, and 95 percent of the located aftershocks were within 14 km of ground zero of the preceding explosion. Engdahl (1972) presented the different results of a similar study following Milrow and Cannikin on Amchitka Island. Each explosion was followed by hundreds of discrete events which were apparently related to cavity collapse. This activity intensified and then terminated within minutes of surface subsidence at the explosion sites (Milrow, 37 hours; Cannikin, 38 hours). The aftershock sequences following these explosions were less extensive and shorter than those observed in Nevada. Israelson, Slunga, and Dahlman (1974) reported an after- shock sequence within a 4-hour period following a large underground nuclear explosion at the southern end of Novaya Zemlya. McEvilly and Peppin (1972) found that the Rayleigh to P_n amplitude relations for aftershocks are different from those for explosions at the Nevada Test Site.

2.9   Thermal Effects

It was noted in preceding sections that shock (inelastic stress) waves involve irreversible thermodynamic processes that deposit thermal energy and produce irreversible deformation in the rock. Analysis of postshot rock temperatures for eight nuclear explosions at the Nevada Test Site indicates that 90 to 95 percent of the total energy released is deposited as residual thermal energy if the explosion is completely contained (Heckman 1964). More detailed thermal analysis of the Salmon explosion indicates that about 90 percent of the total energy released was deposited within 50 m of the explosion (Edwards and Holzman 1968). This percentage does not include the energy dissipated in crushing and fracturing the salt out to 90 m from the edge of the 17-m-radius Salmon cavity (Rawson, Taylor, and Springer 1967).

3.   SPHERICAL EXPLOSIONS: THEORY AND MEASUREMENTS

A contained underground nuclear explosion is, generally, approximately spherical. This sphericity is convenient because it permits one to model an explosion in only two variables: time and a radial spatial coordinate. In this section, it is assumed that the geometry of explosions is spherical; nonspherical geometry is considered in the following section. In the remainder of this lecture, all explosions are understood to be nuclear, except when explicitly identified as chemical.

3.1   Simple Theoretical Models

A number of simple theoretical models have been developed to represent explosions; a selected few are described in the following paragraphs. These models demonstrate some of the physical principles involved in explosions, and certain mathematical solutions define upper or lower limits for some phenomena.

The simplest physical system used to model seismic wave radiation from explosions and earthquakes is a spherical cavity in a homogeneous, isotropic, infinite elastic solid. Solutions for the output (compressional elastic wave motion) as functions of time and the radial spatial coordinate may be obtained as functions of the input (cavity pressure variation with time) and the following parameters: cavity radius, medium density, and the Lamé constants (λ and μ) that define the elastic properties of the solid. Jeffreys (1931) used the assumption that $\lambda = \mu$ in obtaining the first solution of this problem; Kawasumi and Yosiyama (1935) were the first to obtain a general solution in terms of λ and μ. We will use this model in several subsequent parts of this lecture. For the present, we will use the result that, in our assumed

elastic solid, the peak particle velocity in a spherical wave in
the far field varies as

$$u \propto r^{-1},$$ (1)

where u is the peak particle velocity and r is the radial
coordinate in space. This relation is consistent with zero
inelastic energy dissipation because the kinetic energy in the
wave is proportional to $u^2/2$, the surface area of a sphere with
radius r is $4\pi r^2$, u varies inversely with r [Eq. (1)], and the
product $(u^2/2)(4\pi r^2)$ is proportional to the constant 2π.

An ideal elastic medium is a useful mathematical fiction; all
real materials are inelastic to some degree, even at very low
stress levels. This "almost elastic" behavior of solid materials
is often called "anelastic" or "viscoelastic." The mathematical
solutions for wave propagation in anelastic or viscoelastic solids
are more complex than in the elastic case. Viecelli (1973a)
obtained a similarity solution for a spherical compressional wave
in a viscoelastic solid described by the Voigt model. His analytic
result is the asymptotic solution for the wave generated by a
spherical cavity expanding at constant velocity. He found that

$$u \propto r^{-3/2}$$ (2)

in the far field. In this case, the kinetic energy in the wave is
dissipated by viscous effects as indicated by the product
$(u^2/2)(4\pi r^2)$ being proportional to $2\pi r^{-1}$.

In work done during World War II, but published later, Sedov
(1959) in the USSR, Taylor (1950a) in the United Kingdom, and von
Neumann (1963) in the US independently obtained similarity solutions for an intense point explosion, a simple model for a nuclear
explosion. Taylor (1950b) successfully applied his solution to
the analysis of the first nuclear explosion. This was Trinity, an
atmospheric explosion in New Mexico in 1943. The point source
solution is also a good approximation to the initial stage of an
underground nuclear explosion because, at early times, the rock
vapor is a highly ionized gas. The properties of such a high
temperature, high pressure gas are independent of initial chemical
composition and crystalline structure. In the point-source
solution,

$$u \propto r^{-3/2}.$$ (3)

Surprisingly, this result is identical to the viscoelastic
solution given by Eq. (2). Perhaps this is because both are
similarity solutions.

Zel'dovich and Raizer [1967, Ch. XII (4) 21] presented a solution for a strong explosion in an infinite porous medium. They assumed that the porous medium is compressed to the density of the continuous medium, and that the continuous medium is incompressible. In other words, they assumed that the shock is strong with respect to the strength of the porous material, but weak with respect to the compressibility of the continuous material. They found that

$$u \propto r^{-3n/2}, \tag{4}$$

where n is bounded by the limits n > 1 (approaching zero medium porosity) and n < 2 (approaching 100 percent porosity).

The radial variations of the peak radial stress (above equilibrium) for the three cases corresponding to Eqs. (1), (3), and (4) are also of interest. For an elastic wave, corresponding to Eq. (1),

$$\Delta\sigma_r \propto r^{-1}. \tag{5}$$

For a strong shock in a continuous medium, corresponding to Eq. (3),

$$\Delta\sigma_r \propto r^{-3}. \tag{6}$$

For a strong shock in a porous medium, corresponding to Eq. (4),

$$\Delta\sigma_r \propto r^{-3n}, \tag{7}$$

where 1 < n < 2 as in Eq. (4).

3.2 Stress Wave Measurements in Geological Media

Peak-radial-stress measurements in the vicinity of a number of underground nuclear explosions are presented in Fig. 5. These explosions in alluvium, dolomite, granite, salt, and tuff had yields on the order of 10 kt. The data, replotted from a study by Holzer (1966), are scaled to a yield of 1 kt (scaling is discussed in Section 6.2). The shock pressures for vaporization are approximately 100 GPa (1 Mbar) and the radii of vaporization are about $2 \text{ m/kt}^{1/3}$ for these materials (Butkovich 1967). Therefore, the reference curve for the strong-shock solution is not necessarily arbitrarily drawn through the point 100 GPa (1 Mbar), $2 \text{ m/kt}^{1/3}$. As noted in Rodean (1971a), the following conclusions can be made from Fig. 5:

● The data for nuclear explosions in all five materials tend toward a common strong-shock solution at peak radial stresses greater than 10 GPa (100 kbar) and scaled radii of a few meters.

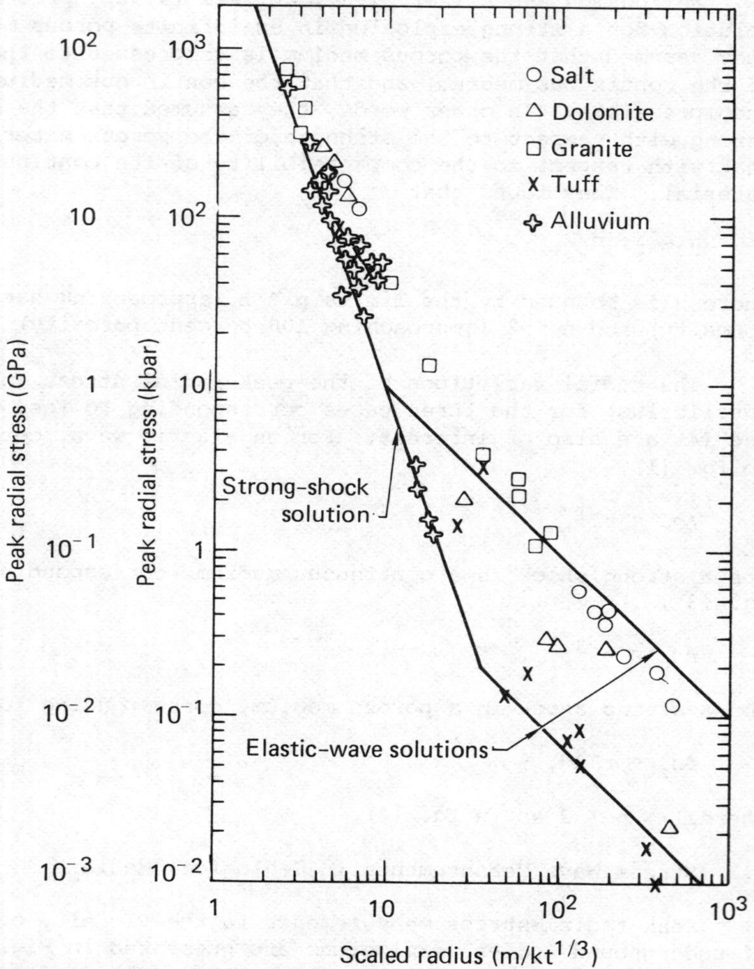

FIG. 5. Measurements and two theoretical solutions for peak (above equilibrium) radial stress produced by underground nuclear explosions (from Rodean 1971a).

● The data for the porous materials, alluvium and tuff, tend to follow the strong-shock solution down to peak radial stresses of about 0.1 GPa (1 kbar).

● The data for granite and salt are so similar that a common curve could be fitted to the points for both materials.

● The data for dolomite are similar to those for granite and salt at scaled radii of a few meters and peak radial stresses above 10 GPa (100 kbar), but the peak radial stresses are intermediate between those in granite and salt and those in tuff at scaled radii of tens and hundreds of meters.

● At scaled radii of hundreds of meters, the data for dolomite, granite, salt, and tuff tend toward (but do not reach) the elastic wave solution.

There are three categories of stress levels surrounding an underground nuclear explosion which must be considered in studying the effects of rock properties on seismic coupling. The first is the very high stress or hydrodynamic region in which the effects of material properties, such as strength, are negligible. This region is illustrated in Fig. 5 at stresses approaching 100 GPa (1 Mbar) and scaled radii of a few meters. The second is the moderate-stress region where material strength has a significant effect and the rock behavior is definitely inelastic. This region corresponds approximately to stresses of 10^{-1} to 10^{1} GPa (1 to 100 kbar) and scaled radii from ten to a few tens of meters in Fig. 5. The third is the low-stress region in which the material response tends toward elastic behavior. In Fig. 5, this region corresponds approximately to stresses less than 10^{-1} GPa (1 kbar) and scaled radii on the order of 100 m and greater. The relative importance of these regions in seismic coupling is a strong function of material properties; however, the lower-stress regions tend to dominate coupling because more of the rock surrounding an explosion is subjected to the lower stresses (the volumetric contribution is proportional to r^3).

In 1970, Larson (1977a) began a comprehensive experimental investigation of the effects of rock properties on seismic coupling efficiency, with emphasis on the moderate- and low-stress regions. He conducted a series of small-scale high-explosive experiments in the 15 geological materials listed in Table 1. Westerly granite and Nugget sandstone were selected to represent high-strength, low-porosity rocks, and Blair dolomite and poly-crystalline salt were chosen to represent moderate- to low-strength, low-porosity rocks. Dry Mt. Helen tuff and Indiana limestone were chosen to represent rocks with large values of dry porosity. Water-saturated samples of the latter two rocks, saturated Tunnel tuff, and water were selected to demonstrate the effects of water saturation. Ice and two frozen soils, Ottawa banding sand and West Lebanon glacial till, were included to represent permafrost and to provide information on the effects associated with the melting-ice phase transition. The data on the Kemmerer coal were available and included for the sake of completeness.

TABLE 1. Properties of the fifteen geologic materials selected
for study by Larson (1977a).

Geologic material	Bulk density (Mg/m^3)	Dry porosity (%)	Longitudinal sound speed[a] (km/s)
Westerly granite	2.65	1	4.80
Nugget sandstone	2.55	4	3.45
Blair dolomite	2.84	1	5.00
Polycrystalline NaCl	2.13	1.6	4.10
Mt. Helen tuff (dry)	1.46	40	2.75
Mt. Helen tuff (saturated)	1.86	0	2.60
Indiana limestone (dry)	2.28	16	4.20
Indiana limestone (saturated)	2.37	8	4.35
Tunnel tuff (saturated)	1.73	14	–
Ice	0.92	0	3.35
Frozen Ottawa banding sand (ice saturated)	2.03	0	4.40
Frozen West Lebanon glacial till (ice saturated)	2.08	5.5(?)	3.45
Frozen West Lebanon glacial till (50% ice saturated)	1.96	19	2.50
Water	1.00	0	1.45
Kemmerer coal	1.30	0	2.25

[a]Or compressional wave velocity.

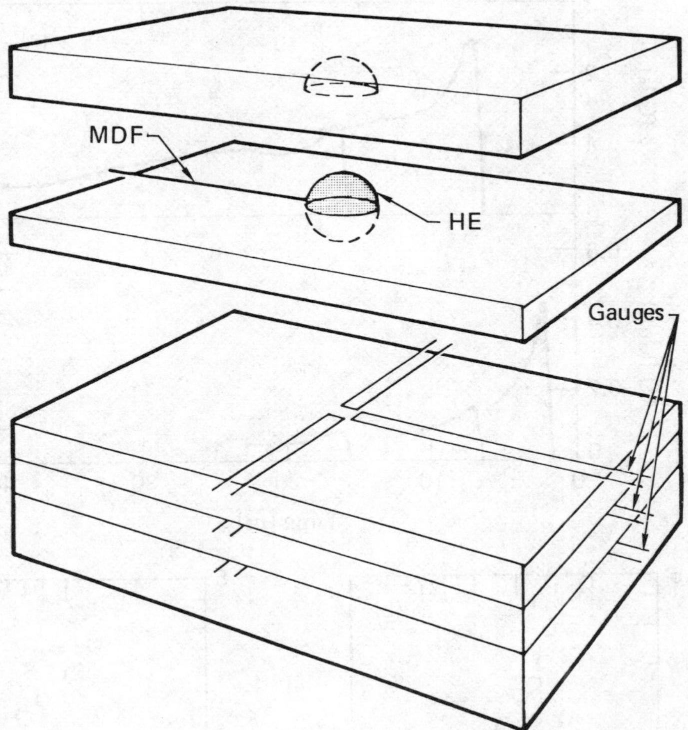

FIG. 6. Schematic drawing of Larson's (1977a)
high-explosive (HE) experimental assemblies for
dynamic stress-wave measurements in geological
materials. The LX04 HE is initiated with a
mild detonating fuse (MDF) attached to a deto-
nator within the sphere of HE. The entire assembly is placed in the center of a large elec-
tromagnet when the experiment is conducted.

Larson's experimental assemblies are schematically illustrated in Fig. 6. The assemblies formed a cube about 0.35 m on a
side. The high-explosive spherical charges had radii of either
1.9×10^{-2} m or 9.5×10^{-3} m. Gauges, placed radially from
the spherical explosive charges as shown in Fig. 6, recorded
particle velocity and radial-stress time histories.

Larson compared seismic coupling efficiency using peak
particle velocity and peak radial stress attenuation as functions
of scaled radius (in this case, the ratio of the radial distance
to the gauge to the radius of the explosive charge). The experimental results for a strong, low-porosity rock (Westerly granite)

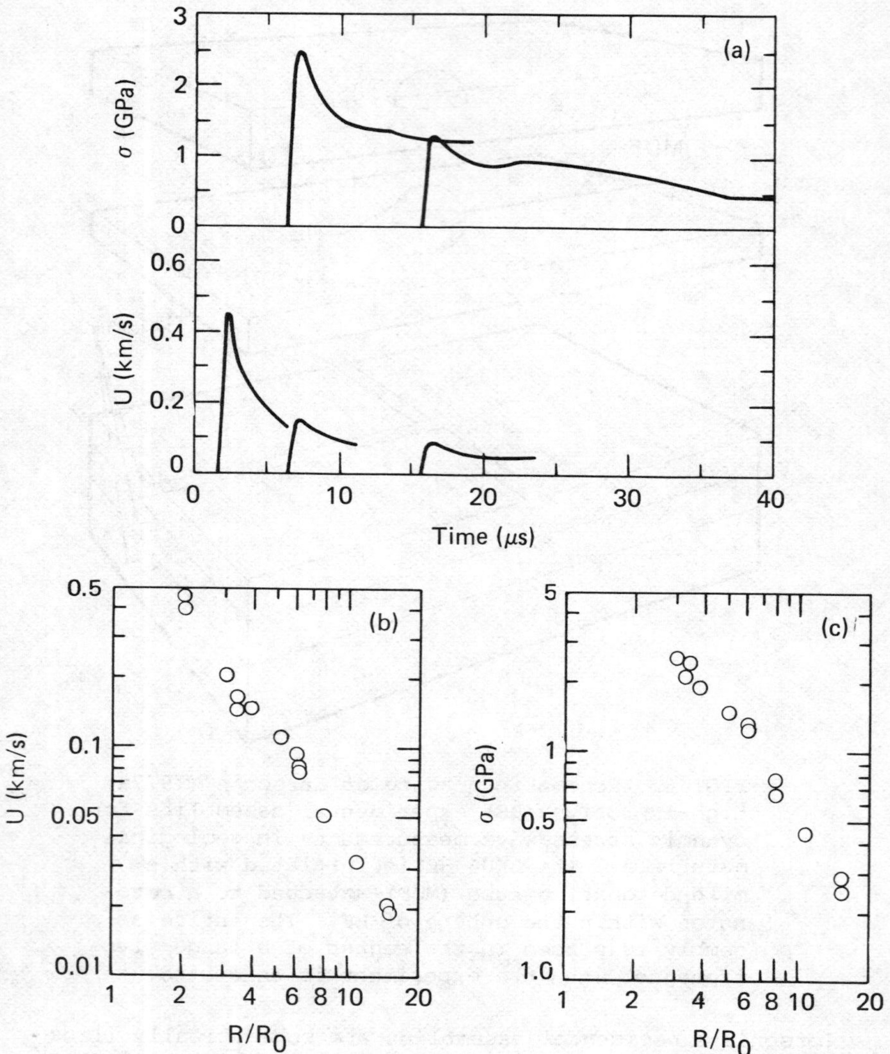

FIG. 7. Experimental results for Westerly granite: (a)
particle-velocity and stress vs time, (b) peak particle-
velocity vs reduced radius, and (c) peak stress vs reduced
radius (from Larson 1977a).

and for a weak, high-porosity rock (dry and saturated Mt. Helen
tuff) are presented in Figs. 7, 8, and 9, respectively. Note the
steep stress-wave fronts in the granite [Fig. 7(a)] and the
dispersive stress waves in the tuff [Figs. 8(a) and 9(a)]. Also
note that water saturation reduces the stress-wave attenuation and

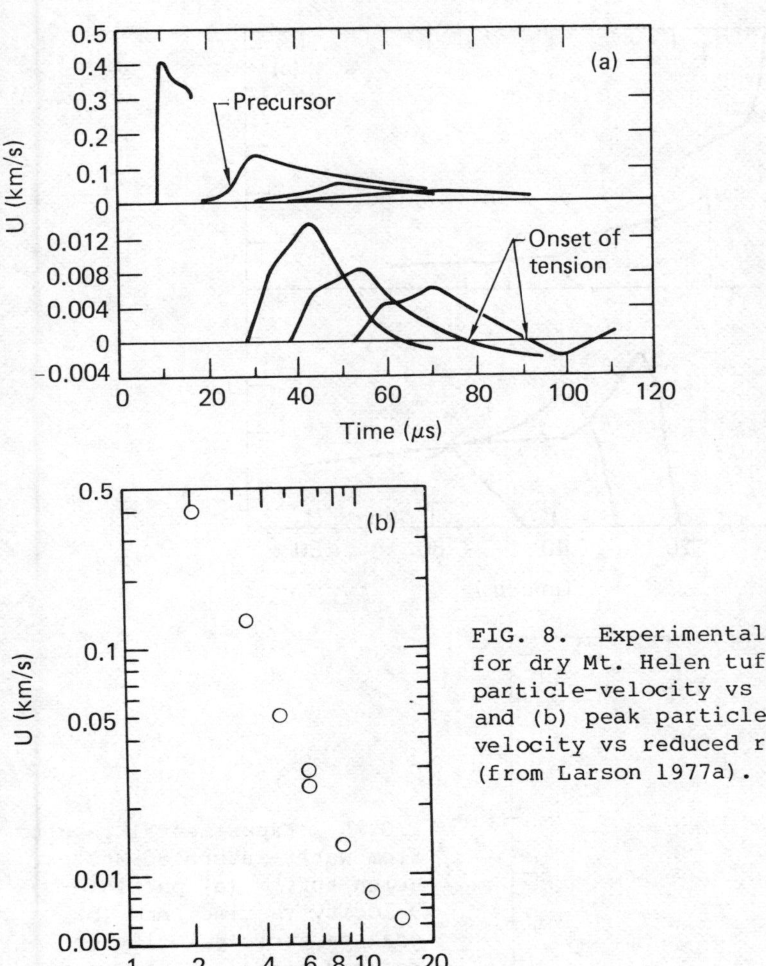

FIG. 8. Experimental results
for dry Mt. Helen tuff: (a)
particle-velocity vs time,
and (b) peak particle-
velocity vs reduced radius
(from Larson 1977a).

dispersion in tuff [Figs. 8(b) and 8(a) for dry vs Figs. 9(b) and
9(a) for saturated tuff]. The data for H_2O in two natural states
are compared in Figs. 10 (water) and 11 (polycrystalline ice).
Note that the stress waves are more dispersive and that the stress-
wave attenuation is greater in ice than in water. The extremes of
stress-wave attenuation, from the lowest to the highest, for these
15 geological materials are summarized in Figs. 12 and 13. The
lowest attenuation was observed in water, the lowest attenuation
observed in rock was in granite, and the highest attenuation
observed in rock was in dry Mt. Helen tuff; these data are
presented in Fig. 12. The attenuation for water and the frozen
materials is presented in Fig. 13; the attenuation in 50% saturated

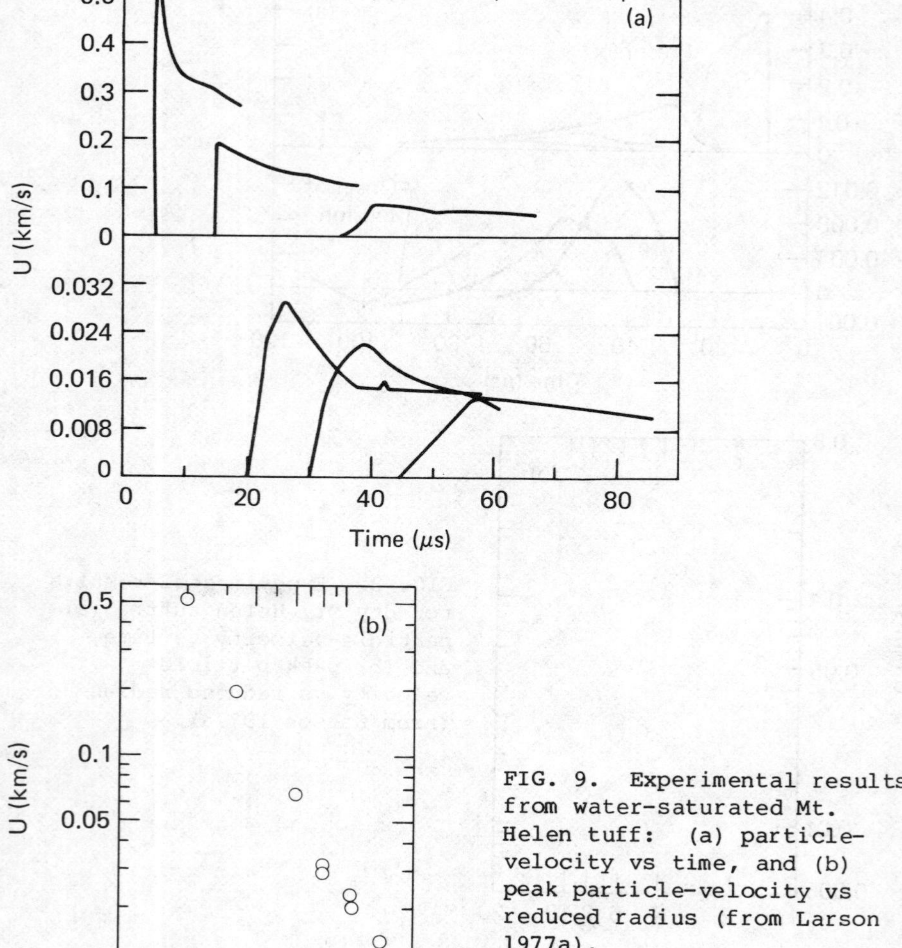

FIG. 9. Experimental results from water-saturated Mt. Helen tuff: (a) particle-velocity vs time, and (b) peak particle-velocity vs reduced radius (from Larson 1977a).

West Lebanon glacial till was the highest observed in all the materials tested.

Larson found that the peak-particle velocity attenuation in the 15 geological materials could be approximated by the relation $u \propto r^{-n}$ where u is peak-particle velocity, r is scaled radius, and n is an empirical constant. His attenuation exponent data are summarized in Table 2 and are compared with attenuation exponents

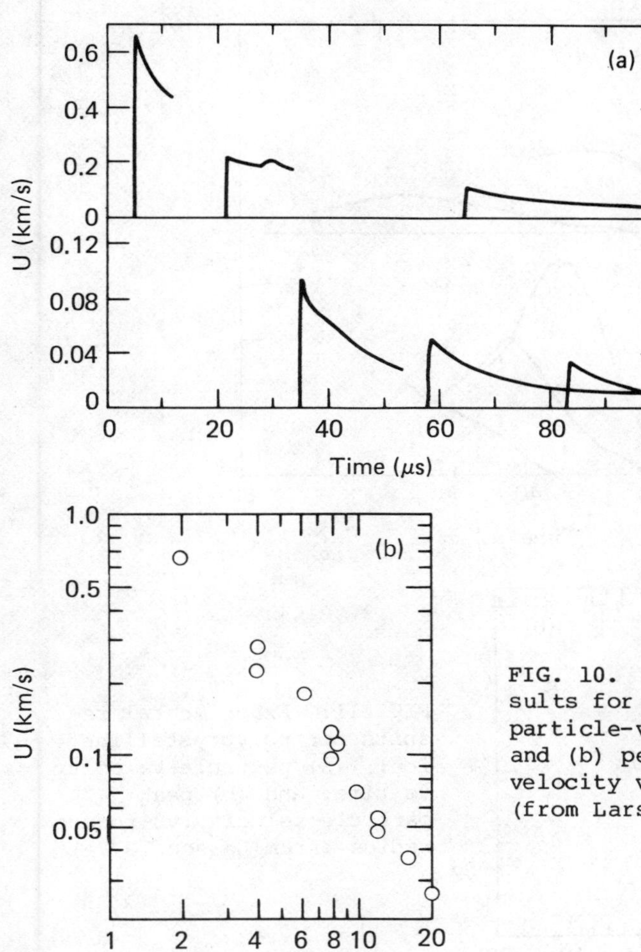

FIG. 10. Experimental results for water: (a) particle-velocity vs time, and (b) peak particle-velocity vs reduced radius (from Larson 1977a).

from the theoretical models represented by Eqs. (1)-(4). Note that none of the materials exhibited ideal elastic behavior (n = 1), that water and the low-porosity rocks approximated visco-elastic or strong-shock behavior (n = 1.5), and that the behavior of the porous rocks is bracketed by the strong-shock response of a porous solid (n = 1.5 to 3). Larson's conclusions include the following:

● The large difference between the peak-particle velocity attenuation relation for water, $r^{-1.4}$, and that for an ideal elastic solid in the far field, r^{-1}, is attributed to viscosity and finite strain. In consolidated rocks, the observed limiting

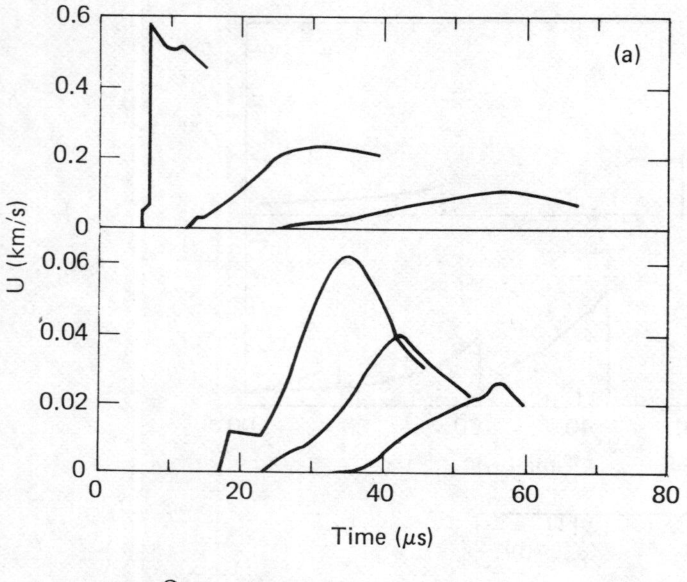

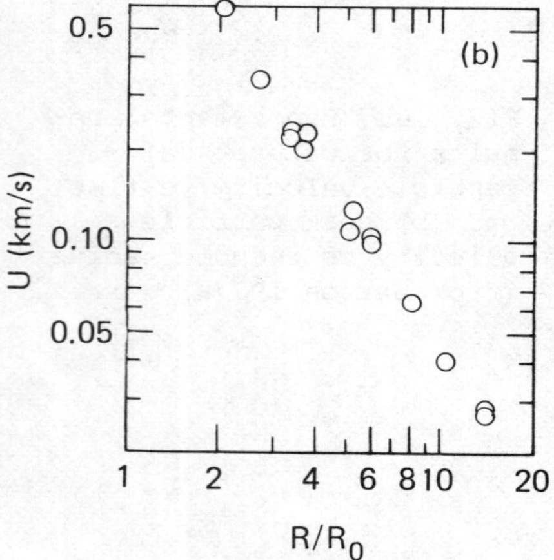

FIG. 11. Experimental results for polycrystalline
ice: (a) particle-velocity
vs time, and (b) peak
particle-velocity vs reduced
radius (from Larson 1977a).

attenuation relation, $r^{-1.5}$, is attributed to viscosity and
finite strain, as well as to the presence of microcracks and other
grain-boundary effects. These attenuation rates appear to be
limiting values for real materials.

● The large difference in coupling between water, $r^{-1.4}$,
and water-saturated Mt. Helen tuff, $r^{-1.9}$, suggests that two-
component interactions lead to both dispersion and dissipation as
the wave propagates through saturated rock. The smaller attenuation observed in water-saturated coal, $r^{-1.6}$, is attributed to
the absence of two-component interactions. This absence is the

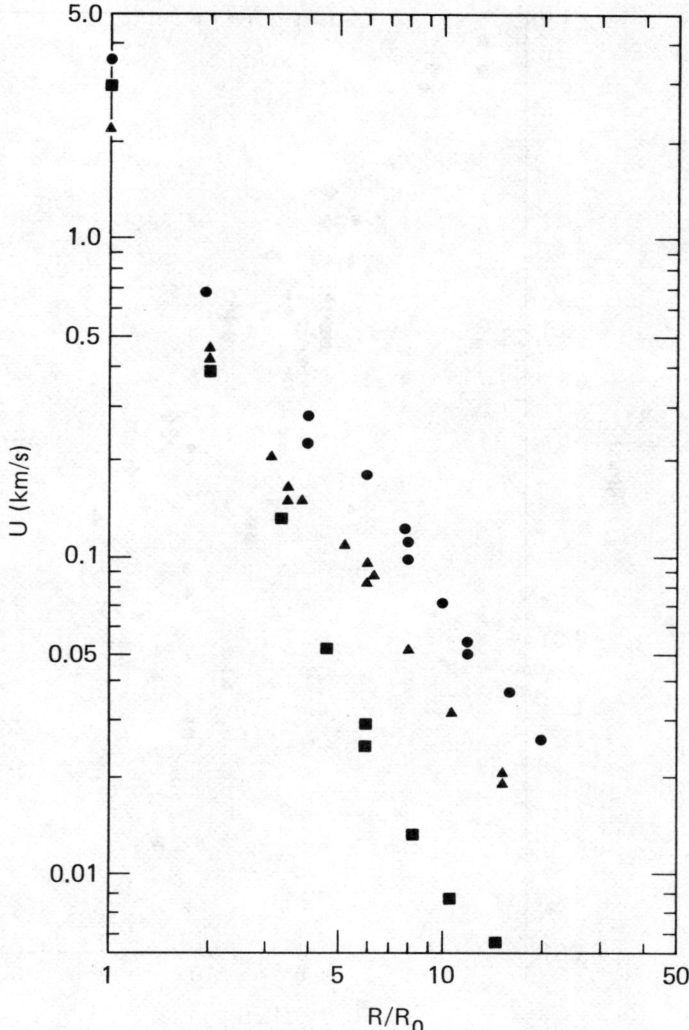

FIG. 12.   A comparison of peak particle-
velocity attenuation for water (●), Westerly
granite (▲), and dry Mt. Helen tuff (■) (from
Larson 1977a).

result of chemical bonding of the water into the coal matrix.   The
attenuation that does occur in coal is caused by grain-boundary
effects that are associated with the highly fractured structure.

● The difference in coupling between water, $r^{-1.4}$, and
ice, $r^{-1.5}$, is attributed to the ice-I melting transition.
Introduction of soil into the ice matrix causes a significant

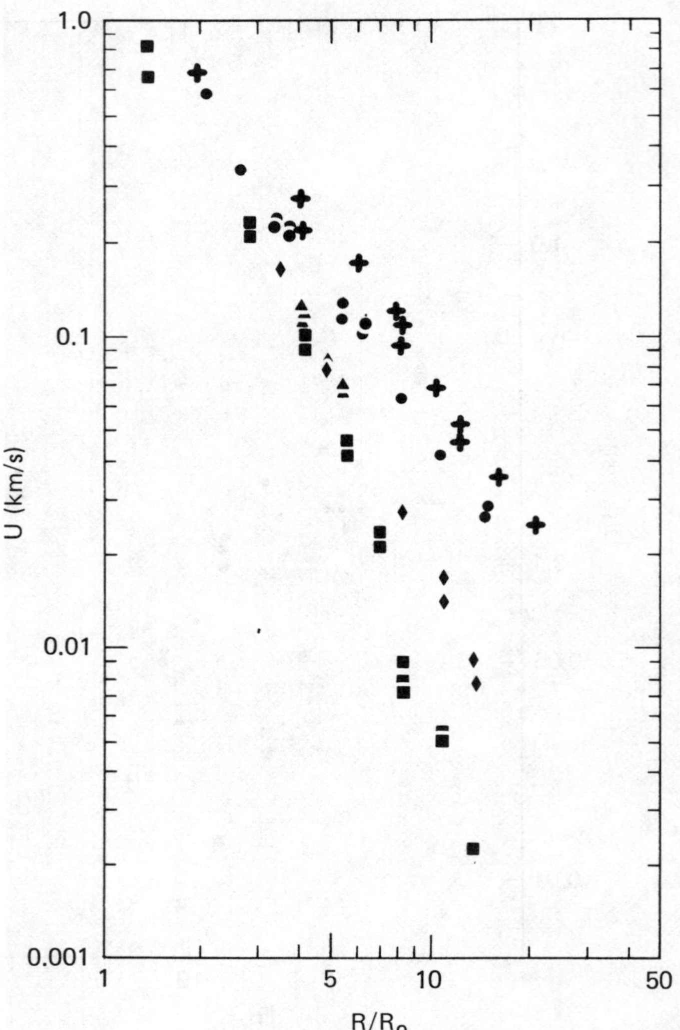

FIG. 13. A comparison of peak particle-
velocity attenuation for water (+) and frozen
materials: ice (●), Ottawa bonding sand (▲),
50% ice-saturated West Lebanon glacial till
(■), and 100% ice-saturated West Lebanon
glacial till (♦) (from Larson, 1977a).

increase in attenuation rates to $r^{-2.0}$. This increased
dissipation is thought to be associated with hysteresis in the
transition that results from the presence of the soil. The
coupling in 50% ice-saturated soil, $r^{-2.7}$, is probably the

TABLE 2. Exponents for radial attenuation of peak particle velocity observed in small-scale high-explosive experiments in 15 geological materials (Larson 1977a). Theoretical attenuation exponents are presented for reference from Eqs. (1)–(4).

	Exponent n for radial attenuation ($u \propto r^{-n}$)		
	Low stresses	Overall average	High stresses
[Far-field elastic wave, Eq. (1)]		(1.0)	
Water		1.4	
[Viscoelastic wave, Eq. (2); strong shock, Eq. (3)]		(1.5)	
Ice		1.5	
Westerly granite		1.5	
Nugget sandstone		1.5	
Blair dolomite		1.6	
Kemmerer coal		1.6	
Polycrystalline NaCl		1.6 to 1.8	
Indiana limestone (saturated)	1.4		2.2
Indiana limestone (dry)	1.6		2.4
Mt. Helen tuff (saturated)		1.9	
Mt. Helen tuff (dry)	1.5		2.5
Tunnel tuff (saturated)		2.1	
Both frozen soils (saturated)		2.0	
Frozen West Lebanon glacial till (50% saturated)		2.7	
[Strong shock in a weak porous solid, Eq. (4)]		(1.5 to 3.0)	

result of the yielding of the weak ice matrix which allows pore
collapse at very low stresses.

● For rocks and soils with significant dry porosity, the
peak-particle velocity attenuation relation ranges from $r^{-2.5}$ to
$r^{-2.7}$. For these materials, dissipation of energy in crushing
pores dominates all other inelastic mechanisms and continues until
a stress is reached that corresponds to matrix strength.

● Therefore, the order of importance for material properties
in determining relative coupling is (1) porosity, (2) strength,
(3) multiple-component interactions, and, in frozen materials, (4)
phase transformation.

One may ask, "What is the quantitative relation between the
data from nuclear explosions and the data from small-scale experiments with high explosives?" Larson successfully scaled his small
scale, high explosive, peak particle-velocity data for salt (NaCl)
to the data for the 5.3-kt Salmon explosion in salt (Rogers 1966).
Trulio (independently and simultaneously) found that the data from
the higher-yield Cowboy experiments with tamped high-explosives
(Murphey 1961) could be scaled with equal success to the Salmon
data (Larson 1977b; Trulio 1977). [See footnote 1, after text.]

A result of their subsequent joint work is presented in
Fig. 14: the data from the latter three sources, scaled over eight
orders of magnitude of explosive energy release (see Section 6.2
for a discussion of scaling), form a common peak-particle velocity
vs scaled radius relation over corresponding large ranges of
velocity and radius. The combined data are approximated by the
relation $u \propto r^{-1.7}$, which is consistent with the exponent values
of -1.6 to -1.8 given in Table 2 for Larson's data alone. [Attempts
to scale data from small-scale high-explosive experiments and
nuclear-explosion data in other materials have not been successful
(Larson 1977b). These failures are attributed to large-scale
inhomogeneities, such as block motion, in the nuclear explosions,
or to differences in material properties for the two sets of
explosions.]

With respect to the above, particle-velocity data that fit a
relation $u \propto r^{-n}$, with $1 < n < 2$, *can* represent elastic behavior.
This is because $u \propto r^{-1}$ in the far field [Eq. (1)], but nearer
the source, the term $u \propto r^{-2}$ also contributes to the particle
velocity in spherical elastic waves [Eq. (30)]. The transition
between the near and far fields for particle velocity, at which
the contributions of the near and far field terms [Eq. (30)] are
equal in amplitude, is defined by the relation $r\omega = \alpha$ (Meyer 1964)
where r is the radial coordinate, ω is the angular frequency, and
α is the compressional wave velocity. The far field effects are
dominant only if $r\omega \gg \alpha$.

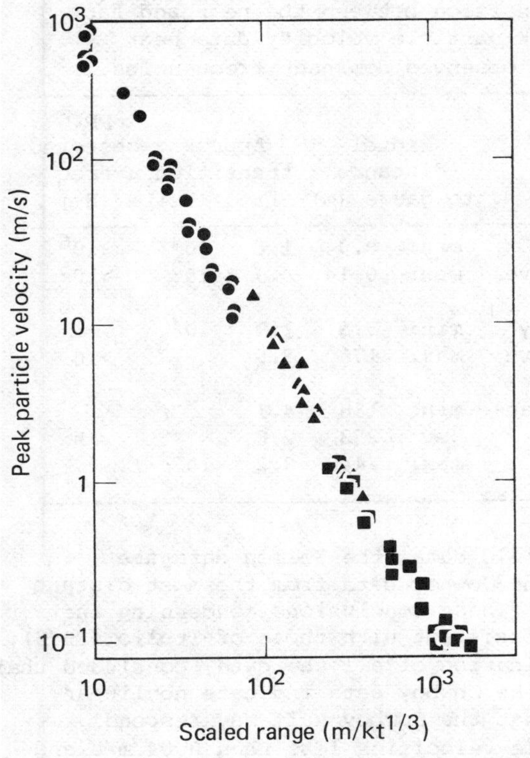

FIG. 14. Peak particle-velocity vs scaled range for chemical
and nuclear explosions in salt: Larson's (1977a) small-scale
high-explosive experiments (LX04), the Cowboy experiments
(low-density TNT), and the Salmon experiment (nuclear explosive) (from Larson 1977b, and Trulio 1977).

Table 3 gives a summary of the conditions for transitions
between the near and far fields for the three sets of peak-
particle velocity data for explosions in salt (Fig. 14), together
with estimates of observed dominant frequencies in the leading
edge of the stress waves. Larson's (1977a) particle-velocity time
histories in salt are similar to those shown in Figs. 7(a) and
9(a) with rise times from zero to peak particle-velocity that
correspond to frequencies of approximately 10^5 to 10^6 Hz. A
gauge record in Murphey's (1961) paper indicates a dominant
frequency greater than 10 Hz for a tamped Cowboy explosion. A
gauge record in Rogers's (1966) paper indicates a dominant
frequency of about 5 Hz. These Cowboy and Salmon observations are
consistent with Trulio's (1979) observation that the Salmon and
Cowboy (scaled to Salmon) particle-velocity data are dominated by
frequencies in the 1- to 10-Hz band. From Table 3, it appears

TABLE 3. Conditions for transition between the near and far fields for three sets of peak-particle velocity data near explosions in salt, and some observed dominant frequencies.

References	Experiments	Radial distance to gauge (m)		Approx transition freq (Hz)	Approx observed freq (Hz)
Larson (1977a)	Small-scale	min:	0.04	1.6×10^4	$\approx 10^6$
	high-explosive	max:	0.14	4.7×10^3	$\approx 10^5$
Murphey (1961)	Tamped Cowboy	min:	5.9	1.2×10^2	
	high-explosive	max:	178	3.8	>10
Rogers (1966),	Salmon nuclear-	min:	168	4.1	
Patterson	explosive		253	2.8	≈ 5
(1966)		max:	744	9.2×10^{-1}	

that Larson's data are far-field data, the Salmon data are transition-field data, and the Cowboy data from the most distant gauge may be far-field data. These conclusions concerning the Salmon and Cowboy data are consistent with those of Trulio (1979). Trulio, after a careful examination of all the data, concluded that the Salmon data and most of the Cowboy data indicate nonlinear inelastic attenuation, and that the Cowboy salt may respond "elastically" at peak-particle velocities less than 0.04 m/s and peak radial stress values less than 4 bars (4×10^5 Pa). These low values of velocity and stress were measured on a "decoupled" Cowboy explosion in a mined cavity.

With respect to the rest of Larson's data summarized in Table 2, let us consider the wave forms illustrated in part(a) of Figs. 7-11. The rise times from zero to peak particle-velocity are approximately 1 µs, corresponding to frequencies on the order of 10^6 Hz for Westerly granite [Fig. 7(a)] and water [Fig. 10(a)]. Clearly, the radial decay exponents of 1.4 (water) and 1.5 (Westerly granite) represent inelastic stress-wave decay in the far field, and can be compared with the ideal elastic decay exponent of unity in the far field. This is also true for the observed decay exponents for Nugget sandstone (1.5), Blair dolomite (1.6), and Kemmerer coal (1.6). On the other hand, the dispersed wave forms for dry Mt. Helen tuff [Fig. 8(a)], saturated Mt. Helen tuff [Fig. 9(a)], and ice [Fig. 11(a)] have dominant frequencies of 10^4 to 10^5 Hz. In an ideal elastic version of the experimental assembly shown in Fig. 6, the near-field term may not be negligible at these frequencies. Therefore, it may be that the attenuation exponents of 1.5 (dry Mt. Helen tuff at low stresses, and ice) and 1.9 (saturated Mt. Helen tuff) should be compared

with an ideal elastic radial decay exponent somewhat greater than
unity. The same caution applies to the decay exponents for dry
(1.6) and saturated (1.4) Indiana limestone, Tunnel tuff (2.1),
and the frozen soils (2.0 and 2.7). A spectral analysis of the
particle-velocity wave-form data would serve to resolve this near-
vs far-field issue.

The conditions for the transition from inelastic to "elastic"
response to stress waves are discussed further in Section 3.4.

3.3 Computer Calculations of Underground Explosions

Calculations with modern digital computers have been used in
the study of underground nuclear explosions beginning with the
first such explosion, Rainier (Johnson, Higgins, and Violet 1959).
Computer calculations have been used to predict and interpret
experimental measurements on explosions such as Salmon (Rogers
1966; Patterson 1966). They have been very useful in the
prediction of inelastic effects for engineering applications of
peaceful nuclear explosions (Cherry and Petersen 1970) and in the
analysis of inelastic phenomena of importance in explosion
containment (Terhune et al. 1977a,b). They are also useful in the
study of explosions as sources of elastic seismic waves, but they
have limitations for this application; the most serious limitation
is discussed in this lecture.

Early examples of one- and two-dimensional codes for finite-
difference calculations of elastic and inelastic stress-wave
motion in solid materials are described by Wilkins (1964) and by
Maenchen and Sack (1964). More recent descriptions of one- and
two-dimensional codes used for underground explosion calculations
at the Lawrence Livermore Laboratory are given by Schatz (1973,
1974) and by Burton and Schatz (1975). Similar codes have been
developed and are being used by other organizations.

The above codes are based on a computational technique that
identifies a finite number of material points and determines the
motion in space-time and the physical state as a function of time
of each point. This technique is the use of Lagrangian coordinates which are fixed in the material instead of being fixed in
space as is the Eulerian system. The Eulerian and Lagrangian
coordinate systems and the relations between them are described by
Zel'dovich and Raizer [1966, Ch. I (1) 1-2]. The Lagrangian
coordinates in a one-dimensional code correspond to a linear array
of discrete points spaced along the radial coordinate axis. The
Lagrangian coordinates in a two-dimensional code correspond to a
two-dimensional array of discrete points in the plane defined by
the longitudinal and radial axes of the cylindrical coordinate
system.

The numerical solution for stress-wave propagation through a material is obtained by means of a step-by-step procedure through the feedback loop illustrated in Fig. 15 for each point (Lagrangian coordinate). The equation of motion (based on the fundamental equations for conservation of mass, linear momentum, and angular momentum) provides a functional relation between the applied stress and the acceleration of each point in the coordinate system. The energy conservation equation does not have to be used if there is no energy transport by heat conduction or radiation from one point to another; this is a valid approximation for stress-wave propagation in rock. However, the calculation of internal plus kinetic energy is useful in checking the accuracy of the calculation. The accelerations determined by the equation of motion are integrated over a discrete time interval (Δt) to obtain velocities; velocities are integrated over a discrete time interval (Δt) to obtain displacements; and displacements (relative to other point displacements), in turn, determine strains or deformations. The constitutive or state equations for the material give the relation between the applied strains and the resulting stress field. At the end of each cycle, time is incremented by $+\Delta t$ and the cycle is repeated.

There are several reasons for the present limitations of finite-difference computer calculations in the study of explosions as sources of seismic waves. The principal reason is discussed in this lecture: the constitutive or state equations relating strain and stress that have been developed do not model the inelastic response of rock with sufficient accuracy over the full range of the conditions generated by explosions.

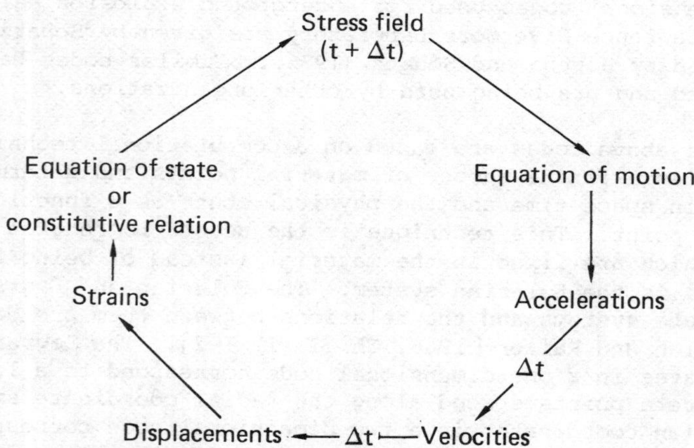

FIG. 15. Feedback loop for finite-difference calculations of stress-wave propagation (from Rodean 1971a).

The constitutive or state relations (hereafter called constitutive relations) are very simple in the case of linear, adiabatic, isotropic, elastic stress-wave propagation: the isentropic bulk modulus for compressional waves in a perfect fluid, and two moduli defined by the Lamé constants for compressional waves $(\lambda + 2\mu)$ and shear waves (μ) in an ideal solid. The problem in calculating the explosion process, including the generation of seismic waves, is that of calculating the propagation of adiabatic stress waves that are initially nonlinear and inelastic, are attenuated by geometric and inelastic effects, and emerge from the explosion as almost-elastic or anelastic waves. The constitutive relations used in present computer programs or codes are very complex; but they approximately model only selected phenomena, not all aspects of stress-wave propagation. For example, let us consider the principal features of the constitutive relations for solids in the one-dimensional SOC73 code (Schatz 1973, 1974). (SOC73 also includes constitutive relations for liquid and gas phases and relations for solid-gas and solid-liquid phase transitions, but they are not considered in this lecture.) SOC73 is based as much as possible on known physical principles and measured rock properties, and as little as possible on nonphysical adjustable parameters.

Before discussing SOC73, it is appropriate to review some fundamentals of stress and strain in isotropic solid materials. The stress tensor Σ has nine components consisting of three normal (σ) and six shear (τ) stresses as shown in the first two terms of Eq. (8):

$$\Sigma = \begin{vmatrix} \sigma_{xx} & \tau_{xy} & \tau_{xz} \\ \tau_{yx} & \sigma_{yy} & \tau_{yz} \\ \tau_{zx} & \tau_{zy} & \sigma_{zz} \end{vmatrix} = \begin{vmatrix} \sigma_{rr} & 0 & 0 \\ 0 & \sigma_{\theta\theta} & 0 \\ 0 & 0 & \sigma_{\theta\theta} \end{vmatrix}. \tag{8}$$

There is a set of orthogonal axes, the principal axes of stress, along which the stresses are purely normal. In the symmetrical spherical coordinate system for an ideal spherical explosion, the stress tensor reduces to the third (right-hand) term in Eq. (8), where

$$\sigma_{\theta\theta} = \sigma_{\phi\phi}, \tag{9}$$

and the maximum shear stress is

$$\tau_m = (1/2)(\sigma_{\theta\theta} - \sigma_{rr}) = (1/2)(\sigma_{\phi\phi} - \sigma_{rr}). \tag{10}$$

The principal stresses can be written in terms of a mean normal stress (by definition, equal to the negative of the hydrostatic or thermodynamic pressure) that determines uniform compression or dilatation,

$$-P = (1/3)(\sigma_{rr} + \sigma_{\theta\theta} + \sigma_{\phi\phi}), \tag{11}$$

and stress deviators that determine distortion,

$$\sigma_{rr} = -P + s_r, \tag{12}$$

$$\sigma_{\theta\theta} = -P + s_\theta, \tag{13}$$

$$\sigma_{\phi\phi} = -P + s_\phi. \tag{14}$$

From Eqs. (9)-(14),

$$s_r = -(4/3)\tau_m, \tag{15}$$

$$s_\theta = s_\phi = (2/3)\tau_m. \tag{16}$$

From these relations, it follows that the stress behavior of a material can be described in terms of a response to hydrostatic stresses and a response to shear stresses. Similarly, the strain can be described as a combination of volumetric strain and distortional strain (Jaeger and Cook 1971, Ch. 2.2, 2.4, 2.8, 2.12).

There are invariants of stress and of stress deviators that are independent of the orientation of orthogonal coordinates; it is obviously desirable to express criteria for material failure in terms of these invariants (Jaeger and Cook 1971, Ch. 2.4, 2.8). The stress invariants are, in spherically-symmetric coordinates,

$$I_1 = \sigma_{rr} + \sigma_{\theta\theta} + \sigma_{\phi\phi}, \tag{17}$$

$$I_2 = -(\sigma_{\theta\theta}\sigma_{\phi\phi} + \sigma_{\phi\phi}\sigma_{rr} + \sigma_{rr}\sigma_{\theta\theta}), \tag{18}$$

$$I_3 = \sigma_{rr}\,\sigma_{\theta\theta}\,\sigma_{\phi\phi}. \tag{19}$$

From Eqs. (11)-(19), the stress invariants can be written as

$$I_1 = -3P + J_1 = -3P, \tag{20}$$

$$I_2 = -3P^2 + 2J_1P + J_2 = -3P^2 + (4/3)\tau_m^2, \tag{21}$$

$$I_3 = -P^3 + J_1P^2 + J_2P + J_3$$

$$= -P^3 + (4/3)P\tau_m^2 - (16/27)\tau_m^3, \tag{22}$$

and the stress deviator invariants are defined as

$$J_1 = s_r + s_\theta + s_\phi = 0, \tag{23}$$

$$J_2 = -(s_\theta s_\phi + s_\phi s_r + s_r s_\theta) = (4/3)\tau_m^2, \qquad (24)$$

$$J_3 = s_r s_\theta s_\phi = -(16/27)\tau_m^3. \qquad (25)$$

Equations (20) and (23) are valid for all sets of principal stress axes, with appropriate substitutions for the subscripts r, θ, ϕ. The first two parts of Eqs. (21), (22), (24), and (25) are similarly valid for all sets of principal stress axes, but the third (right-hand) parts of these equations apply only to the principal stress axes of a spherically symmetric system as defined by Eqs. (8)-(10).

In SOC73 (Schatz 1974), it is assumed that the shear stresses do not affect the volumetric response of a material to changes in hydrostatic pressure, although it is known that this assumption is not correct. Shear stresses can enhance the compaction of a porous material under increasing hydrostatic loading. Shear stresses can also cause dilatation of a compacted material under increasing hydrostatic pressure. These effects are illustrated in Fig. 16. A modification of SOC was developed which included a

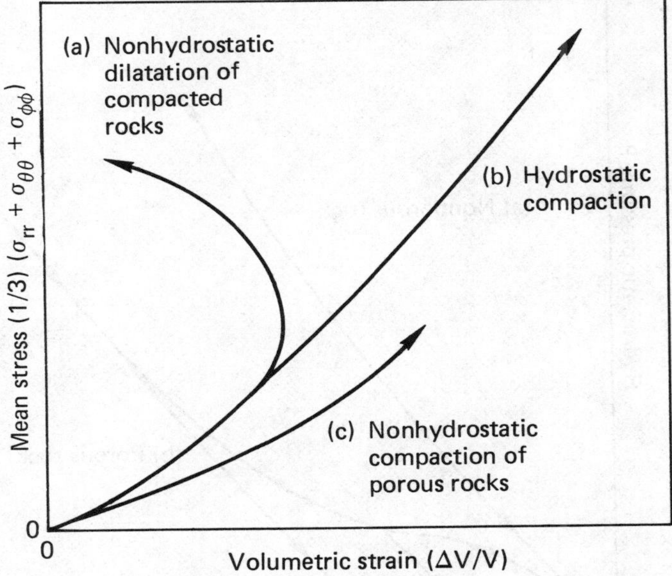

FIG. 16. Three types of quasi-static stress-strain response: (a) dilatation of compacted rocks under increasing nonhydrostatic stress, (b) compression under increasing hydrostatic pressure, and (c) shear-enhanced compaction of porous rock under increasing nonhydrostatic stress.

model for shear-enhanced compaction (Schatz et al. 1977), but the
improved accuracy in modeling this aspect of rock deformation was
not deemed to be worth the large increase of computer time required
for a typical problem. SOC73 uses relations between hydrostatic
pressure (P) and volumetric strain ($\Delta V/V$) that are based on
laboratory measurements of stress and strain in rock samples.
These stress strain relations are illustrated in Fig. 17 for
nonporous and porous rocks. It is assumed that the material is
incrementally elastic with a variable bulk modulus along each
P vs $\Delta V/V$ curve; the shear modulus may be assumed to be constant
or vary with the bulk modulus by assuming that Poisson's ratio is
constant. If the rock is not porous, the loading and unloading
curves are assumed to be identical; that is, there is no inelastic
strain from hydrostatic stress, and there is no hysteresis in the
P vs $\Delta V/V$ plane. If the rock is porous, the pores are crushed at
sufficiently high pressures; this crushing is inelastic, and there
is hysteresis because the inelastic strain is irreversible. The

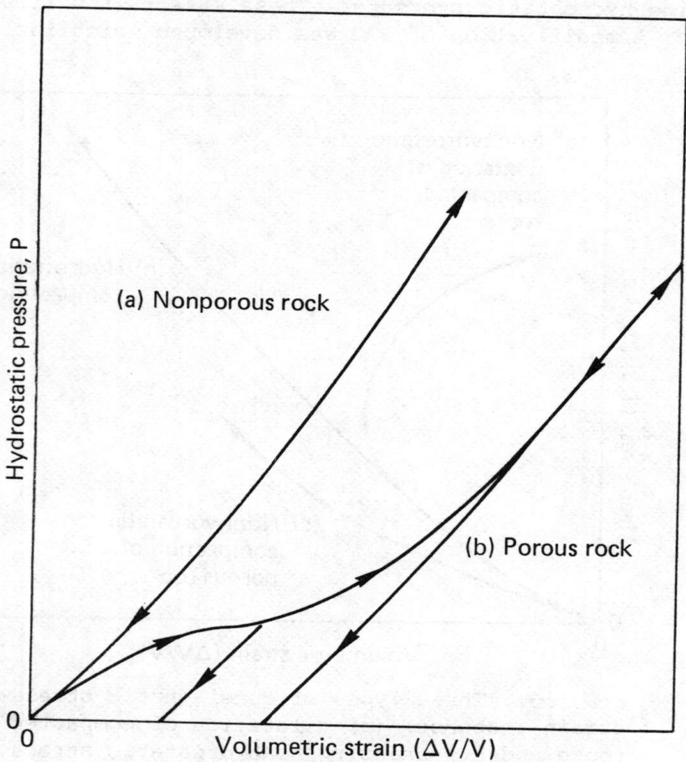

FIG. 17. Hydrostatic loading and unloading of
(a) nonporous rock and (b) porous rock.

inelastic crushing of pores is very effective for dissipation of energy.

In SOC73, it is assumed that distortion in response to shear stress takes place at constant volume. The maximum shear strength of a material is a complex function of the applied stress. Cherry and Petersen (1970) found that the results of various destructive tests (compression, extension, and torsion) on glass, dolomite, limestone, and granite are quite consistent when plotted in terms of the second stress deviator invariant (J_2) vs a combination of the first stress invariant (I_1 or $-P$) and the third stress deviator invariant (J_3) as shown in Fig. 18. In the spherically-symmetric system implicit in SOC73, the invariant functions defining the coordinate axes in Fig. 18 can be simplified to the following, from Eqs. (9)-(16), (24), and (25):

$$|\tau_m| = \left|(1/2)(\sigma_{\theta\theta} - \sigma_{rr})\right| = \left|(1/2)(\sigma_{\phi\phi} - \sigma_{rr})\right| \tag{26}$$

for the vertical axis, and

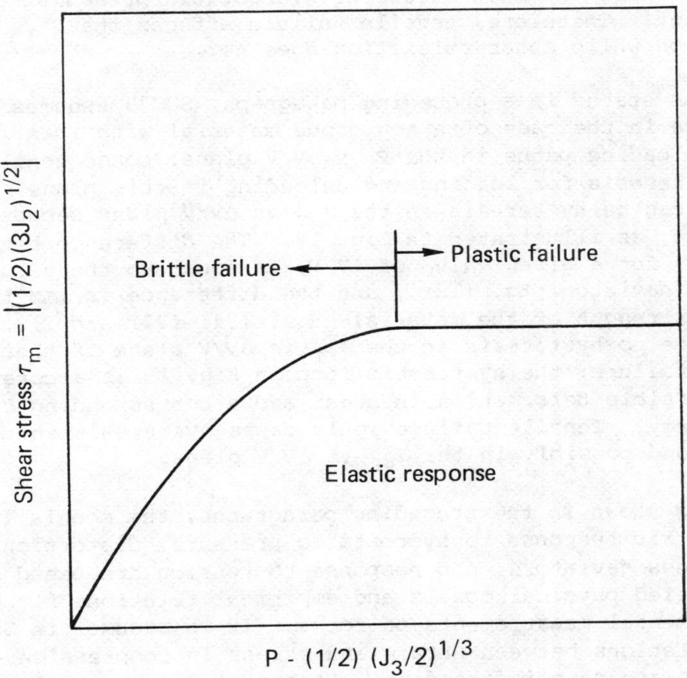

FIG. 18. Shear strength of rocks as a function of confining stress showing elastic response, brittle failure, and plastic failure.

$$P + (1/3)\tau_m = -(1/2)(\sigma_{rr} + \sigma_{\theta\theta}) = -(1/2)(\sigma_{rr} + \sigma_{\phi\phi}) \qquad (27)$$

for the horizontal axis. In SOC73, the maximum shear stess (τ_m) is limited to the values defined by an experimentally determined failure surface as illustrated in Fig. 18. If this maximum value of τ_m is reached at any Lagrangian coordinate, the material at that point is considered to have failed in shear, and a different failure surface for failed material is used for that material point during the remainder of the calculation. The transition between brittle and ductile failure in shear is defined as the condition at which increased loading ($P + (1/3)\tau_m$) does not result in an increase in shear strength (τ_m) as shown in Fig. 18. Strain hardening, which is observed in some materials, is not accounted for in SOC73.

In SOC73, tensile failure is defined as the state in which the shear-stress limit defined in Fig. 18 is reached and any principal stress is in tension. Tensile failure relaxes the tensile principal stresses while other principal stresses remain constant. As previously stated, shear failure relaxes the shear stress while the mean stress or hydrostatic pressure remains constant. Therefore, tensile failure affects the P vs $\Delta V/V$ relation while shear relaxation does not.

As stated in a preceding paragraph, SOC73 assumes no inelastic failure in the case of a nonporous material with identical loading and unloading paths in the P vs $\Delta V/V$ plane; consequently, there is no hysteresis for loading and unloading in this plane. However, there can be hysteresis in the σ_{rr} vs $\Delta V/V$ plane because of shear failure, as illustrated in Fig. 19. The difference between σ_{rr} and $-P$, for a given value of $\Delta V/V$, is equal to the value of the stress deviator [Eq. (12)], and the difference is limited by the shear strength of the material [Eqs. (24)-(27) and Fig. 18]. There would be no hysteresis in the σ_{rr} vs $\Delta V/V$ plane if there were no shear failure; the hysteresis loop in Fig. 19 is a consequence of irreversible deformation in shear and a corresponding dissipation of energy. Tensile failure would cause hysteresis in the P vs $\Delta V/V$ plane and possibly in the σ_{rr} vs $\Delta V/V$ plane.

As shown in the preceding paragraphs, the models in SOC73 for volumetric response to hydrostatic pressure, distortional response to stress deviators, and response to tension are based on simplified physical models and empirical relations fitted to experimental measurements on rocks. It is assumed in SOC73 that the relations between strain and stress in compression, shear, and tension are rate independent. Larson (1980a,b) and Larson and Anderson (1979a,b) conducted shock-wave studies on seven representative geological materials. They found time-dependent behavior in two highly porous rocks (dry and water-saturated tuff and limestone) and in one low-porosity rock (dolomite), but not in two

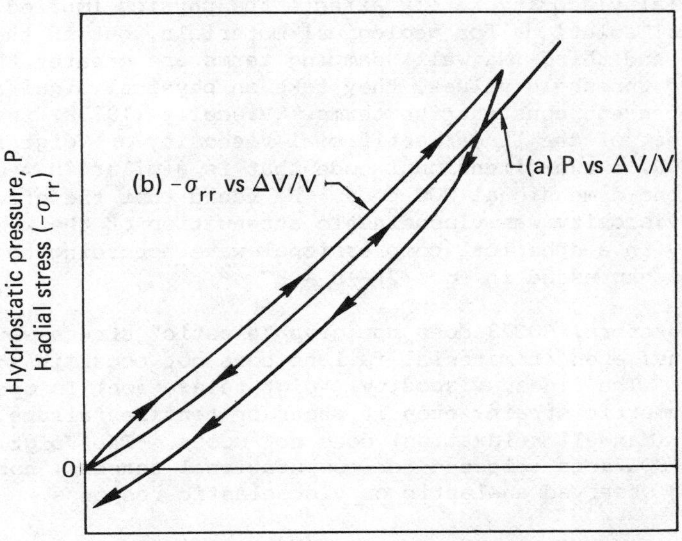

FIG. 19. Stress (P or $-\sigma_{rr}$) vs strain
($\Delta V/V$) loading and unloading cycles for (a)
hydrostatic compression and dilatation, and
(b) uniaxial compression and dilatation. Note
the hysteresis loop in (b) because of shear
failure.

other low-porosity rocks (granite and sandstone). Therefore,
SOC73 cannot model the strain-rate effects that are observed in
some rocks. However, SOC73 includes three strain-rate mechanisms
that are required for the stability and accuracy of the finite-
difference calculations. The first is the standard quadratic
"artificial viscosity" developed by von Neumann and Richtmyer
(1950) for finite-difference calculations of hydrodynamic wave
propagation. This artifical viscosity acts to stabilize the
calculations by "smearing" the shock-wave transition over several
points along the radial Lagrangian coordinate axis. The second is
a linear artificial viscosity that is equivalent to Voigt stress
relaxation (low-pass filtering) that is a function of volumetric
strain rate, and the third is equivalent to Maxwell stress
relaxation (high-pass filtering) that is a function of shear or
tensile strain rate. The Voigt model of a solid consists of a
spring and a dashpot in parallel; the Maxwell model of a solid is
composed of a spring and a dashpot in series (Jaeger and Cook
1971, Ch. 11.3). In shock compression and in shear or tensile
failure, minimum values of these damping mechanisms are necessary
for stable and accurate numerical computations. The quadratic

artificial viscosity rarely affects the physics implied by the
numerical solutions for geological materials, but if the second
(Voigt) and third (Maxwell) damping terms are greater than the
required threshold values, they take on physical significance as
rate-dependent constitutive terms. Viecelli (1973a) investigated
the effect of the linear artificial viscosity or Voigt relaxation
in TENSOR, a two-dimensional code that is similar in many respects
to the one-dimensional SOC code. He found that the effect of the
linear viscosity was viscoelastic attenuation of the peak particle
velocity in a spherical compressional wave according to the
relation expressed in Eq. (2)--$u \propto r^{-3/2}$.

Therefore, SOC73 does not give "elastic" stress-wave
solutions, even if material failure does not occur in shear or
tension. The linear viscosity (Voigt relaxation) is operative for
all volumetric strain, even if shear or tensile failure requiring
damping (Maxwell relaxation) does not occur. The Voigt relaxation
time in SOC73 is selected for computational reasons, not for
modeling observed anelastic or viscoelastic response.

From the above, it is clear that SOC73 calculations represent
an approximation to the physical processes of stress wave propagation. SOC73 solutions for peak particle velocity or peak stress
in the inelastic region can be fairly accurate if the stress-
strain models are based on measured rock properties. However,
SOC73 solutions for stress-wave time histories, particularly at
low stress levels in the transition region between inelastic and
almost-elastic response, are generally inaccurate if based on
measured rock properties. SOC73 can be forced to give a good
approximation to observations at a given gauge location in an
experimental assembly, but the solutions for other gauge locations
will probably be less accurate and the material properties implied
in such forced solutions may be physically unrealistic. As noted
in a preceding paragraph, the modeling in SOC73 could be modified
to take more observed phenomena into account, but at the cost of
increased code complexity and computation time. A comparison of
measurements and calculations [with an earlier version of SOC
(Cherry and Petersen 1970)] for the 5-kt Hardhat explosion is
given in Rodean (1971a, Ch. 2.8); two illustrations are reproduced
in Figs. 20 and 21.

Even if more complete descriptions of material response were
incorporated into SOC73, there would still be substantial limitations to the accuracy of the calculated seismic source function
for a given explosion. Geological inhomogeneities within the
inelastic region of an explosion are more the rule than the
exception. Even if the explosion site were "homogeneous," natural
variability would introduce significant perturbations. Costantino
(1978) conducted a series of 10 similar hydrostatic and triaxial
compression experiments on samples from a single block of granite,

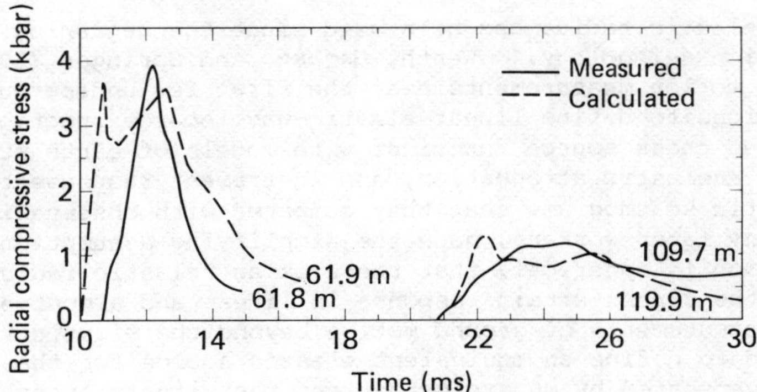

FIG. 20. Calculated and measured radial compressive stresses vs time for the 5-kt Hardhat explosion in granite (from Rodean 1971a).

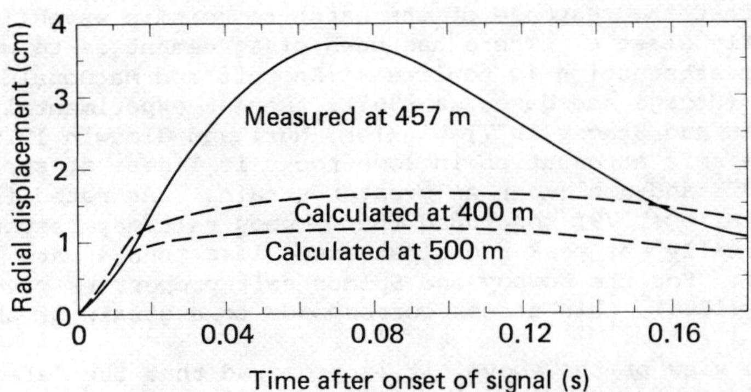

FIG. 21. Calculated and measured radial displacements vs time for the 5-kt Hardhat explosion in granite (from Rodean 1971a).

and concluded that a lower bound for experimental deviation for these physical property measurements in that material is about 20 percent in absolute volume strain.

3.4 Transition from Inelastic to "Elastic" Response

As previously stated, there is some ill-defined boundary called the "elastic radius" beyond which nonlinear inelastic deformation does not occur, but linear anelastic processes continue to attenuate the "elastic" waves from the explosion. The concept of an elastic radius has been used since the beginning of "nuclear-explosion seismology." Werth, Herbst, and Springer (1962) used ground motion measurements near the first few underground nuclear explosions to define linear elastic-wave source functions. They combined these source functions with models of earth structure, linear anelastic attenuation, and instrument response to obtain synthetic seismograms that they compared with observations. They, and many later workers, made the simplifying assumptions that an explosion is spherical, that there is an "elastic radius" beyond which the stress-strain response is linear and almost-elastic, that measurements of ground motion beyond the elastic radius can be used to define an equivalent elastic source for the seismic waves generated by an explosion, and that linear attenuation functions can be used to model anelastic attenuation in the earth.

The term elastic radius implies that there is a distinct transition between nonlinear inelastic and linear elastic stress-strain response. Linear elasticity is a convenient and useful, but fictitious, concept. Linear anelasticity is a similar (perhaps fictitious) concept that is used in recognition of the truth that the response of the earth to seismic waves is not perfectly elastic. There has been disagreement as to whether seismic attenuation is nonlinear (Knopoff and MacDonald 1958) or linear (Savage and Hasegawa 1967). Recent experimental results (Brennan and Stacey 1977; Winkler, Nur, and Gladwin 1979) indicate that seismic attenuation in some rocks is linear at strains less than 10^{-6} and nonlinear at greater strains. As noted in Section 3.2, Trulio (1979) concluded that Cowboy salt may respond "elastically" at peak radial stresses less than 4 bars (4 × 10^5 Pa). For the Cowboy and Salmon salt properties given by Rogers (1966), this stress corresponds to a strain of about 10^{-5}.

In view of the above, it is proposed that the "elastic radius" marking the transition from nonlinear inelastic to linear elastic response be defined as the radial coordinate at which the strain from stress-wave propagation is 10^{-6}. This arbitrary definition could serve until a better one is established as a result of further research on the nature and existence of the transition from nonlinear to linear stress-wave response.

3.5 Reduced-Displacement Potential

The reduced displacement potential is a very useful seismic source function for a spherical wave in an ideal, infinite, homogeneous, isotropic, elastic solid. (Some investigators prefer the reduced-velocity potential, the time-derivative of the reduced-displacement potential. Each has its advantages. The reduced-displacement potential is directly related to the explosion moment [Eqs. (32)-(35)]. The reduced-velocity potential is proportional

to the far-field displacement [Eq. (30)] and its spectrum, as shown
in Fig. 28, Section 5.3.)

The source of the spherical waves may be a finite spherical
cavity with variable pressure or the limiting case of an infinitesimal point soure of dilatation (which may be approached by
letting the cavity radius approach zero and the cavity pressure
approach infinity in such a way that the product of the pressure
and the cube of the radius remain constant and finite).

The first mathematical solution for elastic wave radiation
from a spherical cavity was obtained by Jeffreys (1931) who
assumed equal Lamé constants ($\lambda = \mu$) for the elastic solid. The
first general solution in terms of λ and μ was obtained by
Kawasumi and Yosiyama (1935). Blake (1952) and Selberg (1952)
were the first to use the reduced-displacement potential in
obtaining solutions for spherical wave propagation, but neither of
them used that term for their source functions. Herbst, Werth,
and Springer (1961) were the first to apply the term "reduced-
displacement potential" to the potential function that is a
solution of the spherical wave equation and is not a function of
space but of reduced time alone.

Use of the reduced-displacement potential results in the
following relation between the radial stress σ_{rr} (the driving
function) and the reduced-displacement potential X (the response
function):

$$\frac{\sigma_{rr}(r,\tau)}{4\mu} = \frac{1}{r^3}\left[(r/2\beta)^2\,\frac{\partial^2 X(\tau)}{\partial\tau^2} + (r/\alpha)\,\frac{\partial X(\tau)}{\partial\tau} + X(\tau)\right], \qquad (28).$$

where the reduced time

$$\tau = t - (r - R)/\alpha, \tag{29}$$

and

    r = radial coordinate,
    R = radius of cavity,
    t = time,
    α = compressional-wave velocity,
    β = shear-wave velocity,
    μ = shear modulus.

Selberg (1952) and Yoshiyama (1963) (apparently the Yosiyama of
the Kawasumi and Yosiyama 1935 paper) derived equivalent forms of
Eq. (28). The radial displacement (D_r) is related to the
reduced-displacement potential as follows, where $X(\tau)$ is the input
and $D_r(r,\tau)$ is the output:

$$D_r(r,\tau) = -\left[\frac{1}{\alpha r}\frac{\partial X(\tau)}{\partial \tau} + \frac{X(\tau)}{r^2}\right]. \tag{30}$$

The seismic moment for an earthquake was introduced by Aki (1966):

$$M_q = \mu S_q D_q, \tag{31}$$

where

D_q = average displacement in fault,
M_q = earthquake moment,
S_q = fault area.

Müller (1969) and Tsai and Aki (1971) extended the seismic moment concept to explosions and related the seismic moment to the reduced-displacement potential:

$$M_x(\tau) = 4\pi(\lambda + 2\mu)X(\tau) \tag{32}$$

where

λ = Lamé constant.

The final steady-state value of the reduced-displacement potential $[X(\infty)]$ is often used in defining the static value of the seismic moment of an explosion (Müller, 1973):

$$M_x = (\lambda + 2\mu)S_x D_x \text{ [note the similarity to Eq. (31)]} \tag{33}$$

where

$$D_x = X(\infty)/R_x^2, \tag{34}$$

$$S_x = 4\pi R_x^2, \tag{35}$$

and

D_x = radial displacement at reduced R_x in the elastic zone,
$R_x \geq$ "elastic radius" of explosion,
S_x = surface area of the sphere defined by the radius R_x.

Reduced-displacement potentials have been calculated from free-field ground-motion measurements by assuming spherical symmetry and that the ground motion at that gauge was elastic. Werth and Herbst (1963) published measured reduced-displacement potentials for nuclear explosions in alluvium, granite, salt, and tuff; Patterson (1966) for the Salmon explosion in salt; and Perret (1972a) for the Gasbuggy explosion in shale. Of these measurements, the best are those for Salmon and Gasbuggy because

these explosives were deeply buried; consequently the surface
refection did not perturb the subsurface measurements until
relatively late times. The ideal measurements of reduced-
displacement potentials have never been made. Such ideal
measurements would employ a large number of gauges in a low strain
($<10^{-6}$) region in which the stress-strain relation is most
likely linear, below the explosion along ray-paths from the
explosion to stations at teleseismic distances, and of high
fidelity over the complete seismic spectrum from very short
periods to final steady-state displacements.

4. CAUSES AND EFFECTS OF NONSPHERICAL GEOMETRY

In the preceding discussion, it is assumed that underground
explosions are spherically symmetric. This assumption greatly
simplifies the physics and mathematics of explosions, and it is a
useful and valid approximation although all real underground
explosions are nonspherical to some extent. In this lecture, we
will consider two nonspherical aspects of underground explosions
that are of importance in seismic monitoring: tectonic strain
release and the effects of the free surface above the explosion.

4.1 Tectonic Strain Release

As mentioned earlier, geological and seismological evidence
that some underground explosions released preexisting tectonic
strain began to accumulate early in the history of underground
nuclear testing. Two hypotheses have been advanced to explain
this strain release. Press and Archambeau (1962) proposed that
the strain release occurs around the inelastic fractured zone
produced by the explosion (the tectonic strain field responds to
the creation of a weak fractured zone around the explosion). The
theory of this mechanism was developed further by Archambeau
(1972). Brune and Pomeroy (1963) studied seismic data from the
Hardhat explosion in granite. They concluded that the fractured
zone hypothesis is not sufficient in the case of Hardhat, and that
a dislocation was triggered on a nearby fault plane. They cited
the occurrence of aftershocks as further support of the triggering
hypothesis. Toksöz, Ben-Menahem, and Harkrider (1965) concluded
that the fracture zone hypothesis is satisfactory for the Haymaker
explosion in alluvium and the Shoal explosion in granite, but that
the triggering hypothesis is required to explain the Hardhat
observation. Archambeau and Sammis (1970) and Lambert, Flinn, and
Archambeau (1972) concluded that the seismic data from the Bilby
explosion in tuff and the Shoal explosion are consistent with
stress relaxation around the fractured zone. Aki et al. (1969)
considered both hypotheses in their study of the Benham explosion.
They concluded that Benham triggered an earthquake, in part
because the fracture zone hypothesis requires an unusually high

stress release in order to account for the observed Love wave from
Benham. Aki and Tsai (1972) reviewed the Love wave data from 18
explosions in alluvium and tuff, and concluded that the evidence
is in favor of the earthquake triggering mechanism. All explosions discussed in this paragraph were at the Nevada Test Site,
except Shoal which was at another site in Nevada.

Even though there is disagreement about the mechanism of
tectonic strain release by explosives, there is agreement on the
usefulness of the parameter F that was introduced by Toksöz, Ben-
Menahem, and Harkrider (1965), and has been used by Archambeau and
Sammis (1970) and others. Toksöz, Ben-Menahem, and Harkrider
defined F as the ratio of the strength of the earthquake-like
source to that of the coincident explosion, but did not give an
explicit mathematical definition. Hirasawa (1971) gave a mathematical definition of the parameter q that is the ratio of the
strength of the double-couple component to the strengths of the
explosion component of a composite seismic source. Müller (1973)
found that Hirasawa's q is equal to the ratio of the moments of
the earthquake and explosion components of a composite event; for
example, in the static case,

$$q = M_q/M_x, \tag{36}$$

where M_q and M_x, are defined by Eqs. (31) and (33),
respectively. Müller also gave a mathematical definition of the
parameter F:

$$F = q\alpha^2/2\beta^2, \tag{37}$$

where

α = compressional-wave velocity,
β = shear-wave velocity.

Tectonic strain release is of interest at this Institute
because it may make an underground explosion appear to be "earthquake like." Toksöz and Kehrer (1972) examined the seismic data
for 28 underground nuclear explosions: 23 in the US and 5
(presumed) in the USSR. They estimated F ratios and the orientations of right-lateral, vertical, strike-slip faults that best fit
the data. They found no evidence for tectonic release (F = 0) by
three explosions: Sedan in alluvium, and Gnome and Salmon in
salt. They determined F ratios greater than unity for only three
explosions, all at the Nevada Test Site: Greeley (F = 1.6) in
tuff, and Hardhat (F = 3.0) and Pile Driver (F = 3.2) in granite.
They found no noticeable effect on the $M_s:m_b$ discriminant in
the case of these explosions. The north-northwest orientations of
most of their fault plane solutions for explosions at the Nevada

Test Site are consistent with lineaments and fault planes in that
area.

Toksöz and Kehrer also calculated surface-wave (Rayleigh
wave) radiation patterns for composite sources as a function of F
ratio, assuming vertical strike-slip faulting as the tectonic
strain release mechanism. Patton (1980) made similar calculations
in terms of the q ratio, but assuming dip-slip thrust faulting.
Their results are compared in Table 4. As shown in this table,
the effects on M_S of the two tectonic strain mechanisms are very
different. The effect of a strike-slip mechanism is always
positive as F is increased. In contrast, the effect of a dip-slip
mechanism is first negative and then positive as q (or F) is
increased. This difference in effect on M_S is a consequence of
the different surface-wave radiation patterns produced by the two
tectonic and the explosion mechanisms. The pattern from the
explosion is circular, that from the dip-slip source is two-lobed,
and that from the strikeslip is four-lobed. The phase angle of
the pattern from the explosion differs by π (phase reversal)
from the phase angles in both lobes in the pattern from the dip-
slip source, so the interference between the explosion pattern and
the dip-slip pattern is destructive. The phase angles of adjacent
lobes in the four-lobed strike-slip pattern differ by π, so the
interference between the explosive and tectonic pattern is
alternately constructive and destructive. These mechanisms and
their effects are also discussed in Harkrider's lecture at this
Institute. [See footnote 2, after text.]

TABLE 4. Effect of tectonic strain release on magnitude. The
calculations for the strike-slip case are by Toksöz and Kehrer
(1972) and the dip-slip case are by Patton (1980). For Patton's
calculations, F = 1.5q.

F	ΔM_S (strike-slip)	ΔM_S (dip-slip)	q
0	0	0	0
0.3	0.07		
0.5	0.08		
0.7	0.14	−0.4	0.5
1.0	0.24		
1.5	0.40	−0.4	1.0
2.0	0.53		
3.0	1.08	−0.1	2.0
4.0	1.23		
6.0		0.4	4.0
12.0		0.8	8.0

4.2 Effect of the Free Surface: Spall

The existence of the free surface above an underground explosion perturbs and distorts the spherical wave radiation from an explosion. An ideal spherical underground explosion radiates only compressional (P) waves. In a perfectly elastic half-space, these radiated P waves interact with the free surface and are converted into three types of elastic waves: reflected compressional (pP), vertically-polarized shear (SV), and Rayleigh (R or LR). In this lecture, we will consider the inelastic interaction of P and pP below the free surface (spall) and the possible effects of spall on compressional and Rayleigh waves from the combination of explosion and spall processes.

Spalling is a result of inelastic tensile failure of the rock that is caused by the conversion at the free surface of an upward-propagating compressional (P) wave into a downward propagating tensile (pP) wave. At some depth, the tensile stress in pP exceeds the sum of the compressional stress in P, the lithostatic stress, and the tensile strength of the rock. The tensile failure occurs at a time when the net momentum of the material above the fraction is upward. The spalled material then enters a ballistic trajectory that is terminated when the spall gap is closed. The net momentum of the spalled material is downward at the time of spall closure, and a downward impulse is applied to the underlying material. The seismic signals that may be generated by this impulse are of interest in the context of this Institute.

The spalling mechanism and its local mechanical effects are described in several publications. The dynamics of spalling by contained underground explosions have been investigated theoretically by Chilton, Eisler, and Heubach (1966). Spall measurements in several underground nuclear explosions have been analyzed by Eisler and Chilton (1964), Eisler, Chilton, and Sauer (1966), and Eisler (1967). Toman, Sisemore, and Terhune (1973) reported the spall measurements on the triple Rio Blanco explosion in Colorado and gave a preliminary interpretation of the data.

There have been studies (and speculations) concerning the possible relations between spall and distant seismic signals. Springer (1974) suggested that pP and a spall-closure (slap-down) phase (P_s) may contribute to the P wave and its coda at teleseismic distances. He reported estimates for the delay times corresponding to pP-P and P_s-P based on data from surface-zero accelerometers for a number of US underground nuclear explosions. A number of investigators had previously estimated pP-P delay times from teleseismic data, and Bakun and Johnson (1973) had estimated delay times for pP-P and P_s-P from teleseismic records from the Longshot, Milrow and Cannikin explosions in Amchitka. Springer found that his estimates for pP-P delay times agreed well

with those of the other investigators, but the agreement for P_S-P
delay times was poor. Viecelli (1973b) conducted computer simulations of Rayleigh wave generation with and without spall. He
found that the inelastic spall process generates larger amplitude
Rayleigh waves than the purely elastic process. [Strictly speaking, Viecelli's "elastic" computations are for a viscoelastic
medium because of the effect of the linear artificial viscosity
that is necessary in finite-difference computations; see Section
3.3 on computer calculations and Viecelli (1973a).] Viecelli also
noted that the spall-generated wave is delayed relatively to the
elastically generated wave, but not as much as if the spall closure
were the sole source of the Rayleigh wave. In addition, he noted
that if spall closure were the sole source, the phase of the wave
would be reversed relative to that of the elastically generated
wave. His theoretical calculations, analytic and finite-
difference, indicate that Rayleigh waves are generated by a
combination of the elastic and spall-closure processes associated
with an underground explosion. Viecelli also analyzed spall and
regional Rayleigh wave measurements for underground nuclear tests
at the Nevada Test Site, and concluded that the spall-closure
impulse is sufficient to account for the amplitudes of the
observed Rayleigh wave. [See footnote 3, after text.]

From the above, it appears that the inelastic process of
surface spall has a more significant effect on Rayleigh waves than
on teleseismic P and its coda. Furthermore, spall-closure is one
of three contained underground explosion-related mechanisms that
can generate Rayleigh waves that are reversed in phase at all
azimuths, compared to the waves normally produced by buried
explosions. The other mechanisms are cavity collapse [see Section
2.8 and Smith (1963)], and dip-slip tectonic strain release [see
Section 4.1, Patton (1980), and Harkrider's lecture].

Murphy (1977) found that the observed long-period surface-
wave data are inconsistent with a simple spherically symmetric
source model, and suggested that the spall-closure phenomenon may
be related to this discrepancy.

Rygg (1979) compared the Rayleigh waves from three presumed
explosions of comparable magnitude in eastern Kazakhstan, and
noted that the waves from one explosion were reversed in phase and
delayed, relative to the waves from the other explosions. The
event that generated the anomalous Rayleigh waves also excited
very strong Love waves. Rygg considered three possible explanations for Rayleigh wave phase reversal. The explosion could have
been so shallow that it acted as a downward-surface source, not a
buried source; but a shallow source is not consistent with delayed
Rayleigh waves. Cavity collapse could account for the delay and
phase reversal, but not the observed Rayleigh wave amplitudes.
Spall-closure could be consistent with the observed delay, phase

reversal, and amplitude of the observed Rayleigh waves. Rygg
suggested that spall closure is the likely explanation. He did
not consider dip-slip tectonic strain release. We may conclude
that the role of spall closure in seismic signal generation has
not been established with certainty.

5.   FORWARD AND INVERSE PROBLEMS

This lecture, the other lectures on source theory, and the
lectures on seismic wave propagation emphasize the forward
problem. The forward problem may be defined by the question,
"Given a seismic source, earth structure and properties, and a
seismograph--what is the resulting seismogram at a given station?"
The lectures on inversion theory and source classification
emphasize the inverse problem. In question form, the inverse
problem is, "Given a number of seismograms, the instrument
characteristics, and the station locations--what is the earth
structure; what are the earth properties associated with that
structure; and (of principal interest at this Institute) what are
the source location, origin time, and--especially--classification?"
If the event is classified as an explosion, there is an additional
question, "What was the yield?" In the remainder of this lecture,
we will consider selected aspects of the forward and inverse
problems in explosion seismology.

5.1 The "Equivalent Elastic Source": A Common Solution of
the Forward and Inverse Problems

The physical models used in the numerical computations to
obtain solutions for the forward and inverse problems are equivalent, but not necessarily identical. The numerical computations
in the forward problem proceed in the same time and space directions as the physical processes being modeled, forward from the
source to the receiver. The numerical computations in the inverse
problem (if the data are inverted to obtain a solution) proceed
backward in time and space from the receiver toward the source.
The inverse problem can also be solved by trial-and-error: by
making a series of forward-problem calculations for different
combinations of system parameters.

In principle, the forward problem can be modeled from the
detonation of the explosive to the recording of the seismogram.
However, even if the conditions for a mathematically unique (and
correct) solution from the data inversion obtain, the inverse
problem can be solved only so far and no farther in space and time
(hence the word "toward" in the next-to-last sentence in the
preceding paragraph). The inverse problem can be solved by data
inversion for linear, but not for nonlinear processes. The
impenetrable barrier for data inversion is the so-called "elastic

radius" (more generally, the "elastic boundary") separating linear
and nonlinear stress wave propagation. The data can be inverted
to obtain a description of an "equivalent elastic source," but not
of the nonlinear explosion processes. A solution for an "equivalent elastic source" can be unique in the mathematical sense, but
the word "equivalent" means that this solution is not unique in
the physical sense. This difference between the forward and
inverse problems is illustrated in Fig. 22.

One "equivalent elastic source" is the reduced-displacement
potential defined by Eqs. (28) and (29). As stated in Section
3.5, the reduced-displacement potential can be obtained experimentally by integrating the measured ground motion near an
explosion. Two assumptions are implicit in reduced-displacement
measurements: the ground motion is spherically symmetric and it
is measured at a point where the stress-strain relation is linear
and elastic. In reality, and at best, the ground motion is
approximately spherically symmetric and linearly inelastic or
anelastic. The reduced-displacement potential can be calculated

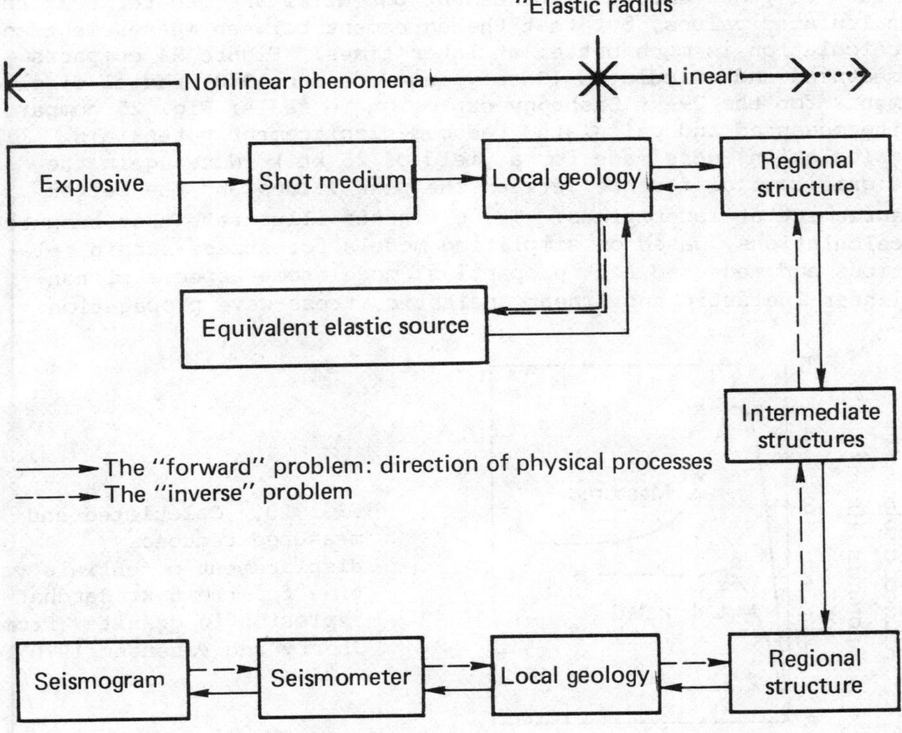

FIG. 22. Schematic diagram for the forward and inverse problems in seismology.

by finite-difference calculations of nonlinear and linear
processes in an explosion (see Section 3.3 on computer calculations). The reduced-displacement potential can also be obtained
from analytic solutions of Eq. (28) for a spherical cavity in an
infinite elastic solid, with the cavity pressure (-P) replacing
the radial stress (σ_{rr}). Therefore, in the context of the
complete forward problem (source-propagation-receiver), the source
component of the problem can be solved to obtain the "equivalent
elastic source."

5.2 Comparison of Measured and Calculated (Forward Problem)
Reduced-Displacement Potentials

In Fig. 21, the radial displacement measured at r = 457 m for
the 5-kt Hardhat explosion in granite is compared with SOC
calculations by Cherry and Petersen (1970) of the displacements
calculated for r = 400 m and r = 500 m. The measured reduced-
displacement potential at 457 m for Hardhat is compared in Fig. 23
with the reduced-displacement potential calculated for 500 m by
Cherry and Petersen. Note that the measured peak displacement and
measured peak reduced-displacement potential are greater than the
calculated values, but that the agreement between measurement and
calculation is much better at later times. Figure 24 compares
measured and calculated (Cherry and Petersen 1970) radial displacements for the 29-kt Gasbuggy explosion in shale; Fig. 25 compares
the measured and calculated reduced-displacement potentials. (The
calculations were made for a yield of 25 kt.) Note again the
significant difference between the peak values and the better
agreement at later times. These figures illustrate that computer
calculations, based on simplified models for stress-strain relations and measured rock properties, model some aspects of nonlinear inelastic and linear anelastic stress-wave propagation

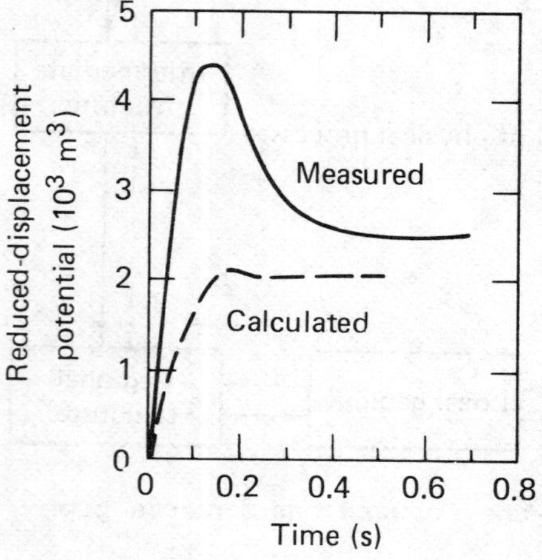

FIG. 23. Calculated and
measured reduced-
displacement potentials vs
time for the 5-kt Hardhat
explosion in granite (from
Cherry and Petersen 1970).

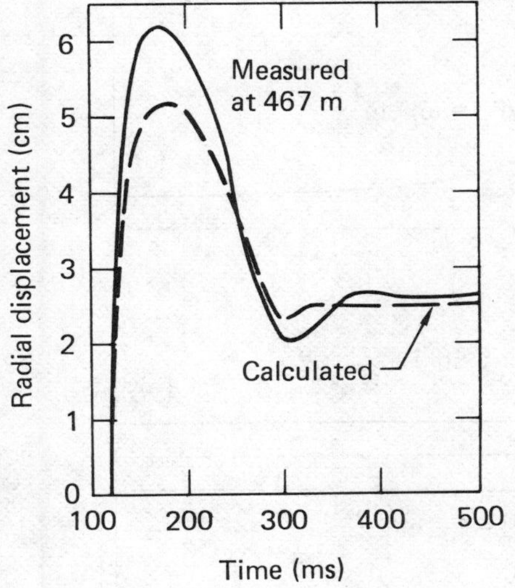

FIG. 24. Calculated and measured displacements vs time at a range of 467 from the 29-kt Gasbuggy in Lewis shale (from Cherry and Petersen 1970).

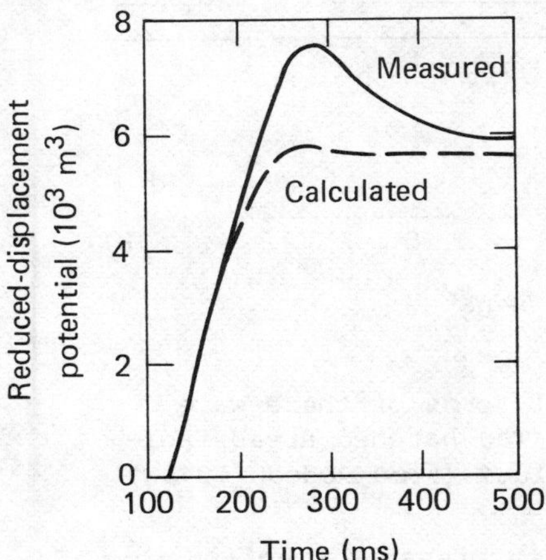

FIG. 25. Calculated and measured reduced-displacement potentials vs time for the 29-kt Gasbuggy explosion in Lewis shale (from Cherry and Petersen 1970).

better than others. The modeling appears to be better for the lower-frequency components of ground motion. As noted in Section 3.3, the accuracy of computer calculations could be improved by the use of more complete modeling of stress-strain relations, but at the cost of increased computer program complexity and computation time.

5.3 Analytic Solutions for the Reduced-Displacement Potential

Figures 26-28 illustrate some analytic solutions for the reduced-displacement potential (Rodean 1971a). Figure 26 shows

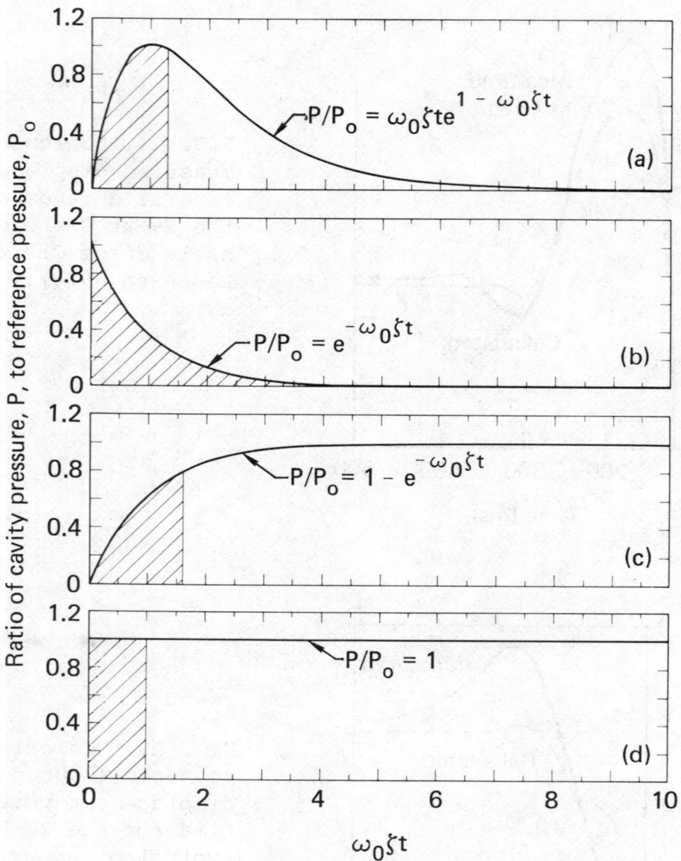

FIG. 26. Four special forms of the P-wave
generating function. The hatched areas indicate equal impulse values (from Rodean 1971a).

four driving functions: (a) a cavity pressure function that
initially increases linearly with time (t is dominant in the
function) and then decays exponentially (exp $-\omega_0\zeta t$ is dominant),
(b) a step in cavity pressure that immediately decays exponentially, (c) an exponential cavity pressure function that asymptotically approaches a constant value, and (d) a step change in
cavity pressure. Functions (a) and (b) are impulse functions, and
function (c) is a modified step function. The sum of the functions in (b) and (c) is equal to the function in (d). The parameters in Figs. 26-28 are defined as follows:

$$\zeta = \beta/\alpha = \text{damping ratio,} \qquad\qquad (38)$$

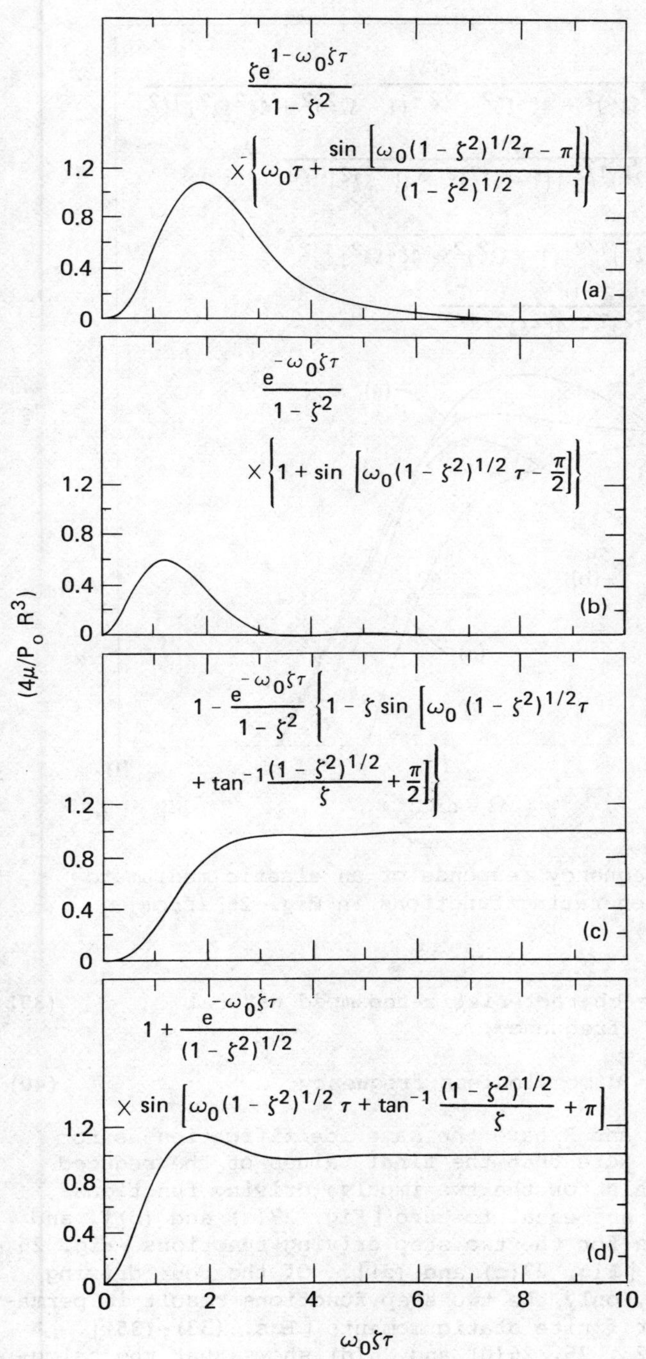

FIG. 27. Reduced-displacement potentials for the P/P_O vs time functions in Fig. 26 (from Rodean 1971a).

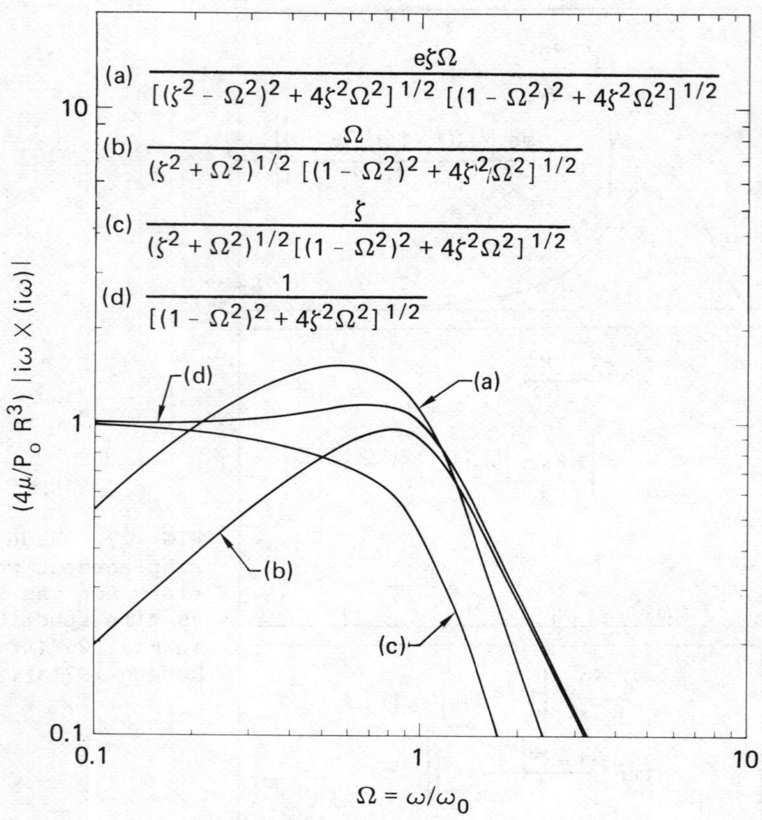

FIG. 28. Frequency response of an elastic medium to
the P-wave generating functions in Fig. 26 (from
Rodean 1971a).

$$\omega_0 = 2\beta/R = \text{characteristic undamped natural frequency,} \tag{39}$$

$$\Omega = \omega/\omega_0 = \text{dimensionless frequency.} \tag{40}$$

The parameters α, β, and R have the same identification as for
Eqs. (28) and (29). Note that the final values of the reduced
displacement potentials for the two impulse driving functions
[Fig. 26(a) and (b)] are equal to zero [Fig. 27(a) and (b)], and
that the final values for the two step driving functions [Fig. 26(c)
and (d)] are nonzero [Fig. 27(c) and (d)]. Of the four driving
functions in Fig. 26, only the two step functions result in permanent displacements or finite static moments [Eqs. (33)-(35)]. A
comparison of Figs. 23, 25, 26(b) and 26(d) shows that the calculated reduced-displacement potentials for Hardhat and Gasbuggy may

be approximated by the reduced-displacement potential for a step
change in cavity pressure, and that the measured reduced-displacement potentials are better approximated by a superposition of the
reduced-displacement potentials for a pressure impulse plus a step
change in pressure. Mueller and Murphy (1971) fitted impulse-plus-
step functions of pressure to close-in free-field data for a
number of underground nuclear explosions. The main point from the
comparison of these figures is that an underground explosion with
its inelastic processes can, for seismological purposes, be
replaced by a spherical cavity in an elastic solid with an appropriate time-variation of cavity pressure.

5.4 Effect of the Driving Function on the Spectrum of the
Reduced-Displacement Potential

     The Fourier amplitudes of the time-derivative of the reduced-
displacement potentials shown in Fig. 28 are proportional to the
Fourier amplitudes of the far-field displacement [Eq. (30)]. The
spectra for the driving functions illustrated in Fig. 26(a) and
(c) (the two with an initial finite rate of cavity pressure
increase) decrease at high frequencies in proportion to ω^{-3}.
The spectra for the driving functions shown in Fig. 26(b) and (d)
(the two with an initial step change in cavity pressure) decrease
in proportion to ω^{-2} at high frequencies. The value of the
exponent n in the function ω^{-n} has other physical significance.
Hanks and Wyss (1972) showed that values of n > 1.5 are required
in order that the energy radiated into the far field be finite.
Randall (1973) discussed this further. He stated that the singularity in the time function with n < 2 would involve an infinite
discontinuity in velocity, and that noninteger values of n > 2
would involve time functions with branch-point singularities which
are not to be expected in physical situations. On the other hand,
n = 2 corresponds to a singularity that has a finite jump in
velocity. Randall did not mention it, but n = 3 implies a finite
jump in acceleration, and acceleration is a continuous function if
n = 4. Randall did state that there are sound theoretical reasons
for expecting ω^{-2} behavior at high frequencies in seismic
spectra.

     There is another approach to approximating reduced-displace-
ment potentials like those shown in Figs. 23 and 25. Haskell
(1967) fitted a combination exponential-polynomial function of
time to the four measured potentials of Werth and Herbst (1963).
He required that displacement, velocity, and acceleration be continuous fuunctions of time, so the Haskell spectra are proportional to ω^{-4} at high frequencies. Von Seggern and Blandford
(1972) relaxed this restriction, requiring that displacement be a
continuous function in time, so their modifications of Haskell's
spectra are proportional to ω^{-2} at high frequencies. They made
this modification because the Haskell model did not match the

teleseismic data from the three explosions on Amchitka: Longshot,
Milrow, and Cannikin.

5.5 Inverse Problem Solutions for Reduced-Displacement Potentials

A number of investigators have inferred equivalent elastic
source time-functions from seismic data. Toksöz, Ben-Menahem, and
Harkrider (1964) corrected observed Rayleigh-wave spectra for
propagation effects to obtain source spectra. They found that the
source spectra for two underground explosions and one cavity
collapse at the Nevada Test Site are consistent with a cavity
pressure function of the form $P = P_0\tau^\xi e^{-\eta\tau}$, with $\xi = 1$ and
$\eta = 0.6$ for Sedan, $\eta = 1.0$ for Haymaker, and $\eta = 1.6$ for the Haymaker collapse. This pressure function is similar to that shown
in Fig. 26(a). Subsequent investigators have favored trial-and-
error solution of the inverse problem for the source function by
assuming source functions and comparing synthetic with observed
seismograms. Helmberger and Harkrider (1972) assumed a reduced
displacement potential function of the form $X = X_0\tau^\xi e^{-\eta\tau}$, and
found that the long- and short-period data from the Boxcar explosion fit this function for $\xi = 0.5$ and $\eta = 0.15$. Von Seggern and
Blandford (1972) found that Mueller and Murphy's (1971) model,
based on close-in free-field data, with its ω^{-2} high-frequency
characteristic, fit the teleseismic P-wave data from Longshot,
Milrow, and Cannikin on Amchitka better than Haskell's (1967)
model with its ω^{-4} high-frequency characteristic. They removed
the quartic and cubic terms from Haskell's model and adjusted parameters in the resultant exponential-polynomial function of time to
approximate the reduced-displacement potentials given by the
Mueller-Murphy impulse-plus-step function for cavity pressure.
Aki, Bouchon, and Reasenberg (1974) compared observed Rayleigh
waves from underground nuclear explosions in the US with synthetic
seismograms for different source models. They found that a
Haskell/von Seggern-Blandford reduced-displacement potential,
modified for ω^{-3} high-frequency behavior and for considerable
overshoot in the time domain, was required to match synthetic with
observed seismograms. Peppin (1977) proposed a very different
model based on local and regional observations of the Jorum and
Handley explosions: a combination of an upward impulse and a
spherical dilatation source. The spectrum of his spherical source
has no overshoot at the corner frequency ($\omega = \omega_0$ in Fig. 28) and
is proportional to ω^{-3} or ω^{-4} at high frequencies. Burdick
and Helmberger (1979) modeled short- and long-period P waves at
teleseismic distances from events on Amchitka and Novaya Zemlya.
They commented that there are problems with pressure and potential
functions of the form $\tau^\xi e^{-\eta\tau}$ (there is no steady-state component consistent with cavity formation, and the far-field functions
have singularities at $\tau = 0$). They used a von Seggern-Blandford
reduced-displacement potential modified to give considerable overshoot (corresponding to a step-function plus a considerable

impulse for the cavity-pressure function in the Mueller-Murphy
model) in matching synthetic seismograms with observations. The
results of these investigations are summarized in Table 5.

There is no general consensus among the above papers, and the
discrepancies among interpretations of observations that Müller
(1973) noted still exist. Burdick and Helmberger (1979) made
important criticisms of the $\tau^\xi e^{-\eta\tau}$ functions used by Toksöz,
Ben-Menahem, and Harkrider (1964), and Helmberger and Harkrider
(1972): these source functions tend toward zero at late times,
and they have singularities in the far field at $\tau = 0$ (also see
the footnote to Table 5 concerning values of ξ). This leaves the
step-plus-impulse functions of cavity pressure (Mueller and Murphy
1971) and the reduced-displacement potentials of the form

$$e^{-\eta\tau} \cdot \left[\sum_{n=0}^{n} a_n \tau^n \right]$$

where η and a_n are adjustable parameters, and $n = 4$ (Haskell
1967), $n = 3$ (Aki, Bouchon, and Reasenberg 1974), or $n = 2$ (von
Seggern and Blandford 1972; Burdick and Helmberger 1979). Since
the von Seggern and Blandford model can be made to closely approximate the Mueller-Murphy model, the two models may be considered
to be equivalent. There is disagreement about the required far-
field high-frequency displacement spectrum: ω^{-2} (Mueller and
Murphy; von Seggern and Blandford; Burdick and Helmberger), or
ω^{-3} (Aki, Bouchon, and Reasenberg; Peppin 1977). There is disagreement about overshoot in the reduced-displacement potential:
none (Peppin), moderate (Mueller and Murphy; von Seggern and
Blandford), considerable (Aki, Bouchon, and Reasenberg; Burdick
and Helmberger), or infinite (Toksöz, Ben-Menahem, and Harkrider;
Helmberger and Harkrider). Peppin alone proposes a nonspherical
source: an upward impulse superimposed on a spherical source of
dilatation.

It is clear that more research is needed to resolve the discrepancies among the explosion source models obtained from solutions of the inverse problem. The high-frequency part of the
displacement spectrum and the overshoot are discussed further in
subsequent sections.

6. DETECTION, IDENTIFICATION, AND YIELD ESTIMATION

We will now apply the preceding material to seismic monitoring; specifically, the problems of detection, discrimination, and
estimating the yields of explosives. We have not discussed the
effects of explosion yield, except in connection with the scaling
of experimental measurements to a common yield (see Figs. 5 and

TABLE 5. Equivalent elastic-source functions based on observations.

Reference	Data	Function	Far-field high-frequency characteristics
Toksoz, Ben-Menahem, and Harkrider (1964)	Rayleigh waves	$P(\tau) = P_0\tau e^{-\eta\tau}$	ω^{-3}
Mueller and Murphy (1971)	close-in free-field	$P(\tau) = P_0 + P_1 e^{-\eta\tau}$	ω^{-2}
Helmberger and Harkrider (1972)	Rayleigh and P waves	$X(\tau) = X_0\tau^{1/2}e^{-\eta\tau}$	$-a$
von Seggern and Blandford (1972)	P waves	$X(\tau) = f_1(e^{-\eta\tau}) \cdot f_2(\tau, \tau^2)$	ω^{-2}
Aki, Bouchon, and Reasenberg (1974)	Rayleigh waves	$X(\tau) = f_1(e^{-\eta\tau}) \cdot f_2(\tau, \tau^2, \tau^3)$	ω^{-3}
Peppin (1977)	Rayleigh and P waves	$X(i\omega)$ flat to corner plus upward impulse	ω^{-3} perhaps ω^{-4}
Burdick and Helmberger (1979)	Long- and short-period P waves	$X(\tau) = f_1(e^{-\eta\tau}) \cdot f_2(\tau, \tau^2)$ with overshoot	ω^{-2}

[a]The noninteger value of the exponent for τ makes the mathematics diffi-cult. In their Appendix C, Gardner and Barnes (1942) list Laplace transforms only for functions of $\tau^\eta e^{\eta\tau}$ with $\eta = -1$, $\eta = 0$, and η = positive integers.

7-14). A key question that we will now try to answer is, "What are the effects of yield on explosion phenomena and related seismic observations?" Or, in other words, "What are the scaling rules for the seismic effects of explosions?" In trying to answer these questions, we will also have to consider the effects of depth of burial and the properties of the explosion medium.

6.1 Detection: Effect of Explosion Environment

Given the capabilities of a seismic net for detecting events at a given location, the principal variables that determine the probability that an explosion will be detected are its yield and the seismic-coupling properties of the surrounding medium. In this section, we consider seismic coupling.

Underground nuclear explosions are a very inefficient means of generating seismic energy. The ratio of the radiated seismic-wave energy to the explosion yield, the seismic-coupling efficiency, is very low for such explosions. As noted in the section on thermal effects, postshot analysis of rock temperatures indicates that 90 to 95 percent of the energy released by contained underground nuclear explosions is deposited as residual thermal energy (Heckman 1964; Rawson, Taylor, and Springer 1967; Edwards and Holzman 1968). Mueller (1969) analyzed close-in free-field and local seismic measurements of ground motion from six underground nuclear explosions: four tamped, one cratering, and one in a decoupling cavity. (Decoupling is discussed in a subsequent paragraph.) The seismic-coupling efficiencies of these explosions, as determined by Mueller, are given in Table 6. Perret (1972b) analyzed close-in free-field ground-motion measurements for 21 contained underground nuclear explosions. His results, in terms of seismic-coupling efficiencies for different rock types, are given in Table 7. Mueller's results show significant seismic coupling differences among contained, cratering, and decoupled explosions. Perret's results illustrate the effects of rock properties on seismic coupling, consistent with the data for explosions in different types of rock in Figs. 5 and 7-14. For a given yield, the seismic signals from an explosion in granite, rhyolite, or salt are much stronger than those from an explosion in dry porous rocks like alluvium or tuff.

The low seismic-coupling efficiency of underground explosions is illustrated by the following simplified model of an underground explosion: a step change in pressure in a spherical cavity in a homogeneous, isotropic, infinite elastic solid. The seismic-coupling efficiency of this system is (Rodean 1972):

$$E_S/E_X = 3(\bar{\gamma} - 1)P/8\mu, \tag{41}$$

where

E_s = radiated seismic wave energy,
E_x = explosion energy,
P = step change in cavity pressure,
$\bar{\gamma}$ = ratio of enthalpy to internal energy of cavity gas,
μ = shear modulus of elastic solid.

P is limited by the shear strength of the elastic solid, and $\bar{\gamma}$ is a function of the cavity gas composition and temperature. It can be shown (Rodean 1972) by means of Eqs. (8)-(16) that the cavity pressure change is limited to

$$P \leq (8/3) \left| \tau_m \right| , \tag{42}$$

TABLE 6. Seismic-coupling efficiencies of underground nuclear explosions determined by Mueller (1969).

Explosion	Type	Medium	Seismic-coupling efficiency (%)
Shoal	Contained[a]	Granite	1.8
Salmon	Contained	Salt	5.8
Boxcar	Contained	Rhyolite	4.6
Benham	Contained	Tuff	6.1
Schooner	Cratering[b]	Tuff	0.32
Sterling	Decoupled[c]	Salt	0.0084

[a]Contained in this table also means "tamped"; that is, the explosive assembly was in close proximity to the surrounding rock.

[b]A Peaceful Nuclear Explosion.

[c]A contained explosion in the cavity formed by the Salmon explosion.

TABLE 7. Seismic-coupling efficiencies of contained underground nuclear explosions determined by Perret (1972b).

Environment	Seismic-coupling efficiency (%)
Alluvium, dry	0.05 to 0.15
Tuff, dry	0.10 to 0.30
Tuff, wet	2 to 3
Shale, deep	2
Dolomite	2
Granite	2
Ryolite	2
Salt, dome	3

where τ_m is the shear strength defined in Fig. 18. From
Eqs. (41) and (42), the upper limit for seismic-coupling
efficiency is

$$E_S/E_X = (\bar{\gamma} - 1)\tau_m/\mu. \tag{43}$$

For representative values of $\bar{\gamma}$ for the cavity gas and measured
values of τ_m and μ, Eq. (43) gives seismic efficiencies consistent with the data in Tables 6 and 7 (Rodean 1972). Equations
(41) and (42) can also be used in the analysis of cavity
decoupling.

The concept of cavity decoupling was introduced by Latter
et al. (1961). Herbst, Werth, and Springer (1961), reporting the
results of the Cowboy experiments in a salt mine with high explosives, showed that seismic signals are significantly reduced in
amplitude if the explosions are detonated in a sufficiently large
cavity. Springer et al. (1968) reported the results of the 380-t
Sterling nuclear explosion in the cavity produced in a salt dome
by the 5.3-kt Salmon explosion. The Sterling reduced-displacement
potential, scaled to the Salmon yield, indicates a decoupling
ratio of 70 ± 20 at 1 to 2 Hz. That is, the Fourier amplitudes at
1 to 2 Hz of the Salmon reduced-displacement potential are 70 ± 20
times those of the Sterling reduced-displacement potential, scaled
to the Salmon yield. The basic reason why decoupling works is
illustrated in Fig. 5: the inelastic explosion-produced stress
wave attenuates more rapidly to a lower stress level in air (the
strong-shock solution) than in rock (Rodean 1971b).

Patterson (1966) made calculations of fully- and partially-
decoupled explosions in salt. (Full decoupling occurs when the
cavity walls respond elastically to the explosion; partial
decoupling when the walls respond inelastically.) He did not make
enough calculations to satisfactorily define the transitions from
full decoupling through partial decoupling to tamped explosions.
Terhune, Snell, and Rodean (1979) made a series of calculations to
define these transitions. They studied the seismic-coupling
efficiency of nuclear explosions in granite by means of computer
calculations as a function of scaled explosion source radius. The
scaled-source radii were varied from 0.1 $m/kt^{1/3}$ (point source)
to 20 $m/kt^{1/3}$ (a nearly full-decoupling cavity). They found
that seismic coupling efficiency is at a maximum when the scaled-
source radius is approximately 2 $m/kt^{1/3}$. The primary cause of
this maximum seismic-source strength is the effect of initial
source radius on peak-particle velocity and the pulse duration of
the outgoing elastic wave. A secondary cause is that rock vaporization (an energy sink) does not occur for scaled-sourced radii
greater than 1 $m/kt^{1/3}$. Therefore, for scaled-source radii
greater than 1 $m/kt^{1/3}$, there is additional energy available for
seismic wave generation. Available data from underground nuclear

explosions at the Nevada Test Site do not provide sufficient
evidence to either support or negate the enhanced coupling that is
indicated by calculations at scaled-source radii of 1 to 2 m/kt$^{1/3}$.
The final (steady-state) value of the calculated reduced-
displacement potential as a function of scaled initial cavity
radius is shown in Fig. 29. Figure 30 shows the final energy
balance (among the cavity gas, rock fractures, and seismic waves)
as a function of scaled initial cavity radius. [Footnote 4, after text

From the above, it is clear that changes in explosion environment can change the seismic signal amplitude from an explosion
of given yield by factors of approximately 10 (variations in
geology) to 100 (cavity decoupling). This corresponds to changes
of 1 to 2 units of seismic magnitude. Therefore, the detection of
an explosion by a seismic network is strongly dependent upon the
environment in which the explosive is emplaced.

6.2 Seismic Scaling Rules for Underground Nuclear Explosions

We will now address seismic scaling rules for nuclear explosions in a given medium. Examples of cube-root of the yield
scaling are given in Figs. 5 and 14 (m/kt$^{1/3}$) for experimental

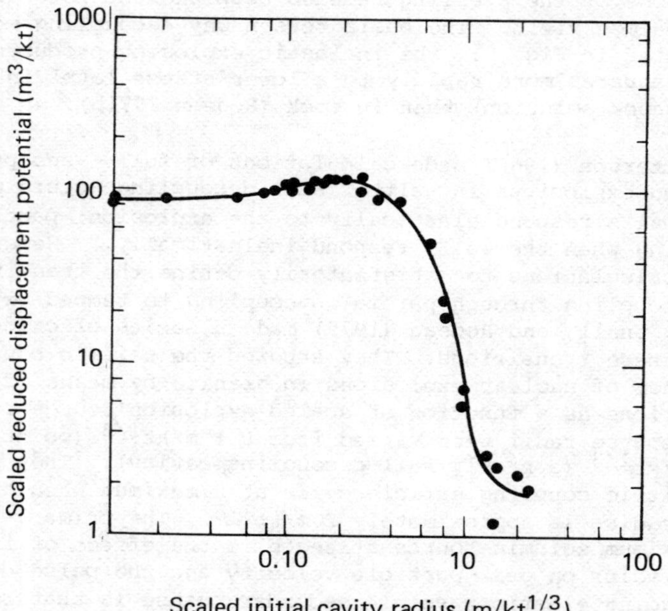

FIG. 29. Final reduced-displacement potential
vs scaled initial source radius (from Terhune,
Snell, and Rodean 1979).

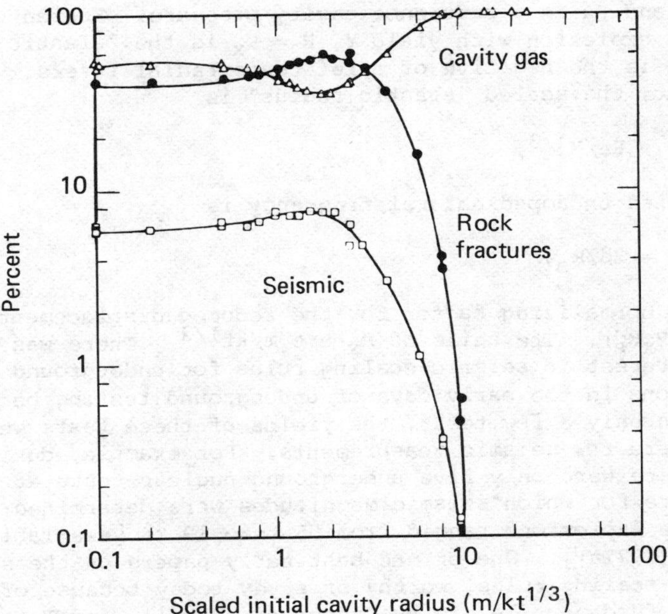

FIG. 30. Final energy balance vs scaled
initial source radius (from Terhune, Snell,
and Rodean 1979).

measurements of explosions with different yields. The rationale
for cube-root scaling is based on the spherical symmetry of an
explosion. During the explosion, the average energy density
varies in inverse proportion to the volume enclosed by the
outward-propagating spherical stress wave. Different explosion
effects are associated with different energy densities: rock
vaporization, rock melting, rock fracture, transition at the
"elastic radius" from nonlinear inelastic to linear anelastic
stress-wave response, etc. Therefore, the characteristic lengths
of an explosion, such as the "elastic radius," in a system with
explosion yield as the only independent variable, are proportional
to the cube-root of the reciprocal of energy density: $m/kt^{1/3}$.
It is assumed that gravity has no effect; its actual effect is
considered in a subsequent paragraph.

In Fig. 27, four normalized solutions for reduced-displacement
potentials are presented, and the corresponding normalized spectra
of the time-derivatives of these potentials are shown in Fig. 28.
The normalizing factor for the potentials is $(4\mu/P_oR^3)$ and for
the frequency, from Eqs. (39) and (40), is $(2\beta/R)$. In Figs. 27
and 28, and in Eq. (39), R is the radius of a cavity in an elastic

medium and P_O is a reference cavity pressure. For an underground
nuclear explosion with yield Y, $R = R_x$ is the "elastic radius" and
$P_O = P_x$ is the negative of a reference radial stress. With P_x
constant, the scaled "elastic radius" is

$$R_s = R_x/Y^{1/3}, \tag{44}$$

the scaled undamped natural frequency is

$$\omega_s = 2\beta/R_s, \tag{45}$$

and the normalizing factor for the reduced-displacement potentials
is $(4\mu/P_x R_s^3)$. The units of R_x are m/kt$^{1/3}$. There was considerable interest in seismic-scaling rules for underground nuclear
explosions in the early days of underground testing because there
had been only a few tests, the yields of these tests were low, and
there were few seismic measurements. For example, during 1957-
1958 there were only five underground nuclear tests at the Nevada
Test Site for which seismic magnitudes were determined; the yield
of these explosions ranged from 55 t to 19 kt [see Table 7.1 in
Rodean (1971a)]. One of the best early papers on the subject of
seismic-scaling rules, worthy of study today because of its treatment of fundamentals, is by Carpenter, Savill, and Wright (1962).
They critically review the cube-root scaling rule and some variations. They show that body-wave amplitude vs yield curves are
strongly dependent upon the seismic source spectrum and the seismometer characteristics, especially narrow band vs broad band.
Werth and Herbst (1963) used cube-root scaling of measured
reduced-displacement potentials for explosions in alluvium,
granite, salt, and tuff to calculate theoretical regional headwave (P_n) amplitude vs yield curves, taking anelastic attenuation in the earth and instrument response into account. Carpenter
(1967) extended this work of Werth and Herbst to P waves at teleseismic distances, and included P-wave amplitude vs yield curves
from scaled source functions for underwater and near-surface
atmospheric explosions. Rodean [see Ch. 7.5 in Rodean (1971a)]
calculated similar regional P_n- and teleseismic P-wave amplitude
vs yield curves for the four reduced-displacement potentials shown
in Figs. 27 and 28. Springer and Hannon (1973) used cube-root
scaling to develop teleseismic P-wave amplitude vs yield relations
for different source functions. In all the above work, the
reduced-displacement potentials were cube-root scaled without
considering possible depth of burial effects. In general, the
theoretical body-wave amplitude vs yield curves obtained by the
above authors are, in log-log coordinates, nonlinear over a yield
range from 1 to 10^3 kt, with slopes of about unity at low yields
and "turnover" to lesser slopes at high yields, as shown in Figs.
31 and 32 for P_n and teleseismic P-waves. This "turnover" is a
consequence of the shift of the source spectrum (see Fig. 28)
toward lower frequencies [see Eqs. (39), (40), (44), and (45)] as

yield increases. The instrument bandpass is centered at about
1 Hz, so the seismometer samples the lower frequencies of a low-
yield spectrum (relative to the corner frequency ω_o) and the
higher frequencies of a high-yield spectrum. Note in Figs. 31 and
32 that the earth-attenuated peak displacement is nearly linear
with yield (in log-log coordinates) even if the seismometer
response to the displacement is not.

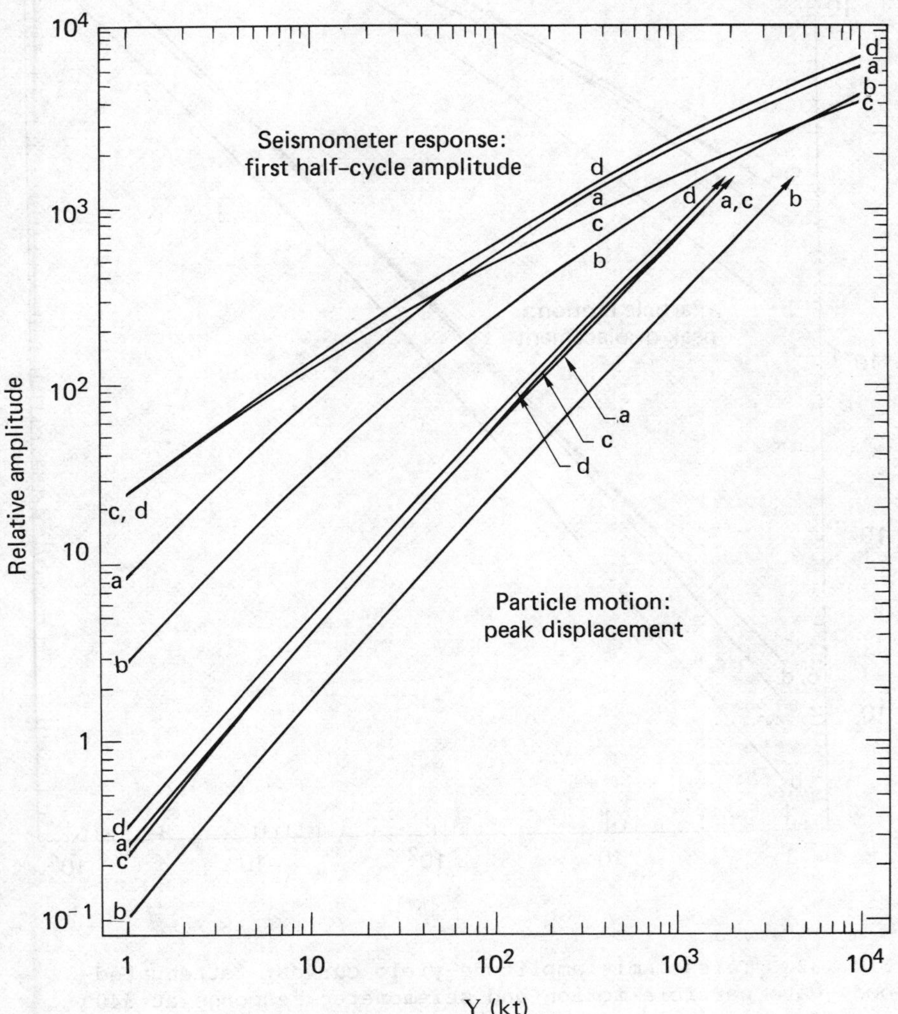

FIG. 31. Regional amplitude-yield curves: attenuated-head-
wave particle motion and seismometer response at 680 km.
Curves a,b,c, and d are based on applied-stress functions of
Fig. 26 a,b,c, and d (from Rodean 1971a).

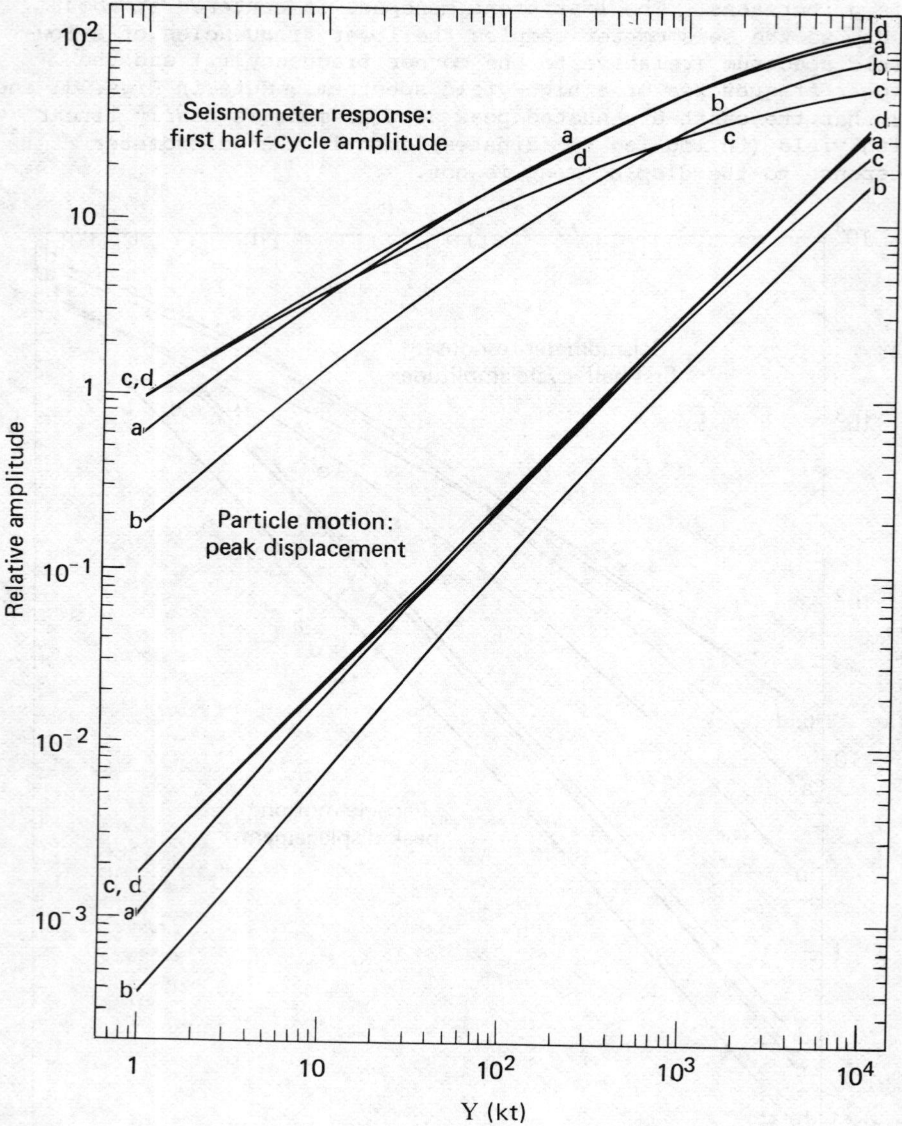

FIG. 32. Teleseismic amplitude-yield curves: attenuated-body-wave particle motion and seismometer response at 3400 km. Curves a,b,c, and d are based on applied-stress functions of Fig. 26 a,b,c, and d (from Rodean 1971a).

After a decade or so of underground testing, there were
sufficient data to derive statistical relations for magnitude vs
yield for yields ranging from about a kiloton to a.little over a
megaton. Ericsson (1971) developed the statistical theory for
linear magnitude vs log yield ($\log_{10}$ Y) relations, considered
the magnitudes to be expected from given yields, and the yields to
be expected for given magnitudes. He demonstrated his approach
with teleseismic m_b and M_s data from a network of Canadian
stations for six tests with announced yields in the Pahute Mesa
testing area at the Nevada Test Site. Marshall, Douglas, and
Hudson (1971) proposed linear M_s vs log yield relations, one for
consolidated and one for unconsolidated rocks, based on data for
31 explosions with announced or presumed yields in the US,
Algeria, and the USSR (20 were at the Nevada Test Site). Basham
and Horner (1973) analyzed teleseismic body and surface-wave data
from a Canadian network for 3 explosions on Amchitka, 32 in the
continental US (28 at the Nevada Test Site), 4 presumed explosions
in eastern Kazakhstan, and 6 presumed explosions on Novaya Zemlya.
They developed linear m_b and M_s vs log yield relations based
on announced and presumed yields for US explosions. Springer and
Hannon (1973) used data from several networks at regional and
teleseismic distances for 17 explosions in the Pahute Mesa testing
area of the Nevada Test Site to develop the equivalent of m_b and
M_s vs log yield relations. The slopes of the m_b and M_s vs
log yield relations for explosions in rhyolite and tuff at the
Nevada Test Site determined in three of these investigations are
summarized in Table 8.

The linear m_b and M_s vs log yield relations that Ericsson
(1971), Basham and Horner (1973), and Springer and Hannon (1973)
developed from statistical analysis of observations are not
consistent with the nonlinear theoretical scaling relations that
Werth and Herbst (1963), Carpenter (1967), Rodean (1971a), and
Springer and Hannon (1973) based on cube-root scaling with yield

TABLE 8. Slopes of linear magnitude vs log yield relations for
explosions in tuff and rhyolite at the Nevada Test Site.

Reference	Regional m_b	Teleseismic m_b		M_s
		Short-period	Long-period	
Ericsson (1971)		0.93		1.19
Basham and Horner (1973)		1.09	0.72	1.23
Springer and Hannon (1973)	0.63	0.99		1.09

of reduced-displacement potentials. Mueller and Murphy (1971)
developed a cube-root scaling rule modified to include empirical
depth-of-burial effects. Murphy (1977) showed that the modified
Mueller-Murphy scaling rule is consistent with the Springer-Hannon
relations for short-period P-wave data at regional and teleseismic
distances, and with the Basham-Horner relations for short- and
long-period P-waves data at teleseismic distances. However, the
Mueller-Murphy scaling rule is not consistent with the Springer-
Hannon and Basham-Horner relations for long-period surface waves.
Murphy concluded that these data are not consistent with a simple,
spherically symmetric model. He suggested that this discrepancy
may be related to the spall-closure phenomenon.

The underground explosion effects are superimposed upon the
underground environment, including the lithostatic or overburden
pressure. Some constitutive relations between stress and strain,
like rock vaporization from shock compression and unloading, may
be assumed to be independent of the overburden pressure associated
with underground explosions. Others, such as the stress-strain
relation for the transition from nonlinear-inelastic to linear-
anelastic response, may be strongly affected by overburden pressure. Therefore, the radius of vaporization may be expected to
vary with the cube-root of the explosion yield, but the "elastic
radius" may be expected to vary in proportion to the cube-root of
the yield and some inverse function of the overburden pressure.
The final value of the reduced-displacement potential or the
static moment of the explosion [Eqs. (32)-(35)] may be a function
of the final cavity volume. According to containment analysis
calculations, the final dynamic cavity pressure can vary between
one-third to twice the initial overburden pressure, depending upon
the mechanical properties of the rock (Terhune et al. 1977a). The
mechanical properties of a given type of rock can vary significantly from site to site; therefore the explosion-produced cavity
volume per kiloton at a given depth can vary from site to site.
The French nuclear explosions in a granite massif in Algeria
formed cavities with scaled volumes (in cubed metres per kiloton)
that averaged about one-fifth the average scaled volumes of
cavities produced by US nuclear explosions in granite rocks in
Nevada (Gauvenet 1970). The scaling relation for these explosions
in Algeria included the effect of rock strength as well as depth
of burial (Michaud 1968). Computer calculations based on quasi-
static rock strength measurements indicate that the relative
absence (Algeria) or presence (Nevada) of water in preexisting
fractures may be responsible for this difference in scaled cavity
size (Rodean 1972). On the other hand, dynamic experiments with
shock waves do not indicate any loss of shear strength because of
the presence of water in Westerly granite (Larson and Anderson
1979b). Therefore, for a given set of rock properties, the final
value of the reduced-displacement potential may be proportional to
the yield and some inverse function of the overburden pressure.

From data obtained at the Nevada Test Site, Mueller and Murphy (1971) give the following respective empirical relations for the cavity radius (R_C) and the "elastic radius" (R_x) as functions of yield and overburden pressure (ρh):

$$R_C \propto Y^{0.29} \rho^{-0.24} h^{-0.11},\tag{46}$$

$$R_x \propto Y^{1/3} (\rho h)^{-0.42},\tag{47}$$

where

    h = depth of burial,
    ρ = average overburden density.

As noted by Mueller and Murphy (1971), yield and depth of burial are generally not independent variables at the Nevada Test Site because of testing practice for containment. For most explosions at that location,

$$h \propto Y^{1/3}.\tag{48}$$

Geological properties (e.g., porosity, water content, strength, etc.) that affect R_C and R_x are also a function of depth. Therefore Eqs. (46) and (47) implicitly describe the variation of R_C and R_x with the overburden effect of depth and with other parameters [material properties and, per Eq. (48), yield] that also vary with depth. From Eqs. (46)-(48),

$$R_C \propto Y^{0.25} \rho^{-0.24},\tag{49}$$

$$R_x \propto Y^{0.19} \rho^{-0.42}.\tag{50}$$

Equation (49) is essentially identical with the theoretical scaling rule that includes the effects of gravity (Chabai 1965):

$$R_C \propto (Y/\rho)^{1/4}.\tag{51}$$

Springer and Denny (1976) studied the broad-band velocity spectra of the complete regional wave trains from about 40 events at the Nevada Test Site. Their spectra therefore contain considerable information about the regional earth structure as well as the source. They found that the "corner frequency" is almost independent of yield, and noted a tendency for it to vary inversely with approximately the sixth power of the yield. From Eqs. (39) and (40) and Fig. 28, this implies that $R_x \propto Y^{1/6}$; this is close to the Mueller-Murphy relation in Eq. (50).

According to Eqs. (40) and (50), the ratio R_x/R_C decreases as yield and depth increase, which is physically reasonable. As a consequence of these and other equations, the impulse component of

the Mueller-Murphy step-plus-impulse pressure functions (see Table
5) increases relative to the step component as yield and depth
increase. Similarly, this overshoot in the time-and-frequency
domains of the reduced-displacement potential increases with yield
and depth. It is this increase in overshoot with yield and depth
that makes this source model approximate the long- and short-period
teleseismic body-wave magnitude vs yield observations of Basham
and Horner (1973), and the short-period regional and teleseismic
body-wave magnitude vs yield observations of Springer and Hannon
(1973). This increase of overshoot with yield and depth is in
contrast to the effect of depth on finite-difference calculations
of the reduced-displacement potential of an explosion. In such
calculations, the overshoot increases with decreasing depth
because of the decrease in shear strength with decreasing overburden pressure (see Fig. 18).

In summary, cube-root scaling is suitable for high-stress
phenomena in the central part of the explosion, but not for
seismic source definition if depth is a variable. In particular,
cube-root scaling does not give results that are consistent with
magnitude and yield data. The Mueller-Murphy modification of
cube-root scaling gives significantly better results for body-wave
magnitude vs yield, but not for surface-wave magnitude vs yield.
No existing spherically symmetric explosion-scaling model accounts
for all observed magnitude vs yield data. Clearly, there is need
for further work on the effects of yield, depth, and material on
seismic source parameters.

6.3 Body vs Surface-Wave Magnitude Relations for Underground Nuclear Explosions

Most of the available unclassified yield data for underground
nuclear explosions are for explosions at the Nevada Test Site, and
most of the explosions at this location have been in alluvium and
volcanic rocks (tuff and rhyolite). For example, Marshall,
Springer, and Rodean (1979) list announced or presumed yields for
46 underground nuclear explosions. Of these, 29 were at the
Nevada Test Site, and 26 of the 29 were in alluvium or volcanic
rocks. The Mueller-Murphy model is based, for the most part, on
data from explosions in alluvium and volcanic rocks at this site.
One way to extend the geological and geographical data base for
underground nuclear explosions is to study the body- vs surface-
wave magnitude data for announced and presumed underground nuclear
explosions in different parts of the world.

In analyzing m_b:M_s data, Liebermann and Pomeroy (1969)
and Basham (1969) noted that the m_b:M_s data for explosions and
earthquakes in the western United States are anomalous with
respect to such data for the Aleutians, Algeria, and the USSR.
Marshall and Basham (1972) confirmed that the m_b:M_s relations

for explosions in the western US differ from those for explosions
in the Aleutians, the USSR, and China, but they did not detect any
regional differences in m_b:M_s data for earthquakes. This is
illustrated in Fig. 33 for explosions at the Nevada Test Site, in
the Aleutians, and in the USSR. There is an offset and a different slope for the m_b:M_s relations for the two populations:
explosions at the Nevada Test Site, and explosions in the
Aleutians and the USSR. Marshall, Springer, and Rodean (1979)
developed an empirical correction to the body-wave magnitude for
anelastic attenuation in the upper mantle, and incorporated this
correction in a new body-wave magnitude, m_Q. The explosions of
Fig. 33 are plotted in terms of m_Q vs M_s in Fig. 34. It is
shown that the change from m_b to m_Q reduces the offset of the
two populations, but there is still a difference in slope. The
present question is, "What is the reason for the difference
between the m_Q vs M_s slope for explosions at the Nevada Test
Site and the slope for explosions in the USSR?" (The two explosions on Amchitka don't have much effect on the statistics.)

Marshall (1979) has suggested that the different slopes for
the two populations of explosions are a consequence of different

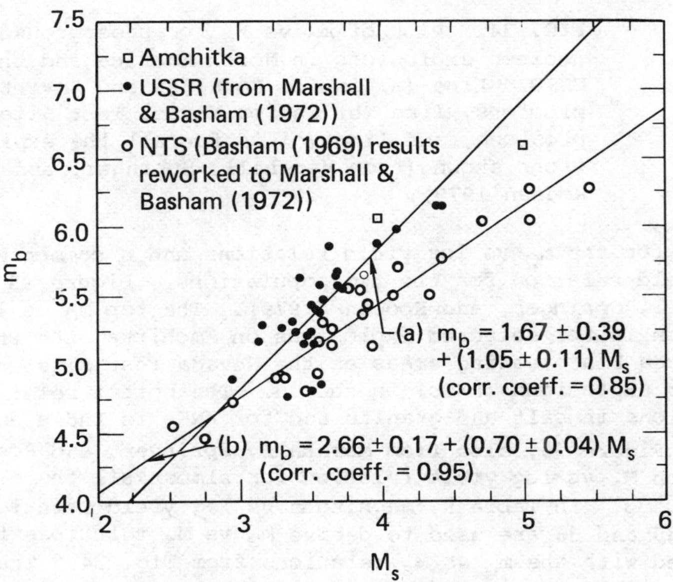

FIG. 33. Plot of m_b vs M_s for underground
nuclear explosions in North America and the
USSR. Line (a) is for Amchitka and Soviet
explosions, line (b) is for Nevada Test Site
explosions (from Marshall, Springer, and
Rodean 1979).

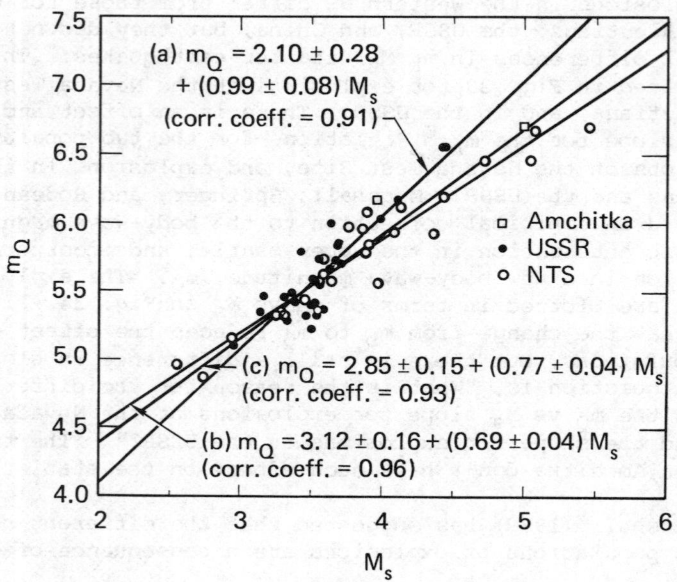

FIG. 34. Plot of m_Q vs M_S for underground
nuclear explosions in North America and the
USSR. Line (a) is for Amchitka and Soviet explosions, line (b) is for Nevada Test Site explosions, and line (c) is for all the explosions shown (from Marshall, Springer, and
Rodean 1979).

slopes for the m_Q vs log yield relations and a common M_S vs
log yield relation for the two populations. Figure 35 is from
Marshall, Springer, and Rodean (1979). The top m_Q vs log yield
relation is for selected explosions on Amchitka, the Pahute Mesa,
and Yucca Flat testing areas at the Nevada Test Site, and peaceful
nuclear explosions (PNEs) in the US. The bottom relation is for
explosions in salt and granite and for PNEs in India and the
USSR. Figure 36, also from Marshall, Springer, and Rodean, gives
a common M_S vs log yield relation for almost all the explosions
in Fig. 33. In Table 9, magnitude vs log yield relations from
Figs. 35 and 36 are used to derive m_Q vs M_S relations that are
compared with the m_Q vs M_S relations from Fig. 34. The comparison of (d) with (e) and of (f) with (g) in Table 9 illustrates
Marshall's hypothesis that the difference between the body-wave
seismic coupling characteristics of (1) the younger volcanic rocks
at Amchitka and the Nevada Test Site, and (2) salt, granite, and
older consolidated rocks in the USSR is responsible for the
different m_Q vs log yield and m_Q vs M_S relations shown in
Figs. 31-34. Data for more explosions in salt, granite, and old

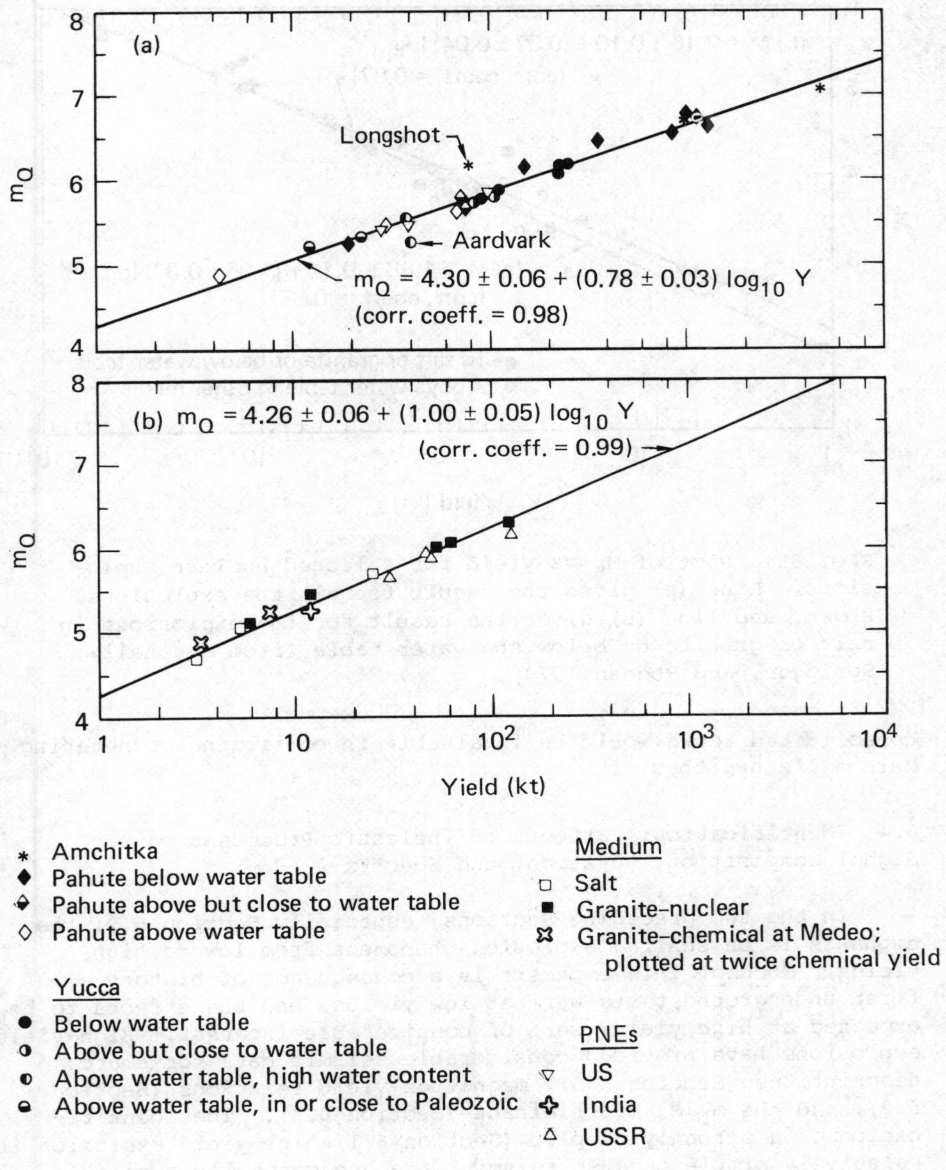

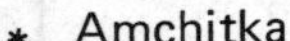

FIG. 35. Plots of m_Q vs yield for selected explosions: (a) at Amchitka, Pahute Mesa, and Yucca Flat at the Nevada Test Site, and PNEs in the Western US; and (b) in salt and granite, and PNEs in India and the USSR. The Medeo explosions (high-explosive) were not included in calculating the line in (b) (from Marshall, Springer, and Rodean 1979).

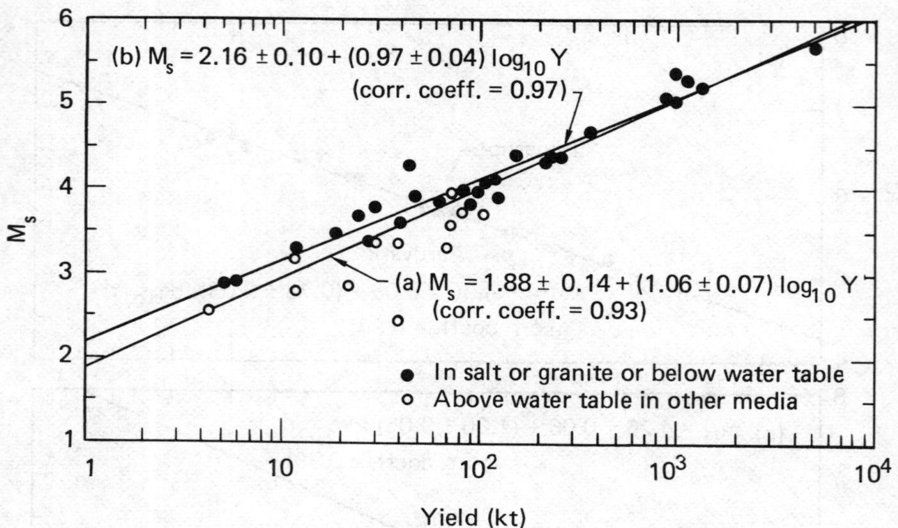

FIG. 36. Plot of M_s vs yield for selected nuclear explosions. Line (a) gives the result for all the explosions shown, and line (b) gives the result for the explosions in salt or granite or below the water table (from Marshall, Springer, and Rodean 1979).

consolidated rocks would be invaluable in confirming or negating Marshall's hypothesis.

6.4 Identification: Effects of Inelastic Processes on Signal Composition, Duration, and Spectra

In the two preceding sections, especially Section 6.2, the emphasis is on scaling explosion phenomena from low to high yields. Much of this emphasis is a consequence of history--the first underground tests were at low yields, and the effects to be expected at high yields were of considerable interest. High-yield explosions have provided considerable seismic data for source descriptions (Section 5.5), magnitude-yield relations (Section 6.2), and the m_b:M_s discriminant (Section 6.3). The identification of a strongly-coupled (Section 6.1) high-yield explosion is relatively simple because seismic data are generally available from a large number of stations. Identification can be difficult for low-yield explosions, especially if they are weakly coupled or decoupled (Section 6.1), because limited seismic data may be available from only a few stations. Detection by a few stations not only reduces the amount of data, but also introduces the problem of magnitude overestimation (Ringdahl 1976) which Christofferson will discuss in his Institute lecture. Therefore,

TABLE 9. Summary of empirical m_Q vs log yield, M_S vs log yield, and m_Q vs M_S relations from Marshall, Springer, and Rodean (1979).

No.	Fig.	Relation	Comments
a	35	$m_Q = 4.30 + 0.78 \log_{10}Y$	Amchitka: volcanic; Nevada Test Site: volcanic; western US: sandstone and shale
b	35	$m_Q = 4.26 + 1.00 \log_{10}Y$	US: salt and granite; Algeria: granite; India: sandstone; USSR: clay, salt, and sandstone
c	36	$M_S = 2.16 + 0.97 \log_{10}Y$	All of the above
d	34	$m_Q = 2.10 + 0.99M_S$	Amchitka, USSR
e	–	$m_Q = 2.03 + 1.03M_S$	From (b) and (c)
f	34	$m_Q = 3.12 + 0.69M_S$	Nevada Test Site
g	–	$m_Q = 2.86 + 0.80M_S$	From (a) and (c)

the emphasis in scaling for purposes of identification should be in scaling from high-yield explosions (which can be exhaustively analyzed) to low-yield explosions (which can be similarly analyzed only if there are extensive local and regional seismic measurements). In Section 6.2 it is noted that yield and depth are generally not independent variables because of testing practices for containment. Clandestine tests, the object of seismic monitoring under a comprehensive test ban, may be expected to be buried at greater than standard depths. Therefore the effects of depth and yield must be separated in scaling data from high-yield explosions to low-yield explosions for seismic identification purposes.

The principal inelastic phenomena in an underground nuclear explosion are associated with cavity formation and the generation of the outward-propagating approximately spherical P wave. As a result of local geological inhomogeneities, no real explosion is perfectly spherical and some S-wave energy is also directly radiated. For explosions with yields of a few kilotons, the inelastic processes associated with the direct-P and direct-S wave generation are completed in 0.1 to 0.2 second. This is the time required for cavity formation (Johnson, Higgins, and Violet 1959;

Rogers 1966), and the radial displacement transient in the almost-elastic zone is completed in a similar time interval (Figs. 21, 23-25). According to the hypothesis that tectonic strain release occurs outside the inelastic fractured zone around the cavity (Archambeau 1972), tectonic strain release is coincident in space and time with the explosion. According to the triggering hypothesis (Brune and Pomeroy 1963; Aki and Tsai 1972), the fault displacement may be triggered outside the immediate vicinity of the shot point. Such tectonic strain release is not coincident in space with the explosion; it may not be coincident in time but delayed a few seconds [e.g., the delayed Rayleigh waves observed by Rygg (1979)]. Spall failure can occur less than a second after detonation, and spall closure can occur 1 to 3 seconds after detonation (Springer 1974). Therefore the principal inelastic processes that are associated with the generation of the main seismic signal from a low-yield explosion (inelastic deformation around the cavity, triggered fault motion, spall closure) are fractional-second processes that take place within a few seconds. Cavity collapse and aftershocks usually generate seismic waves that are distinct from the principal seismic signal.

The explosion is an approximately spherical process; therefore the direct seismic radiation consists of compressional-wave energy and some small fraction of shear-wave energy. Other shear waves are generated by reflection and refraction of compressional waves at the free surface and at interfaces between layers in the local geological formations. Rayleigh waves are generated by interaction of the direct seismic radiation with the free surface. As noted in Sections 4.2 and 6.2, spall closure may make a significant contribution to the generation of Rayleigh waves. Tectonic strain release is associated with perturbations in the Rayleigh-wave radiation pattern and with Love wave generation. The principal differences between explosions and earthquake spectra are at the long periods associated with Rayleigh and Love waves. The reason that the m_b:M_s discriminant works is probably because the primary radiation from explosions is mostly compressional waves. Primary radiation from earthquakes has a much larger fraction of shear-wave energy, and shear waves are more efficient in generating Rayleigh waves than are compressional waves.

6.5 Yield Estimation: Effects of Seismic Coupling Efficiency and Signal Attenuation

Linear magnitude vs log yield relations have been empirically determined for underground nuclear explosions, as noted in Section 6.2. One must make two assumptions in applying such relations developed for one test site to estimate the yields of explosions at another site: the seismic coupling efficiency is approximately

the same at both locations, and the signal propagation characteristics from both sites to their respective receiving networks are
comparable. As noted in Section 6.1, variations in geological
properties around the shot point can cause as much as 1 seismic-
magnitude-unit variation for explosions of a given yield.
Regional variations in attenuation in the upper mantle can cause
magnitude variations to vary from 0.1 to 0.3 units; Marshall,
Springer, and Rodean (1979) developed an empirical magnitude
correction for this effect. Seismic yield estimates for explosions at a given site can therefore have a significant systematic
mean error (or bias) as well as random errors about the mean
(Ericsson 1971; Basham and Horner 1973; Marshall, Springer, and
Rodean 1979).

We will use seismic yield estimates by Dahlman and Israelson
(1977) to illustrate systematic mean errors and random errors
about the mean. They used announced yields for 16 explosions in
tuff and rhyolite in the Yucca Flat and Pahute Mesa testing areas
of the Nevada Test Site to develop a single linear source function
vs log yield relation. [The magnitude vs log yield relations for
explosions in saturated tuff and rhyolite in these testing areas
are essentially identical (Marshall, Springer, and Rodean 1979)].
They used this relation in estimating the yields of 178 explosions
at the Nevada Test Site. They assumed that the coupling of explosions in tuff is three times that of explosions in alluvium. They
used the tuff-rhyolite relation for explosions in other media
(granite, dolomite, limestone). They did not make any corrections
for explosion depth, moisture content of the shot media, or the
depth of the static water table.

Their yield estimates (Y_e) were compared with the official
explosion yields (Y_o) by Rodean (1979). The statistical distributions of the ratios of the estimated to the official yields are
shown in Figs. 37-39 for explosions in the Yucca Flat, Pahute
Mesa, and Rainier and Shoshone Mesa testing areas, respectively.
(The Rainier and Shoshone Mesa testing areas are some distance
apart, but the testing conditions and seismic coupling in those
areas are essentially the same.) The quantitative values of
$\log_{10} (Y_e/Y_o)$ are not shown in these figures because most of
the explosion yields are classified. The main reason for presenting these figures, less numerical scales for the ordinates, is to
show that the ratios of the estimated yields to the official
yields are lognormally distributed for each of these testing
areas. At the Nevada Test Site, the greatest variation in geological properties is found in Yucca Flat; the standard deviation of
$\log_{10} (Y_e/Y_o)$ is somewhat greater for explosions in Yucca Flat
than for explosions in the other testing areas. The means of the
ratios (Y_e/Y_o) for the explosions in Yucca Flat and Pahute
Mesa are approximately unity, as is to be expected because the
estimates are based on calibration data for these testing areas.

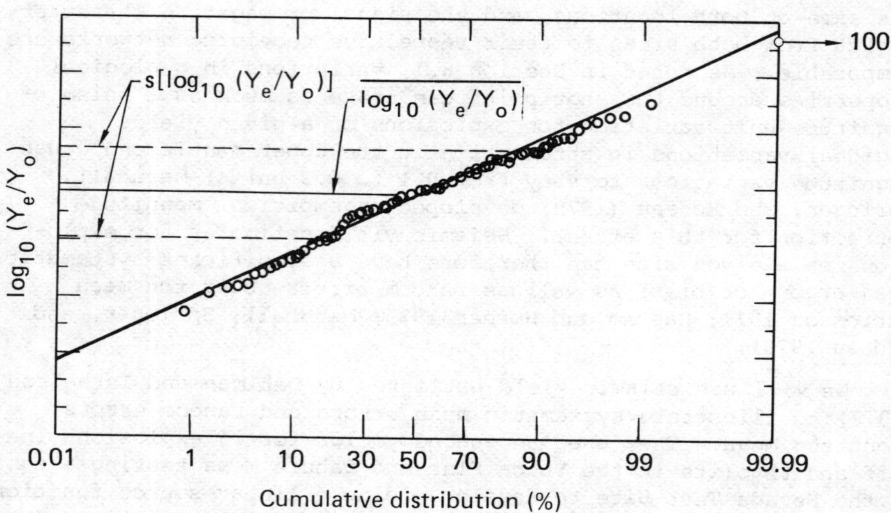

FIG. 37. Cumulative distribution of $\log_{10}\,(Y_e/Y_o)$ for explosions in Yucca Flat (from Rodean 1979, with Y_e values from Dahlman and Israelson 1977).

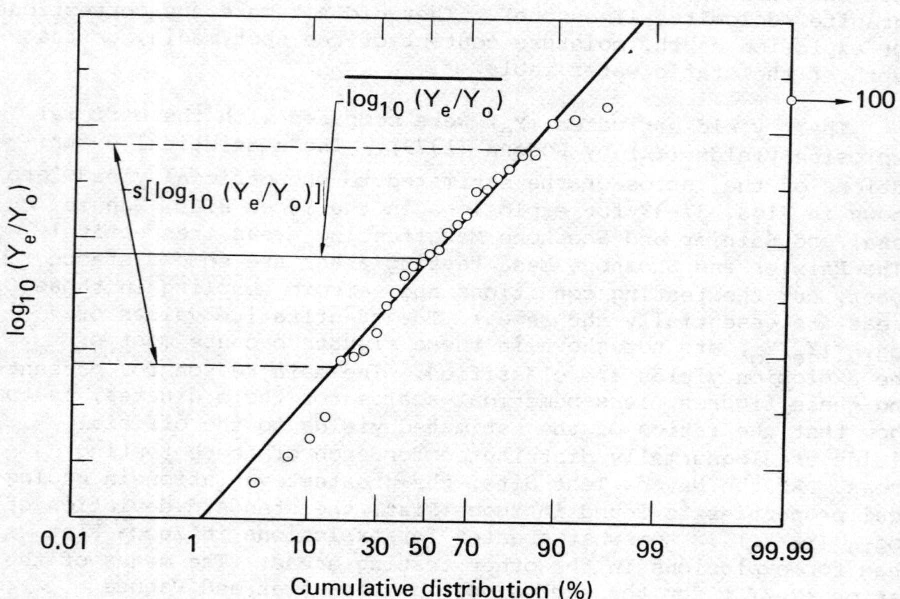

FIG. 38. Cumulative distribution of $\log_{10}\,(Y_e/Y_o)$ for explosions in Pahute Mesa (from Rodean 1979, with Y_e values from Dahlman and Israelson 1977).

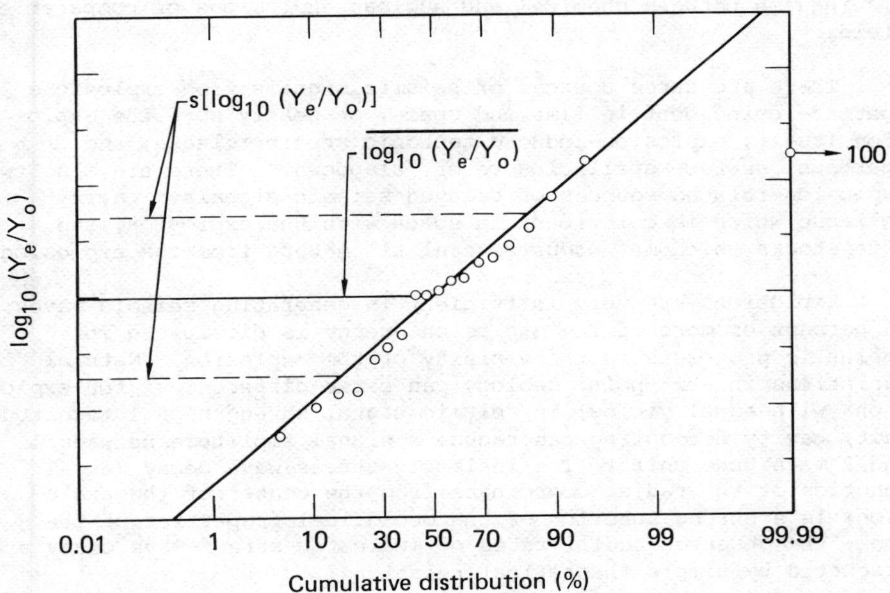

FIG. 39. Cumulative distribution of $\log_{10}$ (Y_e/Y_o) for explosions in Rainier and Shoshone mesas (from Rodean 1971a, with Y_e values from Dahlman and Israelson 1977).

The mean of the ratio (Y_e/Y_o) for Rainier and Shoshone mesas is greater than unity; Dahlman and Israelson did not have any calibration data for these testing areas. As noted by Terhune, Snell, and Rodean (1979), seismic data indicate that the seismic signals from Rainier Mesa are, on the average, a fraction of a magnitude unit greater than those from explosions of comparable yield below the water table in Yucca Flat and Pahute Mesa. This is consistent with the effect of differences in emplacement conditions investigated by Terhune, Snell, and Rodean. (Rainier and Shoshone Mesa explosions generally have greater scaled source radii than explosions in the other testing areas; see Figs. 29 and 30.) However Terhune, Snell, and Rodean concluded that other emplacement and geologic factors that affect signal generation and propagation could be responsible for all or part of the observed seismic magnitude difference.

7.   SUMMARY

The phenomena are different, but there are analogies in the early stages of underground chemical and nuclear explosions. At later times, the inelastic processes in the rock are identical in the two types of explosions. There is no known seismic method to

distinguish between chemical and nuclear explosions of comparable
yield.

There are three sources of seismic signals from explosions
that are coincident in time and space, or nearly so: the explosion itself, explosion-induced tectonic strain release, and
(perhaps) surface spall closure or "slapdown." There are also two
explosion-related sources of delayed seismic signals: cavity
collapse which is coincident in space with the explosion, and
aftershocks which may occur several kilometers from the explosion.

Explosions are very inefficient in generating seismic waves;
95 percent or more of the explosion energy is dissipated in
inelastic processes in the vicinity of the explosion. Natural
variations in shot-point geology can cause differences (for explosions with equal yields) in seismic signal strength of 1 magnitude
unit; cavity decoupling can reduce a signal amplitude between 1
and 2 magnitude units. The inelastic stress-wave decay (as a
function of the radial coordinate from the center of the explosion) is a strong function of the geological properties of the
rock; the observed radial rates of inelastic stress-wave decay are
bracketed by simple theoretical relations.

The transition from a nonlinear inelastic stress wave to a
linear "elastic" seismic wave is not well defined; in part because
ideal elastic stress-wave propagation does not exist in geological
media. Assuming spherical symmetry for explosions, the location
of the transition is often called the "elastic radius." It is
proposed that this transition be defined as the transition between
nonlinear inelastic and linear anelastic behavior. Some data
indicate that linear anelastic behavior exists in some rocks at
strains less than 10^{-6}. More research is needed to define the
transition between nonlinear and linear stress-wave response.

The reduced-displacement potential is a convenient approximate fiction. It is a fiction because no real explosion is
exactly spherical and ideal elastic stress-wave propagation in
rocks does not exist. It is approximate because explosions are
approximately spherical and stress-wave propagation can be almost
elastic. It is convenient because it is a simple mathematical
description of an explosion as a seismic source.

Present programs for the calculation of explosions on large
computers are very useful for predicting phenomena in the high-
stress zone, but are less useful in accurately predicting seismic-
wave radiation from an explosion. This is because the constitutive relations between stress and strain do not model all the
processes that are known to occur at low stress levels at which
nonlinear inelastic response tends toward linear anelastic
response.

In principle, synthetic seismograms can be calculated beginning with the detonation of the explosion and ending with the recording of the seismogram (the forward problem in seismology). However, seismic data can be inverted to obtain only an "equivalent elastic source"; the seismic data cannot be used as inputs to calculate nonlinear inelastic stress-wave propagation backward in space and time (the inverse problem). The reduced-displacement potential is a common solution of the forward and inverse problems. Computer calculations of reduced-displacement potentials for specific explosions indicate that a step function is the principal component of the driving function. Experimental measurements on these explosions indicate the driving function is a step-pulse-impulse function. Seismic data inversion and the matching of synthetic with real seismograms for specific explosions indicate an even greater impulse component in the source function. More research is required to resolve these discrepancies.

A simple spherical explosion model (the reduced-displacement potential) in combination with a theoretical scaling rule (characteristic times and linear dimensions scaled in proportion to the cube-root of explosion yield) is not consistent with observed linear relations between seismic magnitude and the logarithm of the explosion yield. A modified empirical scaling rule in combination with the reduced-displacement potential gives better results for body-wave magnitude vs log yield, but does not account for the observed linear relation between surface-wave magnitude vs log yield. More research is required to better define the effects of yield, depth, and material properties on seismic source parameters. There are speculations and there is some evidence that spall closure or "slapdown" is significant in Rayleigh-wave generation. More research is required to define the role of spall closure in seismic wave generation.

Tectonic strain release makes an explosion appear more "earthquake-like," but apparently not to the extent that the m_b:M_S discriminant between earthquakes and explosions is compromised. The effect on M_S and the Rayleigh wave radiation pattern is significantly different for strike-slip vs dip-slip tectonic release. There are two hypotheses for the release of preexisting tectonic strain by explosions. One is that the strain is released around the weak fractured zone produced by the explosion; a theoretical model has been developed for this phenomenon. The other is that fault motion is triggered near the explosion; a causal theoretical model has not been developed for this phenomenon. Seismic data from many explosions are consistent with both hypotheses, but the large tectonic strain release observed from some explosions is more consistent with the triggering model.

Two final cautions. First, most of the public data (time, yield, depth, medium, etc.) for underground nuclear explosions are

for a limited number of geologic media at one location: alluvium
and volcanic rocks (tuff and rhyolite) at the Nevada Test Site.
Comparable quantities of data for explosions in other media and at
other locations are necessary to provide a balanced perspective
and the answers to some important questions. For example, "Are
the two m_b vs M_s slopes observed for explosions in Nevada and
the USSR a consequence of different slopes for m_b vs log yield
relations for the different rocks at these test sites?" Second,
the best quality and most complete seismic data we now have for
use in seismic discrimination research are from high yield,
strongly-coupled explosions. Under a comprehensive test ban, the
operational problem will concern low yield, weakly coupled
explosions.

ACKNOWLEDGMENTS

This lecture summarizes some of my work, but most of it
concerns the work of many others who are credited in the references. My colleagues at the Lawrence Livermore National Laboratory, especially F. Followill, D. Larson, P. Moulthrop, and
L. Thigpen, made valuable and constructive criticisms of the first
draft of this manuscript. Work performed under the auspices of
the U.S. Department of Energy by the Lawrence Livermore Laboratory under Contract W-7405-Eng-48.

REFERENCES

Ahrens, T. J., and Urtiew, P. A., 1971. The use of shock waves
    in the vaporization of metals, Lawrence Livermore Laboratory,
    Report UCRL-51109, Livermore, California.

Aki, K., 1966. Generation and propagation of G waves from the
    Niigata earthquake of June 16, 1964. Part 2. Estimation of
    earthquake movement, released energy, and stress - strain
    drop from the G-wave spectrum, Bull. Earthquake Res. Inst.,
    Tokyo Univ., 44, 73-88.

Aki, K., Bouchon, M., and Reasenberg, P., 1974. Seismic source
    function for an underground nuclear explosion, Bull. Seism.
    Soc. Am., 64, 131-148.

Aki, K., Reasenberg, P., DeFazio, T., and Tsai, Y.-B., 1969.
    Near-field and far-field seismic evidences for triggering of
    an earthquake by the Benham explosion, Bull. Seism. Soc. Am.,
    59, 2197-2207.

Aki, K., and Tsai, Y.-B., 1972. Mechanism of Love-wave
    excitation by explosive sources, J. Geophys. Res., 77,
    1452-1475.

Archambeau, C. B., 1972. The theory of stress wave radiation
     from explosions in prestressed media, Geophys. J.R., Astr.
     Soc., 29, 329-366.

Archambeau, C., and Sammis, C., 1970. Seismic radiation from
     explosions in prestressed media and the measurement of
     tectonic stress in the earth, Rev. Geophys., 8, 473-499.

Bakun, W. H., and Johnson, L. R., 1973. The deconvolution of
     teleseismic P waves from explosions Milrow and Cannikin,
     Geophys. J. R., Astr. Soc., 34, 321-342.

Basham, P. W., 1969. Canadian magnitudes of earthquakes and
     nuclear explosions in south-western North America, Geophys.
     J. R. Astr. Soc., 17, 1-13.

Basham, P. W., and Horner, R. B., 1973. Seismic magnitudes of
     underground nuclear explosions, Bull. Seism. Soc. Am., 63,
     105-131.

Blake, F. G., Jr., 1952. Spherical wave propagation in solid
     media, J. Acoust. Soc. Am., 24, 211-215.

Boardman, C. R., Rabb, D. D., and McArthur, R. D., 1964.
     Responses of four rock mediums to contained nuclear
     explosions, J. Geophys. Res., 69, 3457-3469.

Boucher, G., Ryall, A., and Jones, A. E., 1969. Earthquakes
     associated with underground nuclear explosions, J.
     Geophys. Res., 74, 3808-3820.

Brennan, B. J., and Stacey, F. D., 1977. Frequency dependence
     of elasticity of rock--test of seismic velocity
     dispersion, Nature, 268, 220-222.

Brune, J. N., and Pomeroy, P. W., 1963. Surface wave radiation
     patterns for underground nuclear explosions and
     small-magnitude earthquakes, J. Geophys. Res., 68,
     5005-5028.

Burdick, L. J., and Helmberger, D. V., 1979. Time functions
     appropriate for nuclear explosions, Bull. Seism. Soc. Am.,
     69, 957-973.

Burton, D. E., and Schatz, J. F., 1975. Rock modeling in
     TENSOR'74, a two-dimensional Lagrangian shock propagation
     code, Lawrence Livermore Laboratory, Report UCID-16719,
     Livermore, California.

Butkovich, T. R., 1967. The gas equation of state for natural
    materials, Lawrence Radiation Laboratory, Report
    UCRL-14729, Livermore, California.

Carpenter, E. W., 1967. Teleseismic signals calculated for
    underground, underwater and atmospheric explosions,
    Geophysics, 32, 17-32.

Carpenter, E. W., Savill, R. A., and Wright, J. K., 1962. The
    dependence of seismic signal amplitudes on the size of
    underground explosions, Geophys. J. R. Astr. Soc., 6,
    426-440.

Chabai, A. J., 1965. On scaling dimensions of craters produced
    by buried explosives, J. Geophys. Res., 70, 5075-5098.

Cherry, J. T., and Petersen, F. L., 1970. Numerical simulation
    of stress wave propagation from underground nuclear explo-
    sions (IAEA-PL-338/15), Peaceful Nuclear Explosions,
    International Atomic Energy Agency, Vienna.

Chilton, F., Eisler, J. D., and Heubach, H. G., 1966. Dynamics
    of spalling of the earth's surface caused by underground
    explosions, J. Geophys. Res., 71, 5911-5919.

Constantino, M. S., 1978. Statistical variation in stress-
    volumetric strain behavior of Westerly granite, Int. J.
    Rock Mech. Min. Sci. and Geomech. Abstr., 15, 105-111.

Dahlman, O., and Israelson, H., 1977. Monitoring Underground
    Nuclear Explosions, Elsevier Scientific Publishing
    Company, Amsterdam, Ch. 11 and Appendix A.2.5.

Edwards, A. L., and Holzman, R. L., 1968. Thermal effects of a
    nuclear explosion in salt, the Salmon experiment,
    Naturwissenschaften, 55, 18-22.

Eisler, J. D., 1967. Near-surface spalling from a nuclear
    explosion in a salt dome, J. Geophys. Res., 72, 1751-1760.

Eisler, J. D., and Chilton, F., 1964. Spalling of the earth's
    surface by underground nuclear explosions, J. Geophys.
    Res., 69, 5285-5293.

Eisler, J. D., Chilton, F., and Sauer, F. M., 1966. Multiple
    subsurface spalling by underground nuclear explosions, J.
    Geophys. Res., 71, 3923-3927.

Engdahl, E. R., 1972. Seismic effects of the Milrow and
    Cannikin nuclear explosions, Bull. Seism. Soc. Am., 62,
    1411-1423.

Ericsson, U., 1971. A linear model for the yield dependence of
    magnitudes measured by a seismographic network, Geophys.
    J. R. Astr. Soc., 25, 49-69.

Gardner, M. F., and Barnes, J. L., 1942. Transients in Linear
    Systems, John Wiley and Sons, Inc., New York.

Gauvenet, A., 1970. Experiments with underground nuclear
    explosions in the Hoggar massif, Proceedings of the
    American Nuclear Society Topical Meeting, Engineering with
    Nuclear Explosives, Las Vegas, Nevada, January 14-16,
    1970, U.S. Atomic Energy Commission, Washington D.C.

Hamilton, R. M., Smith, B., Fisher, F. G., and Papanek, P. J.,
    1972. Earthquakes caused by underground nuclear
    explosions on Pahute Mesa, Nevada Test Site, Bull. Seism.
    Soc. Am., 62, 1319-1341.

Hanks, T. C., and Wyss, M., 1972. The use of body-wave spectra
    in the determination of seismic-source parameters, Bull.
    Seism. Soc. Am., 62, 561-589.

Haskell, N. A., 1967. Analytic approximation for the elastic
    radiation from a contained underground explosion, J.
    Geophys. Res., 72, 2583-2587.

Heckman, R. A., 1964. Deposition of thermal energy by nuclear
    explosives, Lawrence Radiation Laboratory, Report
    UCRL-7801, Livermore, California.

Helmberger, D. V., and Harkrider, D. G., 1972. Seismic source
    descriptions of underground explosions and a depth discri-
    minant, Geophys. J. R. Astr. Soc., 31, 45-66.

Herbst, R. F., Werth, G. C., and Springer, D. L., 1961. Use of
    large cavities to reduce seismic waves from underground
    explosions, J. Geophys. Res., 66, 959-978.

Hirasawa, T., 1971. Radiation patterns of S waves from under-
    ground nuclear explosions, J. Geophys. Res., 76, 6440-6454.

Holzer, F., 1966. Calculation of seismic source mechanisms,
    Proc. Roy. Soc., London, Ser. A, 290, 408-429.

Hudson, J. A., and Douglas, A., 1975. Rayleigh wave spectra
    and group velocity minima, and the resonance of P waves in
    layered structures, Geophys. J. R. Astr. Soc., 42, 175-188.

Israelson, H., Slunga, R., and Dahlman, O., 1974. Aftershocks
    caused by the Novaya Zemlya explosion on October 27, 1973,
    Nature, 247, 450-452.

Jaeger, J. C., and Cook, N. G. W., 1971. Fundamentals of Rock
    Mechanics, Science Paperbacks, London.

Jeffreys, H., 1931. On the cause of osillatory movement in
    seismograms, Mon. Notic. Roy. Astron. Soc., Geophys.
    Suppl., 2, 407-416.

Johansson, C. H., and Persson, P. A., 1970. Detonics of High
    Explosives, Academic Press, Inc., New York.

Johnson, G. W., Higgins, G., and Violet, C., 1959. Underground
    nuclear detonations. J. Geophys. Res., 64, 1457-1470.

Kawasumi, H., and Yosiyama, R., 1935. On an elastic wave ani-
    mated by the potential energy of initial strain, Bull.
    Earthquake Res. Inst., Tokyo Univ., 13, 496-503.

Kedrovskii, O. L., 1970. Primenenie kamufletnyk yadernykh
    vzryvov v promyshlennost. (IAEA-PL-388/20), Peaceful
    Nuclear Explosions, International Atomic Energy Agency,
    Vienna.

Knopoff, L., and MacDonald, G. J. F., 1958. Attenuation of small
    amplitude stress waves in solids, Rev. Mod. Phys., 30,
    1178-1192.

Lambert, D. G., Flinn, E. A., and Archambeau, C. B., 1972. A
    comparative study of the elastic wave radiation from
    earthquakes and underground explosions, Geophys. J. R.
    Astr. Soc., 29, 403-432.

Larson, D. B., 1977a. The relationship of rock properties to
    explosive energy coupling, Lawrence Livermore Laboratory,
    Report UCRL-52204, Livermore, California.

Larson, D. B., 1977b. Lawrence Livermore Laboratory, Livermore,
    California, private communication.

Larson, D. B., 1980a. Shock wave studies in Blair dolomite, J.
    Geophys. Res., 85, 293-297.

Larson, D. B., 1980b. Lawrence Livermore Laboratory, Livermore,
    California, private communication.

Larson, D. B., and Anderson, G. D., 1979a. Plane shock wave
    studies of porous geological media, J. Geophys. Res., 84,
    4592-4600.

Larson, D. B., and Anderson, G. D., 1979b. Plane shock wave
    studies of Westerly granite and Nugget sandstone, Lawrence
    Livermore Laboratory, Report UCRL-82264, accepted for
    publication in Int. J. Rock Mech. Min. Sci. and Geomech.
    Abstr.

Latter, A. L., LeLevier, R. E., Martinelli, E. A., and McMil-
    lan, W. G., 1961. A method of concealing underground
    nuclear explosions, J. Geophys. Res., 66, 943-946.

Liebermann, R. C., and Pomeroy, P. W., 1969. Relative excita-
    tion of earthquakes and underground explosions, J.
    Geophys. Res., 74, 1575-1590.

Maenchen, G., and Sack, S., 1964. The tensor code, Methods in
    Computational Physics, Vol. 3, Academic Press, Inc., New
    York.

Marshall, P. D., 1979. AWRE, Blacknest, England, private
    communication.

Marshall, P. D., and Basham, P. W., 1972. Discrimination be-
    tween earthquakes and underground explosions employing an
    improved M_S scale, J. Geophys. Res., 28, 431-458.

Marshall, P. D., Douglas, A., and Hudson, J. A., 1971. Surface
    waves from underground explosions, Nature, 234, 8-9.

Marshall, P. D., Springer, D. L., and Rodean, H. C., 1979. Mag-
    nitude corrections for attenuation in the upper mantle,
    Geophys. J. R. Astr. Soc., 57, 609-638.

McEvilly, T. V., and Peppin, W. A., 1972. Source characteris-
    tics of earthquakes, explosions, and afterevents, Geophys.
    J. R. Astr. Soc., 31, 67-82.

Meyer, M. L., 1964. On spherical near fields and far fields in
    elastic and visco-elastic solids, J. Mech. Phys. Solids,
    12, 77-111.

Michaud, L., 1968. Explosions nuclaires souterraines etude des
    rayone de cavity, Commissariat A L'Energie Atomique,
    Rapport CEA-R-3594.

Mueller, R. A., 1969. Seismic energy efficiency of underground
    nuclear detonations, Bull. Seism. Soc. Am., 59, 2311-2323.

Mueller, R. A., and Murphy, J. R., 1971. Seismic characteris-
    tics of underground nuclear detonations. Part I. Seismic
    spectrum scaling, Bull. Seism. Soc. Am., 61, 1675-1692.

Müller, G., 1969. Theoretical seismographs for types of
    point-sources in layered media. Part III. Single force
    and dipole sources of arbitrary orientation, _J. Geophys._,
    _35_, 347-371.

Müller, G., 1973. Seismic moment and long-period radiation of
    underground nuclear explosions, _Bull. Seism. Soc. Am._,
    _63_, 847-857.

Murphey, B. F., 1961. Particle motion near explosion in Halite,
    J. Geophys. Res, _66_, 947-958.

Murphy, J. R., 1977. Seismic source functions and magnitude
    determinations for underground nuclear detonations, _Bull.
    Seism. Soc. Am._, _67_, 135-158.

Patterson, D. W., 1966. Nuclear decoupling, full and partial,
    J. Geophys. Res., _71_, 3427-3426.

Patton, H. J., 1980. Lawrence Livermore Laboratory, Livermore,
    California, private communication.

Peppin, W. A., 1977. A near-regional explosion source model for
    tuff, _Geophys. J. R. Astr. Soc._, _48_, 331-349.

Perret, W. R., 1972a. Gasbuggy seismic source measurements,
    Geophysics, _47_, 301-312.

Perret, W. R., 1972b. Seismic-source energies of underground
    nuclear explosions, _Bull. Seism. Soc. Am._, _62_, 763-774.

Press, F., and Archambeau, C., 1962. Release of tectonic
    strain by underground nuclear explosions, _J. Geophys.
    Res._, _67_, 337-343.

Randall, M. J., 1973. The spectral theory of seismic sources,
    Bull. Seism. Soc. Am., _63_, 1133-1144.

Rawson, D. E., 1963. Review and a summary of some Project Gnome
    results, _Trans. Amer. Geophysics Union_, _44_, 129.

Rawson, D. E., Taylor, R. W., and Springer, D. L., 1967. A
    review of the Salmon Experiment, a nuclear explosion in
    salt, _Naturwissenschaften_, _54_, 525-531.

Ringdahl, F., 1976. Maximum-likelihood estimation of seismic
    magnitude, _Bull. Seism. Soc. Am._, _66_, 789-802.

Rodean, H. C., 1971a. _Nuclear-Explosion Seismology_, U.S. Atomic
    Energy Commission.

Rodean, H. C., 1971b. Cavity decoupling of nuclear explosions, Lawrence Livermore Laboratory, Report UCRL-51097, Livermore, California.

Rodean, H. C., 1972. An energy approach to seismic coupling analysis, J. Geophys. Res., 77, 3129-3145.

Rodean, H. C., 1979. Statistical analysis of Swedish yield estimates for U.S. underground nuclear explosions, Lawrence Livermore Laboratory, Report UCRL-52698, Livermore, California. (Title U, Report-Confidential FRD).

Rogers, L. A., 1966. Free-field motion near a nuclear explosion in salt: Project Salmon, J. Geophys. Res., 71, 3145-3426.

Rygg, E., 1979. Anomalous surface waves from underground explosions, Bull. Seism. Soc. Am., 69, 1995-2002.

Savage, J. C., and Hasegawa, H. S., 1967. Evidence for a linear attenuation mechanism, Geophysics, 32, 1003-1014.

Schatz, J. F., 1973. The physics of SOC and TENSOR, Lawrence Livermore Laboratory, Report UCRL-51352, Livermore, California.

Schatz, J. F., 1974. SOC73, a one-dimensional wave propagation code for rock media, Lawrence Livermore Laboratory, Report UCRL-51689, Livermore, California.

Schatz, J., Kusubov, A., Hearst, J., Abey, A., Snell, C., and Thigpen, L., 1977. Rock mechanic project progress and results: rock fracture and pore collapse, Lawrence Livermore Laboratory, Report UCID - 17527, Livermore, California.

Sedov, L. I., 1959. Similarity and Dimensional Methods in Mechanics, 4th Ed., Academic Press, Inc., New York, pp.210-238.

Selberg, H. L., 1952, Transient compressional waves from spherical and cylindrical cavities, Ark. Phys., 5, 97-108.

Smith, S. W., 1963. Generation of seismic waves by underground explosions and the collapse of cavities, J. Geophys. Res., 68, 1447-1483.

Springer, D. L., 1974. Secondary sources of seismic waves from underground nuclear explosions, Bull. Seism. Soc. Am., 64, 581-594.

Springer, D. L., and Denny, M. D., 1976. Seismic spectra of
    events at regional distances, Lawrence Livermore
    Laboratory, Report UCRL-52048, Livermore, California.

Springer, D. L., Denny, M. D., Healy, J., and Mickey, W., 1968.
    The Sterling experiment: decoupling of seismic waves by a
    shot-generated cavity, J. Geophys. Res., 73, 5995-6011.

Springer, D. L., and Hannon, W. J., 1973. Amplitude-yield
    scaling for underground nuclear explosions, Bull. Seism.
    Soc. Am., 63, 477-500.

Springer, D. L., and Kinnaman, R. L., 1971. Seismic source
    summary for U.S. nuclear underground explosions,
    1961-1970, Bull. Seism. Soc. Am., 61, 1073-1098.

Springer, D. L., and Kinnaman, R. L., 1975. Seismic source sum-
    mary for U.S. nuclear underground explosions, 1971-1973,
    Bull. Seism. Soc. Am., 65, 343-349.

Taylor, G., 1950a. The formation of a blast wave by a very in-
    tense explosion. I. Theoretical discussion. Proc. Roy.
    Soc. London, Series A, 201, 159-174.

Taylor, G., 1950b. The formation of a blast wave by a very in-
    tense explosion. II. The atomic explosion of 1945. Proc.
    Roy. Soc. London, Series A, 201, 175-186.

Terhune, R, W., Glenn, H. D., Burton, D. E., and Rambo, J. T.,
    1977a. Containment analysis for the simultaneous detonation
    of two nuclear explosions, Lawrence Livermore Laboratory,
    Report UCRL-52268, Livermore, California.

Terhune, R, W., Glenn, H. D., Burton, D. E., and Rambo, J. T.,
    1977b. Calculational examination of the Baneberry event,
    Lawrence Livermore Laboratory, Report UCRL-52365, Liver-
    more, California.

Terhune, R. W., Snell, C. M., and Rodean, H. C., 1979. Enhanced
    coupling and decoupling of underground nuclear explosions,
    Lawrence Livermore Laboratory, Report UCRL-52806,
    Livermore, California.

Toksöz, M. N., Ben-Menahem, A., and Harkrider, D. G., 1964.
    Determination of source parameters of explosions and
    earthquakes by amplitude equalization of seismic surface
    waves. 1. Underground nuclear explosions, J. Geophys.
    Res., 69, 4355-4366.

Toksöz, M. N., Ben-Menahem, A., and Harkrider, D. G., 1965.
    Determination of source parameters of explosions and
    eathquakes by amplitude equalization of seismic surface
    waves. 2. Release of tectonic strain by underground
    nuclear explosions and mechanisms of earthquakes, J.
    Geophys. Res., 70, 907-922.

Toksöz, M. N., and Kehrer, H. H., 1972. Tectonic strain release
    by underground nuclear explosions and its effect on seismic
    discrimination Geophys. J. R. Astr. Soc., 31, 141-161.

Toman, J., Sisemore, C., and Terhune, R., 1973. The Rio Blanco
    experiment: subsurface and surface effects and measure-
    ments, Lawrence Livermore Laboratory, Report UCRL-51504,
    Livermore, California.

Trulio, J., 1977. Applied Theory Inc., Los Angeles, private
    communication.

Trulio, J., 1979. Applied Theory Inc., Los Angeles, private
    communication.

Tsai, Y.-B., and Aki, K., 1971. Amplitude spectra of surface
    waves from small earthquakes and underground nuclear
    explosions, J. Geophys. Res., 76, 3940-3952.

Viecelli, J. A., 1973a. The linear Q and the calculation of
    decaying spherical shocks in solids. J. Comp. Phys., 12,
    187-201.

Viecelli, J. A., 1973b. Spallation and the generation of sur-
    face waves by an underground explosion, J. Geophys. Res.,
    78, 2475-2487.

von Neumann, J., 1963. Collected works, Vol. VI, Pergamon Press,
    Oxford, pp 219-237.

von Neumann, J., and Richtmyer, R. D., 1950. A method for the
    numerical calculation of hydrodynamic shocks, J. Appl.
    Phys., 21, 232-237.

von Seggern, D., and Blandford, R., 1972. Source time functions
    and spectra for underground nuclear explosions, Geophys.
    J. R. Astr. Soc., 31, 83-97.

Werth, G. C., and Herbst, R. F., 1963. Comparison of amplitude
    of seismic waves from nuclear explosions in four mediums,
    J. Geophys. Res., 68, 1463-1475.

Werth, G. C., Herbst, R. F., and Springer, D. L., 1962. Ampli-
    tude of seismic arrivals from the M discontinuity, J.
    Geophys. Res., 67, 1587-1610.

Wheeler, V. E., Preston, R. G., and Frerking, C. E., 1976.
    Trapped stress waves in underground nuclear explosions,
    Lawrence Livermore Laboratory, Report UCRL-52012, Liver-
    more, California. (Title-U, Report-Confidential FRD).

Wilkins, M. L., 1964. Calculation of elastic-plastic flow, in
    Methods in Computational Physics, Vol. 3, Academic Press,
    Inc. New York.

Winkler, K., Nur, A., and Gladwin, M., 1979. Friction and
    seismic attenuation in rocks, Nature, 277, 528-531.

Yoshiyama, R., 1963. Note on earthquake energy, Bull. Earth-
    quake Res. Inst. Tokyo Univ., 41, 687-697.

Zel'dovich, Ya. B., and Raizer, Yu. P., 1966. Physics of Shock
    Waves and High-Temperature Hydrodynamic Phenomena, Vol. 1,
    Academic Press, Inc., New York.

Zel'dovich, Ya. B., and Raizer, Yu. P., 1967. Physics of Shock
    Waves and High-Temperature Hydrodynamic Phenomena, Vol. 2,
    Academic Press, Inc., New York.

FOOTNOTES

1. Perret (1967) was the first to publish Cowboy and Salmon data scaled to a common yield, but the scaled data sets did not approximate a common relation like that shown in Fig. 14 for reasons that are discussed with the review by Trulio (1978).

2. The effects of tectonic strain release that were calculated by Toksöz and Kehrer and by Patton are not strictly comparable. Toksöz and Kehrer considered the energy but not the phase of the surface waves from the explosive and tectonic components; Patton took both energy and phase into account.

3. According to Bache (1980), the phase difference between explosion- and spall-generated Rayleigh waves is $\pi/2$, not π.

4. As indicated by Terhune, Snell and Rodean, Trulio first noted the phenomenon of enhanced coupling in theoretical calculations.

ADDITIONAL REFERENCES

Bache, T.C., 1980. Systems, Science and Software, La Jolla, California, private communication.

Perret, W.R., 1967. Free-field particle motion from a nuclear explosion in salt, Part I, Project Dribble, Salmon event, Sandia Laboratories, Report VUF-3012, Albuquerque, New Mexico.

Trulio, J.G., 1978. Simple scaling and nuclear monitoring, Applied Theory, Inc., Report ATR-77-45-2, Los Angeles.

ANOMALOUS RAYLEIGH WAVES FROM PRESUMED EXPLOSIONS IN EAST KAZAKH

John R. Cleary

Research School of Earth Sciences, Australian National
University, Canberra, Australia

ABSTRACT

Long-period vertical seismograph recordings at SRO stations
from events in the Shagan River province of eastern Kazakhstan
between 1976 and 1979 show an interesting pattern of Rayleigh
wave anomalies. While Rayleigh waves from some of the events
are normal at all stations, those from other events exhibit
anomalous behaviour, in the form of amplitude perturbations accompanied by phase reversals, in a NE-SW direction, with normal
Rayleigh waves in the other directions. For one event the Rayleigh
waves are phase-reversed in all directions. A remarkable aspect
of the records is that the shapes of the anomalous Rayleigh wave
trains are almost perfectly consistent from event to event, but
their arrival times are delayed by a few seconds compared to the
normal Rayleigh waves. The azimuthal pattern seems to rule out
explanations such as depth of burial and spallation, while the
consistency in wave shape and the associated time delay are difficult to model by combining the radiation patterns of an explosion
and an accompanying earthquake. A reasonable interpretation of
the anomalous signals seems to be that pre-existing strain effectively modifies the source functions of the explosions in
preferred azimuths.

INTRODUCTION

The following preliminary study is based on Rayleigh wave data
recorded by SRO stations from a number of presumed nuclear explosions in the Shagan River province of East Kazakhstan between
November 1976 and August 1979, using locations (Table 1) provided

191

*E. S. Husebye and S. Mykkeltveit (eds.), Identification of Seismic Sources - Earthquake or Underground
Explosion, 191–199.*
Copyright © 1981 by D. Reidel Publishing Company.

TABLE 1

Parameters of E. Kazakh Events

Date	Time* (GMT)	Latitude** (^{o}N)	Longitude** (^{o}E)	Magnitude* (m_b)
23 Nov 76	05.02.57.4	50.014	78.937	5.9
7 Dec 76	04.56.57.4	49.957	78.816	5.9
29 May 77	02.56.57.8	49.968	78.716	5.6
11 Jun 78	02.56.57.7	49.930	78.779	5.9
5 Jul 78	02.46.57.3	49.901	78.836	5.8
29 Aug 78	02.37.06.5	50.023	78.968	5.9
15 Sep 78	02.36.57.3	49.920	78.843	6.0
4 Nov 78	05.05.57.5	50.050	78.918	5.6
29 Nov 78	04.33.02.9	49.979	78.805	6.0
23 Jun 79	02.56.58.6	49.896	78.807	6.2
7 Jul 79	03.46.58.3	50.013	78.947	5.8
4 Aug 79	03.56.58.0	49.887	78.838	6.0

*   From USGS PDE cards

**   From relocations by North and Fitch (pers. comm.)

by North and Fitch (pers. comm.). The SRO stations used are
distributed fairly regularly around the test site in the azimuth
range 80° to 310°, and at distances between 16° and 45° (Figure 1).
The close proximity of the events to each other (cf. Figure 4)
implies that their surface waves travel virtually identical
paths to any particular station.

DATA ANALYSIS AND RESULTS

A remarkable phenomenon associated with the recorded Rayleigh
waves is illustrated here by reference to a series recorded by
the station MAIO in Iran (Figure 2). In this figure the wave
trains marked 'N' are compared with a tracing of the bottom
train (from the event of 9 September 1978), while those marked
'R' are compared with an _inverted_ tracing of the bottom train.
The virtually complete conformity of each recording with the
selected tracing is evident, showing clearly that those marked
'R' are almost perfectly reversed in phase relative to those
marked 'N'.

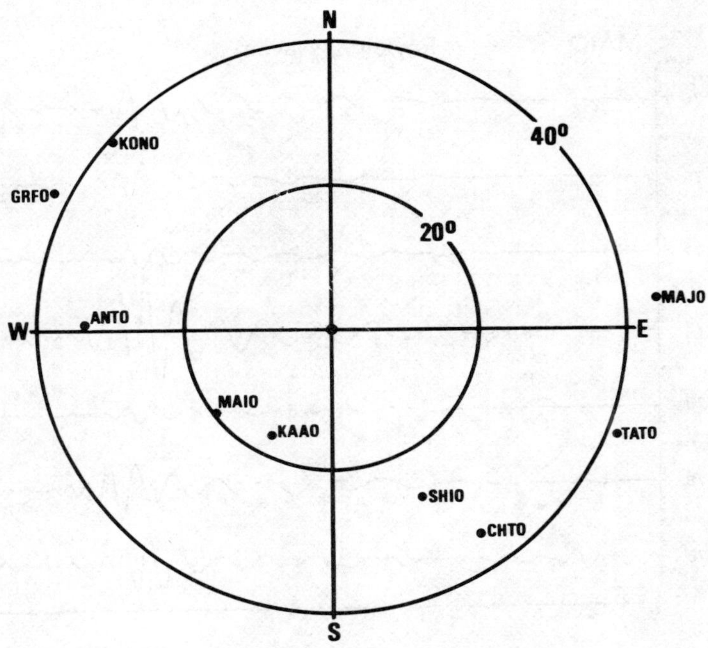

Figure 1.   Positions of SRO stations relative to E. Kazakh test site.

A further aspect of the phenomenon is illustrated in Figure 3, where the arrival times of the largest peaks (indicated by bars) are compared with their _expected_ arrival times relative to that of the bottom event (shown by arrows), after appropriate origin time and distance corrections have been made. It can be seen that whereas the peaks of the 'normal' trains occur within a second of the expected times, those of the 'reversed' trains arrive significantly (2-4 seconds) later than expected. Phase-reversed Rayleigh wave arrivals at other stations are also consistently late; because of the azimuthal coverage of the event of 7 July 1979, it is virtually impossible that this can be an effect of location errors.

Although there is an ambiguity concerning which Rayleigh waves are 'normal' and which are 'reversed', the apparently late trains are here described as reversed because it seems easier to describe 'lateness' rather than 'earliness' of the trains as part of an anomalous pattern. The association of a

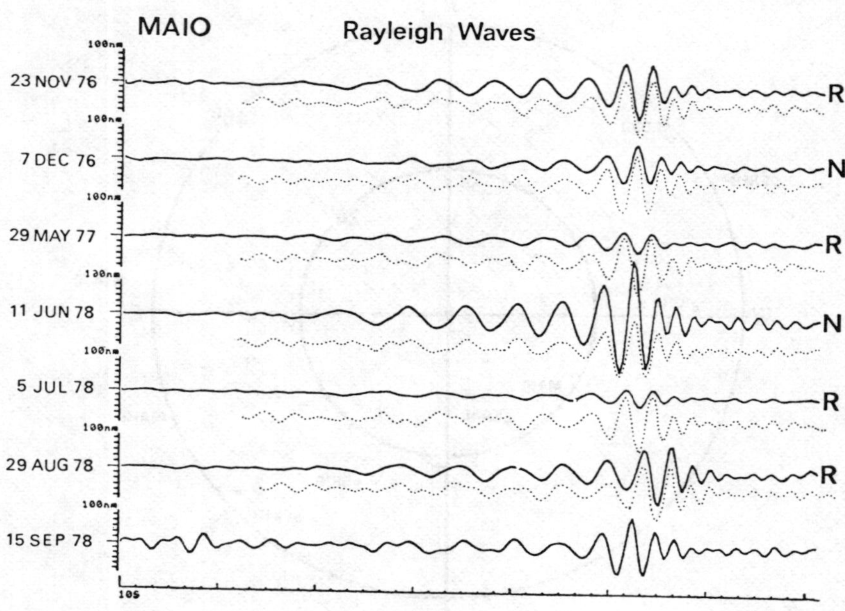

Figure 2. Rayleigh waves from E. Kazakh events recorded at
MAIO, compared with wave train from event of 15 September 1978
(shown dotted). For traces marked 'N' the comparison is direct;
for those marked 'R' the 15 September train has been inverted.

time delay with phase inversion of Rayleigh waves was first
reported by Rygg (1) from observations at KON, CHG and NORSAR
of a Shagan River event which occurred on 10 December 1972.

The amplitude and phase data for the events from all the
stations are summarized in Table 2 and mapped in Figure 4.
In the figure, the direction of the recording station from each
event is shown by a radial bar drawn from a point corresponding
to the NEIS location. Normal and reversed Rayleigh waves (according to the above definition) are indicated by shaded and
unshaded circles at the end of each bar, and the amplitudes of
the largest peaks relative to the amplitudes from the event of
4 August 1979, as indicated in the table, are represented by the
lengths of the bars (where no observation from the 4 August
event was available, a method similar to that described by
Freedman (2) was used to estimate the missing value). In a few

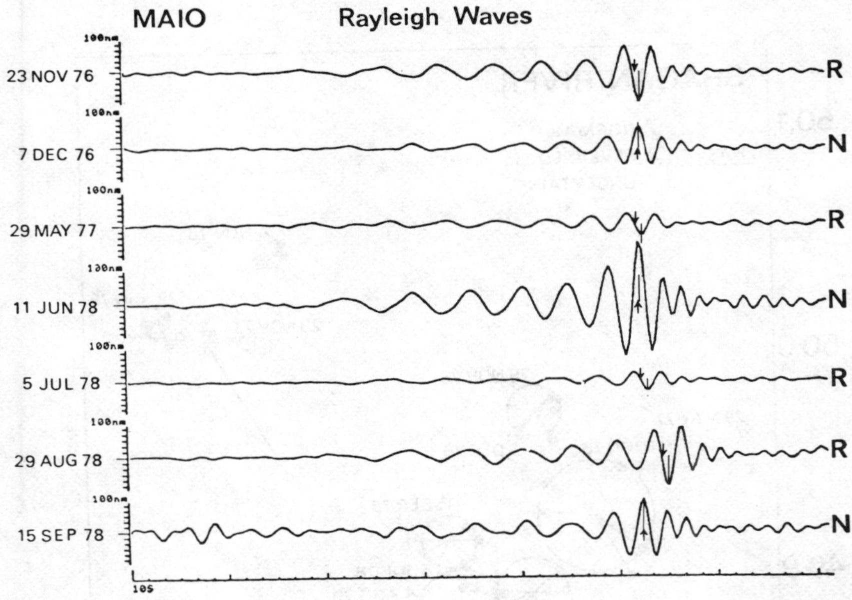

Figure 3. Comparison of arrival times of Rayleigh wave maxima
(indicated by vertical bars) with calculated arrival times (indicated by arrows) based on arrival time of maximum for event of
15 September 1978. Trains with reversed phase (cf. Figure 2)
arrive consistently later than calculated times.

cases it was not possible (usually as a result of low signal-to-
noise ratio) to determine from inspection whether the train
was normal or reversed; these may be resolvable using cross-
correlation methods.

      The figure reveals some additional features associated with
the phase reversal phenomenon. The first is that with the exception of the very anomalous event of 7 July 1979 the reversals
are confined to two stations, MAIO and KAAO, in a southwest direction. Unfortunately, no SRO stations are available to indicate
whether reversals also occurred in the opposite azimuth. A
second feature is that whereas the amplitudes of the normal
Rayleigh waves are reasonably consistent for each event, those
of the reversed waves tend to be anomalous – extremely so, for
the event of 7 July 1979.

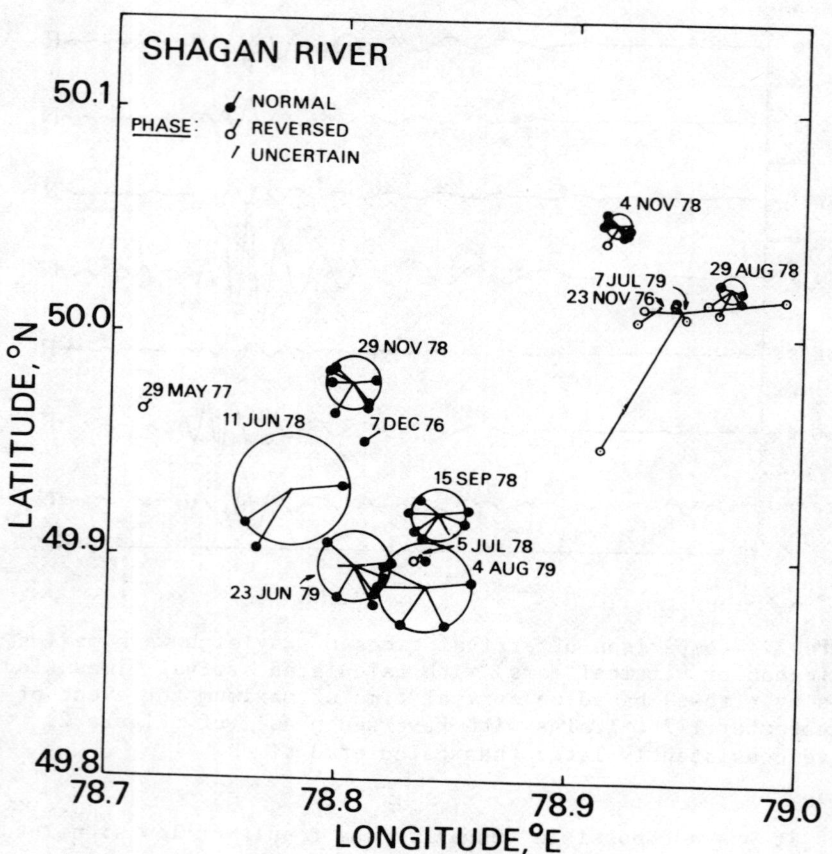

Figure 4. Locations of E. Kazakh events, showing phases and amplitudes of Rayleigh waves at SRO stations (see text). The circles indicate the average amplitudes of 'normal' Rayleigh waves for each event relative to the amplitudes recorded from the event of 4 August 1979 (cf. Table 2).

TABLE 2

Amplitude and Phase of Rayleigh Waves from E. Kazakh*

Date Sta	23 Nov 1976	7 Dec 1976	29 May 1977	11 Jun 1978	5 Jul 1978	29 Aug 1978	15 Sep 1978	4 Nov 1978	29 Nov 1978	23 Jun 1979	7 Jul 1979	4 Aug 1979
SHIO						278 0.24	556 0.48	361 0.31	750 0.64	1083 0.93	222 0.19	1167 1.0
MAJO				587 1.12			348 0.67	174 0.33	261 0.50	435 0.83	1196 2.29	522 1.0
TATO						98 0.22	273 0.62	129 0.30		311 0.71		
CHTO					67 0.16	153 0.37		120 0.29	277 0.55	300 0.72		
KAAO				963 1.50		413 0.64	413 0.64	321 0.50	505 0.79	505 0.79	2294 3.57	642 1.0
MAIO	963 0.62	630 0.40	333 0.21	1926 1.23	296 0.19	1000 0.64	963 0.62					
ANTO							232 0.62	87 0.23	174 0.46	145 0.38	319 0.85	377 1.0
GRFO								177 0.22	442 0.56		139 0.17	795 1.0
KONO							406 0.49	301 0.36	441 0.53	684 0.82	162 0.20	

* Key (for each event/station pair): The top number is the maximum peak-to-peak Rayleigh wave amplitude in nm. If this number is underlined, the phase is reversed; if it is underdotted, the phase is doubtful. The bottom number is the ratio of that amplitude to the amplitude for the event of 4 Aug 1979 at the same station. If no record exists for the latter, the 'missing value' is estimated (see text).

DISCUSSION

The several features noted above may be used to examine
various models which might be constructed to explain the reversals. First of all, it is clear that the essential equivalence
of path to each station eliminates the possibility (in this context) of a transmission-path effect. The strong azimuthal pattern
in the reversals also seems to rule out source-associated models,
such as depth of burial or spallation effects, which do not predict azimuthal dependence and would thus require additional ad
hoc assumptions about local structure in order to conform to
this feature. It seems very likely, therefore, that the occurrence of a phase reversal is an effect of strain release induced
by the explosion, since this would readily account for the consistency in the azimuths of the reversals. On the assumption that
the apparent time delays associated with the reversals are real,
however, the following dilemma then arises. If the explosion and
strain release occur almost simultaneously (so that the components of the two consequent radiation patterns are either in
phase or 180° out of phase), how can a time delay exist? If, on
the other hand, the delay corresponds to a delay in strain release, why is there no evidence of interference between the
time-separated sources?

Figure 5 illustrates an attempt to resolve the dilemma. The
solid curve is intended to represent schematically the source
function of the explosion in the absence of strain release, and
this is presumed to be essentially a step function — at least
as far as Rayleigh waves of these periods are concerned. The
dashed curve is taken to be the source function as modified by
strain release in the direction of maximum pre-stress, which
(according to this model) continues the rebound of the elastic
medium around the source through the initial rest position and
beyond. The model implies that the original explosion step function is converted to something like an impulse function (the
hatched area in the figure), with a consequent reduction in the
energy initiated at Rayleigh-wave periods, so that most of the
observable radiated energy at these periods originates from a
quasi-step function in the opposite direction and with a time
delay corresponding to the rise time, as indicated in the figure.
In addition, the impulse function will (in the ideal case) produce a $\pi/2$ phase delay in the frequency domain of the radiated
energy, such that its associated Rayleigh waves will be shifted
towards a position where they will tend to diminish rather than
distort the waves associated with the step function. At least
in a qualitative sense, therefore, the model seems capable of
accounting for the observed features of the anomalous Rayleigh
waves.

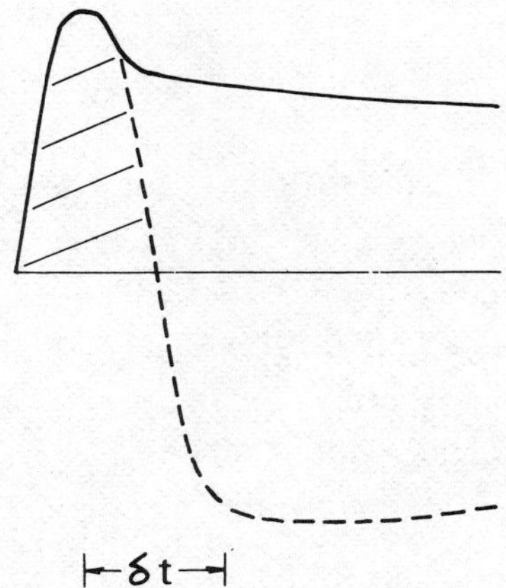

Figure 5. Cartoon illustrating how the explosion source function
(solid curve) is modified by strain release (dashed curve) to
give a quasi-impulse function (hatched area) plus a phase-reversed
step function with a time delay of t.

The above mechanism is based essentially on Archambeau's
model of stress relaxation in the immediate vicinity of the
source (see, for example, ref. 3), which neither requires
nor precludes failure of the medium in the form of an induced
earthquake. It seems likely, however, that such failure has
occurred for the event of 7 July 1979, where the amount of tectonic strain energy released probably exceeds that from the
explosion itself.

ACKNOWLEDGEMENTS

This study was carried out partly under an M.O.D. Research
Fellowship at Blacknest, England, and partly under an NTNF
Fellowship at NORSAR, Norway. I am grateful to Mr. Peter Marshall
at Blacknest, and Dr. Eystein Husebye at NORSAR, for useful discussions.

I am indebted also to North and Fitch of Lincoln Laboratory,
M.I.T., for providing me with their relocations of E. Kazakh events.

REFERENCES

1.   Rygg, E.: 1979, Bull. Seism. Soc. Am. 69, 1995-2002.
2.   Freedman, H.W.: 1966, Bull. Seism. Soc. Am. 56, 677-695.
3.   Archambeau, C.B.: 1972, Geophys. J.R. astr. Soc. 29, 329-366.

P WAVE COUPLING OF UNDERGROUND EXPLOSIONS IN VARIOUS GEOLOGIC
MEDIA

John R. Murphy

Systems, Science and Software
Reston Geophysics Office
11800 Sunrise Valley Drive
Reston, Virginia 22091

Despite the fact that data collection and analyses have been
going on for more than two decades, there is still considerable
controversy concerning the relative seismic coupling efficiency
of underground nuclear explosions in various source media and
the dependence of this coupling on the explosive yield of the
device. In the present study, the available data sources have
been reviewed and evaluated and an attempt has been made to re-
concile them within the framework of an approximate body wave
coupling model.

The data most directly relevant to P wave coupling fall into
two subsets; body wave magnitude (m_b)/yield data and free-field
(i.e. subsurface, near-source) measurements of the seismic
source function. Figure 1 shows some selected m_b/yield data for
explosions in various media, compared with an average m_b/yield
curve derived from a statistical fit to data from a large number
of explosions in saturated tuff. It can be seen that the observed granite data agree very closely with the wet tuff curve
and that the observations from explosions in dry, unconsolidated
materials, such as tuff and alluvium, consistently fall below the
wet tuff curve by 0.50 ± 0.25 magnitude units. However, the observations for the few available explosions in shale, salt and
dolomite are widely scattered, giving evidence to the fact that
the observed m_b value at a fixed yield is not necessarily a good
measure of relative coupling efficiency. This is simply an indication that other factors such as regional differences in propagation path and unaccounted for source variables other than
yield (e.g. depth) can also affect the observed teleseismic m_b
values. In an attempt to resolve some of these remaining uncertainties, the available free-field data from explosions in

E. S. Husebye and S. Mykkeltveit (eds.), Identification of Seismic Sources – Earthquake or Underground
Explosion, 201–205.

                                                      J. R. MURPHY

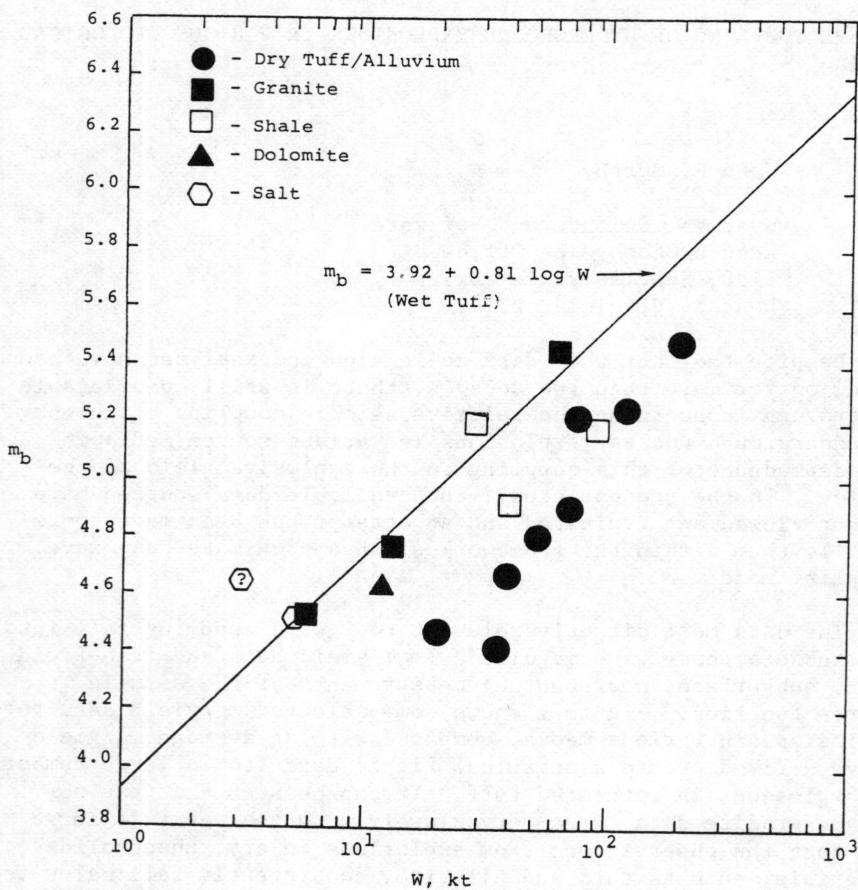

Figure 1. Body wave magnitude data from underground nuclear explosions in various source media.

the various media have also been compared. For example, Figure
2 shows a comparison of observed reduced displacement potentials
(RDP) from four explosions in dry alluvium with those expected
from explosions of the same yield and depth of burial in wet
tuff/rhyolite (T/R) emplacement media. It can be seen that these
data are entirely consistent with the m_b data in that they indicate that explosions in dry alluvium couple significantly less
well than those in saturated T/R media. Corresponding comparisons with data from other media confirm that the body wave
coupling in granite is comparable to that in wet tuff and suggest
that explosions in dolomite couple less well, explosions in shale
about the same and explosions in salt couple better than those in
wet tuff.

These data have been used to calibrate a seismic scaling
model which can be used to estimate theoretical m_b/yield curves
for explosions in various source media. The predicted, theoretical Western U.S. m_b/yield curves for explosions in granite,
salt, shale and wet T/R media are shown in Figure 3 for the
nominal Nevada Test Site depth of burial of $122 \, W^{1/3}$ m. It can be
seen that the predicted m_b/yield curves for granite, wet T/R and
shale are quite similar, differing by less than 0.2 magnitude
units over the yield range from 1 to 1000 kt. The salt curve,
on the other hand, is offset above the other three by an amount
which depends on yield and reaches nearly 0.5 magnitude units
for yields around 10 kt. Moreover, all four predicted curves
are nearly linear in this representation and they can in fact be
approximated very closely by straight lines (m_b versus log W)
with slopes ranging from about 0.8 to 0.9, in good agreement with
the average trends of the observed m_b/yield data.

In summary, the available observed m_b/yield and free-field
data from contained explosions in a variety of source media have
been analyzed in an attempt to define the medium dependence of
the P wave seismic coupling. These data have been interpreted
to indicate that the coupling into the teleseismic P wave transmission path is about the same for explosions in granite, shale
and wet T/R emplacement media. On the other hand, relative to
these three media, explosions in dry unconsolidated material and
dolomite/limestone media appear to couple significantly less
efficiently. In contrast to this, explosions in salt are inferred to couple more efficiently than any other medium studied,
resulting in an estimated m_b/yield curve which is offset above
the granite and wet T/R curve by as much as 0.5 units m_b.

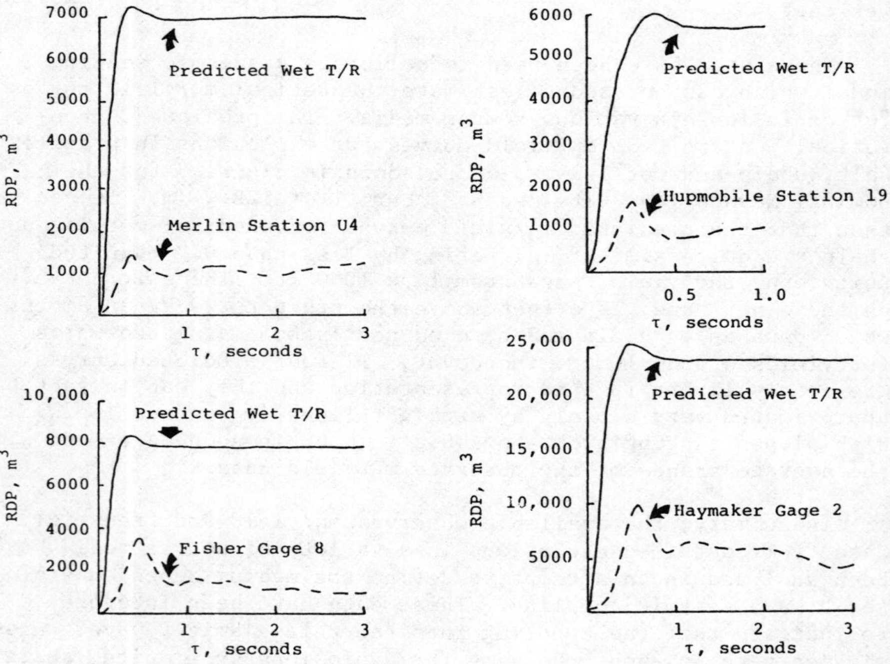

Figure 2. Comparison of observed alluvium RDP's with RDP's predicted for the same yield and depth of burial in a wet tuff/rhyolite emplacement medium.

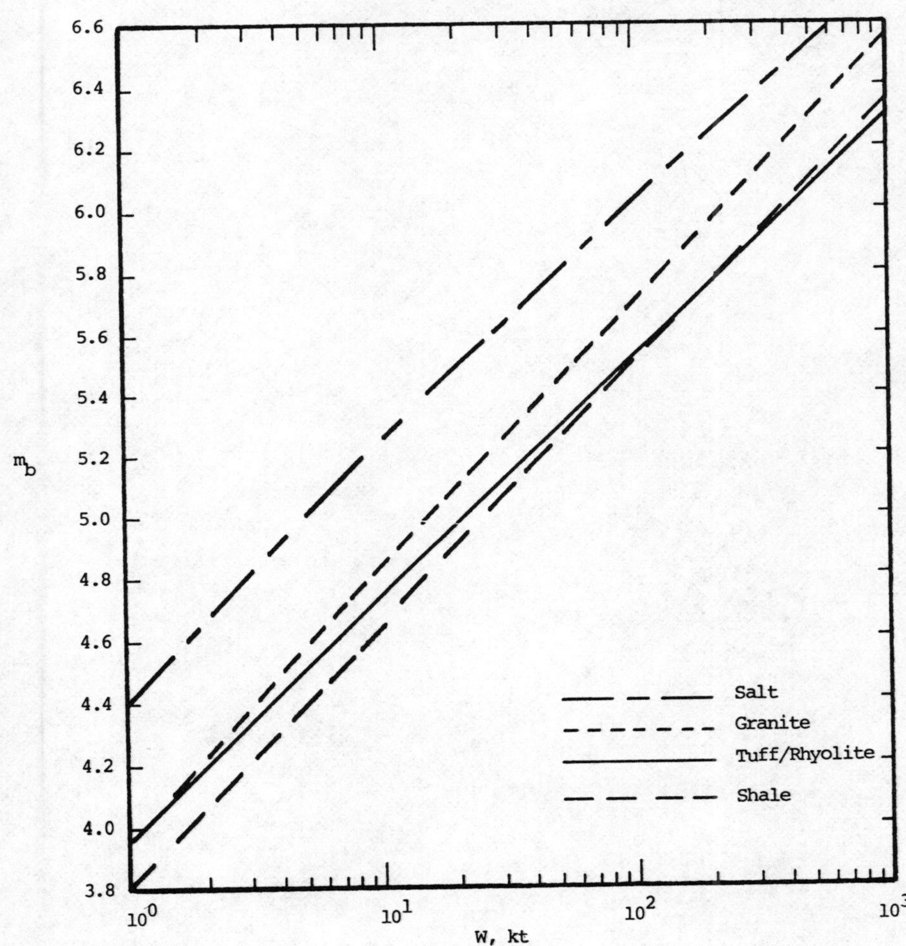

Figure 3. Comparison of theoretical western U.S. m_b/yield curves as a function of source medium, $h = 122W^{1/3}$ m.

SEISMIC MOMENT TENSORS

D.J. Doornbos

Vening Meinesz Laboratory, University of Utrecht,
The Netherlands
Present address: NTNF/NORSAR, P.B. 51, Kjeller, Norway

Originally introduced as a physically meaningful tool to obtain
the excitation coefficients of the normal modes of the earth, the
moment tensor has since been appreciated as a concept of more
general utility. This paper aims to review this development, and
at the same time to obtain simple approximations which are hoped
to bridge a gap between abstract formalism and practical requirements. Emphasis is on those practical aspects of moment tensors
which are likely to be useful in future applications, and on
those theoretical aspects which have led to some confusion in
the past.

1. INTRODUCTION

It is now about 10 years since Gilbert (1) published a note on
the excitation of the normal modes of the earth. That paper
emphasized the utility of a normal mode representation of the
response of the earth, but besides it introduced the concept
of a moment tensor to represent the seismic source; it now
appears that the utility of this last concept is appreciated as
much as was the main topic of the paper (see, e.g., the report
of a recent conference on earthquake parameters (2), or the
book of Aki and Richards (3)). Kostrov (4) introduced the concept in the Russian literature. From a practical point of view,
the utility of this type of source representation lies in the
fact that it is general (that is, it can accomodate different
types of sources) and that the seismic response is linearly
related to it. Once this fact was appreciated (5), it was exploited to investigate the source of two deep earthquakes (6,7).
The results of this investigation were both exciting and puzzling, since it would imply precursive compression in the source

*E. S. Husebye and S. Mykkeltveit (eds.), Identification of Seismic Sources - Earthquake or Underground
Explosion, 207–232.*

regions of these events. It is not surprising therefore to see
that a reappraisal of the moment tensor representation, and of
the inversion procedure, were the topic of several papers in the
years thereafter. In the course of these studies it became apparent that the term 'moment tensor' could mean different things.
Initially, the moment tensor was introduced and understood in
terms of a stress tensor, since stress is released on an earthquake fault. On the other hand, it was known since the work of
Burridge and Knopoff (8) that a dislocation source can be represented by a system of equivalent forces (specifically, double
couples for a shear dislocation), and it is the equivalent forces
that determine the excitation of seismic waves. From this point
of view, one would expect a moment tensor to be associated with
the equivalent force system at the source. In a major contribution Backus and Mulcahy (9,10) clarified the issue; they
demonstrated the existence of and the relation among several
types of moment tensors.

Two other problems had to be faced. First, the moment
tensor is a point source representation, reasonably accurate
only for relatively long waves. In case of a finite region one
can define a moment tensor density, but if the source region is
'reasonably' small (compared to the wave lengths), the advantages of a point source representation can be retained by means
of a limited number of moment tensors of higher degree (9).
Extending this work, Backus (11,12) discussed the interpretation
of this representation in some detail; several aspects of his
analysis still remain to be exploited.

The second problem is common to other types of source analyses and concerns the wave propagation between source and receiver.
Inversion of the seismic records to retrieve the moment tensor
implies a deconvolution of this 'path effect'. Systematic differences between wave propagation in earth models and in the
real earth are thus projected onto the source. Several authors
have tried to estimate this effect (13,14,15). It seems that
the effect can be significant, but suitable calibration procedures could possibly reduce it; attempts to this effect may
be found in (16), (17) and (18).

The moment tensor was originally introduced as a physically
meaningful tool to obtain the excitation coefficients of the
normal modes of the earth, in analogy to the excitation by a
distribution of body forces. If the normal modes $u_n(\xi)$ are
normalized and $\underline{f}(\xi,t)$ is the excitation function in a region V,
then the excitation coefficients are, in the frequency domain
(1):

$$b_n(\omega) = \int_V \underline{u}_n^\dagger(\underline{\xi}) \ \underline{f}(\underline{\xi},\omega) \ dV/(\omega_n^2-\omega^2) \qquad (1.1)$$

where n is the mode number. The coefficients b_n represent the displacement response in terms of the normal modes:

$$\underline{u}(\underline{x},\omega) = \sum_n b_n(\omega) \ \underline{u}_n(\underline{x}) \qquad (1.2)$$

In principle these coefficients represent the excitation of surface waves and body waves as well since these are contained in $\underline{u}(\underline{x},\omega)$, but the ensuing procedure is often not practical. Another summary of the excitation of various types of waves is provided by means of Green's functions (19,20,3):

$$u_i(\underline{x},t) = \int_{-\infty}^{t} \int_V G_i^j(\underline{x},\underline{\xi},t-\tau) \ f_j(\underline{\xi},\tau) \ dV \ d\tau \qquad (1.3)$$

where a summation over the common component index j is understood. In matrix notation

$$\underline{u}(\underline{x},t) = \int_{-\infty}^{t} \int_V \underline{\underline{G}}(\underline{x},\underline{\xi},t-\tau) \ \underline{f}(\underline{\xi},\tau) \ dV \ d\tau \qquad (1.4)$$

and in the frequency domain:

$$\underline{u}(\underline{x},\omega) = \int_V \underline{\underline{G}}(\underline{x},\underline{\xi},\omega) \ \underline{f}(\underline{\xi},\omega) \ dV \qquad (1.5)$$

The relation to the normal mode representation is seen from eqs. (1.1) and (1.2):

$$\underline{\underline{G}}(\underline{x},\underline{\xi},\omega) = \sum_n \underline{u}_n(\underline{x}) \ \underline{u}_n^\dagger(\underline{\xi})/(\omega_n^2-\omega^2) \qquad (1.6)$$

and this form illustrates one of the most useful properties of the Green's function, known as the reciprocity theorem:

$$G_i^j(\underline{x},\underline{\xi},t-\tau) = G_j^i(\underline{\xi},\underline{x},t-\tau) \qquad (1.7)$$

which is valid under rather general conditions, including self-gravitation and prestress (21).

In discussing developments related to the seismic moment tensor, it is appropriate to also consider developments related to the associated Green's functions. After all, one of the reasons why progress in applications has been relatively slow is, I believe, that clearly interpretable and computationally convenient formulas for Green's functions were not, or at least not widely, available. Useful excitation formulas were known for some time (22,23), where the source was taken a combination of forces and couples, so specific types of sources, including dislocations and explosions, could be modelled. Only recently these works have been adapted to the moment tensor representation (24,25,26,27,28). An extensive summary of results in terms of Green's functions is contained in (3). In section 2 I will emphasize the application of recent asymptotic approximations to the Green's functions, partly because of the computational advantages, but also because the resulting formalism allows a clear identification of: 1) the excitation effect at the source, 2) the propagation effect between source and receiver, and 3) the receiver-oriented effect.

Taking into account the extensive treatment in (9,10,11,12), the then following part of the paper will be concerned with those practical aspects of moment tensors which I believe are likely to be useful in future applications, and with those theoretical aspects which have led to some confusion in the past. These aspects include moment tensors of higher degree, the form of the required spatial and temporal derivatives of Green's functions to be associated with these moment tensors, the nature of equivalent forces inside and outside the source region, the interpretation of moment tensors of zero-th and higher degree and the role of physically plausible constraints, and moment tensors of particular types of sources including sources with volume change.

The moment tensor should find its ultimate use in studies of source classification and tectonics, but although some applications with obvious tectonic implications have been published, most efforts to date have been spent on the inversion procedure itself. The last section of this paper contains a brief summary of different methods, and an attempt to compare them. A particular application of source classification is discrimination between explosions and earthquakes, and I will find occasion to include a few comments on the role and possible prospects of the moment tensor in this field.

2. GREEN'S FUNCTIONS AND ASYMPTOTIC APPROXIMATIONS

It may be noted that the formal representations, eqs. (1.1) – (1.5), do not tell how a solution should be obtained. To actually

compute it, one usually assumes the medium vertically or radially
inhomogeneous. Then the Fourier transformed wave field may be
decomposed in vector surface harmonics (e.g., ch. 13 of ref. 19,
and many applications may be found in the seismological literature). For the displacement

$$\underline{u}(\underline{x},\omega) = \sum_n \underline{\underline{Y}}_n(x,y) \; \underline{\tilde{u}}_n(z,\omega) \tag{2.1}$$

where (x,y) are coordinates in the surface, $z = $ constant, and
the matrix $\underline{\underline{Y}}_n$ has as column vectors the 3 vector surface harmonics of degree n. In the notation these have been taken dis-
crete (e.g., the surface spherical harmonics), but depending
on the infiniteness in one or both of the horizontal coordinates,
the sum may be replaced by an integral and a sum (the Fourier-
Bessel transform in cylindrical coordinates), or by a double
integral (the two-dimensional wavenumber transform in cartesian
coordinates). The symbol ~ denotes functions in the 'transformed'
domain, i.e., the coefficients of the vector surface harmonics.
For each surface harmonic, propagation of displacement-stress,
and gravitational potential if required, may now be described by

$$\partial_z \underline{\tilde{D}} = i\omega \, \underline{\underline{A}} \, \underline{\tilde{D}} \tag{2.2}$$

where all quantities depend on ω and on the index n and,
separately for P-SV and SH, $\tilde{D}$ is the stress-displacement vector
(plus gravitational potential if required in the P-SV case).
Equations like (2.2) (with the factor $i\omega$ absorbed in $\underline{\underline{A}}$), together with numerical considerations of its inversion, have been
discussed in (29) and (3). A recent review of solutions in a
layered model is given in (26), and asymptotic solutions in
a layer were obtained in (30). In these methods, continuity and
boundary conditions are imposed at horizontal interfaces and
surfaces, so following Hudson (23) it is convenient to represent the source as a dislocation in displacement and stress
across a horizontal interface z_s: $[\underline{\tilde{D}}_s]_-^+$.

By applying the stress-free surface condition, conditions
at the bottom interface, and 'jump conditions' at the source
level, various solutions of (2.2) have been obtained; for
recent versions see, e.g., (26,28,31). They may all be represented in the form

$$\underline{\tilde{u}}(z) = \underline{\underline{H}}(z,z_s)[\underline{\tilde{D}}_s]_-^+ = \left(\underline{\underline{T}}(z,z_s) \; \middle| \; -\underline{\underline{G}}(z,z_s) \right) \begin{bmatrix} \underline{\tilde{u}}_s \\ \hline \underline{\tilde{\tau}}_s \end{bmatrix}_-^+ \tag{2.3}$$

where the dislocation across z_s has been partitioned in one of displacement $\tilde{\underline{u}}$ and stress $\tilde{\underline{\tau}}$, and the elements of $\underline{\underline{T}}(z,z_s)$ and $\underline{\underline{G}}(z,z_s)$ are related:

$$T_i^j(z,z_s) = c_{i3kl} \; \partial_l \; G_k^j(z,z_s) \tag{2.4}$$

Here c_{imkl} is the elastic tensor in z_s, and the index m=3 denotes the vertical. So if $\underline{\underline{G}}(z,z_s)$ were representing a displacement in z_s, then $\underline{\underline{T}}(z,z_s)$ would be the associated stress on z_s. In fact, $\underline{\underline{G}}(z,z_s)$ is the Green's tensor in the transformed domain, and eq. (2.3) is a special case of Burridge and Knopoff's representation of the displacement due to a dislocation (8). To get the relation to eqs. (1.5) and (1.6), we decompose a point force $\underline{f}(\omega) \; \delta(\underline{\xi})$ in surface harmonics similar to eq. (2.1), and represent it as a stress discontinuity across a horizontal plane. The components for the vector surface harmonics of degree n are

$$[\tilde{\underline{\tau}}_s]_-^+ = -\underline{\underline{Y}}_n^{\dagger}(\xi_x,\xi_y) \; \underline{f}(\omega) \tag{2.5}$$

where we have taken $\xi_z = z_s$. From eqs. (2.1), (2.3) and (2.5) we then find the Green's tensor

$$\underline{\underline{G}}(\underline{x},\underline{\xi},\omega) = \sum_n \underline{\underline{Y}}_n(x,y) \; \underline{\underline{G}}_n(z,z_s,\omega) \; \underline{\underline{Y}}_n^{\dagger}(\xi_x,\xi_y) \tag{2.6}$$

Note the similarity between this representation in terms of eigen-functions in the horizontal plane and that in terms of normal modes, eq. (1.6).

For all but the longest waves, gravitational perturbations may be neglected so $\tilde{\underline{D}}$ in eq. (2.2) is a stress-displacement vector. Let a surface harmonic correspond to a fixed ray parameter p, then a horizontal wavenumber is $\omega q = \omega p/r$ in spherical geometry, $\omega q = \omega p$ in flat geometry. Then the zero-th order approximations to the matrix $\underline{\underline{A}}$ in eq. (2.2), see e.g. (30), is identical in the spherical and flat geometry (apart from a r^{-1} factor in the radial wave functions). Therefore, in the following we make no distinction between these two cases, unless explicitly stated. In an isotropic medium, the zero-th order approximation implies decoupling of P and SV away from interfaces. The decoupled waves may then be represented in terms of scalar potentials, or equivalently in terms of vertical wave functions. The vertical wave functions $U_{u/d}$, $V_{u/d}$ will be taken to denote up/downgoing waves for P and S, respectively. Cormier (28) has discussed various representations in terms of standing or travelling waves. If the wave functions are normalized, coefficients $\tilde{A}_{u/d}$, $\tilde{B}_{u/d}$ must be attached to them and these are connected, through the wave functions, with the stress-displacement vector $\underline{D}$. Separately for SH and P-SV:

$$\tilde{\underline{D}} = \underline{\underline{F}} \; \tilde{\underline{B}} \tag{2.7}$$

where $\tilde{\underline{B}}$ contains the wave coefficients, and the matrix $\underline{\underline{F}}$ depends
on ω and p, on density and elastic constants, and on the wave
functions and vertical derivatives. The explicit form of $\underline{\underline{F}}$ depends also on the arrangement and normalization of stress-
displacement and wave coefficients; here the conventions and
forms in (31) will be followed. Incidentally, $\underline{\underline{F}}$ is equivalent
to the eigenvector matrix of $\underline{A}$ in eq. (2.2) (26), and it has
been called a matrizant, or fundamental matrix (32).

On the basis of (2.7), the excitation problem can now be
restated as a problem to find the wave coefficients $\tilde{A}_{u/d}$ and
$\tilde{B}_{u/d}$. This approach was followed in (26) and (31). Related
results were obtained in (28). The advantages of this approach
are that 1) excitation coefficients can be identified with
wave types (P, SV, SH) that are actually observed, at least in
teleseismic body waves, and 2) in a smoothly varying layer the
wave coefficients are constant whereas stress-displacement
is not. Applying eq. (2.7) on opposite sides of the source
level we have

$$[\tilde{\underline{D}}_s]_-^+ = \underline{\underline{F}}(z_s) \; [\tilde{\underline{B}}_s]_-^+ \tag{2.8}$$

where all functions are evaluated at the source level, and
$[\tilde{\underline{B}}_s]_-^+$ denotes the jump in wave coefficients across this level.
This equation can be inverted, and the wave coefficients at
the source level may be related to those at any other level
through a matrix of generalized reflection coefficients $\underline{\underline{R}}(z,z_s)$:

$$\tilde{\underline{B}}(z) = \underline{\underline{R}}(z,z_s) \; \tilde{\underline{B}}(z_s^{\pm}) \tag{2.9}$$

The structure of these reflection coefficients is closely related to the solution for the displacement response (31). An
iterative procedure to obtain these coefficients was developed
in (33). From eqs. (2.7), (2.8) and (2.9) we have the representation

$$\tilde{\underline{D}}(z) = \underline{\underline{F}}(z) \; \underline{\underline{R}}(z,z_s) \; \underline{\underline{F}}'^{-1}(z_s) \; [\tilde{\underline{D}}_s]_-^+ \tag{2.10}$$

where $\underline{\underline{F}}'^{-1}$ is given by the inverse of $\underline{\underline{F}}$ but with the 2nd and 4th
row of $\underline{\underline{F}}^{-1}$ reversed in sign. The Green's tensor may be identified
with the part of $\underline{\underline{F}}(z) \; \underline{\underline{R}}(z,z_s) \; \underline{\underline{F}}'^{-1}(z_s)$ connecting the stress

part of $[\tilde{\underline{D}}_s]^+$ to the displacement part of $\tilde{\underline{D}}(z)$ (c.f. eq. 2.3).
We will give this result in more explicit form and towards that
end use some properties of asymptotic wave functions. These satisfy the wave equation, e.g. for P:

$$\partial_z^2 U + \omega^2(\alpha^{-2}-q^2)U = 0 \tag{2.11}$$

where α is the wave velocity and ω_q the horizontal wavenumber.
Then

$$U_d\,\partial_z^2 U_n - U_u \partial_z^2 U_d = 0$$

which may be integrated to give

$$U_d\,\partial_z U_u - U_u\,\partial_z U_d = \frac{i\omega}{\alpha}\,\{U_d\,C_u\,U_u + U_u\,C_d\,U_d\} = c \tag{2.12}$$

where C_u, C_d are so-called generalized cosines for up- and downgoing waves, respectively, as defined in (34), and the constant
c depends on the normalization of the wave functions. We will
choose $c = 4i/\pi$, appropriate for a normalization of wave functions
as in eq. (23) of ref. (34).

To apply this to the representation (2.10), it is to be
noted that each element of $\underline{\underline{F}}$ contains a vertical wave function
$U_{u/d}$ for P, or $V_{u/d}$ for S, see eg. (28) or (31). Each element
of $\underline{\underline{F}}^{-1}$ contains

$$\alpha/\{U_{u/d}(C_u + C_d)\} = \frac{\pi}{4}\,\omega\,U_{d/u} \tag{2.13a}$$

or

$$\beta/\{U_{u/d}(D_u + D_d)\} = \frac{\pi}{4}\,\omega\,U_{d/u} \tag{2.13b}$$

with $C_{u/d}$, $D_{u/d}$ generalized cosines for P and S_1 respectively.
Thus from eq. (2.10), substituting for $\underline{\underline{F}}$ and $\underline{\underline{F}}^{-1}$ and using
eq. (2.12), we obtain the asymptotic Green's tensor consistent
with the form in eq. (2.3). For SH:

$$\underline{\underline{G}}(z,z_s) = i\ \frac{\pi}{4}\mu^{-\frac{1}{2}}(z)\ \mu^{-\frac{1}{2}}(z_s)(V_u,V_d)_z\ \underline{\underline{R}}(z,z_s)\begin{pmatrix} V_d \\ V_u \end{pmatrix}_{z_s} \tag{2.14}$$

For P-SV:

$$\underline{\underline{G}}(z,z_s) =$$
$$i\ \frac{\pi}{4}\ \rho^{-\frac{1}{2}}(z)\rho^{-\frac{1}{2}}(z_s)\begin{pmatrix} q\ U_u,\ q\ U_d,\ D_u V_u/\beta,\ -D_d V_d/\beta \\ C_u U_u/\alpha,\ -C_d U_d/\alpha,\ -qV_u,\ -qV_d \end{pmatrix}_z$$
$$\underline{\underline{R}}(z,z_s)\begin{pmatrix} q\ U_d, & C_d U_d/\alpha \\ q\ U_u, & -C_u U_u/\alpha \\ D_d\ V_d/\beta, & -q\ V_d \\ -D_u V_u/\beta, & -q\ V_u \end{pmatrix}_{z_s} \tag{2.15}$$

where ρ and μ are density and rigidity, and α,β are P and S
velocities. The result is a sum of terms, each representing
the contribution of a particular wave type at the receiver (up-
or downgoing P, SV or SH), arising, through a reflection coefficient, from another particular wave type at the source. At the
receiver, U_u,V_u are associated with upgoing waves but at the
source, U_u,V_u must be associated with downgoing waves, due to
the factors of the form (2.13) in the inverse fundamental matrix
$\underline{\underline{F}}^{-1}$. A similar interpretation can be given for U_d, V_d. The parameter q in eq. (2.15), which is related to the ray parameter p
and to the horizontal wavenumber ωq, arises by horizontal differentiation of the corresponding surface harmonic. Use can be
made here of a familiar asymptotic approximation to the surface
harmonic (35, 36) and a splitting in travelling components,
$\exp(\pm i\omega px)$ in plane geometry, $\exp(\pm i\omega p\theta)$ in spherical geometry.
If, in the near field or near the antipode, the asymptotic ap-
proximation is invalid, the representation (2.15) is still useful
if q is identified with an operator, i.e., $i\omega q = \partial_x = \partial_\theta/r$. The
operator applies to the corresponding surface harmonics in the
inverse transformation, eq. (2.6). The inverse transformation involves a summation over azimuthal order, and an integral over
ray parameter or a summation that can be transformed into such
an integral (the Watson transformation). The representations
(2.14) and (2.15) do not depend on azimuthal order, so it is
convenient to do this summation first. With the source (or

receiver) at the pole of the coordinate system, the only contributing azimuthal orders are m = 0, $\pm$1. For each of the terms
in eqs. (2.14) and (2.15) we then get a result of the form

$$\underline{\underline{G}}(\underline{x},\underline{\xi},\omega) = \frac{1}{8} \exp(i\pi/4)(\frac{\omega^3}{2\pi x})^{\frac{1}{2}} (\rho v^2 \rho_s v_s^2)^{-\frac{1}{2}}$$

$$\int_\Gamma p^{\frac{1}{2}} \underline{s}(\underline{x}) \underline{s}^T(\underline{\xi}) U_u(z) U_u(z_s) R(z,z_s) \exp(i\omega p\theta) dp \qquad (2.16)$$

where in this example, the contribution to $\underline{\underline{G}}$ comes from an upgoing
P wave at the receiver due to a downgoing P wave at the source.
Γ is a path in the complex ray parameter plane, V is the local
wave velocity (of P in this example) and, in an asymptotic approxi-
mation, $\underline{s}(\underline{\xi})$, $\underline{s}(\underline{x})$ are unit vectors in the direction of particle
displacement at source and receiver, respectively. They depend
on ray parameter and may involve generalized cosines, and they
determine the radiation pattern for the particular wave type
under consideration. Apart from these vectors, the form (2.16)
is rather similar to Richards' representation of the field from
a potential point source (37). In fact, the latter result may be
used to obtain the Green's tensor equivalent to (2.16), by
requiring that in the appropriate circumstances it reproduces
the geometrical spreading of ray theory, and the excitation
for a homogeneous medium (38). Somehwat less restrictive in
usage than geometrical ray theory is the WKBJ approximation.
In situations where this approximation may be applied to the
vertical wave functions in the integrand of (2.16) including
the wave functions used in the generalized reflection coefficient
R), the integrand becomes a simple harmonic function of frequency
and the inverse Fourier transform to the time domain is trivial.
It is the basis of the 'WBKJ seismogram', which is obtained by
evaluating an integration over ray parameter directly in the time
domain (35). Finally, it should be mentioned that although the
effect of anelastic damping has not been explicitly considered,
it is conventionally included either by specifying complex
velocities (39, 40), or by applying to the results a damping
operator which can be made such that causality is preserved
(41).

3.   MOMENT TENSORS AND EQUIVALENT FORCES

Let us now return to the response given by eq. (1.3). If the
excitation force acts in a finite volume V, the volume integral
can be replaced by a point source representation by expanding
the Green's function in a Taylor series about a suitably chosen
reference source point $\underline{\xi}^o$, and requiring that there be no net

force acting (11, 42). We also apply the reciprocity theorem:

$$u_i(\underline{x},t) = \partial \xi_k \; G_j^i(\underline{\xi}^O,\underline{x},t) * F_{j,k}(\underline{\xi}^O,t) +$$

$$+ \frac{1}{2} \partial \xi_k \partial \xi_\ell \; G_j^i(\underline{\xi}^O,\underline{x},t) * F_{j,k\ell}(\underline{\xi}^O,t) + \dots \qquad (3.1)$$

The force moment tensor of spatial degree one is

$$F_{j,k}(\underline{\xi}^O,\tau) = \int_V (\xi_k - \xi_k^O) \; f_j(\underline{\xi},\tau) \; dV \qquad (3.2)$$

which has the symmetry $F_{j,k} = F_{k,j}$, due to the requirement that there be no net torque. The force moment tensor of spatial degree two is

$$F_{j,k\ell}(\underline{\xi}^O,\tau) = \int_V (\xi_k - \xi_k^O)(\xi_\ell - \xi_\ell^O) \; f_j(\underline{\xi},\tau) \; dV \qquad (3.3)$$

which has the obvious symmetry $F_{j,k\ell} = F_{j,\ell k}$. Moment tensors of higher degree, with their symmetries, follow analogously.

For asymptotic wave functions the spatial derivatives in eq. (3.1) are simple to evaluate. Towards this end the integral representation of the Green's function, as in eq. (2.16), is transformed to the time domain, and it will be written

$$G_j^i(\underline{\xi}^O,\underline{x},t-\tau) = \int_\Gamma \; g_j^i(\underline{\xi}^O,\underline{x},t-\tau) \; dp$$

and

$$\partial \xi_k \; G_j^i(\underline{\xi}^O,\underline{x},t-\tau) = -\int_\Gamma \frac{c_k}{v} \; \dot{g}_j^i(\underline{\xi}^O,\underline{x},t-\tau) \; dp =$$

$$= \dot{G}_{j,k}^i (\underline{\xi}^O,\underline{x},t-\tau) \qquad (3.4)$$

where $\bullet$ denotes a temporal derivative of g, v is the wave velocity and c_k the direction cosine of the wave in $\underline{\xi}^O$. The latter is a function of ray parameter and it is actually a generalized cosine (c.f. eq. (2.12) for the introduction of a generalized cosine with the vertical direction). $\dot{G}_{j,k}^1$ is defined in eq. (3.4)

only for notational convenience in the following equations. Higher
derivatives of the Green's functions follow analogously. The response (3.1) may now be written:

$$u_i(\underline{x},t) = G^i_{j,k}(\underline{\xi}^o,\underline{x},t) * \dot{F}_{j,k}(\underline{\xi}^o,t) +$$

$$\frac{1}{2} \dot{G}^i_{j,k\ell}(\underline{\xi}^o,\underline{x},t) * \dot{F}_{j,k\ell}(\underline{\xi}^o,t) + \ldots \tag{3.5}$$

This representation is in terms of temporal derivatives of the
source function and these are transient even though the source
function itself is usually not; it merely confirms that the far
field response to a step-like source is impulse-like (42). This
makes it possible for the temporal variation of the source to
be described by moments, by expanding $G^i_{j,k}$, $G^i_{j,k\ell}$, etc. in a
Taylor series about a suitably chosen reference time τ^o. The
procedure is analogous to that for spatial moments; in order to
avoid complicated notation, we will drop indices and give a
result for scalar functions:

$$u(t) = G(t) * \dot{F}(t) = G(t-\tau^o) F^{(o)} - \dot{G}(t-\tau^o)F^{(1)}(\tau^o) +$$

$$+ \frac{1}{2} \ddot{G}(t-\tau^o) F^{(2)}(\tau^o) + \ldots \tag{3.6}$$

where the force moment tensor of temporal degree n is

$$F^{(n)}(\tau^o) = \int_{-\infty}^{+\infty} (\tau-\tau^o)^n \dot{F}(\tau) \, d\tau \tag{3.7}$$

and it is assumed that $t \gg \tau^o$ so that $\dot{F}(t) = \dot{F}(\infty) = 0$. Each
term of eq. (3.5) can thus be expanded in temporal moments ac-
cording to eq. (3.6). This procedure appears to be meaningful;
Backus (11) demonstrated that in seismic faulting, a temporal
force moment of degree n would be of roughly the same size as
a spatial force moment of degree n+1, and in general only a few
degrees are needed to specify the long-period response. This
is an important consideration in inversion for the source parameters with only a limited number of data.

The moments defined by eq. (3.1) and other previous equations, are associated with excitation forces; they are called
force moments. Usually, seismic excitation is caused not by
external forces, but by failure of the material inside, i.e.,

in the source region the usual stress-strain relations (Hooke's law) break down, temporarily. The seismic effect of this failure is equivalent to the excitation by a distribution of forces; these would be equivalent forces. In (9) it was demonstrated that these forces arise when replacing, in the equations of motion, the true physical stress τ_{ij} by the model stress σ_{ij} which satisfies Hooke's law:

$$\sigma_{ij} = c_{ijk\ell} \, \partial_k \, u_\ell \tag{3.8}$$

plus additional terms if prestress is present (21). When the material changes constitution, the elastic tensor applies to the material after the transformation. Backus and Mulcahy termed the difference between σ_{ij} and τ_{ij} the stress glut:

$$m_{ij} = \sigma_{ij} - \tau_{ij} \tag{3.9}$$

and the equivalent force is

$$f_j = -\partial_i \, m_{ij} \tag{3.10}$$

which acts in the finite source region and which may be described by generalized functions, in case of faulting (10). When eq. (3.10) is substituted in (1.3) and the result integrated by parts:

$$u_i(\underline{x},t) = \int_{-\infty}^{t} \int_V \partial\xi_k \, G_j^i(\underline{\xi},\underline{x},t-\tau) \, m_{jk}(\underline{\xi},\tau) \, dV \, d\tau \tag{3.11}$$

and this form can be replaced by a point source representation with the aid of moments. The equivalent of eq. (3.1) becomes

$$u_i(\underline{x},t) = \partial\xi_k \, G_j^i(\underline{\xi}^o,\underline{x},t) * M_{jk}(\underline{\xi}^o,t) +$$

$$\partial\xi_\ell \partial\xi_k \, G_j^i(\underline{\xi}^o,\underline{x},t) * M_{jk,\ell}(\underline{\xi}^o,t) + \ldots \tag{3.12}$$

where $M_{jk,\ell}$ are spatial moment tensors of the stress glut. Due to the symmetry in the stress glut they have the symmetry $M_{jk,\ell\ldots} = M_{kj,\ell\ldots}$. Additional symmetries for moments of degree

2 and higher follow, as before, from the definition (c.f. eq. 3.3).

Of the moment tensors discussed here, the glut moments determine the source but the force moments determine the observed response. If the stress glut is known, the equivalent force and hence the motion is uniquely determined. However, it is the equivalent force that can be determined from the motion, and from eq. (3.10) it is clear that the stress glut is not uniquely determined by the equivalent force. The same is true in general for the associated moments, but for moment tensors of low spatial degree (glut moments of degree 0 and 1), the symmetry properties provide sufficient constraints so that the relation is unique. For glut moments of spatial degree two, one additional constraint would be needed, and this must be obtained from a priori information (or by assumption) on the source. For example, the assumption that the source does not involve volume change would be a sufficient constraint. We will take up this point again in section 5, when discussing moment tensors in terms of fault parameters. For completeness, the general relations as given in (9) are:

$$F_{j,k} = M_{jk}$$
$$F_{j,k\ell} = M_{jk,\ell} + M_{j\ell,k} \qquad (3.13)$$
$$F_{j,k\ell m} = M_{jk,\ell m} + M_{j\ell,mk} + M_{jm,k\ell}$$

In using eq. (3.10) we have assumed that $\sigma_{ij} = \tau_{ij}$ outside the physical source region. However, the true physical stress τ_{ij} involves also the true values for elastic constants and if these are incorrectly used in some regions of the earth (i.e., in an anomaly), then $\sigma_{ij} \neq \tau_{ij}$ there. Similarly, the use of incorrect values for density may give rise to equivalent forces. If we put

$$\rho = \rho^o + \rho' , \qquad\qquad c_{ijk\ell} = c^o_{ijk\ell} + c'_{ijk\ell}$$

where the model values have the superscript ^{o}, and substitute in the equation of motion, the equivalent force is

$$f_i = -\rho' \ddot{u}_i + \partial_j c_{ijk\ell} \partial_\ell u_k + \rho' \partial_i \psi \qquad (3.14)$$

where ψ is the gravitational potential. Calculation of these functions is complicated because they depend on the displacement

itself. It is in fact the usual expression for excitation of
scattered waves by inhomogeneities (43), but in the latter case
the Born approximation is often invoked, i.e., u_i is replaced by
u_i^o which is the solution in the unperturbed model ρ^o, $c_{ijk\ell}^o$. The
approximation is invalid for strong perturbations, and this is
likely to be a case of particular interest in the source region.
By analogy, equivalent stresses on a boundary or interface can
be used to replace the effect of perturbations of that interface.

4. DISLOCATIONS AND SOURCES WITH VOLUME CHANGE

The source region is often conceived as a fault plane Σ
across which displacements are discontinuous. Then the model
stress and the stress glut are singular there and these tensors
may be described by generalized functions (10). Backus and
Mulcahy find for the stress glut of a shear dislocation (and
isotropic prestress, if present):

$$m_{jk} = c_{jk\ell m} \, [u_m]_-^+ \, n_\ell \, \delta_\Sigma \tag{4.1}$$

where n_ℓ are components of the unit vector normal to Σ, and

$$\delta_\Sigma = \int_\Sigma \delta(\underline{\xi}-\underline{\xi}') \, d\Sigma$$

The quantities in eq. (4.1) depend on time and space, although
this has not been brought out in the notation. The response
follows from eq. (3.11):

$$u_i(\underline{x},t) = \int_{-\infty}^t \int_\Sigma c_{jk\ell m} \, [u_m]_-^+ \, n_\ell \, \partial\xi_k \, G_j^i(\underline{\xi},\underline{x},t-\tau) \, d\Sigma \, d\tau \tag{4.2}$$

which is a classical result, first obtained from a form of
Betti's reciprocity relation (8) and verified in the case of
self-gravitation and isotropic prestress (21). Particular cases
are well-known; in an isotropic medium with shear modulus μ:

$$m_{jk} = \mu([u_j]_-^+ \, n_k + [u_k]_-^+ \, n_j) \, \delta_\Sigma \tag{4.3}$$

which can be represented by a double couple in each surface
element of Σ. The moment tensor of spatial degree zero is

$$M_{jk} = \int_{\Sigma} \mu([u_j]_-^+ n_k + [u_k]_-^+ n_j) \, d\Sigma \tag{4.4}$$

which represents a double couple if Σ is a plane.

Shear (tangential) dislocations do not involve volume change. If such changes occur they are associated with the isotropic component of the moment tensor, and this provides a means to investigate the possibility of volume change, at least in principle (6,7). A different systematic procedure was advanced in (44). It should be noted however that an isotropic compoment in the moment tensor does not necessarily imply a source with volume change. In (10) several realistic possibilities of faulting were considered which, unlike the ideal case represented by eqs. (4.3) and (4.4), might give an isotropic component. Most important in this respect would be faulting in an anisotropic medium, and bumpiness of the fault (with the effect of plastic fault gouge in the space left between the bumps). At this stage it is not yet clear if these circumstances do indeed arise to the extent that the effects are signficant. The rest of this section will be a brief discussion of the implications and complications in models of sources with volume change. A more comprehensive discussion in terms of moment tensors is given in (45); some specific types of sources with volume change are considered in (46).

The volume change ΔV associated with the isotropic component of the moment tensor $1/3 \, \mathrm{tr} \, \underline{\underline{M}}$, depends on the source model. A change

$$\Delta V = (\lambda + \frac{2}{3})^{-1} \cdot \frac{1}{3} \, \mathrm{tr} \, \underline{\underline{M}} \tag{4.5}$$

where $(\lambda + \frac{2}{3} \mu)$ is the incompressibility, occurs if a cavity collapses (in an isotropic medium). This model is readily incorporated in the dislocation model of eqs. (4.1) and (4.2), by allowing the displacement jump on Σ to have a normal component, so eqs. (4.3) and (4.4) are modified. In a self-gravitating earth however, a complication arises due to the action of hydrostatic pressure during collapse of the cavity. In first-order approximation this can be represented by a temporary moment tensor; for spatial degree zero:

$$N_{jk} = \int_{\Sigma} -P_o \, \delta_{jk} \, [u_j]_-^+ \, n_k \, d\Sigma \cdot H(t) \tag{4.6}$$

where p_o is the hydrostatic pressure, $H(t)$ the unit step function

and it is assumed that the initial displacement discontinuity at
$t=0$ is $[u_j(0)]_-^+$, and the final displacement discontinuity is
zero. Hence N_{jk} is a transient and the static moment tensor is
still obtained from the stress glut of eq. (4.1), by setting
$[u_m]_-^+ = [u_m(0)]_-^+$. The drop in potential energy is however (47):

$$\Delta W = \frac{1}{2} \int_\Sigma [u_j(0)\ \sigma_{jk}(0)]_-^+\ n_k\ d\Sigma\ -p_o\ \Delta V \qquad (4.7)$$

The first term on the right hand side is to be associated with
the static moment tensor obtained from eq. (4.1), but the second
term is associated with (4.6). Part of this energy will not
become available as seismic energy but will be dissipated during
volume change or faulting. One can argue (45) that in realistic
cases, due to the transient nature of the process (4.6), the
seismic energy associated with the action of hydrostatic pressure
is at the high frequencies which usually are not observable.

The above model with associated volume change is not adequate
to describe the effect of a phase transformation at depth. Then
it must be assumed that the material changes constitution in a
finite volume V. The stress glut becomes

$$m_{ij} = c_{ijk\ell}\partial_k u_\ell - \tau_{ij} = c_{ijk\ell}\varepsilon_{k\ell}^F \qquad (4.8)$$

where τ_{ij} is the initial static stress (before the phase transformation), $_Fc_{ijk\ell}$ the elastic tensor after the phase transformation, and $\varepsilon_{k\ell}^F$ is termed the stress-free strain (48). The stress
glut for a dislocation (4.1) may be obtained from (4.8) by a
limiting procedure. Since τ_{ij} is finite, its contribution to
u_i in eq. (3.11) vanishes if V shrinks to a vanishing volume
V_Σ^1 and the contribution to the excitation by (4.1) is related
to the first term on the right hand side of eq. (4.8).

The dilatation in a point $\underline{\xi}$, associated with the isotropic
part of the stress glut $1/3$ tr $\underline{\underline{m}}\ \delta_{ij}$, may be expressed (45):

$$\partial_i u_i(\underline{\xi}) = -\ \frac{1}{4\pi(\lambda+2\mu)}\ \int_V\ \frac{1}{3}\ \text{tr}\ \underline{\underline{m}}\ \partial_i\partial_i\left(\frac{1}{|\underline{\xi}-\underline{\xi}'|}\right)\ dV' \qquad (4.9)$$

(in an isotropic medium). Thus, if the stress-free strain is
a pure dilatation, the volume change becomes

$$\Delta V = \int_V \partial_i u_i(\underline{\xi}) dV \simeq \frac{1}{(4\pi(\lambda+2\mu)} \int_V \{\frac{1}{3} \, tr \, \underline{\underline{m}} \, \partial_i \partial_i \int_V \cdot \frac{dV}{|\underline{\xi}-\underline{\xi}'|}\} \, dV$$

$$= (\lambda+2\mu)^{-1} \cdot \frac{1}{3} \, tr \, \underline{\underline{M}} \tag{4.10}$$

which differs from eq. (4.5) for a dislocation. In this model, the energy associated with the static stress glut equals the drop in potential energy which does not depend on hydrostatic pressure:

$$\Delta W = \frac{1}{2} \int_V m_{ij} \cdot \partial_i u_j \, dV = \frac{1}{2} \int_V c_{ijk\ell} \, \partial_k u_\ell \, \varepsilon_{ij}^F \, dV \tag{4.11}$$

which equals the total strain energy in an inclusion, as derived in (4.8). If the transformation of the volume V were instantaneous, all of ΔW would become available as seismic energy. In practice, however, part of ΔW is dissipated or transferred to internal energy of the precipitating phase, and the transformation of V is not instantaneous. Since excitation of seismic energy can be represented in terms of temporal derivatives of the source function, as in eq. (3.5), one may infer that a slower source process implies less seismic radiation. In fact, we expect for the seismic energy in this case

$$E_{seismic} = \Delta G \simeq -\delta p \, \Delta V \tag{4.12}$$

where ΔG is the change in the Gibbs free energy between the products and reactants of the transformation, ΔV is the volume change and δp is the amount of superpressure in the metastable zone. We may note that in this form, the superpressure in a phase transformation replaces the total pressure on a cavity, but the mechanism of seismic radiation is rather different in the two cases.

5.   MOMENT TENSORS AND FAULT PARAMETERS

In this section two related questions will be posed, some aspects of which have been discussed in previous sections:
1) How to interpret moment tensors of the stress glut in terms of fault parameters. To answer this question, the type of faulting should be known or assumed a priori; as argued by Backus (12), this assumption may then be tested a posteriori. Information on the type of faulting is precisely what is needed to answer the second question: 2) How to obtain the glut moments

of spatial degree two uniquely from the seismic response. As
argued in section 3, moments of degree zero and one can always
be obtained uniquely, at least in principle. These moments
completely specify a simple point source, and the interpretation
for point sources was discussed during the early development of
the moment tensor concept (1,5). The discussion will be confined
here to the most simple type of faulting which is also the most
popular model, represented by the shear dislocation in an isotropic medium with isotropic prestress, eq. (4.3). A more extensive discussion including other types of faulting may be found
in (11,12).

The tensor in eq. (4.3) has zero trace, and this property
then applies to all moment tensors derived from it. In particular

$$\sum_{j=1}^{3} M_{jj} = \sum_{j=1}^{3} M_{jj,\ell} = \sum_{j=1}^{3} M_{jj,\ell m} = 0 \tag{5.1}$$

Two different sources with the same force moments may have different glut moments of spatial degree two and higher. However,
it the constraint (5.1) is imposed they also have the same glut
moments of spatial degree two (11). It follows that, in principle,
these constrained moments can be determined uniquely from the
seismic response; we can use eq. (3.12) together with (5.1).

A special case is that where the fault surface Σ is a
plane. Then the moment tensor (4.4) can be rewritten

$$M_{jk} = M(n_j \nu_k + \nu_j n_k) = M(a_j a_k - b_j b_k) \tag{5.2}$$

where M is 'the seismic moment' and $\underline{\nu}, \underline{n}$ are unit vectors in the
direction of slip and in the direction normal to the fault plane.
Moment tensors of higher spatial degree have similar constraints
on their first two indices. The structure (5.2) implies that one
of the eigenvalues of $\underline{\underline{M}}$ is zero, and the eigenvectors (say $\underline{a}$ and
$\underline{b}$) corresponding to the two nonzero eigenvalues constitute the
compression and tension axis of the source (the P and T axis).
The fault parameters $\underline{n}$ and $\underline{\nu}$ are simply related to $\underline{a}$ and $\underline{b}$
(e.g. 5), but $\underline{n}$ and $\underline{\nu}$ can be freely interchanged without violating that relation; this is the 'fault plane ambiguity'. The
ambiguity can be removed with the aid of moments of higher
degree, because from them one can estimate, among other things,
the orientation of the fault and the direction of slip. Of
course, finiteness of the source is explicitly considered then,
and moment tensors up to at least spatial degree two should be
included in the representation (3.12). It is appropriate then to

also include temporal moments up to degree two (see section 3).
Taking account of the symmetries in the moment tensors, the
number of unknowns amounts to 90, which may be impractical in
case of limited number of data (which is a common case). The
number of unknowns can be considerably reduced by means of
'smoothing assumptions'; the assumptions made here are that the
6 components of the stress glut have a common space and time
history. Moment tensors are then approximated as follows:

$$M_{jk,\ell}(\xi^o) \simeq M_{jk}\,\Delta\xi_\ell \ , \quad M_{jk}^{(1)}(\tau^o) \simeq M_{jk}\,\Delta\tau \ , \text{ etc.} \tag{5.3}$$

Specifically, if the final static value of the stress glut (e.g.,
in terms of the final offset on the fault) is almost the same
throughout the source region V (end effects are ignored), and
the source function is well approximated by a ramp with rise
time T, then

$$\Delta\xi_\ell = \int_V (\xi_\ell - \xi_\ell^o)\,\frac{dV}{V} \ , \quad \Delta\tau = \int_V (\tau - \tau^o)\,\frac{d\tau}{T}$$

This leaves the 20 unknowns:

$$M_{jk}, \ \Delta\tau, \ \Delta\xi_\ell, \ \Delta(\tau^2), \ \Delta(\tau\xi_\ell), \ \Delta(\xi_\ell\xi_m); \quad \ell,m = 1,2,3$$

and these can be estimated from a system of equations related
to (3.5), (3.6) and (3.12):

$$u_i(\underline{x},t) = G^i_{j,k}\,M_{jk} - \dot{G}^i_{j,k}\,M_{jk}\,\Delta\tau + \dot{G}^i_{j,k\ell}\,M_{jk}\,\Delta\xi_\ell \tag{5.4}$$

$$+ \frac{1}{2}\,\ddot{G}^i_{j,k}\,M_{jk}\,\Delta(\tau^2) - \ddot{G}^i_{j,k\ell}\,M_{jk}\,\Delta(\tau\xi_\ell) + \frac{1}{2}\,\ddot{G}^i_{j,k\ell m}\,M_{jk}\,\Delta(\xi_\ell\xi_m)$$

where $G^i_{j,k} \equiv G^i_{j,k}(\underline{\xi}^o,\underline{x},t-\tau^o)$.

A solution to (5.4) can be obtained by iteration; constraints
are imposed by (5.1), and positivity constraints to $\Delta(\tau^2),\Delta(\xi_\ell^2)$
could also be added. Our best estimate of the source location
$\underline{\hat{\xi}}^o$ and source time $\hat{\tau}^o$ are such that, referred to these new
coordinates:

$$\Delta\tau = \Delta\xi_\ell = 0 \ , \quad \text{and} \quad M_{jk,\ell}(\underline{\hat{\xi}}^o) = M_{j,k}^{(1)}(\hat{\tau}^o) = 0 \tag{5.5}$$

$\overset{\wedge o}{\tau}$ is the time at which the source is roughly halfway during rising from the initial to the final static state, and $\overset{\wedge o}{\xi}$ is roughly the 'centre of gravity' of the source region. With the approximations (5.3) and the reference coordinates $\overset{\wedge o}{\xi}$, $\overset{\wedge o}{\tau}$, we also find the parameters defined in (11):

The 'rise time'

$$T \simeq 2\{\Delta(\tau^2)\}^{\frac{1}{2}} \tag{5.6}$$

an estimate of 'rupture velocity' (in unidirectional rupture propagation)

$$v_\ell \simeq \Delta(\tau\xi_\ell)/\Delta(\tau^2) \tag{5.7}$$

a tensor describing the size, shape and orientation of the final static source region:

$$W_{\ell m} = \Delta(\xi_\ell \xi_m) \tag{5.8}$$

If the source region is a plane, then eq. (5.7) or (5.8) can be used to remove the 'fault plane ambiguity', since the rupture velocity vector $\underline{v}$ should be tangent to the fault plane, and the normal $\underline{n}$ to that plane should satisfy

$$\underline{\underline{W}} \; \underline{n} = 0$$

so $\underline{n}$ corresponds to the eigenvector of $\underline{\underline{W}}$ with eigenvalue zero. This interpretation may also be used to impose additional constraints to the solution of (5.4).

6.   DISCUSSION

Besides (6,7), several more recent results on inversion for the moment tensor of earthquakes have now become available (e.g. 16,17,18). These results may have interesting tectonic implications, but most efforts so far have been spent on experimenting with the inversion procedure itself. Various procedures have been advocated and it may be instructive to see in how far they are really different, and what the relation is to more 'classical' methods of source analysis. The inversion procedure in most cases uses the classical least squares method, although it was sometimes communicated under a different name (7,49). The incorporation of linear constraints in such a method is a simple procedure, and if a solution is obtained by iteration, it is possible to also incorporate non-linear constraints (50).

Most results have been obtained in the frequency domain; the system of equations can be obtained by Fourier transforming eq. (3.12). Although this procedure appears to be simpler than inversion in the time domain, the source time function which results after inverse Fourier transforming the solution, is often unjustifiably detailed compared to the spatial information of the source that is retrieved. In fact, in all these applications a simple point source is assumed, although results with moments of spatial degree one have recently been reported (51); these results were used to relocate earthquake hypocentres. To meet the above objections and also to obtain sensible results with relatively few stations, smoothing assumptions have been invoked. In (52) a smooth source spectrum was assumed. A similar assumption can be made to simplify the inversion procedure in the time domain, but this implies a representation by temporal moments of higher degree, and these can be retrieved together with the spatial moments, as described in section 5. To date, experiments with this procedure have not yet been reported. In its simplest form, the representation is by just the moment tensor of spatial and temporal degree zero, and this inversion has been done (18); the representation is that of a point source with step function time dependence.

The question of the effect of anomalous earth structure (which may also be held responsible for mislocations) has been addressed in most of the practical applications, and attempts have been made to reduce this effect (16,17,18). In this respect one might also ask which seismic parameters are least sensitive to perturbations of an earth model. No parameter is completely reliable, but clearly observed polarities appear to be more reliable than are amplitude and phase, and they are well documented (unfortunately, polarities are not always clearly observable). If polarities are included in the inversion procedure, then application of the least squares (L2) method is not trivial. In (53), polarities were combined with amplitude data in an inversion based on the L1 norm (54), and results of this procedure were reported in (55). This procedure perhaps most clearly reminds one of the relation to classical fault plane solutions: The moment tensor not necessarily entails more information (especially if a double couple constraint is imposed), but the information is obtained in a more systematic and objective way. In particular, the trial and error procedures which are common in other methods are avoided here. The same statement applies to moment tensors of higher degree; their interpretation is in terms of fault parameters that can also be retrieved by more conventional methods (3).

An important but difficult problem in inversion for the moment tensor is that of model testing. Based on physical considerations, tests of various types of faulting were advanced in (12). Statistical aspects of model testing were emphasized in (50). The statistical aspects pertain to the fact that, when a priori assumptions as to the type of source are made, the number of free parameters decreases and the least squares error increases; a χ^2-test was proposed to compare models with different number of free parameters.

Model testing is closely related to source classification, a particular application of which is discrimination between earthquakes and explosions. Several comprehensive reviews of discrimination are available, e.g. (56,57,58). It is not the purpose here to discuss this material, but rather to identify the role of the moment tensor. Ideally, the moment tensor of an explosion is isotropic, and of an earthquake it is predominantly deviatoric. It may be noted that the moment tensor of an explosion is not predominantly due to a stress glut, but to an imposed stress pulse; for a consideration of the physics involved, see (59). Since the moment tensor discriminates different excitation patterns at the source, it should be related to discriminants which are also based, perhaps in part, on differences in excitation pattern. This is obviously true for polarity, and it is also true for the m_b/M_s ratio although this ratio is also affected by source time function and source size. A simple demonstration of how different excitation, as represented by the isotropic and deviatoric parts of the moment tensor, contributes to the m_b/M_s discriminant, was given in (60). Another demonstration of this effect, at a time that the moment tensor was not yet in use, was given in (61). In that work the sensitivity to source depth and to orientation of the fault plane with respect to the receiver was also shown. Of course, the moment tensor itself is not affected by these circumstances, and so it would appear to be a potential discriminant. The point has been made before, e.g. (49), but only some occasional results have been reported (62). One of the difficulties in application is almost certainly the effect of anomalous earth structure, as noted previously. Another, perhaps less obvious, problem is that whereas explosions can be treated as point sources, this is not always permitted for earthquakes, even with long-period data. Of course, source size itself can be put to use as a discriminant. This would suggest the use of moments of higher degree as an alternative to more common observational parameters like spectral ratio. Thus, moment tensors incorporate several commonly used discriminants and in this sense the philosophy is similar to that of a pattern recognition approach (63); in a physical sense however, the use of moment tensors would be preferable.

ACKNOWLEDGEMENTS

The present work was initiated during a visit to NTNF/
NORSAR in 1979. I thank Drs. Husebye and Sandvin for their
help in preparing this paper.

REFERENCES

1.   Gilbert, F.: 1970, Geophys. J.R. astr. Soc. 22, pp 223-226.
2.   Engdahl, E.R. and Kanamori, H.: 1980, Trans. Am. Geophys.
     Un. 61, pp. 62-64.
3.   Aki, K. and Richards, P.G.: 1980, "Quantitative Seismology",
     Freeman and Co., San Francisco.
4.   Kostrov, B.V.: 1970, Izv. Earth Physics 4, pp. 84-101.
5.   Gilbert, F.: 1973, Phil. Trans. R.Soc. Lond. A274, pp.
     369-371.
6.   Dziewonski, A.M. and Gilbert, F.: 1974, Nature 247, pp.
     185-188.
7.   Gilbert, F., and Dziewonski, A.M.: 1975, Phil. Trans. R.
     Soc. London A278, pp. 187-269.
8.   Burridge, R. and Knopoff, L.: 1964, Bull. seism. Soc.
     Am. 54, pp. 1875-1888.
9.   Backus, G. and Mulcahy, M.: 1976, Geophys. J.R. astr. Soc.
     46, pp. 341-362.
10. Backus, G. and Mulcahy, M.: 1976, Geophys. J.R. astr. Soc.
     47, pp. 301-329.
11. Backus, G.: 1977, Geophys. J.R. astr. Soc. 51, pp. 1-25.
12. Backus, G.: 1977, Geophys. J.R. astr. Soc. 51, pp. 27-45.
13. Mendiguren, J.A. and Aki, K.: 1978, Geophys. J.R. astr.
     Soc. 55, 539-556.
14. Okal, E.A. and Geller, R.J.: 1979, Phys. Earth Planet.
     Int. 18, pp. 176-196.
15. Patton, H. and Aki, K.: 1979, Geophys. J.R. astr. Soc. 59,
     pp. 479-495.
16. Patton, H.: 1980, J. Geophys. Res. 85, pp. 821-848.
17. Strelitz, R.A.: 1980, Phys. Earth Planet. Int. 21, pp. 83-96.
18. Ward, S.N.: 1980, Bull. seism. Soc. Am. 70, pp. 717-734.
19. Morse, P.M. and Feshback, H.: 1953, "Methods of Theoretical
     Physics", McGraw-Hill, New York.
20. Ben-Menahem, A. and Singh, S.J.: 1972, in "Methods in
     Computational Physics" 12, pp. 299-375 (ed. Bolt, B.A.).
21. Dahlen, F.A.: 1972, Geophys. J.R. astr. Soc. 28, pp. 357-383.
22. Saito, M.: 1967, J. Geophys. Res. 72, pp. 3689-3699.
23. Hudson, J.A.: 1969, Geophys. J.R. astr. Soc. 18, pp. 233-249.
24. McCowan, D.W.: 1976, Geophys. J.R. astr. Soc. 44, pp. 595-
     599.
25. Mendiguren, J.A.: 1977, J. Geophys. Res. 82, pp. 889-894.

26. Kennett, B.L.N. and Kerry, N.J.: 1979, Geophys. J.R. astr.
     Soc. 57, pp. 557-583.
27. Ward, S.N.: 1980, Geophys. J.R. astr. Soc. 60, pp. 53-66.
28. Cormier, V.F.: 1980, Bull. seism. Soc. Am. 70, pp. 691-716.
29. Takeuchi, H. and Saito, M.: 1972, in "Methods in Computa-
     tional Physics" 11, pp. 217-295 (ed. Bolt, B.A.).
30. Woodhouse, J.H.: 1978, Geophys. J.R. astr. Soc. 54, pp. 263-
     281.
31. Doornbos, D.J.: 1980, Geophys. J.R. astr. Soc., in press.
32. Gilbert, F. and Backus, G.E.: 1966, Geophys. 31, pp. 326-
     332.
33. Kennett, B.L.N.: 1974, Bull. seism. Soc. Am. 65, pp. 1643-
     1651.
34. Richards, P.G.: 1976, Bull. seism. Soc. Am. 66, pp. 701-717.
35. Chapman, C.H.: 1978, Geophys. J.R. astr. Soc. 54, pp. 481-
     518.
36. Nussenzweig, H.M.: 1965, Ann. Phys. 34, 23-95.
37. Richards, P.G.: 1973, Geophys. J.R. astr. Soc. 35, pp. 243-
     264.
38. Doornbos, D.J. and Mondt, J.C.: 1980, Pure Appl. Geophys. 18,
     pp. 1291-1307.
39. Cormier, V.F. and Richards, P.G.: 1976, J. Geophys. Res. 81,
     pp. 3066-3068.
40. O'Neill, M.E. and Hill, D.P.: 1979, Bull. seism. Soc. Am.
     69, pp. 17-25.
41. Sipkin, S.A. and Jordan, T.H.: 1980, J. Geophys. Res. 85,
     pp. 853-861.
42. Stump, B.W. and Johnson, L.R.: 1977, Bull. seism. Soc. Am.
     67, pp. 1489-1502.
43. Hudson, J.A.: 1977, J. Geophys. 43, pp. 359-374.
44. Randall, M.J.: 1972, in "Methods in Computational Physics" 12,
     pp. 267-298 (ed. Bolt, B.A.).
45. Doornbos, D.J.: 1977, Geophys. J.R. astr. Soc. 51, pp. 465-
     474.
46. Knopoff, L. and Randall M.J.: 1970, J. Geophys. Res. 75,
     pp. 4957-4963.
47. Dahlen, F.A.: 1973, Geophys. J.R. astr. Soc. 31, pp. 469-
     484.
48. Eshelby, J.D.: 1957, Proc. R. Soc. London A241, pp. 376-396.
49. Buland, R. and Gilbert, F.: 1976, Geophys. Res. L. 3, pp. 205-
     206.
50. Strelitz, R.A.: 1978, Geophys. J.R. astr. Soc. 52, pp. 359-
     364.
51. Chou, T.A. and Dziewonski, A.M.: 1980, Trans. Am. Geophys.
     Un. 17, pp. 296.
52. Gilbert, F. and Buland, R.: 1976, Geophys. J.R. astr. Soc. 47,
     pp. 251-255.
53. McCowan, D.W.: 1978, Semiann. Techn. Summ. Seism. Discr.
     M.I.T. Lincoln Lab, pp. 21-23.

54. Claerbout, J.F.: 1976, "Fundamentals of Geophysical Data
    Processing", McGraw-Hill, New York.
55. Fitch, T.J., McCowan, D.W. and Shields, M.W.: 1980, Trans.
    Am. Geophys. Un. 17, pp. 295.
56. Bolt, B.A.: 1976, "Nuclear Explosions and Earthquakes: The
    Parted Veil", Freeman and Co., San Francisco.
57. Dahlman, O. and Israelson, H.: 1977, "Monitoring Underground
    Nuclear Explosions", Elsevier, Amsterdam.
58. Blandford, R.: 1977, Ann. Rev. Earth Planet. Sci. 5, pp. 111-
    122.
59. Rodean, H.C.: 1971, "Nuclear Explosion Seismology", U.S.
    Atomic Energy Commission.
60. Gilbert, F.: 1973, Geophys. J.R. astr. Soc. 33, pp. 487-488.
61. Douglas, A., Hudson, J.A. and Kembhavi, V.K.: 1971,
    Geophys. J.R. astr. Soc. 23, pp. 451-460.
62. Fitch, T.J. and Shields, M.W.: 1978, Semiann. Techn. Summ.
    Seism. Discr. M.I.T. Lincoln Lab., pp. 20-21.
63. Tjøstheim, D. and Husebye, E.S.: 1976, Geophys. Res. L. 3,
    pp. 499-502.

DETERMINATION OF SOURCE MECHANISM AND HYPOCENTRAL COORDINATES FROM WAVEFORM DATA

A.M. Dziewonski, T.-A. Chou and J.H. Woodhouse

Department of Geological Sciences
Harvard University, Cambridge, Mass. 02138

1. INTRODUCTION

We describe a procedure that could become an important tool in the utilization of the recently established global digital seismograph network towards routine estimation of the principal seismic source parameters (hypocentral coordinates, seismic moment tensor) for earthquakes of moderate size. Considering the usual level of global seismic activity, the analysis could eventually be applied to as many as several hundred events per year, rapidly transforming our level of understanding of the process of stress accumulation and release.

We show that it is possible to use the wave-form data not only to derive the source mechanism of an earthquake, but also to establish the "best point source" (the centroid of the stress glut density) at a given frequency. Thus two classical problems of seismology are combined into a single procedure.

Given an estimate of the origin time, epicentral coordinates and depth, an initial moment tensor is derived using one of the variations of the method described in detail by Gilbert and Dziewonski [1]. This set of parameters represents the starting values for an iterative procedure in which perturbations to the elements of the moment tensor are found simultaneously with changes in the hypocentral parameters. In general, the method is stable and convergence rapid.

The motivation of this work was to develop an approach to the quantitative analysis of earthquakes that would satisfy the following requirements:

(i) relative insensitivity to the earth model used in the generation of the excitation kernel functions ("matching filters" in the terminology of Buland and Gilbert, [2]);

E. S. Husebye and S. Mykkeltveit (eds.), Identification of Seismic Sources - Earthquake or Underground Explosion, 233–254.

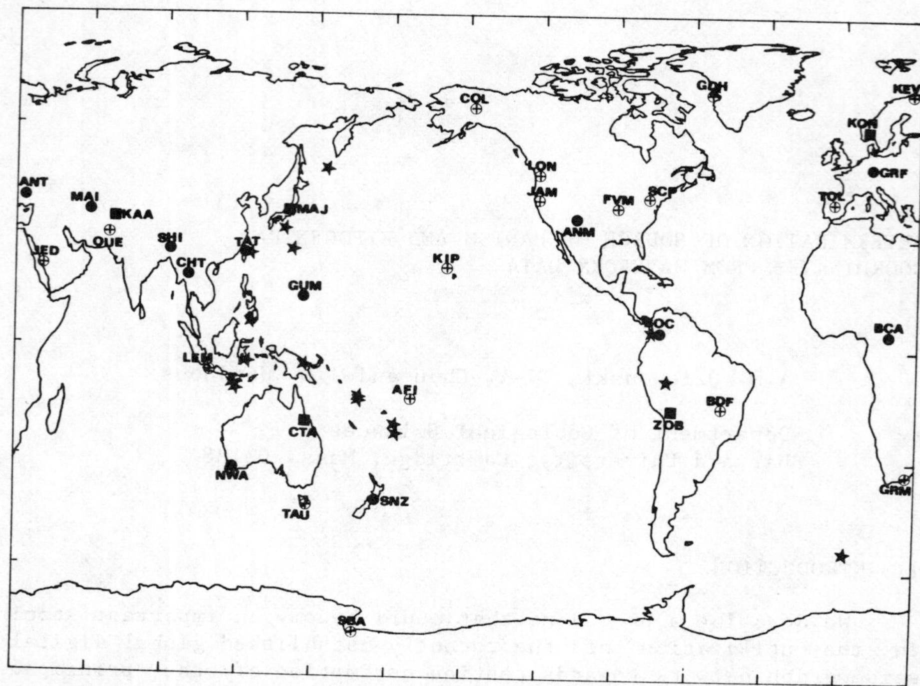

Figure 1. Stars show locations of 14 of the events analyzed
in this study. Squares show the ASRO and filled circles the SRO
stations operational at any time during the period from August
1977 to November 1979. Crossed circles are the proposed locations
of WWSSN stations which are to be equipped with digital data acquisition system. Addition of those stations would greatly improve the azimuthal coverage; for all events analyzed here this
coverage was rather poorly balanced.

(ii) ability to investigate earthquakes of moderate magnitudes ($M_b \sim 5.5$), rather than only exceptionally large events
which are rare and unevenly distributed on the earth's surface;

(iii) "objectivity" of the method; by this we mean that relatively few and preferably no subjective judgements need to
be made by an operator;

(iv) reasonable numerical efficiency of the method, so that
calculations can be carried out on a powerful mini-computer
and data for a substantial number of events could be
analyzed in a relatively short time.

At present there are in operation approximately 15 Seismic
Research Observatories (SRO) and Abbreviated Seismic Research Observatories (ASRO); Figure 1 shows their world-wide distribution.
The sensors (for details, see Peterson, et al. [3]) are placed
either in a borehole (SRO) or in a pressurized vessel in a seism-

ic vault (ASRO). The long-period channels are sampled at a rate of 1 sample per second; the peak magnification of the ground motion is at a period of 25 seconds.

Because of the sensitivity of these instruments (nominal dynamic range of SRO stations is 120 db), they yield recordings with an adequate signal-to-noise ratio for long period (T > 45 sec) body waves for events as small as $M_b \sim 5.5$. This property is important, as in this work we intend to study the source mechanism using the body-wave part of a seismogram. For a typical epicentral distance of about 90°, the interval between the emergence of the P-wave and the arrival of the fundamental Rayleigh mode has duration of, roughly, 30 minutes; for the transverse component this time is somewhat shorter, as the waves of the fundamental Love mode travel with higher group velocities.

Such an interval of a record contains in addition to the P- and S-wave arrivals also multiple reflections of these waves, converted phases, core reflections and, of course, all the phases including the primary "p" and "s" surface reflections. Thus the excitation of seismic waves, contained in our hypothetical recording, projected back onto the focal sphere would yield information not about a single point on that sphere, as is usually the case in focal mechanism studies, but about the entire meridional section.

Little seems to be gained, unless excitation at very long periods is the subject of study, from extending the analysis to greater times. Dispersion and the amplitudes of fundamental mode surface waves in a period range from 40 to 100 seconds are very sensitive to lateral variations in the earth's structure; body waves that arrive along the major arc are not likely to contribute much to the resolution of the properties of the source, and the effect of differences between the real earth and the model is likely to be magnified.

The most straightforward method for construction of the kernel functions (synthetic seismograms) is that of superposition of normal modes. If all normal modes within the appropriate frequency range and group velocity window are used in construction of a synthetic seismogram, we are assured that no part of the theoretically predictable seismic signal has been omitted. Although the number of normal modes that are used in the summation is large, the computation times involved can be quite reasonable if a proper strategy is used.

The additional advantage of the normal mode method is its objectivity: the person processing records need not identify the phases in order to achieve correct results. This makes the approach suitable for essentially routine operations without the need for intervention by a skilled seismologist.

2. DERIVATION OF EXCITATION KERNELS

Following Gilbert and Dziewonski [1], a component of ground

motion excited by a point source may be expressed as:

$$u_k(\underline{r},t) = \sum_{i=1}^{6} \psi_{ik}(\underline{r},\underline{r}_s,t) * f_i(t): \qquad (2.1)$$

where u_k is the k-th record in a set of seismograms, with the receiver at position $\underline{r}$ and the source at r_s. The excitation kernels ψ_{ik} depend also on the properties of the earth, both in terms of the displacement functions (eigenfunctions) of individual normal modes as well as their eigenvalues (complex frequency). Functions $f_i(t)$ represent the six independent components of the moment rate tensor, and we follow the notation used by Gilbert and Dziewonski: $f_1 = M_{rr}$, $f_2 = M_{\theta\theta}$, $f_3 = M_{\phi\phi}$, $f_4 = M_{r\theta}$, $f_5 = M_{r\phi}$, $f_6 = M_{\theta\phi}$; the asterisk signifies convolution.

Buland and Gilbert [2] discussed the solution of eq. (2.1) for $f_i(t)$ in the frequency domain representation, and there is no need to repeat their argument here. Dziewonski [4,5] pointed out that significant efficiency in evaluation of the excitation kernels can be achieved by summing up first contributions from all the modes of the same angular order number, as such a sum does not depend on the position of the receiver.

In Figure 2 we show a set of the excitation kernels, ψ_{ik}, computed for a deep source (580 km) and a three-component receiver at a distance of 98.8° and azimuth of 294.1°. The catalog of normal modes computed by Buland [6] for an earth model 1066B [1] was used in calculaton of the traces in Figure 2 and throughout this paper. Attenuation coefficients were computed using model FSQMK of Dziewonski [7]. It is quite clear that the excitation kernels associated with the individual elements of the moment tensor have a distinctly different character and that it would not be unreasonable to expect that, with a high signal-to-noise ratio and the reference earth model close to the real earth, a dependable point source mechanism could be derived from recordings at a single station [8].

Application of the procedure proposed by Buland and Gilbert to the system of equations of condition immediately following from eq. (2.1) can lead to reliable solutions for the moment tensor if the estimated position of the source $\underline{r}_s$ and the origin time t_0 are correct. But this need not always be the case.

3. ITERATIVE INVERSION OF WAVE-FORM DATA FOR SOURCE MECHANISM AND HYPOCENTRAL PARAMETERS.

There are three reasons why the position of the source and the origin time estimated from the first arrivals of the P-waves may not be the best parameters to use in inversion for the moment tensor:

(i) representation of the source as a point in space and time is essentially correct, but the standard hypocentral parameters obtained from arrival times of P-waves are in er-

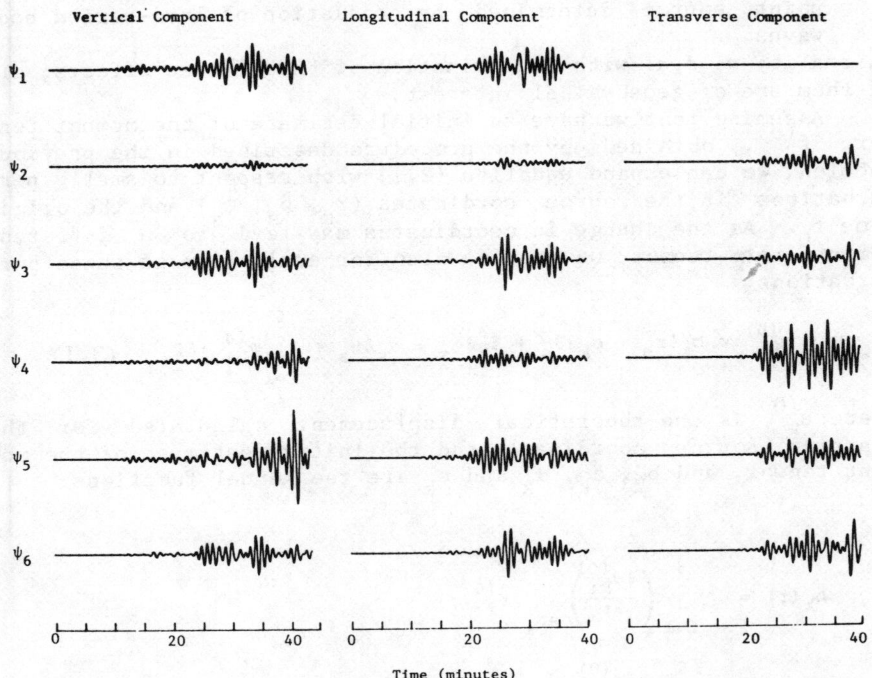

Figure 2. Excitation kernels for a deep earthquake (580 km) computed for the three components of the ground motion for a point on the earth's surface at a distance of 98.8° and azimuth 294.1° from the epicenter. The scale is common for all the traces in the figure. The traces represent synthetic seismograms excited by the individual components of the moment tensor of unit amplitude. A synthetic seismogram for an arbitrary point source is a linear combination of the six traces for a given component of the ground motion. The objective of our inversion procedure is to find the weights that give the best match between the observed and synthetic seismograms. Notice that there is a great diversity of shapes among the kernel functions. Intuitively, this indicates that for deep earthquakes good resolution of the source properties could be achieved with relatively few recordings.

ror;
(ii) the hypocentral parameters are correct for the point of origin of rupture but the source is distributed in time and space, consequently the position of the centroid of the stress glut density [9,10], the best point source, is incorrect;
(iii) errors in the reference earth model or lateral heterogeneities, in conjunction with a particular distribution of the station network, may lead to an apparent shift of the

point source determined by radiation of long-period body waves.

In reality we deal with a combination of these three effects; all of them are of geophysical interest.

Assuming that we have an initial estimate of the moment tensor, $f^{(0)}$, obtained by the procedure described in the previous section, we can expand equation (2.1) with respect to small perturbations in the source coordinates (r_s, θ_s, ϕ_s) and the origin time t_0. As the change in coordinates may lead to a different moment rate tensor, we provide also for evaluation of these perturbations:

$$u_k - u_k^{(0)} = b_k \delta r_s + c_k \delta \theta_s + d_k \delta \phi_s + e_k \delta t_s + \sum_{i=1}^{6} \psi_{ki}^{(0)} * \delta f_i \qquad (3.1)$$

where $u_k^{(0)}$ is the theoretical displacement calculated for the starting source coordinates and the initial estimate of the moment tensor, and b_k, c_k, d_k and e_k are the kernel functions:

$$b_k(t) = \sum_{i=1}^{6} \left(\frac{\partial \psi_{ki}^{(0)}}{\partial r} \right)_{r=r_s} * f_i^{(0)} ;$$

$$c_k(t) = \sum_{i=1}^{6} \left(\frac{\partial \psi_{ki}^{(0)}}{\partial \theta} \right)_{\theta=\theta_s} * f_i^{(0)} ;$$

$$d_k(t) = \sin\theta_s \sum_{i=1}^{6} \left(\frac{\partial \psi_{ki}^{(0)}}{\partial \phi} \right)_{\phi=\phi_s} * f_i^{(0)} ; \qquad (3.2)$$

$$e_k(t) = -\sum_{i=1}^{6} \frac{\partial \psi_{ki}^{(0)}}{\partial t} * f_i^{(0)} .$$

Equation (3.1) without the last term is conceptually equivalent to the usual expression for perturbation of hypocentral coordinates needed to achieve the best fit to the observed travel times. Here, however, we are attempting to match the wave-forms of extended duration.

The location kernels are obtained by summing the contributions from all relevant normal modes using the same strategy as described in Section 2. The expressions for perturbations of amplitude of individual normal modes with respect to changes in hypocentral coordinates were derived by Woodhouse [11].

The improved estimate of the moment rate tensor is again substituted into eq. (3.1) after the kernel functions have been

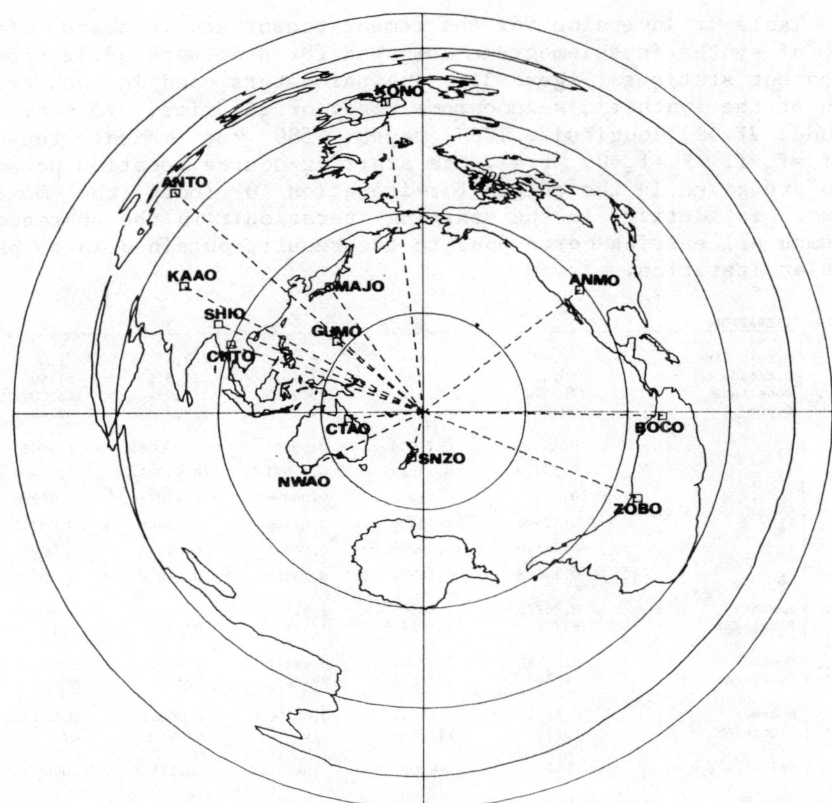

Figure 3. Equidistant projection centered on the coordinates 20.84S, 178.64W showing distribution of the stations used in our experiment with the synthetic data. This location and station coverage corresponds to that for the event of April 24, 1979; the results of the analysis of the actual data for this earthquake are presented later in the text.

recalculated for the new location of the source, and the process is repeated until convergence is achieved.

We have designed the following numerical experiment to test the procedure. Synthetic seismograms were computed for a point source with the components of a moment tensor representing superposition of an explosion with a strike–slip earthquake ($f_1 = f_2 = f_3 = 1$; $f_4 = f_5 = 0$; $f_6 = 1$), placed at a depth of 580 km and geographical coordinates 21.50S and 178.00W. Also, the origin time has been delayed by 7.5 sec with respect to an arbitrary t_0. The starting coordinates were h = 600 km, latitude 20.84S, longitude 178.64W, $t_0 = 0$ (these hypocentral coordinates correspond to those reported by N.E.I.S. for an actual earthquake on April 24, 1979, which will be analyzed later in this report). Thus the

Table 1. Inversion for the moment tensor and location of a set of synthetic seismograms computed for a network of 12 three-component stations (Figure 3). The parameters used in construction of the synthetic seismograms were: origin time, 7.5 sec; latitude, 21.5S; longitude, 178W; depth, 580 km; moment tensor, $f_1=f_2=f_3=1$, $f_4=f_5=0$, $f_6=1$. The starting source location parameters are given in the column for iteration 0 (only the moment tensor is derived in the starting iteration); in the subsequent columns all entries correspond to the results obtained in a particular iteration.

		ITERATION	0	1	2	3	4
Location		Origin Time	0.00	5.53	6.93	7.47	7.50
		Latitude	20.84S	21.10S	21.40S	21.51S	21.50S
		Longitude	178.64W	178.04W	178.23W	177.98W	178.00W
		Depth	600.00	584.27	587.31	579.75	580.00
Moment Tensor		f_1	0.69844	0.71056	0.99344	1.00006	1.00000
		f_2	0.33230	0.35757	0.97671	0.99992	1.00000
		f_3	0.47704	0.51578	0.98099	1.00022	0.99998
		f_4	-0.13696	-0.15316	-0.01460	0.00048	-0.00003
		f_5	-0.02306	-0.09278	-0.00845	0.00034	-0.00001
		f_6	0.14698	0.14584	0.93683	0.99946	1.00000
Principal Axes	I	Moment	0.76273	0.83205	1.91597	1.99995	2.00000
		Plunge/Az	63/145	55/133	1/135	0/135	0/135
	II	Moment	0.52791	0.49440	0.99317	1.00006	1.00000
		Plunge/Az	24/293	32/287	89/300	90/-	90/-
	III	Moment	0.21714	0.25246	0.04200	0.00061	-0.00001
		Plunge/Az	13/28	12/25	0/45	0/45	0/45
		Relative r.m.s.	.84706	.74561	.16420	.00175	.00000

starting location of the source is some 100 km away from the true position. The distribution of the stations used in this experiment is shown in Figure 3.

The results are listed in Table 1. The zeroth iteration yields a moment tensor that is very remote from the proper solution; even so, the first iteration brings about a major correction of the epicentral coordinates but the moment tensor shows very little improvement. With the new location some 45 km from the true coordinates, the moment tensor after the second iteration is within 93% of its expected value, and after four iterations the input parameters and the inversion results agree to five significant figures.

The conclusions from this experiment are the following:

(i) results of inversion for the moment tensor can be sensitive to the assumed location of the hypocenter;

(ii) errors in estimates of the moment tensor are highly non-linear functionals of the magnitude of mislocation of the source;

(iii) the process shows rapid convergence and stability, although there are obviously limits to how far the starting solution can be removed from the true one. Our example is close to that limit; from subsequent tests we have found out that the solution does not converge if the mislocation vector is increased by 50%.

4. APPLICATION TO THE ANALYSIS OF THE EARTHQUAKE DATA

In this study we use the data from the SRO/ASRO network. Theoretical seismograms are obtained using the normal mode catalog computed by Buland [6] for an Earth model 1066B [1]. Since this catalog does not extend to periods below 45 sec., we apply a $\cos^2$ taper to the spectra of observed and theoretical seismograms in the period range from 45 to 60 seconds, in order to avoid the ringing associated with an abrupt truncation. Because of the response of the instruments used, this leads to a rather narrow-band signal with the maximum energy at about 60 seconds. For this reason, our solution for the moment tensor will be characteristic of the source spectrum at a period of approximately 1 minute.

The signals used span the time from the arrival of P-wave until the arrival of the fundamental mode surface waves: Rayleigh for the vertical and longitudinal components and Love for the transverse component. An average Earth model would not be useful in the analysis of surface waves, as in the period range considered here their dispersion and amplitudes are too much perturbed by lateral heterogeneity.

The segments of records with obvious glitches are eliminated and a preliminary inversion is performed to examine the data for consistency; this allows us to discover, for example, reversed polarities of components and erroneous multiplexing. After these corrections are made, the data are inverted again to obtain the starting solution, f_i^0, and then we proceed with the iterative procedure for the simultaneous determination of the moment tensor and hypocentral parameters.

The solutions have been constrained to have a vanishing trace of the tensor ($f_1+f_2+f_3= 0$); in our experiments we have found no evidence that the isotropic component is necessary to satisfy the data. This does not mean that this component is not present; our data are so clearly dominated by shear energy that, at least at this stage, attempts to recover the monopole component of the source do not seem warranted.

In Figure 4 we compare the observed and computed seismograms for several stations characteristic of the azimuthal coverage for a deep (600 km) earthquake of April 24, 1979 under the Fiji islands. Distribution of all stations is shown in Figure 3; we have used the location of the event for this synthetic experiment described in the previous section. The azimuthal distribution of stations for this event is not well balanced: 8 out of 13 stations are in a single quadrant. This is the case for most of the earthquakes in the circum-Pacific belt. Addition of the proposed digital WWSSN stations (see Figure 1) would significantly

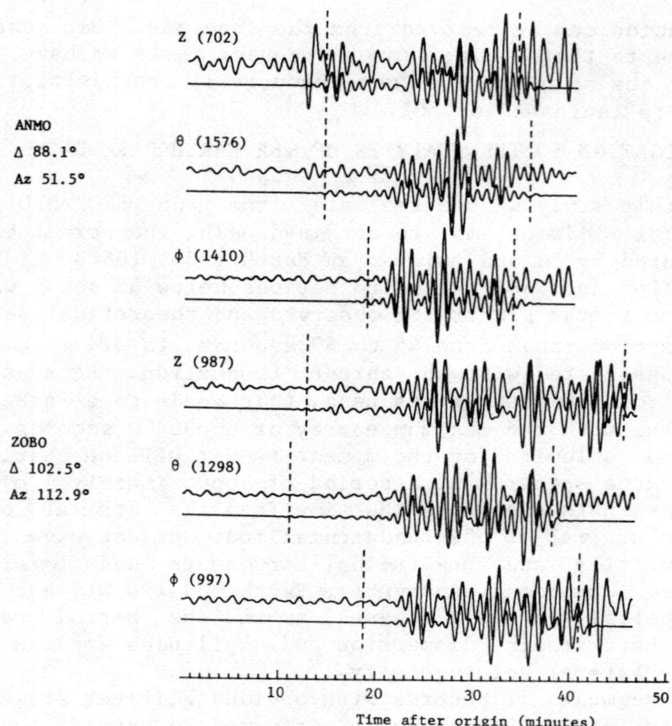

Figure 4. Comparison of the observed and synthetic seismograms for a deep (600 km) event of April 24, 1979 (Event #10 in Table 2). The top trace for each pair represents the observed seismogram, the bottom trace is the synthetic computed for the source mechanism obtained by the iterative inversion of the data from 12 SRO/ASRO stations. The scale is common for each pair of traces. The numerical factor corresponds to the maximum amplitude for a given pair of traces. It represents the digital count of the SRO instruments; the ASRO response has been appropriately normalized. Only the data between the vertical broken lines were used in the analysis. Notice the glitch associated with the P-wave arrival on the vertical component of station ANMO. This is a rather distant station ($\Delta = 88^{\circ}$) and the only one that exhibited the non-linear behavior, even though station SNZO is only 21° from the epicenter. This unpredictability of the non-linear response of the SRO stations is both puzzling and worrisome.

improve the azimuthal coverage; we consider early installation of these stations to be of significant benefit for scientific research.

Even though the moment of the earthquake was rather small (2.3×10^{25} dyne-cm), the SRO records are not free of nonlinearities: there is a large glitch at station ANMO, at an epicentral

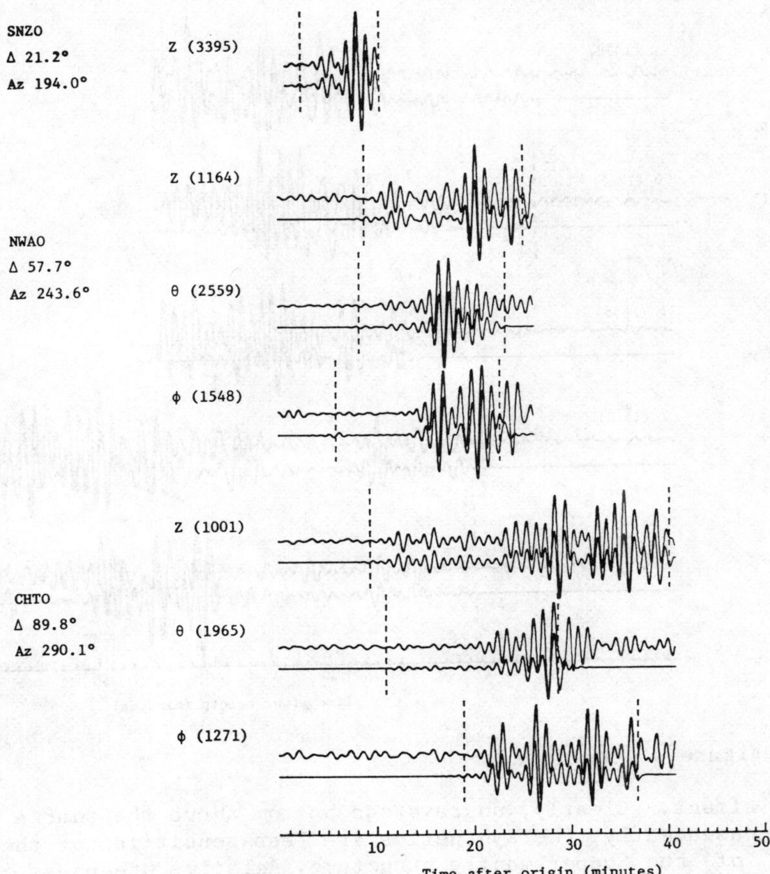

Figure 4 (continued)

distance of $88°$. This station yielded apparently good quality records for events many times larger and at comparable distance. The unpredictability of the nonlinear behavior of the SRO stations is rather puzzling.

The agreement between the observed and synthetic seismograms for this event is better than for any other case analyzed. We would like to draw the reader's attention to the vertical component records for stations SNZO and KONO; in the former case the observed and synthetic traces are identical, for all pratical purpose, but good reproduction of observed records for relatively small epicentral distances ($\Delta < 50°$) is fairly common. However, for the station KONO ($\Delta = 141°$) the duration of the analyzed record is approximately 45 minutes, yet the observed and synthetic records agree in all important details.

We think that the depth of an earthquake has much to do with

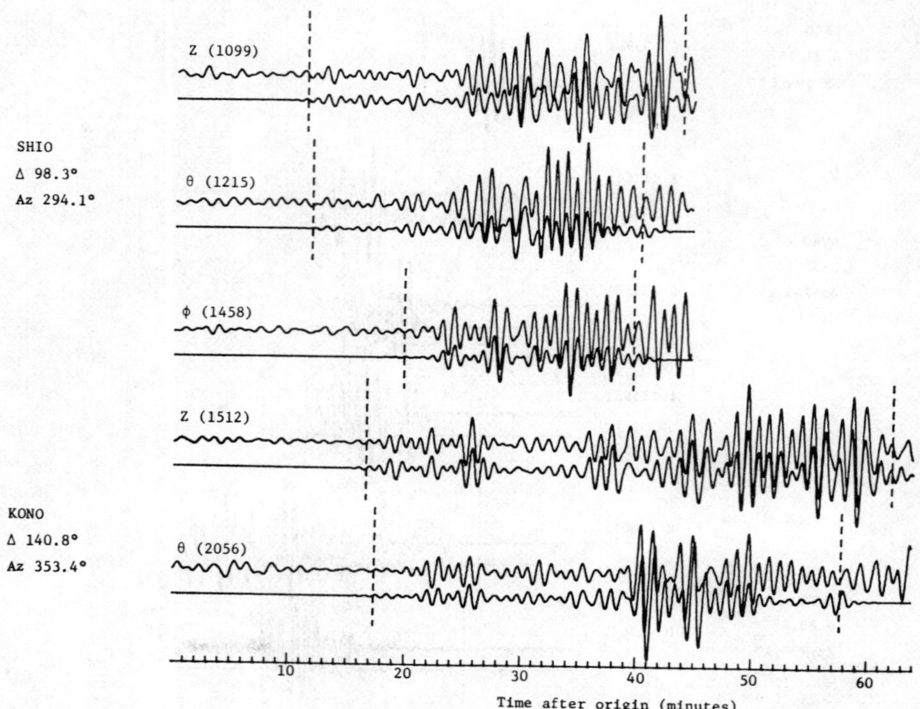

Figure 4 (continued)

this effect. Clearly, no rays can bottom above the source depth and, consequently, the synthetics are less sensitive to the details of the upper mantle structure. Relative steepness of the incidence angles at the surface also reduce the importance of P-S_V conversions which, we feel, are primarily responsible for the discrepancies between the observed and synthetic seismograms for earthquakes at shallower depths.

There is, however, a systematic misfit that seems to be common to all records of longitudinal components observed at distant stations ($\Delta > 80°$), regardless of the source depth. The synthetics predict amplitudes systematically too low at late times. This cannot be explained by an error in the source mechanism or the Q-structure of the model, since we would not then expect satisfactory agreement for the vertical and transverse components. We think that the explanation must be related to some signifcant difference between the elastic properties of the upper mantle or the crust of the real earth and the model used in this study. This observation supports the view that the investigation of source mechanism and the earth's structure cannot, and should not, be decoupled.

An interesting aspect of the result of the analysis for this event is the very small perturbation in the hypocentral parameters: only about 6 km in location and 3 sec in the origin time.

Table 2 summarizes results of our analysis for 14 earthquakes, arranged in a chronological order; one has already been discussed in some detail. Now we shall present a brief summary of the important characteristics of all the entries in the table.

Event #1. This deep event (439 km; M_b=6.9; N.E.I.S.) was large enough to be well recorded by the instruments of the IDA network (International Deployment of Accelerometers, [12]). Thus it provides an opportunity to compare solutions observed using an entirely different set of instruments, differnet range of periods and different time span of analyzed records.

Our moment is some 20 percent less than that obtained from IDA data by Masters and Gilbert [13]. This may be an effect of the finite duration of the source; 20 seconds would be sufficient to explain the difference. The plunges of the tension and compression axes are nearly identical, but their azimuth differ by 15°, on average; the solutions may thus be considered highly compatible. Professor H. Kanamori (personal communication) informed us that his solution for this event, also based on IDA data but obtained using a different procedure, is similar.

This means that the combined capability of the global digital seismograph network for the high quality analysis of seismic sources extends from events of seismic moment on the order of 10^{24} dyne-cm upwards.

Relatively large perturbations in hypocentral parameters should be considered with some caution as the network coverage has not yet achieved its full strength; the average number of stations used to analyze earthquakes in 1979 was roughly twice that used in the analysis of the Honshu event.

Event #2. An intermediate depth event under the Bonin Island (M_b=6.1; N.E.I.S.); approximately 30 km shift in epicentral location; depth decreased by 5 km. The source mechanism represents a superposition of a strike-slip (dominant) with a vertical dip slip; the M_{rr} component of the moment tensor is, essentially, zero. Relatively large intermediate eigenvalue.

There appears to be a tendency in the literature to decompose the set of principal values in terms of simpler mechanisms such as a "major" and "minor" double couple or superposition of a double couple with a "linear vector-dipole" [14,15]. This is a highly non-unique procedure which may tend to obscure the physical meaning of the solution.

We do not feel that such attempts at decomposition are particularly useful. One simple physical explanation of a non-vanishing intermediate eigenvalue is that the shear failure takes place on a curved surface and, subsequently, all three eigenvalues of the moment tensor assume finite values.

In the absence of the isotropic component, deviation from a simple double couple mechanism can be effectively characterized by a parameter $\varepsilon = \lambda_{min} / \lambda_{max}$; for a double couple, $\varepsilon = 0$; for

Table 2. Results of inversion for the moment tensor and hypocentral parameters for 14 earthquakes. The entries 7a and 7b represent results of the analysis for the same earthquake but using different time windows. A finite duration of the source (24 sec) has been assumed for Event #14; accordingly, 12 sec should be added to the value of t_0 to obtain a result comparable to that for all other events in the table. See the text for details.

	EVENT·NO.	1	2	3	4	5
	DATE	03/07/78	03/15/78	07/11/78	09/02/78	09/06/78
N.E.I.S.	Origin time (h:m:s)	2:48:47.6	22:04:40.1	12:17:7.8	1:57:33.4	11:07:43.1
	Latitude	32.01N	26.42N	7.89S	24.90N	13.32S
	Longitude	137.61E	140.56E	71.42W	121.99E	167.14E
	Depth (km)	439	263	645	109	198
Our Results	δt_0 (sec)	7.7±0.3	3.5±0.2	3.0±0.4	7.6±0.3	9.3±0.2
	Latitude	31.52N±0.02	26.34N±0.01	7.91S±0.05	24.50N±0.02	13.31S±0.02
	Longitude	136.99E±0.03	140.88E±0.02	71.22W±0.06	122.16E±0.03	166.67E±0.02
	Depth (km)	420±2.0	258±0.8	652±0.4	101±1.0	190±0.8
	Scale Factor (dyne-cm)	10^{26}	10^{25}	10^{24}	10^{25}	10^{26}
Moment Tensor	f_1	0.87±0.08	-0.07±0.07	-3.99±0.30	1.35±0.09	1.27±0.03
	f_2	0.07±0.10	-2.33±0.12	0.74±0.42	3.06±0.15	-0.36±0.05
	f_3	0.94±0.10	2.40±0.13	3.25±0.62	-4.41±0.16	-0.91±0.04
	f_4	1.53±0.09	-2.32±0.10	-1.25±0.21	1.71±0.10	-0.28±0.03
	f_5	-3.52±0.11	-2.27±0.09	4.42±0.28	-1.59±0.09	-0.79±0.03
	f_6	0.47±0.07	-3.77±0.11	-1.83±0.35	-1.50±0.14	0.80±0.05
T-axis	Moment	3.9	4.7	6.2	4.7	1.6
	Plunge/AZ	52/66	13/65	24/246	32/16	64/127
N-axis	Moment	0.2	1.3	-0.1	0.2	0.0
	Plunge/AZ	0/157	59/178	6/339	55/171	24/333
P-axis	Moment	-4.1	-6.0	-6.1	-4.9	-1.6
	Plunge/AZ	37/247	28/329	65/83	12/278	11/239
	No. of stations used	7	7	7	9	12

a linear vector-dipole, ε =0.5, the maximum for a moment tensor with no istropic component. For this particular event, ε =0.22.

Event #3. The deepest of the analyzed events (652 km; M_b=5.8; N.E.I.S.), but of rather small moment (6×10^{24} dyne-cm) and relatively low signal-to-noise ratio. Our solution is similar to nodal plane solutions obtained by Isacks and Molnar [16] for earthquakes in the same region and comparable depths. Even though data for only seven stations were available for the analysis, the perturbation in the epicentral coordinates is on the order of 20 km and 7 km in depth.

Event #4. North Taiwan; M_b = 6.1 (N.E.I.S.). Approximately 50 km shift in epicentral coordinates and an 8 km decrease in depth. A nearly perfect double couple solution with a large strike-slip component.

Event #5. An intermediate depth in the New Hebrides area (M_b=6.0; N.E.I.S.); a simple double couple earthquake.

Table 2 (continued)

	EVENT NO.	6	7a	7b	8	9
	DATE	09/15/78	09/23/78	09/23/78	10/01/78	12/18/78
N.E.I.S	Origin time (h:m:s)	11:39:25.1	16:32:11.1	16:32:11.1	13:23:50.1	10:15:53.1
	Latitude	48.25N	13.92S	13.92S	6.68N	54.48S
	Longitude	154.28E	167.21E	167.21E	123.98E	2.07E
	Depth (km)	44	201	201	46	10
Our Results	δt_0 (sec)	2.6±0.4	9.4±0.3	8.2±0.2	3.0±0.3	8.2±0.3
	Latitude	47.73N±0.03	13.94S±0.02	13.88S±0.01	6.71N±0.04	54.69S±0.03
	Longitude	155.77E±0.05	166.90E±0.02	166.93E±0.02	124.34E±0.03	2.63E±0.04
	Depth (km)	31±1.5	200±0.7	201	15±1.3	10
	Scale Factor (dyne-cm)	10^{25}	10^{26}	10^{26}	10^{24}	10^{26}
Moment Tensor	f_1	1.15±0.03	1.34±0.03	1.46±0.02	4.56±0.13	-0.08±0.05
	f_2	-0.27±0.03	-0.15±0.04	-0.25±0.04	-0.14±0.16	1.40±0.06
	f_3	-0.88±0.03	-1.19±0.04	-1.21±0.04	-4.42±0.18	-1.32±0.06
	f_4	0.09±0.05	-0.22±0.03	-0.24±0.03	0.64±0.38	-0.08±0.05
	f_5	0.23±0.06	-0.13±0.03	-0.16±0.03	-2.72±0.76	0.30±0.05
	f_6	-0.44±0.03	0.50±0.04	0.46±0.04	1.03±0.14	0.07±0.05
Principal Axes — T-axis	Moment / Plunge/AZ	1.1 / 84/286	1.4 / 79/154	1.5 / 80/152	5.4 / 74/76	1.4 / 3/179
N-axis	Moment / Plunge/AZ	0.0 / 1/27	0.0 / 11/333	-0.1 / 10/338	0.0 / 0/67	0.0 / 77/281
P-axis	Moment / Plunge/AZ	-1.1 / 6/117	-1.4 / 0/248	-1.4 / 1/248	-5.4 / 15/257	-1.4 / 13/88
	No. of stations used	15	14	14	14	12

	EVENT NO.	10	11	12	13	14
	DATE	04/24/79	05/13/79	06/25/79	08/05/79	11/23/79
N.E.I.S	Origin time (h:m:s)	1:45:10.0	17:30:56.8	5:29:5.6	0:53:54.8	23:40:29.8
	Latitude	20.84S	4.05S	4.98S	22.72S	4.80N
	Longitude	178.64W	123.15E	145.58E	177.51W	76.22W
	Depth (km)	599	615	189	250	108
Our Results	δt_0 (sec)	3.2±0.2	2.9±0.3	10.0±0.2	2.9±0.2	3.5
	Latitude	20.79S±0.02	4.05S±0.02	4.93S±0.02	22.93S±0.03	4.86N±0.02
	Longitude	178.61W±0.02	122.95E±0.03	145.60E±0.03	177.63W±0.02	75.59W±0.03
	Depth (km)	600±1.7	616±1.8	191±0.8	225±1.0	116±1.2
	Scale Factor (dyne-cm)	10^{25}	10^{25}	10^{26}	10^{25}	10^{26}
Moment Tensor	f_1	-1.83±0.04	0.07±0.02	-0.97±0.03	-3.99±0.12	-4.11±0.13
	f_2	1.45±0.07	-0.39±0.02	0.50±0.03	1.46±0.24	-3.30±0.24
	f_3	0.38±0.07	0.32±0.02	0.47±0.04	2.53±0.20	7.41±0.26
	f_4	-1.59±0.06	0.26±0.02	0.61±0.02	-4.87±0.13	-2.69±0.12
	f_5	-0.61±0.05	-0.98±0.02	-0.22±0.03	-4.27±0.14	-3.92±0.13
	f_6	0.11±0.05	0.10±0.02	-0.48±0.05	0.28±0.16	1.18±0.18
Principal Axes — T-axis	Moment / Plunge/AZ	2.1 / 23/168	1.2 / 42/85	1.1 / 16/40	5.9 / 33/131	8.7 / 18/95
N-axis	Moment / Plunge/AZ	0.4 / 5/76	-0.3 / 17/339	0.0 / 11/307	2.3 / 5/224	-1.6 / 31/196
P-axis	Moment / Plunge/AZ	-2.5 / 66/334	-0.9 / 43/233	-1.1 / 71/185	-8.2 / 57/321	-7.1 / 53/339
	No. of stations used	12	12	13	11	12

Event #6. Kurile Islands (M_b=6.0; N.E.I.S.). A major shift in epicentral coordinates; approximately 140 km to the south-east. A double-couple solution with the horizontal axis of compression perpendicular to the trench axis and nearly vertical axis of the tension; the null axis is also horizontal and parallel to the trench axis. The focal depth of this earthquake is 31 km (reduced from 44, reported by the N.E.I.S.). If our relocation of this event is correct, then it would be associated with the outer rise, and its mechanism (reverse faulting) would be consistent with the stress regime in the lower part of the buckled lithosphere (c.f. Chen and Forsyth, [17]).

Event #7(a and b). This intermediate depth earthquake (M_b= 6.3) is a "carbon copy" of the Event #5. Both of these earthquakes have similar moments, orientation of the principal axes and, what is perhaps the most important, very similar perturbation in the epicentral coordinates. Thus even if our relocation procedure is biased by the lateral heterogeneities or erroneous features in the reference earth model, the relative locations appear to be highly reliable if the network coverage is good. The entry (a) corresponds to the analysis which included late segments of seismograms with, on occasion, an obvious disagreement between the observed seismograms and synthetics; these segments have been truncated in the inversion for which the results are presented as the entry (b). The results are very similar, including the estimates of the standard errors; the smaller average deviation obtained from the analysis (b) has been offset by the reduced number of the data points.

Event #8. A shallow event off the south-west of the Mindanao Island (M_b=5.6; N.E.I.S.). Major decrease in depth from the estimate given by the N.E.I.S. (reduction from 46 to 15 km). Notice that because of the shallowness of the event the errors associated with the f_4 and f_5 ($M_{r\theta}$ and $M_{r\phi}$) elements of the moment tensor are several times greater than those for other components. Even so, the normal fault mechanism is clearly dominant. The seismic moment (5×10^{24} dyne-cm) is the smallest of all events analyzed in this study.

Event #9. This event near Bouvet Island in the South Atlantic can be characterized as a "slow" earthquake, because of a large difference between the M_b (5.5) and M_s (6.5) magnitudes. This earthquake has also been analyzed by Kanamori [18], who used the IDA data in his study of the excitation of the long period surface waves (200-300 sec.). His estimate of the seismic moment (2×10^{26} dyne-cm) is higher than ours: 1.4×10^{26} dyne-cm, which is consistent with the "infra-red" character of earthquakes associated with spreading centers [19]. There was a major change in epicentral coordinates reported by N.E.I.S. in the P.D.E. cards (55.4S, 1.1E) and in the Monthly Bulletin (54.47S, 2.07E). Our inversion procedure, in which we started with the P.D.E. coordinates, converged to a location at 54.69S and 2.63E, some 40 km from the improved estimate but about 150 km from the initial

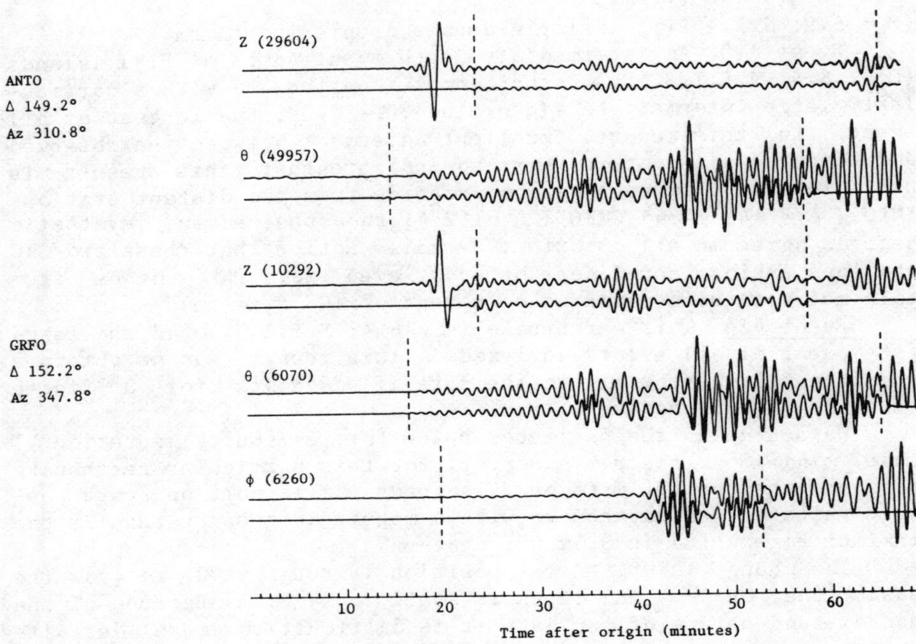

Figure 5. Comparison of observed and synthetic seismograms for stations ANTO and GRFO for an intermediate depth earthquake of August 5, 1979 (Event #13 in Table 2). The top trace for each pair represents the observed seismogram, the bottom trace is the synthetic computed by the iterative solution. For additional details see the caption to Figure 4. Our source mechanism for this event shows a particularly large deviation from a simple double couple mechanism. The agreement between the observed and synthetic seismograms is very good; at these large distances (150°) the traces contain significant contributions from waves travelling along the major arc. Notice that even for PKP waves the SRO stations exhibit non-linear response.

P.D.E. location. The mechanism is dominated by the strike-slip component. Despite a very shallow depth, fixed at 10 km, the f_4 and f_5 components remained stable.

$\underline{\text{Event #10}}$. Deep event under Fiji Islands discussed above (see $\underline{\text{Figure 4}}$).

$\underline{\text{Event #11}}$. Deep event under the Banda sea (M_b=5.7; N.E.I.S.). As for all deep earthquakes analyzed by us, the change in the hypocentral coordinates is rather small. This event had, we believe, a resolvable intermediate eigenvalue: $\varepsilon = 0.22$. The dominant component of the moment tensor is f_5, indicating that the slip occurred either in nearly vertical or horizontal plane with the null axis pointing $20°$ NW.

$\underline{\text{Event #12}}$. An intermediate depth earthquake under New Guinea

(M_b = 6.2; N.E.I.S.); a simple double couple mechanism.

Event #13. An intermediate depth event south of Fiji Islands (M_b = 6.4; N.E.I.S.). A rather complex earthquake with a particularly large intermediate eigenvalue : ε = 0.28, the largest of all events in this report. Excellent agreement between the observed and computed seismograms seems to indicate that this result is reliable. In Figure 5 we show records from two distant stations ANTO (Δ = 149°) and GRFO (Δ =152°); the observed and synthetic records agree in all important details. Notice that these two SRO stations exhibit non-linear behavior even for PKP phases from this moderate size earthquake (moment 7×10^{25} dyne-cm).

Event #14. This earthquake (m_b = 6.4; N.E.I.S.) had the largest moment of all events analyzed in this report, and we obtain a significantly better fit to the data if we allow for a finite duration, T, of the event.

We search in the parameter space for a source duration T that minimizes the r.m.s. error. For this particular earthquake we found the optimal T to be 24 seconds. It is not an unreasonable value for an earthquake with a magnitude greater than 7: the maximum eigenvalue is 8.7×10^{26} dyne-cm.

The change in epicentral position is roughly 70 km to the east, the change in depth is small: only 8 km. Because of the narrow-band nature of our data it is difficult to make definite statements with respect to the faulting history. The relatively large value of ε (0.18) may be indicative of the complexity of faulting geometry. Undoubtedly analysis of the IDA data for this earthquake should be very helpful, as it was certainly large enough to excite a rich normal mode spectrum.

5. DISCUSSION AND CONCLUSIONS

In the introduction we have stated several objectives that should be met by a method of quantative analysis of the data from a global digital seismograph network. We think that the examples of application shown in this paper demonstrate that the approach proposed by us does indeed satisfy these requirements.

The time window and the frequency band chosen for the analysis lead to repeatable solutions for the moment tensor even when there is a marked disagreement between the observed and synthetic seismograms for a significant part of the total data set used in the inversion.

An earth model obtained by inversion of a large body of reliable seismological data, such as model 1066B used throughout this study, is bound to match well the early part of observed traces; the level of agreement at late times and large distances depends strongly on the source depth. For very deep earthquakes the agreement is generally excellent as the waveforms are weaker functionals of the upper mantle structure. The disagreement at late times may be related in part to the effect of lateral heterogeneities and in part to discrepancies between the reference earth model and the optimal "average earth" structure. Systematically too small amplitudes predicted for the longitudinal components at late times and large distances indicate that this may be the case. Our results demonstrate that we have a good "bootstrap" for a future joint inversion for the earthquake parameters and perturbation in the Earth's structure.

The seismic moments of the events analyzed in this report ranged from 5×10^{24} dyne-cm to 8×10^{26} dyne-cm. The lower range could be extended by incorporating higher frequencies in the analysis, although this would increase markedly the required expenditure of the computer time. The other possibility is to calibrate the phase velocity dispersion and apparent attenuation of the fundamental mode surface waves for a particular source-receiver path using the moment tensor obtained by the method described in this paper.

In their present configuration, the SRO instruments are too non-linear to allow the analysis of the early parts of the recordings of the events with moments significantly exceeding 10^{27} dyne-cm. These can be studied using later parts of the recordings, sometimes beginning hours after the origin time [13,18,20], but the loss of the information on the properties of the source spectrum at periods below, say, 100 seconds is irreversible. It is regrettable that the recommendations of the Panel on Global Seismograph Network [21] with respect to the need for on-scale recording have not been followed.

For the set of 14 earthquakes our solutions for the moment tensor are highly consistent with those obtained by other researchers using different methods and data sources (nodal plane solutions, inversion for the moment tensor using ultra-long period data from IDA network). In this sense, there were no surprises, but rather we have demonstrated that our approach gives highly reliable solutions even at shallow depths, at which inversions based on long period data encounter difficulties with respect to the $M_{r\theta}$ and $M_{r\phi}$ components.

Our most important finding with respect to the nature of source mechanisms is that a substantial number of earthquakes of intermediate and great depth have significant non-double couple components. Of nine events in this depth range, four have substantial values of ε ($\varepsilon = |\lambda_{min} / \lambda_{max}|$) ranging from 0.16 to 0.28, and for five ε is very small (from 0 to 0.05). Admittedly the population is small, but the indication is that the frequency of deviation from the double couple mechanism is quite high. The only pair of earthquakes in the immediate vicinity of each other (Events #5 and #7) available in this study yielded nearly identical solutions, and in both cases ε was essentially zero (pure double couple). At the same time, nearly all shallow earthquakes yielded essentially pure double couple solutions.

The novel element of this study has been the expansion of the inverse problem to include perturbations in the hypocentral parameters. Ideally, we would like to distinguish between the location and time of the initiation of rupture and the centroid of the stress glut; such information would be invaluable in studies on the nature of earthquakes, particularly at depths and locations that make other methods of estimating the source dimension inapplicable.

Unfortunately, the absolute locations obtained by the analysis of the arrival times of P-waves have uncertainty that may reach several tens of kilometers, and on occasion our procedure shows changes in location that seem excessive and, in the case of poor station coverage, very sensitive to the particular distribution of receivers. Thus, at this stage the results are of a purely exploratory nature. However, two of the parameters that are determined by the inverse procedure seem to yield reliable and useful information even now.

Hypocentral depth. The energy radiated upward from a source and reflected from the surface is approximately equal to that radiated downward. Thus our excitation kernels are highly sensitive to the source depth as they contain the complete suite of depth phase. On the basis of comparison with the standard depth determinations, we infer that our solutions are reliable. In instances of a major revision in depth estimates between the PDE and Monthly Bulletins issued by the N.E.I.S., our results show better agreement with the improved estimate, even if we use the PDE parameters as the starting location. With few exceptions, our results show a systematic shift towards shallower depths; this is particularly apparent for the shallow earthquakes. Thus, the ability of the method to provide an objective depth determination should be considered one of its most important immediately applicable features. There is no reason to think that the depth determination would be potentially as sensitive to the influence of lateral heterogeneities as the epicentral coordinates and, perhaps, the origin time might be .

Origin time. Perturbations in the origin time, δt_0, are positive for all events analyzed in this report, indicating the finite duration of the analyzed sources. There is an overall increase of δt_0 with the increase in the moment, as should be expected.

In this study we have used an earth model that was derived before the frequency dependence of the elastic parameters due to anelastic attenuation was widely recognized [22]. In the future, it will be desirable to use a model in which this effect is considered, as this may affect the estimates of δt_0.

Epicentral coordinates. This is perhaps the most uncertain aspect of our results. In some cases, the perturbation with respect to the standard locations is very small: for event #10 it was aproximately 6 km. It is also reasonable ($\sim$20 km) for two other deep events (#3 and #11). However, for several shallow earthquakes it appears to be excessive: see Event #6, in particular. The relative locations seem to be more consistent. Two close intermediate depth events (#5 and #7) show very similar perturbation in their location.

We expect that the effect of lateral heterogeneities is the main source of error. This, in part, can be compensated for by the improved network coverage. The planned deployment of WWSSN stations with digital recording should be extremely helpful in this respect, particularly in terms of a better balanced azimuthal distribution of receivers.

Although we admit that some of our relocations should be considered with caution, we believe that it is very important that the problem be studied further. As the standard locations are not free from systematic errors due to lateral heterogeneities, results obtained using our method offer an opportunity to look at the source through a different filter; the wavelengths used in our study are one to two orders of magnitude greater. The effect of lateral heterogeneities associated, for example, with subducted slabs may be different and, perhaps, reduced if one looks at the source with wavelengths of several hundred kilometers.

Even though elimination of some of the technical problems in the performance of the SRO/ASRO network would allow us to save much effort in the data editing stage, this research would not have been possible without the high quality data base that this network provides. The IDA network is essential for monitoring large magnitude events; there is a range of overlap between this source of data for ultra-long periods and our application of the SRO/ASRO data that is sufficient to establish that we are able to consistently study source mechanism of earthquakes with moments ranging from 10^{24} to 10^{30} dyne-cm. We hope that the global digital seismograph network will be maintained in the future; addition of some 20 WWSSN stations with digital recording capability would be extremely valuable, as for studies of events of moderate magnitudes there now exists a rather severe imbalance in the geographical distribution of receivers with digital recording.

REFERENCES

1. Gilbert, F. and Dziewonski, A.M.: 1975, Phil. Trans. R. Soc., Lond., A 278, pp. 187-269.
2. Buland, R. and Gilbert, F.: 1976, Geophys. Res. Lett., 3, pp. 205-206.
3. Peterson, J., Bulter, H.M., Holcomb, L.G., and Hutt, C.R.: 1976, Bull. Seism. Soc. Am., 66, pp. 2049-2068.
4. Dziewonski, A.M.: 1978, EOS, 59, pp. 325.
5. Dziewonski, A.M.: 1978, Seismic Discrimination, Semiannual Technical Summary, September 30, 1978, Lincoln Laboratory, M.I.T., pp. 40-42.
6. Buland, R.P.: 1976, Ph.D. Thesis, University of Califorina, San diego.

7. Dziewonski, A.M.: 1977, Seismic Discrimination, Semiannual Technical Summary, March 31, 1977, Lincoln Laboratory, M.I.T., pp. 94-98.
8. Gilbert, F. and Buland, R.P.: 1976, Geophys. J. R. astr. Soc., 50, pp. 251-255.
9. Backus, G.E. and Mulcahy, M.: 1976, Geophys. J. R. astr. Soc., 46, pp. 341-361.
10. Backus, G.E.: 1977, Geophys. J. R. astr. Soc., 51, pp. 1-25.
11. Woodhouse, J.H.: 1980, Seismic Discrimination, Semiannual Technical Summary, March 31, 1980, Lincoln Laboratory, M.I.T., in press.
12. Agnew, D., Berger, J., Buland, R., Farrell, W. and Gilbert, F.: 1976, Trans. Am. Geophys. Un., 57, pp. 180-188.
13. Masters, G. and Gilbert, F.: 1979, EOS, 60, pp. 879.
14. Knopoff, L. and Randall, M.J.: 1970, J. Geophys. Res., 75, pp. 4957-4963.
15. Randall, M.J. and Knopoff, L.: 1970, J. Geophys. Res., 75, pp. 4965-4976.
16. Isacks, B. and Molnar, P.: 1971, Rev. Geophys. Space Phys., 9, pp. 103-174.
17. Chen, T. and Forsyth, D.W.: 1978, J. Geophys. Res., 83, pp. 4995-5003.
18. Kanamori, H.: 1980, EOS, 61, pp. 296.
19. Kanamori, H. and Stewart, G.S.: 1976, Phys. Earth Planet. Interiors, 11, pp. 312-332.
20. Dziewonski, A.M. and Gilbert, F.: 1974, Nature, 247, pp. 185-188.
21. Engdahl, E.R.: 1977, Report of the Panel on Seismograph Networks, Committee on Seismology, National Academy of Sciences, Washington, D.C.
22. Liu, H.-P., Anderson, D.L., and Kanamori, H.: 1976, Geophys. J. R. astr. Soc., 47, pp. 41-58.

THE EFFECT OF GREEN'S FUNCTIONS ON THE DETERMINATION OF SOURCE
MECHANISMS BY THE LINEAR INVERSION OF SEISMOGRAMS

Brian W. Stump

Air Force Weapons Lab/NTE, Kirtland AFB NM 87117

Lane R. Johnson

Dept of Geology and Geophysics, University of
California, Berkeley CA 94720

The moment tensor representation of seismic sources is applied
to some close-in (2-13 km) accelerometer data of contained
nuclear explosions. The two explosions studied are Handley and
Pipkin, both detonated on Pahute Mesa at the Nevada Test Site.
A variety of Green's functions are used in the inversions for
the moment tensor to test the sensitivity of the procedure to
assumed propagation path effects. The simplest model consisted
of a half-space structure with the most complicated a three-
layer over a half-space. The resulting source functions are
dominated by the isotropic component of the moment tensor with
the deviatoric component as much as a factor of twenty-five
smaller. The time function of the source appears to have a
double peaked nature. The fit of the calculated seismograms to
the observed is not greatly affected by the assumed Green's
function.

E. S. Husebye and S. Mykkeltveit (eds.), Identification of Seismic Sources - Earthquake or Underground
Explosion, 255–267.

I.   Seismic Source Representation

Given a set of observed waveforms from an explosion or earthquake and knowing the propagation path effects of the geological material, a quantitative procedure for determining the source function in space and time is desirable. The moment tensor source representation is such a formulation (Gilbert, 1970; Gilbert and Dziewonski, 1975; Backus and Mulcahy, 1976 a and b; Stump and Johnson, 1977; Backus 1977 a and b; Strelitz, 1977).

If an explosion or earthquake source can be represented as a set of equivalent body forces, then the source can be written as a series of moments. For small sources or large wavelengths only the first term of the series is retained and the displacements at any point and time can be written as:

$$u_k(\underline{x}',t') = G_{ki,j}(\underline{x}',t';\underline{0},o) \otimes M_{ij}(\underline{0},t') \tag{1}$$

Where u_k is the displacement in the k direction, G_{ki} is the Green's function, M_{ij} is the moment tensor, ,j indicates derivative with respect to x_j, and $\otimes$ represents temporal convolution. A more complete derivation of these results is given in Stump and Johnson (1977).

In the frequency domain, the equation becomes:

$$u_k(\underline{x}',f) = G_{ki,j}(\underline{x}',f;0,o) \; M_{ij}(\underline{0},f) \tag{2}$$

Knowing the propagation path effects ($G_{ki,j}$), one can determine the source (M_{ij}) from a set of observational data (u_k).

In this study, the moment tensor formulation will be applied to a set of data from contained nuclear explosions. The sum of the diagonal elements of the moment tensor, the isotropic component, is proportional to volume changes in the source region and will be interpreted as the explosion component of the source. This component can then be compared to the remaining portion of the moment tensor, its deviatoric component, to analyze that portion of the source not accounted for by a symmetric explosion.

II.   Data Set:

The two underground nuclear explosions from NTS used in this study were the Handley event detonated on 26 March 1970 with a Wood-Anderson magnitude at the U.C. Berkeley Seismographic Station of 6.3 and the Pipkin explosion of 8 October 1969 with a magnitude of 5.5. The locations of these two events are given in Figure 1. The depth of the Handley shot was 1.2 km while that for Pipkin was 0.6 km.

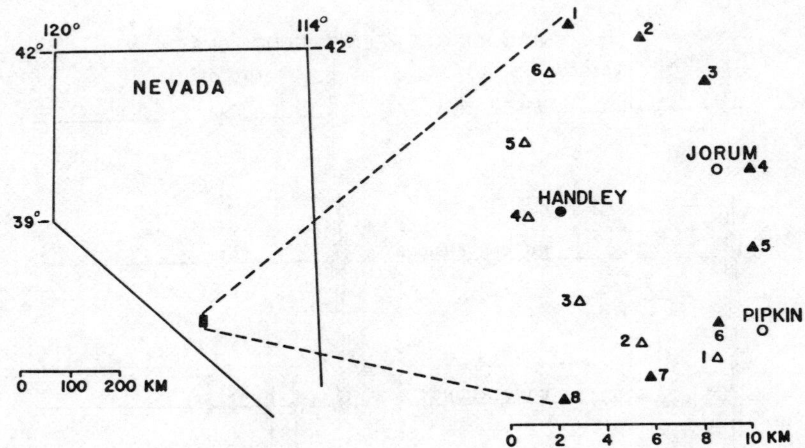

Figure 1. The location of the Nevada Test Site within the state
of Nevada. The location of the Handley, Jorum, and Pipkin tests.
The solid triangles were recording the Handley event while the
open triangles recorded Jorum and Pipkin.

Two arrays of three-component accelerometers were installed
to record the underground tests Jorum and Handley (Figure 1).
While the instruments were recording for the Jorum event, the
Pipkin test was detonated and thus recorded. The instruments
were all placed at eight kilometers from Handley and Jorum.
Three-component data from stations at the same distance, but a
variety of azimuths, was obtained from Handley, and data at a
variety of distances and azimuths was obtained from Pipkin. A
total of twelve components of data were recorded for Handley and
eleven for Pipkin.

The instruments used in these experiments were accelerometers
with a flat acceleration response between 0.02 to 50 Hz and a
clipping level of $\pm \frac{1}{2}$ g. The seismograms were recorded on analog
tapes. A low pass anti-alias filter (four-pole Butterworth) at
10 Hz was applied to the data prior to digitization at 54 samples/
second.

III. Green's Functions:

The purpose of this study was not only to investigate the
explosion source mechanism, but to also determine the sensitivity
of the moment tensor procedure to the assumed propagation path
effects.

Figure 2 illustrates the vertical and radial explosion
Green's functions for the models tested. The responses have

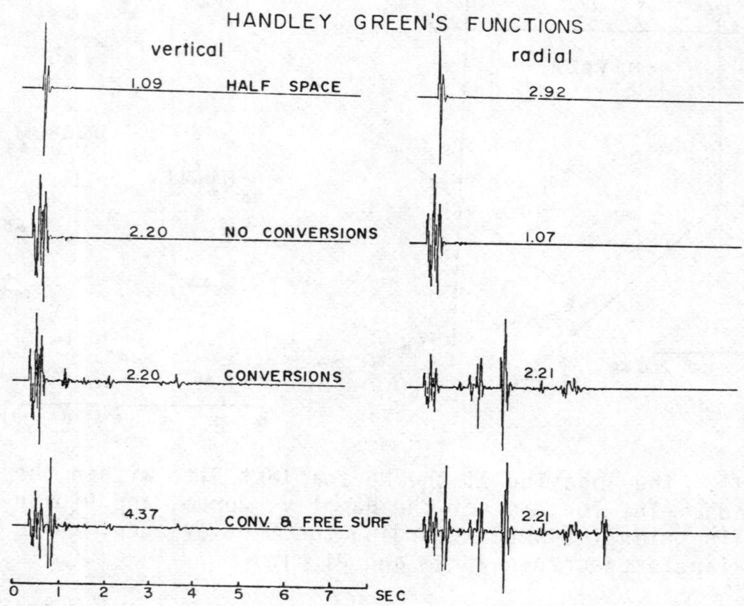

Figure 2. The radial and vertical explosion Green's functions
for the four models considered. The accelerometer response has
been convolved with the Green's function.

been convolved with the accelerometer. The simplest synthetic
was that of an elastic half-space (Johnson, 1974) with the velocities chosen to match the arrival times observed in the data
(Table 1). Since the initial compressional pulse from the source
is traveling nearly horizontally in this model, the radial component of motion is approximately a factor of three larger than the
vertical. In the observational data, the vertical motion is larger
than the radial indicating that the approach angle of the compressional waves is more nearly vertical. In order to model this
phenomena, a layered structure was introduced (Table 1). This
velocity model is a result of well logs in the Pahute Mesa area,
first arrival data for shots in the mesa and results of other
researchers (Hamilton and Healy, 1969; Hadley and Helmberger,
1980).

        The synthetics for these layered structures are calculated
using the generalized ray method (Helmberger, 1968; Pao and
Gajewski, 1977). The no conversions model includes only rays
which make no P to S or S to P conversions for this layered structure. Both up-going and down-going energy are included in the
calculation. The inclusion of the down-going energy yields a

TABLE 1

HALF-SPACE MODEL:

   V_p = 3.34 km/sec V_s = 1.93 km/sec ρ = 2.67 gm/cc

LAYERED MODEL:

H_1 = 0.6 km	V_p = 2.7 km/sec	V_s = 0.8 km/sec	ρ = 2.6 gm/cc
H_2 = 0.8 km	V_p = 3.4 km/sec	V_s = 1.2 km/sec	ρ = 2.7 gm/cc
H_3 = 1.6 km	V_p = 3.8 km/sec	V_s = 1.8 km/sec	ρ = 2.8 gm/cc
H_4 = ∞	V_p = 4.4 km/sec	V_s = 2.5 km/sec	ρ = 2.82 gm/cc

vertical to radial ratio in the synthetic explosion quite close to the data.

The third model utilized the same layered structure, but the conversions at the interfaces were included. The effect of the added rays can be seen primarily on the radial component of motion. The final model was the same as the third except the effect of the free surface was added.

The four sets of Green's functions have a wide range of complexity from the single pulse half-space response to the complete response of the layered material which has significant energy past four seconds.

IV.   Synthetic Fits:

Not only will the inversion of the Handley data set allow one to study explosion source functions, but it will give some feel for our ability to fit observational accelerograms at relatively close distances. Approximately five seconds of data was inverted in this exercise. The results of the inversion using the half-space model are shown in Figure 3. The initial P pulse on the vertical component of motion is well matched with the radial and transverse components yielding somewhat degraded fits. The large transverse components of motion not expected from a spherically symmetric explosion are modeled.

For purposes of comparison of the effects of the various Green's functions on the fits attention is focused on the vertical accelerogram at station 4, Z4. Figure 4 illustrates the observed seismogram Z4 and the fits to this data using the four Green's functions. There is little change in the calculated seismograms from the four inversions. The largest change occurs between the half-space and the layered model with no conversions.

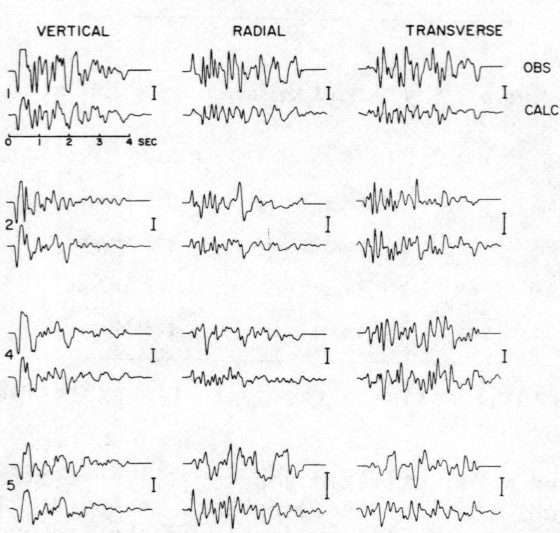

Figure 3. The observed seismograms (top) and fitted seismograms (bottom) using the moment tensor from the inversion of the Handley data using the half-space Green's function. The bars on each pair of seismograms are equal to 250 cm/sec^2.

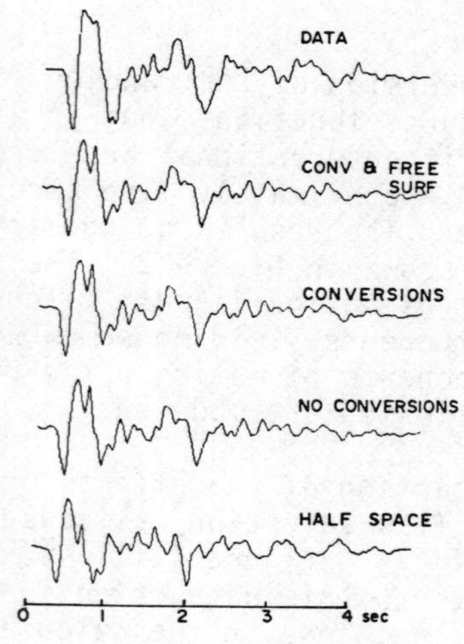

Figure 4. The observed and fitted Z4 accelerograms using the four different Green's functions.

Utilizing the degree of fit of the calculated accelerograms to the observations, one cannot distinguish the adequacy of the four propagation path models. The Green's functions are quite different for the four models; thus, the source and propagation path effects are trading off with one another while the fits to the data remain stable. The conclusion for this data set is that once the observations are well modeled one cannot assume that the source and propagation path effects are understood. One can fit complex, near-source accelerograms with only a simple half-space Green's function.

V.   Source Functions:

In order to judge the adequacy of each of the inversions, one must look at the Green's functions, source functions, and the fit to the observational data. It is these three factors, in conjunction with peripheral data such as reflection/refraction surveys in the area and field observations of the source, which allow one to judge the uniqueness of the solution.

The isotropic source spectra (far-field displacement) for the four Green's functions are given in Figure 5. The solid lines in each figure are the source spectra, while the dashed lines are the estimates of the variances. All four source models have high frequency slopes of three; although as one approaches the spectral corner at 1-3 Hz, these roll-offs become gentler. The Green's functions for the half-space model have an interference hole in the spectrum at 4 Hz. Since the data exhibits no such hole, a large peak at 4 Hz results in the moment tensor. This phenomenon is not seen in the more complex models where additional rays are included. Because of this spectral peak, the half-space result yields a 3 Hz corner while the other models have corners closer to 1 Hz.

The long period level of the spectra decreases as more rays are added to the response functions. In the half-space result, only up-going energy can reach the receiver. Both up-going and down-going energy are included in the layered structures. Since more energy can reach the receivers in the layered structure, the size of the source term needed to match the observed waveforms is reduced. A reduction in the isotropic component of the moment tensor of four is seen in comparing the half-space and the con-versions free surface models.

It appears in Figure 5 that, as more rays are included in the response functions, the spectra become peaked. Signal to noise studies of the data indicate that the energy at the long periods (5 seconds) is close to the noise level. Conclusions concerning the long period of the source spectra are, therefore, tentative.

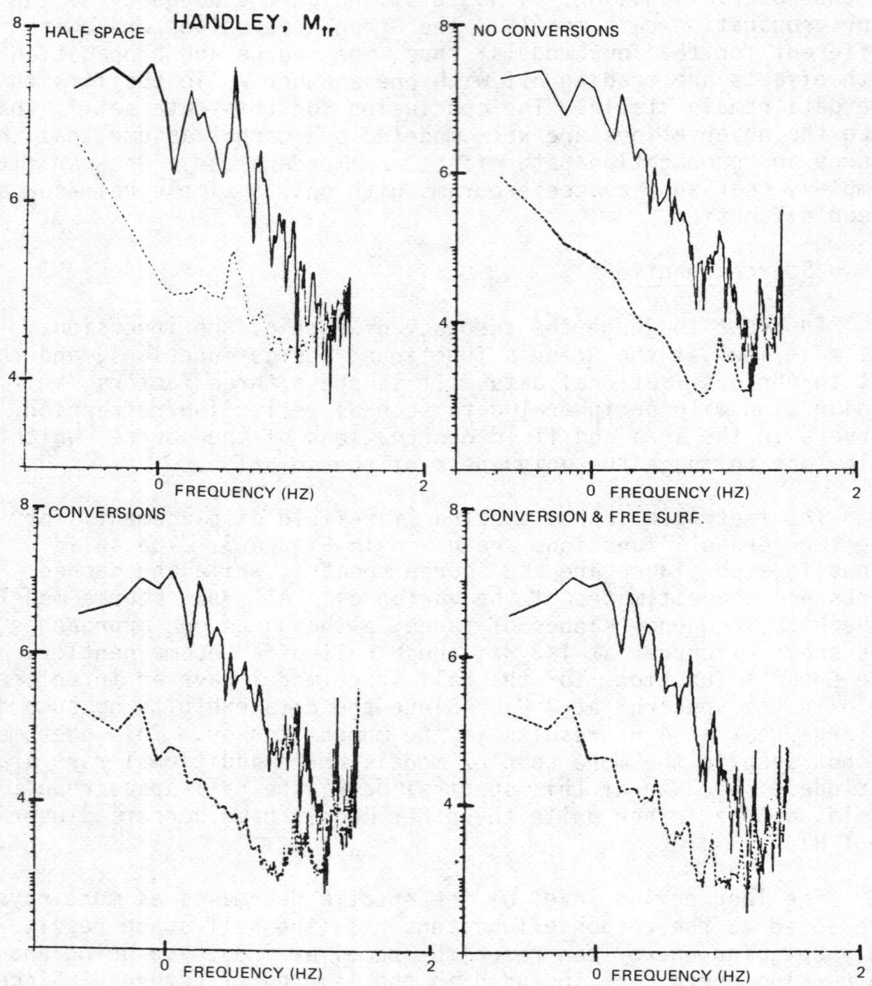

Figure 5. The Isotropic far-field moment tensors in the frequency domain resulting from the four different Green's functions. All modulus values should be multiplied by 8.84×10^{16} dyne-cm/sec.

The importance of the deviatoric component of the moment tensor and the symmetry of the explosion source are summarized in Table 2. The values in the table are determined from the long period level of the moment tensor spectrum. In the half-space Green's function, the symmetric explosion results in a much larger radial component of motion than vertical. The large

TABLE 2

FAR FIELD MOMENTS

HANDLEY

	M_{TR}	M_{TR}/MAX DEV	% ERROR IN SYMMETRY
Half Space	4.8×10^{24} $\frac{\text{dyne-cm}}{\text{sec}}$	3.7	48.3%
No Conversions	2.0×10^{24}	25.0	1.7%
Conversions	1.3×10^{24}	25.0	3.7%
Conversions and Free Surface	1.2×10^{24}	23.0	5.0%

vertical accelerations observed are matched in the inversion by introducing a large vertical dipole. The 48.3 percent error in the source symmetry reflects this fact. The error was the largest deviation of any of the diagonal elements of the moment tensor from the isotropic component. Inclusion of turning rays below the source depth in the layered Green's functions greatly improved the source symmetry. The layered Green's functions also reduced the sizes of the deviatoric components of the moment tensor. With the deviatoric component a factor of 25 smaller than the isotropic, the transverse accelerograms are modeled.

The far-field and near-field isotropic displacement time functions resulting from the four models are given in Figure 6. The effect of the 4 Hz spike in the half-space results can be seen. It is felt that this noncausal time signal is a result of interference effects between the limited number of arrivals in the half-space Green's functions.

The near-field time functions for all four models indicate a double pulsed nature to the source, the second pulse following the first by 1.2 seconds. The time functions have no permanent displacement, but, because the data used to determine the source function were accelerograms, no information about this part of the source is available. The long period motion between the first and second pulse in the waveforms is similarly suspect.

VI.   Pipkin Data:

In studying the effects of the Green's functions, primary emphasis has been placed on the Handley data set. Similar results have come out of the Pipkin data. A comparison of the isotropic component of the source for Handley and Pipkin is given in Figure 7. The layered Green's function including conversions and the free surface was used.

HANDLEY M_{TR}

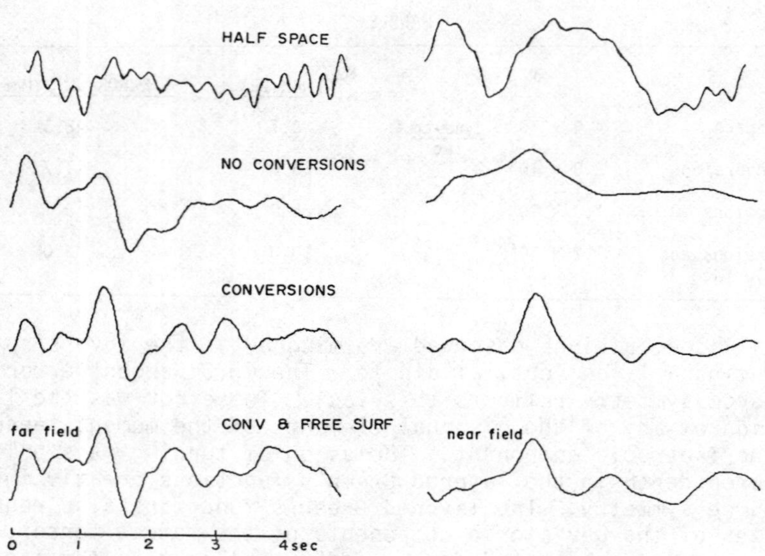

Figure 6. Far-field and near-field moment tensors in the time domain for the four different Green's functions.

The source functions are nearly identical except for the scale factor which is reflective of the yield variation between the two events. The Handley explosion gives a near-field moment of 3.4×10^{23} dyne-cm and that for Pipkin is 3.1×10^{22} dyne-cm. The corner frequencies, high frequency roll-off, and spectral shape are quite similar. The double source is retained.

VII. Conclusions:

The results of this study indicate that one can fit the observed close-in accelerograms from underground contained explosions. The fits of the observational data are relatively insensitive to the Green's function chosen to model propagation path effects. The match of the calculated waveforms to the observations is not a good measure of the adequacy of either the source or propagation path model for this data set.

The source models indicate that a spherically symmetric explosion can explain the data with deviatoric components of the source as much as a factor of 25 smaller. These small deviatoric components yield good fits to the transverse data.

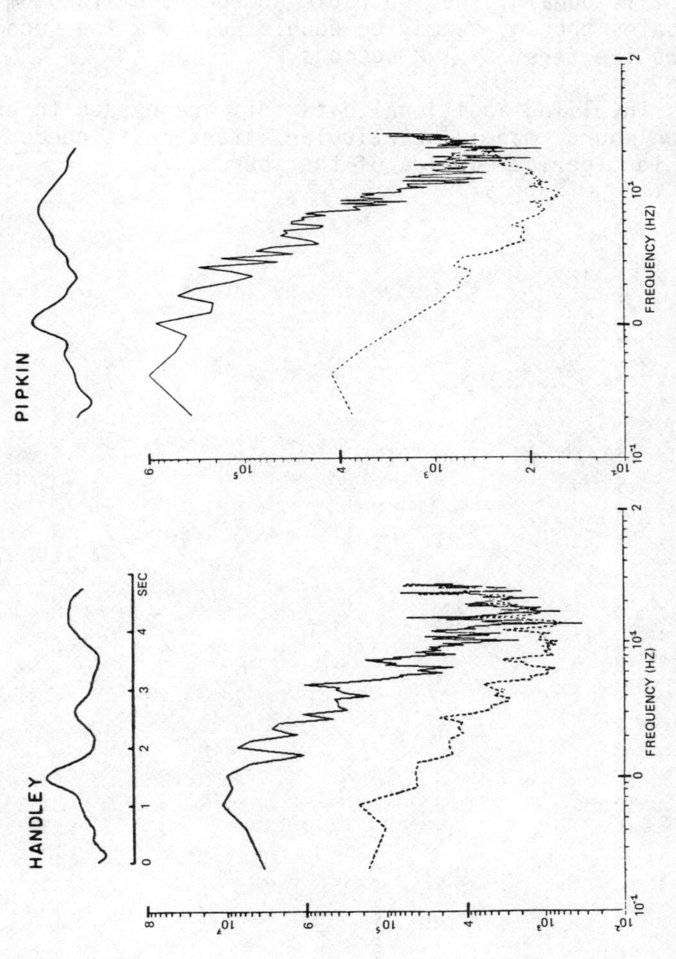

Figure 7. Comparison of the Handley and Pipkin isotropic moment tensors. The spectra are the far-field source function and should be multiplied by 8.84×10^{16} dyne-cm/sec. The time functions are for the near-field source.

The isotropic spectral model indicates a 1 Hz corner frequency and a 3-slope high frequency roll-off. Because of the limited signal to noise ratio in the data at the long periods (5 sec), questions concerning spectral peaking cannot be addressed.

In the time domain, the isotropic component of the moment tensor indicates the source may be double pulsed. The second pulse follows the first by 1.2 seconds.

Studies involving additional data sets are needed to extend the explosion source model. Particular attention is needed concerning the long period portion of the source.

REFERENCES

Backus, G.E., 1977a. Interpreting the seismic glut moment of total degree two or less, Geophys. J. Roy. Ast. Soc. 51, 1-25.

Backus, G.E., 1977b. Seismic sources with observable glut moment of spatial degree two, Geophys. J. Roy. Astr. Soc. 51, 27-45.

Backus, G.E. and M. Mulcahy, 1976a. Moment tensors and other phenomenological descriptions of seismic sources - I. Continuous displacement, Geophys. J. Roy. Astr. Soc. 46, 341-361.

Backus, G.E. and M. Mulcahy, 1976b. Moment tensors and other phenomenological descriptions of seismic sources - II. Discontinuous displacements, Geophys. J. Roy. Astr. Soc. 47, 301-329.

Gilbert, F., 1970. Excitation of normal modes of the earth by earthquake sources, Geophys. J. Roy. Astr. Soc. 22, 223-226.

Gilbert, F., and A.M. Dziemonski, 1975. An application of normal mode theory to the retrieval of structural parameters and source mechanisms from seismic spectra, Phil. Trans. Roy. Soc. London, Ser. A, 278, 187-269.

Hadley, D.M. and D.V. Helmberger, 1979. Seismic source functions and attenuation from local and teleseismic observations of the NTS events Jorum and Handley, Abstract Fall AGU, EOS, 60, 881.

Hamilton, R.M. and J.H. Healy, 1969. Aftershocks of the Benham nuclear explosion, Bull. Seism. Soc. Am. 59, 2271-2281.

Helmberger, D.V., 1968. The crust-mantle transition in the Bearing Sea, Bull. Seism. Soc. Am. 58, 179-214.

Johnson, L.R., 1974. Green's function for Lamb's problem, Geophys. J. Roy. Astr. Soc. 37, 99-131.

Pao, Y-H. and R.R. Gajewski, 1977. The generalized ray theory and transient responses of layered elastic solids, Phys. Acous. 13, 183-265, Academic Press.

Strelitz, R.A., 1977. Moment tensor inversions and source models, Geophys. J. Roy. Astr. Soc. 52, 359-364.

Stump, B.W. and L.R. Johnson, 1977. The determination of source properties by the linear inversion of seismograms, Bull. Seism. Soc. Am. 67, 1489-1502.

SIMPLIFIED BODYWAVE SOURCE TERMS WITH ONE APPLICATION IN MOMENT TENSOR RECOVERY

Steven N. Ward

Department of Geological Sciences, Harvard University, Cambridge, MA 02138

It is well known that the elastic equations of motion in spherically symmetric, inhomogeneous media can be transformed into

$$\frac{\partial}{\partial r} \underline{v}_\ell^{\,m}(r,\omega) = \underline{\underline{m}}_\ell(r,\omega) \, \underline{v}_\ell^{\,m}(r,\omega) + \underline{F}_\ell^{\,m}(r,\omega) \tag{1}$$

which is formally solved using propagator matrices $\underline{\underline{\Lambda}}$

$$\underline{v}(\underline{r},\omega) = \sum_{\ell,m} \underline{\underline{\Lambda}}_\ell^{\,m}(\underline{r},r',\omega) \, \underline{v}_\ell^{\,m}(r',\omega)$$

$$+ \int_{r'}^{r} \underline{\underline{\Lambda}}_\ell^{\,m}(\underline{r},\bar{r},\omega) \, \underline{F}_\ell^{\,m}(\bar{r},\omega) \, d\bar{r} \tag{2}$$

The $\underline{v}_\ell^{\,m}$, of course, are traction-displacement vectors and $\sum_{\ell,m} = \sum_{\ell=0}^{\infty} \sum_{m=-\ell}^{\ell}$. Any scheme for the recovery of the seismic moment tensor requires the calculation of absolute scale synthetic seismograms. Equation (2) forms the basis for such calculations. Unfortunately, obtaining the source vector $\underline{F}_\ell^{\,m}(r,\omega)$ by direct

evaluation for moment tensor forces $(-\nabla \cdot \underline{M}(\underline{r},\omega))$ involves considerable algebra, and the possibility of erring in sign or normalization during its derivation is a concern. Obviously, for moment tensor recovery purposes it is desirable to recast (2) into a simpler and more intuitive form.

The recasting begins by decomposing the propagator using

E. S. Husebye and S. Mykkeltveit (eds.), Identification of Seismic Sources – Earthquake or Underground Explosion, 269–272.

fundamental matrices $\underline{\underline{L}}_\ell(r,\omega)$

$$\underline{\underline{\Lambda}}_\ell^{\,m}(\underline{r},r',\omega) = \underline{\underline{L}}_\ell^{\,m}(\underline{r},\omega)\ \underline{\underline{L}}_\ell^{\,-1}(r',\omega) \tag{3}$$

The columns of $\underline{\underline{L}}$ are the displacements and tractions of six in-dependent solutions $\underline{v}_\ell^{\,m(i)}$ of system (1) with $\underline{F}_\ell^{\,m} = 0$. Using the Hermitian properties of the equations of motion, Chapman and Woodhouse (1) proved that for exact $\underline{\underline{L}}_\ell(r,\omega)$

$$\underline{\underline{L}}_\ell^{\,-1}(r,\omega) = r^2\ \underline{\underline{N}}_\ell^{\,-1}(\omega)\ \underline{\underline{\hat{L}}}_\ell^{\,-1}(r,\omega) \tag{4}$$

where $\underline{\underline{N}}_\ell^{\,-1}(\omega)$ is constant. $\underline{\underline{\hat{L}}}_\ell^{\,-1}(r,\omega)$ is obtained by a simple re-arrangement of the complex conjugate elements of $\underline{\underline{L}}_\ell(r,\omega)$.[1] Upon substituting (3) and (4) into (2) we find that only terms in dis-placement appear in the product $\underline{\underline{\hat{L}}}_\ell^{\,-1}(r,\omega)\ \underline{F}_\ell^{\,m}(r,\omega)$ because the traction elements in $\underline{\underline{\hat{L}}}_\ell^{\,-1}$ always multiply zeros in the vector $\underline{F}_\ell^{\,m}$. This elimination of traction terms permits (2) to be written

$$\underline{v}(\underline{r},\omega) = \sum_{\ell,m} \underline{\underline{L}}_\ell^{\,m}(\underline{r},\omega)\ \underline{\underline{N}}_\ell^{\,-1}(\omega)\left[{}_0\underline{v}_\ell^{\,m}(r',\omega)\right.$$
$$\left. - \int_{\hat{V}} \underline{\underline{u}}_\ell^{\,*m}(\hat{\underline{r}},\omega)\ \cdot\ \nabla\ \cdot\ \underline{\underline{M}}\ (\hat{\underline{r}},\omega)\ d\hat{v}\right] \tag{5}$$

The ij-th element of the displacement tensor $\underline{\underline{u}}$ is defined to be the j-th component of displacement from the i-th solution $\underline{v}_\ell^{\,m(i)}$. After shuffling terms, applying the divergence theorem, and as-suming that the source is completely buried in a smoothly vary-ing region, we reduce (5) to its final appearance

$$\underline{v}(\underline{r},\omega) = \sum_{\ell,m} \underline{\underline{L}}_\ell^{\,m}(r,\omega)\ \underline{\underline{N}}_\ell^{\,-1}(\omega)\left[{}_0\underline{v}_\ell^{\,m}(r',\omega)\right.$$
$$\left. + \int_V \underline{\underline{\varepsilon}}_\ell^{\,*m}(\hat{\underline{r}},\omega)\ :\ \underline{\underline{M}}(\hat{\underline{r}},\omega)\ d\hat{v}\right] \tag{6}$$

The ijk-th element of the strain tensor $\underline{\underline{\varepsilon}}$ is the jk-th component of strain for the i-th solution $\underline{v}_\ell^{\,m(i)}$.

The calculation of body wave excitation using (6) is much to

be preferred over (2). First, the source terms in (6), simply
being components of strain, are straightforward and immediately
expressible for any particular choice of $\underline{\underline{L}}_\ell(r)$. Secondly, the

excitation expressed in (6) clearly shows the similarities and
differences among body wave and normal mode expansions. Displacement solutions to (1) using normal mode eigenfunctions
$\underline{U}_\ell{}^m(\underline{r})$ (3) are

$$\underline{u}(\underline{r},\omega) = \sum_{\ell,m} \frac{\underline{U}_\ell{}^m(\underline{r})}{(\omega_\ell^2-\omega^2)\tau_\ell} \left[\int_{\hat{v}} \underline{\underline{\varepsilon}}_\ell{}^{*m}(\hat{\underline{r}}) : \underline{\underline{M}}(\hat{\underline{r}},\omega)\ d\hat{v} \right] \tag{7}$$

Clearly (6) and (7) are similar, as both contain a term giving
the field strength at the receiver location $\underline{r}$; both have a normalization dependent upon angular order but not radius; and both
have a source term involving only strain and moment tensors in a
volume integral. Two fundamental differences between (6) and (7)
exist, however: 1) The source term in (7) is a scalar whereas in
(6) it is a vector. This difference speaks of the contrasting
ways that the two expansions partition the total displacement
field. For normal mode solutions only two independent fields are
available (spheroidal and toroidal). For body wave solutions,
six independent fields are available (upward and downward travelling P, Sv and Sh). 2) Regularity and free surface boundary
conditions are prebuilt into (7) through $\underline{U}_\ell(r)$, whereas for body
waves a wide variety of boundary conditions (not necessarily
true-life) can be met with different selections of the starting
vector ${}_0\underline{v}_\ell{}^m(r',\omega)$. The finer partitioning of the displacement

field and the added flexibility to remove unwanted wave contributions through judicious application of boundary conditions are
the fundamental characteristics discriminating body wave and normal mode formulations.

     Using a version of equation (6) appropriate for a point
source, a simplified moment tensor inversion program employing
teleseismic P phases from the SRO network was developed. The
program was applied to a large shallow focus Oaxaca, Mexico
earthquake of November, 1978. We found that a relatively small
number (8-10) of P phases were sufficient to form a statistical
basis for testing various solutions. One result of the investigation was that unconstrained source models, which include a minor double couple with or without an isotropic component, did not
describe this event significantly better than a moment tensor
constrained to be a single double couple. Complete results of
this experiment are presented in (4).

[1]   We do not give explicit expression for $\hat{\underline{\underline{L}}}_\ell^{-1}$ here. Fuller
details can be found in (2).

REFERENCES

(*1*) Chapman, C. H. and J. H. Woodhouse: 1981, *Symmetry of the wave equation excitation of body waves*, Geophys. J. R. astr. Soc. (in press).

(*2*) Ward, S. N.: 1981, *On elastic wave calculations in a sphere using moment tensor sources*, Geophys. J. R. astr. Soc. (in press).

(*3*) Gilbert, F.: 1970, Geophys. J. R. astr. Soc. 22, pp. 223-226.

(*4*) Ward, S. N.: 1980, Bull. Seism. Soc. Am. 70, pp. 717-734.

THE INTERPRETATION OF MOMENT TENSOR INVERSIONS

Richard A. Strelitz

University of Southern California
Department of Geological Sciences
University Park
Los Angeles, California 90007

The moment tensor formalism provides a concise means of expressing the body force equivalents for a variety of simple point sources; consequently, it is ideally suited for use in the forward problem of generating synthetic seismograms for most sorts of natural and artificial sources. Furthermore, this straightforward approach for the computation of synthetic seismograms would indicate a relatively simple inverse problem. However, it is the very generality of the method which poses potential problems when trying to interpret the relationship between the elements of the moment tensor determined from the inverse problem and the physical reality or model of the source. The very same a priori knowledge and assumptions that permitted a simple construction of the moment tensor will provide the stumbling block that precludes a facile implementation of an automated analysis system. The root of the problem lies in the inability of a rote technique to duplicate the complex chain of reasoning which led to the imposition of the a priori constraints and the adoption of an idealized earth model. In those cases where the inversion has been limited to determining the magnitude and orientation of the best double couple, the difficulties alluded to above are not restrictive. In the more complex situation wherein one seeks to compare source models or to investigate the existence of a particular source type, these constraints may present quite formidable obstacles.

The first consideration must be to realize that not all source models are equally likely to be solutions of an inversion. In particular, the double couple represents a specific statement as to the distribution of errors, especially in the sense of nodal planes or null amplitudes. Furthermore, the competing source

273

E. S. Husebye and S. Mykkeltveit (eds.), Identification of Seismic Sources - Earthquake or Underground Explosion, 273–275.
Copyright © 1981 by D. Reidel Publishing Company.

models - the pure double couple, the isotropic source and the
compensated linear vector dipole have distinct numbers of degrees
of freedom. Therefore, to truly be able to make a statement that
a certain event is the result of a specific model type, one must
solve for each source type in question separately and then compare the goodness of fit (however defined) with explicit reference
to the differing numbers of degrees of freedom, a consequence of
the varying numbers of constraints that characterize each source
type.

The next phase of the analysis should be a study of the uncertainties in the derived solution. Since each model is characterized by the eigenvectors of the associated moment tensor (and
here we restrict discussion to the lowest order solution), the
uncertainty in the principal components will also describe how
well a particular model is constrained.

Finally, there is the more intractable problem of the effect of
deviations of the true earth from the idealized form of the model.
This is very succintly illustrated by the results of Fitch McCowan
and Shields (1980) in their inversion of the body wave data for
the shallow focus event of October 23, 1964 (M=6.2; Dep=43 km);
in this study, a straightforward inversion led to a solution that
was only 65% double couple; by altering the model, they were able
to increase the percentages to 89%. The moral of this is that
the solution is dependent on the model. The solution is to
realize that no finite parameter problem (moment tensor inversion
or epicentral location) can be done without considering the
effect of the infinite number of model parameters, in effect
converting the least squares problem to a true inverse problem.
A corollary might be that just because you do not solve for
parameters does not mean that they are zero. For an unbiased
set of unmodeled variables (the changes i.e. to the model which
are not solved for have zero mean), the solution will not change
although the estimate of the variance will be significantly
different; in the more realistic case such as the one cited above,
the solution will vary. This method of solution is called the
"method of unmodeled variables" (Bush, 1971) and the solution
hinges on being able to characterize the uncertainty and bias in
the model. It should be noted that this source of error is
distinct from measurement error in the data, and does not have
the same effect on the solution. As is immediately clear, the
chief hurdle to implementing this technique lies in the determination of the estimate of the variance in the model parameters;
for the Earth, this is possible equivalent to finding an estimate
of the variance in the model parameters; for the Earth, this is
possibly equivalent to finding an estimate of the degree of
inhomogeneity. One possible approach is to use Kalman filtering
which permits concurrent updating of solution and model covariance.

There is no need to panic and dismiss use of the moment tensor, the above cavils notwithstanding. Instead, what I have sought to outline is the source of problems when one seeks to force the results of an inversion to coincide with a priori statements or preconceived notions and prejudices.

REFERENCES

Bush, N., Unmodeled Error Analysis on Trajectory and Orbital Estimation, Technometrics 13, 303-314 (1971).

Fitch, T., D. McCowan and M. Shields, Estimation of the Seismic Moment Tensor from Teleseismic Body Wave Data with Applications to Intraplate and Mantle Earthquakes, Journal of Geophysical Research 85, 3817-3828 (1980).

COUPLING NEAR SOURCE PHENOMENA INTO SURFACE WAVE GENERATION

David G. Harkrider

Seismological Laboratory
California Institute of Technology

ABSTRACT

Theoretical explosion mechanisms for an all azimuth
reversal of Rayleigh waves are presented. Linear code methods
for modeling nonlinear source region phenomena and techniques
for coupling near source phenomena into surface wave generation,
such as vertical separation of source radiation, mixed path
approximations, and the elastodynamic representation theorem
are reviewed. Qualitative and quantitative results obtained
using these algorithms, in particular source medium scaling,
are discussed.

1. INTRODUCTION

Non-isotropic and "reversed" surface waves from underground
events in Russia have emphasized the need for estimating the
magnitude and mechanism of these anomalous seismic source region
effects. At present the favored explanation is the superposition of a thrust-like double couple earthquake or its
tectonic release equivalent.

There are a number of feasible geological and tectonic
settings, where the one double couple like source plus explosion
assumption is too restrictive. Volume tectonic release in a
region of plane compression or tension is an example of where
the source must be modeled by two right angle quadrupoles plus
isotropic source. This mechanism is disturbing in that there
is an isotropic component which can not be separated from the
explosion by moment tensor inversion.

E. S. Husebye and S. Mykkeltveit (eds.), Identification of Seismic Sources – Earthquake or Underground Explosion, 277–326.

Because of the various parameters in the source region which can effect the amplitude and polarity of Rayleigh waves even when the signals have traversed relatively well calibrated propagation paths, this presentation will review some of these effects and present linear methods for modeling nonlinear source regions and techniques for coupling near source phenomena into surface wave generation. In the remainder of this introduction, we will discuss two possible explosion mechanisms for an all azimuth reversal of Rayleigh wave signature.

1.1 Tectonic release or triggering

A substantial amount of research has been done on the effect of superimposing a vertical pure strike slip double couple like mechanism on explosion generated surface waves. This tectonic generation model was felt to be most appropriate to NTS and the French Sahara Test site (Harkrider, 1977, and Rodi et al., 1978).

Another and potentially more troublesome tectonic mechanism is the 45° dipping pure thrust fault or tectonic release due to elastic failure in an equivalent orientation of a pre-stressed field. If strong enough the additional seismic radiation can reverse the Rayleigh wave signature which would be observed for the explosion alone.

In Figure 1, we show the synthetic fundamental Rayleigh waves for an explosion and a pure thrust 45° dipping double couple at a range of 4000 km in a western U.S. model, CIT 109. The two mechanisms are synthesized separately for enough azimuths to determine the radiation pattern for the double couple. The phase of the double couple is found to be reversed with respect to the explosion at all azimuths. The bottom synthetic is a composite seismogram of the two mechanisms. The amplitude reversed double couple (dashed) is plotted on top of the explosion. They are shown with the same initial time, i.e., no shift in time, and demonstrate the nearly perfect reversal in polarity. The peak Airy phase amplitudes are plotted in Figure 2 as a function of azimuth. The amplitudes are scaled such that the P wave moment for both the explosion and the double couple are the same, $M_O = 10^{25}$ (dyne-cm). This does not imply that their P waves have the same amplitude since this is a strong function of source rise time and was not investigated here.

This thrust double couple mechanism can be considered the sum of two orthogonal vector dipoles of opposite sense. The vertical dipole contributes the azimuthally symmetric part of the radiation pattern and the horizontal dipole normal to the strike contributes a symmetric plus quadrupole part. The relative weights of the two dipoles to the radiation pattern is a function of frequency and source depth. The illustrated

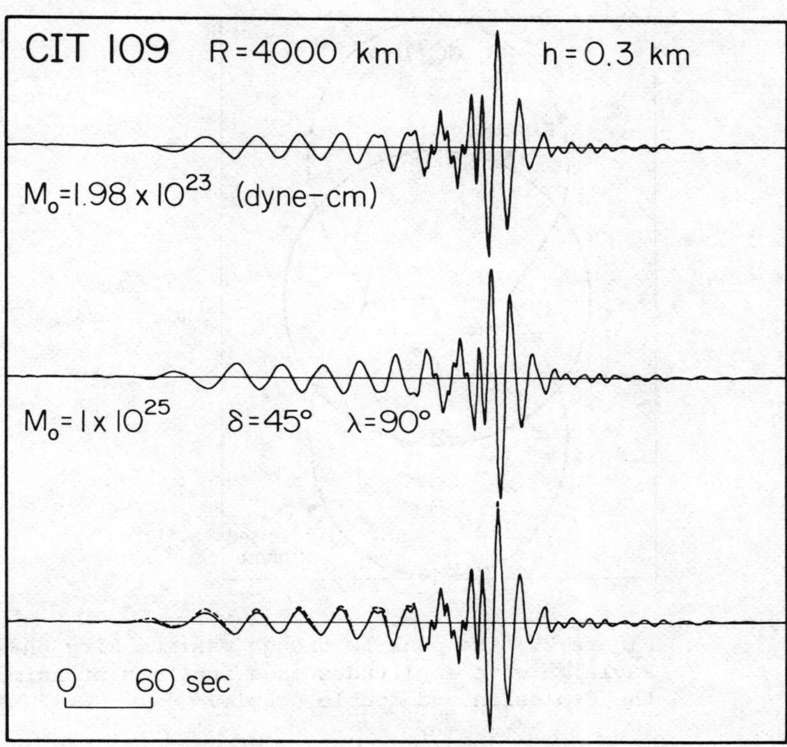

Figure 1. Vertical Rayleigh waves from an explosion and a 45° dipping pure thrust double couple fault singly and superimposed in the bottom trace. The azimuth from the fault strike is 22.5°.

synthetics and radiation pattern are for a source depth of 0.3 km. As source depth is increased from 0.01 km to 5 km in this continental model, the high frequency component of the explosion synthetic is reduced compared to the longer periods but the phase remains unchanged. At source depths less than 1 km, the double couple is out of phase with the explosion at all azimuths. On the other hand, at a source depth of 5 km, the early arriving long periods are always out of phase with the explosion but the later arriving short periods, which include the Airy phase, vary from out of phase at an azimuth of 90° to in phase at 0° which is along the fault strike.

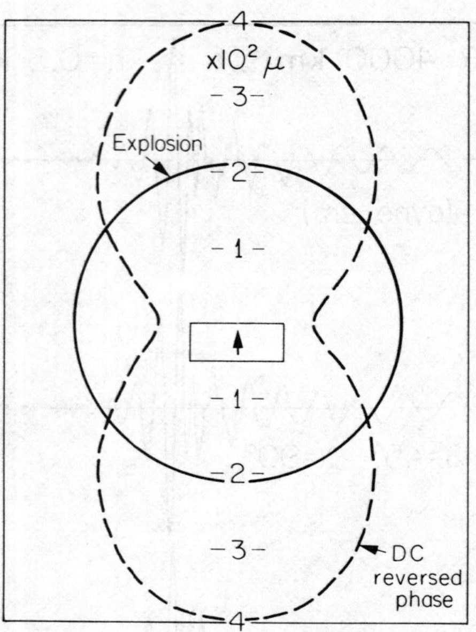

Figure 2. The peak to trough maximum Airy phase
Rayleigh wave amplitudes as a function of azimuth for
the explosion and double couple.

At shallow source depths, the combination of an explosion
with this pure thrust 45° dipping double couple like mechanism
can result in a reversal of phase at all azimuths, a quadrupole
typical of pure strike slip double couple, or an ideal explosion
pattern depending on the relative strengths of the two sources.
Local variations away from a pure thrust or the 45° dip can also
similarly modify the resultant radiation pattern.

1.2 Degradation of upgoing and downgoing source radiation

In order to investigate the effects of nonlinear regions
above and below the source such as spall and cracking, Harkrider
and Bache (1981) developed a wave field separation in terms of
upoing and downgoing waves from the source that can be used in
most computer programs based on residue theory, steepest descent
approximation or wave number integration. This technique allows
one to model the resultant source wave in the linear homogeneous
regime by source radiation whose history, amplitude, or travel
time differs above and below the source point.

When comparing theoretical Rayleigh wave generation for two source models of NTS, Bache _et al_. (1977) found that the polarity of the Rayleigh waves generated by the downgoing radiation in the soft tuff over hard basement model were opposite from those from a source region in which the soft rock-hard rock contrast below the source is absent. If the Rayleigh waves from the upgoing source radiation or the source wave itself were degraded by surface or near surface irregularities or momentum traps then it would be possible to reverse the polarity of Rayleigh waves at all azimuths from explosions in the same general source area. This effect would be similar to that due to the superposition of the tectonic surface wave radiation from a 45° dipping pure thrust fault like mechanism mentioned previously.

In the next section we give a brief derivation of the theoretical aspects of the technique applied to explosions and the depth effect on Rayleigh wave amplitudes when the vertically suppressed explosion sources are used to model directional source effects in simple and realistic earth models.

2.   VERTICAL SUPPRESSION OF SEISMIC SOURCE RADIATION

The seismic wave propagation problem is commonly formulated in terms of an integral in the wavenumber domain. Ray theory techniques for solving the problem naturally separate the contribution of waves along different takeoff angles. However, this separation is not obvious in methods where the wavenumber integral is solved directly.

Harkrider and Bache (1981) present a wavenumber formulation where the source discontinuity condition is separated into discontinuity conditions due to up and downgoing source radiation. This separated source "jump" condition can be conveniently used in most computer programs based on residue theory or wavenumber integration. With minor modification it can also be used in codes for free oscillation or normal mode synthetics. The following is an example of the formulation applied to explosion generated surface waves.

2.1 Rayleigh wave explosion source suppression theory

From Harkrider (1964) equation (43), an azimuthally symmetric compression source is given by

$$
\bar{\Phi}(r,z) \;=\;
\begin{cases}
\displaystyle\int_0^\infty S_{01}^{+}\, e^{-ikr_\alpha(z-h)}\, J_0(kr)\, dk \quad , \quad z > h \\[2em]
\displaystyle\int_0^\infty S_{01}^{-}\, e^{-ikr_\alpha(h-z)}\, J_0(kr)\, dk \quad , \quad z < h
\end{cases}
\tag{2.1}
$$

where S_{01}^{+} is the strength of the downgoing and S_{01}^{-} is the strength of the upgoing source radiation in wave number space. For a spherically symmetric explosion source

$$
S_{01}^{+} \;=\; S_{01}^{-} \;=\; i\,\frac{\bar{\Psi}(\omega)}{r_\alpha}
\tag{2.2}
$$

with $\bar{\Psi}(\omega)$ the reduced displacement potential i.e.,

$$
\Phi(r,z,t) \;=\; -\,\frac{\Psi(t - R/\alpha)}{R}
\tag{2.3}
$$

The surface vertical displacement for Rayleigh waves is given in Harkrider (1964), equation (79), as

$$
\left\{ \bar{w}_0 \right\}_R \;=\; \frac{\pi}{k}\,\frac{N_R^{(1)}}{\left(\dfrac{\partial F_R}{\partial k}\right)_\omega}\, H_0^{(2)}(kr)
\tag{2.4}
$$

where

$$
N_R^{(1)} \;=\; [GN - LH]\,\left[Y + \frac{K}{L} Z\right]
\tag{2.5}
$$

and

$$
\left[Y + \frac{K}{L} Z\right] \;=\; \left\{ \delta\sigma\, \bar{y}_1^{R} + \delta w\, \bar{y}_2^{R} - i\delta\tau_R\, \bar{y}_3^{R} - i\delta u\, \bar{y}_4^{R} \right\}
\tag{2.6}
$$

The correspondence between the Saito (1967) and Haskell (Harkrider, 1964) notation is given by

$$
\bar{y}_3^{R} \;=\; \left[\frac{\dot{u}^{*}}{\dot{w}_0}\right]
\tag{2.7}
$$

$$
\bar{y}_1^{R} \;=\; \left[\frac{\ddot{w}}{\dot{w}_0}\right]
\tag{2.8}
$$

$$
\bar{y}_2^{R} \;=\; k\left[\frac{\sigma^{*}}{\dot{w}_0/c}\right]
\tag{2.9}
$$

$$\bar{y}_4^R = -k \left[\frac{\tau_R}{\dot{w}_0/c} \right] \qquad (2.10)$$

and

$$\dot{x}/c = ikx \qquad (2.11)$$

For a dilatation source

$$\delta u = -ik (S_{01}^+ - S_{01}^-) \qquad (2.12)$$

$$\delta w = -ikr_\alpha (S_{01}^+ + S_{01}^-) \qquad (2.13)$$

$$\delta \sigma = k^2 c^2 \rho (\gamma - 1)(S_{01}^+ - S_{01}^-) \qquad (2.14)$$

$$\delta \tau_R = -2k^2 \mu r_\alpha (S_{01}^+ + S_{01}^-) \qquad (2.15)$$

where

$$\gamma = 2 \frac{\beta^2}{c^2} \qquad (2.16)$$

and

$$r_\alpha = (c^2/\alpha^2 - 1)^{1/2} \qquad (2.17)$$

(Harkrider, 1964, equations 37)

In order to eliminate the Rayleigh wave contribution from the downgoing source wave, we set the downgoing source coefficient, S_{01}^+, equal to zero and obtain

$$\left\{ \bar{w}_0 \right\}_R = -2\pi \mu kr_\alpha S_{01}^- \left[K_R + \frac{i}{2\mu r_\alpha} M_R \right] \underline{A}_R H_0^{(2)}(kr) \qquad (2.18)$$

where

$$\underline{A}_R = \frac{[G^* N - L^* H]}{\left(\dfrac{\partial F_R}{\partial k} \right)_\omega} \qquad (2.19)$$

$$K_R = \bar{y}_3^R(h) - \frac{1}{2\mu k} \bar{y}_2^R(h) \qquad (2.20)$$

$$M_R = \rho c^2 (\gamma - 1) \bar{y}_1^R(h) - \frac{1}{k} \bar{y}_4^R(h) \qquad (2.21)$$

or in terms of Saito (1967)

$$A_R = \frac{1}{2cUI_1^R} \tag{2.22}$$

and

$$I_1^R = \int_0^\infty \rho[\,(\bar{y}_1^{-R})^2 + (\bar{y}_2^{-R})^2\,]\,dz \tag{2.23}$$

Similarly suppressing the upgoing source radiation by setting $\bar{S}_{01}$ equal to zero and combining the two results we have

$$\left\{\bar{w}_0\right\}_R = -2\pi\mu ki\bar{\Psi}(\omega)\ A_R\ H_0^{(2)}(kr)\left\{(2 - |\Gamma|)K_R - i\,\frac{\Gamma}{2\mu r_\alpha}\,M_R\right\} \tag{2.24}$$

where $\Gamma = 1$: All upgoing source wave suppressed
$\Gamma = 0$: Total source
$\Gamma = -1$: All downgoing source wave suppressed
Γ can take any value in the range $[-1, 1]$ and represents the proportion of up or downgoing source radiation suppressed. For $\Gamma = 0$ we have the result given by Harkrider, 1964, Hudson, 1969, Douglas et al., 1971 and Bache et al., 1978a, among others.

Even though this formulation has been restricted to Rayleigh waves generated by explosion sources, it is easily extended to more general seismic sources imbedded in vertically inhomogeneous media. Expressions have also been obtained in terms of the multipole source coefficient representation of finite sources (Bache and Harkrider, 1976, and Harkrider and Archambeau, 1981).

From equation (2.24), we see that separating the elastic field into that generated by the upgoing and downgoing source radiation introduces a square-root singularly, $1/r_\alpha$, into the frequency domain solution which is not found in the total source field relation $\Gamma = 0$. This is also true for the more general source type expressions. For residue or normal mode solutions, the integratable singularity occurs when the horizontal phase velocity is equal to the source depth body wave velocity corresponding to the type of source radiation. For our explosion source it is the compressional velocity at the source depth.

For the upgoing source wave formulation the jump or source condition in a layer over a half-space is exactly equivalent to satisfying the free surface with the layer homogeneous solutions plus the source term and the lower interface with only the homogeneous solutions. For the downgoing source wave formulation, the source term is incorporated only with the lower interface boundary conditions. When viewed in this light, it is not as apparently straight forward as identifying directional source terms in the multilayered formulation. Because of this we devised a few tests as to the validity and effectiveness of this technique.

The first is analytic and demonstrates that for downgoing source waves from a source located in the half-space below the layers, no Rayleigh or Love waves are generated. On the other hand, the upgoing waves in the same example, yield the total surface wave solutions. The details are given in Harkrider and Bache (1981).

The second is numerical. Bache and Harkrider (1976) presented expressions for computing the steeply emergent body waves exiting from the base of a model in which a generalized source is imbedded in a stack of plane elastic layers. The modified multipole source coefficient relations make it possible to use this technique to calculate the P-wave displacements at a depth of 40 km in the mantle for an explosion imbedded in a simple one layer crust-mantle model. The results are shown in Figure 3. The synthetics due to the upper, lower and total source radiation for a homogeneous half-space are shown in the top box. In the middle box, we show the synthetics for a one layer model. At the bottom we have sketched the ray paths corresponding to the arrivals labeled on the synthetics. The separation of arrivals into those initially leaving the source in upward or downward directions is excellent.

Since the steepest descent method is essentially a plane wave calculation at the dominant wave number it is not surprising that the implicit nulling of source coefficients in the wave number integrands result in good separation. It is primarily a test of the algebra and resulting numerical code. The third test is a numerical comparison of the mode theory separation using generalized ray seismograms as standards (Harkrider and Helmberger, 1981).

Isolating the contribution from upgoing and downgoing source radiation to the displacement field for SH waves trapped in a plane layered half-space is straight forward using generalized rays. SH displacement fields are especially appropriate since their total contribution is from homogeneous shear waves and not interface or surface waves. If the source and receiver are separated by distance large enough for the major contribution to be interpreted as from multiple reflected rays beyond critical angle, then there should be a close correspondence between generalized rays and summed locked mode solutions.

In Figure 4, we show comparisons between the summed modes and the generalized ray techniques for azimuthal seismograms

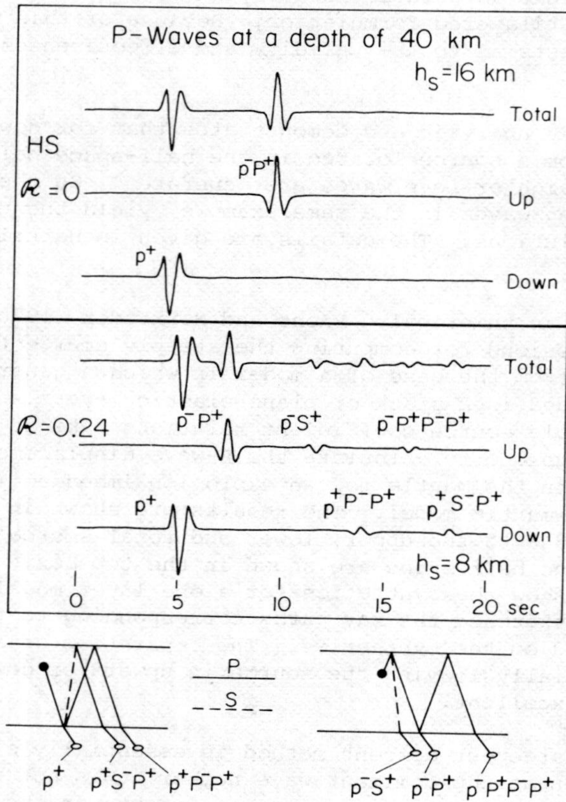

Figure 3. Steepest descent P-wave arrivals in a
half-space and a one layer half-space at a depth of
40 km. Raypath diagrams at figure bottom.

excited by the upgoing, downgoing and total SH source radiation
in a single layer over a half-space. The layer thickness is
32 km and the source is a horizontal pure shear dislocation at a
depth of 24 km and a range of 800 km from the surface receiver.
The farfield source time history is a trapazoid with $\delta t_1 = \delta t_2 =
\delta t_3 = 0.5$ sec. The seismograms on the left are mode sums and on
the right are generalized rays. The top, middle and bottom
traces are seismograms excited by the upgoing downgoing, and
total source radiation, respectively. All the seismograms in
Figure 4 are on the same time and displacement scales. Considering that only six modes were used in the sums and that the time
spacing was different for the two techniques, the agreement is

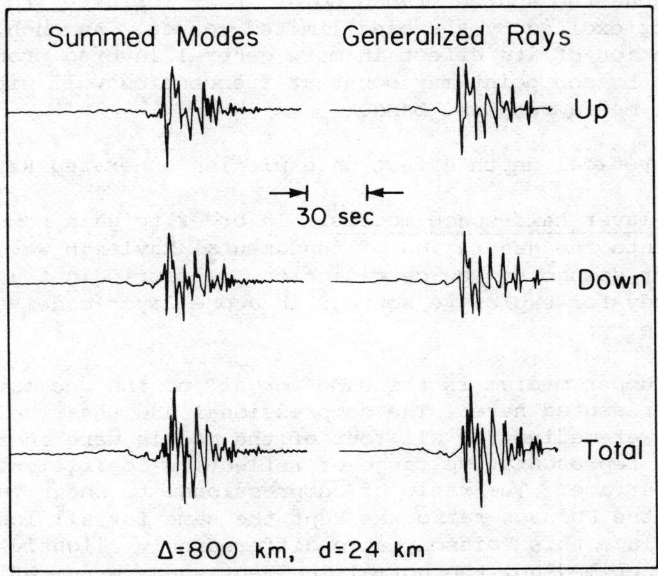

Figure 4. SH wave comparison between summed modes and generalized rays demonstrating vertical separation of source radiation in mode sum.

exceptionally good especially in phase arrival time and overall amplitude.

An interesting feature of this particular decomposition is that the SH amplitudes due to the upgoing and downgoing source radiation are essentially equal to those due to the total source even though their respective seismic energies are one half that of the total source radiation. The partition of energy is evidenced by the difference in apparent frequency in the seismograms. The effect of summing the two components is to approximately increase the number of peaks and troughs by a factor of two. Thus a better estimate of the relative seismic energies in this example is obtained by dividing the peak amplitude of each seismogram by its apparent period.

For this SH source decomposition, the branch point singularity occurs at a phase velocity equal to the shear velocity of the upper or source layer. Thus a limitation of this particular

test is that the source branch point is at infinite frequency
and is not excited by the band limited source. As such, we have
no indication of its effect in more general layered problems
where the branch point may occur at frequencies well within the
source or media response band.

2.2 Theoretical depth effect on explosion generated Rayleigh waves

One layer half-space models. In order to gain some physical
insight into the generation of fundamental Rayleigh waves by the
upgoing or downgoing source radiation, we carried out a parametric
model study for explosive sources in a one layer model over a
half-space.

The upper medium is the same for all of the one layer model
results presented here. The compressional and shear velocities
of the lower medium for all four of the models were chosen so as
to have a representative range of reflection coefficients at the
lower interface. The ratio of compressional to shear velocities
and thus the Poisson ratio was kept the same for all lower elastic
media. Since this Poisson ratio differed only slightly from that
of the upper medium, the normal incident shear wave reflection
coefficient, R, distinguishing each model, is essentially the
same as the compressional wave coefficient.

The layer thickness is 32 km in all models and the R = 0.24
model has been frequently used as a one layer continental crust
over mantle model. The lower medium density for all but the
negative reflection coefficient model was 3.4 gm/cm^3. In order
to obtain a negative reflection coefficient model without
drastically changing the disperson and thus the shape of the
Rayleigh wave, we use the same velocities as the R = 0.24 model
but reduced the lower medium density to 1.3 gm/cm^3. The elastic
model parameters are given in the Table.

For comparison purposes, we include the simple half-space
results for each of the five elastic materials used in the
parametric reflection coefficient models. The spectral characteristics for all nine simple propagation models are shown in
Figures 5 and 6. The phase, c, and group, U, velocities as a
function of period, T, for the fundamental Rayleigh mode of the
four dispersive models are shown in Figure 5. Their long period
asymptotes corresponding to their lower half-space Rayleigh
velocities are notated as HS1, HS2 and HS3 at the right hand edge
of the figure. Since only their half-space densities differ,
models R = 0.24 and R = -0.24 both have the same long period
asymptote HS2. For the limited period range used, the R = -0.24
model is still significantly different from its asymptotic value.
All four layer models have the same short period asymptote, HS0,
the Rayleigh velocity of their common upper layer medium.

Table. Elastic model parameters

R	D	α km/sec	β km/sec	ρ gm/cm^3	σ	V_R km/sec
HS0	∞	6.2	3.5	2.7	.266	3.227
0.15	32.	6.2	3.5	2.7	.266	3.227
HS1	∞	6.83	3.75	3.4	.285	3.47
0.24	32.	6.2	3.5	2.7	.266	3.227
HS2	∞	8.2	4.5	3.4	.285	4.16
0.40	32.	6.2	3.5	2.7	.266	3.227
HS3	∞	11.83	6.49	3.4	.285	6.00
-0.24	32.	6.2	3.5	2.7	.266	3.227
HSR	∞	8.2	4.5	1.3	.285	4.16

The amplitude responses $\underline{A}_R$ of the four dispersive models are shown as solid lines in Figure 6. The asymptotic model values are shown as dashed curves. These half-space responses are inversely proportional to period, T, and differ only by a multiplicative constant. A closed form expression for this factor as a function of density, Rayleigh and body velocities can be found in Harkrider (1970) and Harkrider et al. (1974).

For two materials with the same Poisson ratio, the ratio of their amplitude responses is given by

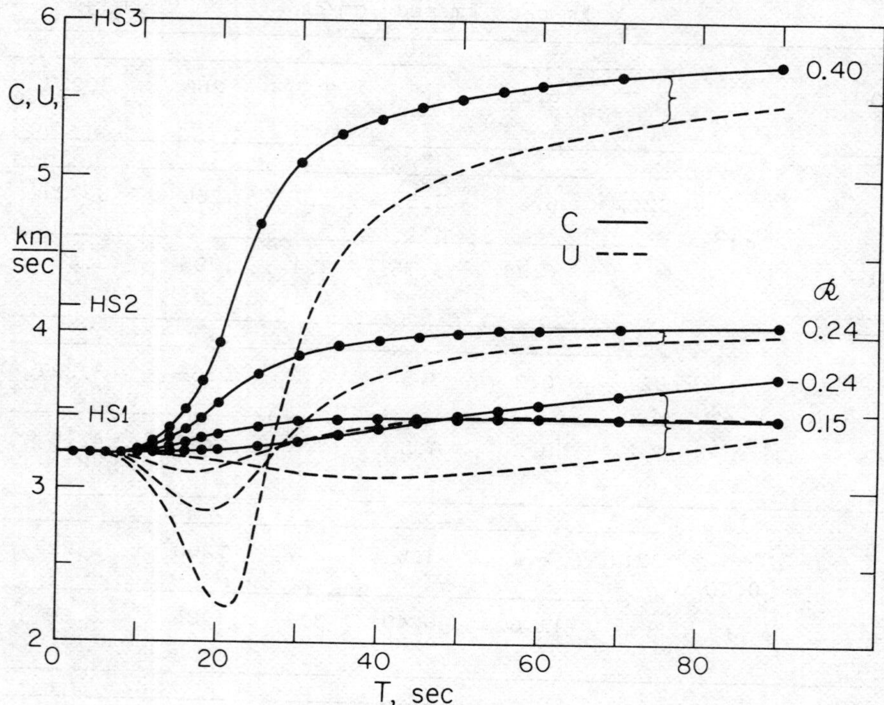

Figure 5. Phase, c, and group, U, velocity dispersion
versus period, T, for the one layer half-space models.
The half-space Rayleigh velocities for HS1, HS2, and
HS3 are shown at left margin.

$$(A_{R1}/A_{R2}) = (\rho_2 V_2^3)/(\rho_1 V_1^3) \tag{2.25}$$

where V is either a body or Rayleigh velocity. Since the Poisson
ratio of HS0 is nearly the same as the rest of the materials, all
of the dashed curves in Figure 6 are approximately related by the
above. Since half-spaces HSR and HS2 have the same elastic
velocities their curves differ only by the ratio of their
densities.

As noted for the velocities, the R = -0.24 model is still
appreciably different from its long period asymptote HSR at

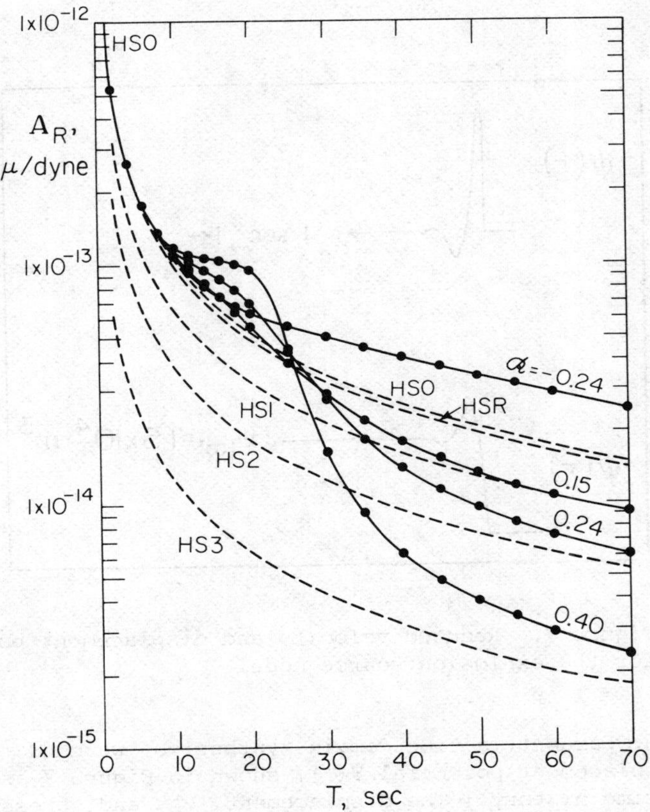

Figure 6. Amplitude response, $\underline{A}_R$, for the half-space and one layer half-space models as a function of period, T, in microns/dyne.

70 seconds. For periods shorter than 6 seconds, all the dispersive models are essentially the same as their short period asymptotic model, HSO.

Even though the source induced branch point singularity is integratable, we used the FFFT algorithm for generating our time domain synthetics and we were not sure of its applicability to integrands of this type. For these models the Rayleigh wave phase velocities are all less than the layer surface P wave

velocity and thus there is no frequency domain singularity
for their fundamental Rayleigh wave mode.

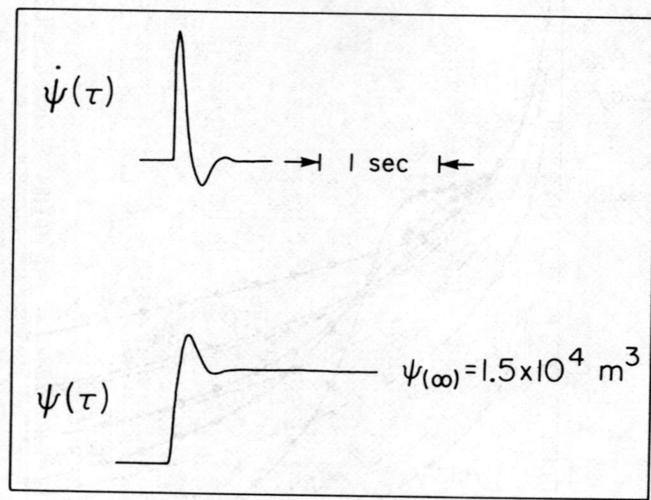

Figure 7. Reduced velocity and displacement potentials
for the explosion source model.

For our parametric time domain synthetics, we used the
reduced displacement potential, $\Psi(\tau)$, shown in Figure 7 as the
explosion time history. The displacement, U_R, and stress σ_{RR},
fields for the direct source P wave at a spherical distance, R,
are given by

$$U_R = \frac{1}{\alpha R} \dot{\Psi}(\tau) + \frac{1}{R^2} \Psi(\tau) \qquad ; \ \tau = t - R/\alpha \qquad (2.26)$$

$$\sigma_{RR} = -\rho \frac{\ddot{\Psi}(\tau)}{R} - \frac{4\mu}{R} U_R(\tau) \qquad\qquad (2.27)$$

where

$$\dot{\Psi}(\tau) = \frac{d}{d\tau} \Psi(\tau) \qquad , \qquad \ddot{\Psi}(\tau) = \frac{d^2}{d\tau^2} \Psi(\tau) \qquad (2.28)$$

and the reduced velocity potential, $\dot{\Psi}(\tau)$, is also shown in Figure
7. This particular $\Psi(\tau)$ is one of many resulting from finite

difference code calculations for source materials appropriate for the Nevada Test Site (Bache et al., 1975). For the periods of interest, the amplitude spectra of the reduced velocity potential, $\dot{\Psi}(\omega)$, is essentially equal to a constant $\Psi(\infty)$, equal to 1.5×10^4 m^3. The far-field whole space displacements and stresses are given by the first terms in equations (2.26) and (2.27). The far-field displacement is then given by $\dot{\Psi}(\tau)$ in the upper layers, at a distance of 160 km, the peak amplitude is 250 microns (μ).

Since $\Psi(\tau)$ is the same for all the parametric depth and model calculations, materials with the same velocities have the same far-field source displacement histories and materials with the same density have the same far-field source stress histories. The latter is the case for all the one layer calculations at a given depth with the one exception of a source in the half-space of the R = -0.24 model.

Since we were interested in determining the amplitude as well as signal character of Rayleigh waves generated by upgoing and downgoing source radiation as a function of source depth, we wanted the shape of our Rayleigh signals to remain reasonably unchanged over a wide range of depths. For meaningful absolute amplitude comparisons between our parametric models, we decided to use the amplitudes measured at ranges near enough to reduce the effects of the different degrees of geometric dispersion in the layer models. The distance used for the depth effect figures is 160 km, which is well beyond critical distance for a critically reflected P wave.

Since the dominant periods for the surface Rayleigh displacement are also a strong function of depth, it was decided to include the effect of an instrument in order to realistically band limit the Rayleigh waves. For our amplitude-depth figures the amplitudes are the maximum peak to trough values (PP) of Rayleigh waves measured through a long period LRSM seismograph response. Unless stated otherwise, the following figures in this section are for synthetics which include this instrument.

The depth effect on the dominant period of the surface vertical Rayleigh wave in a homogeneous half-space is shown in Figure 8 with and without a long period WWSN instrument response. The effect with the LRSM-LP response is shown at the left of Figure 10. The half-space, HS0, is of the same material as the upper layer medium in the one layer model. In some figures this half-space is labeled R = 0. For all illustrated Rayleigh synthetics, upward displacement is toward the bottom of the figure with the exception of those which include a WWSSN-LP response.

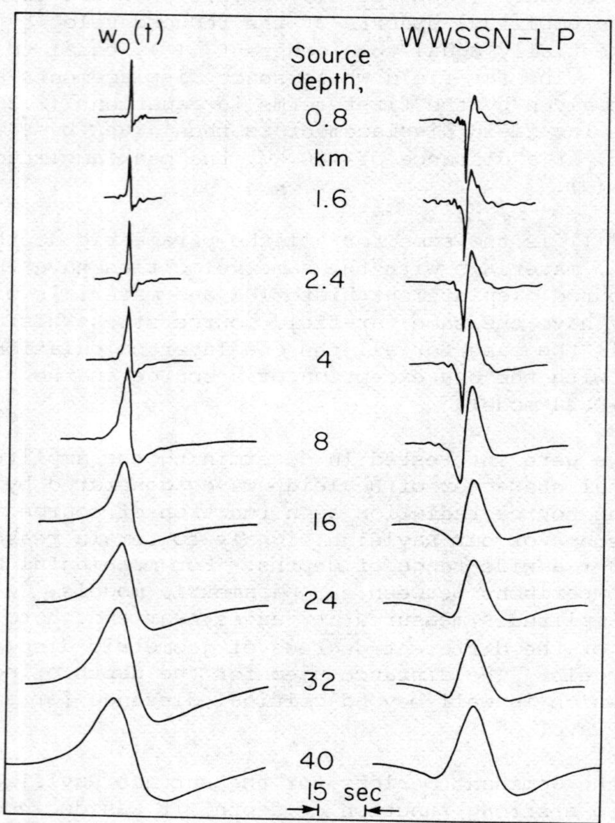

Figure 8. Vertical Rayleigh wave displacements on a half-space with and without WWSSN-LP instrument for selected source depths.

The explosion spectral source depth, h, dependence for a half-space is given by

$$K_R(h) = \frac{(\gamma - 1)}{\gamma r_\alpha^*} \exp[\omega r_\alpha^* h / V_R] \qquad (2.29)$$

where

$$r_\alpha^* = - (1 - V_R^2/\alpha^2)^{1/2} \qquad (2.30)$$

which acts as a low pass exponential filter. The attenuation of
high frequencies increases with increasing source depth. For a
surface source, the filter is flat for all frequencies. The
change in shape with increasing depth is most discernable for the
Rayleigh wave synthetics without an instrument, e.g., the left-
hand side (LHS) of Figure 8, and is most rapid between the near
surface, 0.8 and 8.0 km, depths.

The change of dominant period and shape at the near surface
depths is slightly less rapid as seen through the WWSSN-LP
instrument and is negligible for the R = 0, LRSM-LP synthetics in
Figure 10. The characteristics of the LRSM-LP instrument were
deemed desirable in that the maximum peak to trough value occurred
at or near the same phase as the source depth was varied. This
led to less scatter in amplitude versus depth plots for all the
models including the one layer models.

Rayleigh waves excited by step RDP explosions in half-spaces
of the same Poisson ratio are invariant when scaled in time as
Vt, where V is one of the three intrinsic velocities. Their
shapes are identical and differ only in absolute amplitude.
Except for the spectral depth factor $K_R(h)$, this would be true
for all half-spaces when scaled by $V_R t$. The inclusion of the
instrumental responses changes this, but it was observed for all
our half-space synthetics that the dominant period decreased as
the model Rayleigh velocity V_R was increased. The maximum
Rayleigh amplitudes for the LRSM-LP half-space synthetics as a
function of source depth are plotted in Figure 9.

For half-spaces with the same Poisson ratio, the explosion
spectra is proportional, ($\sim$), to the elastic constants as

$$\left\{\bar{w}_0\right\}_R \quad \sim \quad \frac{\mu}{\rho} \frac{\Psi(\infty)}{V_R^3} K_R(h) \ (\omega/V_R)^{1/2} \tag{2.31}$$

Thus for near surface sources and by keeping $\Psi(\infty)$ the same in
each source media

$$\left\{\bar{w}_0\right\}_R \quad \sim \quad (1/\beta)^{3/2} \tag{2.32}$$

This dependence is essentially the same as that deduced by
Marshall (1970) and is the approximate relation between time
domain amplitudes shown in Figure 9 at a source depth of 0.8 km.
This also predicts the shallow depth values of half-space HS0
which has only a slightly different Poisson ratio. Since half-
spaces HS2 and HSR have the same Poisson ratio and body velocities,
their explosion seismograms are identical at all ranges and

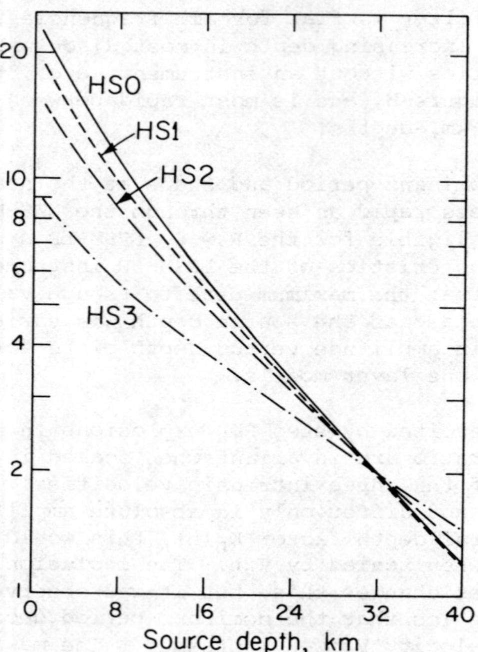

Figure 9. Maximum peak to trough, PP, Rayleigh wave
amplitude on the half-space models in microns as a
function of source depth.

depths. All the depth curves in Figure 9 show the expected
almost exponential relation predicted by spectral equation (2.29).
The curves deviate from straight lines because of the changing
dominant period with changing depth. The use of the narrow band
LRSM-LP instrument reduces this deviation.

Rayleigh wave LRSM-LP synthetics for the standard one layer
crust-mantle model, R = 0.24 at the two distances 160 and 1000 km
are shown in Figure 10 for a variety of source depths. The
corresponding synthetics for the extreme dispersion model R =
0.40 can be seen on the LHS of Figures 14 and 15. The synthetics
for both models at the 160 km distance are quite similar. However,
at the distance of 1000 km, their signatures are very dissimilar
as one would expect because of the difference in dispersion. As
in the half-space models, the effect of increasing source depth
is to progressively filter out the higher frequencies.

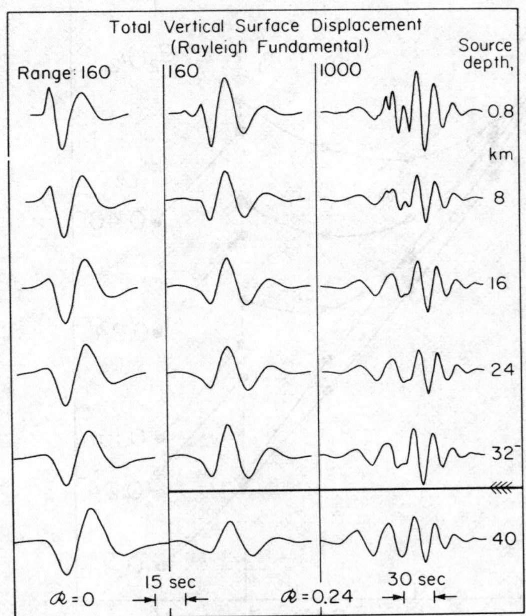

Figure 10. Rayleigh wave LRSM-LP seismograms on the one layer half-space, R = 0.24, and the half-space of the same surface material at two distances and six source depths.

The LRSM-LP maximum PP values for the one layer models as a function of depth are shown in Figure 11. For comparison purposes we show the half-space values of model R = 0 or HS0. As previously mentioned this is a half-space of the same material as the upper layer in all the one layer models. Referring to Figure 9, HS0 has the largest amplitudes of all the half-space models for source depths from the surface down to 32 km, which is the layer thickness of the one layer models. Thus all the one layer models have larger explosion generated LRSM-LP amplitudes than any individual half-space model. This is also generally true for WWSSN synthetics if we ignore the scatter at source depths less than 16 km.

The increase in amplitude with source depth in the lower part of the layer for the positive reflection coefficient models is due to the contribution of the downgoing source radiation to the generation of the fundamental Rayleigh wave and, of course,

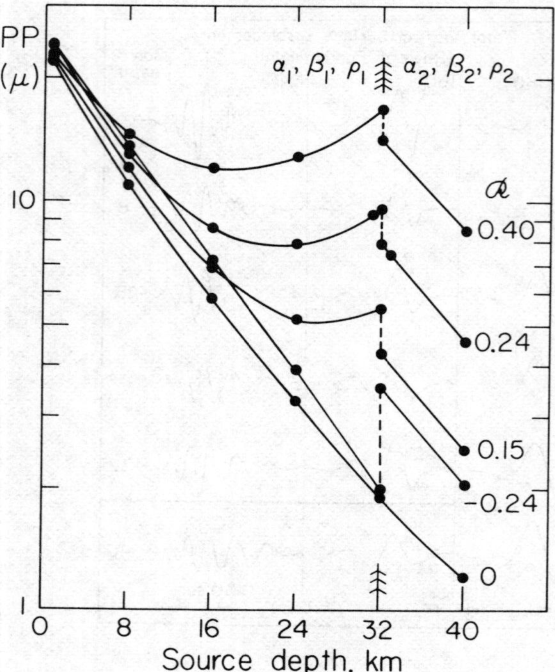

Figure 11.   Maximum LRSM-LP Rayleigh amplitudes, PP,
for the one layer half-space models as a function of
source depth.

is not present in the half-space model. This effect will be
discussed in more detail later. Below the layer interface the
amplitudes show the expected exponential decay of a half-space
as source depth is increased. This decay rate is essentially
the same for all one layer models and is practically indistinguishable from that of HS0 or HS1 at those depths.

The discontinuity in source depth effect as the source
crosses the bottom interface is due to the spectral discontinuity
in $\mu K_R(h)$ at this interface. The magnitude and sign of this
jump is frequency dependent and in the synthetics depends on the
dominant spectral frequencies. The dashed discontinuities in
Figure 11 roughly correspond to spectral values between 20 and
30 sec.

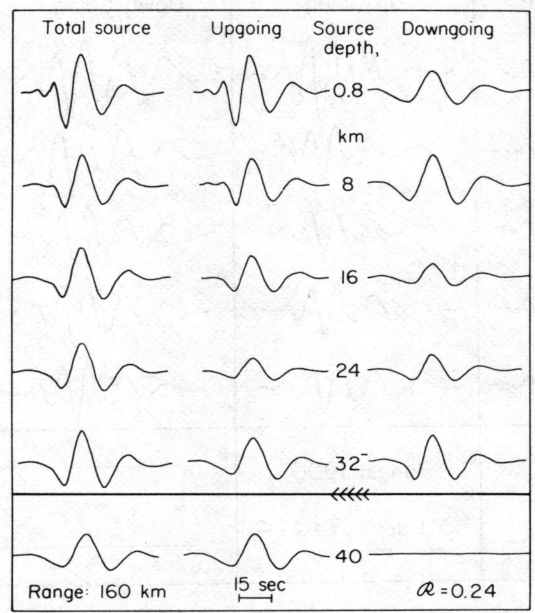

Figure 12. Rayleigh wave LRSM-LP seismograms at a range of 160 km for the R = 0.24 model demonstrating the contribution of the upgoing and downgoing source radiation at a sequence of increasing source depths.

The LRSM-LP synthetics generated by the total, the upgoing, and the downgoing compressional source radiation at the selected depths and ranges are shown in Figures 12 and 13 for the standard model, R = 0.24, and in Figures 14 and 15 for the extreme dispersion model, R = 0.40. The amplitude scales are relative and vary between synthetics.

The downgoing source radiation synthetics below 32 km are shown as straight lines. They are several orders of magnitude less than the total and the upgoing synthetics which can't be distinguished when plotted on the same amplitude scale. The 32^- and 32^+ synthetics are calculated using the eigenfunctions $\bar{y}_1^R$, $\bar{y}_2^R$, $\bar{y}_3^R$ and $\bar{y}_4^R$ for the depth h = 32 kms. These eigenfunctions are continuous across interfaces and the difference in synthetics is due to the difference in elastic constants on the (+) and (−) negative sides of the interface. Since theoretically the upgoing and the total contribution are identical for sources infinitesimally below the interface, the numerical differences between

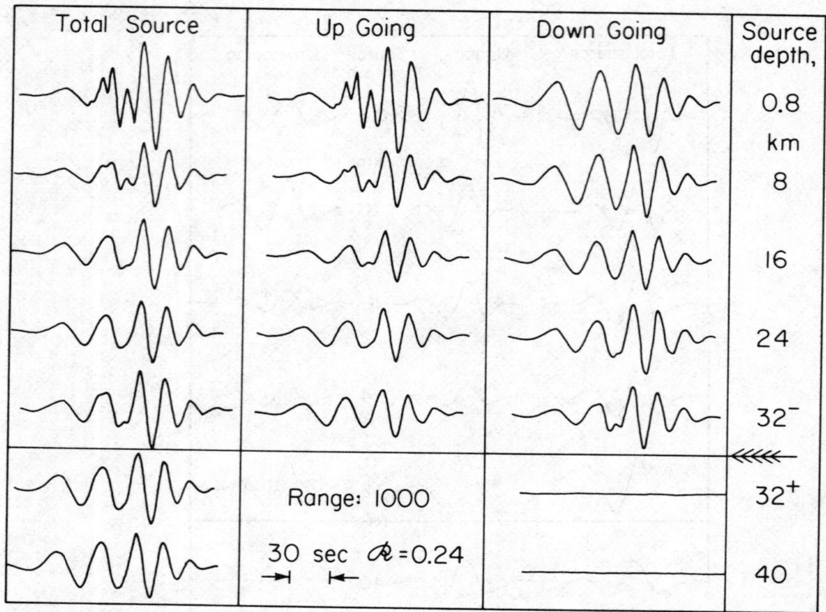

Figure 13.   Rayleigh wave LRSM-LP seismograms at a range of 1000 km for the R = 0.24 model demonstrating the contribution of the upgoing and downgoing source radiation at a sequence of increasing source depths.

the synthetics are a measure of the accuracy in computing the four eigenfunctions at the top of the terminating half-space.

From Figures 12-15, we see that for models R = 0.24 and R = 0.40, the greater the source depth the less the high frequency content in the "total" and "upgoing" synthetics. Conversely, an increase in source depth increases the high frequencies in the "downgoing" synthetic with a maximum in high frequency content for a source at the bottom interface. This maximum high frequency content in synthetics for the downgoing source at the interface appeared to be the same as that for the upgoing source at some intermediate depth. This observation was true for all our one layer models.

The maximum PP amplitudes due to the total and separated source radiation for the three positive reflection coefficient models are shown in Figures 16, 17 and 18 as a function of source depth.

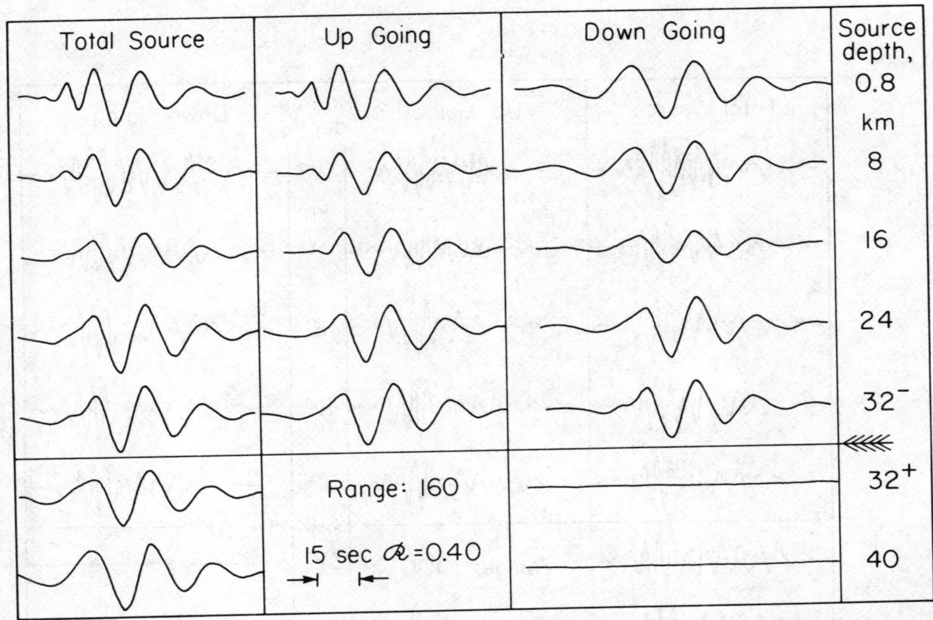

Figure 14. Rayleigh wave LRSM-LP seismograms at a range of 160 km for the R = 0.40 model demonstrating the contribution of the upgoing and downgoing source radiation at a sequence of increasing source depths.

In all three figures the amplitudes excited by the upgoing source radiation, labeled U, from a source in the upper layer are very close to those observed for sources in a half-space with the same elastic properties of the upper layer, i.e., HS0. These half-space amplitudes are shown in Figures 16, 17, 18 and 20 as dashed curves and labeled HS. Even though the difference is slight, the amplitude enhancement due to the lower half-space appears to be proportional to the value of the reflection coefficient, R.

For the contribution due to the downgoing source radiation, labeled D, the effect of R is very noticeable. Even though this is not basically a multiply reflected trapped wave phenomena like higher mode P-SV generation; the larger the positive reflection coefficient, the greater the contribution of the downgoing source wave to the excitation of the fundamental P-SV or Rayleigh wave mode. It is the contribution of the downgoing source radiation which cause the total source excitation, labeled T, to eventually

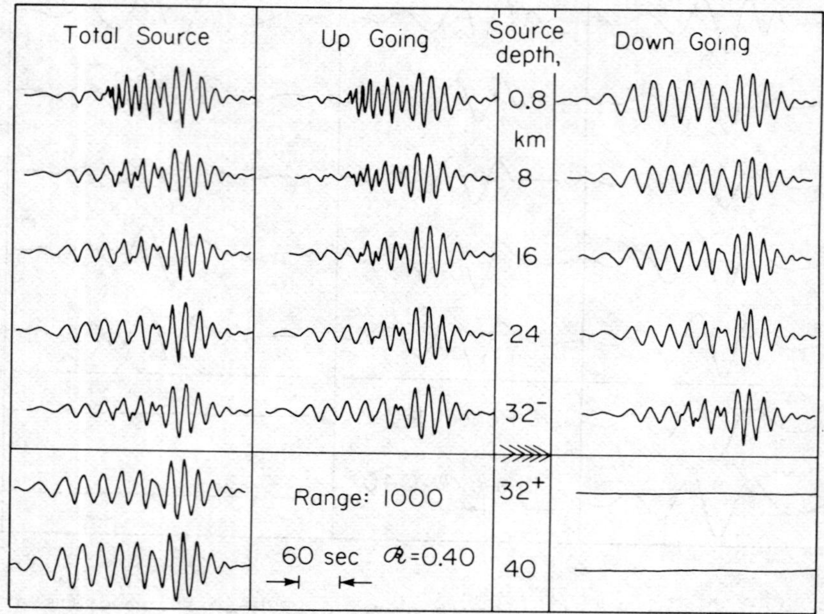

Figure 15. Rayleigh wave LRSM-LP seismograms at a
range of 1000 km for the R = 0.40 model demonstrating
the contribution of the upgoing and downgoing source
radiation at a sequence of increasing source depths.

increase with increasing source depth in the upper layer.

As mentioned earlier, the high frequency content in the
"downgoing synthetics" and the "upgoing synthetics" appeared to
be similar at some intermediate source depth in the upper layer.
This depth corresponds to the crossover depth of equal amplitude
excitation of upgoing and downgoing source radiation, which is
determined by the intersection of the U and D labeled curves in
Figures 16, 17 and 18.

Since the positive reflection coefficients enhanced the
excitation of the Rayleigh wave in one layer models over the
half-space model and this enhancement appeared to be proportional
to its value, we wondered what the effect of a negative reflection
coefficient would be. In particular, we wanted to know if a
negative value would reverse the polarity of the downgoing

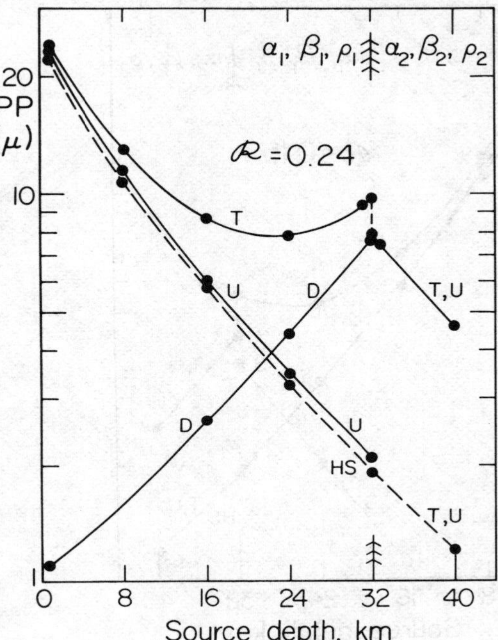

Figure 16. Maximum LRSM-LP Rayleigh amplitudes, PP, for the one layer half-space model, R = 0.24, versus source depth due to the upgoing, U, downgoing, D, and total, T, source radiation.

contribution so that the combination of the two Rayleigh waves would result in a completely different source depth behavior than the positive coefficient models.

As mentioned earlier, the negative R model has the same velocity structure as the standard one layer model, R = 0.24. The negative R value is obtained by a density reversal at depth. The upper medium density is the same as the positive models. The lower half-space density is reduced from the positive model values of 3.4 to a new value of 1.3 gm/cm^3.

Rayleigh wave synthetics for the two models at a range of 160 km and for our standard source depth set as observed with the WWSSN-LP and LRSM-LP seismograph systems are shown in Figure 19. The Rayleigh wave signatures are different for the two models but not drastically so. Without the side by side

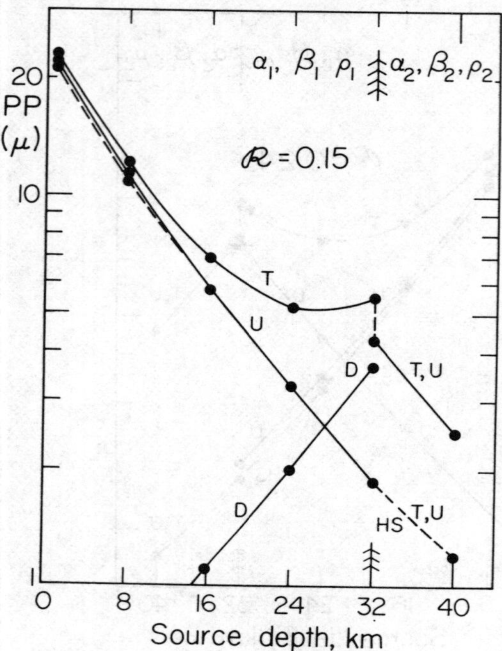

Figure 17. Maximum LRSM-LP Rayleigh amplitudes, PP, for the one layer half-space model, R = 0.15, versus source depth due to the upgoing, U, downgoing, D, and total, T, source radiation.

comparison, one would be hard pressed to delineate the differences.

The major differences in the two models is in the amplitude excitation for the total and separated source radiation as a function of source depth (Figure 20). The R = −0.24 model has a common feature with the positive R models in that the upgoing source radiation contribution decreases and the downgoing contribution increases with the increasing source depth in the layer. There are two notable differences. One is in the magnitude of the separated contributions and the other is that the upgoing contribution is greater than the total source contribution. The downgoing contribution is at least an order of magnitude smaller than the downgoing contribution to the positive R model. The upgoing contribution is significantly greater than the upgoing contribution to half-spaces and positive reflection coefficient models.

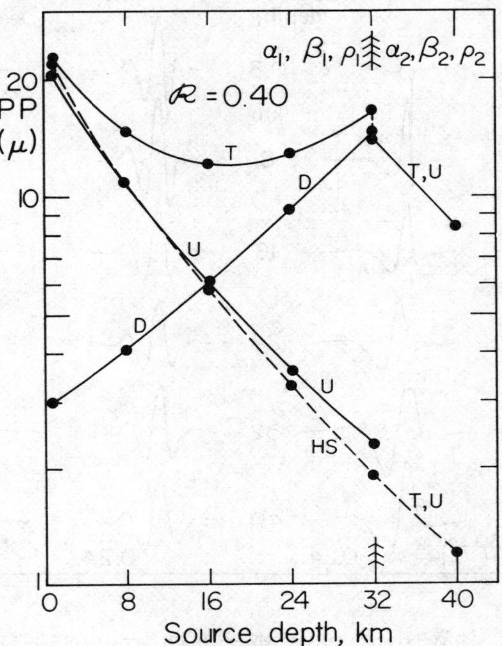

Figure 18.   Maximum LRSM-LP Rayleigh amplitudes, PP,
for the one layer half-space model, R = 0.40, versus
source depth due to the upgoing, U, downgoing, D, and
total, T, source radiation.

The postulated reversal in polarity between the separated
contributions indeed occurred for this negative R model. This
reversal caused the total source contribution to be less than the
upgoing with a resulting difference in amplitude trend for
increasing source depth with respect to the positive models. The
two Rayleigh synthetics due to the vertically separated source
radiation are shown in Figure 21. As before, the synthetics are
scaled for best illustration.

The negative reflection coefficient model was used only to
determine if it was possible to reverse the polarity between the
two separated source Rayleigh waves. No further investigation of
this class of models was undertaken since one layer negative R
models are unrealistic earth models.

Because of the source depth behavior of all models,

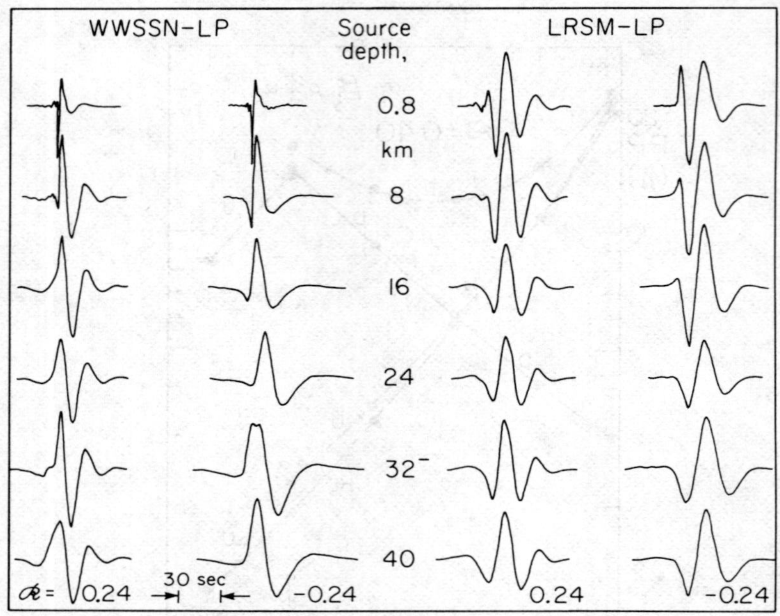

Figure 19. Rayleigh wave WWSSN-LP and LRSM-LP seismograms for models R = 0.24 and R = -0.24 at a range of
160 km and selected source depths.

especially the amplitude and frequency content of the Rayleigh
synthetics as the separated sources approached the interface from
above and below, we concluded that in one layer models at short
ranges there are two fundamental Rayleigh waves. These upper
layer surface waves are primarily generated by the interaction of
the free surface with the upgoing source wave and of the returning
downgoing source wave after reflection from the lower interface.
The strength and polarity of the interface controlled Rayleigh
wave depends on the sign and magnitude of the reflection coefficient. Although their effect was not isolated, it was felt that
the contribution due to later multiple reflections would be
reduced by their decrease in wave front curvature at the top and
bottom of the wave guide.

NTS source region models. In the previous section, we
restricted our discussion to that part of the Rayleigh wave train
most influenced by the near surface. This was done by our choice
of frequency band, by varying only the properties of the half-

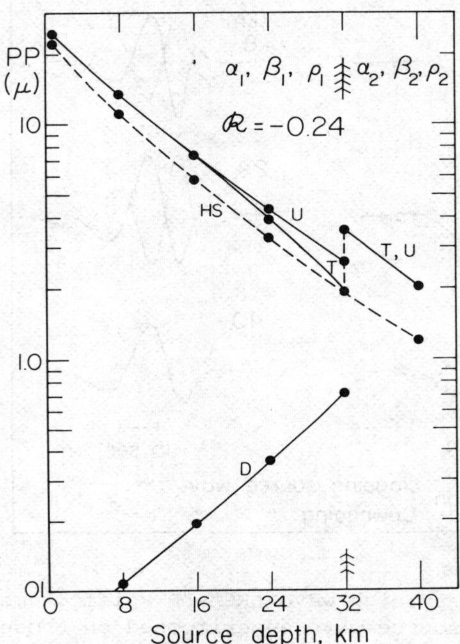

Figure 20. Maximum LRSM-LP Rayleigh amplitudes, PP,
for the one layer half-space model, R = -0.24, versus
source depth due to the upgoing, U, downgoing, D, and
total, T, source radiation.

space and by measurements at near ranges where dispersion effects
had not yet dominated the wave signature. The reversal of
Rayleigh wave polarity resulted from a negative reflection
coefficient which was obtained with an unrealistic density depth
contrast.

Reversal can also occur when the upper media are realistically varied and the Rayleigh wave is primarily controlled by the
lower media. In fact, a reversal can occur in source structure
models as reasonable as that of the NTS tuff region.

Earlier we mentioned several reasons to suspect that the
upgoing waves from explosion sources might make a substantially
smaller contribution to the far-field Rayleigh waves than is
predicted by elastic wave theory. If so, what are the effects on
the dependence of surface wave amplitude on source material and

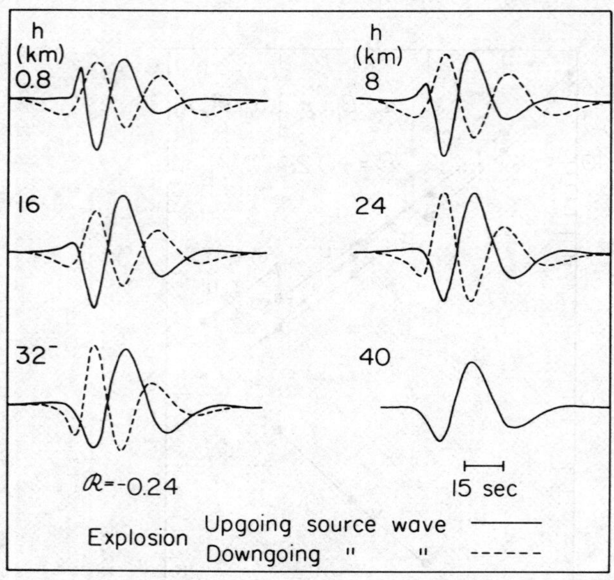

Figure 21. Rayleigh wave LRSM-LP seismograms due to
the upgoing source wave superimposed on those due to
the downgoing source wave at a range of 160 km and
selected source depths.

explosion yield? To address these questions we use an approximate Rayleigh wave synthesis technique described by Bache et al.,
(1978a), that separates the contribution of the source region
geology from that of the average propagation path. Separate
crustal models are used for the source region and average path,
with the vertical displacements computed from

$$\left\{ \bar{w}_0 \right\}_R = -4\pi i \mu k_1 \bar{\Psi}(\omega)\, K_{R1} \underline{A}_{R1} T(\omega)\, H_0^{(2)}(k_2 r) \qquad (2.33)$$

where quantities computed for the source region and average path
are denoted by subscripts 1 and 2, respectively. The transition
between the two structures is accounted for by the approximate
transmission coefficient

$$T(\omega) = (c_2\, \underline{A}_{R2})^{1/2} / (c_1\, \underline{A}_{R1})^{1/2} \qquad (2.34)$$

The derivation of this equalized horizontal energy flux expression
is given in a later section.

Bache et al. (1978a) computed synthetic seismograms for explosions in three distinct source regions encountered at NTS, Climax Stock, where the source is in granite, Yucca Flat and Pahute Mesa, where the source materials are tuff or rhyolite. The Climax Stock and Yucca Flat source regions represent the extremes in material properties, and so will be used for our examples. The average path model is the western United States model 35-CM2 of Alexander (1963). The GRTM attenuation model and LRSM-LP instrument response were used in all the computations to be presented.

The remainder of this section is essentially a reevaluation by T.C. Bache of Bache et al. (1977 and 1978b). First, we construct theoretical amplitude-yield curves for explosions in the two different source regions. Synthetic seismograms were computed at a range of 3000 km for several yields (W) with the source depth (in meters) varying as $107 \ W^{1/3}$. All seismograms were computed with the same reduced displacement potential which was cube-root scaled to the yield. At the dominant periods, this source function is nearly constant and so can be characterized by its static level, $\Psi(\infty)$, at a particular yield. For these examples $\Psi(\infty) = 91 \ m^3$ at 20 KT.

Selected synthetic seismograms are shown in Figure 22. We see that the waveform is essentially independent of yield and source depth and the source structure has only a minor effect. Consistent amplitude measurements are easily made and the M_s is computed from

$$M_s = \log A + 1.656 \log \Delta - 0.18 \qquad (2.35)$$

where A is peak-to-peak amplitude in nanometers and Δ is the range in degrees.

Now let us look at the effect of suppressing all or part of the upgoing waves from the explosion source, which is illustrated by the examples in Figures 23 and 24. When the source is in granite the Rayleigh waves from the up and downgoing waves are almost exactly in phase and simply sum to make up the total seismogram. For the source in tuff (Figure 24) the situation is entirely different. The contributions from the up and downgoing waves are nearly opposite polarity (the phase difference is about 225 degrees at long period). Therefore, suppression of the upgoing waves actually increases M_s.

Synthetic seismograms analogous to those in Figure 22 were computed with the upgoing waves totally suppressed. As with the total source seismograms, the waveform is essentially independent of explosion yield and depth and looks like the "downgoing" examples in Figures 23 and 24. For each of the four cases we plot

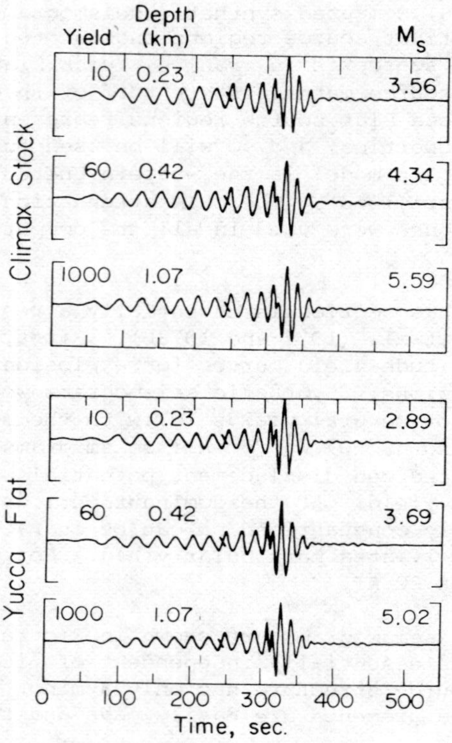

Figure 22. Typical vertical component synethetic
seismograms at 3000 km for a range of yields at fixed
scaled depth in structures representing NTS Climax
Stock and Yucca Flat. The cycle at which amplitude
was measured is indicated with a bar. The period of
this cycle is 19.7 seconds for the Climax Stock seismograms and 18.8 seconds for the others.

M_S versus yield in Figure 25. The linear least squares fit to
these data is shown with each set. The interesting features are
the slope of the M_S on the source material and on the degree of
suppression of the upgoing waves.

     In all cases the slope is slightly greater than unity. This
is due to two factors. First, the source level at 20 seconds is
not precisely proportional to W, but increases slightly faster.
If this were the only factor, the slope would be 1.02. The

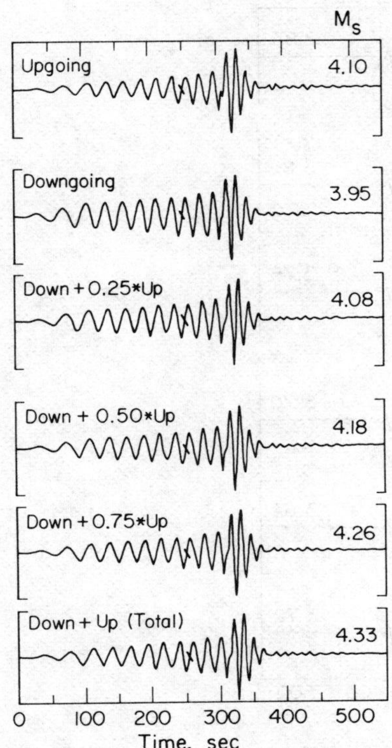

Figure 23. The relative contribution of the upgoing
and downgoing source waves to one of the Climax Stock
seismograms of Figure 22. Note that positive vertical
is down for this set.

second factor is the depth dependence of the eigenfunctions; for
example, K_{R1} is a slightly increasing function of depth for the
tuff events. Of course, we have ignored any depth-dependent
source coupling effects which are likely to be much more
important. As an indication of these effects, we refer to
Murphy (1977), who points out that the semi-empirical source
model of Mueller and Murphy (1971) gives

$$\psi(\infty) \sim W^{0.76} \qquad (2.36)$$

for NTS explosions at a constant scaled depth of burial. The
M_S-log yield curves obtained using this model would have slopes
that are 0.24 smaller than those in Figure 25.

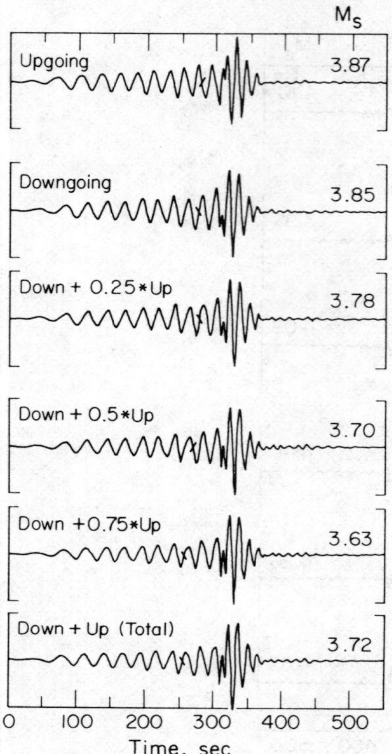

Figure 24. The relative contribution of the upgoing
and downgoing source waves to one of the Yucca Flat
seismograms of Figure 22. Note that positive vertical
is down for this set.

The dependence of M_S on the source material is strongly
influenced by the degree of suppression of upgoing waves. Bache
et al. (1978a) discussed this dependence for a total RDP source.
For a fixed path and source, the Rayleigh waves from explosions
in different source materials are proportional to

$$G_S = \mu_S \frac{K_{Rl} A_{Rl}}{c_l} T(\omega) \qquad (2.37)$$

This quantity was plotted versus period for the Climax Stock,
Yucca Flat and Pahute Mesa source regions discussed in that paper.
Comparing the fairly similar Yucca Flat and Pahute Mesa regions,
Bache et al., pointed out that $G_S \sim \mu_S$, leading to $M_S =$

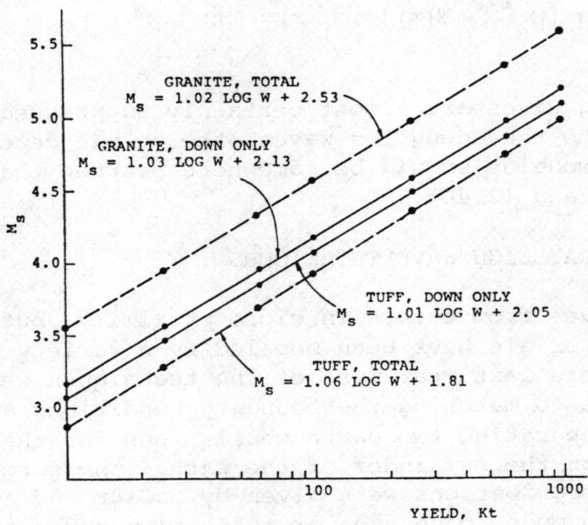

Figure 25. M_S versus yield for four data sets. The
results of a linear least squares regression, assuming
all error in M_S, are shown for each data set.

$\log [\mu_S \Psi(\infty)]$. However, for the dissimilar Climax Stock and
Yucca Flat regions the dependence on μ_S is weaker. For the path
models in that earlier paper or the 35-CM2 used here, this
dependence is approximately

$$M_S \sim \log [\mu_S^{3/4} \Psi(\infty)] \tag{2.38}$$

It is interesting to compare this to the elementary scaling

$$M_S \sim \log [\mu_S^{-3/4} \Psi(\infty)] \tag{2.39}$$

derived from the half-space scaling in (2.31). This is inappropriate because it assumes that the entire path is changed with
the local source medium and is valid only when the source layer
path length and thickness are large enough to be effectively
considered a uniform half-space. Limited source region results,
e.g., an embedded transparent spherical source medium, predict
an approximate dependence of $M_S \sim \log [\mu_S \Psi(\infty)]$ as above.

If the upgoing wave is suppressed, the dependence on the
local shear modulus is much weaker. From the M_S values in
Figures 23 and 24 we compute that total suppression of the

upgoing waves leads to the approximate scaling

$$M_s \sim \log [\mu_s^{1/8} \Psi(\infty)] \tag{2.40}$$

Since the upgoing waves are almost certainly suppressed to some extent compared to the downgoing waves, the actual dependence on the local shear modulus should be somewhere between that in equations (2.39) and (2.40).

3. MIXED PATH RAYLEIGH WAVE TECHNIQUES

Rayleigh waves from events in close proximity, but in different source materials have been modeled by a variety of techniques. Before last year most of the techniques were based on approximations to matching the boundary conditions across a vertical plane separating two earth models, one for the source region and one for the remainder of the path. The formulation and suggested approximations were given by McGarr and Alsop (1967). If one assumes no mode conversion or reflection and normal incidence to the boundary, the approximation is equivalent to assuming that the total horizontal energy flux remains constant during the transmission of each mode across the boundary.

Recently a major effort has been placed on using the Knopoff-de Hoop elastodynamic integral representation, given by Burridge and Knopoff (1964), to couple nonlinear and laterally inhomogeneous source regions to laterally homogeneous propagation models. In the following sections we will give a brief derivation of the first technique and finish with a representation theorem application to Rayleigh waves generated by a vertical point force in a multilayered media.

3.1 Equal horizontal energy flux method

The assumption that there is no mode conversion is probably most appropriate for the fundamental mode Rayleigh wave since it is a true surface wave and only requires a free surface for its existence. On the other hand, higher mode Rayleigh waves and all Love modes are primarily constructive interfering multiply reflected and refracted body waves trapped in the crust upper mantle wave guide. In order for there to be minimal mode conversion for these waves, constructive interference angles should be nearly the same on both sides of the transition zone. Of course, in the presence of low velocity layers at periods for which the phase velocity of the fundamental Rayleigh wave is large enough to represent body waves in these layers, there may be considerable mode conversion even for the fundamental mode. Finite element results indicate that for an oceanic-continental transition zone this approximation is reasonable for the fundamental Rayleigh wave and that there is considerable mode conversion for all Love wave modes.

The energy flux or transport is given by Biot (1956) as

$$W = 2UE\bar{w}_0^{-2} \qquad (3.1)$$

where U is the group velocity and E is total potential or kinetic mode energy at a particular frequency normalized to the square of the surface vertical spectral displacement or in the case of Love waves surface horizontal displacement. Equating the energy flux across the boundary we have

$$\bar{w}_{02} = (U_1 E_1)^{1/2} / (U_2 E_2)^{1/2} \, \bar{w}_{01} \qquad (3.2)$$

The expression can be further reduced to

$$\bar{w}_{02} = (c_2 \underline{A}_{R2})^{1/2} / (c_1 \underline{A}_{R1})^{1/2} \, \bar{w}_{01} \qquad (3.3)$$

using the relation in Harkrider and Anderson (1966)

$$EU = \omega^2 (2c\underline{A}_R)^{-1} \qquad (3.4)$$

where ω is the angular frequency, c is the phase velocity and $\underline{A}_R$ is the media spectral amplitude response. Specializing to the case of Rayleigh waves generated by an explosion, i.e., equation (2.24), $\Gamma = 0$ we have on the receiver side of the boundary at r_1

$$\left\{ \bar{w}_0 \right\}_{R2} = -4\pi\mu k_1 \, i\bar{\Psi}(\omega) \, K_{R1}\underline{A}_{R1} \, T(\omega) \, H_0^{(2)} (k_1 r_1) \qquad (3.5)$$

and as before

$$T(\omega) = (c_2 \underline{A}_{R2})^{1/2} / (c_1 \underline{A}_{R1})^{1/2} \qquad (3.6)$$

In order to approximate the correct wavelength spreading in phase and amplitude in the receiver medium, we modify the distance factor to

$$H_0^{(2)} (k_1 r_1 + k_2 r_2) \qquad (3.7)$$

where

$$k = \frac{\omega}{c} \qquad (3.8)$$

In modeling events at NTS where source region structural differences are spatially limited compared to the propagation path length r_1 is usually set to zero and we have the relation used in (2.2).

Other mixed path approximations, which have been used, are to
take the expression for an explosion in a uniform receiver type
medium and replace $\mu \, \Psi(\omega) \, K_R(h)$ with values appropriate for the
source medium. Even the best of these approximations is based
upon assumptions which are questionable for limited source regions,
e.g., that the source region wave front and the boundary are
parallel.

3.2 Elastodynamic integral representation method

The Knopoff-de Hoop integral representation theorem can be
used to extend near source calculations in laterally and vertically
inhomogeneous regions to regional and teleseismic ranges using
standard laterally homogeneous half-space codes (Harkrider et al.,
1981). We define a surface which encloses the heterogeneity of
the source region and exterior to which the medium can be represented by a plane stratified medium. As long as the possible
interaction of back reflected exterior radiation with the source
region including the source itself can be neglected, i.e., a
transparent source region, the Green's functions are those
appropriate for the exterior region with no heterogeneities
present. For a plane stratified half-space terminated at depth
by a homogeneous half-space, the source containing surface can be
a cylinder, i.e., canister, extending from the free-surface down
into the terminating half-space. Various numerical techniques can
then be used to determine the stress and displacement at a finite
number of nodes on the surface. By spatially integrating the
convolution of these nodal quantities with the appropriate surface
wave Green's functions for ring sources in the laterally homogeneous region, we can obtain synthetic Rayleigh and Love waves for
arbitrarily complex sources.

The vertical spectral displacement at the free surface of a
multilayered elastic half-space can be expressed in terms of a
surface integral of observed displacements and stresses as

$$\bar{w}_0 = \int_{\Sigma_{SES}} \{ \bar{G}_{3j} \, \bar{\sigma}_{jk}(\xi) - \bar{T}_{3kj} \, \bar{u}_j(\xi) \} \, n_k \, dS(\xi) \qquad (3.9)$$

where

$$\bar{T}_{ikj} = \lambda \, \delta_{kj} \, \bar{G}_{ip,p} + \mu \, (\bar{G}_{ik,j} + \bar{G}_{ij,k}) \qquad (3.10)$$

$\bar{G}_{3j}$: Spectral displacement in the vertical, x_3, direction
on the free surface due to a point force in the j
direction at ξ.

$\bar{u}_j$: Components of the spectral displacements observed
at ξ.

$\bar{\sigma}_{ij}$: Components of the stress tensor at $\underline{\xi}$.

n_k : Unit vector normal to the surface Σ_{SES}, positive into the nonlinear source region.

Σ_{SES}: Surface excluding the nonlinear source region from the linear propagation media

and

$$\bar{G}_{ij,k} \equiv \frac{\partial}{\partial \xi_k} \bar{G}_{ij} = -\frac{\partial}{\partial x_k} \bar{G}_{ij} \qquad (3.11)$$

Assuming an azimuthally symmetric distribution of displacements and stresses on Σ_{SES}, we integrate the integrand of equation (3.9) over a ring of radius r_0 and at a depth z_0. The G_{ij}'s are the wave number integral solutions found in Harkrider (1964). Taking the residue contribution to obtain the Rayleigh waves, we obtain the following ring integration contributions to $\{\bar{w}_0\}_R$:

$$\{\bar{w}_0\}_R = \int_0^h (\bar{u}_z\{\bar{T}^o_{zr}\} + \bar{u}_r\{\bar{T}^o_{rr}\} - \bar{\sigma}_{rz}\{\bar{G}^o_z\} - \bar{\sigma}_{rr}\{\bar{G}^o_r\})\, a\,dz_0$$

$$+ \int_0^a (\bar{u}_z\{\bar{T}^o_{zz}\} + \bar{u}_r\{\bar{T}^o_{zr}\} - \bar{\sigma}_{zz}\{\bar{G}^o_z\} - \bar{\sigma}_{rz}\{\bar{G}^o_r\})\, r_0\,dr_0 \qquad (3.12)$$

with the ring Green's functions given as

$$\{\bar{T}^o_{zz}\} = -2\pi(\tfrac{i}{2})\, \bar{y}^R_2(z_0)\, J_0(kr_0)\, \underline{A}_R\, H^{(2)}_0(kr) \qquad (3.13)$$

$$\{\bar{T}^o_{zr}\} = -2\pi(\tfrac{i}{2})\, \bar{y}^R_4(z_0)\, J_1(kr_0)\, \underline{A}_R\, H^{(2)}_0(kr) \qquad (3.14)$$

$$\{\bar{T}^o_{rr}\} = -2\pi(\tfrac{i}{2})\, \{\mu\, \bar{y}^R_3(z_0)\, [J_0(kr_0) - J_2(kr_0)]$$

$$+ 2\, \frac{\lambda\mu}{(\lambda + 2\mu)}\, K_R(z_0)\, J_0(kr_0)\}\, k\, \underline{A}_R\, H^{(2)}_0(kr) \qquad (3.15)$$

$$\{\bar{G}^o_z\} = -2\pi(\tfrac{i}{2})\, \bar{y}^R_1(z_0)\, J_0(kr_0)\, \underline{A}_R\, H^{(2)}_0(kr) \qquad (3.16)$$

and

$$\{\bar{G}^o_r\} = -2\pi(\tfrac{i}{2})\, \bar{y}^R_3(z_0)\, J_1(kr_0)\, \underline{A}_R\, H^{(2)}_0(kr) \qquad (3.17)$$

with $r_0 = a$ in the first integral which is over the canister's side and $z_0 = h$ in the second integral which is over the canister's bottom.

An example of the technique is shown in Figure 27. The nodal displacement-stress values were calculated at Del Mar Technical Associates and the results were for a joint AFOSR project. The synthetics shown in the figure are the vertical

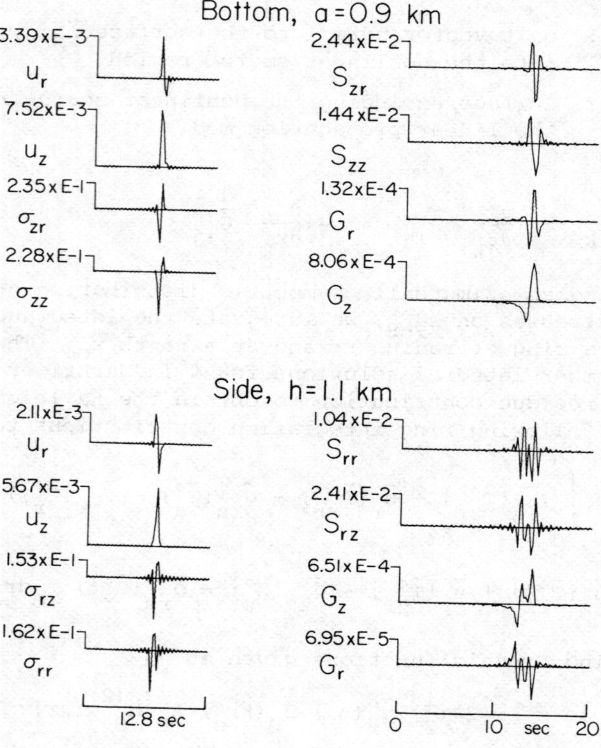

Figure 26. Nodal forcing functions and corresponding
Green's functions for a ring on the bottom and a ring
on the canister bottom for the HSO model.

Rayleigh wave displacements due to a vertical point source at a
depth of 0.4 km in a homogeneous half-space at a range of 300.1 km.
The nodal displacements and stresses are calculated for the
bottom and sides of a cylindrical canister of radius 2.1 and
vertical extent of 2.1 km. The vertical force is located on the
axis with a band limited delta function history. The frequency
range is 0.01 - 2.0 hertz.

The nodal values were calculated by two very different
techniques. The SWIS code is a finite element code capable of
calculating displacements and stresses in a vertically and
laterally inhomogeneous medium. For practical purposes the
lateral range is restricted to no more than five (5) source depths
in this problem. The PROSE code is a full (ω-k) numerical
integration solution to a source and receiver in a plane layered
viscoelastic half-space. The two nodal value results are combined

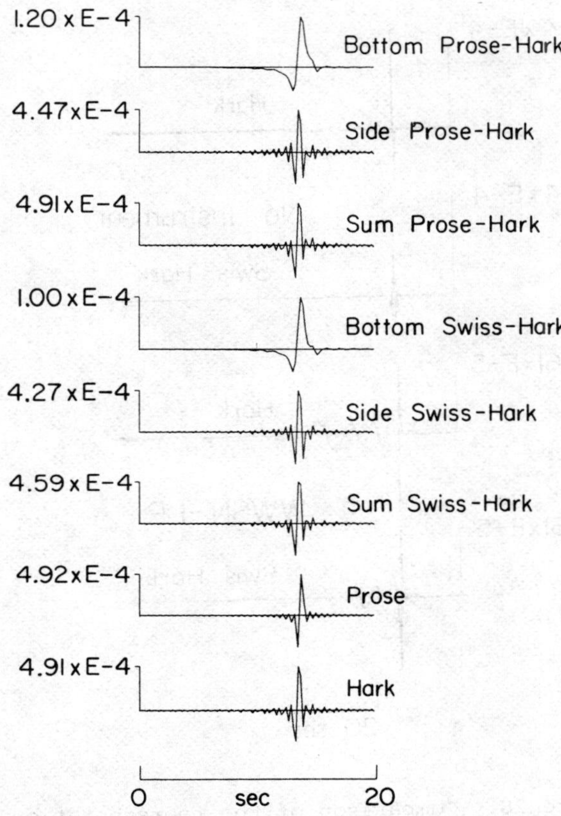

Figure 27. Contributions to the final solution of the
summed convolutions from the bottom and side of the
canister for the HSO model.

with ring source surface wave Green's functions, which propagate
the values from the canister to receiver and are calculated by
the plane layered surface wave code HARK.

Nodal displacement and stress forcing histories calculated
by SWIS and their associated surface wave Green's functions from
HARK are shown in Figure 26. The upper traces are for a ring of
width 0.2 km and radius 0.9 km located on the bottom of the
canister. The lower traces are for a ring of radius 2.1 km,
depth 1.1 km, and width 0.2 km located on the side. These results
are for the HSO homogeneous half-space discussed in section 2.2.
The S's in the figure are the time domain equivalents of the
spectral T's in the previous equations. The units are such that

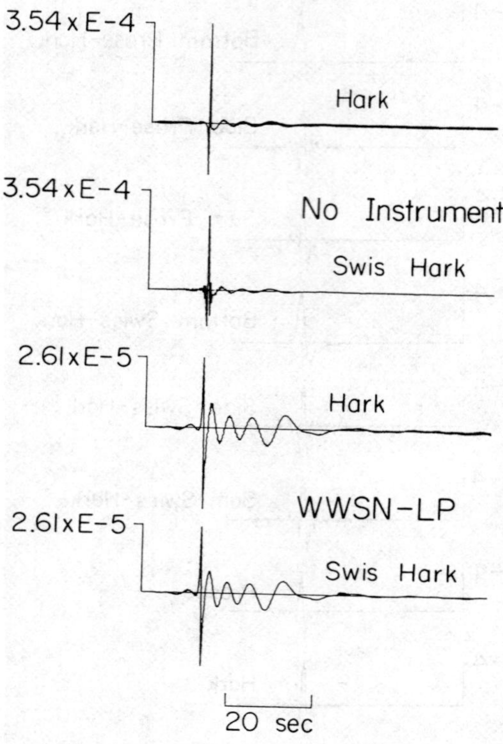

Figure 28. Comparison of the representation theorem results (SWIS-HARK) with the direct solution (HARK) for model CIT109 with and without WWSSN-LP instrument.

the nodal forcing - Green's function convolutions when scaled by the ring areas, represent ground displacement at the receiver in microns due to a vertical point force of one dyne located at the source.

The resultant Rayleigh waves due to the bottom, side and total canister nodes are shown in Figure 27. The top three traces are formed using forcing functions generated by PROSE. The next three by the finite-element code SWIS. The bottom two traces are PROSE and HARK vertical displacement calculations at a range of 300.1 km from the vertical point force. HARK generates only the Rayleigh wave while PROSE includes the direct body waves which at this range are negligible compared to the Rayleigh wave. The comparison between signatures of PROSE-HARK, SWIS-HARK, and HARK is very good. This is due to the fact that propagation over the major part of the path is given by HARK calculations. The peak

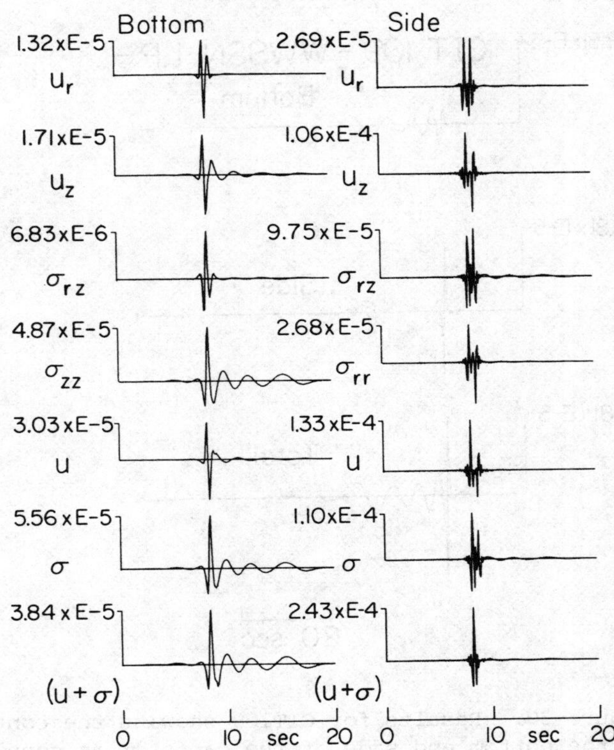

Figure 29. Contributions to the final solution for each
forcing function on the side and bottom of the canister
due to a vertical point force in model CIT109. Traces
represent ground motion in microns.

amplitude values of PROSE-HARK, PROSE and HARK are in excellent
agreement. The choice of a band limited delta function source
history is a very severe test of the technique in that a slight
shift of phase can give differences on the order of 10% in time
domain side lobe amplitudes, i.e., the time spacing misses the
narrower side lobe peak values. The differences between SWIS and
PROSE as seen in SWIS-HARK and PROSE-HARK amplitude comparisons
are not unexpected considering the difference in technique.

Earlier we stated that as long as the exterior reflected
energy interaction with the source region is negligible, then the
exterior Green's functions should be those appropriate for the
propagation medium. As a test of this, we coupled the above
canister driving functions, i.e., nodal displacements and stresses

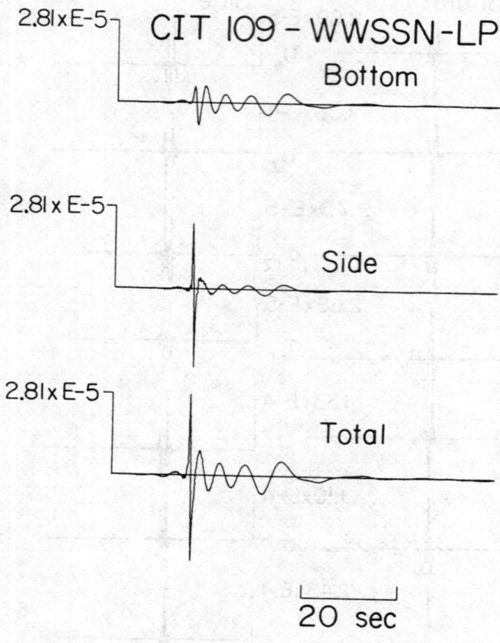

Figure 30.   Results for CIT109 showing the contributions
of the bottom and side of the canister as seen through
a WWSSN-LP instrument.

for a vertical point force in a simple half-space, to the Rayleigh
Green's functions for a multilayered crust-mantle model of the
western U.S., CIT109. The negligible reflection criteria to be
satisfied in this case since the surface layer of CIT109 was the
same elastic medium as the interior source region.

The results are shown in Figure 29 at a range of 1500 km with
a WWSSN-LP instrument and without. The comparison of the direct
calculation HARK with the SWIS forcing or driving function result
as seen through the long period instrument is outstanding both in
absolute amplitude and shape expecially before and after the
maximum peak to trough amplitude. The two signals are virtually
indistinguishable when overlayed.

The contribution of each forcing function on the bottom and
side of the canister is shown in Figure 29. The traces represent
surface vertical displacement at the receiver. The high frequency
and maximum amplitude contribution comes from the canister side
and the long period from the bottom. One might expect this type
of frequency partition because of the mean depth of the contrib-

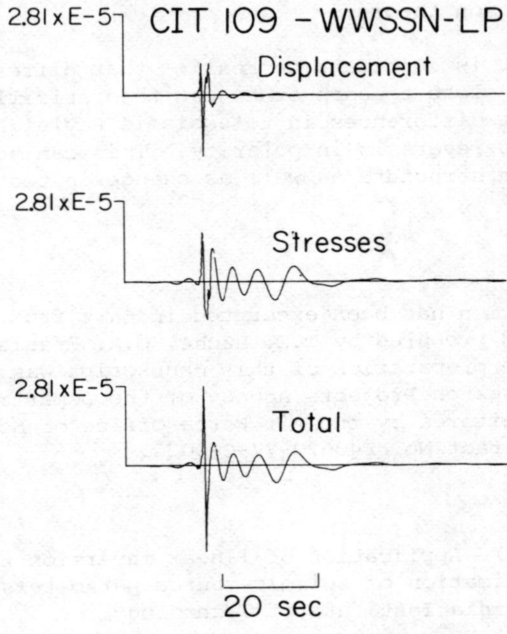

Figure 31. Contributions of the displacement and stress forcing functions to the CIT109 WWSSN-LP seismogram.

uting canister surfaces. On the bottom, the stresses make a larger contribution than the displacements and on the side the displacements are more important although neither dominate by more than a factor of two.

In Figure 30, the bottom, side, and total canister contributions to the Rayleigh wave WWSSN-LP seismograms are shown. As in the synthetic without an instrument (Figure 29), the high frequency and maximum amplitude contribution is from the side of the canister. From Figure 31, we see that the high frequency portion of the WWSSN-LP seismogram requires both displacement and stress values on the canister, whereas the long period energy comes predominately from the stresses.

4.   SUMMARY

In the preceeding sections, the need for the theoretical inclusion of different source regions, various techniques for this, and some of the results from coupling near source phenomena into surface wave generation were presented. The techniques reviewed were vertical separation of source radiation, mixed path

approximations, and source region coupling through the elastodynamic representation theorem.

In closing, it is important to realize that different source
regions relatively close to each other can theoretically not only
exhibit significant differences in teleseismic Rayleigh wave
amplitudes but also reversals in polarity. This can be caused by
local source region structure as well as change in tectonic
environment.

ACKNOWLEDGEMENTS

This presentation has been excerpted in part from papers and
reports written and prepared by T.C. Bache, G.A. Frazier, P.
Glover and me. The preparation of this manuscript was supported
by the Advanced Research Projects Agency of the Department of
Defense and was monitored by the Air Force Office of Scientific
Research under Contract No. F49620-77-C-0022.

REFERENCES

Alewine, R.W. (1974). Application of linear inversion theory
    toward the estimation of seismic source parameters, Ph.D.
    Thesis, California Institute of Technology.

Alexander, S.S. (1963). Crustal structure in the western United
    States from multi-mode surface wave dispersion, Ph.D. Thesis,
    California Institute of Technology.

Bache, T.C., J.T. Cherry, N. Rimer, J.M. Savino, T.R. Blake, T.G.
    Baker, and D.G. Lambert (1975). An explanation of the
    relative amplitude generated by explosions in different test
    areas at the Nevada Test Site, Systems, Science and Software
    Final Contract Report, DNA 3958F.

Bache, T.C. and D.G. Harkrider (1976). The body waves due to a
    general seismic source in a layered earth model: 1. Formu-
    lation of the theory, Bull. Seism. Soc. Am., 66, 1805-1819.

Bache, T.C., P.L. Goupillaud, and B.F. Mason (1977). Seismic
    studies for improved yield determination, Systems, Science
    and Software Quarterly Technical Report SSS-R-77-3345.

Bache, T.C., W.L. Rodi, and D.G. Harkrider (1978a). Crustal
    structures inferred from Rayleigh wave signatures of NTS
    explosions, Bull. Seism. Soc. Am., 68, 1399-1413.

Bache, T.C., W.L. Rodi, and B.F. Mason (1978b). Source amplitudes
    of NTS explosions inferred from Rayleigh waves at Albuquerque
    and Tucson, Systems, Science and Software Topical Report
    SSS-R-78-3690.

Biot, M.A. (1956). General theorems or the equivalence of group
    velocity and energy transport, Phys. Rev., 105, 1129-1137.

Burridge, R. and L. Knopoff (1964). Body force equivalent for
    seismic dislocation, Bull. Seism. Soc. Am., 54, 1875-1888.

Douglas, A., J.A. Hudson, and V.K. Kembhavi (1971). The analysis
    of surface wave spectra using a reciprocity theorem for
    surface waves, Geophys. J., 23, 207-223.

Harkrider, D.G. (1964). Surface waves in multilayered media I.
    Rayleigh and Love waves from buried sources in a multilayered
    elastic half-space, Bull. Seism. Soc. Am., 54, 627-679.

Harkrider, D.G., and D.L. Anderson (1966). Surface wave energy
    from point sources in plane layered earth models, J. Geophys.
    Res., 71, 2967-2980.

Harkrider, D.G. (1970). Surface waves in multilayered media II.
    Higher mode spectra and spectral ratios from point sources
    in plane-layered earth models, Bull. Seism. Soc. Am., 60,
    1937-1987.

Harkrider, D.G., C.A. Newton, and E.A. Flinn (1974). Theoretical
    effect of yield and burst height of atmospheric explosions
    on Rayleigh wave amplitudes, Geophys. J., 36, 191-225.

Harkrider, D.G. (1977). Preliminary report on the French Sahara
    Event Saphire, Semi-annual Technical Report, April-September
    1977, California Institute of Technology, AFOSR No.
    F49620-77-C-0022.

Harkrider, D.G. and C.B. Archambeau (1981). Theoretical Rayleigh
    and Love waves from an explosion in the prestressed source
    regions (to be submitted for publication).

Harkrider, D.G. and T.C. Bache (1981). Rayleigh waves generated
    by the suppression of upgoing or downgoing explosion source
    waves (to be submitted for publication).

Harkrider, D.G., P. Glover, and G.A. Frazier (1981). A represen-
    tation theorem approach to modeling the excitation of
    Rayleigh waves by complex near source phenomena (to be
    submitted for publication).

Harkrider, D.G. and D.V. Helmberger (1981). Ray and mode modeling of seismic waves at regional distances. SH waves (to be submitted for publication).

Hudson, J.A. (1969). A quantitative evaluation of seismic signals at teleseismic distances - I. Radiation from point sources, Geophys. J., 18, 353-370.

Marshall, P.D. (1970). Aspects of the spectral differences between earthquakes and underground explosions, Geophys. J., 20, 397-416.

Mueller, R.A. and J.R. Murphy (1971). Seismic characteristics of underground nuclear detonations, Part I. Seismic spectrum scaling, Bull. Seism. Soc. Am., 61, 1675-1692.

Murphy, J.R. (1977). Seismic source functions and magntiude determinations for underground nuclear detonations, Bull. Seism. Soc. Am., 67, 135-158.

McGarr, A. and L.E. Alsop (1967). Transmission and reflection of Rayleigh waves at vertical boundaries, J. Geophys. Res., 72, 2169-2180.

Rodi. W.L., J.M. Savino, T.G. Barker, S.M. Day, and T.C. Bache (1978). Analysis of explosion generated surface waves in Africa, results from the discrimination experiment and summary of current research, Systems, Science, and Software Quarterly Technical Report SSS-R-78-3653.

ON SEISMIC SYNTHESIS

B.L.N. Kennett

Department of Applied Mathematics & Theoretical Physics
University of Cambridge, Silver Street
Cambridge, CB3 9EW, England

ABSTRACT

The response of a stratified Earth model to excitation by a
seismic source may be described in terms of the moment tensor of
the source and the reflection and transmission properties of
portions of the stratification. This approach is well suited to
generating approximations to the response and the associated
computational schemes have good numerical behaviour. This
representation of the seismic field allows the role of the source
and the structural model to be clearly seen and indicates the
character of the corresponding inverse problems.

INTRODUCTION

One of the most successful methods for constructing the
response of a medium to excitation by a seismic source is to work
in terms of the reflection and transmission properties of
portions of the medium. This approach yields valuable physical
insight into the nature of the seismic wave propagation and also
has desirable characteristics for computational applications.

The detailed theory for the reflection matrix method has
been presented in a set of recent papers [1-5] and we will
restrict our attention here to the application of this approach
to the interpretation of the seismic wavefield. This will enable
us to justify a number of common approximations for the seismic
response. We will also look at the way in which the seismic
response can be used to study the inverse problems of reconstruc-
ting the nature of the seismic source.

327

*E. S. Husebye and S. Mykkeltveit (eds.), Identification of Seismic Sources – Earthquake or Underground
Explosion, 327–345.*

For simplicity in exposition we will confine our attention to the excitation of a horizontally stratified elastic half space by a source within the stratification. The generalisation of these results to spherical stratification may be readily made [14].

RESPONSE OF A STRATIFIED EARTH

Basic Equations

We consider a horizontally stratified isotropic elastic half space in which the elastic properties depend only on the depth coordinate z (measured downwards). The seismic response of the half space then depends on the P wave speed (α) , S wave speed (β) and density (ρ) distributions with depth z ; and the depth of the source.

Since we have lateral uniformity of properties within the model we make use of the powerful method of integral transforms to solve the governing coupled partial differential equations. We introduce a cylindrical set of coordinates (x , φ , z) with the depth axis passing through the source region, and then make a Fourier-Bessel expansion of the seismic displacement field within the half space in terms of vector surface harmonics. In terms of the unit vectors $\hat{x}$, $\hat{\varphi}$, $\hat{z}$ we may represent the displacement vector $\underset{\sim}{w}$ and the associated traction vector, across a horizontal plane, $\underset{\sim}{t}$ as

$$\underset{\sim}{w}(x,\varphi,z,t) = \frac{1}{2\pi}\int_{-\infty}^{\infty}dw\,e^{-iwt}\int_{0}^{\infty}dk\,k\sum_{m=-\infty}^{\infty}(U\underset{\sim}{R}_{k}^{m} + V\underset{\sim}{S}_{k}^{m} + W\underset{\sim}{T}_{k}^{m}),$$

$$\underset{\sim}{t}(x,\varphi,z,t) = \frac{1}{2\pi}\int_{-\infty}^{\infty}dw\,e^{-iwt}\int_{0}^{\infty}dk\,k\sum_{m=-\infty}^{\infty}(P\underset{\sim}{R}_{k}^{m} + S\underset{\sim}{S}_{k}^{m} + T\underset{\sim}{T}_{k}^{m}),$$

where the orthogonal surface harmonics are given by

$$\underset{\sim}{R}_{k}^{m} = \hat{z}\,Y_{k}^{m}(x,\varphi), \qquad Y_{k}^{m}(x,\varphi) = J_{m}(kx)e^{im\varphi},$$

$$\underset{\sim}{S}_{k}^{m} = k^{-1}\underset{\sim}{\nabla}_{1}Y_{k}^{m}(x,\varphi), \qquad \underset{\sim}{\nabla}_{1} = \hat{x}\,\partial_{x} + \hat{\varphi}\,x^{-1}\partial_{\varphi}.$$

$$\underset{\sim}{T}_{k}^{m} = k^{-1}\hat{z}\wedge\underset{\sim}{\nabla}_{1}Y_{k}^{m}(x,\varphi),$$

which depend on the mth order Bessel function $J_{m}(kx)$ and the angular term $\exp(im\varphi)$. The traction quantities P , S and T are related to the displacement terms by

$$P = \rho\alpha^{2}\partial_{z}U - k\rho(\alpha^{2}-2\beta^{2}V), \quad S = \rho\beta^{2}(\partial_{z}V + kU), \quad T = \rho\beta^{2}\partial_{z}W$$

For each frequency ω the Fourier-Bessel expansion corresponds to an expansion of the wavefield into a set of angularly modulated cylindrical waves. The part of the field associated with the R_k^m and S_k^m harmonics lie in a vertical plane through the source and correspond to the coupled P and SV wave parts of the seismic field. The remaining portion depending on T_k^m corresponds to SH waves since its motion is confined to a horizontal plane.

For a point source on the z axis the angular summation is restricted to $|m| < 2$, but for an extended source we have a (possibly) infinite expansion over angular order m. Since we can simulate extended sources by the superposition of point source fields we will confine attention at present to the case of a general point source. For ease of handling the conditions of continuity of displacement and traction with the stratified medium , it is convenient to introduce stress-displacement vectors containing the displacement and traction terms corresponding to the P-SV and SH wave systems. Thus for each frequency ω , horizontal wavenumber k and angular order m we introduce

$$\underset{\sim}{b}_P = [\, U,\ V,\ \omega^{-1}P,\ \omega^{-1}S \,]^T,$$

$$\underset{\sim}{b}_H = [\, W,\ \omega^{-1}T \,]^T,$$

and then these vectors satisfy coupled sets of ordinary differential equations of the form

$$\partial_z \underset{\sim}{b}\,(\omega, p, m, z) = \omega\, A\,(p, z)\, \underset{\sim}{b}\,(\omega, p, m, z),$$

in terms of the horizontal slowness p ($= k\,/\,\omega$). This reduction of the partial differential equations to first order ordinary differential equations was introduced by Alterman,-Jarosch & Pekeris [6] in the spherical case.

The stress-displacement vectors $\underset{\sim}{b}$ at two different levels within the medium may then be related by

$$\underset{\sim}{b}\,(z_1) = P(z_1, z_2)\, \underset{\sim}{b}\,(z_2),$$

where the matrix $P(z_1, z_2)$ is termed the propagator matrix for the elastic equations [7]. The propagator satisfies the matrix differential equation

$$\partial_z P = \omega\, A\, P$$

with the initial condition

$$P(z_1, z_1) = I,$$

so that its columns are independent solutions of the vector differential equations. In a uniform medium the propagator takes a rather simple form

$$P(z_1, z_2) = exp[\,\omega A\,(z_1 - z_2)],$$

which is frequently referred to as a Haskell matrix [8] This propagator matrix has a useful multiplicative property so that the propagator between levels z_1 and z_3 may be represented in terms of the propagators for the intervals (z_1, z_2) and (z_2, z_3) as

$$P(z_1, z_3) = P(z_1, z_2)\,P(z_2, z_3).$$

We may introduce a point source into the medium by forcing a jump in the stress-displacement vector across the source plane z_s so that

$$b(z_s +) - b(z_s -) = \mathcal{S}$$

and the source discontinuity $\mathcal{S}$ will vary with angular order m . The source discontinuity can be specified in terms of a moment tensor representation of the source [2].

Reflection and Transmission

In a uniform region the stress displacement vector b may be expressed in terms of up and downgoing waves via the transformation

$$b\,(\omega, p, m, z) = D\,(p)\,v\,(\omega, p, m, z),$$

where the columns of the transformation matrix are the eigenvectors of the matrix A . These eigenvectors can be recognised as stress-displacement vectors for up or downgoing P,SV or SH waves so that

$$D_P = (\,b_U^P \;\; b_U^S \;\; b_D^P \;\; b_D^S\,) \;, \quad D_H = (\,b_U^H \;\; b_D^H\,)$$

and the elements of the wave vector $\underset{\sim}{V}$ may be identified with up and downgoing wave amplitudes

$$\underset{\sim}{V} = [\underset{\sim}{V}_U , \underset{\sim}{V}_D]^T .$$

The decomposition may be written in partitioned form as

$$\left(\frac{\underset{\sim}{w}}{\underset{\sim}{t}} \right) = \left(\begin{array}{c|c} M_U & M_D \\ \hline N_U & N_D \end{array} \right) \left(\frac{\underset{\sim}{V}_U}{\underset{\sim}{V}_D} \right) ,$$

where M_U , M_D are the displacement transformations from up and downgoing waves and N_U , N_D the corresponding traction transformations.

For a varying medium it is not always possible to make an unambiguous separation of the up and downgoing parts of the wave field within the medium. However, if we isolate a portion (z_1 , z_2) of the stratification by sandwiching it between two uniform half spaces with continuity of properties at the limits, we may decompose the seismic wavefield into up and downgoing parts in these uniform media. We may then define Reflection and Transmission matrices for the region (z_1 , z_2) by relating the wave vectors at $z_1 -$, $z_2 +$. With incident downgoing waves at z_1 we have reflection and transmission matrices R_D^{12} , T_D^{12} whose entries are the reflection and transmission coefficients for the region. Thus for P–SV waves

$$R_D^{12} = \left(\begin{array}{cc} r_{PP} & r_{PS} \\ r_{SP} & r_{SS} \end{array} \right)_D^{12} ,$$

and for SH waves

$$R_D^{12} = \left(r_{HH} \right)_D^{12} .$$

For incident upgoing waves at the base z_2 we get the corresponding matrices R_U^{12} , T_U^{12} .

The wave vectors at $z_1 -$ and $z_2 +$ are related by the wave propagator

$$\underset{\sim}{V}(z_1 -) = Q(z_1 -, z_2 +) \underset{\sim}{V}(z_2 +) ,$$

where

$$Q(z_1 -, z_2 +) = D(z_1 -)^{-1} P(z_1, z_2) D(z_2 +) .$$

In partitioned form the wave propagator depends on the reflection and transmission matrices for the region (z_1 , z_2) and

$$\begin{pmatrix} \underset{\sim}{V}_U(z_1-) \\ \hline \underset{\sim}{V}_D(z_1-) \end{pmatrix} = \left(\begin{array}{c|c} T_U^{12} - R_D^{12}(T_D^{12})^{-1}R_U^{12} & R_D^{12}(T_D^{12})^{-1} \\ \hline -(T_D^{12})^{-1}R_U^{12} & (T_D^{12})^{-1} \end{array} \right) \begin{pmatrix} \underset{\sim}{V}_U(z_2+) \\ \hline \underset{\sim}{V}_D(z_2+) \end{pmatrix}.$$

If we now consider the reflection and transmission properties of a composite region (z_1 , z_3) in terms of the properties of (z_1 , z_2) and (z_2 , z_3), we may proceed by introducing at z_2 a 'fictional' uniform medium with the properties at z_2 . Then we may write, by an extension of our treatment above,

$$\underset{\sim}{V}(z_1-) = Q(z_1-, z_3+) \underset{\sim}{V}(z_3+),$$

$$= Q(z_1-, z_2) Q(z_2, z_3+) \underset{\sim}{V}(z_3+),$$

and each of the wave propagators $Q(z_1-, z_2)$, $Q(z_2, z_3+)$ can be expressed in terms of the corresponding reflection and transmission matrices e.g. R_D^{12}, R_D^{23}. After performing the multiplication of the partioned matrices we find that we have a set of addition rules for the reflection matrices. For the downgoing matrices for example

$$R_D^{13} = R_D^{12} + T_U^{12}R_D^{23}[I - R_U^{12}R_D^{23}]^{-1}T_D^{12},$$

$$T_D^{13} = T_D^{23}[I - R_U^{12}R_D^{23}]^{-1}T_D^{12}.$$

The physical interpretation of these expressions becomes apparent if we make a series expansion of the matrix inverse in the form

$$[I - A]^{-1} = I + A + A^2 + A^3 + \dots$$

so that the matrices R_D^{13}, T_D^{13} may be expressed as

$$R_D^{13} = R_D^{12} + T_U^{12}R_D^{23}T_D^{12} + T_U^{12}R_D^{23}R_U^{12}R_D^{23}T_D^{12} + \dots$$

$$T_D^{13} = T_D^{23}T_D^{12} + T_D^{23}R_U^{12}R_D^{23}T_D^{12} + \dots$$

and the higher terms in these expansions have further powers of the compound matrices $R_U^{12}R_D^{23}$ introduced. These expansions are illustrated in fig 1.

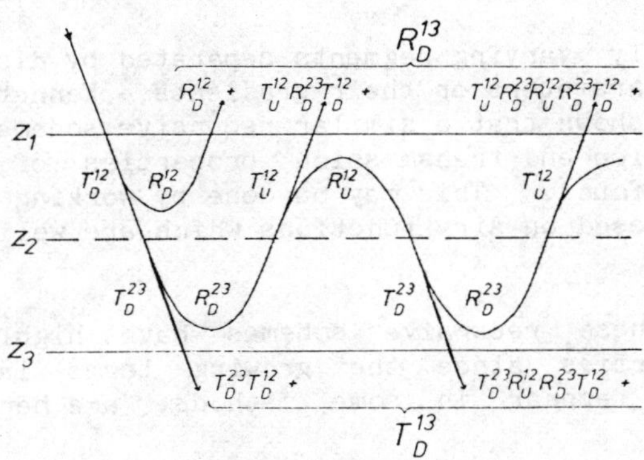

Fig 1. Interaction series expansion for reflection and transmission matrices

The total response to some incident downgoing wavefield $V_D(z_1)$ can be considered as the sum of contributions from each term in the series. The action of any individual term may be seen by reading it from right to left.

If we consider the reflected part of the field then R_D^{12} corresponds to reflection from the region (z_1 , z_2). The second term $T_U^{12} R_D^{23} T_D^{12}$ arises from transmission down through the upper region (12), reflection by the lower part (23) and transmission back up through the upper zone (12). In $T_U^{12} R_D^{23} R_U^{12} R_D^{23} T_D^{12}$ an additional internal multiple within (z_1 , z_3) is introduced via $R_U^{12} R_D^{23}$. The total response includes all such internal multiples. Truncated forms of these interaction series for reflection and transmission matrices limit the number of possible internal reflections; this can be very convenient when one wishes to examine the nature of a reflection process [9-11]

So far we have not indicated how we may construct the wave propagator and thus find the reflection and transmission matrices.

For a model composed of a stack of uniform layers one of the most effective methods is to use the addition rule for reflection matrices developed above in a recursive manner [1]. Thus one starts at the deepest interface and finds the interfacial reflection and transmission matrices and then adds in the phase delays through the overlying layer and then the effect of the next interface. The calculation can then be stepped up to the next higher interface using the reflection and transmission matrices for all the material below the current level.

For smoothly varying segments separated by discontinuities in the elastic parameters or their gradients, Kennett & Illingworth [4] have shown that a similar recursive scheme may be used once the reflection and transmission properties of a gradient zone have been found. This may be done by working with uniform approximations based on Airy functions which are well behaved at turning points.

Both of these recursive schemes have highly desirable numerical properties since the growing terms in evanescent regions, which appear in some methods, are here explicitly excluded.

The reflection matrix method sketched above provides a convenient means of handling coupled P and S wave propagation, but to what extent are P and S waves coupled within the Earth? Over much of the Earth's mantle there is very little coupling [12], but interconversion between P and S waves is particularly marked at the free surface and the core-mantle boundary and to a slightly lesser degree at the Moho discontinuity and in the gradient zone down to 900 km or so in the upper mantle [4]. Conversion may also occur within the crust.

Response of the half space

We consider a general point source at the level z_S within the half space, which below the level z_L we take to be uniform (fig 2).

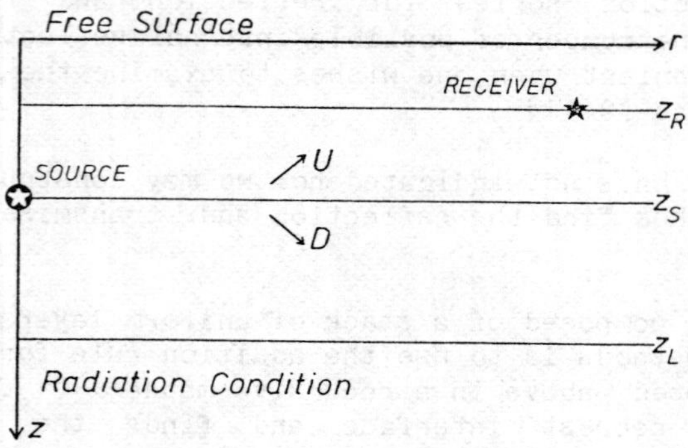

Fig 2. Source in elastic half space

At the free surface we require the traction on the plane $z = 0$ to vanish and so the stress-displacement vector must take the partitioned form

$$\underset{\sim}{b}(0) = [\underset{\sim}{W_0}, \underset{\sim}{0}]^T.$$

In the underlying half space we wish there to be only downgoing or decaying evanescent waves and so from the eigenvector decomposition of the stress-displacement field we require

$$\underset{\sim}{b}(z_L) = D_L [\underset{\sim}{0}, \underset{\sim}{V_{DL}}]^T.$$

Now across the source level we have a jump $\underset{\sim}{\mathcal{S}}$ in the stress- displacement vector. Thus we may relate the surface displacement $\underset{\sim}{W_0}$ to the source term $\underset{\sim}{\mathcal{S}}$ and the radiation into the lower half space $\underset{\sim}{V_{DL}}$ by

$$\begin{pmatrix} \underset{\sim}{W_0} \\ \underset{\sim}{0} \end{pmatrix} = P(0, z_L) \, D_L \begin{pmatrix} \underset{\sim}{0} \\ \underset{\sim}{V_{DL}} \end{pmatrix} - P(0, z_S) \, \underset{\sim}{\mathcal{S}} ,$$

$$= F(0, z_L) \begin{pmatrix} \underset{\sim}{0} \\ \underset{\sim}{V_{DL}} \end{pmatrix} - \begin{pmatrix} \underset{\sim}{S_W} \\ \underset{\sim}{S_T} \end{pmatrix} ,$$

where we have introduced the equivalent surface source $[\underset{\sim}{S_W}, \underset{\sim}{S_T}]^T$ to the buried jump $\underset{\sim}{\mathcal{S}}$. In terms of the partitions of the matrix F

$$\underset{\sim}{W_0} = F_{WD} \, F_{TD}^{-1} \, \underset{\sim}{S_T} - \underset{\sim}{S_W} ,$$

F may be found from the propagator $P(0, z_L)$ for the entire stratification or alternatively in terms of the reflection properties of the whole medium beneath the free surface. In this form the surface displacement $\underset{\sim}{W_0}$ is given by [2]

$$\underset{\sim}{W_0} = (M_D + M_U \, R_D^{0L})(N_D + N_U \, R_D^{0L})^{-1} \underset{\sim}{S_T} - \underset{\sim}{S_W} ,$$

in terms of the displacement and traction transformations M , N . The reflection matrix for the free surface arises from the vanishing traction condition at the surface and is given by

$$\tilde{R} = - N_D^{-1} N_U .$$

We may therefore display the free surface reverberation operator within the surface displacement by writing

$$\underset{\sim}{W}_0 = \left(M_D + M_U R_D^{OL} \right)\left(I - \tilde{R} \, R_D^{OL} \right)^{-1} N_D^{-1} \underset{\sim}{S}_T - \underset{\sim}{S}_W \, .$$

This form is very suitable for shallow sources [13], but the surface source term involves a propagator

$$\left[\underset{\sim}{S}_W, \ \underset{\sim}{S}_T \right]^T = P(0, z_S) \, \underset{\sim}{\mathcal{J}},$$

and has poor numerical behaviour for evanescent fields. For deep sources it is therefore advantageous to seek representations of the seismic field which depend only on the reflection and transmission matrices and so acheive good numerical behaviour.

Hitherto we have represented the source via the jump $\underset{\sim}{\mathcal{J}}$ in the stress–displacement vector $\underset{\sim}{b}$ across the source plane, alternatively we may work in terms of a jump in the wave vector across this plane

$$\underset{\sim}{V}(z_S+) - \underset{\sim}{V}(z_S-) = D_S^{-1} \underset{\sim}{\mathcal{J}} = \underset{\sim}{\Sigma},$$

and $\underset{\sim}{\Sigma}$, like $\underset{\sim}{\mathcal{J}}$, will depend on angular order m. If we now partition $\underset{\sim}{\Sigma}$ in the form

$$\underset{\sim}{\Sigma} = \left[\underset{\sim}{\Sigma}_U, \ \underset{\sim}{\Sigma}_D \right]^T$$

then we can identify $-\underset{\sim}{\Sigma}_U$ with the upward radiation produced by the source and $\underset{\sim}{\Sigma}_D$ with the downward radiation.

With this representation we may show [2,3] that the surface displacement from a buried source is given by

$$\underset{\sim}{W}_0 = \left(M_U + M_D \tilde{R} \right)\left(I - R_D^{OS} \tilde{R} \right)^{-1} T_U^{OS} \left(I - R_D^{SL} R_U^{FS} \right)^{-1} \cdot$$
$$\left(R_D^{SL} \underset{\sim}{\Sigma}_D - \underset{\sim}{\Sigma}_U \right)$$

and the elements appearing in this solution are represented schematically in fig 3. The reflection matrix R_U^{FS} includes the effect of the stratification and the free surface above the source. The expression $\left(M_U + M_D \tilde{R} \right)$ is a conversion factor which takes the amplitude of a upgoing wave into the displacement appropriate to a free surface.

$$(I - R_D^{SL} R_U^{FS})^{-1} \qquad (I - R_D^{RS} \tilde{R})^{-1}$$

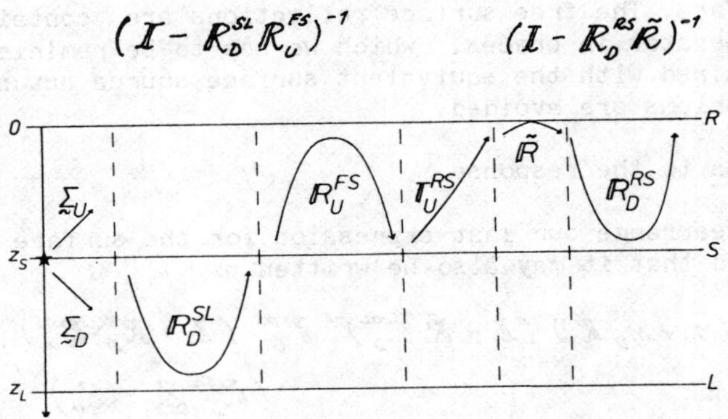

Fig 3. Elements in surface displacement solution

The physical meaning of the displacement representation is, as before,   obtained by reading from right to left.   The term $(R_D^{SL} \Sigma_D - \Sigma_U)$ corresponds to the entire upward radiation effect of the source at the plane $z = z_s$. This arises in part from direct upward radiation $(- \Sigma_U)$ and in part by reflection of energy which initially set off downwards from the source $(R_D^{SL} \Sigma_D)$.   The term $(I - R_D^{SL} R_U^{FS})^{-1}$ corresponds to a reverberation operator through the entire half space, including the free surface, coupling the upper and lower parts at the level z_s . Free surface waves (Rayleigh or Love waves) correspond to the singularities of this operator and thus to the vanishing of the secular function

$$det\ (I - R_D^{SL} R_U^{FS}) = 0 .$$

The remaining terms $(I - R_D^{OS} \tilde{R})^{-1} T_U^{OS}$ transmit the waves from the source level up to the surface , allowing for surface reflections near the receiver.

The effect of the free surface may be isolated by using an alternative expression for the surface displacement [5]

$$W_0 = [\ M_U - (M_D + M_U R_D^{OL})(I - \tilde{R} R_D^{OL})^{-1} \tilde{R}\ ]\ G(z_s),$$

where

$$G(z_s) = T_U^{OS} [I - R_D^{SL} R_U^{OS}]^{-1} (R_D^{SL} \Sigma_D - \Sigma_U).$$

and includes reverberations within the stratification but not from the surface. The free surface reflections are contained in the first operator in braces, which we see to be reminiscent of the form obtained with the equivalent surface source but now the numerical problems are avoided.

Approximations to the response

If we rearrange our last expression for the surface displacement we find that it may also be written as

$$\underset{\sim}{W}_0 = (M_U + M_D \tilde{R})(I - \tilde{R} R_D^{OL})^{-1} T_U^{OS}(I - R_D^{SL} R_U^{OS})^{-1}$$
$$(R_D^{SL}\underset{\sim}{\Sigma}_D - \underset{\sim}{\Sigma}_U)$$

and we may now generate approximations to this full response by expanding out the free surface and stratification reverberation operators

$$(I - \tilde{R} R_D^{OL})^{-1} = I + \tilde{R} R_D^{OL} + \tilde{R} R_D^{OL}\tilde{R} R_D^{OL} + \dots$$

$$(I - R_D^{SL} R_U^{OS})^{-1} = I + R_D^{SL} R_U^{OS} + R_D^{SL} R_U^{OS} R_D^{SL} R_U^{OS} + \dots$$

Then if we extract the simplest contributions which involve no internal multiples in the stratification

$$\underset{\sim}{W}_0 = (M_U + M_D \tilde{R})\left[T_U^{OS} R_D^{SL}(\underset{\sim}{\Sigma}_D - T_D^{OS}\tilde{R} T_U^{OS}\underset{\sim}{\Sigma}_U) - T_U^{OS}\underset{\sim}{\Sigma}_U \right.$$
$$+ T_U^{OS} R_D^{SL} T_D^{OS}\tilde{R} T_U^{OS} R_D^{SL}\underset{\sim}{\Sigma}_D$$
$$\left. + R_D^{OS}\tilde{R} T_U^{OS}(R_D^{SL}\underset{\sim}{\Sigma}_D - \underset{\sim}{\Sigma}_U) + Rem \right]$$

The remainder (Rem) includes all those contributions with further multiples of the types $\tilde{R} R_D^{OL}$, $R_D^{SL} R_U^{OS}$.

We now examine the nature of these terms. Direct upward propagation gives $(- T_U^{OS}\underset{\sim}{\Sigma}_U)$, whilst all energy which has returned once from beneath the level of the source occurs in $T_U^{OS} R_D^{SL}(\underset{\sim}{\Sigma}_D - T_D^{OS}\tilde{R} T_U^{OS}\underset{\sim}{\Sigma}_U)$. These two terms constitute the main arrivals characterised as P and S and through $T_U^{OS}(R_D^{SL}\underset{\sim}{\Sigma}_D - \underset{\sim}{\Sigma}_U)$ include the energy which left the source through both the upper and lower focal hemispheres. The contribution $T_U^{OS} R_D^{SL} T_D^{OS}\tilde{R} T_U^{OS}\underset{\sim}{\Sigma}_U$ represents waves which started off upwards but which is reflected by the free surface before being returned by the structure beneath the source. This term thus

includes   the surface reflected phases pP,sP,pS,sS generated near
the source (fig 4a).

The next contribution with just a single surface reflection
$T_U^{os} R_D^{SL} T_D^{os} \tilde{R} T_U^{os} R_D^{SL} \underset{\sim}{\Sigma}_D$ has been reflected back twice below
the level of the source (fig 4b). The first passage to the
surface is similar to that in straight P and S propagation and
then at the free surface conversion can occur to give PP,PS,SP,SS
phases.

The last class of contribution $R_D^{os} \tilde{R} \, T_U^{os} (R_D^{SL} \underset{\sim}{\Sigma}_D - \underset{\sim}{\Sigma}_U)$
arises from reverberations near the receiver between the free
surface and the layering above the source.

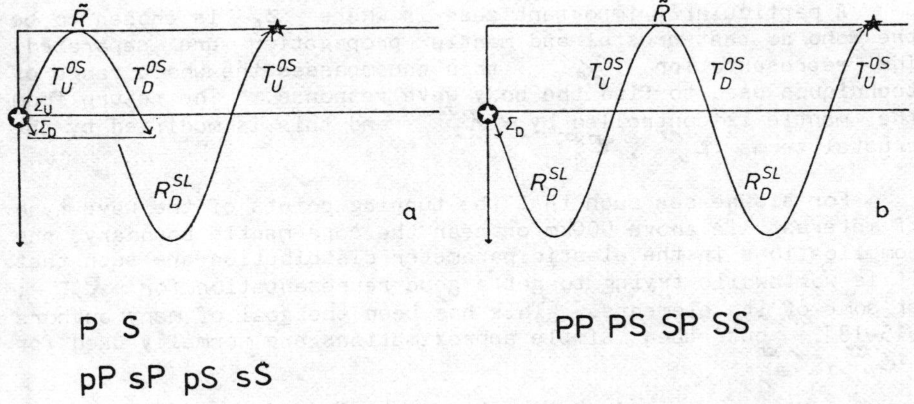

Fig 4. Portions of the seismic response

For long range propagation from a shallow source, as is of
interest in explosion studies,   the time separation between
arrivals such as P and PP is such that that we do not need to
retain a full allowance for free surface effects. We would
however like to include surface reflections near the source, and
thus pP,sP phases and their crustal reverberations and also
crustal reverberations near the receiver.

We fix a level Z_J as the base of the crustal zone and
then after some manipulation of the expression for the surface
response we can extract a portion depending on only simple
reflection below Z_J :

$$\underset{\sim}{W}_0' = (M_U + M_D \tilde{R})(I - R_D^{oJ} \tilde{R})^{-1} T_U^{oJ} R_D^{JL} T_D^{SJ} (I - R_U^{FS} R_D^{SL})^{-1} (R_D^{SL} \underset{\sim}{\Sigma}_D - \underset{\sim}{\Sigma}_U).$$

The structure of our approximation is

$$W_0' = C_U^{RJ} \, R_D^{JL} \, C_D^{SJ},$$

where C_U^{RJ} represents upward transmission effects, from z_J, in the neighbourhood of the receiver allowing for local surface effects and C_D^{SJ} represents the total effect of the source at z_J including near source reverberations.

This approach may be employed in a wide variety of problems where the elastic waves speeds increase rapidly with depth so that there is a large time separation between deeply penetrating waves and those confined to the near surface zone and surface reflected phases.

A particularly important case is where z_J is chosen to be the Moho so that crustal and mantle propagation are separated. The representation W_0' then encompasses the whole range of techniques used to find the body wave response. The return from the mantle is controlled by R_D^{JL} and this is modified by the crustal terms C_U^{RJ}, C_D^{SJ}.

For slownesses such that the turning points of the wave type of interest lie above 900km or near the core-mantle boundary, the complications in the elastic parameter distribution are such that it is worthwhile trying to get a good representation for R_D^{JL}, or some of its elements. This has been the goal of many authors [15-18], but then simple approximations are normally used for C_U^{RJ}, C_D^{SJ}.

On the other hand for turning points in the lower mantle below 900 km but away from the core-mantle boundary, the effect of mantle propagation is relatively simple and the crustal effects at source and receiver become rather more important. Douglas, Blamey and Hudson [19] have used the approximation W_0', with full calculation of source and crustal reverberations, to calculate body wave seismograms . They have made a simple allowance for lateral variation in crustal structure by taking different crustal structures on the source side and receiver side. This scheme provides a good model for P wave arrivals for epicentral distances from $30°$ to $85°$ for which R_D^{JL} can be approximated by simple ray tracing. Improved approximations to R_D^{JL} based, for example, on the reflection properties of piecewise smooth stratification [4], enable the approximation W_0' to be used with success from epicentral distances of $15°$ outwards.

In shorter range studies it may be appropriate to use a similar approximation with the level z_J lying near the surface,   and then the surface reflections within O , z_J could be suppressed as in the Reflectivity technique of Fuchs and Muller [20].

Inversion of the Integral Transforms

We have shown how to represent the surface displacement in a half space in terms of reflection and transmission matrices and source terms as a function of frequency, slowness and azimuthal order.   We now consider the inversion of the transforms to give time series for the displacement at stations at the surface of the half space.

With our assumption of a point source the azimuthal summation is restricted to angular orders $|m|<2$ and this presents no complication once the integrals over frequency and slowness have been performed for each m.   We therefore consider as an example azimuthal symmetry and construct the vertical component of the surface displacement

$$W_{oz}(x,t) = \frac{1}{2\pi} \int_{-\infty}^{\infty} d\omega\, e^{-i\omega t} \int_{0}^{\infty} dk\, k\, U_0(k,\omega)\, J_0(kx).$$

In terms of horizontal slowness p :

$$W_{oz}(x,t) = \frac{1}{2\pi} \int_{-\infty}^{\infty} \omega^2 d\omega\, e^{-i\omega t} \int_{0}^{\infty} dp\, p\, U_0(p,\omega)\, J_0(\omega p x),$$

for real ω .

We now have a choice of the order in which the slowness and frequency integrals are undertaken.   When the slowness integral is calculated first the intermediate result is the complex frequency spectrum $\breve{W}_{oz}(x,\omega)$ at a particular location, and so this approach may be designated the spectral method.   If alternatively the frequency integral is evaluated first, we have a final integral over slowness and we follow Chapman [21] by calling this approach the slowness method.   Within each of these broad classes a variety of techniques are used to evaluate the slowness integral.

The nature of the technique to be employed is principally governed by the character of the seismic response retained in the expression for $U_0(p,\omega)$ which is to be integrated.

The spectral method has mostly been used with either the full response [3,22] or with an approximation to the response, which for some portion of the model includes a complete treatment

of all reflection phenomena [18,20].   In these cases the contour
of integration is generally taken along the real axis and
physical attenuation is introduced to move the poles in the
response, due to surface waves, off the contour of integration.
Such an spectral integration scheme has been widely used for
theoretical seismogram calculation in the interpretation of
seismic refraction experiments on land and at sea.   The approach
is also now being applied to the interpretation of seismic
records from near earthquakes.   In fig 5 we illustrate a record
section of three component seismiograms for a shallow earthquake
(45 dip slip at 2.5 km) in a simple crustal model, calculated by
the spectral approach [3].

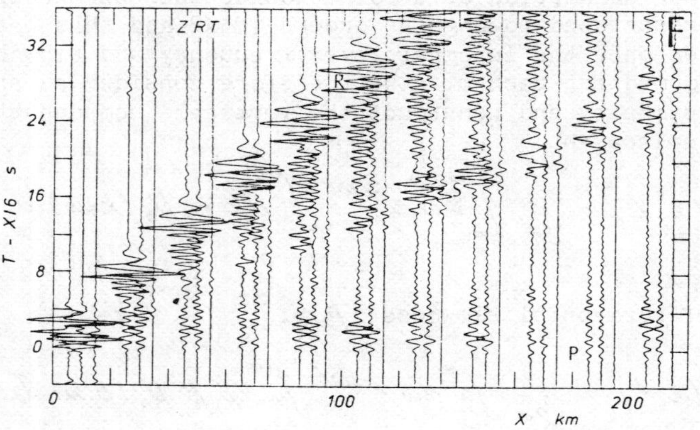

Fig 5.   Composite record section of three component seismograms

Richards [23] introduced complex slowness contours in the
spectral method ao that he could concentrate attention on a
particular phase in teleseismic propagation and this approach has
been extended to deal with many core phases [16,17].

The slowness method may be applied to the full response of a
medium by direct numerical integration [24], but has commonly
been used in a 'generalised ray' treatment.   In this case
$U_0(p,\omega)$ is represented as a sum of ray contributions with
a common source-time dependence

$$U_0(p,\omega) = s(\omega) \sum_I G_I(p) \exp[i\omega\tau_I(p)],$$

where $G_I(p)$ represents amplitude modulation along the Ith
path from source to receiver and $\tau_I(p)$ the corresponding
phase delay. The displacement is approximated by taking a finite.

set of rays from the infinite expansion representing the full response. For each ray the displacement contribution may then be constructed using a complex p contour as in the Cagniard method [25] or by an integral along the real p axis [21]. The Cagniard technique is most effective for a model of the stratification in terms of uniform layers. The method recently introduced by Chapman ('WKBJ seismograms') [21,26], may be used for a smoothly varying medium and gives good results in the presence of turning points.

For all ray expansion methods the fidelity of synthesis of the final seismograms is only as good as the choice of rays employed

THE NATURE OF THE SEISMIC SOURCE

The dependence of the seismic wavefield on the source and on the velocity structure are rather different. This is well illustrated by the surface displacement form

$$\underset{\sim}{W}_0 = (M_U + M_D \tilde{R})(I - \tilde{R} R_D^{OL})^{-1} T_U^{OS} (I - R_D^{SL} R_U^{OS}).$$

$$. (R_D^{SL} \underset{\sim}{\Sigma}_D^S - \underset{\sim}{\Sigma}_U)$$

The source contribution $\underset{\sim}{\Sigma}$ depends only on the nature of the source and the elastic parameters at the source level. The remainder of the response depends in a very involved way on the elastic parameter distribution in the model. To understand observed seismograms and to synthesise realistic seismograms we must get good information on both the source and the structure through which the waves propagate.

A convenient description of a seismic source is provided by the seismic moment tensor [27]. The elements of this tensor provide the weighting functions for the set of nine dyad elements (couples and dipoles) relative to fixed axes. For explosion and earthquake sources , conservation of angular momentum requires that the moment tensor be symmetric. The model of an explosion is the scaled unit tensor

$$M_{ij} = M_0 \, \delta_{ij} \, ,$$

for moment M_0 . For a double couple with fault plane specified by a normal $\underset{\sim}{n}$ and the auxiliary plane by normal $\underset{\sim}{e}$

$$M_{ij} = M_0 (n_i e_j + n_j e_i).$$

The source representations $\underset{\sim}{\mathcal{A}}$ or $\underset{\sim}{\mathcal{E}}$ may be constructed in terms of the moment tensor components [2]. The particular components involved depend on angular order m.

If all elements of the moment tensor have the same time dependence $\bar{s}(t)$ then the surface displacement may be represented as a weighted sum of the contributions for the six independent moment tensor components

$$\underset{\sim}{W}_0 \ (x,t) \ = \ \sum_{j=1}^{6} \ \bar{s}(t) * \underset{\sim}{W}_j \ (x,t) \ M_j$$

where $\underset{\sim}{W}_j$ is the seismogram generated by just the jth tensor component. If the structural model is known then this relation defines a linear inverse problem for the weights M_j , and this forms the basis of much current research. A similar but more involved treatment is possible if each moment tensor component may have a separate time dependence.

The success of this approach depends on having a good structural model , since errors in the propagation characteristics introduced by inaccuracies in the model will contaminate source mechanism recovery.

The problem of recovering the structure is a non-linear inverse problem, and at present is attempted by model refinement using travel-time methods supplemented by amplitude and waveform studies using theoretical seismograms, which can be calculated by the methods we have described.

REFERENCES

[1] Kennett, B.L.N.: 1974, Bull. Seism. Soc. Am. 64, pp 1685-1696
[2] Kennett, B.L.N., and Kerry, N.J.: 1979, Geophys. J. 57, pp 537-553
[3] Kennett, B.L.N.: 1980, Geophys. J. 61, pp 1-10
[4] Kennett, B.L.N., and Illingworth, M.R.: 1981, Geophys. J. (in press)
[5] Kennett, B.L.N.: 1981, Geophys. J. 63, (in press)
[6] Alterman, Z.S., Jarosch, H., and Pekeris, C.L.: 1959, Proc. R. Soc. Lond. 252A, pp 80-95
[7] Gilbert, F., and Backus G.: 1966, Geophysics 31, pp 326-332
[8] Haskell, N.A.: 1953, Bull. Seism. Soc. Am. 43, pp 17-34
[9] Kennett, B.L.N.: 1975, Bull. Seism. Soc. Am. 65, pp 1643-1651
[10] Stephen, R.A.: 1977, Geophys. J. 51, pp 169-182
[11] Hughes, V.J., and Kennett, B.L.N.: 1981, Geophysics (submitted)
[12] Richards, P.G.: 1974, Bull. Seism. Soc. Am. 64, pp 1575-1588

[13] Kennett, B.L.N.: 1979, Geophys. Prospect. 27, pp 301-321

[14] Illingworth, M.R.: 1981, Geophys. J., (submitted)

[15] Helmberger, D.V., and Wiggins, R.A.: 1971, J. Geophys. Res. 76, pp 3229-3245

[16] Choy, G.L.: 1977, Geophys. J. 51, pp 273-312

[17] Cormier, V.F., and Richards, P.G.: 1976, J. Geophys 43, pp 3-31

[18] Kind, R., and Muller, G.: 1975, J. Geophys. 41, pp 149-172

[19] Douglas, A., Hudson, J.A., and Blamey, C.: 1973, Geophys. J. 28, 285-410

[20] Fuchs, K., and Muller, G.: 1971, Geophys. J. 23, pp 417-433

[21] Chapman, C.H.: 1978, Geophys. J. 54, pp 481-518

[22] Kind, R.: 1978, J. Geophys. 44, pp 603-612

[23] Richards, P.G.: 1973, Geophys. J. 35, pp 243-264

[24] Fryer, G.: 1980, Geophys J. (in press)

[25] Wiggins, R.A., and Helmberger, D.V.: 1971, Geophys. J. 37, 73-90

[26] Dey-Sarkar, S.V., and Chapman, C.H.: 1978, Bull. Seism. Soc. Am. 68, pp 1577-1593

[27] Backus, G., and Mulcahy, M.: 1976, Geophys. J. 46, pp 341-361; 47, pp 301-329

SOME RECENT EXTENSIONS OF THE REFLECTIVITY METHOD

Gerhard Müller

Institut für Meteorologie und Geophysik
Universität Frankfurt
Feldbergstr. 47
6000 Frankfurt

Wolfgang Schott

Institut für Geophysik
Westfälische Berggewerkschaftskasse
Herner Str. 45
4630 Bochum

Abstract. Different aspects of the reflectivity method
are illustrated for the case of SH-wave propagation in
a layered half-space. Starting with the original form
of the reflectivity method, the combination with
generalized ray theory is described which allows calculations of many interesting seismic phases, such as
surface reflections and surface multiples, for only
little additional computing time. Then the calculation
of complete seismograms is described, first with wavenumber integrals whose integrands have little or
nothing to do with reflectivities, and second with
integrands consisting mainly of the reflectivities of
different parts of the layered medium. Each case is
illustrated with an example of theoretical seismograms.
Finally, a few remarks are made on aliasing in the
time domain, on fast Hankel transforms and on causal
absorption with frequency-dependent Q-factor.

INTRODUCTION

Since the original development of the reflectivity

*E. S. Husebye and S. Mykkeltveit (eds.), Identification of Seismic Sources - Earthquake or Underground
Explosion, 347–371.*
Copyright © 1981 by D. Reidel Publishing Company.

method by Fuchs (1968) this method for theoretical seismograms has undergone several modifications. It is the purpose of this report to describe some of these. Concentration naturally is on extensions that have been developed mainly at the Geophysical Institute at Karlsruhe; the reflectivity method, however, is in widespread use and thus has seen other modifications which will not be covered in the following.

The extensions which are discussed here can be divided into two groups. The first group includes combinations of the reflectivity method in its basic form (i.e. for a <u>homogeneous</u> half-space on top of a layered reflecting half-space) with generalized ray theory in order to allow for modifications of the incident and/or reflected wave field by a <u>layered</u> medium on top of the reflecting half-space. The principle of such combinations has been described some time ago by Fuchs and Müller (1971). It has recently been applied by Schott (1979) in the SH-wave case to the mantle S wave, the surface reflection sS and the free-surface multiples of the wavesystem S+sS, and by Faber and Müller (1980) to the corresponding P waves (including sP) and to conversions from S to P and from P to S at discontinuities above the reflecting half-space. In all these cases the basic feature of the original version of the reflectivity method, namely calculation of reflectivities or plane-wave reflection coefficients, is preserved. The theoretical seismograms are not complete seismograms, but display only those body-wave types which one wants to calculate.

The second group of extensions of the reflectivity method has the more ambitious aim of calculating complete seismograms. In principle the problem has been solved analytically some time ago in connection with surface-wave studies (e.g. Harkrider 1964), but only recently Kind (1978) calculated complete seismograms numerically along these lines. The integrands in the corresponding wavenumber integrals in this case include <u>all</u> wave types of the medium, not only the body waves, and thus are not reflectivities. Kennett and Kerry (1979) have shown, however, that the integrands can be expressed by reflectivities of different sections of the layered medium; Kennett (1980) has calculated theoretical seismograms with this method.

Since the purpose of this presentation is mainly tutorial,   these extensions will be described in some detail for the simple case of SH-wave propagation. The

principles for P-SV waves are the same, the corresponding formulas, however, are much more complicated and
numerical calculations much more time consuming. The
examples shown in the following are for both cases and
should illustrate the potential of the different extensions of the reflectivity method.

SH-WAVE THEORY

The model that we investigate in the following is
a half-space, consisting of homogeneous, horizontal
layers and having a free surface where the receivers
are assumed. A point source, e.g. a single force or a
double couple, acts inside of one of the layers. Such
a model is sufficiently general to be suited for
different seismological applications. The curvature
of the earth can be taken into account, if necessary,
by a flat-earth transformation (e.g. Müller 1977)
which is accurate enough for all practical purposes.
Fig. 1 shows the geometry, layer parameters and the
indexing   of layers and interfaces. Cylindrical

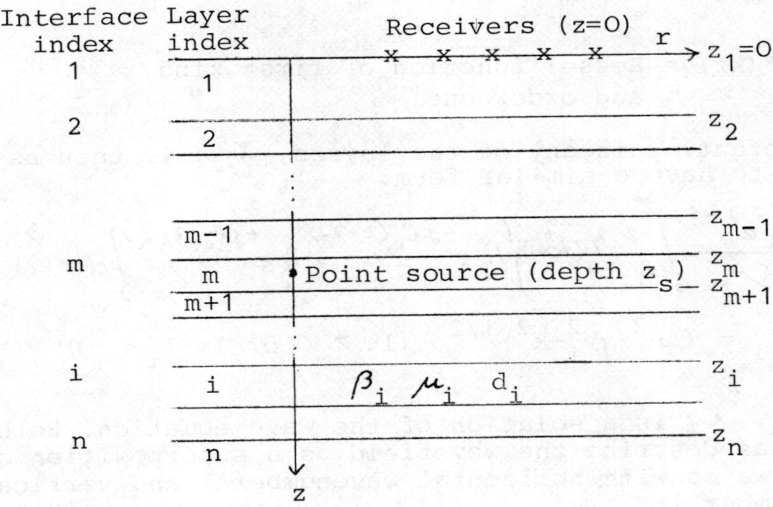

Fig. 1   Layered half-space, consisting of n-1 homo-
         geneous layers on top of a homogeneous half-
         space.
         β_i = S velocity, μ_i = rigidity, d_i = thick-
         ness, z_i = depth to top of layer i. The point
         source is located in layer m at depth z_s.

coordinates (r, φ, z) are used, and we work with an SH-wave potential $\chi(r, \varphi, z)$ from which the SH-wave displacement vector $\vec{u}$ follows by $\vec{u} = \mathrm{rot}(\vec{e}_z \chi)$ ($\vec{e}_z$ = unit vector in z direction). χ satisfies the wave equation, and the boundary conditions require continuity of χ and $\mu \, \partial \chi / \partial z$ at the interfaces within the half-space; at the free surface $\partial \chi / \partial z$ = O holds. The point source orientation is described with the aid of a Cartesian coordinate system related in the usual way to the cylindrical coordinates. We start with a harmonic single force, directed along the unit vector $\vec{f} = (f_1, f_2, f_3)$. The potential of this point force in a full-space with the properties of the source layer is (see, e.g., Müller, 1969, eq. (10)):

$$\chi_0 = \frac{\gamma_2 e^{j\omega t}}{4\pi \mu_m} \int_0^\infty \frac{j}{l_m} J_1(kr) e^{-jl_m|z-z_s|} \, dk \qquad (1)$$

$$\gamma_2 = -f_1 \sin\varphi + f_2 \cos\varphi$$

$$l_m = (\omega^2/\beta_m^2 - k^2)^{1/2} \qquad (l_m > O \text{ or } \mathrm{Im}\, l_m < O)$$

$$J_1(kr) = \text{Bessel function of first kind and order one}$$

The potential in any of the layers, χ_i, is then assumed to have a similar form:

$$\chi_i = \frac{\gamma_2 e^{j\omega t}}{4\pi \mu_m} \int_0^\infty \frac{j}{l_m} J_1(kr) \left\{ A_i e^{-jl_i(z-z_i)} + B_i e^{+jl_i(z-z_i)} \right\} dk \qquad (2)$$

$$l_i = (\omega^2/\beta_i^2 - k^2)^{1/2} \qquad (l_i > O \text{ or } \mathrm{Im}\, l_i < O)$$

As χ_0, χ_1 is a solution of the wave equation. Both formulas describe the wavefield as a superposition of plane waves with horizontal wavenumber k and vertical wavenumber l_i.

In the following we use the denotation $\bar{\chi}_i(k, \omega, z)$ for the expression in the curly brackets of (2). The first term of $\bar{\chi}_i$ represents all downgoing waves in layer i and the second term all upgoing waves. The k- and ω-dependent coefficients A_i and B_i are determined with the aid of the boundary conditions for the interfaces. Considering interface i, we have

$$\overline{\chi}_i = \overline{\chi}_{i-1} \quad \text{and} \quad \mu_i \frac{\partial \overline{\chi}_i}{\partial z} = \mu_{i-1} \frac{\partial \overline{\chi}_{i-1}}{\partial z}$$

for $z = z_i$. This leads to 2 equations relating the co-efficients A_i and B_i with A_{i-1} and B_{i-1}, which can be written in matrix form:

$$\begin{pmatrix} A_i \\ B_i \end{pmatrix} = \frac{e^{+jl_{i-1}d_{i-1}}}{2\mu_i l_i} \underbrace{\begin{pmatrix} (\mu_i l_i + \mu_{i-1} l_{i-1})e^{-2jl_{i-1}d_{i-1}} & (\mu_i l_i - \mu_{i-1} l_{i-1}) \\ (\mu_i l_i - \mu_{i-1} l_{i-1})e^{-2jl_{i-1}d_{i-1}} & (\mu_i l_i + \mu_{i-1} l_{i-1}) \end{pmatrix}}_{\text{layer matrix } m_i} \begin{pmatrix} A_{i-1} \\ B_{i-1} \end{pmatrix} \quad (3)$$

The <u>reflectivity method in its original form</u> assumes
that source and receivers are located in the same
medium, i.e., that m = 1 and that this layer actually
is a homogeneous half-space which extends infinitely
in negative z-direction. The source can be at any
depth $z_s < z_2$, and d_1 is the distance of the receivers
from the reflecting half-space which occupies the
region $z \gtrless z_2$. In this case there is only one downgoing wave in the upper homogeneous half-space (i = 1),
namely the wave from the source which is represented
by (1); here $|z-z_s| = z-z_s$ has to be chosen. Compari-
son with (2) for i = 1 (hence z_1 = 0) yields

$$A_1 = e^{+j l_1 z_s} . \quad (4)$$

The upgoing wave in the upper half-space, i.e., the
reflection from the half-space $z \gtrless z_2$, is the wave of
interest. Hence, B_1 is the coefficient which has to be
calculated. Applying (3) iteratively one obtains:

$$\begin{pmatrix} A_n \\ B_n \end{pmatrix} = m_n \cdot m_{n-1} \cdot \ldots \cdot m_3 \cdot m_2 \begin{pmatrix} A_1 \\ B_1 \end{pmatrix} = M \begin{pmatrix} A_1 \\ B_1 \end{pmatrix} = \begin{pmatrix} M_{11} & M_{12} \\ M_{21} & M_{22} \end{pmatrix} \begin{pmatrix} A_1 \\ B_1 \end{pmatrix} \quad (5)$$

The matrix M is obtained by multiplication of the
layer matrices m_i, as defined in (3). Since in the
lower homogeneous half-space there is no upgoing wave,

B_n vanishes and (5) can be solved for A_n and B_1 in terms of A_1, as given in (4). The ratio $T^{(1)} = A_n/A_1$ is the transmission coefficient and the ratio $R^{(1)} = B_1/A_1$ the reflection coefficient (or reflectivity) of the layered medium:

$$T^{(1)} = \frac{1}{M_{22}}\left(M_{11}M_{22} - M_{12}M_{21}\right), \qquad R^{(1)} = -\frac{M_{21}}{M_{22}} \tag{6}$$

The superscript (1) indicates that all layers below layer 1 constitute the reflecting and transmitting medium. The potential of the reflection follows by inserting $B_1 = R^{(1)}A_1$ into (2):

$$\chi_1^{refl} = \frac{r_z e^{j\omega t}}{4\pi\mu_1} \int_0^\infty \frac{\tilde{\phi}}{\ell_1} e^{j\ell_1(z+z_s)} J_1(kr)\, R^{(1)}(\omega,k)\,dk \tag{7}$$

The non-zero displacement components of the reflection are

$$u_r^{refl} = \frac{1}{r}\frac{\partial \chi_1^{refl}}{\partial \varphi}$$

in r-direction (a near-field term), and

$$u_\varphi^{refl} = -\frac{\partial \chi_1^{refl}}{\partial r}$$

in φ-direction. Considering only far-field displacements, one has $u_r^{refl} = 0$ and

$$u_\varphi^{refl} = -\frac{r_z e^{j\omega t}}{4\pi\mu_1} \int_0^\infty \frac{jk}{\ell_1} e^{j\ell_1 z_s} J_0(kr)\, R^{(1)}(\omega,k)\,dk. \tag{8}$$

Here, $z = 0$ has been assumed; $J_0(kr)$ is the Bessel function of first kind and order zero.

For numerical calculations of theoretical SH-wave seismograms in the case of an impulsive single force the wavenumber integral in (8) is transformed into an integral over the angle of incidence γ, related to k by $k = (\omega/\beta_1)\sin\gamma$, or over the slowness $s = k/\omega$, and the integration range is suitably restricted. The exponential factor $\exp(j\omega t)$ is replaced by the spectrum of the single force. Inverse Fourier transformation finally gives the theoretical seismograms.

SIMPLE EXTENSIONS OF THE REFLECTIVITY METHOD

These extensions take account of the fact that one often is not interested in the response of <u>all</u> depth ranges in the medium. For instance, in studies of the earth's mantle and core with the aid of earthquake waves the effect of crustal reverberations both on the source and the receiver side can often be disregarded, i.e., only the wave represented by the direct ray through the crust is considered. The effect of the crust in this case consists in time shifts and amplitude changes due to transmission across the crustal interfaces. In the case of downward propagation the amplitude coefficient A_i depends only on A_{i-1} (and not additionally on B_{i-1} as in (3)), and in the case of upward propagation B_{i-1} depends only on B_i:

$$A_i = t_i^{\downarrow} e^{-j l_{i-1} d_{i-1}} A_{i-1} \quad , \qquad t_i^{\downarrow} = \frac{2\mu_{i-1} l_{i-1}}{\mu_{i-1} l_{i-1} + \mu_i l_i} \tag{9}$$

$$B_{i-1} = t_i^{\uparrow} e^{-j l_{i-1} d_{i-1}} B_i \quad , \qquad t_i^{\uparrow} = \frac{2\mu_i l_i}{\mu_{i-1} l_{i-1} + \mu_i l_i} \tag{10}$$

$t_i^{\downarrow}$ and $t_i^{\uparrow}$ are the transmission coefficients of interface i for downward and upward transmission, respectively, and the exponential terms describe the phase shift across layer i-1.

We may thus subdivide the layered medium into, say, l layers where (9) and (10) are used ($m \leqslant l < n$) and n-1 layers below these l layers where the exact relation (3) is used. The reflection coefficient $R^{(l)}(\omega,k)$ of these n-1 layers follows from (5) with A_l and B_l replaced by A_l and B_l, respectively. The final SH-wave displacement for z = O follows from (8) by using $R^{(l)}(\omega,k)$, μ_m and l_m instead of $R^{(l)}(\omega,k)$, μ_1 and l_1 and by replacing $\exp(j l_1 z_s)$ by the product

$P(\omega,k)$ of all transmission coefficients $t_i^{\downarrow}$ and $t_i^{\uparrow}$ and all exponential phase terms of the ray from the source to the top $z = z_{l+1}$ of the reflecting layer stack and back to the surface $z = 0$:

$$u_{\varphi}^{refl} = -\frac{\delta_z e^{i\omega t}}{2\pi\mu_m} \int_0^{\infty} \frac{ik}{l_m} P(\omega,k)\, J_0(kr)\, R^{(l)}(\omega,k)\, dk \tag{11}$$

Here, an additional factor of 2 has been introduced to represent the effect of the free surface at the receivers. Double-couple formulas, corresponding to (11), have been assembled by Kind and Müller (1975); they include also P-SV waves.

The procedure that leads to (11) is a combination of the reflectivity method in its original form with generalized ray theory. It can, of course, also be applied to other rays. For instance, if the ray leaves the source in the upward direction, is reflected at the free surface and subsequently at the layer stack, we are modelling the phase sS in seismological terms. Compared with the case above, which corresponds to the calculation of the S phase, $P(\omega,k)$ includes some more transmission coefficients and phase terms and in addition the reflection coefficient of the free surface (which is equal to 1). More complicated ray paths in the upper layers are treated similarly. Reflections at interfaces between these layers introduce into $P(\omega,k)$ reflection coefficients of the general form

$$r_i^{\downarrow} = \frac{\mu_{i+1} l_{i+1} - \mu_i l_i}{\mu_{i+1} l_{i+1} + \mu_i l_i} \quad \text{or} \quad r_i^{\uparrow} = -r_i^{\downarrow}. \tag{12}$$

The rays considered so far have only one interaction with the reflecting zone. Multiple interaction takes place in the case of phases such as SS, sSS, SSS etc. The corresponding modifications of (11) are obvious: $P(\omega,k)$ has to be changed according to the ray path above the reflecting zone, and the reflectivity $R^{(l)}(\omega,k)$ is replaced by its squared, cubed etc. value.

The calculation of phases such as sS and those just mentioned in addition to S does not require much extra computing time, since the reflectivity calculation which is the most time consuming step has to be done only once. Thus, the method described is an effective method for the calculation of many interesting seismic SH phases. An example of such calculations, taken from Schott (1979), is shown in Fig. 2. The seismograms are for a double-couple which models the Tonga-Fiji deep-focus earthquake of 9 October, 1967. They include the phases S and sS and their free-surface multiples of first and second order, i.e., SS, sSS, SSS and sSSS; the corresponding core reflections are automatically included. The average layer thickness was about 50 km, and the reflecting zone started at a depth of 1000 km. Observed transverse-component seismograms for this earthquake are shown in Fig. 3.

The procedure described above for SH waves can, of course, also be used for the compressional phases P, pP, sP and their P multiples, produced by interaction with the earth's surface. All these phases are represented by wavenumber integrals over the P-P reflectivity or its powers. Another application of the procedure is the calculation of compressional precursors Sp to the main SV-wave phases S, ScS and SKS, produced by wave conversion at interfaces in the lithosphere or upper mantle below the receivers. In this case the SV-SV reflectivity is used both for the main phases and the conversions; the functions $P(\omega,k)$ of the latter include SV-P transmission coefficients of all interfaces for which conversion is desired and exponential phase terms both for S and P waves. An observational study of Sp phases has been performed by Faber and Müller (1980). This study showed that, depending on focal depth, the precursors to the main SV phases are not only Sp conversions, but also the higher order P multiples of the wave system P+pP+sP mentioned above. Fig. 4, taken from that paper, shows several theoretical seismograms covering the time span from the mantle P wave to the mantle S wave and including Sp phases from the two transition zones at depths of 400 and 670 km. For a comparison with observations and discussions of the potential of Sp precursors in structural investigations the reader is referred to Faber and Müller (1980).

Our experience with these relatively simple extensions of the reflectivity method is that for little additional computing time the possibilities to calculate body-wave phases are greatly expanded. A disadvantage is that seismograms are synthesized in sort of a patchwork manner. The calculation of complete seismograms (to be described in the next section) with subsequent investigation of special time windows is

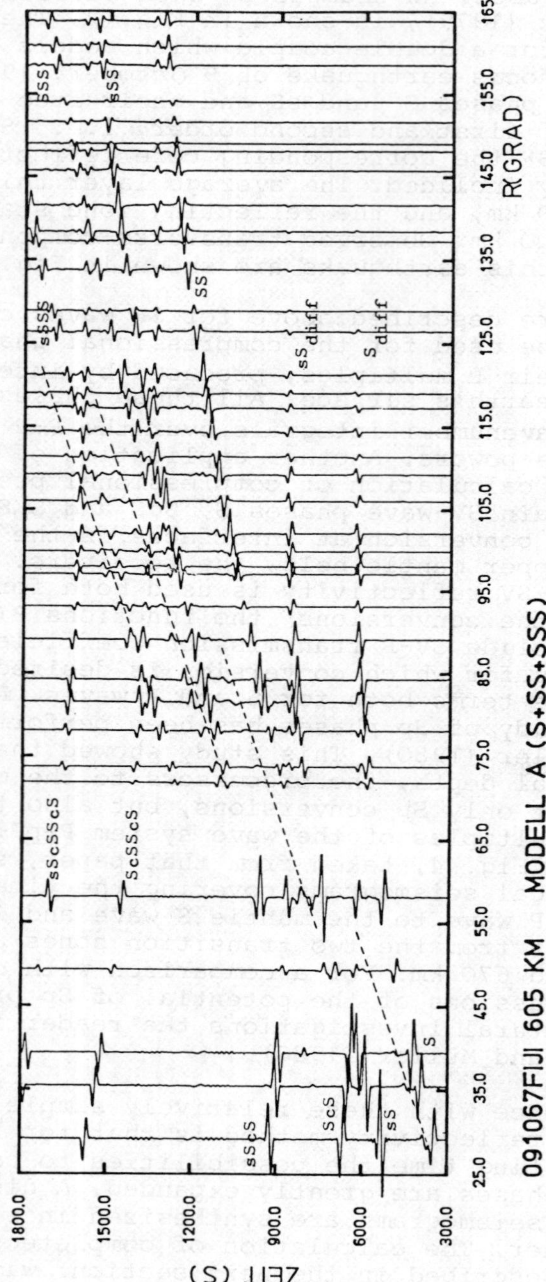

Fig. 2   Theoretical SH-wave seismograms for the Tonga-Fiji earthquake of 9 October, 1967. The arrivals along the dashed lines are numerical phases which propagate with the highest slowness (lowest phase velocity) included in the slowness integration.

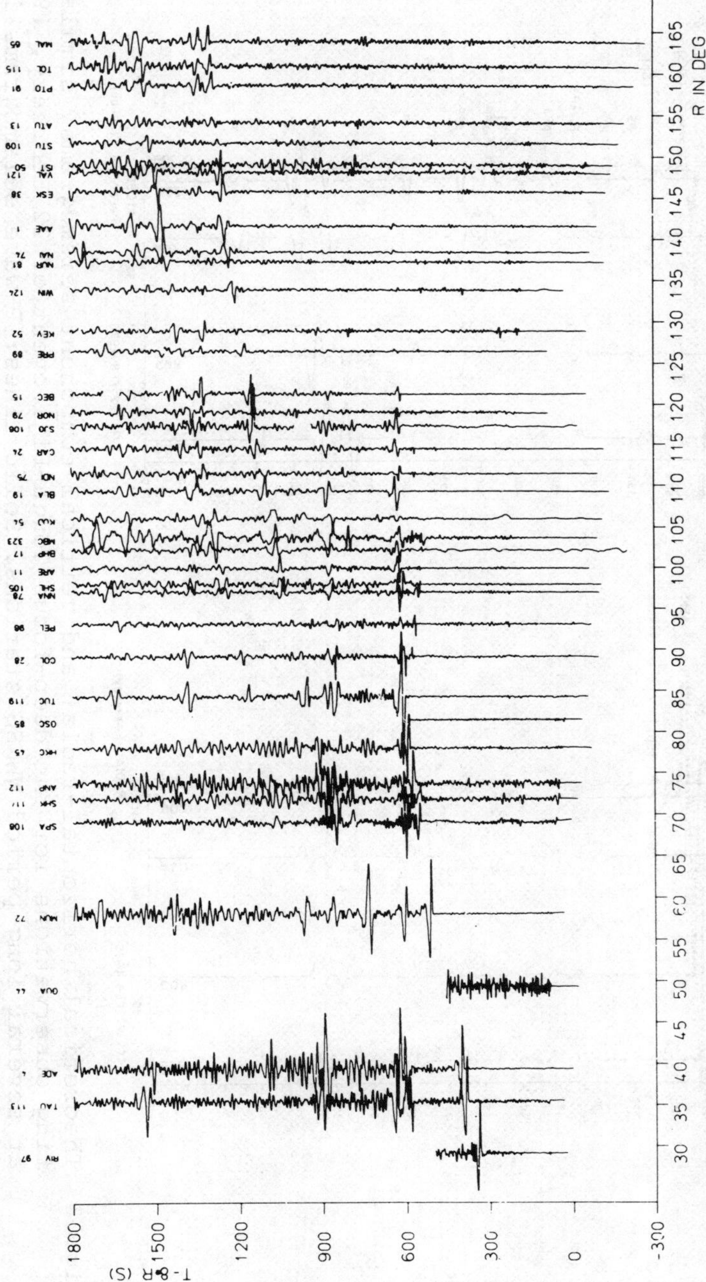

Fig. 3 Transversal-component seismograms for the earthquake of 9 October, 1967, assembled from WWNSS long-period records.

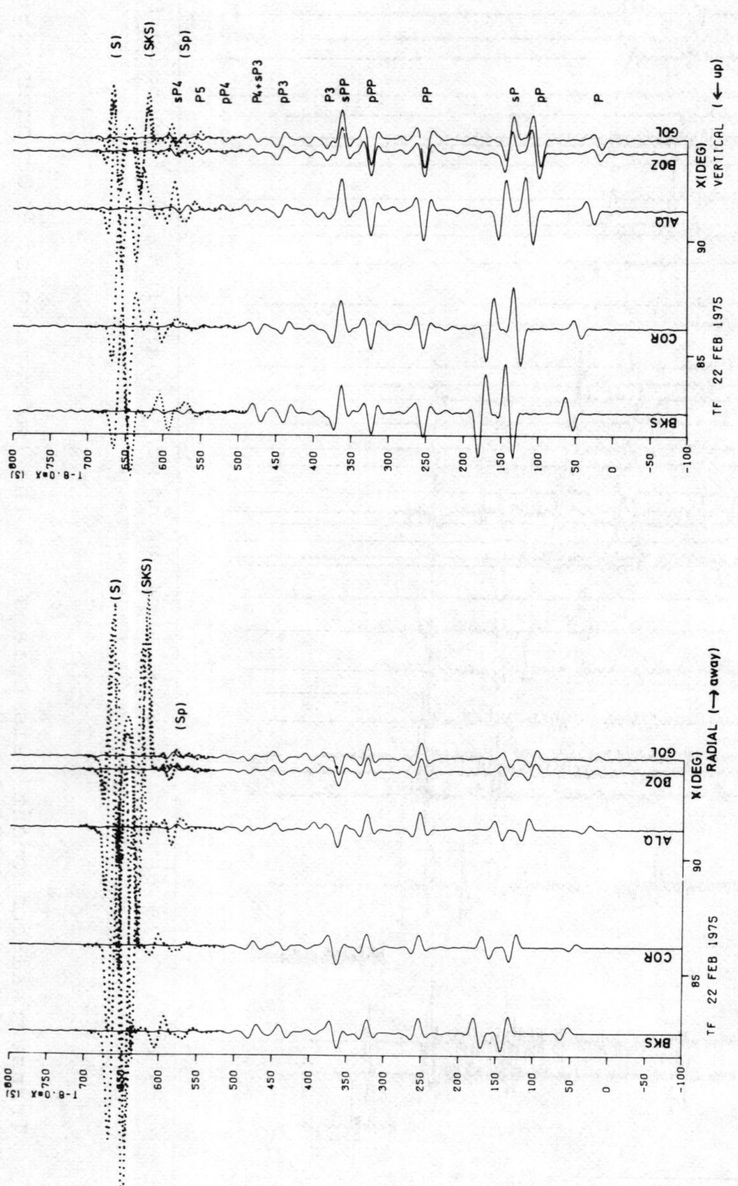

Fig. 4   Theoretical horizontal–radial and vertical component seismograms, simulating observations for the deep-focus Tonga-Fiji event of 22 February, 1975 at several long-period WWNSS stations. Solid lines: P-wave seismograms including high-order surface multiples. Dotted lines: S-wave seismograms including Sp precursors.

<u>in principle</u> the superior method. Whether this is also true <u>in reality</u> is currently an open question; applications of this method to cases such as the one illustrated in Fig. 4 are required before more definite conclusions can be drawn.

COMPLETE SEISMOGRAMS

The calculation of complete seismograms, including all body and surface waves, requires the use of formula (3), connecting the potential coefficients in neighbouring layers, throughout the whole layered half-space. This procedure has to be modified when the source layer m is traversed, since here the source radiation has to be included. We make a distinction between $\overline{\chi}_m^+$ (for $z_m \leqslant z \leqslant z_s$) and $\overline{\chi}_m^-$ (for $z_s \leqslant z \leqslant z_{m+1}$):

$$\overline{\chi}_m^+ = A_m e^{-jl_m(z-z_m)} + B_m e^{+jl_m(z-z_m)} + e^{+jl_m(z-z_s)}$$

$$\overline{\chi}_m^- = A_m e^{-jl_m(z-z_m)} + B_m e^{+jl_m(z-z_m)} + e^{-jl_m(z-z_s)}$$

The third term in these formulas represents the source radiation and follows from (1). Writing these expressions in the form

$$\overline{\chi}_m^+ = A_m^+ e^{-jl_m(z-z_m)} + B_m^+ e^{+jl_m(z-z_m)}$$

$$\overline{\chi}_m^- = A_m^- e^{-jl_m(z-z_m)} + B_m^- e^{+jl_m(z-z_m)},$$

$$\tag{13}$$

we obtain the relation between A_m^+ and B_m^+ on the one hand and A_m^- and B_m^- on the other:

$$\begin{pmatrix} A_m^- \\ B_m^- \end{pmatrix} = \begin{pmatrix} A_m^+ \\ B_m^+ \end{pmatrix} + \begin{pmatrix} A_s \\ B_s \end{pmatrix} , \quad A_s = e^{jl_m(z_s-z_m)} , \quad B_s = -1/A_s \tag{14}$$

After successive application of (3) we have

$$\begin{pmatrix} A_m^+ \\ B_m^+ \end{pmatrix} = m_m \cdot m_{m-1} \cdot \ldots \cdot m_2 \begin{pmatrix} A_1 \\ B_1 \end{pmatrix} = M^+ \begin{pmatrix} A_1 \\ A_1 \end{pmatrix}, \tag{15}$$

where M^+ is the layer matrix product for the layers above the source, and $B_1 = A_1$ has been used as a consequence of the stress-free condition at the surface $z = 0$. Similarly

$$\begin{pmatrix} A_m \\ B_m \end{pmatrix} = m_m \cdot m_{m-1} \cdot \ldots \cdot m_{m+1} \begin{pmatrix} A_m^- \\ B_m^- \end{pmatrix} = M^- \begin{pmatrix} A_m^- \\ B_m^- \end{pmatrix}, \tag{16}$$

where M^- is the layer matrix product of the layers below the source. With the aid of (14) and (15) we obtain the connection between the coefficients of the half-space with index n and those of layer 1:

$$\begin{pmatrix} A_m \\ B_m \end{pmatrix} = M^- M^+ \begin{pmatrix} A_1 \\ A_1 \end{pmatrix} + M^- \begin{pmatrix} A_s \\ B_s \end{pmatrix} = M \begin{pmatrix} A_1 \\ A_1 \end{pmatrix} + M^- \begin{pmatrix} A_s \\ B_s \end{pmatrix}$$

$$= \begin{pmatrix} M_{11} & M_{12} \\ M_{21} & M_{22} \end{pmatrix} \begin{pmatrix} A_1 \\ A_1 \end{pmatrix} + \begin{pmatrix} M_{11}^- & M_{12}^- \\ M_{21}^- & M_{22}^- \end{pmatrix} \begin{pmatrix} A_s \\ -1(A_s) \end{pmatrix},$$

where the matrix M is the layer matrix product of the whole layered half-space, as defined in (5). Since $B_n = 0$, we have

$$A_1 = B_1 = \frac{1}{M_{21} + M_{22}} \left(\frac{M_{22}^-}{A_s} - M_{21}^- A_s \right), \tag{17}$$

and the far-field displacement for $z = 0$ is

$$u_\varphi = -\frac{\gamma_2 \, e^{i\omega t}}{2\pi \mu_m} \int_0^\infty \frac{ik}{l_m} J_0(kr) A_1(\omega, k) \, dk. \tag{18}$$

In numerical calculations only matrix M (or its lower row) has to be computed; matrix M^- is an intermediate result of this computation.

Surface-wave, i.e. Love-wave contributions in u_φ are due to poles of the integrand, which are zeros of $M_{21} + M_{22}$. In a purely elastic medium these poles would

be located on the integration path along the positive
real k axis and thus produce serious difficulties in
numerical calculations. These problems can be circum-
vented by introducing absorption via complex shear velocities whereby the poles are shifted away from the
integration path.

Kind (1978) has treated the much more cumbersome
case of P-SV waves. Fig. 5 shows one of his seismogram
sections which display both P and SV body waves and
Rayleigh waves. In this calculation the phase velocity
range was chosen sufficiently broad to include all
these wave types. Of course, one can also use more restricted ranges and thus study only selected phases,
as mentioned at the end of the previous section.

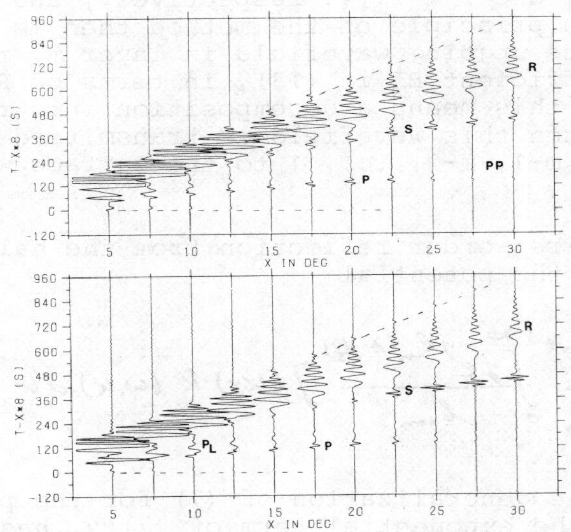

Fig. 5   Complete P-SV seismograms, calculated by Kind
         (1978) for a realistic model of the earth's
         crust and uppermost mantle. The source is an
         explosion at a depth of 0.3 km below the sur-
         face. Top: vertical component seismograms.
         Bottom: horizontal component seismograms.
         Arrivals along the dashed lines are numerical
         phases due to the restriction of the phase ve-
         locities to the window from 2.8 to 15 km/sec.

In the remainder of this section we discuss an alternative way of calculating complete seismograms (Kennett and Kerry, 1979), in which the kernel $A_1(\omega,k)$ in (18) is represented with the aid of the reflectivity $R^-(\omega,k)$ of the half-space $z \geq z_S$ and the reflectivity $R^+(\omega,k)$ of the layers $z \leq z_S$. From (15) we find

$$R^+ = \frac{A_m^+ e^{-jl_m(z_S - z_m)}}{B_m^+ e^{+jl_m(z_S - z_m)}} = \frac{M_{11}^+ + M_{12}^+}{M_{21}^+ + M_{22}^+} e^{-2jl_m(z_S - z_m)} \tag{19}$$

and from (16) and $B_n = 0$

$$R^- = \frac{B_m^- e^{+jl_m(z_S - z_m)}}{A_m^- e^{-jl_m(z_S - z_m)}} = - \frac{M_{21}^-}{M_{22}^-} e^{+2jl_m(z_S - z_m)}. \tag{20}$$

Both coefficients refer to the plane $z = z_S$, i.e., they include the phase shift due to travel from $z = z_S$ to $z = z_{m-1}$ and $z = z_{m+1}$, respectively, and back to $z = z_S$. The principle of the method then is to express the complete upgoing wavefield in layer m, represented by the coefficient B_m^+ in (13), in terms of R^+ and R^-; physically this means a decomposition into ray contributions. Then this wavefield is transmitted through the layers m-1, m-2, ..., 1 to the surface with the aid of (15).

The first-order reflection from the half-space $z \geq z_S$ has the potential

$$\frac{r_m e^{j\omega t}}{4\pi\mu_m} \int_0^\infty \frac{j e^{jl_m(z-z_S)}}{l_m} J_n(kr) R^-(\omega,k) \, dk, \tag{21}$$

which is the generalization of (7) for the present case. (In the exponential term of (7) z_S has to be set equal to zero and z replaced by $z-z_S$.) The first-order reflection (21) produces multiples between the layers $z \leq z_S$ and the half-space $z \geq z_S$. Upgoing multiples of order $p = 1,2,3,...$ have potentials which follow from (21) by the substitution

$$R^- \longrightarrow R^-(R^+R^-)^p.$$

The direct upgoing wave follows from (1), using $|z-z_S| = -(z-z_S)$. This is the same as substituting in (21)

$$R^- \longrightarrow 1.$$

This wave also produces multiples, and upgoing multiples of order $p = 1,2,3,\ldots$ are obtained from (21) by the substitution

$$R^- \longrightarrow (R^+ R^-)^p.$$

Summation of all upgoing waves gives the potential

$$\frac{\dot{r}_z e^{j\omega t}}{4\pi\mu_m} \int_0^\infty \frac{je^{jl_m(z-z_s)}}{l_m} J_n(kr) \frac{1+R^-}{1-R^+R^-} \, dk.$$

From this result we can determine by inspection the coefficient B_m^+ in (13). Recalling the relation between the potential χ and the function $\bar{\chi}$ which was given in connection with formula (2), we obtain:

$$B_m^+ = \frac{1+R^-}{1-R^+R^-} e^{jl_m(z_m-z_s)}$$

From (15) we find the relation between B_m^+ and A_1

$$A_1 = \frac{B_m^+}{M_{21}^+ + M_{22}^+}, \tag{22}$$

which has to be inserted into (18) to give the far-field displacement for $z = 0$.

There is no essential difference in the numerical calculation of the kernel A_1 between the formulas (17) and (22), provided that the matrix multiplication giving M, M^+ and M^- is done properly. This means to avoid the possibility of overflow of the exponential term $\exp(+jl_{i-1}d_{i-1})$ in the layer matrix m_i, as defined in (3). Overflow can occur if the wavenumber l_{i-1} is (negative) imaginary, i.e., whenever the waves in a layer are inhomogeneous, provided that the frequency and/or layer thickness is large enough.

However, by careful inspection of (17), (19), (20) and (22) it can be seen that the appearance of these exponential terms can be avoided completely or replaced by the appearance of their reciprocal values in the numerator of A_1.

An example of seismogram calculations with the aid of (18) and (22) in the case of a horizontal single force at the surface ($m = 1$, $z_m = z_s = 0$, $M_{21}^+ = 0$, $M_{22}^+ = 1$, $R^+ = 1$) is shown in Fig. 6. The model is a layer of coal of 1 m thickness on top of a half-space of rock with higher velocity and density. (This model is equivalent to a 2 m thick coal layer sandwiched in a full-space of rock with wave excitation in the middle of the layer.) The seismograms are dominated by the fundamental Love mode. The Airy phase at the end of the traces is relatively weak because absorption has been assumed with Q values of 50 in coal and 100 in the surrounding rock.

SOME OTHER ASPECTS OF THE REFLECTIVITY METHOD

Aliasing in the time domain

In numerical calculations of theoretical seismograms the seismogram length strongly influences the computing time, since usually the spectra are calculated for equidistant frequencies and the frequency interval is the reciprocal value of this length. Thus, the number of frequencies in a specified frequency window is proportional to the seismogram duration. For complicated models this duration can be very long, especially if complete seismograms are calculated. If the seismogram duration chosen is shorter than the response of the medium, those parts of the response beyond the end of the seismograms will be superposed on the early parts, a situation which is completely analogous to aliasing in the frequency domain and which therefore is called aliasing in the time domain. A consequence often is unwanted interference of late and early arrivals. The most straightforward although most time consuming way to avoid this situation is to increase the seismogram duration until the complete response fits into this time interval. An alternative method of which one could think (Kind, 1979) is to work with non-equidistant frequencies, such that also frequency intervals occur whose reciprocal value is

Fig. 6   Fundamental-mode Love waves in a coal seam of thickness 2 m, surrounded by rock of higher velocity and density. Source and receivers are located on the seam axis. Frequencies up to 600 Hz are included. Each seismogram is normalized with respect to its maximum amplitude.

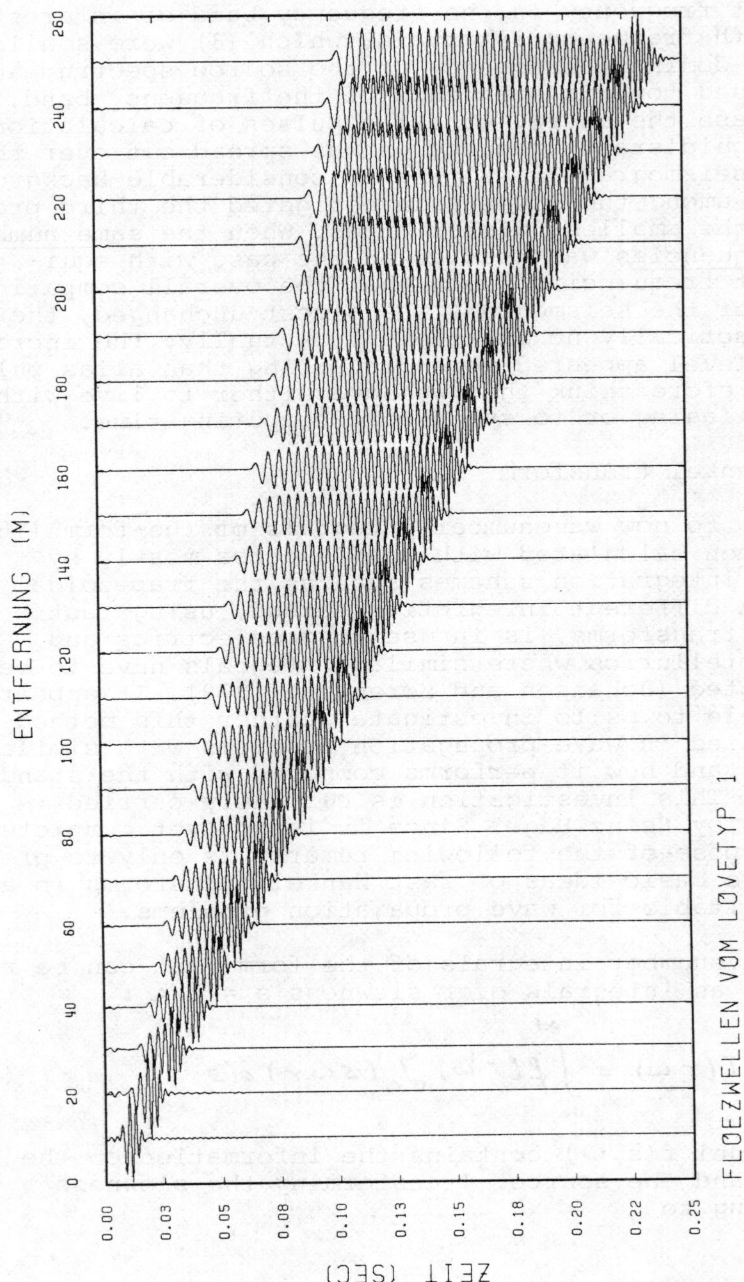
ENTFERNUNG (M)
ZEIT (SEC)
FLOEZWELLEN VOM LOVETYP 1

longer than the response of the medium. We have ex-
perimented with frequency intervals, which (1) increased or (2) decreased from the lowest to the
highest frequency in the frequency band of interest,
and with frequency intervals, which (3) were smallest
at the dominant frequency of the source spectrum and
increased towards both ends of the frequency band. In
each case the energy of alias pulses of calculations
with equidistant frequencies was spread out over the
whole seismogram, resulting in considerable background
noise; among the 3 cases investigated the third produced the smallest disturbances. When the <u>same number
of frequencies</u> was used as in the case with equidistant frequencies, such that the overall computing
time for the seismograms was almost unchanged, there
was practically no improvement. Actually, the increased
noise level appeared more disturbing than alias pulses.
We therefore think that we have either to live with
time aliasing or to spend more computing time.

Fast Hankel transform

Up to now wavenumber integrals of the form (18)
have been calculated with good success mostly by
simple integration schemes such as the trapezoidal
rule. A different integration method, using fast
Hankel transforms, is in use in geoelectrics and
magnetotellurics where similar integrals have to be
calculated (Johansen and Sørensen, 1979). It appeared
desirable to us to investigate whether this method can
be applied in wave propagation problems with similar
success and how it performs compared with the standard
method. This investigation is currently carried
through by Heinz Häge. Since it is not yet completed,
the purpose of the following remarks is only to present the basic ideas of fast Hankel transforms in a
form suitable for wave propagation problems.

Wavenumber integrals of the form (18) can be rewritten as integrals over slowness $s = k/\omega$:

$$I(r, \omega) = \int_0^\infty f(s, \omega) \, J_0(s\omega r) \, ds \tag{23}$$

The kernel $f(s, \omega)$ contains the information on the
medium and the source. Transforming the slowness
according to

$$s = s_0 e^{-u},$$

where s_0 is a reference slowness and u a dimensionless logarithmic slowness, and introducing

$$\omega r = \frac{1}{s_0} e^{v}, \tag{24}$$

where v is a dimensionless logarithmic phase velocity, (23) is changed to

$$I\left(\frac{1}{s_0 \omega} e^{v}, \omega\right) = s_0 \int_{-\infty}^{+\infty} f(s_0 e^{-u}, \omega) J_0(e^{v-u}) e^{-u} du.$$

The function

$$\overline{I}(v, \omega) = \frac{1}{s_0} e^{v} I\left(\frac{1}{s_0 \omega} e^{v}, \omega\right)$$

$$= \int_{-\infty}^{+\infty} f(s_0 e^{-u}, \omega) J_0(e^{v-u}) e^{v-u} du = F(v, \omega) * H_0(v) \tag{25}$$

is a convolution integral of the function

$$F(v, \omega) = f(s_0 e^{-v}, \omega)$$

and the filter operator (here of order zero)

$$H_0(v) = e^{v} J_0(e^{v}). \tag{26}$$

$F(v, \omega)$ is assumed to be band-limited not only in ω, but also in v, the limits being given by the limits in slowness. $\overline{I}(v, \omega)$ is calculated for equidistant values of both ω and v. The original spectrum (23) for constant distance r,

$$I(r, \omega) = \frac{1}{\omega r} \overline{I}(ln(\omega r s_0), \omega),$$

is found from $\bar{I}(v, \omega)$ by interpolation along the curve $\omega(v)$, defined by (24).

The filter operator in the form (26) is badly behaved, since for $v \rightarrow \infty$ it oscillates with increasing frequency and amplitude. However, since the kernel $F(v, \omega)$ is sampled at equidistant values of v, all that is needed is a low-pass filtered version $H_O^*(v)$ of $H_O(v)$, the cutoff-frequency being determined by the increment Δv. Johansen and Sørensen (1979) discussed in detail analytical low-pass filtering; we found numerical low-pass filtering to be suited equally well.

The calculation of (25) is more favorable than the calculation of (23) only if the number of points on the v-axis necessary to represent $F(v, \omega)$ is smaller than the number of points on the s-axis necessary to represent the original kernel function $f(s, \omega)$. An improvement can be expected, if $f(s, \omega)$ has a dispersive appearence with high "frequencies" at small slownesses and low "frequencies" at high slownesses. The increment Δs in the integration with (23) is determined by the high "frequencies", but also applied for the low "frequencies", although here one could work with a larger Δs. Going over to logarithmic slownesses gives a more balanced "frequency" distribution in $F(v, \omega)$, such that working with one value of Δv is not connected with oversampling. In this case one may need less samples in v than in s, and the fast Hankel transform would deserve its name. It is a matter of tests to find out to what extent this situation prevails in wave-propagation problems.

Causal absorption with frequency-dependent Q

Absorption can be incorporated into the reflectivity method by making the wave velocities complex. In terms of the complex, frequency-dependent rigidity $\mu(\omega) = A(\omega)\exp(j\varphi(\omega))$ $(A(\omega) = $ modulus, $\varphi(\omega) = $ phase) the S velocity is

$$\beta(\omega) = \left(\frac{\mu(\omega)}{\rho}\right)^{1/2} = \left(\frac{A(\omega)}{\rho}\right)^{1/2} e^{j\frac{\varphi(\omega)}{2}} . \tag{27}$$

The quality factor $Q(\omega)$ is defined as the ratio of the real and the imaginary part of $\mu(\omega)$, hence

$$\tan \varphi(\omega) = 1/Q(\omega) . \tag{28}$$

In weakly dissipating media absorption can be treated as a linear phenomenon. Causality then requires that $\mu(\omega)$, which is the transfer function relating strain and stress, obeys the Kramers-Krönig relations. From these follows a relation between $A(\omega)$ and $\varphi(\omega)$:

$$\ln A(\omega) = \ln A(\infty) - \frac{1}{\pi} P \int_{-\infty}^{+\infty} \frac{\varphi(\omega')}{\omega' - \omega} \, d\omega' \qquad (29)$$

In the case of power-law dependence of Q on ω,

$$Q(\omega) = Q(\omega_r) \left(\frac{\omega}{\omega_r}\right)^{\gamma}, \quad 0 \le \gamma < 1, \qquad (30)$$

the principal-value integral in (29) can be calculated to give

$$A(\omega) = A(\infty) \exp\left\{ -\frac{1}{Q(\omega)} \cot\left(\gamma \frac{\pi}{2}\right) \right\}. \qquad (31)$$

This result is valid for Q greater than about 10, in which case (27) can be replaced by

$$\beta(\omega) = \left(\frac{A(\omega)}{\rho}\right)^{1/2} \left(1 + \frac{i}{2Q(\omega)}\right). \qquad (32)$$

We then assume that $\beta(\omega)$ is known at the reference frequency ω_r which was introduced in (30):

$$\beta(\omega_r) = \beta_r \left(1 + \frac{i}{2Q(\omega_r)}\right) \qquad (33)$$

With the aid of (31) and (33) we obtain from (32) the final expression for $\beta(\omega)$:

$$\beta(\omega) = \beta_r \exp\left\{ \frac{1}{2} \cot\left(\gamma \frac{\pi}{2}\right) \left(\frac{1}{Q(\omega_r)} - \frac{1}{Q(\omega)}\right) \right\} \left(1 + \frac{i}{2Q(\omega)}\right) \qquad (34)$$

In the case of weak absorption considered here the exponential term in (34) describes the phase-velocity dispersion due to absorption. In the limit $\gamma \to 0$ which means frequency-independent Q we have

$$\beta(\omega) = \beta_r \exp\left\{ \frac{1}{\pi Q} \ln \frac{\omega}{\omega_r} \right\} \left(1 + \frac{i}{2Q}\right) \approx \beta_r \left(1 + \frac{1}{\pi Q} \ln \frac{\omega}{\omega_r} + \frac{i}{2Q}\right) \qquad (35)$$

O'Neill and Hill (1979) have used this complex velocity law in reflectivity calculations of theoretical seismograms. (34) is the generalization of (35) for frequency-dependent Q.

It remains open at present to what extent this generalization is of practical importance. Probably, it is not relevant for amplitude investigations. However, for values of γ as large as 1/3, as they have recently been suggested, dispersion effects in pulse propagation may become significantly different from those at $\gamma = 0$.

ACKNOWLEDGMENTS

We are grateful to Heinz Häge and Paul Temme for discussions on fast Hankel transforms and aliasing in the time domain, and to Ingrid Hörnchen for typing the manuscript.

REFERENCES

Faber, S., and Müller, G., 1980, Sp phases from the transition zone between the upper and lower mantle, Bull. Seism. Soc. Am. 70, pp. 487-508.

Fuchs, K., 1968, The reflection of spherical waves from transition zones with arbitrary depth-dependent elastic moduli and density, J. Phys. Earth 16, Special Issue, pp. 27-41.

Fuchs, K., and Müller, G., 1971, Computation of synthetic seismograms with the reflectivity method and comparison with observations, Geophys. J.R.A.S. 23, pp. 417-433.

Harkrider, D.G., 1964, Surface waves in multilayered elastic media, 1. Rayleigh and Love waves from buried sources in a multilayered elastic halfspace, Bull. Seism. Soc. Am. 54, pp. 627-679.

Johansen, H.K., and Sørensen, K., 1979, Fast Hankel transforms, Geophys. Prospecting 27, pp. 876-901.

Kennett, B.L.N., 1980, Seismic waves in a stratified medium II-theoretical seismograms. Geophys. J.R. A.S. 61, pp. 1-10.

Kennett, B.L.N., and Kerry, N.J., 1979, Seismic waves in a stratified half-space, Geophys. J.R.A.S. 57, pp. 557-584.

Kind, R., 1978, The reflectivity method for a buried source, J. Geophys. 44, pp. 603-612.

Kind, R., 1979, Extensions of the reflectivity method, J. Geophys. 45, pp. 373-380.

Kind, R., and Müller, G., 1975, Computation of SV
    waves in realistic earth models, J. Geophys. 41,
    pp. 149-172.

Müller, G., 1969, Theoretical seismograms for some
    types of point-sources in layered media, Part III:
    Single force and dipole sources of arbitrary
    orientation, Z. Geophys. 35, pp. 347-371.

Müller, G., 1977, Earth-flattening approximation for
    body waves derived from geometric ray theory - im-
    provements, corrections and range of applicability,
    J. Geophys. 42, pp. 429-436.

O'Neill, M.E., and Hill, D.P., 1979, Causal absorption:
    its effect on synthetic seismograms computed by the
    reflectivity method, Bull. Seism. Soc. Am. 69,
    pp. 17-25.

Schott, W., 1979, Die Reflektivitätsmethode für SH-
    Wellen in Theorie und Anwendung, Diploma thesis,
    University of Karlsruhe, 97 pp.

Addendum

A more effective method to avoid or reduce aliasing in
the time domain than the one mentioned above has been
sketched by Rosenbaum (1974); Kennett (1979) and Kind
(personal communication) have drawn attention to this
paper. The basic idea is to calculate in a first step
the _damped_ seismogram $u_\tau(t)=u(t)e^{-t/\tau}$ instead of $u(t)$
by evaluating the Fourier transform $\bar{u}(\omega)$ of $u(t)$ at
the complex frequencies $\omega-i/\tau$ and using the damping
theorem of Fourier transforms: $\bar{u}_\tau(\omega)=\bar{u}(\omega-i/\tau)$. Evidently, $u_\tau(t)$ is less disturbed by aliasing in the time
domain than $u(t)$. Multiplication of $u_\tau(t)$ by $e^{+t/\tau}$ in
a second step gives the desired seismogram $u(t)$. First
experiments with this method, with τ equal to 20 to 40
per cent of the desired length of $u(t)$, show promising
results.

Kennett, B.L.N., 1979, Theoretical reflection seismo-
    grams for elastic media, Geoph. Prospecting 27,
    pp. 301-321.

Rosenbaum, J.H., 1974, Synthetic microseisms: logging
    in porous formations, Geophysics 39, pp. 14-32.

ISOCHRONAL FORMULATION OF SEISMIC DIFFRACTION

P.W. Buchen and R.A.W. Haddon

Department of Applied Mathematics, University of Sydney, Australia

1.   INTRODUCTION

Consideration of the interaction of both P and S wave motions in boundary-value problems requires the solution of the full elastodynamic equation of motion. There are, however, a number of situations in which such phenomena can be ignored, like for example (i) near-vertical reflection seismics extensively used in oil exploration and (ii) certain seismological P-wave phenomena. In these circumstances, seismic wave propagation can be studied via the scalar-wave equation. Diffraction theory in terms of the scalar-wave equation can be tackled from a number of different viewpoints. The most rigorous approach is in the setting of a full boundary-value problem, though only very simple geometries can be handled. On the other hand, Kirchhoff theory, though not exact, does have the advantage that complex media can be satisfactorily treated. The purpose of this study is to demonstrate synthetic seismograms calculations for a number of structural configurations utilising an efficient computational procedure derived from the Kirchhoff integral solution of the scalar-wave equation. The method, which is briefly described in the text, is an extension of the works of Hilterman (1) and Trorey (2). Three applications are specifically considered. First, a study is made of the effects that the angle of a plane wedge or corner has on diffraction. Second, a study of a hypothetical reef structure of some complexity is made and thirdly, the method is applied to a study of the diffraction of PKP waves due to shadowing by the Earth's core.

E. S. Husebye and S. Mykkeltveit (eds.), Identification of Seismic Sources - Earthquake or Underground Explosion, 373–381.

2.   ISOCHRONAL FORMULATION OF THE KIRCHHOFF SOLUTION

Kirchhoff's integral solution of the scalar-wave equation can be expressed in the form

$$u(P,t)= \int_{S} \left\{ u*\frac{\partial G}{\partial n} - G*\frac{\partial u}{\partial n} \right\} dS_Q \tag{1}$$

where $u(P,t)$ denotes an appropriate scalar field variable at point P and time t while G is the Green's function. Under the integral we have $u = u(Q,t)$, $G = G(P,Q,t)$ and * denotes time-domain convolution. For a homogeneous medium characterized by wave velocity c, we choose $G = (-4\pi r)^{-1}\delta(t-r/c)$ where $r = PQ$. In a weakly-heterogeneous medium of slowly varying physical properties, we shall choose the ray-theory approximation $G \approx -(4r\pi R)^{-1}\delta(t-T)$ where R^{-1} represents geometrical spreading and T travel time for a ray from Q to P. Kirchhoff diffraction theory approximates solutions of the wave equation by specifying u and its normal derivative $\partial u/\partial n$ on some arbitrary closed S surrounding P. Usually their specific values are determined from an assumed incident wave field on S. It may happen that part of S (say S') is in the geometrical shadow (vis-a-vis the incident wave), then both u and $\partial u/\partial n$ are taken to vanish on S'. These are referred to as Kirchhoff boundary conditions. They are, of course, not exact because secondary interactions between the assumed incident waves and the boundary are ignored. However, good approximations to the exact solution can be obtained when (i) dominant wavelengths are small compared with diffractor dimensions, (ii) sources and receivers are many wavelengths removed from diffracting surfaces, and (iii) angular deviations from diffracting elements are small. We remark that in terms of classical Kirchhoff diffraction theory, seismic problems will almost always belong to the class of Fresnel diffraction rather than Fraunhoffer diffraction.

Our approach differs from others mainly in the parameterization of the surface integral appearing in Eq. (1). Let τ denote the total travel time $(T_{OQ}+T_{QP})$ along two ray segments from a source at O to a diffracting element Q of S and from Q to a sensor location at P. Further, let θ denote an appropriately chosen curvilinear coordinate on S. For example, in the case of a plane diffractor (say, representing a dipping crustal horizon), we may conveniently choose θ to be the polar angle in the plane with the geometrical reflection point as its pole. In general, the surface element dS_Q can be represented mathematically by

$$dS_Q = J(\tau,\theta)d\tau d\theta \tag{2}$$

where J is the surface Jacobian. For simplicity S will be assumed
either plane or spherical so that J can easily be determined
analytically. We first integrate Eq. (1) with respect to θ,
keeping τ constant. Curves of constant τ on S are called <u>isochrons</u>.
Clearly, all elements of a particular t-isochron constructively
interfere at P to contribute to the ultimate signal at time
t. For plane diffractors, isochrons are families of confocal
ellipses being the curves of intersection of the diffracting plane
with the two-way travel time ellipsoid associated with the
given source-receiver pair. It transpires that the integrand
of Eq. (1) varies slowly along isochrons at all but a number of
isolated points so that a robust integration scheme can be used.
At these isolated points, which ultimately give rise to diffracted
pulses, the integrand can be shown to have $t^{\frac{1}{2}}H(t)$ behaviour.
The final integration with respect to τ is transformed into a
time-domain convolution in the of the form

$$u(P,t) = \int W(\tau)\ \dot{F}(t-\tau)\ d\tau \tag{3}$$

where $\dot{F}$ denotes the derivative of the source-time function. The
weight function W(t) contains all the geometrical information
of the diffraction model as 'viewed' from the given source-
receiver locations. The convolution can be readily performed
numerically, either directly in the time domain, or via the FFT-
algorithm. We have found the former to be quite satisfactory so
that the synthetic traces shown have all been computed without
any reference to the frequency domain.

3.   APPLICATIONS

We consider in this section details of the three applications
mentioned in the Introduction. The first two refer to plane
diffractors, while the third regards the CMB as a spherical
diffractor. The source-time function F(t) is taken in all cases
to be a single swing sine wave of period specified in the text.

   (a)   Figure 1 shows a plane horizontal wedge with corner
angle designated by α. The model is assumed to lie at a depth of
1000 m, and the source function has a period of 10 millisecs.
Ten reflection points, corresponding to ten surface receivers,
are indicated. Receiver no. 1 coincides with the projection of
an assumed surface shot point. Several features of the reproduced
waveforms are worth elaborating upon. Traces 1 to 5 correspond to
the shadow zone of the wedge, and so represent pure diffraction effects. Amplitudes increase to a maximum at the corner and are in
phase with the source function. Traces 6 to 10 correspond to the
illuminated zone of the wedge and should contain both diffracted
and reflected pulses. For clarity only the diffracted pulses have
been shown. Away from the shadow boundary, reflected pulse ampli-

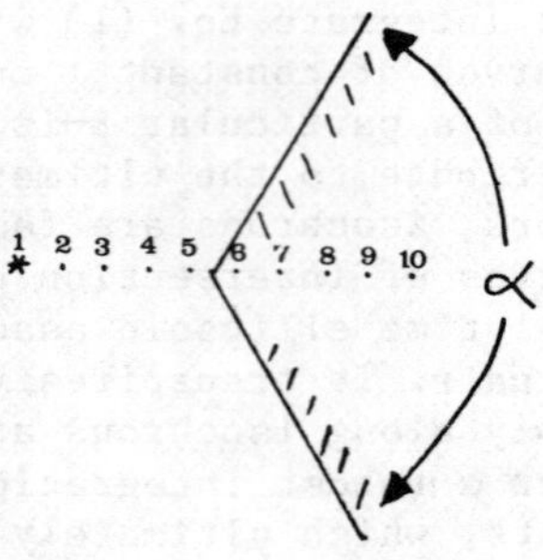

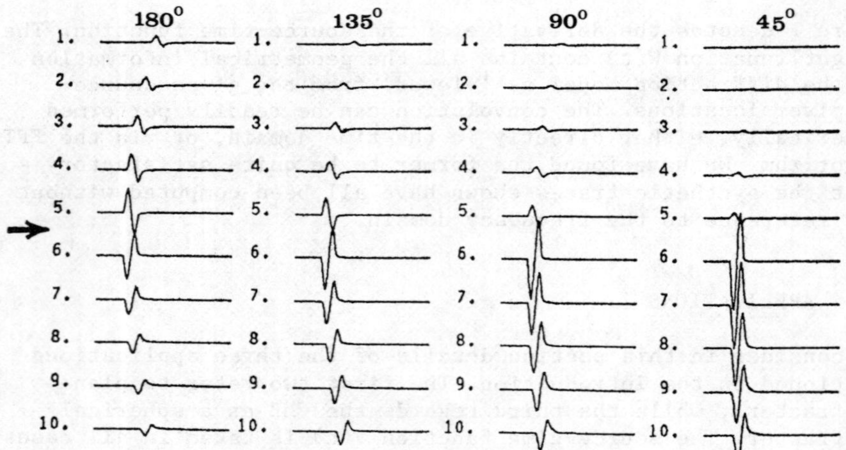

180°   135°   90°   45°

Figure 1. Synthetic waveforms of corner diffraction for varying corner angles α.

tudes would dominate, but at the shadow boundary diffracted and reflected pulse amplitudes are both equal to half the total amplitude. Diffracted pulses in the illuminated region are seen to have amplitudes which decrease away from the corner and are 180° out of phase with the source function. The discontinuity in the diffracted pulse across the shadow boundary is well known and is

exactly compensated by a corresponding discontinuity in the reflected pulse. The total wave field is thus continuous across
the shadow boundary. We also observe that proportionally more
diffracted energy is propagated into the illuminated zone at the
expense of that in the shadow zone as the wedge angle α is decreased. The case $\alpha = 180^\circ$ corresponds to diffraction by a
straight edge.

(b)   In Figure 2 we demonstrate how the method works for
calculating diffracted pulse forms for complex structures of arbitrary shape. Two hypothetical planar reef structures are modelled
at a depth of 1000 m. The line of reflection points shown, corresponding to a surface geophone spread and single-shot point,
has been deliberately chosen so that only diffracted waves are
generated without complications from the reflected pulses. The
first geophone coincides with the assumed shot point. The source
function period is again taken as 10 millisecs. Both the weight
function and its convolution with $\dot{F}(t)$ (see Eq. (3)) are shown.
The calculated synthetics exhibit several interesting features
worth commenting upon. For example, the first arrival cross-over
at about trace 13 corresponds to the interference of diffracted
pulses from both structures. Many other distinct diffracted
pulses can be noticed. Each one corresponds to a configuration when a t-isochron is tangential to a diffracting edge
of the model. The phase of the resulting diffraction depends on
whether the t-isochron cuts into or out of an edge as t increases
through the critical value. The former produces a diffracted
pulse which is in phase with the source function; the latter one
which is out of phase. Amplitudes depend on the geometry in a
complicated way, but generally decrease away from shadow boundaries. Exceptions may occur when two or more diffractions interfere constructively.

(c)   Figures 3 and 4 demonstrate the extension of the
method to slowly varying media. Synthetic seismograms for one
second period core phases are shown in Figure 4 for a spherically symmetric Earth model. Three distinct phases are evident.
Two of these are predicted by ray theory and correspond to the
AB and BC branches of PKP, which 'meet' at the caustic B. The
third diffracted phases which pass through B and C are due to
shadowing by the Earth's outer and inner cores, respectively.
For this application ordinary ray theory for source - diffractor
- receiver paths breaks down. Rays emerging from the source
side of the CMB are strongly focused so that a caustic surface
arises within the core and also throughout the mantle. Ordinary
ray theory is invalid in such circumstances. This problem is overcome by applying the Kirchhoff formalism twice over two different
surfaces. The first surface, S_1, coincides with part of the
CMB on the source side while the second surface, S_2, with part

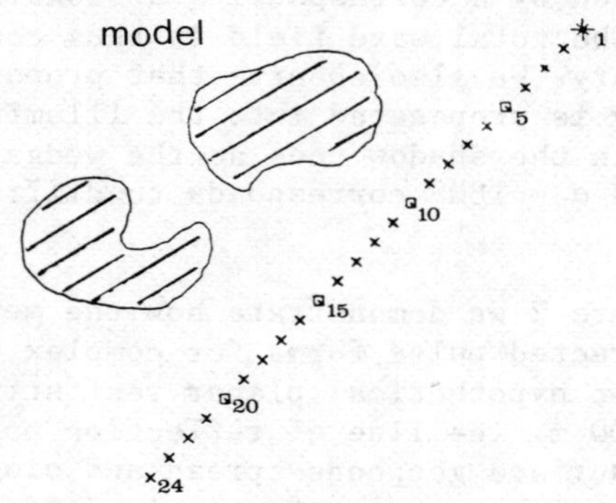

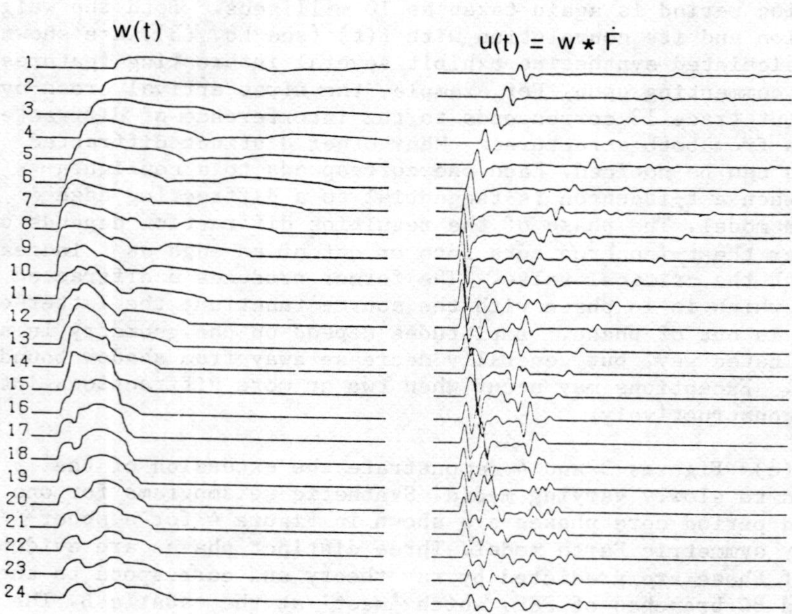

Figure 2.   Complex reef model diffraction study. Depth = 1000 m, geophone spacing = 50 m. Both the weight function W(t) and its convolution with Ḟ(t) are shown.

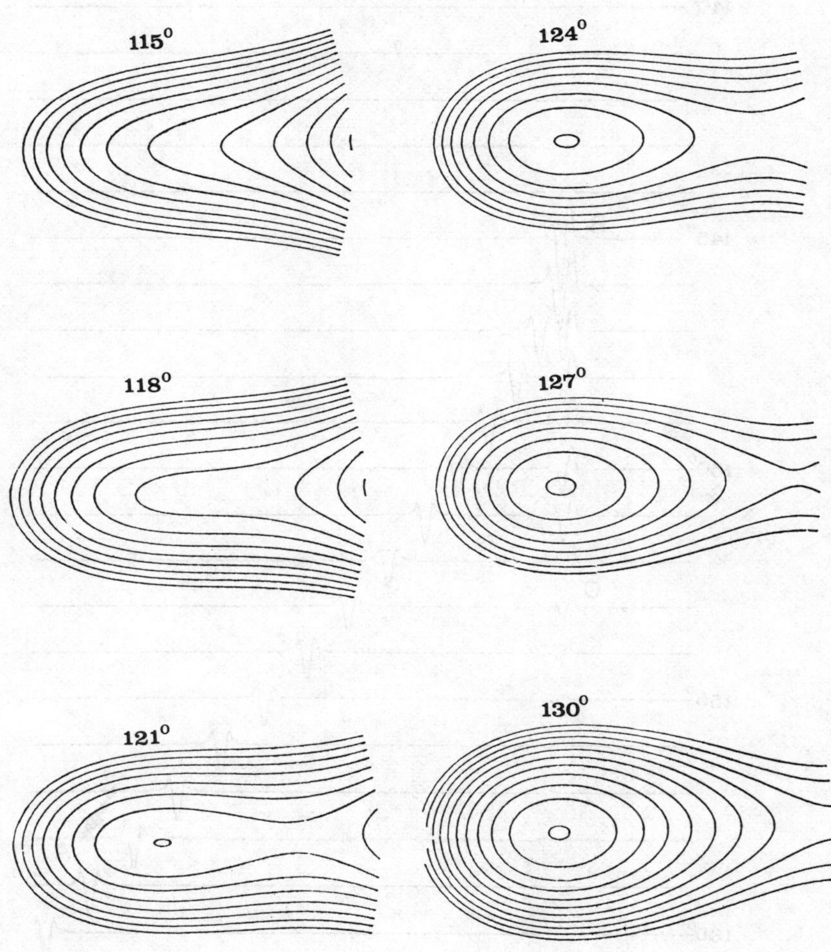

Figure 3.   Isochronal curves on the source side of the CMB for receivers at various angular distances on the opposite side of the CMB. The caustic is at 118.5° for this model. These curves are used in calculating the waveforms shown in Figure 4.

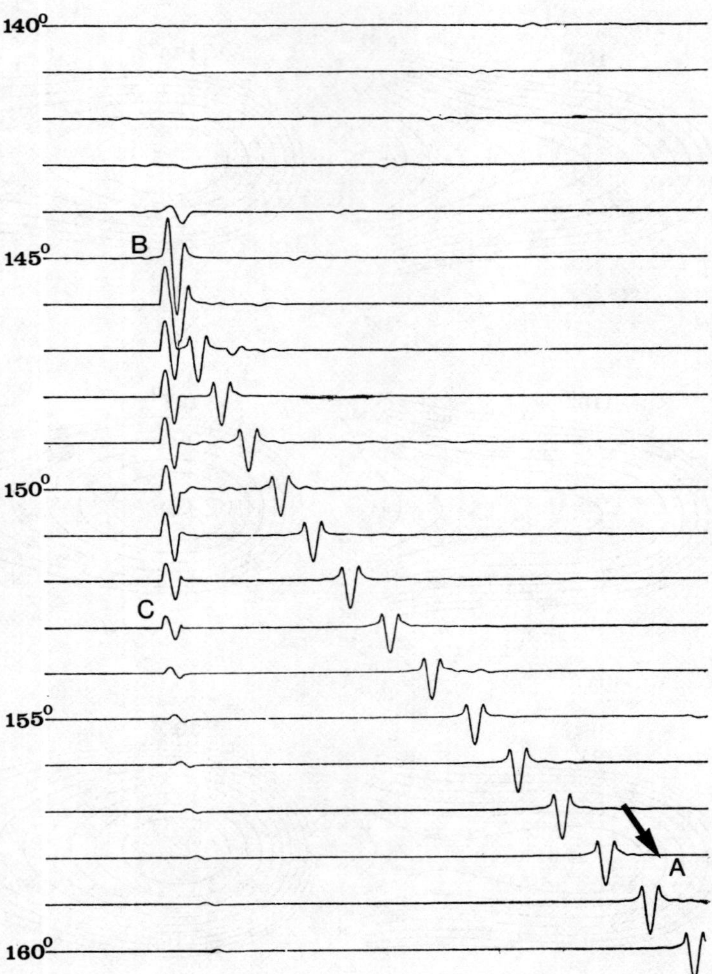

Figure 4. Synthetic seismograms for 1 sec period PKP-waves. The AB and BC phases are indicated. Diffracted waves due to screening by the outer and inner cores are also evident.

of the CMB on the receiver side. If Q_1 and Q_2 denote generic points of S_1 and S_2, then ray paths of the type $O - Q_1 - Q_2 - P$ where O is the source and P the receiver are adequately represented by ray-theory. Figure 3 shows families of isochrons on S_1 corresponding to ray segments of the type $O - Q_1 - Q_2$; that is, for field points on S_2. For the particular Earth model used, the caustic surface intersects S_2 at 118.5°. This is indicated in Figure 3 by the confluence of a local minimum and saddle-point associated with the isochronal contours. The minimum is directly related to the BC phase, while the saddle-point gives rise to the AB phase. At greater distances the minimum and saddle-point become well separated resulting in distinct BC and AB phases. Evidently the minimum and saddle-point generate different pulse shapes as seen in Figure 4. The right-hand terminator on the isochronal contours represents the shadow boundary on the CMB for a surface source. It gives rise to an observed diffraction event in Figure 4. Similarly, the left-hand terminator seen at about 130° in Figure 3 gives rise to diffraction due to shadowing by the Earth's inner core. Figure 4 shows calculated wave forms of the various core phases mentioned for surface receivers. The caustic surface intersects the Earth's surface for the model at 145°.

4. CONCLUSIONS

The isochronal formulation of Kirchhoff diffraction theory entirely in the time domain leads to an efficient computational procedure for the calculations of synthetic seismograms. As demonstrated, a wide class of seismological problems can be treated by this method. In particular, this approach can provide a valuable aid in three-dimensional modelling of complex crustal structures. In addition, although we have only considered the spherically symmetric case here, synthetics can be obtained by the same method for the scattering of waves by assumed heterogeneities on the CMB. This is an important consideration because such seismic waves have been observed as precursors to core phases and the evidence overwhelmingly points to a scattering origin on the CMB.

REFERENCES

1. Hilterman, F.J.: 1975, Geophysics <u>40</u>, 745-762.
2. Trorey, A.W.: 1977, Geophysics <u>42</u>, 1177-1182.
3. Buchen, P.W. and Haddon, R.A.W.:1980, J. Acous. Soc. Amer. <u>68</u>, 309-313.

HIGH FREQUENCY TOROIDAL MODES

T.J. Clarke

Department of Applied Mathematics & Theoretical Physics
University of Cambridge, Silver Street
Cambridge, CB3 9EW, England

ABSTRACT

The nature of asymptotic toroidal mode dispersion is discussed,
both for smooth and for discontinuous Earth models, making use of
a fixed slowness approach. The information content of the set of
high frequency modes is assessed, and a simple procedure developed
to compute the perturbations produced by the Moho. Finally
Brune's method of extracting toroidal mode data from body wave
pulses is discussed, and in particular the systematic error
introduced by the presence of the Moho.

INTRODUCTION

This paper is concerned with the problem of trying to
retrieve the shear velocity structure within the Earth from
toroidal mode data. In particular, the question "How much
information may be extracted from the set of observed toroidal
mode frequencies of the Earth?" is posed.

The approach adopted is to consider sets of modes of fixed
phase velocity, or slowness, since by this means it is possible
to work gradually down through the Earth, uncovering the structure
a little at a time. In addition, an analytic inversion technique
is available, in the form of the Tau method [3]. The major
disadvantage of this approach is that it makes use only of those
modes of sufficiently high frequency, that they may be considered
in terms of asymptotic solutions of the mode equation.

E. S. Husebye and S. Mykkeltveit (eds.), Identification of Seismic Sources – Earthquake or Underground
Explosion, 383–390.

The first part of the paper will be concerned with the asymptotic behaviour for a smooth Earth model; we will then consider a more realistic structure.

ASYMPTOTICS FOR A SMOOTH MODEL

For an Earth model with no discontinuities in material parameters, the set of toroidal modes at fixed slowness, Ω_ρ, shows a simple asymptotic behaviour governed only by the shear velocity structure within the model. Thus, for modes corresponding to a ray with no turning point in the mantle, that is, ScS modes, the asymptotic dispersion relation is given by:

$$\omega_n(\rho) \sim \frac{n\pi}{H(\rho)} + 0\left(\frac{1}{n}\right) \qquad H(\rho) = \tfrac{1}{2}\tau_\beta(\rho) = \int_a^b \left(\frac{1}{\beta^2(r)} - \frac{\rho^2}{r^2}\right)^{\frac{1}{2}} dr \qquad (1)$$

where ρ = slowness, n = radial order.

For modes which do have a turning point in the mantle, say at $r = R_\beta$, the equation becomes:

$$\omega_n(\rho) \sim (n+\tfrac{1}{4}) \frac{\pi}{H(\rho)} \qquad\qquad (2)$$

where in this case the lower limit of integration is R_β . In future only the first case will be written out explicitly; the second may always be produced by adding in the appropriate number of $\pi/4$'s, associated with the turning point.

Thus, for a smooth model, observations of the set Ω_ρ , in the asymptotic regime will give only one piece of information, τ_β . Such limited information is, however, adequate to infer the velocity structure, by means of Tau inversion.

The process of inferring velocity structure for a smooth model is as follows:

a)   Observation of set Ω of toroidal modes
b)   Interpolation to obtain sets Ω_ρ for $\beta(b)^{-1} < \rho < \beta(a)^{-1}$
c)   Extraction of best estimate of τ for each ρ value
d)   Tau inversion.

There are problems associated with each stage of this scheme; I shall mention only that arising in the third step.

Given a large, infinitely accurate set of modes, the value
of τ would by preference be deduced from the large n part of
the spectrum, to ensure that the behaviour were as close as
possible to that predicted by asymptotic theory. In practice,
however, this part of the spectrum is increasingly difficult to
resolve as n increases, and the frequencies become more liable
to observational error. It is thus necessary to find a compromise frequency, where neither effect is intolerably large.

ASYMPTOTICS FOR A DISCONTINUOUS EARTH

The asymptotic behaviour of the mode frequencies is complicated significantly by the existence of discontinuities in the
elastic parameters, and thus in the shear velocity structure,
within the Earth. The asymptotic dispersion relation becomes:-

$$\omega_n(\rho) \sim \frac{n\pi}{H(\rho)} + \delta_n + O(\frac{1}{n}) \tag{3}$$

where the quantities δ_n do not tend to zero as $n \to \infty$, but
represent a periodic disturbance of constant amplitude. This is
the solotone effect [5]-[9].

The nature of this effect may be seen most clearly in the
case of a single discontinuity. For ScS equivalent modes the
asymptotic dispersion relation may be obtained readily by a
Liouville transformation. It is:

$$\sin(\omega H) - R \sin(\omega\tilde{H}) = 0 \qquad \tilde{H} = H - 2H^+ \tag{4}$$

where H^+ is H evaluated for the region above the discontinuity,
and R is the reflection coefficient for the discontinuity, given
by the equation:

$$R = \frac{1-Q}{1+Q} \qquad Q = \frac{\mu(\frac{1}{\beta^2} - \frac{\rho^2}{r^2})^{\frac{1}{2}} \; |\text{below}}{\mu(\frac{1}{\beta^2} - \frac{\rho^2}{r^2})^{\frac{1}{2}} \; |\text{above}}$$

$|R|$ is typically less than 0.25, but may be as large as this for
the Moho, particularly for modes with turning points high in the
mantle.

The eigenfrequencies are thus given by the values of ω at
which the curve $y = \sin(\omega H)$ intersects the curve $y = R\sin(\omega H)$.
The second curve is contained entirely within the region
$-|R|<y<|R|$, and there is a root associated with each passing of
the first curve through this region. For small R, it is clear

that there is a root in the neighbourhood of each point $n\pi/H$, but displaced by an amount δ_n. The largest values of this displacement are obtained when the point of intersection is close to one of the lines $y = \pm\,|R|$, and the amplitude of the solotone of effect, say D, is thus given by:

$$D = \frac{\sin^{-1}(|R|)}{H} \ .$$

Although periodic, the solotone effect is not a simple sine function, but for small R, and for $H^+ \ll H$, an assumption valid in the case of the Moho, we may write an approximate dispersion relation thus:

$$\omega_n = \frac{n\pi}{H} + \delta_n \ ; \qquad \delta_n \approx \frac{\sin^{-1}|R|}{H}\sin(\frac{2\pi H^+}{H}n) \ . \tag{5}$$

It is easily shown that this relation tends to exactness as $R \to 0$.

For a model with two or more discontinuities the solotone effect is more complicated. It is possible for two and three interfaces to produce an analytic expression for the asymptotic dispersion relation, although in the case of three discontinuities it is rather unwieldy. For two discontinuities the dispersion relation is given by:

$$0 = \sin(\omega H) - R_1\sin(\omega\tilde{H}_1) - R_2\sin(\omega\tilde{H}_1) + R_1 R_2 \sin(\omega[H + 2(H_1^+ - H_2^+)]) \tag{6a}$$

where $\tilde{H}_1$, $\tilde{H}_2$ are the quantities $\tilde{H}$ for the two discontinuities. This may be simplified if $R_1, R_2 \ll 1$, by ignoring the last term, to produce the equation:

$$\sin(\omega H) - R_1\sin(\omega\tilde{H}_1) - R_2\sin(\omega\tilde{H}_2) = 0 \ . \tag{6b}$$

The effects of the two discontinuities thus add together almost linearly in this case. The same will hold for more than two, although less accurately as the number increases. In particular, it will hold to reasonable accuracy in the case of the Earth, where there is one major discontinuity, the Moho, and other smaller ones. If we assume there to be two others, with reflection coefficients R_2, R_3, a measure of the accuracy obtained by taking into account the Moho phase shift only is given by η, where η satisfies

$$\eta = 1 - (R_1 R_2 + R_1 R_3 + R_2 R_3)/R_1 \quad .$$

For typical values of R_2, R_3 , this produces a value of approximately 0.8 for η . We would thus expect the Moho alone to account for approximately eighty percent of the observed solotone effect.

A reasonably good approximation to the asymptotic dispersion relation for the Earth should therefore be given by:

$$\omega_n \sim \frac{n\pi}{H} + \frac{\sin^{-1}(|R|)}{H} \sin(2\pi \frac{H^+}{H} n) \quad . \tag{7}$$

Since $H^+ \ll H$, the disturbance will have a long period.

The set Ω_ρ for a particular ρ value now contains more than the single piece of information τ_β . In addition it contains information about the τ value for the crust, τ_β^+ , as well as the reflection coefficient for the Moho. Thus given an accurate set Ω_ρ , containing modes up to a large n value, the inversion procedure would be as follows:-

a), b) As above - observation and interpolation
c) Extract a mean τ value for the set
d) Check this by plotting the phase shifts δ_n . These should oscillate symmetrically about zero
e) Measure D (phase shift amplitude) and derive R
f) Measure T (period) and derive τ_β^+
g) Use τ_β in a Tau inversion·

In practice there are two objections to this approach. The first is that the τ values for the crust, as well as the Moho reflection coefficients, are known to reasonable accuracy from other methods, and any results obtained would simply duplicate known information.

The second, more powerful objection, is that such accurate mode information is simply not available. Directly observed toroidal mode data is available accurately only for $n < 8$ approximately, which is only just within the asymptotic regime. From such a limited set it is clearly not feasible to extract the amplitude and period of the solotone effect, since the information covers typically only a quarter cycle or less. For modes with turning points in the middle of the mantle, a complete cycle would take until approximately $n = 20$, with correspondingly very high angular order, typically 250 or 300. Such modes are not resolvable by conventional techniques.

There are two contrasting approaches which may be adopted in the face of this difficulty. The first, due to Brune [4], consists of a method of extracting high frequency mode data, which avoids the problem of resolution of separate modes. Unfortunately the frequencies derived by this approach are not genuine toroidal mode frequencies at all, and should really be classed as body wave data.

The second approach is to assume that the structure of the Earth is known down at least as far as the Moho, and to attempt to extract from the mode data which is available accurate values for the rest of the mantle. This approach will be described now.

Consider the set Ω_ρ , and a particular member ω_N. A rough estimate of H is given by:

$$H^O = N\pi/\omega_N \ .$$

We know also that the dispersion relation is given approximately by:

$$\sin(\omega H) - R\sin(\omega \tilde{H}) = 0 \ .$$

Assuming we know R and H^+ , and hence an approximation to $\tilde{H}$, we can look for the root of this equation near ω_N. The phase shift for these approximate H values will be an approximation to the actual phase shift. This quantity may then be subtracted from ω_N, and revised estimates for the H values obtained. The process converges rapidly, to produce an accurate value for τ_β .

Given an original set Ω_ρ of modes, we have calculated a new set Ω_ρ . The ρ value is unchanged, and we have maintained the radial order n for each mode. Thus the L value must have changed, and is given by the usual equation:

$$\rho = \frac{\sqrt{L(L+1)}}{\omega}$$

The set of corrected mode frequencies for integer L, if required, may be calculated by interpolation. In general, however, we are interested only in the quantity τ_β .

The success of this method may be seen by considering a particular case. The mode frequencies were calculated for a model with a Moho discontinuity at 30 km, but otherwise smooth, and the fixed ρ sets of data extracted. These show the predicted solotone effect, but when the iterative correction method above is applied to each mode in turn, the τ values obtained show very good agreement with each other. (See Table below.)

TAU VALUES FROM TOROIDAL MODE DATA AT FIXED SLOWNESS

(phase velocity [flattened] = 11.00km/s)

n	Tau (uncorrected)	Tau (corrected)
5	506.06	511.40
6	505.70	510.74
7	505.96	510.66
8	506.06	510.32
9	506.68	510.40
10	506.98	510.24
11	507.60	510.24
12	508.24	510.22

BRUNE METHOD

The alternative approach, due to Brune, is to attempt to
extract large n toroidal mode data from body wave pulses, so
avoiding the problem of high frequency resolution.

This method consists of measuring the phase velocity and τ
value for a doubly reflected S wave, and is based on the
principle of ray-mode duality. In its simplest form, this states
that the ray with velocity V is built up of modes with the
same phase velocity. The equation:

$$\omega_N = \frac{2\pi N}{\tau_\beta} \tag{8}$$

derived by the principle of constructive interference, is then
used to calculate a set of mode frequencies at fixed slowness,
one for each n value. Equation 8 may be recognised as the
simple asymptotic equation introduced earlier in the smooth case.

It is clear from the preceding analysis that the effect of
discontinuities in the Earth, and in particular of the Moho, will
be to introduce errors into the procedure. An extra phase shift
must be introduced into 8, and the frequency predicted will be
altered. The data extracted using this method is actually body
wave data, and not toroidal mode. It is, however, possible to
make a correction for the presence of the Moho, and thus to bring
the values closer to the real mode frequencies. To do this, the
process used previously to remove the effect of the Moho is
reversed. The phase shifts are calculated for the inexact modes

extracted by the Brune method, and are added back onto these
frequencies.

The resulting change in frequency is equivalent to an error
in the Tau value of between two and four seconds, which although
not large is clearly not negligible. Such an error may also be
important when the mode frequencies are used in an iterative
inversion scheme, or in any other inversion procedure. It is
particularly significant in this respect that most of the mode
data tabulated is for small n , (n<12 approximately) and since
for most phase velocities this restricts us to the first half
cycle of the solotone oscillation, the errors in frequency will
be consistently of the same sign.

REFERENCES

[1]   Kennett, B.L.N., and Nolet, G.: 1979, Geophys. J. 56,
        pp283-303

[2]   Woodhouse, J.H.: 1978, Geophys. J. 54, pp263-280.

[3]   Bessonova, E.N., Fishman, V.M., Ryaboyi, V.Z., Sitnikova,
        G.A.: 1974, Geophys. J. 36, pp377-398

[4]   Brune, J.N.: 1964, Bull. Seism. Soc. Am. 54, pp2099-2128

[5]   Lapwood, E.R.: 1975, Geophys. J. 40, pp453-464

[6]   McNabb, A., Anderssen, R.S., and Lapwood, E.R.: 1976, J.
        Math. Anal. Appl. 54, pp741-751

[7]   Anderssen, R.S., and Cleary, J.R.: 1974, Geophys. J. 39,
        pp241-268

[8]   Anderssen, R.S.: 1977, Geophys. J. 50, pp303-309

[9]   Wang, C.: 1978, Ph.D. Thesis, A.N.U. Canberra.

ELASTIC AND ELECTROMAGNETIC WAVE PROPAGATION IN HORIZONTALLY LAYERED MEDIA

Bjørn Ursin

SINTEF, Division of Petroleum Technology,
N-7034 Trondheim NTH, Norway

The mathematical similarity between electromagnetic and elastic wave propagation in layered media is well known (1, 2,3). By applying a combination of Fourier, Laplace and Bessel transforms to the different partial differential equations, we obtain a system of 2n linear ordinary differential equations where the coefficients are functions of the depth coordinate and the transformed time and horizontal space coordinates. The 2n x 2n coefficient matrix may be partitioned into 4 nxn submatrices. By a proper choice of variables the two diagonal submatrices are zero, and the off-diagonal submatrices are symmetric. This technique has been used by Richards (4) for plane P-SV waves. A similar equation has been considered by Reid (5) for an electromagnetic transmission line problem. By considering this general type of equation a number of results are derived. The wavefield is decomposed into upgoing and downgoing waves by an eigenvalue decomposition, which is much simplified compared with the general case of a full 2n x 2n coefficient matrix. The propagator matrix (6) for a stack of inhomogeneous layers is computed by a simplified method. For a stack of homogeneous layers we obtain recursive equations for the computation of the submatrices of the propagator matrix. This extends previous results given by Thomson (7), Haskell (8), Frasier (9), Goupillaud (10), Treitel and Robinson (11) and Claerbout (12). Specific computational schemes have been discussed by Dunkin (13), Richards (4), Foster (14), Burdick and Orcutt (15) and Kennett (16,17). Due to the symmetry of the off-diagonal submatrices two propagation invariants are derived. One of these is only valid for lossless media and corresponds to the conservation of energy.

E. S. Husebye and S. Mykkeltveit (eds.), Identification of Seismic Sources - Earthquake or Underground Explosion, 391–393.

For a stack of inhomogeneous layers transmission and
reflection matrices for upward and downward propagation may be
defined. Using the two propagation invariants we may derive a
number of symmetry properties for the transmission and reflection
matrices as demonstrated by Kennett, Kerry and Woodhouse (18)
for P-SV waves. The computation of the reflection and transmission matrices for two inhomogeneous layers is done by Redheffer's
star product. Redheffer (19) gives many interesting properties
of the star product in a discussion on transmission line theory.
This composition rule has also been derived by Kennett (20) and
Kennett and Kerry (21) for elastic waves. The effect of an inhomogeneous layer with known parameters may be removed by applying the inverse of the star product. This could be used in geophysical inverse schemes. It should be added that in an inhomogeneous medium the reflection and transmission matrices can be
computed by solving a matrix Riccati equation (4, 22, 23, 24).
The coefficient matrices in the Riccati equation are derived
from the simplified equations for the propagator matrix.

For a stack of layers bounded above by a free surface we
obtain modified transmission and reflection matrices. By considering the propagation invariants we obtain a number of symmetry properties for the transmission and reflection matrices.
These include results derived by Kunetz and d'Erceville (24),
Claerbout (12) and Frasier (9). The relationship between the
usual transmission and reflection matrices and the modified
transmission and reflection matrices is also obtainable.
In case the response of a buried receiver from a buried point
source is desired, we may follow the same strategy as outlined in (21). More general sources may easily be considered by
applying the superposition principle.

Finally it should be remarked that a comprehensive discussion of elastic and electromagnetic wave propagation in
horizontally layered media is recently given by Ursin (25)

REFERENCES

1.   Brekhovskikh, L.M.: 1960, Waves in Layered Media, Academic
      Press, New York.

2.   Szaraniec, E.: 1976, Geophys. Prosp. 24, 528-548.

3.   Szaraniec, E.: 1979, Geophys. Prosp. 27, 576-583.

4.   Richards, P.G.: 1971, Geophysics 36, 798-809.

5.   Reid, W.T.: 1972, Riccati Differential Equations, Academic
      Press, New York.

6.   Gilbert, F. and Backus, G.E.: 1966, Geophysics 31, 326-332.

7.   Thomson, W.T.: 1950, J. Appl. Phys. 21, 89-93.

8.   Haskell, N.A.: 1953, Bull. Seism. Soc. Am. 43, 17-34.

9.   Frasier, C.W.: 1970, Geophysics 35, 197-219.

10.   Goupillaud, P.L.: 1961, Geophysics 26, 754-760.

11.   Treitel, S. and Robinson, E.A.: 1966, Geophysics 31, 17-32.

12.   Claerbout, J.F.: 1968, Geophysics 33, 264-269.

13.   Dunkin, J.W.: 1965, Bull. Seism. Soc. Am. 55, 335-358.

14.   Foster, M.: 1975, Geophys. J.R. astr. Soc. 42, 519-527.

15.   Burdick, L.J. and Orcutt, J.A.: 1979, Geophys. J.R. astr.
      Soc. 58, 261-278.

16.   Kennett, B.L.N.: 1979, Geophys. Prosp. 27, 301-321.

17.   Kennett, B.L.N.: 1979, Geophys. Prosp. 27, 584-600.

18.   Kennett, B.L.N., Kerry, N.J. and Woodhouse, J.H.: 1978,
      Geophys. J.R. astr. Soc. 52, 215-230.

19.   Redheffer, R.: 1961, Difference equations and functional
      equations in transmission line theory, in: Beckenbach, E.F.
      (ed.), Modern Mathematics for the Engineer (second series),
      McGraw-Hill, New York.

20.   Kennett, B.L.N.: 1974, Bull. Seism. Soc. Am. 64, 1685-1696.

21.   Kennett, B.L.N. and Kerry, N.J.: 1979, Geophys. J.R. astr.
      Soc. 57, 557-584.

22.   Friedlander, B., Kailath, T. and Ljung, L.: 1976, J. Franklin
      Inst. 301, 71-82.

23.   Ljung, L., Kailath, T. and Friedlander, B.: 1976, Proc. IEEE 64,
      131-139.

24.   Kunetz, G. and d'Erceville: 1962, Ann. de Geophys. 18, 351-359.

25.   Ursin, B.: 1980, Elastic and electromagnetic wave propagation
      in horizontally layered media, SINTEF, Trondheim.

CALCULATION OF WAVE FIELDS IN MANTLE VELOCITY MODELS

M.R. Illingworth and B.L.N. Kennett

Department of Applied Mathematics and Theoretical
Physics,
Silver Street, Cambridge CB3 9EW
England.

A reasonable approximation to the seismic velocity distribution in the earth's mantle and crust is to take a model composed of regions with smoothly varying properties separated by discontinuities in velocity or velocity gradient. Within each region, at each frequency and slowness, the wave functions are constructed from a Langer uniform approximation in terms of Airy functions supplemented by an expansion in the elastic parameter gradients which represents multiple internal reflections. This procedure allows an efficient treatment of turning point problems and in general it is an adequate approximation to retain only the leading order term in the interaction series, provided the parameter gradients are not too large.

We may then construct the reflection properties of each gradient zone sandwiched between two uniform half spaces with continuity of properties at the limits of the layer. The effects of interfaces are introduced by using the reflection matrix between two uniform half spaces with the properties at the interface. The overall reflection properties of the stratification are obtained by a recursive procedure incrementing successively on the effects of gradient zones and interfaces which is simply an extension of the corresponding recursive development for uniform layers.

This computational scheme gives good results for crustal and upper mantle models with only about a dozen subdivisions of the stratification required down to 950 km. Checks on the accuracy of the computation show that for periods less than 20 seconds the error associated with using only the Langer approximation in the upper mantle monotonic velocity distributions is

395

E. S. Husebye and S. Mykkeltveit (eds.), Identification of Seismic Sources - Earthquake or Underground Explosion, 395–396.

less than 0.1 per cent, except for a few cases where higher
order reflection processes are important. The error may reach
2 per cent when a turning point is just at a structural boundary.
The leading order approximation allows no conversion of wave
type except at a first order discontinuity and the error introduced by neglecting conversion at steep gradient zones may reach
a few per cent. For a low velocity zone, rather larger errors
may occur for slownesses such that two turning points are close
together but these do not affect waves which turn well above or
below the zone.

The details of this approach are to be found in Kennett,
B.L.N., and Illingworth, M.R. (1981): Geophys. J. (accepted for
publication).

THE EFFECT OF FOCAL DEPTH AND SOURCE TYPE ON SYNTHETIC SEISMOGRAMS

Keith K. Nakanishi

Earth Sciences Division, Lawrence Livermore National
Laboratory, L-205, Livermore, CA 94550

The effects of various parameters such as focal depth,
source type, and source-to-receiver distance on the vertical
components of Pn, Pg, Lg, and Rg are investigated using synthetic
seismograms formed by the summation of P-SV normal modes. The
normal mode method of seismogram synthesis is particularly well
suited for studies of this type. The calculation of the normal
modes, the most time-consuming part of the computations, depends
only on the earth model and can be done once. The computation of
the synthetic seismograms then requires summation of the individual modes as excited by the source function. The cost of
computing the suite of seismograms in this step is due mainly
to the inverse Fourier transformation. Summation of normal modes
has been successfully used previously to synthesize crustal
phases such as the transverse component of Lg (1,2), the transverse component of Sn (3,4) and the vertical component of Pg
(5). In this study, we will investigate the source and depth
dependence of the vertical component of all of the major crustal
phases.

In this experiment an earth model consisting of a crustal
layer (Vp = 6.1 km/s, Vs = 3.55 km/s, 30 km thick) over a mantle
(Vp = 7.8 km/s, Vs = 4.5 km/s, 500 km thick) is used in order to
generate a simple seismogram. The structure is terminated by a
high velocity halfspace (Vp = 20 km/s, Vs = 10 km/s) representing
the deeper parts of the earth. The model is a flat, elastic
representation of a continental crust and upper mantle.

The P-SV normal modes are computed using the Knopoff-Schwab
algorithm (6,7,8). The solutions are found by a grid-search

E. S. Husebye and S. Mykkeltveit (eds.), Identification of Seismic Sources - Earthquake or Underground
Explosion, 397–399.

method due to D. Harvey (9). All of the possible normal modes
from zero to 1.28 Hz are computed at equally-spaced intervals of
0.005 Hz. For this model there are 387 modes for phase velocities
ranging from 0 to 10 km/s.

Three different source types are used: an isotropic explosion, a strike-slip on a vertical fault, and a reverse thrust
on a 45° dip fault. All three are point sources in space, have
a step source time function, and have equal low-frequency scalar
moments. Displacement time series are computed at an azimuth of
30° off strike for focal depths that range from 0 to 20 km and
epicenter-to-receiver distances of 150 to 500 km.

Results using this structure and source model indicate
that Pn, the first arrival on all but the closest stations, is
small, but recognizable in most of the time series. The amplitude and frequency content of Pn are relatively independent of
source type and source depth for this crustal model.

The dominant arrivals on the seismogram are Pg, the compressional wave propagating in the crust, and Rg, the fundamental
Rayleigh mode. Comparisons between the different phases on the
synthetic seismograms indicate that Pg is insensitive to source
type as well as depth. Rg is extremely depth-dependent in both
amplitude and frequency content. In addition, Rg is excited less
for explosions than for thrust or strike-slip sources. Lg,
however, shows a very strong decrease of amplitude with depth
for explosion sources. Lg from thrust and strike-slip sources
is less sensitive to depth, with about the same amplitude, but
a more complicated waveform for the deeper sources. All of the
above observations are independent of epicenter-to-receiver
distances.

Work performed under the auspices of the U.S. Department of
Energy by the Lawrence Livermore Laboratory under contract number
W-7405-ENG-48.

REFERENCES

1.   Knopoff, L., Schwab, F, and Kausel, E.: 1973, Geophys. J.R.
     astr. Soc., 33, 389-404.
2.   Knopoff, L., Schwab, F., Nakanishi, K., and Chang, F.: 1974,
     Geophys. J.R. astr. Soc., 39, 41-70.
3.   Stephens, C., and Isacks, B.L.: 1977, Bull. Seism. Soc. Am.,
     67, 69-78.
4.   Mantovani, E., Schwab, F., Liao, H., and Knopoff, L.:1977,
     Geophys. J.R. astr. Soc., 51, 709-726.
5.   Harvey, D.J.: 1980, EOS, Trans. Am. Geophys. Un., 61, 308.

6. Knopoff, L.: 1964, Bull. Seism. Soc. Am., 54, 431–438.
7. Schwab, F.: 1970, Bull. Seism. Soc. Am., 60, 1491–1520.
8. Schwab, F., and Knopoff, L.: 1972, in: Methods in Computational Physics (B. Bolt, ed.), 11, 87–180.
9. Harvey, D.: 1979, Personal communication.

SURFACE WAVE PROPAGATION ACROSS DIFFERENT TECTONIC REGIONS.

Michel CARA

L.A. 195 – I.P.G. – Tour 14, 4 place Jussieu,
75230 PARIS Cedex 05, FRANCE
present address : I.P.G., 5 rue R. Descartes,
67084 Strasbourg Cedex.

Abstract. Separation of propagation and source effects in
observed seismic signals is a central problem in any attempt
to identify seismic sources. Surface-waves play a major role in
tele-seismic signals at periods greater than 10-20 sec. They
have been investigated in great details for numerous tectonic
regions and different types of seismic sources. As propagation
parameters (dispersion, attenuation) are sensitive to regional
structures crossed by the waves, they can be used to infer
regional models of the Earth. Such models can, in turn, be used
1) to predict the actual dispersion and attenuation of surface
waves in a given region and 2) to compute the model-dependent
terms in the source-excitation problem.
   After reviewing basic theoretical results and data
processing techniques which are pertinent to surface wave
analysis, we describe propagation parameters in different
tectonic provinces and we present some inferred regional mantle
models. Different types of surface wave data are used in this
paper. Some of them, now well known, have been extensively
studied during the last two decades (e.g. fundamental Rayleigh
waves of 15-100 sec. period) and others obtained more recently,
are currently the object of vigorous research (very long period,
overtones ...).

I - INTRODUCTION

   A typical long period record is presented in figure 1a.
For such a large earthquake, several surface wave groups R_1, R_2,
... Rn are observed. A Fourier transform of this record gives
a discret spectrum which can be analysed in terms of stationary

401

E. S. Husebye and S. Mykkeltveit (eds.), Identification of Seismic Sources - Earthquake or Underground
Explosion, 401–420.

oscillations of the Earth. Surface wave studies are, on the other
hand, generally restricted to the "direct path" wavetrain (e.g.
R.1 in figure 1). For smaller seismic sources (magnitude <6), the
"direct path" wavetrain is often the only one observed and it
can provide useful constraints on source parameters.

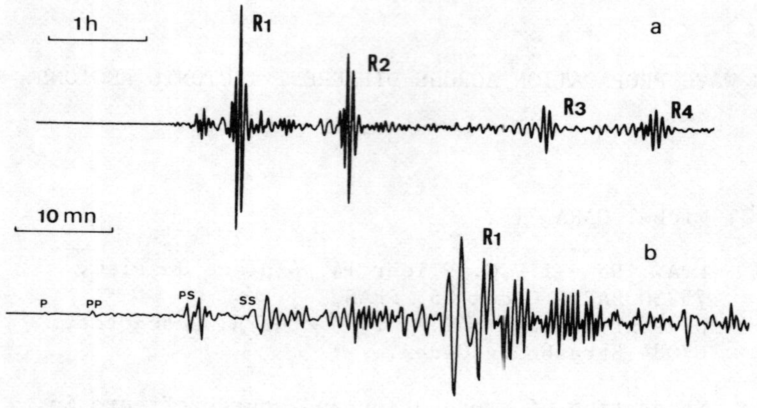

Figure 1. Indonesian earthquake of August 1977 recorded with
a wide dynamic range instrument near Paris. Figure 1 a displays
the same record as 1 b after low pass filtering.

     Regional studies of surface wave propagation are commonly
described in the framework of flat Earth models. Sphericity
effects can generally be neglected at short period. They remain
small up to about 100 sec period for the fundamental mode of
both Love and Rayleigh modes. Due to the improvement of long
period instruments, more accurate long period data become
available and at shorter period, application of wavenumber
analysis techniques to surface waves has recently provided
overtone data. These two types of surface waves can sample the
Earth over more than 10 % of its radius and sphericity effect
must be taken into account. Direct computations for a standard
oceanic model show for instance that sphericity effect can
reach several percents for the first Rayleigh overtones (table I).

Table I . Sphericity effect (in percent) for Rayleigh modes 0,1 and 2 (fundamental and first two higher modes) [1] on phase velocity ($\delta C/C$) and group velocity ($\delta U/U$) at different periods T (in sec.). The numbers in parentheses are from Brune&Dorman[2].

Mode 0			Mode 1			Mode 2		
T	$\delta C/C$	$\delta U/U$	T	$\delta C/C$	$\delta U/U$	T	$\delta C/C$	$\delta U/U$
50	1.0(0.8)	0.0	40	3.9	1.8	30	4.6	4.1
75	1.3(1.2)	0.0	60	4.7	2.8	50	6.5	0.1
100	1.6(1.6)	0.0	80	5.7	1.8	70	8.2	5.9
150	2.2(2.4)	0.1	100	6.9	2.7	90	9.9	-1.4

II – NORMAL MODE THEORY AND DATA PROCESSING TECHNIQUES.

Normal mode excitation of an isotropic vertically heterogeneous sphere by a realistic earthquake model, a point double couple, was derived by Saito [3]. The transversely isotropic case is given in [4]. Saito's results were extensively used in the past [5-6] . An independent approach is that of Gilbert [7] who introduced a general representation of point sources by the seismic moment tensor.

II-a. An expression for surface wave signal

The motion due to a point source in a spherically symetric, and isotropic earth model can always be written as the real part of complex functions of spherical coordinates r, θ, ϕ and time t, representing a sum of free oscillations :

$$U_r = \sum_{n,\ell} \omega_{n,\ell}^{-2} \, {}_nU_\ell(r) \exp[i\omega_{n,\ell}t - \alpha_{n,\ell}t] \sum_{m=o}^{2} {}_nA_\ell^m \, P_\ell^m(\cos\theta)\exp(im\phi)$$

$$(1) \quad U_\theta = \sum_{n,\ell} \omega_{n,\ell}^{-2} \, {}_nV_\ell(r) \exp[i\omega_{n,\ell}t - \alpha_{n,\ell}t] \sum_{m=o}^{2} {}_nA_\ell^m \, \frac{d}{d\theta} P_\ell^m(\cos\theta)\exp(im\phi)$$

$$U_\phi = \sum_{n,\ell} \omega_{n,\ell}^{-2} \, {}_nW_\ell(r) \exp[i\omega_{n,\ell}t - \alpha_{n,\ell}t] \sum_{m=o}^{2} {}_nC_\ell^m \, P_\ell^m(\cos\theta)\exp(im\phi)$$

These expressions are obtained for sources varying as step time functions. The first two relations give spheroidal modes (corresponding to Rayleigh waves) and the latter one gives toroidal modes (Love waves). These relations involve a sum over angular order ℓ, radial order n (mode branch order hereafter), and azimuthal order m. The functions $_nU_\ell$, $_nV_\ell$ and $_nW_\ell$ give the depth variation for each mode. $P_\ell^m(\cos\theta)$ is the associate Legendre function, $\alpha_{n,\ell}$ the attenuation coefficient and $\omega_{n,\ell}$ the circular eigenfrequency. The functions $_nA_\ell^m$ and $_nC_\ell^m$ involve linear combinations of the seismic moment tensor elements weighted by the values of $_nU_\ell$, $_nV_\ell$ and $_nW_\ell$ and their first derivatives at the source level. Detailed expressions for $_nA_\ell^m$ and $_nC_\ell^m$ can be easily inferred from [8]. Note that the static term arising from Gilbert's approach [7] is neglected in (1).

There are two steps in developing the expressions (1) into a surface wave representation. A first step is to use an asymptotic expansion of $P_\ell^m(\cos\theta)$. For $\ell>>m$, $\varepsilon<\theta<\pi-\varepsilon$ and $\ell>>\varepsilon$ it can be written :

$$P_\ell^m(\cos\theta)e^{i\omega_{n,\ell}t}=[1/(2\pi\sin\theta)]^{1/2}n^{m-1/2}[e^{i\psi^+(\theta,t)}+e^{i\psi^-(\theta,t)}]$$

where :

$$\psi^\pm(\theta,t)=\pm[(\ell+1/2)\theta-m\pi/2-\pi/4]+\omega_{n,\ell}t.$$

The two terms $\pm$ correspond to waves propagating in opposite directions at phase velocity $C_{n,\ell}=\omega_{n,\ell}/k_\ell$ where :

$$k_\ell=\frac{1}{a}\left|\frac{\partial\psi}{\partial\theta}\right|=\frac{1}{a}\left(\ell+\frac{1}{2}\right), \tag{3}$$

" a " being the Earth's radius. To this degree of approximation, the phase velocity does not depend on θ . Such an assumption, fundamental for Fourier integral representation of the surface wave field, is generally implicitly made except near a pole ($\theta=0$ or $\theta=\pi$) where a $\frac{\pi}{2}$ polar phase shift occurs [9] . At very long period this "phase shift" cannot be considered as punctual and a θ-dependent phase velocity should be considered [10], for example by using more terms in the asymptotic expansion (2).

A second step is necessary to get expressions for the individual surface wave groups like R_n in figure 1 a. The Poisson sum formula transforming the sum over ℓ in expressions (1), into a sum of integrals over ω can be used [11] . The "direct path" surface wave can be obtained more simply from the transformation :

$$\Sigma_\ell \rightarrow a\int\frac{d\omega}{_nU(\omega)}$$

where $_nU(\omega)$ is the group velocity $d\omega/d_nk$ along a fixed mode branch n [3,12] . A multimode surface wave observed at epicentral distance $x = a\theta$ can then be written as the real part of a complex function :

$$f(x,t) = \sum_n \int_o^\infty I(\omega)\,_nS(\omega)\ e^{i(\omega t-_nk(\omega)x)}e^{-_n\alpha(\omega)x}d\omega \qquad (4)$$

where $I(\omega)$, the complex transfer function of the instrument has been added, and $_nS(\omega)$ is the complex source function. Continuous functions $_nS(\omega)$, $_nk(\omega)$ and $_n\alpha(\omega)$ have to be defined from the discrete functions in (1). In particular $_nk(\omega)$ has to be defined as a continuous curve along a mode n branch from $\omega_{n,\ell}$ by using the relation (3) between the wavenumber and the angular order ℓ.

The general expression (4) can be applied to either Love or Rayleigh waves, derived respectively from toroidal and spheroidal oscillations of a spherical, non-rotating and isotropic Earth model. Rotation affects only the very long periods and ellipticity has marginal effects on long period surface waves. Lateral heterogeneities cause for example much greater perturbations on Rayleigh wave great-circle phase velocity at 200-300 sec period, than ellipticity [13] . At longer period (low ℓ order) when the involved wavelengths become comparable to the Earth dimensions, these effects are not negligeable and can be taken into account by perturbation methods [14].

The main inadequacy of (4) to fit observed surface waves is thus due to lateral heterogeneities. To deal with this problem, it is usually assumed that optical geometry approximation holds and the average phase slowness along a propagation path is considered as a linear average of local phase slowness [15-17] (i.e. $_nk$ is a function of x in (4)). It is furthermore generally assumed that the propagation paths do not differ from great circles. These approximations are in some cases certainly not valid. A breakdown of optical geometry can for example occur when lateral heterogeneities present scale dimensions close to the involved wavelengths. This has been questionned for instance at very long period for surface waves circling several times around the Earth, like R_n in figure 1 a [18]. Experimentally, departure from results predicted in the framework of optical geometry have been observed beyond 300 sec period, for fundamental mode Rayleigh waves [19]. In laterally varying structures, another question often addressed is the possibility of mode conversions, particularly for sharp structural transitions like ocean/continent in case of crustal surface waves. Lateral heterogeneities are sources of secondary waves (overtones, diffracted body waves) which can cause errors in surface wave measurements [20] . However, in most cases (for mantle surface waves in particular) secondary wave amplitudes are small enough to cause no severe perturbations to the primary wave field. We must finally mention the possibility of multipathing which can be source of important perturbations in the surface wave signal. Direct observations of secondary signals due to multipathing have for example been obtained from two dimensional wavenumber analysis on arrays of seismometers [21].

II-b. Data processing techniques

Due to expression (4) Fourier transform is a basic tool in surface wave studies. In case of a <u>single mode</u> surface wave (e.g. fundamental mode), direct Fourier transform of time-windowed records should yield all the quantities of interest. In fact, multipathing, seismic noise and in some cases overtones, can be responsible for perturbations in the records. A wide class of time-frequency data processing techniques have been developed to deal with these problems. The most used ones are *time variable filtering techniques* (an improvement of the rough time windowing) and *moving-window analysis* or its equivalent *multiple filtering techniques* [22,23].

A central question in these techniques is the time-frequency resolution problem in presence of dispersion. For high resolution in frequency (e.g. narrow band filters) the classical uncertainty relationships hold while in case of poor resolution (e.g. wide band filters), the first and higher order derivatives of group velocity with respect to frequency play a major role [24]. Because time resolution can be improved by reducing dispersion, *residual dispersion methods* [25] are sometimes used[26]. In time-frequency methods of analysis, a special problem arises when amplitude variations with frequency are important [26,27] . In these cases, it is desirable to correct the Fourier transform of the records, not only for dispersion [25], but also for amplitude variations prior to perform the analysis [28].

More recently, several techniques have been developed 1) to retrieve both source and propagation parameters from wide sets of records, and 2) to isolate overtone parameters. These techniques, initially applied on a global scale to free oscillations [5,29] are now applied to surface waves. In source studies, the concept of moment tensor permits the linearization of the inverse problem [30,31] and in overtone studies, one-dimensional wavenumber techniques can be applied to arrays of long period instruments with appropriate geometry. These latter techniques allow us [32,28], or should allow us [33], to retrieve overtone parameters. An example of one-dimensional wavenumber analysis of Rayleigh overtones observed in the continental United States, is presented figure 2.

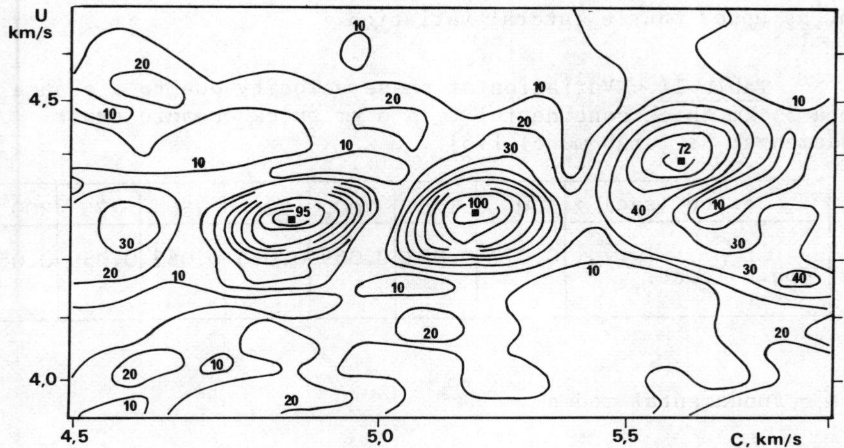

Figure 2 – U(group velocity) – C (phase velocity) amplitude
diagram, at 35 sec period, of 14 WWSSN records of a New Hebrides
earthquake (from [28]). Rayleigh overtones 1,2 and 3 (full squares
at 95 %, 100 % and 72 % of maximum amplitude) are clearly
isolated in this diagram.

III – CRUSTAL SURFACE WAVES

The depth of penetration of a surface wave mode increases
with both the rank of the mode branch (n in equation (1) and (4))
and the horizontal wavelength ($2\pi/_n k(\omega)$ in equation (4)). In
surface wave litterature, the most used propagation parameters
are : 1) phase velocity C $=\omega/k$; 2) group velocity U $= d\omega/dk$;
3) spatial attenuation coefficient α (or apparent quality factor
$\pi/U\alpha$). These parameters are generally described as functions
of the period T $= 2\pi/\omega$. The maximum of sensitivity of C to
S-velocity (the most significant model parameter) occurs at a
depth which is approximately equal to 0.3-0.4 wavelengths for
fundamental mode Rayleigh waves [13,27,34,35]. In continental
structures, a transition period between "mantle" and "crustal"
Rayleigh waves can be set near 30 sec (maximum of sensitivity
at 30-35 km depth). In oceanic region, of course, the upper
mantle plays a major role at such a period and for overtones,
the transition period is much smaller (a few seconds for Love
or Rayleigh overtones with rank > 2). It is important to note
that, although convenient for presentation purpose, the crustal/
mantle surface wave concept is a rather vague notion and the
transition between both types is by no means sharp. Crustal
properties can, for example, affect fundamental mode Rayleigh
waves up to very long period, the effect shown in table II is
approximately of the same order of magnitude, but opposite in

sign, as upper mantle lateral variation.

Table II - Variation of phase velocity due to a change
from a 35 km thick continental to a 6 km thick oceanic crust
(fundamental Rayleigh mode) [13] .

Period (sec) :	150	175	200	225	250	275	300
$C_{oc.} - C_{cont.}$ (km/s)	0.077	0.073	0.069	0.066	0.063	0.059	0.055

III a - Fundamental modes

Crustal properties can thus significantly affect "mantle"
waves. In the same way, shallow superficial layers can also greatly
affect the dispersion and attenuation of "crustal" waves. Figure
3 displays an example of perturbation of phase and group velocity
by a water layer and a low velocity sedimentary layer (S-velocity :
1.9 km/sec.). This shows that care must be taken when searching
for crustal thickness, for example, by using narrow band "crustal"
surface wave data. This also indicates that a laterally heterogeneous layer close to the surface, can cause strong perturbations
in a 20 sec. period Rayleigh wave (multipathing, diffraction
etc...).

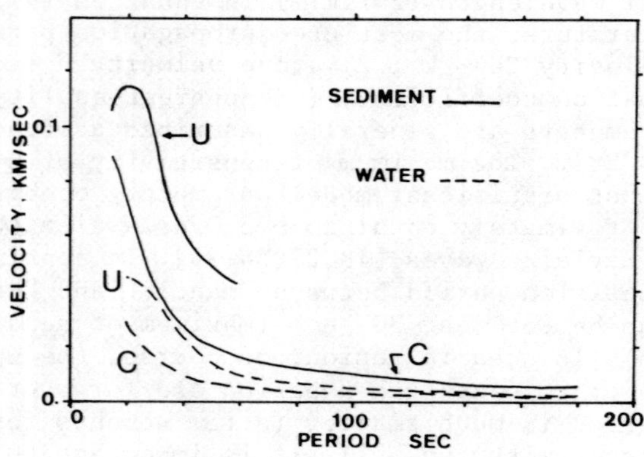

Figure 3 -   Effect of a 1 km thick sedimentary layer and
a 1 km thick water layer on group (U) and phase (C) velocity
(fundamental Rayleigh mode). Computations are made on a continental model, after Souriau [27] .

The most important parameters at periods between 15 and
40 sec are average properties of the crust (i.e. crustal thickness, average crustal velocity, ...). In that sense, the most
spectacular lateral variation in surface wave propagation is the
ocean/continent variations due to strong differences in crustal
thickness between both types of structure. The S or P velocity
contrast between the upper mantle and the crust reaches more
than 25 % and Moho depth varies from 6 km in an oceanic bassin
down to more than 50 km under some mountain ranges. 25 % lateral
variations of velocity are then typically expected in the upper
50 km of the Earth.

In some stations located on a continental border, the
differences in dispersion are so important for surface waves
coming from ocean and continent that the nature of the propagation
path can be determined at first sight on teleseismic long period
seismograms.

Figure 4 displays examples of group velocity curves observed
in such a situation.

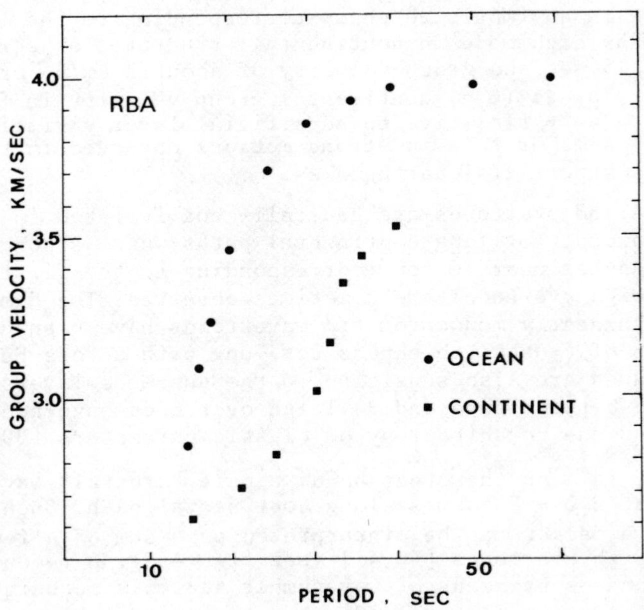

Figure 4 - Rayleigh wave group velocities from records obtained
on WWSSN-type long period instrument in Rabat (Morocco). The
oceanic data correspond to a 1010 km path in the Atlantic while
continental data correspond to a 1440 km path across North Africa.
Data from [36].

Crustal structure has been studied in most part of the world by using fundamental Rayleigh mode and numerous papers have been published during the last two decades on this subject. In regions where accurate crustal structures have been obtained by P seismic refraction data, surface waves remain useful to obtain S-velocity and Poisson ratio [37] .

Love waves are much less in use than Rayleigh waves for crustal structure investigations, mainly due to instrumental and data processing difficulties. However interesting constraints have been obtained from Love waves for crustal models (e.g.[36]).

Attenuation of crustal surface waves are mainly determined by the general existence of a highly attenuating superficial layer (in particular when sediments are involved) overlying a generally weakly attenuating medium (crust + uppermost part of the mantle). Lateral heterogeneities can greatly affect the propagation of crustal surface waves and separation of intrinsic attenuation from geometrical or diffraction effects is not obvious.

III b - Rg, Lg, overtones

Rg phase is an impulsive phase corresponding to the fundamental Rayleigh mode in continental structure, at periods smaller than 15 sec and group velocity of about 3 km/s, or less in sedimentary areas (e.g. continental group velocity in figure 4). Such waves are very sensitive to superficial layer variations. They can play a major role in strong motions observed in the near field of superficial earthquakes.

Short period overtones are generally not isolated in a seismogram, except for long continental paths where strongly dispersed branches near 10 sec (corresponding to the first Love or Rayleigh overtone) are sometimes observed. The dominant period of such nearly monochromatic wavetrains have been used as a measurement of crustal thickness for long path across Eurasia [38]. Such waves are also sensitive to the sub-Moho layers and some coupling between Love and Rayleigh overtones have been interpreted as due to anisotropy of elastic parameters [39] .

Lg phase is , on the other hand, a pure "crustal" wave, propagating at 3.3 - 3.5 km/s along continental path. Such an impulsive wave packet can be interpreted as a sum of a few Love or Rayleigh overtones [40,41] (see figure 5). A recent study of Lg phases by means of wavenumber analysis techniques mentionned in section II, has indicated that actual Lg wavefields observed across southern California were more complicated than what is predicted by superposing a few overtones in a laterally homogeneous model [33]. It is probable that in this region, lateral heterogeneities play a major role in the observed short period Lg. On the other hand on the Sierra Nevada batholite (California), preliminary results indicates that Lg wave field is simpler and a mode representation is more justified [33].

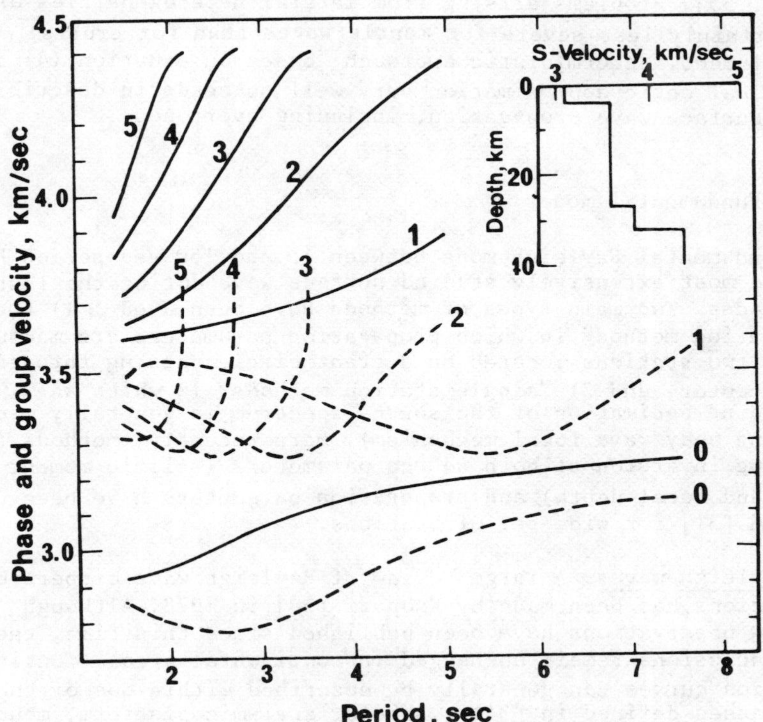

Figure 5 – Phase velocity (solid line) and group velocity (dashed line) for short period Rayleigh overtones (from [33]). Computations are made for the southern California model given in the inset.

IV - MANTLE SURFACE WAVES.

Lateral variations of seismic velocity can reach 25 % (or more if superficial layers are taken into account) in the upper 50 km of the Earth. On the other hand, between 50 and 250 km, an approximate upper bound for S-velocity lateral variations is 10 % [34, 53, 55]. Problems arising from lateral heterogeneities are thus certainly less severe for mantle waves than for crustal waves. Indeed, deterministic approach based on equation (4) and geometrical optic approximation very well succeeds in describing mantle surface wave propagation, including overtones.

IV a - Fundamental modes.

Fundamental Rayleigh mode between 15 and 150 sec-period has been the most extensively studied surface wave during the last two decades. Two main types of methods have been used : 1) the "two-station method" in which propagation parameters are measured between two stations located on a great circle passing through the epicenter, and 2) "single station methods" in which an independent estimation of the source spectrum is generally made (by using body wave focal mechanism). More recently, methods involving inversion of both source parameters (seismic moment tensor and focal depth) and propagation parameters have been proposed [31] for wide set of stations.

Rayleigh waves. A large review of Rayleigh wave dispersion observations has been made by Knopoff [34] in 1972. Although many new observations have been published since this time, the main conclusions remain unchanged in continental areas. Continental dispersion curves can generally be described within one of the four classes defined in [34] : shield, aseismic platform, mountains, rift zones. Extreme variations of phase velocity are observed between shields and rift zones. They reach 10 % at 80 sec period. In oceanic regions important lateral variations are also observed. They are well correlated with the age of the lithosphere [42,43]. At 80 sec.period, phase velocity variations reach 5 % between a young oceanic area (ridge) and an old (>100 M.y.) oceanic basin.
Upper mantle structures can be inferred down to at least 150-200 km from fundamental mode Rayleigh wave observations at periods between 30 and 150 sec. Interpretation of phase velocity data has generally been made in the framework of flat isotropic Earth models by using empirical sphericity corrections. The proposed models show that S-velocity lateral variations up to 10 % are expected near 150 km depth (lowest velocity in rift or ridge areas, highest velocity in shields). In oceanic areas, the correlation between phase or group velocity and lithospheric age is generally interpreted as a thickening of the lithosphere which is associated with the high velocity "lid". Another general

feature of surface wave S-velocity models is the presence of a
well developed low velocity zone in oceanic and young continental
area [34,35].

Love waves. Similar results have been obtained from Love
waves in oceanic regions but Love and Rayleigh waves models
inferred under assumption of isotropy are generally not compatible.
Although observational errors play probably a role in such
discrepancies, a slightly anisotropic upper mantle (a few percent)
is probable in regions where relatively recent tectonic processes
have been involved [43-45].

Great circle data. All the above results have been obtained
from "direct path" surface waves. We now briefly turn to great
circle phase velocity observations which provide most of very
long period data. At 200 sec period, maximum variations of phase
velocity observed between different great circle paths reach 6%
for the fundamental Rayleigh mode [13]. Local variations, of
course, are greater. Recent great circle regionalizations
based on geometrical optic   [46,47] and a few direct measurements
between couples of stations [48] lead to the conclusion that,
at 200 sec period, maximum lateral variations of phase velocity
do not exceed 0.06 - 0.09 km/s. A recent inversion of one of
these regionalized great circle data in four tectonic provinces
[47] strongly suggests that S-velocity lateral variations mainly
occur in the uppermost 250 km. Deeper but smaller lateral variations
are also suggested from this study. These latter variations might
be in connection with recent tectonic processes (ridges, subduction
zones)[47,49].

Attenuation. Regional attenuation data present much larger
errors than dispersion data : "direct path" Rayleigh and Love
waves have been used to infer regional attenuation models[50,51].
A review of surface wave attenuation properties was made by
B. Mitchell [51] in 1977, for the United States and the Pacific
Ocean. As for dispersion studies, "direct path" surface wave
data allow us to constrain attenuation models down to about 150 km
depth and further data are needed to obtain information at greater
depth. Attempts to regionalize great circle attenuation data have
provided regional values exhibiting considerable scatter. The most
convincing lateral variations are obtained from "direct path"
measurements, high attenuation being correlated with low velocity
[51].

Despite of their obvious interest for mantle rheology,
surface wave attenuation data are of primary importance for
interpretation of dispersion because an attenuating medium is
a priori dispersive. This physical dispersion has until recently,
generally been neglected in surface wave studies. It should be
taken into account, particularly when the highly attenuating
asthenosphere is involved. Some simple attenuation models proposed

for this correction[52], have been recently used in surface wave
inversion [45].

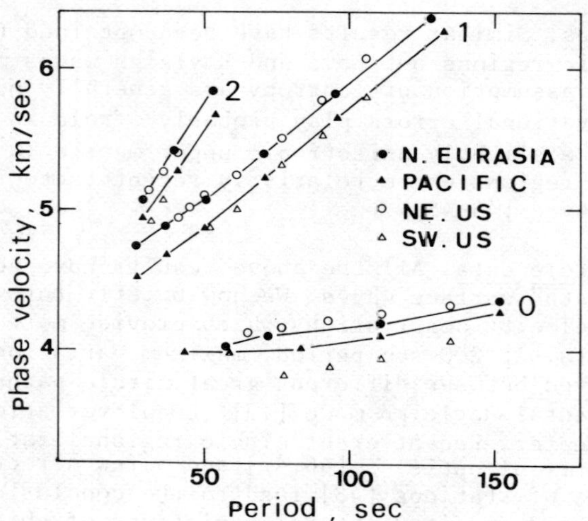

Figure 6 -   Phase velocity for Rayleigh modes 0 (fundamental mode),
             1 (first overtone) and 2 (second overtone) in different
             tectonic provinces (data from [53]).

IV b - Overtones.

        "Direct path" measurements can provide information at depth
greater than 150-200 km if use is made of 30-100 sec period
overtones. Group velocity plateaus at about 4.4 km/s for Love or
Rayleigh overtones lead to an impulsive "S_a" phase which, if
excited at the source, can be well observed with classical long
period instruments (i.e. WWSSN seismograms). Wavenumber analysis
techniques applied to wide arrays of long period seismometers
have recently provided accurate mantle overtone data in several
parts of the world [28, 32, 53] . Figure 6 and 7 display phase
and group velocity curves obtained from "direct path" Rayleigh
overtones for different regions [53] . A clear "first order"
variation is observed for overtones between typical continental
(N.Eurasia) and oceanic (Pacific Ocean) paths. Overtone phase
velocities exhibit similar lateral variations within the continental
United States : use of a block regionalization algorithms [28]
has shown that strong lateral variations could be resolved between
a central-northeastern block and a southwestern block (figure 5b).
Somewhat intermediate overtone curves have also been obtained
for western Europe [32,53]. Inversion of fundamental mode and
overtone dispersion curves, including those presented in figure 6,
has provided constraints on upper mantle S-velocity down to at

at least 500 km depth, the deepest parts being controlled by
overtone data and the upper 150 km by fundamental mode. A set of
"smooth" S-velocity models obtained from such sets of data [1]
are presented in figure 8 (i.e. only 4 largest eigenvalues are
kept in the linear inversion algorithm).

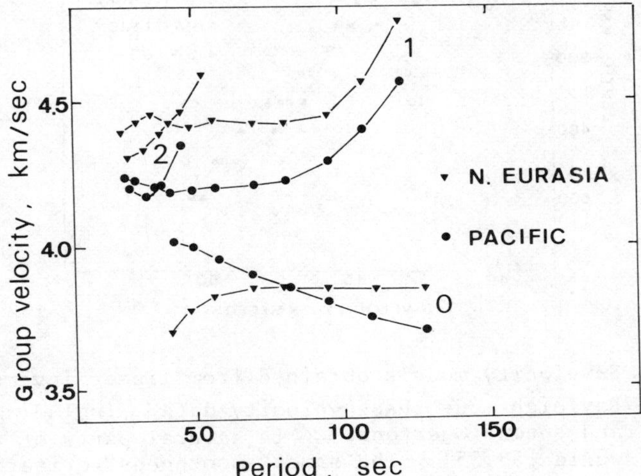

Figure 7 - Group velocity for Rayleigh modes 0 to 2
(data from [53]).

Attenuation data are rather difficult to obtain for "direct
path" mantle overtones. In addition to all sources of error
encountered in fundamental mode amplitude measurement, overtone
amplitudes are furthermore contaminated by interferences with
other modes. It has been however recently demonstrated [54] that
some constraints on upper mantle models can be obtained by using
several data processing   techniques, including band-pass
wavenumber filtering (or "spatial filtering" [28]). For the
Pacific Ocean, it seems in particular that the upper mantle
attenuation is much less important in the depth range 250-500 km
between 50 and 250 km depth [54] .

Overtone data reviewed above are obtained from Rayleigh
wave overtones. Few studies have been made on Love wave overtones
[43,53]. Difficulties to interpret both Love and Rayleigh
fundamental mode data in terms of isotropic elastic parameters
[43-45] have also been encountered for overtones in western
Europe [53] where observed Love wave phase velocities are
systematically greater than those predicted from Rayleigh wave
models. It is thus likely that, at least in some regions, the
usual isotropy hypothesis should be dropped.

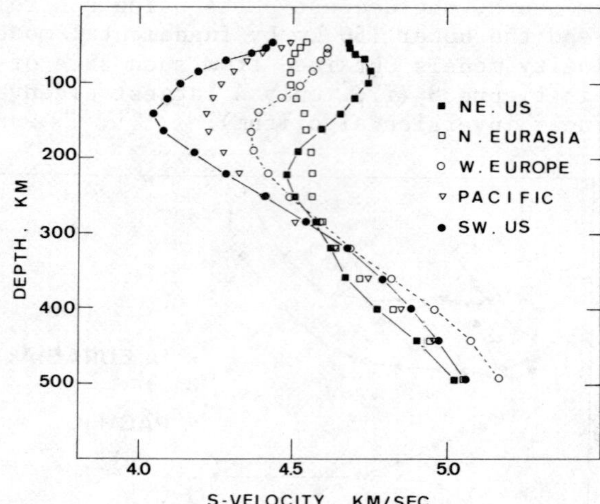

Figure 8 -   S-velocity models obtained from linear inversion of
Rayleigh mode phase velocity data — including first
and second overtones —— in several parts of the
world [53,55] . The same "smoothness" criterion is
used for all models (i.e. 4 largest eigenvalues kept
in the linear inversion scheme).Models for the
continental US are obtained by a regionalization in
two blocks, the boundary between "NE US" and "SW US"
being set just East of the Rocky Mountains.

V - CONCLUSION

This paper is an attempt to briefly review surface wave
observations, including multimode surface waves, at periods lying
between a few seconds (Lg) and a few hundred seconds (very long
period fundamental modes). Several assumptions, generally implicitly
made, are commonly used in surface wave studies. These include
approximations made for Fourier integral representation of the
surface wave field and geometrical optic approximation. Interpretation of propagation parameters in terms of regional earth models
are also generally based on simplified models (i.e. isotropy,
non-dispersive medium ...). For the most extensively studied type
of surface wave-fundamental Rayleigh mode at periods between 15
and 150 sec - most of these assumptions are, in general, satisfactory. On the other hand, as new independant types of surface
waves become available (e.g., Love waves, mantle overtones ...)
internal inconsistencies (e.g. Love/Rayleigh modes) or inadequacies
of the classical surface wave approach to describe the observed
wavefield soon appear. In addition to extensively studied types

of surface waves we have presented in this paper more recent
observations, some of them still uncertain, which could provide
important data in the future for both structural and source
investigations.

Acknowledgements

        Part of this paper was written while the author was a
Research Fellow of the California Institute of Technology (supported
in part by fellowships from C.N.R.S./N.S.F. and N.A.T.O.). I
benefited from numerous discussions for this paper and, in
particular I would like to thank Don Anderson, Hiroo Kanamori,
Nelly Jobert and Bernard Minster. Annie Souriau and Jean-Jacques
Lévêque provided some material presented in this paper, they
are fully acknowledged.

Contribution IPG n°418.

REFERENCES

1   CARA, M., 1978 : Thèse de doctorat d'Etat, Paris.

2   BRUNE, J.N. & DORMAN, J., 1963 : Bull. Seism. Soc. Am.
         53, pp 167-209.

3   SAITO, M., 1967 : J. Geophys. Res. 72, pp 3689-3699.

4   TAKEUCHI, H. & SAITO, M., 1972 : Mehtods in computational
         Physics, 11, pp 217-295, Acad. Press, London.

5   MENDIGUREN, J.A., 1973 : Geophys. J. 33, pp 281-321.

6   KANAMORI, H. & CIPAR, J.J., 1974 : Phys. Earth Planet.
         Int. 9, pp 128-136.

7   GILBERT, F., 1970 : Geophys. J. 22, pp 223-226.

8   DESCHAMPS, A., LYON-CAEN, H. & MADARIAGA R., 1980 :
         Ann. Geophys. in press.

9   BRUNE, J.N., NAFE, J.E. & ALSOP, L.E., 1961 : Bull. seism.
         Soc. Am., 51, pp 247-257.

10   SCHWAB, F. & KAUSEL, E.G., 1976 : Geophys. J. 45, pp 407-435.

11   GILBERT F., 1976 : Geophys. J. 44, pp 275-280.

12   KANAMORI, H. & STEWART, G.S., 1976 : Phys. Earth Planet.
         Int. 11, pp 312-332.

13   LEVEQUE, J.J., 1978 : Thèse de 3ème cycle, Paris.

14   DAHLEN, F.A. & SAILOR, R.V., 1979 : Geophys. J. 58,
         pp 609-623.

15   BACKUS, G.E., 1964 : Bull. seism. Soc. Am. 54, pp 571-610.

16   KNOPOFF, L., 1969 : J. geophys. Res. 74, p 1701.

17   JORDAN, T.H., 1978 : Geophys. J. 52, pp 441-455.

18   MADARIAGA, R. & AKI, K., 1972 - J. geophys. Res. 77,
         pp 4421-4431.

19   JOBERT, N. & LEVEQUE, J.J. & ROULT, G., 1978 : Geophys.
         Res. Lett. 5, pp 569-572.

20   CISTERNAS, A. & JOBERT, G., 1977 : J. Geophys. 43, pp 59-74.

21   CAPON, J., 1971 : Bull. seism. Soc. Am. 61, pp 1327-1344.

22   LANDISMAN, M. DZIEWONSKI, A.M. & SATO,Y., 1969 :
          Geophys. J. 17, pp 369-403.

23   DZIEWONSKI, A.M., BLOCH, S. & LANDISMAN, M., 1969 : Bull.
          seism. Soc. Am. 59, pp 427-444.

24   CARA, M., 1973 : Geophys. J. 33, pp 65-80.

25   DZIEWONSKI, A.M., MILLS, J. & BLOCH, S., 1972 : Bull. seism.
          Soc. Am. 62, pp 129-139.

26   CARA, M. & HATZFELD, D., 1976 : Ann. Geophys. 32, pp 85-91.

27   SOURIAU, A., 1978 : Thèse de Doctorat d'Etat, Paris.

28   CARA, M., 1978 : Geophys. J. 54, pp 439-460.

29   GILBERT, F. & DZIEWONSKI, A.M., 1975 : Phil. Trans. R. Soc.
          London Ser. A 278, pp 187-269.

30   MENDIGUREN, J.A., 1977 : J. Geophys. Res. 82, pp 889-894.

31   PATTON, H., J., 1978 : Ph. D. thesis, M.I.T., Cambridge.

32   NOLET, G., 1975 : Geophys. Res. Lett. 2, pp 60-62.

33   CARA, M. & MINSTER, J.B., 1980 : submitted to Bull. seism.
          Soc. Am.

34   KNOPOFF, L., 1972 - Tectonophys. 13, pp 497-519.

35   KOVACH, R.L., 1978 : Rev. Geophys. Space Phys. 16, pp 1-13.

36   CARA, M. & HATZFELD, D., 1977 : Bull. Soc. géol. France
          19, pp 757-764.

37   HADLEY, D. & KANAMORI, H., 1979 : Geophys. J. 58, pp 655-666.

38   CRAMPIN, S., 1966 : Bull. Seism. Soc. Am. 56, pp 1227-1239.

40   KNOPOFF, L., SCHWAB, F., NAKANISHI, K. & CHANG, F., 1974 :
          Geophys. J. 39, pp 41-70.

41   PANZA, G.F. & CALGAGNILE, G., 1975 : Geophys. J. 40,
          pp 475-487.

42   KAUSEL, E.G., LEEDS, A.R. & KNOPOFF, L., 1974 : Science
          186, pp 139-141.

43   FORSYTH, D.W., 1975 : Geophys. J. 43, pp 103-162.

44   SCHLUE, J.W. & KNOPOFF, L., 1977 : Geophys. J. 49,
         pp 145-165.

45   YU, G.K. & MITCHELL, B.J., 1979 : Geophys. J. 57,
         pp 311-341.

46   NAKANISHI, I. 1978 : Geophys. J. 58, pp 35-59.

47   LEVEQUE, J.J., 1980 : Geophys. J. 63, pp 23-44.

48   OKAL, E., 1978 : Geophys. J. 53, pp 663-668.

49   CARA, M. & LEVEQUE, J.J., 1979 : XVII IUGG General Assembly
         Canberra (abstract).

50   TSAI, Y.B. & AKI, K., 1969 : Bull. seism. Soc. Am. 59,
         pp 275-287.

51   MITCHELL, B.J., YACOUB, N.K. & CORREIG, A.M., 1977 - AGU
         monograph 20, pp 405-425.

52   KANAMORI, H. & ANDERSON, D.L., 1977 : Rev. Geophys. Space
         Phys. 15, pp 105-112.

53   CARA, M., NERCESSIAN, A. & NOLET, G., 1980 : Geophys. J. 61,
         pp 459-478.

54   CARA, M., 1979 : AGU monograph, in press.

55   CARA, M., 1979 : Geophys. J., 57, pp 649-670.

Lg WAVE PROPAGATION IN EURASIA

Svein Mykkeltveit and Eystein S. Husebye

NTNF/NORSAR, Post Box 51, N–2007 Kjeller, Norway

ABSTRACT

WWSSN records for explosions and earthquakes in Eurasia are
analyzed for propagation characteristics of regional seismic phases, especially Lg. An overall prominence of the commonly reported phases Sn, Lg and Rg is not evident in view of
the large scatter in our observed group velocities from reading
all clear, wavelike onsets in the seismograms. When only the most
energetic secondary arrival for each seismogram is included,
however, Lg stands out as a reasonably stable and consistent
phenomenon. Still the propagation efficiency of Lg is less than
that reported for eastern U.S., even for the Western Russia/
Baltic Shield region. Propagation characteristics of Sn, Lg and
Rg phases are complex and the associated amplitude scatter is of
the order of one m_b magnitude unit even for nearly identical
travel paths. The tectonic barrier concept occasionally introduced for explaining strong Lg attenuation observations is not
consistently valid even across the Himalayas. The potential of
regional phases in a source identification context is not found
particularly promising.

1.   INTRODUCTION

A longstanding problem in seismology has been the proper physical
understanding of seismic wave propagation at regional distances,
and of particular interest in this context is the prominent Lg
wave. This phase, often characterized by sharp onsets, amplitudes
larger or comparable to any of the conventional phases for distances say less than 15°, was first identified by Press and

*E. S. Husebye and S. Mykkeltveit (eds.), Identification of Seismic Sources - Earthquake or Underground
Explosion, 421–451.*

Ewing (1) as a surface shear wave with group velocity around 3.5 km s^{-1} and periods in the range of 0.5-6.0 sec. In the notation adopted, L refers to Love waves and the g-index to presumed propagation in the granitic part of the crust. Initially Lg and associated phases were interpreted in terms of channel wave propagation effects which in turn led to many suggestions of low velocity channels in the crust and upper mantle. Further evidence, both observational and theoretical, for the Lg-waves being higher mode Love and Rayleigh waves, was provided by many seismologists during the following years. See (2-9) for references to early work on Lg waves. In a series of papers Knopoff and his colleagues (10-16) succeeded in modelling Lg waves by incorporating the effects of source mechanisms, anelastic attenuation and the selective frequency response of the recording instrument in addition to the involved propagation features of higher modes. In essence, they demonstrated that the propagation modes in question in addition to having the 'correct' group velocities, are associated with sufficiently large amplitudes to provide a satisfactory explanation to the Lg observations. Results here are that no low velocity layer is required for Lg propagation, and also that the virtually total elimination of Lg over a distance of 100 km of oceanic crust is probably due to strong shear attenuation.

Remaining problems are handling of Lg wave propagation in heterogeneous media, particularly across so-called tectonic boundaries, and our understanding of the role of scattering which besides formidable computational problems is difficult to model realistically due to the structural details required for the short wavelength sampling of these waves.

There is still a considerable interest in Lg studies on account of the prominence of this phase at local and regional distances. The Lg phase is of major interest to seismic risk assessments, local structure and magnitude studies and not least to problems associated with differentiating between earthquakes and underground nuclear explosions. The topic of this paper is tied to the latter type of problems. We will provide evidence on the relative detectability of Lg waves for various tectonic regimes in Eurasia and then discuss qualitatively the earthquake/explosion discrimination potential of this phase. For the sake of completeness, data bearing on the propagation efficiency of Sn are also included. The observational data are taken from WWSSN records at regional distances for events in Western Russia and Central Asia.

This contribution is organized as follows: The next section describes our observational material and also gives details on analysis procedures used. The results are given in a separate section and comparison is made with similar investigations for

the same regions and elsewhere. Then the results are discussed
and finally some digital NORSAR recordings are presented that
allow a more in-depth analysis of Lg wave propagation characteristics.

2.   OBSERVATIONAL DATA AND ANALYZING PROCEDURES

The main scope of this work was to analyze Lg and to some extent
also Sn, Li and Rg wave propagation efficiencies and characteristics across various tectonic provinces of the Eurasian plate,
in particular Central Asia and Western Russia/Baltic Shield.
This constraint together with that of short period Lg observations being mainly confined to regional distances, limited the
choice of easily available data to records from WWSSN stations
in southern and western Asia and Fennoscandia. Likewise, the

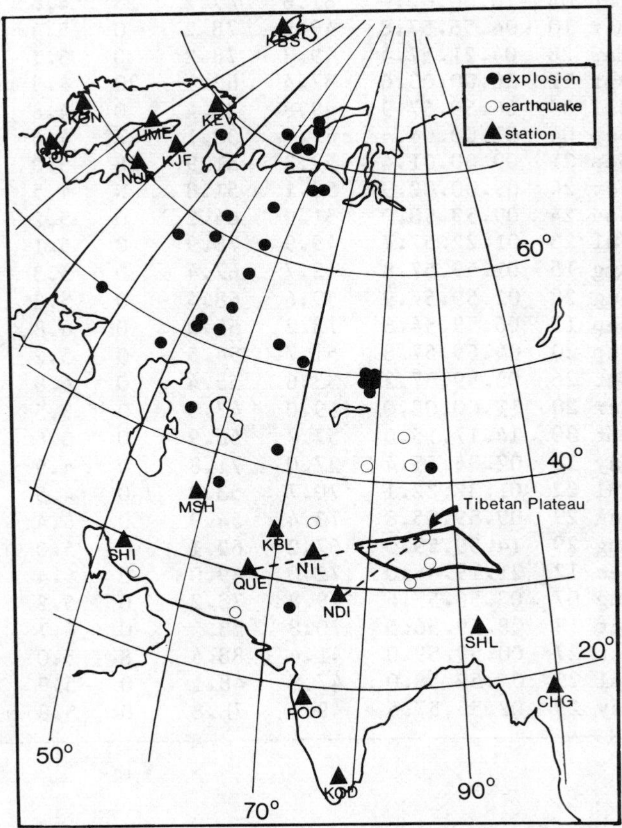

Figure 1.   Location of events and stations used in this study.

events subjected to analysis are mainly presumed underground
explosions in Kazakh, Caspian Sea area, Western Russia and Novaya
Zemlya, in addition to Central Asian earthquakes. For further
details we refer to Figure 1 and Tables 1 and 2. Noteworthy, it

Table 1.   Presumed explosions used in this study

No.	Origin Time (GMT)			Lat.	Long.	H	m_b	Agency
1	1970 Dec 23	07.00.57.3		43.8	54.8	0	6.0	ISC
2	1971 Mar 23	06.59.56.4		61.4	56.2	0	5.5	ISC
3	1971 Jul 02	17.00.01.9		67.7	62.0	0	4.7	ISC
4	1971 Jul 10	16.59.59.6		64.2	54.8	0	5.2	ISC
5	1971 Sep 19	11.00.06.9		57.8	41.4	N	4.5	ISC
6	1971 Oct 04	10.00.02.0		61.6	47.2	13	4.6	ISC
7	1972 Mar 10	04.56.57.8		49.8	78.2	0	5.4	ISC
8	1972 Mar 28	04.21.57.4		49.7	78.2	0	5.1	ISC
9	1972 Apr 11	06.00.03.0		37.4	62.1	20	4.9	ISC
10	1972 Jul 09	06.59.57.9		49.8	35.4	0	4.8	ISC
11	1972 Sep 04	07.00.04.4		67.7	33.1	7	4.6	ISC
12	1972 Sep 21	09.00.01.4		52.2	51.9	28	5.0	ISC
13	1972 Nov 24	09.00.02.9		52.1	51.8	N	4.5	ISC
14	1972 Nov 24	09.59.58.0		51.9	64.2	0	5.2	ISC
15	1973 Jul 23	01.22.57.7		49.9	78.9	0	6.1	ISC
16	1973 Aug 15	01.59.57.8		42.7	67.4	0	5.3	ISC
17	1973 Aug 28	02.59.57.9		50.6	68.4	0	5.2	ISC
18	1973 Sep 12	06.59.54.6		73.3	55.0	0	6.8	ISC
19	1973 Sep 30	04.59.57.8		51.7	54.5	0	5.2	ISC
20	1973 Oct 26	05.59.57.2		53.6	55.4	0	4.8	ISC
21	1974 Mar 20	11.00.08.0		59.0	47.5	0	3.5	NAO
22	1974 Apr 30	14.17.15.0		57.7	56.9	0	3.6	NAO
23	1974 May 18	02.34.55.4		27.0	71.8	0	4.9	ISC
24	1974 Jul 22	01.32.22.1		70.7	53.3	0	4.5	ISC
25	1974 Aug 29	09.59.55.8		73.4	54.9	0	6.4	ISC
26	1974 Aug 29	14.59.59.0		67.2	62.1	0	5.0	ISC
27	1974 Dec 12	21.19.46.0		72.0	49.0	0	4.4	ISC
28	1975 Aug 07	03.56.57.6		49.8	78.2	0	5.2	ISC
29	1975 Oct 18	08.59.56.5		70.8	53.5	0	6.7	ISC
30	1975 Oct 27	00.59.59.0		41.4	88.4	8	5.0	ISC
31	1976 Jul 29	04.59.58.0		47.8	48.1	0	5.9	ISC
32	1977 May 29	02.56.57.5		49.7	78.8	0	5.8	ISC

Table 2.   Presumed earthquakes used in this study.

No.	Origin Time (GMT)			Lat.	Long.	H	m_b	Agency
1	1971 Mar 23	20.47.16.0		41.4	79.2	14	5.8	ISC
2	1971 Nov 08	03.06.34.0		27.0	54.5	11	5.6	ISC
3	1972 Mar 15	06.00.30.2		30.5	84.4	12	5.1	ISC
4	1972 Apr 09	04.10.48.9		42.1	84.6	21	5.8	ISC
5	1972 Sep 03	23.03.53.6		36.0	73.2	46	5.6	ISC
6	1973 Jun 02	23.57.02.4		44.1	83.6	12	5.7	ISC
7	1973 Jul 14	13.39.29.4		35.2	86.5	29	5.7	ISC
8	1973 Aug 14	18.24.16.0		25.4	65.6	2	4.9	ISC
9	1973 Sep 08	07.25.41.0		33.3	86.8	11	5.5	ISC

was decided to separate the observational data into two populations. To have a unified analysis of a rather homogeneous sub-
region of Eurasia, all short-period (and most long-period) records
of the events (all explosions) numbers 2, 3, 4, 5, 6, 10, 11,
13, 14, 17, 19, 20, 24 and 26 of Table 1 were read consistently
for the Finnish stations NUR, KJF and KEV, comprising a Western
Russia/ Baltic Shield data base with the remainder of the data
constituting a Eurasian data base.   Tables 3 and 4 give for the
respective events the actual records analyzed. It should here be
emphasized that the WWSSN analog records have definite shortcomings vis-à-vis complex high frequency signals, which is a
typical feature of the Lg wave train.   These kinds of problems
are illustrated in Figure 2, which shows group velocity curves
for the fundamental and first five higher Love wave modes for a
continental type structure without velocity inversions. The
small group velocity bracket for the first five modes immediately
suggests that the Lg wave train would exhibit a complex interference pattern.   We find extrema in the group velocity curves
around 4.5 km s^{-1} and 3.5 km s^{-1}, that is, at group velocities
typical of Sn and Lg phases. Dispersion curves for Rayleigh
modes (e.g., see (14)) show the same main features for a wide
range of models with and without low velocity channels.

Returning to the analog record analysis, the efforts were
concentrated upon reading onset times, amplitudes and periods
of all prominent phase arrivals within the group velocity window of approximately 4.8–2.7 km s^{-1} for all three components on
both short and long period records. Observations bearing on P
waves were read from the vertical component as this information
was used for estimating the relative propagation efficiency of
secondary seismogram phases. Any analysis based on reading analog
seismograms involves some personal judgement, say, whether the
presumed first onset or the maximum amplitude time of the Lg
train should be used in estimating group velocities. Many authors

Table 3. Presumed explosion records analyzed in this study. 1=SPZ, 2=SPNS, 3=SPEW, 4=LPZ, 5=LPNS, 6=LPEW. Event number refers to Table 1.

Event No.	COP	KBS	KEV	KJF	KON	NUR	UME	CHG	KBL	KOD	MSH	NDI	NIL	POO	QUE	SHI	SHL
1						1			1 4							1	
2			123 456	123 456		123 456											
3			123 456	123 456		123 456											
4			123 456	123 456		123 456											
5			123 456	123 456		123 456											
6			123 456	123 456		123 456											
7											123		123				
8									123								
9			1		1	1	1					1	1		1	1	1
10			123 456	123 456		123 456											
11	123	123 456	123	123 456	123	123 4	123										
12			1		1	1						2	2			1	

Table 3. Cont.

Event No.	COP	KBS	KEV	KJF	KON	NUR	UME	CHG	KBL	KOD	MSH	NDI	NIL	POO	QUE	SHI	SHL
13			123 456	123 456		123 456											
14			123 456	123 456		123											
15			1		1 456	123	456		456			1	456		1	1	
16			1		1	1						1			1	1	
17			123 456	123 456		123 456											
18						4		4									
19			123 456	123 456		123 456											
20			123 456	123 456		123 456											
21					1				2							1	1
22					1	1	1		1								
23						1	1								1		
24			123 456	123 456		123 456											
25	123	456	456	456		456											

Table 3. Cont.

Event No.	COP	KBS	KEV	KJF	KON	NUR	UME	CHG	KBL	KOD	MSH	NDI	NIL	POO	QUE	SHI	SHL
26			123	123		123											
27			123														
28									123			123					
29			4		4	4									456	123	
30						1							1		1		
31			1		1	1										1	
32			123			1	1										1

Table 4. Presumed earthquake records analyzed in this study. For codes see caption of Table 3. Event number refers to Table 2.

Event No.	COP	KBS	KEV	KJF	KON	NUR	UME	CHG	KBL	KOD	MSH	NDI	NIL	POO	QUE	SHI	SHL
1							1										4
2			1		1	1	1					1					1
									4			4					
3						1		123		123	123	123	12		123	1	123
											4	4	456		456		4 6
4			1		1	1	1					1				1	
									4			4			4		4
5			1		1	1	1								1	1	
												4	4		4	4	
6					1	1	1					1			1	1	1
							4					4			4		
7			1			1	1					123			1	1	1
						4	4		456			4 6	4		4	4	4
8									123			123		123			123
									456						456		456
9			1			1						1			1	1	
			4						4						4	4	

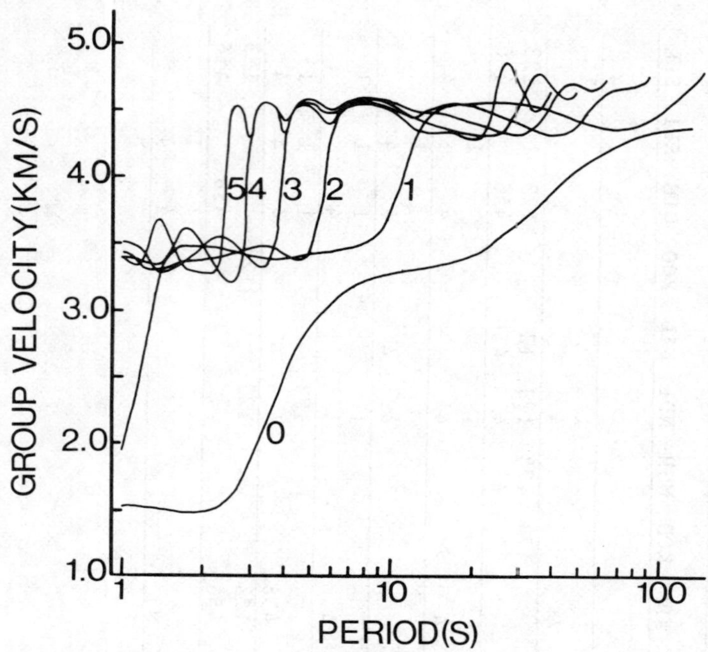

Figure 2. Spherical group velocity of fundamental and first
five higher Love wave modes for a continental model with a
35 km thick crust and Sn velocity 4.54 km/s. Redrawn after (13).

report Lg and Sn velocities without measurement comments but the
associated group velocity brackets are very narrow and in good
agreement with what is expected from theoretical dispersion
curves, like the ones in Figure 2. Indeed, in cases where these
waves appear as concentrated energy bursts, a simple listing of
group velocities is adequate. However, it soon became clear from
our seismogram analysis that the Lg wave does not necessarily
exhibit 'sharp commencements' as initially reported by Press and
Ewing (1) but might be rather outdrawn in time and may be compared
to P waves in relatively strongly inhomogeneous media where scattering, multipathing and mode conversions are important. For
this reason, we chose to read all prominent phases within the
4.8–2.7 km s^{-1} velocity window. In practice, this meant that
all wavelike arrivals that stand out reasonably clearly from the
general noise/coda level were read, with arrival times corresponding to the wavepacket onset time. Amplitudes and periods,
however, were read for the maximum amplitude-to-period ratio
within the same wavepacket but never more than three full cycles
behind the onset time.

To summarize so far: with the data extracted from the WWSSN
records we can calculate travel times, group velocities, event
magnitudes, amplitude decays and also which component(s) is most
prominent as regards higher mode surface wave recordings. As mentioned in the previous section, analog seismograms are not the
most efficient nor the best way of extracting signal parameters
from complex wave trains. For our purpose of mapping average Lg
propagation efficiencies across various parts of Eurasia, this
kind of data is, however, adequate, especially as high-quality
digital records having comparable regional sampling are for
practical purposes not available.

3.   ANALYSIS RESULTS

In this section we will present results bearing on group velocity
observations, dominant signal periods and relative propagation
efficiencies in Western Russia and Eurasia. A comparison is also
made between our results and those reported by others for these
tectonic regions.

3.1 Group Velocity Observations

All our group velocity measurements for 'local' amplitude maxima
found in the vertical component records are displayed in Figures
3 and 4 for the two data sets under consideration. We start the
discussion here with the Western Russia/Baltic Shield data of
Figure 3 and immediately note that the few LP observations presented reflect that most of the events available for analysis
were of moderate magnitude. Anyway, the LP observations here
appear to be mainly of the Sn and Rg type. On the other hand,
the SP data are slightly dominated by Lg and Sn type of group
velocity observations in the respective ranges of 3.35-3.60 km/s
and 4.30-4.60 km/s. There are also some observations in the
interval 3.80-4.00 km/s which is somewhat above Båth's (17) Li
observations (Love waves in the intermediate part of the crust).
That this should be a prominent phase is not evident from the
theoretical dispersion curves for Rayleigh waves, but on the
other hand this may reflect the fact that the current concepts
of the lower crustal strctures are somewhat crude. Perhaps the
most likely explanation here is in terms of scattering, multipathing and possibly lags introduced by mode conversions. The
most striking aspect of Figure 3 is the more or less continuous
distribution of observed wave velocities. For the short period
data, practically any velocity within the window (3.10, 4.70)
km/s is represented, and from this viewpoint, the significance
of any phase is questionable. This situation is altered, however,
by considering only the most energetic arrival for each seismogram.

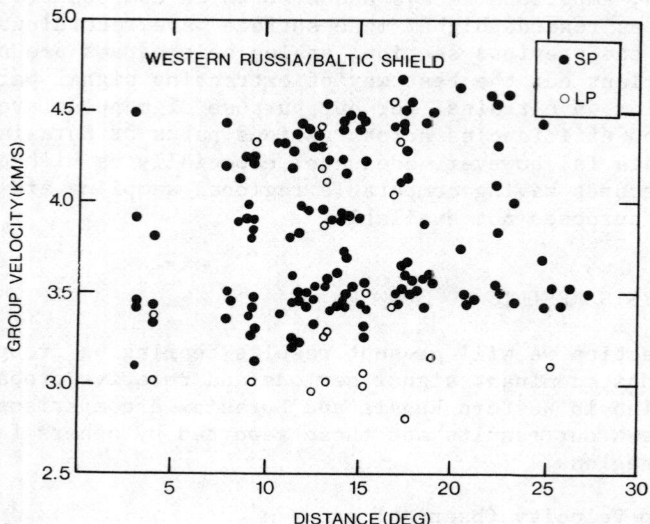

Figure 3. Group velocity as a function of epicentral distance
for all 'local' amplitude maxima found in the vertical component
seismograms belonging to the Western Russia/Baltic Shield data
base.

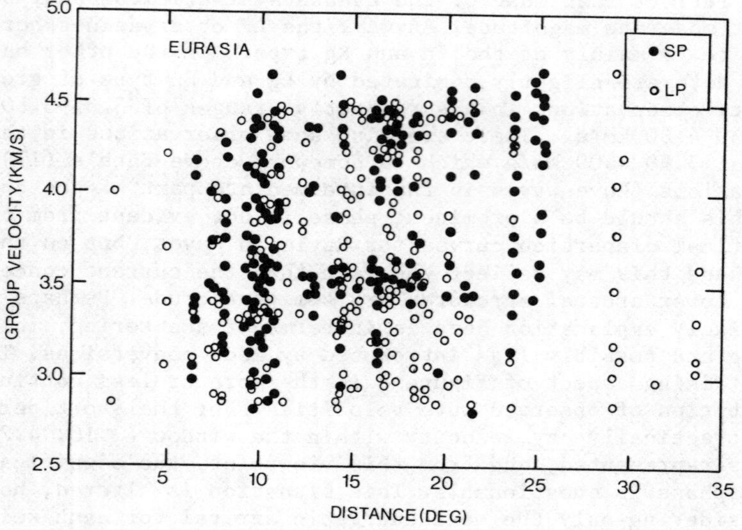

Figure 4. Same as Figure 3 for the Eurasian data base.

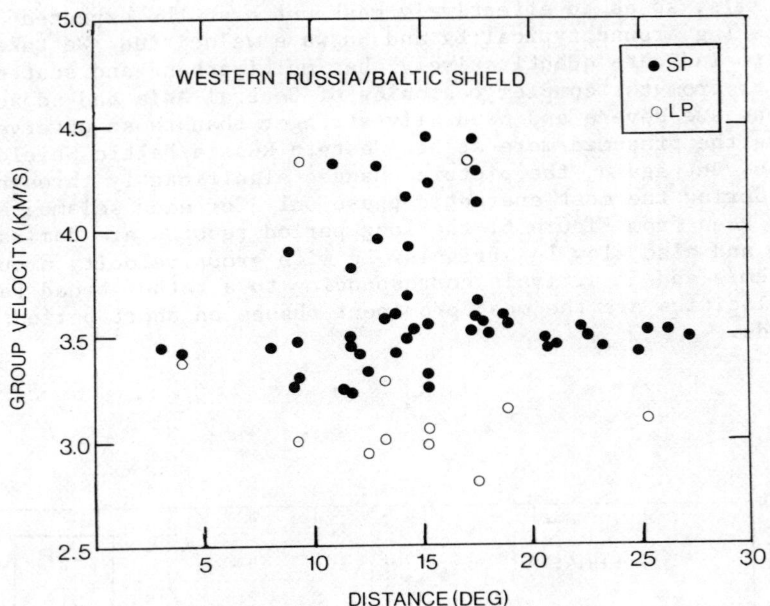

Figure 5.   Group velocity as a function of epicentral distance
for the strongest phase in the group velocity window 4.8-2.7 km/s
for each vertical component seismogram in the Western Russia/
Baltic Shield data base.

This has been done in Figure 5 with the result that Lg stands
out rather clearly at a velocity of approximately 3.50 km/s
as the most likely observable phase, especially for distances
exceeding about 18°. This means that Lg is a rather stable phenomenon for propagation in Western Russia/Baltic Shield. Rg appears to be the dominating phase on long period instruments,
with group velocity around 3.0 km^{-1}, which compares favorably
with the fundamental mode dispersion curve for Rayleigh waves
at a period of 5-10 s.   Li or Sn is only occasionally the
strongest phase. Finally, beyond epicentral distances of 25-30°,
reasonably clear phases of the types considered here are very
seldom observable.

The Eurasian group velocity data displayed in Figure 4 exhibit an even greater scattering than that observed for the Western Russian/Baltic Shield area. Indeed, the observations are 'smeared' over the whole group velocity interval of 2.80-4.60 km/s, so as to effectively mask out even the expected clustering around typical Lg and Sn wave velocities. We take this to indicate quantitatively that multipathing and scattering effects from the complex tectonics of Central Asia and adjacent regions are severe and naturally stronger than those observed across the presumed more stable Western Russia/Baltic Shield region. But again, the picture changes significantly through considering the most energetic phase only for each seismogram. As is seen from Figure 6, the long period records are dominated by Rg and also slow Lg arrivals. Lg with group velocity around 3.50 km/s and Sn arrivals corresponding to a rather broad band of velocities are the most prominent phases on short period records.

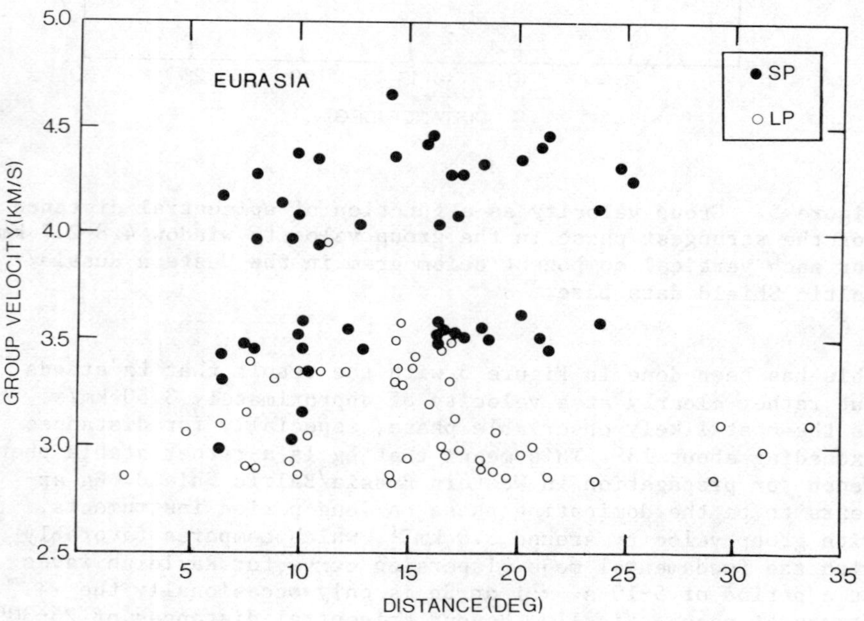

Figure 6.   Same as Figure 5 for the Eurasian data base.

3.2 Dominant Periods of Lg and Sn Observations

The distribution of periods for the readings included in Figures
3 and 4 is given both for short and long period records in
Figure 7. From this figure and previous ones it can be concluded that the Lg wave with period in the range 0.8-1.2 s is a
rather prominent feature in WWSSN short period vertical component
seismograms from Eurasian events. Other phases propagating at
group velocities between 4.8 and 2.7 km/s are not consistently
observable throughout the Eurasian plate as a whole on short
period records. For our kind of modest magnitude events, Rg
type of waves of typical period 4-12 s are most prominent on
WWSSN long period vertical component records, with other phases
only occasionally being stronger.

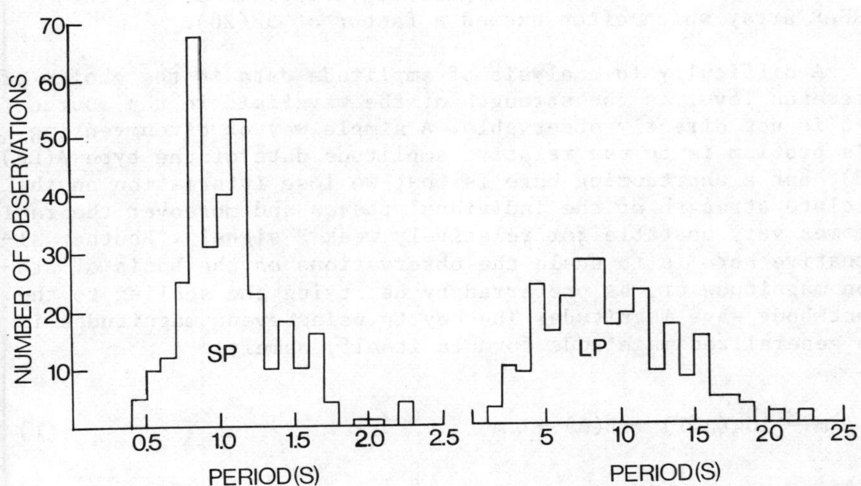

Figure 7. Distribution of periods for all arrivals included in
Figures 3 and 4.

3.3 Propagation Efficiency/Event Detectability

Analysis of short period phase amplitudes is always difficult
and particularly so when the data at hand are a mix of different
event/receiver combinations. In our case, the problem is twofold, namely: i) average propagation efficiency of Lg and Sn
as a function of distance and ii) event detectability or how to
compare Lg and Sn amplitudes with those of P. Appropriate analyzing techniques for handling this kind of problems have been

demonstrated by Nuttli (18) and Bollinger (19), but before giving details on the procedures chosen by us, we want to briefly comment on the major factors affecting the observed phase amplitudes. These factors are instrument response, geometrical spreading, attenuation/scattering, source mechanism and finally the strength of the source (event magnitude). Wave propagation in the real earth would include besides more grand structural differences, also multipathing and scattering effects which are not easily modelled. Our main concern is average amplitude decay with distance, however, so simple physical modelling schemes will be used. This in turn implies that source radiation effects, multi-pathing effects, and so on, are interpreted statistically in terms of scatter in the observations themselves. We note in passing that pronounced radiation effects have not been recognized in short period Lg observations in the 0.5-5.0 s range. In case of underground explosions, these comments apply to possible release of tectonic stresses. The above assumptions are justified from experience with P-wave amplitude variations across the NORSAR array which often exceed a factor of 5 (20).

A difficulty in analysis of amplitude data is the choice of reference level as the strength of the wavefield in the source area is not directly observable. A simple way of circumventing this problem is to use relative amplitude data of the type A(Lg)/A(P), but a shortcoming here is that we lose information on the absolute strength of the individual phases and moreover the ratio becomes very unstable for relatively weak P signals. Another alternative here is to scale the observations on the basis of station magnitude or, as preferred by us, tying the scaling to the event body wave magnitude. The key to using event magnitudes is the generalized magnitude formula itself, namely:

$$m = \log(A/T) + B(\Delta) + \delta m \tag{1}$$

where m = magnitude, A/T = amplitude-period ratio and δm = bias term (ignored here). The B(Δ)-term directly reflects amplitude-distance decay (the depth index omitted) given that both the event magnitude m and the station-observed A/T-values refer to the same phase, commonly P. On the other hand, if the event m_b magnitude is coupled to A(Lg)/T then the $B_{Lg}(\Delta)$ parameter in comparison with that for P waves gives the relative propagation efficiency of Lg, namely:

$$B_{Lg}(\Delta) = B_P(\Delta) + \log\left[(A/T)_P/(A/T)_{Lg}\right] \tag{2}$$

Eq. (2) gives that the Lg amplitude distance factor is equal
to that of P waves plus another relatively slowly varying
distance term. Also, the Lg detectability can easily be de-
ducted if the corresponding P-wave detectability is known before-
hand.

Since we tie our amplitude analysis to the common body wave
formula of eq. (1), our work in the following essentially amounts
to an estimation of the $B_{Lg}(\Delta)$ term for regional distances. As
for established $B_P(\Delta)$ tabulations, which of course are needed,
the widely used Gutenberg-Richter tables (21) are considered
unreliable at regional distances. For example, ISC ignores magnitude estimates for stations at such a distance interval. More
recently Veith and Clawson (22) and Booth et al (23) have presented $B_P(\Delta)$ tabulations which appear to be far more reasonable
than the Gutenberg-Richter ones. Such curves have been generated
also by J. Fyen of NORSAR from the ISC data tapes. The latter
results are presented in Figure 8 together with the Gutenberg-
Richter and Veith Clawson curves. Obviously, beyond $\Delta \sim 20°$ the
agreement between different tabulations is quite good, while
for shorter distances the Gutenberg and Richter (21) results are
clearly inconsistent with the others.

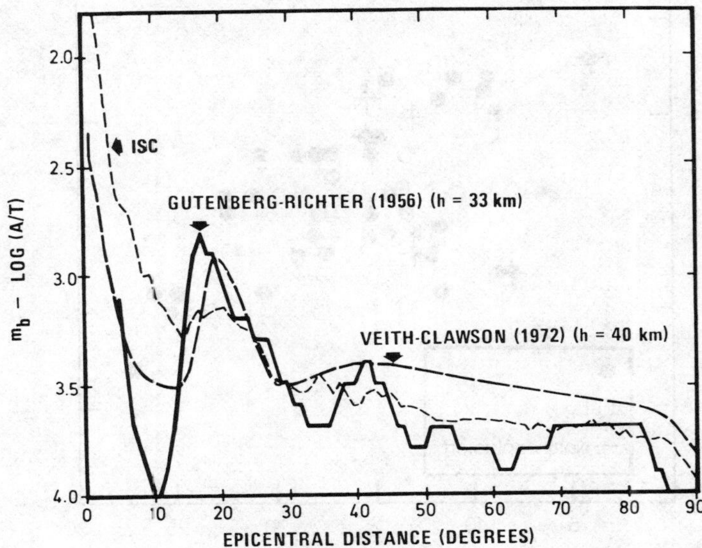

Figure 8. Average observed magnitude correction term based on
ISC data for 23198 events and 136 stations together with
Gutenberg-Richter and Veith-Clawson 'standard' curves.

3.4 Amplitude Decay across Western Russia/Baltic Shield

Using eq. (1) and the ISC-reported body wave magnitudes for the
events in question, the corresponding estimates for the magnitude
distance factor $B(\Delta)$ for both P and Lg waves are plotted in
Figure 9, and estimates of $B(\Delta)$ for Sn in comparison with $B(\Delta)$
for Lg are given in Figure 10. The analysis was restricted to
max. amplitudes of wavelets in the rather broad Sn and Lg group
velocity windows of 4.15-4.70 km/s and 3.25-3.70 km/s, respectively. Furthermore, all maxima were read on the vertical short
period component. Returning to Figures 9 and 10 we see that P
is generally the strongest phase and especially so for distances

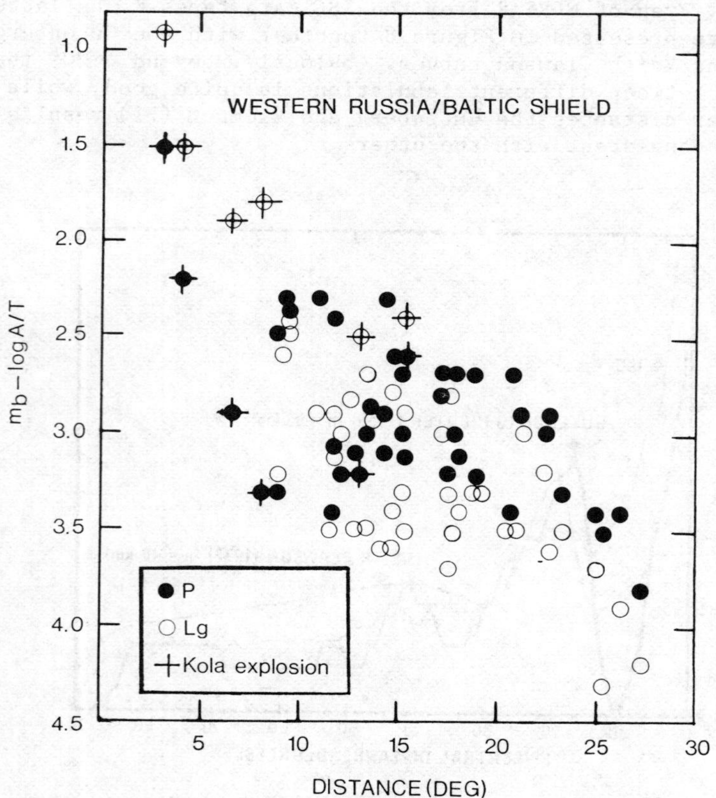

Figure 9. Magnitude distance factor $B(\Delta)$ for P and Lg waves
for events in the Western Russia/Baltic Shield data base.

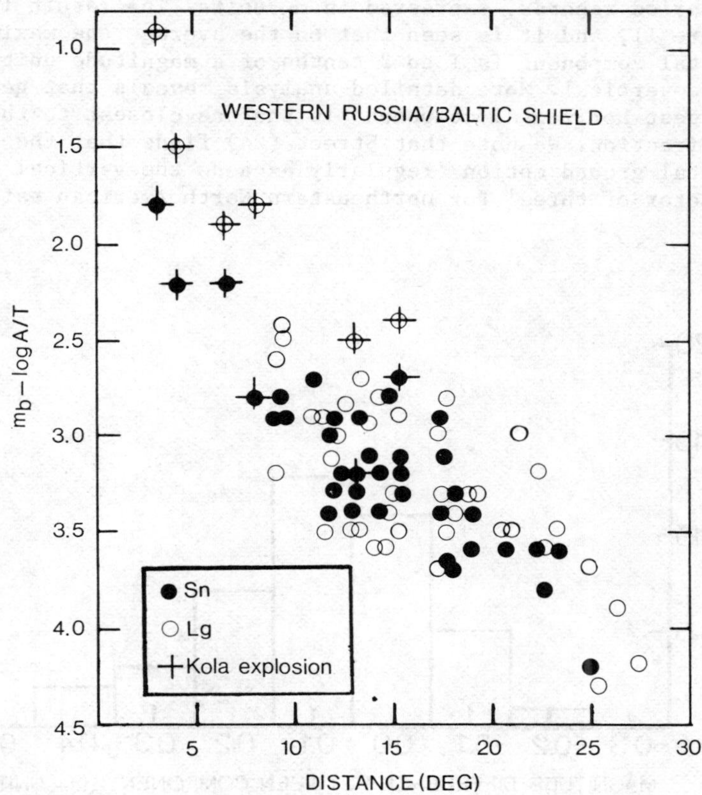

Figure 10.   Magnitude distance factor B(Δ) for Sn and Lg waves
for events in the Western Russia/Baltic Shield data base.

beyond 10⁰. A notable exception here is a presumed explosion
on the Kola peninsula – event no. 11 in Table 1 – where Lg waves
are significantly stronger than the P waves, and in this respect
resembles attenuation characteristics of eastern U.S. For
this particular event, we have also included readings from 3
additional Fennoscandian stations (COP, KON and UME) in Figures
9 and 10. The presumed Kola explosion is clearly an anomalous
event as compared to other events in Western Russia. This difference may be explained in terms of sedimentary overburden
along the paths to Fennoscandia from Western Russia events.
Tectonic barriers like the Baltic Sea appear to be 'nonactive'
vis-à-vis the data at hand.

For Lg waves in our Western Russia/Baltic Shield data base,
we have computed amplitude differences between components in the
short period records, expressed in m_b units. The result is given
in Figure 11, and it is seen that on the average the maximum
horizontal component is 1 to 2 tenths of a magnitude unit larger
than the vertical. More detailed analysis reveals that generally
the largest horizontal component is the one closest to the transverse direction. We note that Street (24) finds that the resultant
horizontal ground motion 'regularly exceeds the vertical component
by a factor of three' for northeastern North American earthquakes.

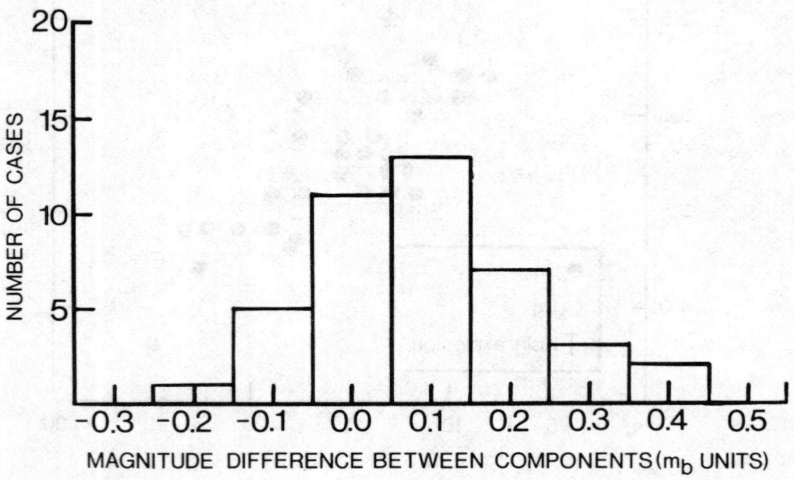

Figure 11. Magnitude difference between the largest horizontal
and the vertical component for events in the Western Russia/
Baltic Shield data base.

3.5 Amplitude Decay across Central Asia and the Himalayas

In conformity with Figure 9, Figure 12 shows the magnitude distance factor for P and Lg waves as a function of distance for
events in the Eurasian data base. As only Asian stations have
been included, the observations are representative of the amplitude decay across Central Asia and the Himalayas based on both
earthquake and explosion records. The down-pointing arrows indicate a maximum possible value of log(A/T) for Lg. In fact, for

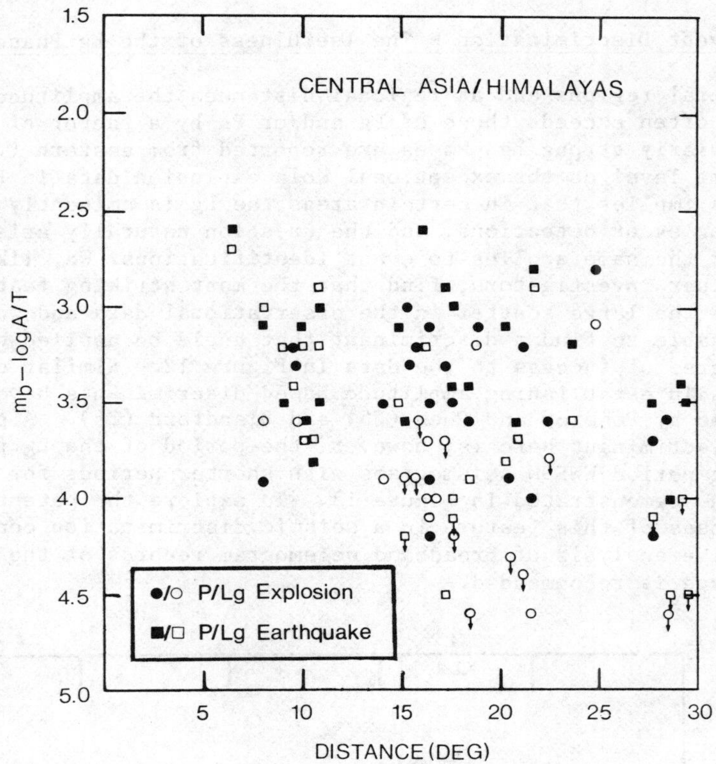

Figure 12. Magnitude distance factor B(Δ) for P and Lg waves for
stations in Central Asia. Symbols with down-pointing arrows
represent estimated maximum amplitude values (see text).

these cases there is no sign of clear wave onsets so the corresponding (A/T)-value is extracted from the coda at the appropriate time. A study of Figure 12 reveals a very high attenua-
tion/scattering level for the Central Asia/Himalayas region
in comparison with the Western Russia/Baltic Shield area both
for P and Lg waves. The average amplitude difference amounts
to about half a magnitude unit for both wave types and the
above remarks on nondetections are liable to increase this
difference for Lg waves. Further analysis shows that also for
this region, Lg motion is largest on the transverse horizontal
component, but the observations do not differ significantly

from what is depicted in Figure 11 for Western Russia/Baltic
Shield.

3.6 Event Discrimination - The Usefulness of the Lg Phase

In several regions and at regional distances the amplitude of Lg
phases often exceeds those of Pg and/or Pn by a factor of 5 to 10.
Particularly strong Lg phases are reported from eastern U.S., on
the same level as the exceptional Kola explosion data in Figure
9. This implies that in certain areas the Lg is eminently suitable for event detections, and the question naturally arises
whether the same applies to event identifications. We, like
most other investigators, find that the most striking feature
here is the large scatter in the observational data and we have
been unable to find a discriminant that could be applied with
any degree of success to the data in Figure 12. Similar difficulties in establishing amplitude-based discriminants have been
reported by Pomeroy and Chen (25) and Blandford (26). A potential discriminant here is, however, the period of the Lg phase
on long period WWSSN seismograms with shorter periods for explosions as demonstrated in Figure 13. To explore the potential
usefulness of this feature in a seismic discrimination context,
extensive analysis of broadband seismogram records of the SRO
(27) type is recommended.

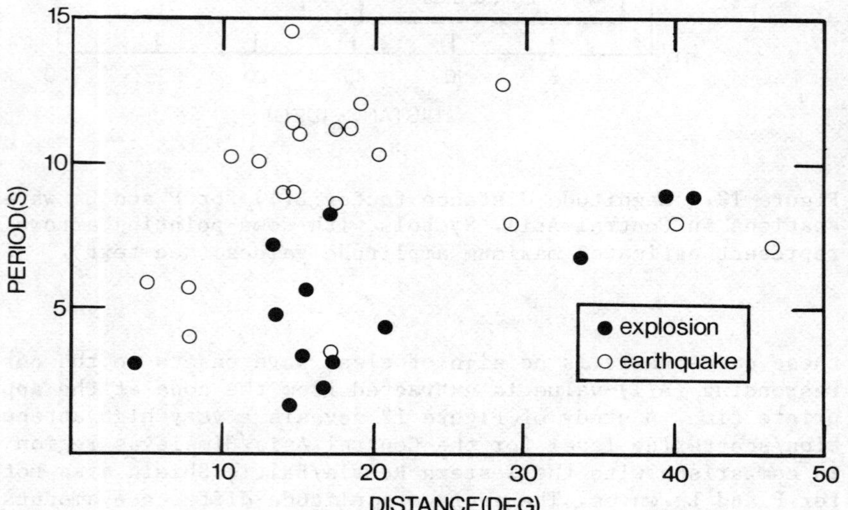

Figure 13. Period of largest Lg amplitude (group velocity window
3.30-3.70 km/s) in long period vertical component records, both
data sets.

3.7   Other Lg and Sn Investigations – Summary of Observational Results

Already in 1954 Båth (2) published a comprehensive set of observational data bearing on Lg propagation characteristics across Eurasia and later on supplemented it with Li observations (17). These and other early studies all testified to the consistency of the Lg group velocity around 3.5 km s^{-1}. For a comprehensive review here, reference is made to Piwinskii and Springer (28), who also tabulate the various observations.

During the later years there has been a renewed interest in Lg propagation across various parts of Eurasia which has partly been motivated by the use of Lg phases as a tool for 'mapping' tectonic boundaries (29, 30, 31) and partly for using these phases for detecting and identifying seismic events at regional distances (25, 26, 28, 32, 33). In general there is good agreement between the reported results, in particular if due regard is given to differences in instrumentation and tectonically inferred variabilities in the observations. The essence of these results is as follows:

- Lg is a dominating phase in the distance range 1°–15° with periods on WWSSN records of 0.8–1.2 s.

- Its group velocity is around 3.5 km s^{-1} though Båth (2) using Weichert pendulum instruments reported Lg$_1$ at 3.35 km s^{-1} and Lg$_2$ at 3.60 km s^{-1}.

- Other prominent phases are Sn having group velocities around 4.50–4.70 km s^{-1}, though our data favor the relatively lower velocities. Li observations mainly based on analysis of Weichert and similar seismogram records are not clearly observed on WWSSN records.

- Lg propagation efficiency varies considerably and very few areas in Eurasia or Western Russia match the observations from eastern U.S. Notwithstanding considerable scattering in the observational data, relatively poor transmission efficiencies are found for propagation paths transecting more recent tectonically active areas like the Himalayas, Tibet and the Iran–Afghanistan region.

- In a discrimination context the potential of Lg and related phases is not rated particularly promising. There is too much overlap in earthquake/explosion parameter population which is not surprising in view of relatively large scattering in the observational data even in case of modest path changes for the same source type.

4. DISCUSSION

Observational results and theoretical studies have clearly
demonstrated that Lg waves represent a sum of several higher
modes of Love and Rayleigh waves travelling in the upper parts
of the crust with group velocities close to 3.5 km s^{-1}. Although the Lg group velocity depends on crustal thickness
and associated velocity structure, such effects are not easily
observable in view of the consistencies of Lg velocities of
3.5 km s^{-1} for most regions as reported here (Figures 5 and 6)
and elsewhere. Turning to Sn waves, group velocities are reported
to be around 4.7 km s^{-1} (32) for Eurasian paths while our observations are mainly below 4.5 km s^{-1} with very few observations
exceeding 4.55 km s^{-1}.

A conspicuous feature of Figures 3 and 4 are secondary
arrivals with group velocities ranging from 3.6 - 4.3 km s^{-1}
including Båth's Li phases of 3.8 km s^{-1}. Such arrivals have
no counterparts in group velocity extrema for realistic earth
models. Our somewhat generalized explanations are in terms of
mode conversion (Sn to Lg as reported in (34)), scattering and
multipathing.

We have expressed the amplitude-distance decay rate, including anelastic attenuation and geometrical spreading, in
terms of magnitude-distance factor for P waves. In most other
studies Lg amplitude decay rates are commonly expressed in terms
of attenuation coefficients (18). Our choice of Lg amplitude
display was motivated by the large scatter in the observational
data which is of the order of one magnitude unit and makes a
fit to a specific attenuation coefficient impossible. Specifically, the scatter in the amplitude data is also large for
presumed underground explosions which implies that source effects like mechanism and focal depth are secondary compared to
the path effect.

We find a stronger Lg attenuation for Central Asia/Hima-
layas than for Western Russia/Baltic Shield (Figures 9 and 12).
This result, which is similar to what is obtained by others
studying relative Lg propagation efficiency across Eurasia, is
quantitatively explained in terms of recent tectonic activity
and consequently more complex structures in Central Asia/Himalayas.
This explanation is not entirely satisfactory as the Lg propagation across Western Russia/ Baltic Shield is with the exception
of the presented Kola explosion significantly less efficient
than that observed for eastern U.S., which exhibits a rather
similar crustal structure. More detailed explanations forwarded
for observations of relatively inefficient Lg propagation are
in terms of tectonic barriers like mountain chains, intermediate

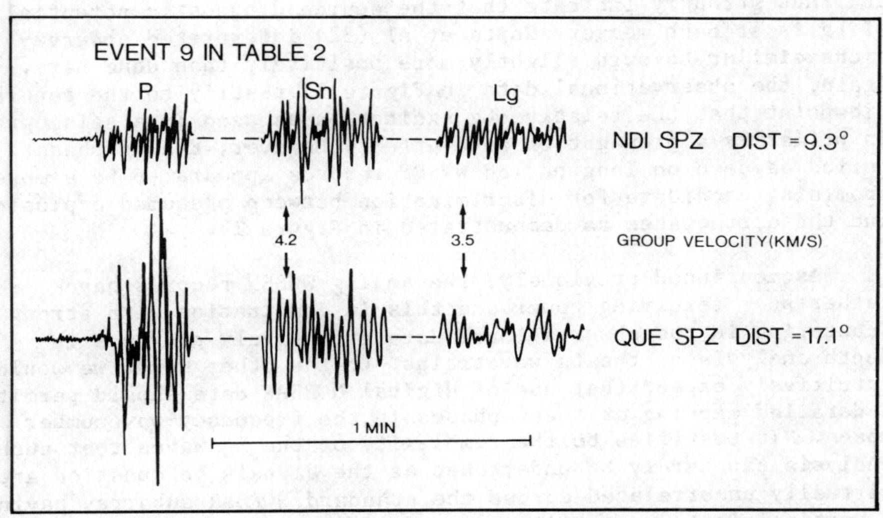

Figure 14.   Recorded waveforms at NDI and QUE for event 9 in
Table 2. The station magnifications are 50000 and 200000 for
NDI and QUE, respectively. The corresponding paths and the
location of the Tibetan Plateau are indicated in Figure 1.

oceanic stretches, sedimentary basins, etc. In particular,
Ruzaikin et al (30) made a very detailed study of this kind for
parts of Central Asia and found that the Tibetan Plateau was a
very effective barrier to Lg propagation. Using somewhat different event/station combinations we found that occasionally Lg
waves propagate reasonably efficiently across this plateau as
demonstrated in Figure 14.   Even if Lg is of smaller amplitude
than P and also Sn for these two records, it stands out clearly
from the background noise, with readable wavelike onsets. The Lg
phase contrasts with many readings in Figure 12, which give only
estimates of maximum Lg amplitudes and do not refer to readable
onsets (down-pointing arrows). We take this to indicate that Lg
propagation is indeed complex, and only broadly can be correlated
even with prominent tectonic features.

      Initially it was mentioned that many of the recent Lg
studies were undertaken to explore the event discrimination
potential of this phase.   The results in Figure 12 clearly

demonstrate that the relative P and Lg excitation for earthquakes and presumed explosions exhibit a considerable overlap and thus strongly indicate that the source diagnostic potential of Lg is at best meager. Gupta et al (32) interpreted observations similar to ours slightly more positively than done here. Again, the observational data in Figure 12 testify to the general viewpoint that the relative Lg excitation as seen in a seismogram is primarily a propagation path effect. However, the Lg phase period as seen on long period WWSSN records appears to be a more promising candidate for discrimination between presumed explosions and the earthquakes as demonstrated in Figure 13.

As mentioned previously, the analog WWSSN records have rather poor resolving power and this in combination with strong scattering in the observational data effectively prevents in-depth analysis of the Lg wavetrains. On the other hand, we would intuitively expect that use of digital NORSAR data should permit a detailed mapping of these phases in the frequency–wavenumber space. It testifies to the complexity of the Lg waves that such analysis can hardly be undertaken as the signals in question are virtually uncorrelated across the standard NORSAR subarray having 6 vertical instruments within an area of radius 5 km (see Figure 15). However, for sensor separations in the order of 1 km and less, signal correlation is good and consequently the Lg wavetrain can be subjected to frequency–wavenumber analysis as actually demonstrated in Figure 16, where a phase velocity of 4.4 km/s at 2.0 Hz is found for the Lg phase in Figure 15. This result agrees well with phase velocity dispersion curves for higher mode Rayleigh waves (14). Still, whether it is possible to advantageously use high quality Lg observations as diagnostics for crustal structures and different types of signal sources remains an open question in view of the apparent sensitivity of Lg waves to small-scale crustal heterogeneities as demonstrated in Figure 15.

5.   CONCLUDING REMARKS

Extensive analysis of WWSSN records at regional distances from earthquakes and presumed nuclear explosions in Western Russia and Central Asia is summarized as follows:

1.   On the short period records the most prominent phases are P, Lg and Sn and respective group velocities for Lg and Sn are around 3.5 km s^{-1} and 4.4 km s^{-1}. On long period records Rg is often observed with a group velocity around 2.9 km s^{-1}.

2.   Lg is generally the strongest secondary phase in the seismogram, but only occasionally stronger than P. The Sn phase is normally weaker than Lg.

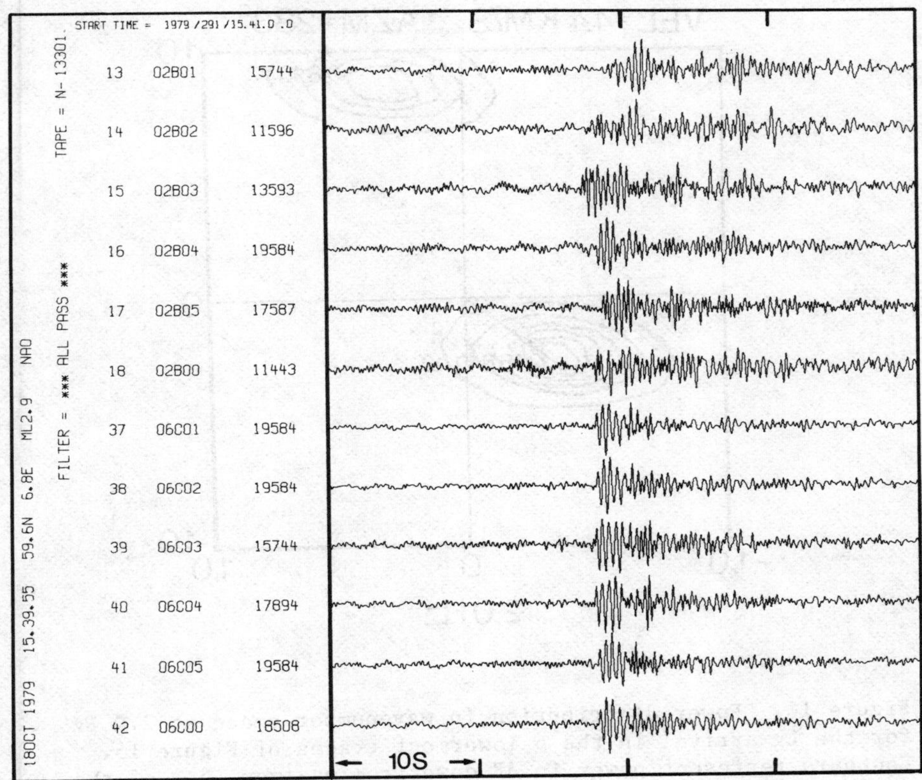

Figure 15. NORSAR data showing the Lg phase from an explosion at distance 285 km. The upper 6 traces are from a standard subarray with radius 5 km, the lower six from an experimental subarray with radius 1 km. The distance between the two subarrays is about 40 km.

3.   There is a considerable scatter in the observed P, Lg and Sn amplitudes which roughly amounts to one m_b magnitude unit. Another feature testifying to the complexity of regional phases is that Lg signals are virtually uncorrelated across arrays unless the sensor separation is about 2 km or less.

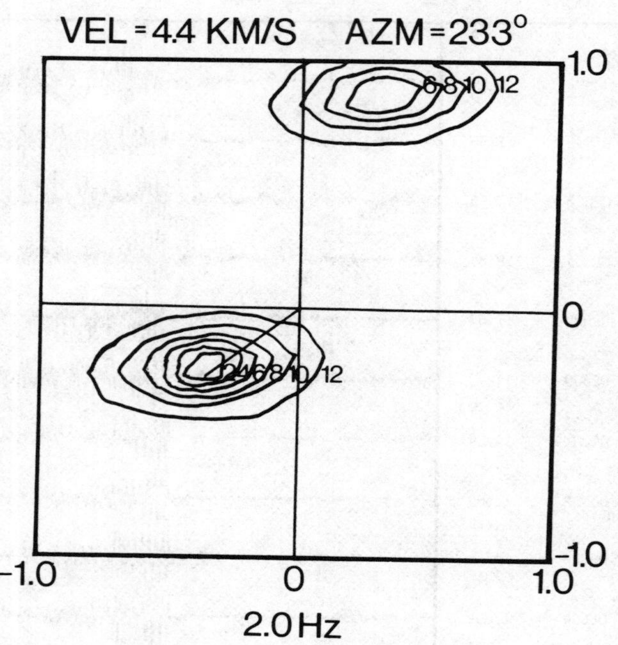

Figure 16. Power distribution in wavenumber space at 2.0 Hz
for the Lg arrival in the 6 lowermost traces of Figure 15.
Contours represent power in dB down from maximum. 5 s of the
Lg phase is included in the computation of the power distribution. The upper power peak is a side lobe effect.

4.   Recent theoretical developments in seismogram synthesis
     are difficult to use profitably in a diagnostic context
     of crustal structure and source types in view of the
     large scattering in the observational data.

5.   Lg propagation efficiency is clearly tectonically dependent
     being best for Western Russia/Baltic Shield paths and
     relatively poor for paths crossing the Himalayas. However,
     local scattering is considerable, so pinpointing of clear,
     specific 'tectonic barriers' is a very difficult task –
     if at all possible.

6. Lg and P/Lg observations do not appear to be an efficient diagnostic for source identification – there is too much overlap between earthquake and explosion populations. However, the dominant period of Lg on long period records appears to be a promising diagnostic in this context.

7. To take advantage of recent developments in seismogram synthesis, better observational data must become available – say digital recordings from mini-arrays where sensor separation should be limited to about 1 km. In this respect, analog WWSSN records are clearly inadequate.

Finally, the Lg wave is a fascinating phase which requires rather extensive and detailed analysis if its information potential on crustal structure and source type is to be fully utilized.

REFERENCES

1. Press, F. and Ewing, M.: 1952, Bull. Seism. Soc. Am. 42, 219–228.

2. Båth, M.: 1954, Arkiv för Geofysik 2, 295–342.

3. Båth, M.: 1956, Ann. Geofis. 9, 411–450.

4. Gutenberg, B.: 1955, Geophys. 20, 283–294.

5. Oliver, J.E. and Ewing, M.: 1957, Bull. Seism. Soc. Am. 47, 187–204.

6. Oliver, J.E. and Ewing, M.: 1958, Bull. Seism. Soc. Am. 48, 33–49.

7. Romney, C., Brooks, B.G., Mansfield, R.H., Carder, D.S., Jordan, J.N. and Gordan, D.W.: 1962, Bull. Seism. Soc. Am. 52, 1057–1074.

8. Brune, J. and Dorman, J.: 1963, Bull. Seism. Soc. Am. 53, 167–210.

9. Kovach, R.L. and Anderson, D.L.: 1964, Bull. Seism. Soc. Am. 54, 161–182.

10. Schwab, F. and Knopoff, L.: 1971, Bull. Seism. Soc. Am. 61, 893–912.

11. Panza, G.F., Schwab, F.A. and Knopoff, L.: 1972, Geophys. J.R. astr. Soc. 30, 273–280.

12. **Knop**off, L., Schwab, F. and Kausel, E.G.: 1973, Geophys. J. R. astr. Soc. 33, 387–402.

13. Knopoff, L., Schwab, F., Nakanishi, K. and Chang, F.: 1974, Geophys. J.R. astr. Soc. 39, 41–70.

14. Panza, G.F. and Calcagnile, G.: 1975, Geophys. J.R. astr. Soc. 40, 475–487.

15. Calcagnile, G., Panza, G.F., Schwab, F. and Kausel, E.G.: 1976, Geophys. J.R. astr. Soc. 47, 73–81.

16. Knopoff, L., Mitchel, R.G., Kausel, E.G. and Schwab, F.: 1979, Geophys. J.R. astr. Soc. 56, 211–218.

17. Båth, M.: 1957, Geofis. Pura et Appl. 38, 19–31.

18. Nuttli, O.W.: 1973, J. Geophys. Res. 78, 876–885.

19. Bollinger, G.A.: 1979, Bull. Seism. Soc. Am. 69, 45–63.

20. Berteussen, K.-A.: 1975, J. Geophys. 41, 595–613.

21. Gutenberg, B. and Richter, C.F.: 1956, Ann. di Geofis., Vol. IX, 1–15.

22. Veith, K.F. and Clawson, G.E.: 1972, Bull. Seism. Soc. Am. 62, 435–452.

23. Booth, D.C., Marshall, P.D. and Young, J.B.: 1974, Geophys. J.R. astr. Soc. 39, 523–537.

24. Street, R.L.: 1976, Bull. Seism. Soc. Am. 66, 1525–1537.

25. Pomeroy, P.W. & T.C. Chen: 1980, Regional Seismic Wave Propagation, Roundout Assoc. Final Technical Report, 15 December 1977 – 30 September 1980.

26. Blandford, R.: 1981, Ibid., Seismic discrimination problems at regional distances.

27. Peterson, J.: 1981, Ibid., The global digital seismograph network. A status report.

28. Piwinskii, A.J. and Springer, D.L.: 1978, Propagation of Lg waves across eastern Europe and Asia, Lawrence Livermore Laboratory, Univ. of California.

29. Antonova, L.V., Aptikayev, F.F., Kurochkina, R.I., Nersesov,
     I.L., Sitnikov, A.V., Tregub, F.S., Fedorskaya, L.D. and
     Khalturin, V.I.: 1978, Experimental seismic investigations
     of the earth's interior, Inst. of Phys. of the Earth, Moscow.

30. Ruzaikin, A.I., Nersesov, I.L., Khalturin, V.I. and Molnar,
     P.: 1977, J. Geophys. Res. $\underline{82}$, 307–316.

31. Shishkevish, C.: 1979, Propagation of Lg seismic waves
     in the Soviet Union, Rand Corp., Santa Monica, California.

32. Gupta, I.N., Barker, B.W., Burnetti, J.A. and Der, Z.A.:
     1980, Bull. Seism. Soc. Am. $\underline{70}$, 851–872.

33. Nuttli, O.W.: 1981, Bull. Seism. Soc. Am. $\underline{71}$, 249–261.

34. Isacks, B.L. and Stephens, C.: 1975, Bull. Seism. Soc. Am.
     $\underline{65}$, 235–244.

SEISMOGRAMS OF EXPLOSIONS AT REGIONAL DISTANCES IN THE WESTERN
UNITED STATES: OBSERVATIONS AND REFLECTIVITY METHOD MODELING

K. H. Olsen[1] and L. W. Braile[2]

[1] Geosciences Division, Los Alamos National
    Laboratory, Los Alamos, New Mexico 87545, U.S.A.
[2] Geoscience Department, Purdue University,
    West Lafayette, Indiana 47907, U.S.A.

ABSTRACT. Seismic energy propagating through vertically and
laterally varying structures of the earth's crust and lower
lithosphere–uppermost mantle is responsible for the numerous and
complex seismic phases observed on short-period seismograms at
regional distance ranges (100 to 2000 km). Recent advances in
techniques for computing synthetic seismograms make it practical
to calculate complete seismograms that realistically model many
features of regional phases. A modified reflectivity method
program is used to interpret some details of record sections of
Nevada Test Site (NTS) underground explosions that were observed
700 to 800 km from the sources.

I.   INTRODUCTION

Regional seismic phases recorded by high-gain, short-period or
broadband instruments are likely to play an increasingly important role in seismic source location and identification as
acceptable magnitude thresholds are pushed to lower levels.
From the standpoint of complexity of seismograms, the epicentral
distance range between ~200 km and the transition to simpler
teleseismic waveforms around 2000 km presents many challenges to
the seismic analyst. In this range, propagation paths can
traverse the crust, the lower lithosphere, and the uppermost
mantle where both vertical and lateral heterogeneities strongly
influence waveform characteristics. Good observational data are
rare for testing analysis techniques developed for regional
problems. In contrast to the numerous detailed crustal refrac-
tion/reflection profiles that have been obtained from many parts
of the world out to distances ~200 km, relatively few long-range

*E. S. Husebye and S. Mykkeltveit (eds.), Identification of Seismic Sources - Earthquake or Underground
Explosion, 453–466.*

profiles exist where station spacing is sufficiently tight to facilitate a clear interpretation of the onset, development, and amplitude vs. distance behavior of the many observable phases. Thus, although signals from sources of interest may be easily observable at regional distances, derivation of source parameters from observations at sparsely located observatories or arrays will require careful analysis and modeling of the intricacies of wave propagation at these scales.

Phases of interest in regional identification studies fall into two main categories: large amplitude, long duration, but somewhat indistinct wave groups such as Lg and $\bar{P}$; and body waves (mainly compressional) that appear either as first arrivals or closely following as possible wide angle reflections/near-critical refractions from interfaces and/or steep velocity gradients in the deep crust, lower lithosphere, and uppermost mantle. The Lg and $\bar{P}$ phases are often the largest amplitude features on regional short-period seismograms, but a clear explanation of how Lg and $\bar{P}$ propagate is still lacking [1]; this lack perhaps is reflected in the fact that seismologists frequently use the notations $\bar{P}$ or P_g interchangeably in reference to a broad, large amplitude phase following P_n. We adopt the $\bar{P}$ notation here. The phase in question propagates very well in the western United States, but attenuates rapidly in the eastern U.S. A group velocity around 6 km/s implies $\bar{P}$ propagates as compressional waves multiply reflected within the crust--which may thus act as a waveguide. Similarly, the ~3.5 km/s group velocity for Lg suggests shear waves multiply reflecting within the crustal layers. Some authors [2] prefer to treat Lg as a superposition of higher mode Love and Rayleigh waves propagating in a nearly laterally homogeneous, vertically layered crust. In any case, the propagation physics is complicated and will require quite sophisticated synthetic seismogram codes to properly model and interpret observed waveforms.

Record sections of long-range seismic refraction profiles often show one or more nearly parallel travel time (T) vs. distance (Δ) branches following within several seconds of first arrivals [3, 4, 5]. Each secondary branch may be traceable only over a distance interval of 50 to 200 km before being replaced in a "shingle-like" fashion with another branch or set of arrivals [5, 6, 19]. These are usually interpreted as parts of cusp phases arising from critical refractions and/or wide-angle reflections from first order discontinuities or steep velocity gradients in the upper mantle. Archambeau et al. [7] and Burdick and Helmberger [8], for example, have derived velocity vs. depth models for the major features of the upper mantle beneath the U.S. by a joint analysis of travel times, amplitude vs. distance variations, and waveform fitting of the first few compressional arrivals observed at widely separated seismograph stations throughout the U.S. These and similar models by others are most valid for depths greater than about 250 km. Although these analyses suggest that the main features of mantle structure at depths below about 300 km (corresponding to compressional first arrivals at epicentral ranges beyond ~1500 km) may be more uniform over a global scale [8], it is known that significant lateral variations in lower lithosphere and uppermost mantle properties occur beneath the continents on regional and perhaps even finer scales [8, 9, 10, 11]. In the depth range between the Moho and ~300 km, several types of structural variations have been suggested in the literature that would give rise to wide angle reflections, converted phases, and similar closely spaced arrivals on seismograms at regional ranges. These include the presence or absence of the S-wave and/or the P-wave low velocity zone (LVZ) in the asthenosphere, high velocity mantle lids [12, 13], alternating lamellae of positive and negative velocity gradients [6, 19], etc. These early arriving phases often have better defined onsets than the $\bar{P}$ and Lg phases and, since they are observed at distances beyond that where a true head wave Pn arrival can be expected, they may be useful in regional source location and identification. In order to make use of the information contained in these arrivals (especially the amplitude vs. distance behavior for particular paths of interest), it will be necessary to use modern sophisticated synthetic seismogram techniques to derive localized fine scale details from generalized crust-mantle models.

The purpose of this paper is to explore a few of the problems in modeling regional short-period seismograms by means of a modified reflectivity method [14] computer program developed by R. Kind [15]. This numerical program accounts for the effects of a buried source and is thus capable of computing 'complete' seismograms—including refracted waves, surface reflected body waves such as the pP phase, and surface waves. The effects of anelastic attenuation (Q) for each layer are included as an integral part of the method [15]. The most severe limitation of the technique for studies of regional seismograms is the assumption of lateral homogeneity (this is also a limitation for normal modes summation techniques). An item of interest will be the extent synthetics can be made to match observed waveforms under this restriction.

Two problems are considered. The first, labeled the B-3 model for brevity, employs a simple model consisting of three layers in the crust without velocity gradients and an almost uniform velocity mantle. A large range of apparent surface phase velocities is used in order to display S phases and surface waves. The second calculation, the A-10 model, treats the mantle structure in detail, but confines attention to compressional phases near ther start of the seismogram. The more

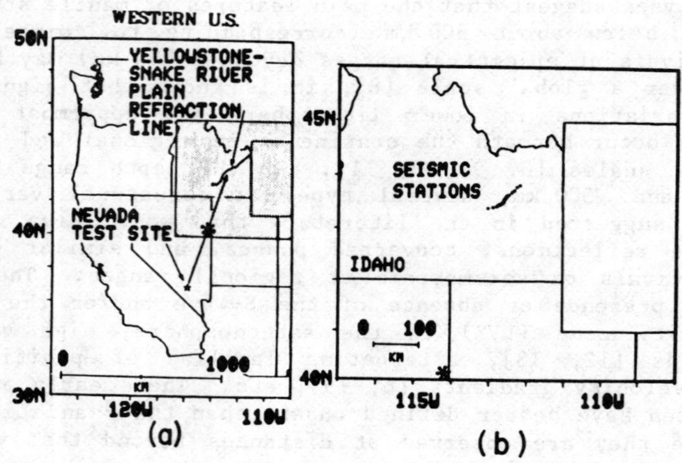

Fig. 1. (a) Location map of the western United States with relative positions of the Nevada Test Site and the Y-ESRP recording line. (b) Enlargement showing positions of stations that recorded the 27 September 1978 RUMMY explosion. Asterisk denotes approximate area for mantle ray turning points from NTS explosions.

important conclusions of the A-10 model are summarized here--a fuller discussion of this calculation and the implications for uppermost mantle structure beneath the western U.S. can be found in a previous publication [16].

A comparison of the synthetic seismogram calculations has been made with a 100-km-long record section of short-period vertical component seismograms obtained in eastern Idaho during the 1978 Yellowstone-Eastern Snake River Plains (Y-ESRP) seismic profiling experiment. For these observations, the sources were underground nuclear explosions at the Nevada Test Site (NTS) at distances between 720 and 820 km from the nearly radially oriented linear station array (Fig. 1). Only the records from the largest NTS explosion, the m_b = 5.7 RUMMY event at 1720:00.076 GMT, 27 September 1978, are reproduced here since they have the best signal-to-noise ratio of the three NTS explosions observed during the experiment. Additional details of the Y-ESRP instrumentation, experiment, and data can be found elsewhere [16].

2.   COMPUTATIONAL TECHNIQUE

As discussed by Kind [15] and by Fuchs and Müller [14], the reflection coefficient and time shift calculations in the reflectivity method are carried out in the frequency domain and then Fourier transformed to plot seismograms. We included Müller's [17] earth flattening approximation in both of our problems to account for earth curvature effects. Both P and S velocities are independently specified in all calculations, since the reflection coefficients are functions of both P and S velocity contrasts at non-normal incidence angles and are required even when only computing P phases over a narrow time window. In the A-10 calculation, for example, the departure of the P/S velocity ratio in a layer from that given by Poisson's ratio = 1/4 is an important factor in our interpretation [16]. Densities are given by a Birch's Law relation (density = 0.252 + 0.3788*P velocity). The attenuation factor Q_α for P waves was chosen as 25 in the source layers, 200 in the upper crust, and 1000 in the lower crust and the uppermost mantle layers; for the LVZ modeling of the A-10 model, Q_α in the asthenospheric layers was adjusted as part of the fitting procedure (see Fig. 5). The attenuation factor for S waves was always assumed to be $4Q_\alpha/9$ [20]. The explosive source algorithm [16] was used with the source buried at a depth of 0.640 km in a layer of P velocity = 3.55 km/s. These were close to actual field values for the NTS RUMMY explosion. Time intervals, number of samples, and computed lengths of seismograms were chosen so that the dominant frequency of the source spectrum was 1.6 Hz for the A-10 calculation--again close to the observed value. In order to save computer time for the extended duration B-3 seismogram sections, the parameters were chosen so that the dominant frequency of the source was shifted to 0.25 Hz; although this was low compared to observed frequencies, we felt it was adequate for the puposes of this initial study. To avoid long computer runs, the wave field was only computed within a limited phase velocity window: 1 km/s to 20 km/s for B-3, and 6.5 km/s to 1000 km/s for A-10. These integration limits sometimes introduced spurious single cycle "phases" at these apparent velocities in the computed record sections. The limit velocities were chosen so as to not overlap or interfere with arrivals of interest in the observations. In the record section plots, the amplitudes of each trace have been multiplied by station distance to maintain a convenient scaling of the amplitudes of the phases which are subject to geometrical spreading and attenuation due to anelasticity.

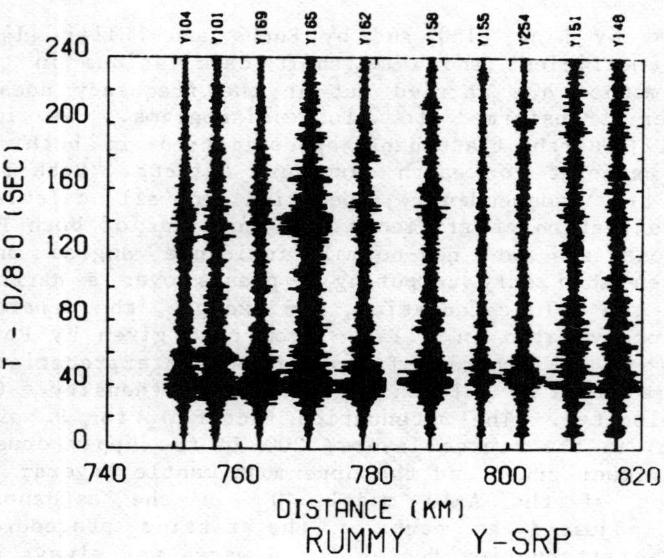

Fig. 2. Vertical component low time resolution seismic record
section of the RUMMY explosion as recorded at Snake River Plains
stations. The time scale is compressed to show envelope behavior; individual waveforms not readily seen. The $\bar{P}$ phase is
the broad feature at reduced times between 30 and 60 s. Upward
ground motion to the left.

3.   DISCUSSION

3.1 The Extended Time Seismograms: B-3 Model

Figure 2 is a true relative amplitude vertical component record
section of the RUMMY explosion recorded on ten matched short-
period (1 Hz natural frequency) instruments deployed in the
eastern Snake River Plains (Fig. 1). Although the time scale is
too compressed to reveal many details of the waveforms, several
important overall features can be noted. The broad (~40-second-
long) envelope of the $\bar{P}$ phase appears at reduced times between
approximately 30 to 60+ seconds, and is the largest amplitude
feature on the record. In contrast, the Lg phase expected at
reduced times of ~130+ seconds (an average velocity of about 3.5
km/s) is poorly developed on these unfiltered records; it is
only obvious at the 770-km station. A few impulsive arrivals
can be seen (such as the first arrivals at reduced time ~10
seconds, which will be discussed in Sec. 3.2, and perhaps an
Sn [?] phase at t_{red}~80 seconds and ~780 km), but the

impression one gets by viewing this observed section is that the correlations seem to be better described as broad energy correlations rather than phase correlations. A similar conclusion is suggested by seismograms from central Asia shown in the paper of Ruzaikin et al. [1]. A coherent structure in the $\bar{P}$ and Lg phases is difficult to trace from station to station even though the stations are only separated by 8 km on the average.

The results of an attempt to model late time arrivals over a regional distance range is shown in Fig. 3. A rudimentary, almost trivial, crust/mantle velocity structure was assumed that consisted of three constant velocity layers in the crust overlaying a nearly constant velocity halfspace. (A slight negative gradient in P velocity was introduced just below the Moho in order to suppress the Pn amplitudes as required by the observations; see Sec. 3.2.) We note several points.

(a) The seismogram section from 100 to 900 km and the enlarged individual record for 800 km shows a surprising amount of complexity at times beyond the first arrivals even though an extremely simple earth model and source function is used. Groups corresponding to the $\bar{P}$ and Lg phases can be identified.

(b) There appears to be a considerable amount of S-wave energy although none is present in the explosion source algorithm. This is probably due to P-to-S and S-to-P, etc., conversions at interfaces and to multiples which the program adequately includes.

(c) The calculated dispersed fundamental mode Rayleigh wave is very large. There are at least two reasons this Rayleigh wave is not representative of the observations. First, no corrections for the short-period bandpass response of the seismometers were included in the synthetics. Second, the assumed source spectrum has too much energy at the longer periods as compared with a near point-source representative of a NTS explosion, thus over enhancing the Rayleigh waves. Long-period Rayleigh waves from actual underground explosions are probably generated or modified and enhanced by mechanisms such as spall closure and/or tectonic strain release; these mechanisms are not adequately treated by the explosion algorithm used for the present calculation.

(d) Because the calculated seismogram sections are quite complicated even for this simple earth model, they give the impression that broad "packets of energy" can be more readily correlated than any well defined phases--for at least the $\bar{P}$ and Lg phases. This was the case with the observations in Fig. 2. In order to better understand the gross behavior of these phases with distance and to identify the origin of obscure features, it will be necessary to include calculations of the horizontal

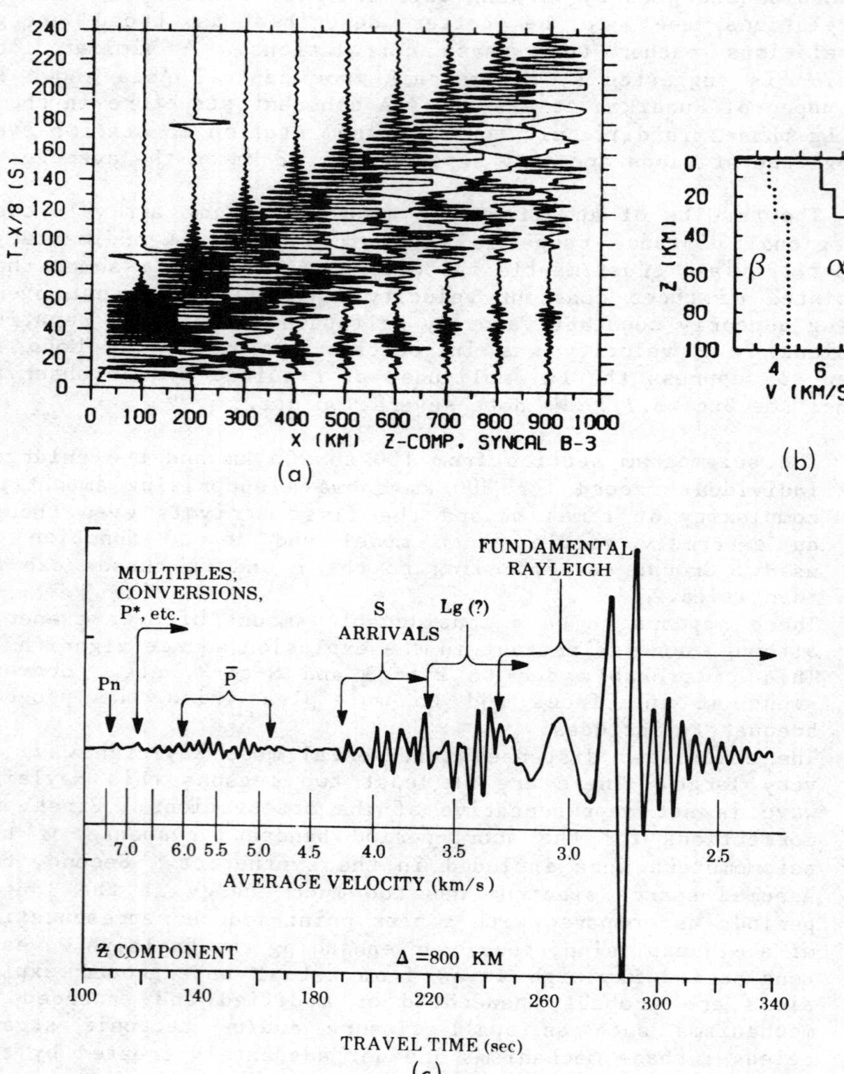

Fig. 3. (a) Synthetic seismogram vertical component record section calculated from the P and S velocity vs. depth structure (Model B-3) shown in (b). (c) Expanded plot of the synthetic seismogram at the 800-km distance. Approximate arrival time and average velocity windows for different phases or groups are indicated; the phase velocities of the different wave types are equal to or slightly greater than the average velocities. The Rayleigh waves on plot (a) are arbitrarily clipped in plotting to avoid large overlays in the seismograms.

(radial) component and to perform calculations at small station separation to increase recognizability of phase correlations.

These results suggest that the modified reflectivity method, even with the restrictive assumption of lateral homogeneity, can be a useful technique in understanding the intricacies of Lg and P phases and the types of earth structures that most affect them. In addition, these studies suggest that observations of complex and apparently-incoherent seismic phase arrivals--even over short distances--do not necessarily imply strong lateral heterogeneity in crustal structure. Parameter studies would help identify those aspects where refinements due to lateral heterogeneity and/or scattering need to be considered in order to better match observations.

3.2 Early Time Arrivals: A-10 Model

Figures 4a and 4b are enlarged portions of the first few seconds of the digitized RUMMY vertical component seismograms (see also Fig. 2) that show details of the earliest arrivals. We have interpreted [16] this record section in terms of three different compressional phases, all having apparent velocities close to 8 km/s: (a) an extremely weak leading arrival labeled Pn, which was lost in the background noise for the two other, lower yield, NTS shots that were also recorded during the Y-ESRP experiments; (b) a stronger phase labeled P_{lid} follows Pn by about two or three seconds for epicentral distances between 700 and 780 km; (c) beyond 780 km, the P_{lid} phase appears to be overtaken and overwhelmed by a low-frequency phase, P_l, whose amplitude increases rapidly with distance out to at least the farthest station of the linear array. The detailed reasons for these labels and identifications are discussed in [16]; they can be summarized as follows.

The phase labeled Pn could be a wide angle reflection from a weak P-velocity contrast in the lower lithosphere below the Moho rather than a true headwave (in the strict sense of the mathematical definition) that travels along the M-discontinuity interface over the entire 800-km path. However, the sub-Moho P velocity (7.7 to 7.9 km/s) in this region of the Great Basin is known to be close to both the average and the apparent velocity observed in Figs. 2 and 4. This, plus the fact that other travel time arguments [16] suggest there is no evidence for mantle lids or other thin but fairly high gradient zones down to a depth of about 100 km, argues that the most straight-forward explanation for this arrival is that it is a P_n-type phase. We calculate that the energy at 800 km is greatly reduced because the wave travels in a region beneath the Moho that has a slight negative velocity gradient.

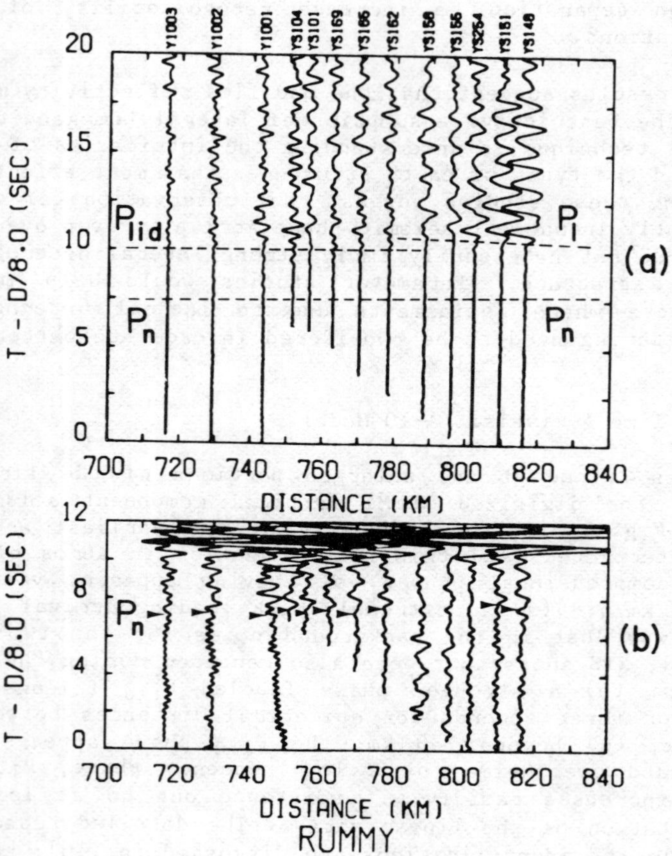

Fig. 4. (a) True relative amplitude record section of early compressional arrivals from the RUMMY explosion. (b) Same as (a) with increased amplitudes to show weak P_n phase. Upward motion to the left. All traces low pass filtered at 3 Hz.

The sudden onset at about 780 km and subsequent rapid amplitude growth of the P_1 phase indicates it is the cusp of the critically refracted P-waves from the steep velocity gradient at the base of the asthenospheric low velocity zone. The observed dominant low frequency content is then explained by the attenuation of the high frequency components as the energy travels first downward and then back up through the very low-Q region of the LVZ. The notation of P_1 for this phase follows the convention established by Archambeau et al. [7].

The travel times, moderate amplitudes, and relatively high frequency content imply the phase identified as P_{lid} is a wide angle reflection from a discontinuity near the base of the mantle lid (= top of LVZ) in this area.

The conclusions concerning these three early arriving compressional phases summarized above were confirmed by using the modified reflectivity program to quantitatively model the arrival times, amplitudes, and waveforms in the first 15 seconds of the record sections. The procedure was to begin with a generic P-velocity vs. depth model for the western U.S. (the T-7 model) derived from a wider data set by Burdick and Helmberger [8] and then to perturb the model to achieve a better fit [16]. Because of the influence of S-velocity contrasts on the P-wave reflectivity calculations, an S-velocity vs. depth model derived by Priestly and Brune [18] from an analysis of Rayleigh and Love wave dispersion on paths crossing the area of interest in the Great Basin of Eastern Nevada was incorporated into the synthetic seismogram modeling. The starting T-7 and Priestly-Brune (P/B) velocity models are shown by dotted lines in Fig. 5. The generic T-7 P-wave model has a pronounced mantle lid with a strong positive P-velocity gradient beneath the Moho for depths from 33 to 65 km. Calculation of synthetics for this lid structure gave very large amplitudes for the "P_n" arrival, which was superimposed on a strong reflection from the base of the lid at 65 km [16]. Thus, the T-7 + P/B starting model gave results very different from observations. However, as seen in Fig. 5, only small changes to the initial model were necessary to match the observations. To bring the calculated synthetic seismograms into agreement with observations, the gradient at the base of the LVZ had to be raised to shallower depths and the positive gradient lid replaced with a smooth but gradual negative gradient starting at the M-discontinuity. The final model, A-10, that matches observations is shown by the solid lines in Fig. 5. Figure 6 is the comparison between the observed and synthetic record sections. Interestingly, no discontinuity in P-velocity is necessary to explain the P_{lid} reflections; the reflections can be adequately modeled by a small negative step in S velocities at a depth of about 100 km. The synthetics, however, do not seem to adequately model the long oscillatory trains following the P_1 phase onset. This is probably due to interference effects caused by fine structure in the lower LVZ velocity gradient that we have not yet modeled by thin enough layers in the calculation [16].

These calculations illustrate that synthetic modeling techniques can be helpful in phase identification and in quantitative calculations of amplitude vs. distance behavior and waveform characteristics. With a sophisticated reflectivity method calculation we were able to model several important features of

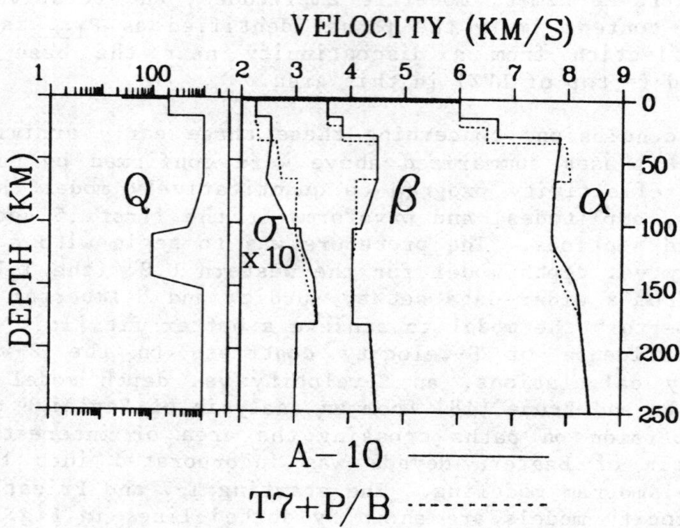

Fig. 5. P-velocity (α) and S-velocity (β) vs. depth plots for the T-7/Priestly-Brune and A-10 models. Assumed Q structure at left: σ (dimensionless) is Poisson's ratio.

regional short-period seismograms. The technique appears promising in advancing knowledge of wave propagation and source identification at regional distance ranges.

ACKNOWLEDGMENTS

We especially thank Rainer Kind for making available to us the modified reflectivity method computer program that we have adapted for our analyses. Paul A. Johnson was responsible for the computer runs. We thank Terry C. Wallace and Mike Shore for discussions and for reviewing the manuscript. The calculational and data reduction efforts for this research were supported by the U.S. Department of Energy and partially by ONR Earth Physics Program grant N00014-75-C-0972 to L.W.B. The Snake River Plains data were collected during research partially funded by the U.S. National Science Foundation (grant EAR-77-23707 to the University of Utah and EAR-77-23357 to Purdue University) and by the U.S.G.S. Geothermal Research Program grant 14-08-0001-G-532 to Purdue.

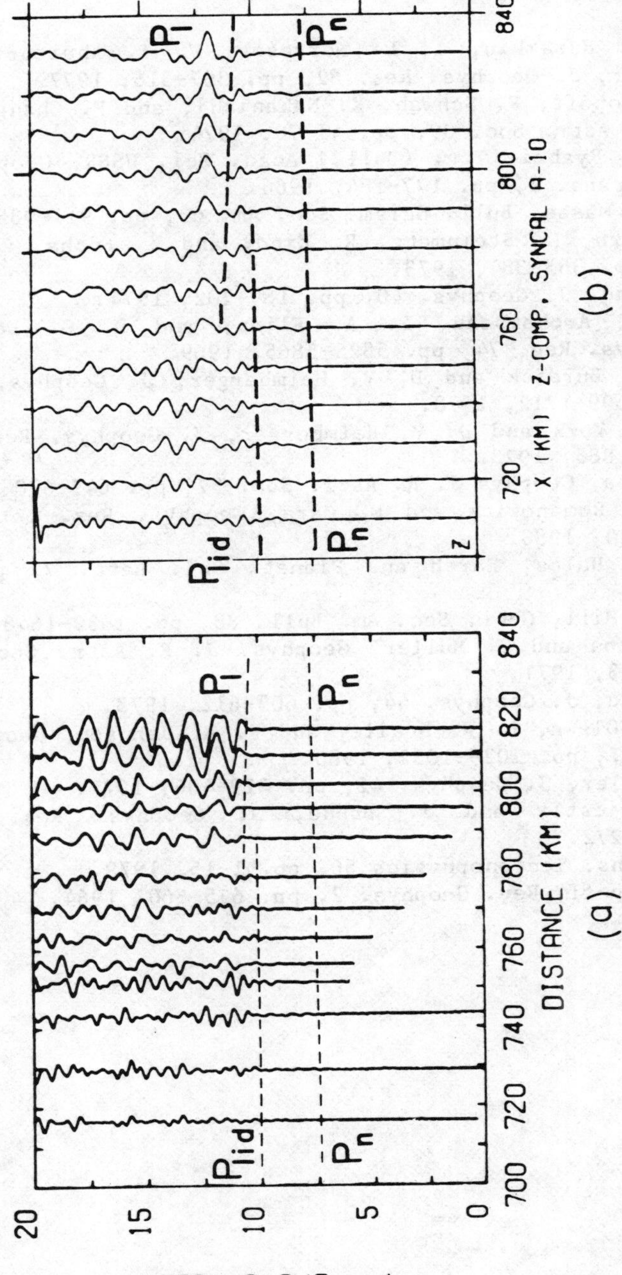

Fig. 6. Comparison of the observed (a) record section with the synthetic section calculated from the A-10 model (b).

REFERENCES

1. A. I. Ruzaikin, I. L. Nersesov, V. I. Khalturin, and P. Molnar, J. Geophys. Res. 82, pp. 307-316, 1977.

2 L. Knopoff, F. Schwab, K. Nakanishi, and F. Chang, Geophys. J. R. Astr. Soc. 39, pp. 41-70, 1974.

3. V. Z. Ryaboi, Izv. (Bull.) Acad. Sci. USSR, Geophys. Ser., AGU Trans. 3, pp. 177-184, 1966.

4. R. P. Masse, Bull. Seism. Soc. Am. 63, pp. 911-935, 1973.

5. A. Hirn, L. Steinmetz, R. Kind, and K. Fuchs, Z. Geophys. 39, pp. 363-384, 1973.

6. R. Kind, J. Geophys. 40, pp. 189-202, 1974.

7. C. B. Archambeau, E. A. Flinn, and D. G. Lambert, J. Geophys. Res. 74, pp. 5825-5865, 1969.

8. L. J. Burdick and D. V. Helmberger, J. Geophys. Res. 83, pp. 1699-1712, 1978.

9. J. E. York and D. V. Helmberger, J. Geophys. Res. 78, pp. 1883-1886, 1973.

10. M. Cara, Geophys J. R. Astr. Soc. 57, pp. 649-670, 1979.

11. B. A. Romanowicz and M. Cara, Geophys. Res. Lett. 7, pp. 417-420, 1980.

12. A. L. Hales, Earth and Planet. Sci. Lett. 7, pp. 44-46, 1969.

13. D. P. Hill, Geol. Soc. Am. Bull. 83, pp. 1639-1648, 1972.

14. K. Fuchs and G. Müller, Geophys. J. R. Astr. Soc. 23, pp. 417-433, 1971.

15. R. Kind, J. Geophys. 44, pp. 603-612, 1978.

16. K. H. Olsen, L. W. Braile, and P. A. Johnson, Geophys. Res. Lett. 7, pp. 1029-1032, 1980.

17. G. Müller, J. Geophys. 42, pp. 429-436, 1977.

18. K. Priestly and J. Brune, J. Geophys. Res. 83, pp. 2265-2272.

19. K. Fuchs, Tectonophysics 56, pp. 1-15, 1979.

20. L. Knopoff, Rev. Geophys. 2, pp. 625-660, 1964.

PHASE IDENTIFICATION AND EVENT LOCATION AT REGIONAL DISTANCE
USING SMALL-APERTURE ARRAY DATA

Svein Mykkeltveit and Frode Ringdal

NTNF/NORSAR, Kjeller, Norway

ABSTRACT

As part of ongoing research aimed at a better understanding of
seismic wave propagation at regional distances, one of the NORSAR
subarrays has been modified to a small aperture array with station
distances from 125 to 2051 meters. Regional phases including Pn,
P, Sn, Lg and Rg from reported events within 10° have been analyzed with respect to phase velocity and azimuth. This is done
by computing frequency-wave number spectra for time series of
100 samples each (20 Hz data), which for most phases include both
onset and amplitude maximum. Derived values for phase velocity
and azimuth are in general accordance with expected values, even
at frequencies of 6-7 Hz. Discrimination between fast (P) and
slow types of phases (Sn, Lg) is possible using data from this
array. A location alogrithm for the small aray data is described
that enables us to locate most regional events with a difference
from the location by the whole Fennoscandian network that is
within the uncertainty limits of the latter.

1.   INTRODUCTION

One of the principles governing the design of the NORSAR array
was that distances between instruments should be such that
ordinary beamforming should give near to optimum gain in signal-
to-noise ratio for teleseismic events. This implies distances
large enough to ensure a low noise coherence (at around 1 Hz)
and yet sufficiently small to maintain a high signal coherence.
This resulted in inter-station separations of about 3 km. It was
soon discovered, however, that this gave very low signal coher-

*E. S. Husebye and S. Mykkeltveit (eds.), Identification of Seismic Sources – Earthquake or Underground
Explosion, 467–481.*

ences for regional and local events, having spectral peaks at
typically 2-3 Hz or above.

With the recent increased interest in regional and local
events in a discrimination context, it was decided to implement
- within the NORSAR system - a test subarray with small station
distances. The purpose of the NORESS (NORSAR Experimental Small
Subarray) experiment then has been to obtain data from closely
spaced sensors in order to

-       Investigate noise and signal coherences at high frequencies
-       Identify P, S and Lg phases from regional seismic events
-       Detect and locate regional events.

This paper deals with results obtained from processing of local
and regional events recorded by NORESS during its first half
year of operation.

2.   NORESS CONFIGURATION AND DATA

The NORESS array consists of 6 closely spaced vertical seismometers and was implemented in October 1979 through an altera-
tion of the geometry of standard NORSAR subarray 06C. The configuration of subarray 06C before and after the change is shown
in Figure 1. Data are sampled at 20 Hz rate and all sensors
are equipped with 8 Hz anti-aliasing filters.

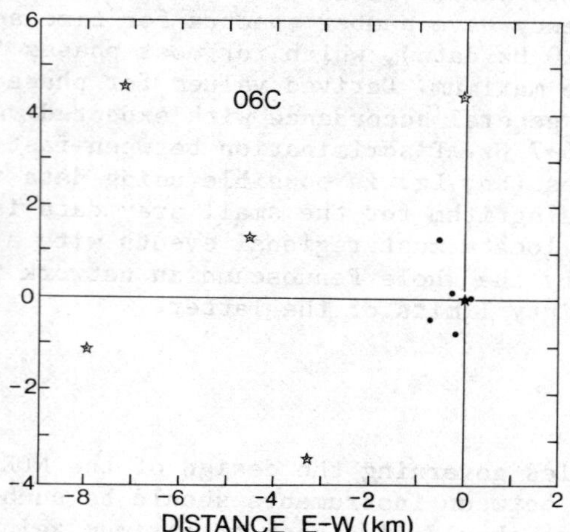

Figure 1.   Geometry of the NORESS array. Dots represent NORESS
instrument locations while asterisks indicate the positions of
the 'old' 06C instruments. NORESS inter-station distances range
from 125 to 2051 m.

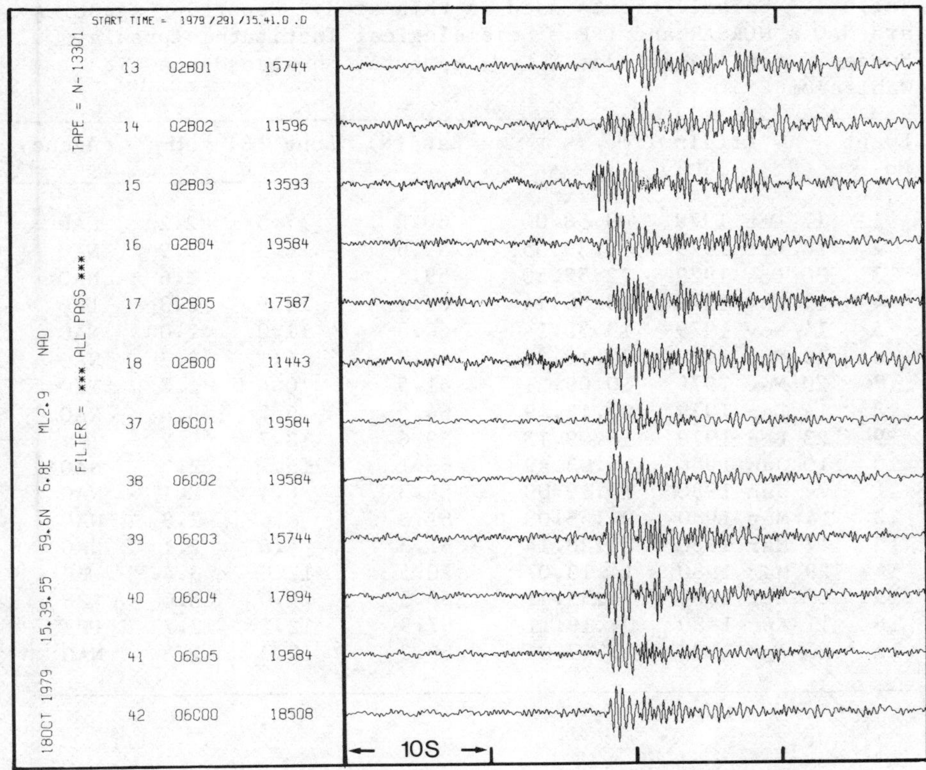

Figure 2. Lg waves from event no. 2 in Table 1. Epicentral distance is 285 km. The top six traces are from standard subarray 02B and the lower six traces are from NORESS.

Figure 2 gives an example of how NORESS records compare with those of a standard NORSAR subarray. The Lg phase in Figure 2 shows visual lack of coherence across sensors within an ordinary subarray at NORSAR (6 top traces of Figure 2 from subarray 02B), with sensor-to-sensor distances of 3 to 10 km. The NORESS array is seen to provide a significant improvement in this respect, as can be seen from the bottom 6 traces. As good coherence is an implicit prerequisite for most processing schemes, dense spatial sampling as offered by a NORESS-type array seems necessary for successful mapping of regional propagation characteristics.

Table 1 provides information on 17 regional events recorded by NORESS in the time period October 1979–April 1980. The events are both earthquakes and explosions, the latter mostly associated with mining activity and hydroelectric power plant construction.

Table 1. Seismic events used in this study. Reporting agencies are NAO = NORSAR and UPP = Seismological Institute, Uppsala, Sweden. Local magnitudes M_L are computed in accordance with Wahlstrøm (1).

Event No.	Origin Time (GMT)		Lat (N)	Long (E)	M_L	Agency
1	17 Oct 1979	09.58.00	60.3	7.5	2.2	NAO
2	18 Oct 1979	15.39.55	59.6	6.8	2.9	NAO
3	30 Oct 1979	12.52.50	59.5	6.9	2.6	NAO
4	11 Nov 1979	23.58.14	61.1	16.9	2.3	UPP
5	14 Nov 1979	13.30.14	60.8	11.0	2.0	NAO
6	19 Nov 1979	23.15.28	59.7	6.1	2.3	NAO
7	20 Nov 1979	10.09.21	61.5	10.4	1.7	NAO
8	14 Dec 1979	03.13.49	64.5	6.2	3.4	NAO
9	23 Dec 1979	14.09.13	59.6	18.7	3.2	UPP
10	10 Jan 1980	16.03.49	66.1	15.7	2.3	NAO
11	24 Jan 1980	14.19.00	58.3	6.4	2.1	NAO
12	14 Mar 1980	04.56.08	61.3	4.6	2.9	NAO
13	17 Mar 1980	23.28.14	61.3	4.6	2.1	NAO
14	29 Mar 1980	14.13.07	70.5	17.0	3.4	UPP
15	02 Apr 1980	14.55.41	56.5	21.0	3.0	UPP
16	11 Apr 1980	04.19.11	57.9	12.1	2.7	UPP
17	23 Apr 1980	15.49.06	59.2	10.3	1.8	NAO

Locations in Table 1 are as determined from the regional seismic network of Scandinavia. The NORESS records from these events are dominated by P and Lg arrivals. Quite often two clear P arrivals are found; a typical situation is the one with Pn as a rather weak first arrival followed by a stronger P phase. In addition, some records show a clear Sn between the P and Lg phases. No other S phases were observed for these events. Event no. 5 (epicentral distance 30 km) provided a very strong Rg phase.

3. DATA PROCESSING

3.1 Coherence measurements

Coherence as a function of interstation separation within NORESS was investigated for both signal and noise. For the noise, the block-averaging method of direct spectral estimation was used, with 20 blocks each of 512 samples of 20 Hz data, amounting to a total of 8.53 minutes. With that much data the bias is relatively small, and we should expect 90% of the uncorrelated values (observed coherence for true zero coherence) to fall in

the range 0.05 to 0.35 (2). The time interval was carefully
selected to avoid detectable signals within it. Results are
given in Figure 3 for two frequencies of interest. At 2 Hz
the coherence is maintained above the random level out to distances of 0.7 km and for 4 Hz to about 0.4 km. For a 12 second
interval of an Lg phase very similar to the one in Figure 2,
the coherence was computed for the same two frequencies and
results plotted in Figure 3. It is seen that signal coherence
at 4 Hz is maintained at a level above 0.5 to about 1 km and
then drops to the random level. For lower signal frequencies,
like 2 Hz, the coherence is maintained at a relatively high
level throughout the range of NORESS station distances. In comparing the coherences, one sees that at 4 Hz there is still

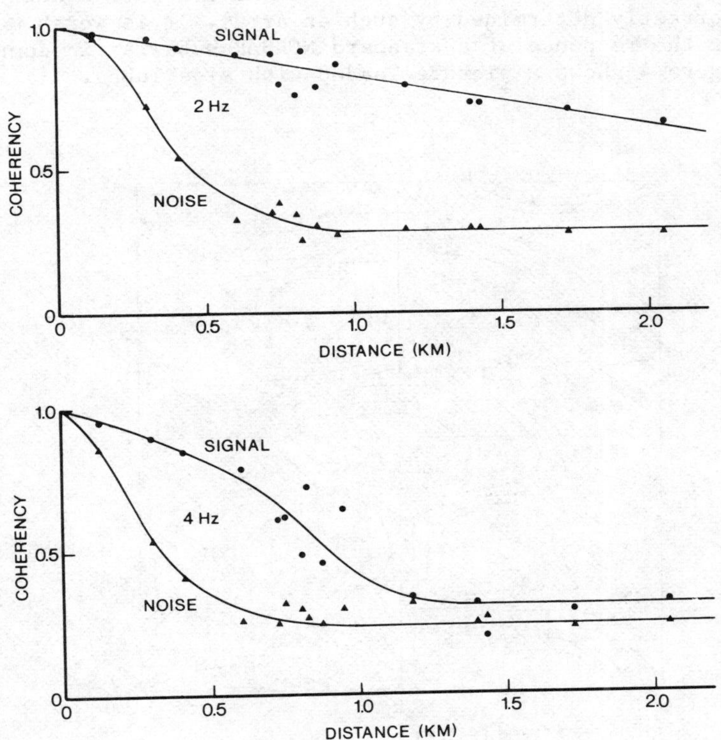

Figure 3. Signal and noise coherences at 2 Hz (top) and 4 Hz
(bottom) as a function of sensor distances within NORESS.

some 'space' between the signal and noise curves. This in turn
indicates the possibility of success for beamforming-oriented
procedures when applied to regional phases recorded at a
NORESS type array. But again one realizes the necessity of having
an array with station separations in the order of 1 km and less
when dealing with high-frequency regional phases.

3.2   F-k analyses

The power distribution in wave number domain was computed for
all clearly recorded phases from the events in Table 1. Of major
interest here is the single frequency response pattern of the
NORESS array which is shown in Figure 4. For slow phases (phase
velocity in the order of 3.5–4.0 km s^{-1}) and high frequencies
(say 5 Hz), the corresponding wave number is larger than 1 km^{-1}
and side lobe effects related to array geometry might become important. On the other hand, P phases of all frequencies (8 Hz
lowpass filters on all channels) should have a fair chance of
being correctly determined by such an array. It is worth mention-
ing that the response of a standard NORSAR subarray in comparison
with Figure 4 shows a picture loaded with side lobes.

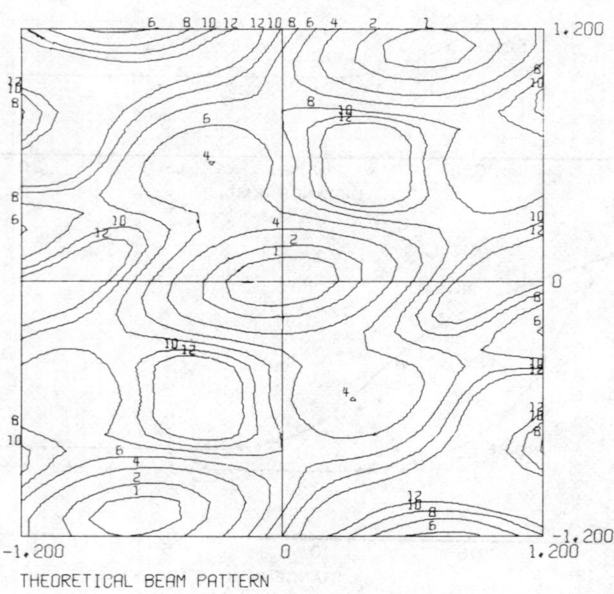

Figure 4.   Single frequency theoretical response pattern for
NORESS. Contours give power in dB down from maximum.

The high resolution as well as the conventional method
for estimating the frequency-wave number (F-k) spectrum (3)
was used. The high resolution or maximum-likelihood array processing has previously been applied to transients by several
authors (e.g., 4,5). For our regional phases, we have consistently
used 5 sec (100 samples at 20 Hz sampling) of data for each F-k
plot. The blocking of the data represents a problem in the case
of short time segments. We have chosen to treat the data in one
block only, thus violating one of the underlying assumptions
for the high resolution method and, strictly speaking, the associated F-k plots should not be termed 'high resolution'. Nevertheless, we chose this approach for the only reason that it
produced plots with more rapid breakdowns around power peaks.
The actual position in k-space of the power maximum is, however,
the same as for the conventional method. Therefore, all results
in the following analyses on phase velocities and azimuths would
have remained the same, even if the conventional method for estimating F-k spectra had been used.

Figure 5 shows records from the 6 NORESS sensors for event
no. 9 in Table 1. For each of the phases denoted Pn, P and Lg
in Figure 5, 5 sec of data from the phase onset was subjected to

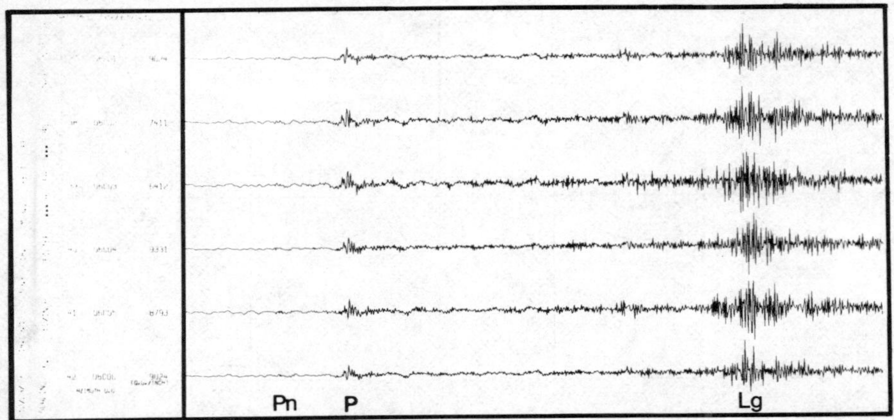

Figure 5.   NORESS records for event 9 in Table 1. The time
covered is 90 seconds. The Pn phase is also shown in Figure 6.

F-k analysis. For the Pn phase the results are shown in Figure
6. According to the reporting agency (UPP), the 'true' azimuth
from NORESS to the epicenter is 104°; the three frequencies
shown give values of 98°, 97° and 103°. The power distribution
is computed on a 31 x 31 grid, resulting in an azimuth resolution of about 3-4° for this Pn phase. The phase velocity values

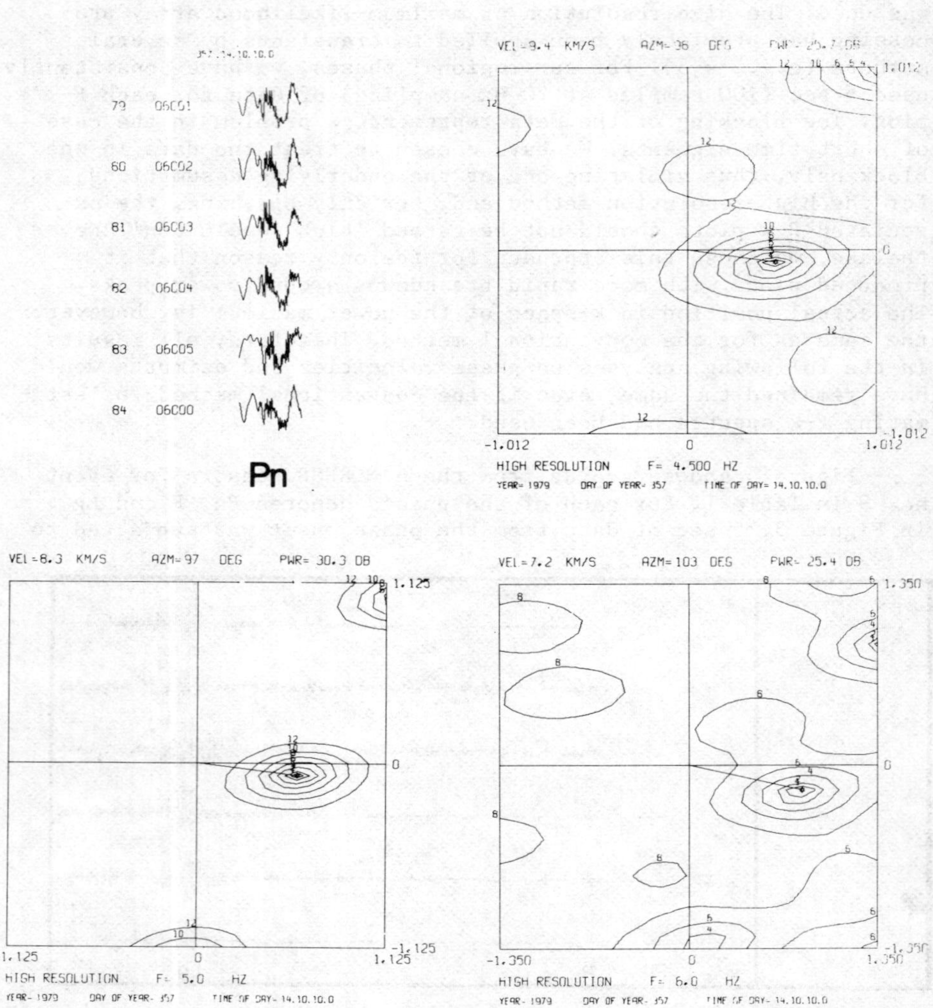

Figure 6.   F-k plots at three frequencies for the 5 seconds
of the P_n phase shown in the upper left corner.

are seen to fluctuate around 8 km s^{-1}. The coarseness of the
k-space grid accounts for part of these fluctuations. Also
for the two other phases in Figure 5 we get reaonsable results
from the F-k analysis; the phase termed 'P' shows typical P-
type velocities and reasonable azimuths (Figure 7), while the
Lg phase has velocities in the range 4-5 km s^{-1} and again the
azimuths are 'good' (Figure 8).

4.   PHASE VELOCITY RESULTS

All distinct phase arrivals from the events in Table 1 were subjected to the kind of analysis described above for event no. 9.
For each phase F-k plots were produced for 5 frequencies centered
around the spectral peak. A high degree of consistency among
results for various frequencies were found for nearly all phases
analyzed. The only difficulties encountered were a few cases
where the point of maximum power in the F-k plots were caused by
side lobes resulting from the array geometry. Most of these cases
of spatial aliasing could, however, be easily identified from corresponding unrealistically low phase velocity values.

Results from the F-k analysis as regards phase velocities
are summarized in Figure 9. For each phase we chose the frequency with maximum power and plotted the corresponding phase
velocity. For some events more than one P phase was considered
and for these cases only the phase velocity of the strongest
P phase is included in Figure 9. Some of the scatter in Figure 9
is explained by the resolution in the k-space grids. If the
maximum power were shifted to the neighboring point in k-space,
the corresponding change in phase velocity would roughly amount
to 0.4 km s^{-1}. Despite these kinds of limitations, the following
conclusions are readily reached:

-     The P and Lg/Sn phases separate into nonintersecting phase
      velocity bands and a discrimination between P and Lg/Sn
      is possible from NORESS data.

-     It is hardly possible to separate Lg from Sn on the basis
      of phase velocity measurements alone.

The phase velocities found for Sn are for all 4 cases higher than
those for Lg for the same event. If only one of these phases is
present in the seismogram, however, F-k analysis alone cannot tell
whether it is an Lg or Sn phase.

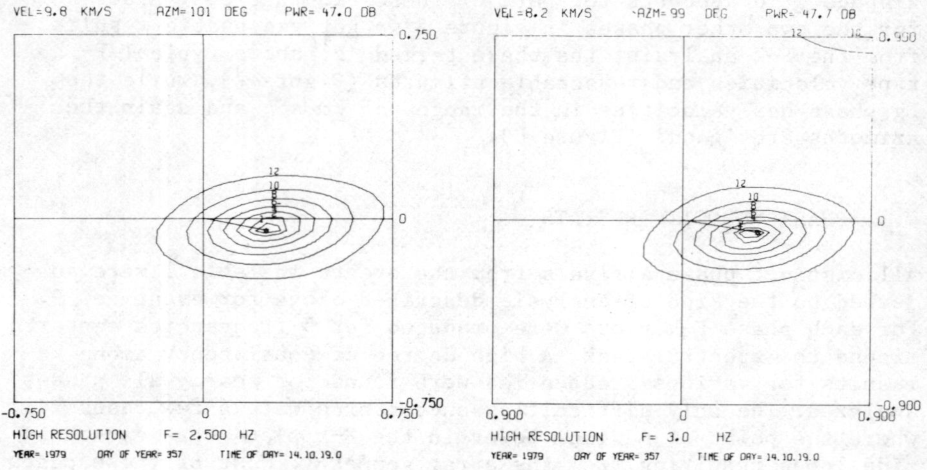

Figure 7. F-k plots at two frequencies for 5 seconds of the
P phase in Figure 5.

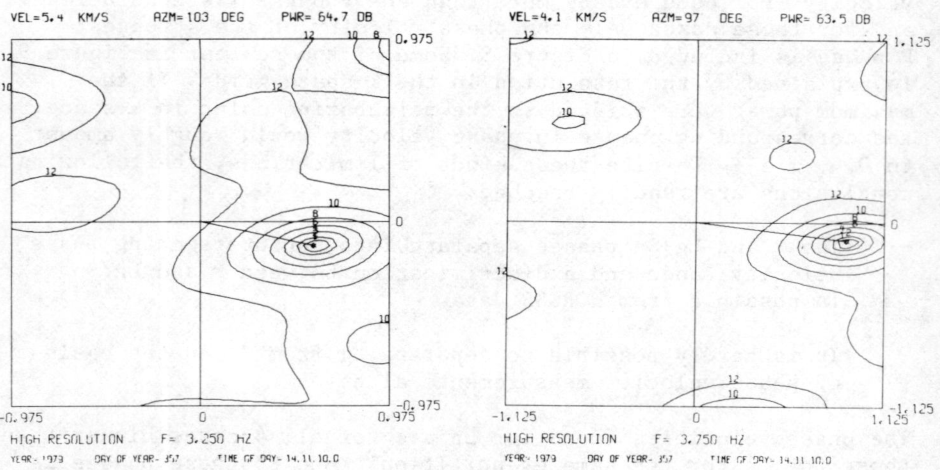

Figure 8. F-k plots at two frequencies for 5 seconds of the
Lg phase in Figure 5.

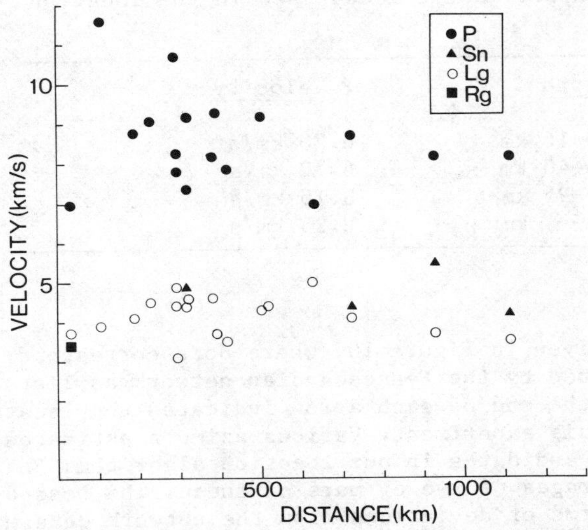

Figure 9. Phase velocities determined from F-k analysis of
NORESS data from events in Table 1. Results for one event at
epicentral distance 900 km is included which does not appear
in Table 1, as it remained undetected by the reporting agencies.

5.   A LOCATION EXPERIMENT

Based on the results from F-k analyses of P and Lg phases, we
have developed a simple location procedure and applied it to the
events in Table 1. The location algorithm consists of the following steps:

i)   Identification of P and Lg phases for each event using phase
      velocity results from F-k analyses.

ii)   Manual pick of onset times for P and Lg.

iii) Epicenter azimuth determination from F-k analyses.

iv)   Epicenter distance determination using Lg-P travel time
      differences and a simple crust/upper mantle model.

The crustal model adopted is given in Table 2 and represents
a simplified average over various models derived for parts of
Fennoscandia. For Lg group velocity we have used 3.50 km s^{-1}
which is also consistent with findings from explosion seismology
investigations in the same area. The result of the location

Table 2. Crust/upper mantle model used in the location
experiment.

Depth	P Velocity
0–16 km	6.20 km/s
16–40 km	6.70 km/s
40–95 km	8.15 km/s
95– km	8.25 km/s

experiment is given in Figure 10, where dots correspond to epicenters determined by the Fennoscandian network as listed in
Table 1, while the end of each arrow indicates the location as
determined by this experiment. Various azimuth estimates were
considered for candidates in our location algorithm: Pn, P, Lg
azimuths or averages of two or more of these. The best overall
results – in terms of deviations from the network determinations
– are achieved using the azimuth for the strongest P phase in
cases where more than one P phase was processed. Only two events
show substantial location differences: For event no. 8 there is
an azimuth difference of 13°. Our Lg azimuth in this case,
however, points directly to the epicenter as determined by the
network. The location capability of the regional network in the
area of event 14 is not very good due to poor station coverage.
This fact may explain the location difference for this event.
Alternatively, our distance estimate is wrong. The main results
from this location experiment can be summarized as follows:

-   The median 'error' of our distance estimates using Lg–P
    times is 11 km.

-   The median 'error' of our azimuth estimates from the
    strongest P phase is 3°.

-   The median location 'error' is 30 km.

It should be noted here that the above 'errors' are relative to
the location results from the Fennoscandian network, and that
these locations are uncertain by typically 10–30 km or more.
Thus the majority of the processed events can be located with
a deviation from the network location that is within the uncertainty limits of the latter. From Figure 10 we also see that
the best agreement between solutions is in southeastern Norway
and southern Sweden, which are also the regions of best location
capability of the Fennoscandian network.

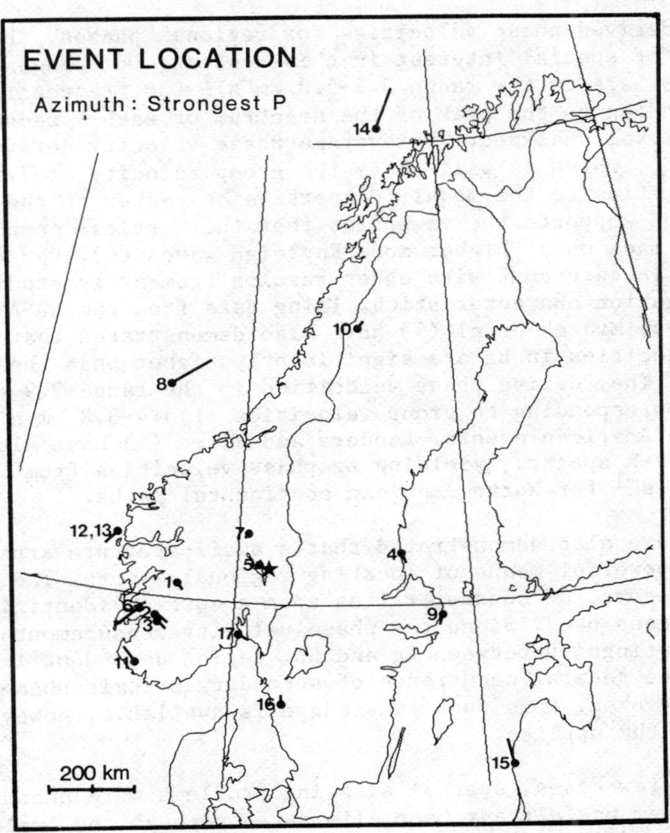

Figure 10. Location 'errors' using NORESS (with azimuths determined from the strongest P phase for each event) for regional event location. Dots indicate epicenters determined by the Fennoscandian station network, while the ends of the arrows give location from NORESS processing. The location of NORESS is indicated by an asterisk. Event numbers refer to Table 1.

6.   DISCUSSION

We have derived phase velocities for regional phases. The Lg
phase is of special interest in this context. Our phase
velocities are in the range 3.2–5.0 km s^{-1} for frequencies
corresponding to the peak of the spectrum of each 5 second
time interval analyzed. The average phase velocity derived is
4.1 km s^{-1}, which is well above the group velocity of 3.5 km s^{-1}.
This testifies to the modal, dispersive character of the Lg
phase, and supports the viewpoint that the vertical component
of Lg is made up of higher mode Rayleigh waves (6). Our findings
are well in agreeemnt with other results from array studies of
Lg propagation characteristics. Using data from the LASA and
CPO arrays, Mrazek et al (7) have also demonstrated that the
phase velocities in Lg are significantly higher than the group
velocity. They derive phase velocities in the range 3.9–4.7
km s^{-1} corresponding to group velocities of 3.4–3.8 km s^{-1}
for North American events. Landers and Fitch (5) have also
computed F-k spectra, yielding Lg phase velocities from 3.9
to 4.6 km s^{-1} for North American continental paths.

We have also demonstrated that a small–aperture array
can be a powerful means of locating regional events. The location procedure, of course, relies upon a correct identification
of the phases used. Since the phase velocity measurements alone
cannot distinguish between Lg and Sn, region–dependent information on the general occurrence of secondary seismic phases
must be invoked. Once such knowleddge is available, however,
our algorithm applies.

In a few cases, spatial aliasing problems were encountered.
This problem has already been eliminated through the implementation of an additional 6 instruments in the NORESS array. Work
now continues on finding the optimum layout for NORESS, both
with respect to sensor number and geometry.

REFERENCES

1.   Wahlstrøm, R.: 1978, Magnitude–scaling of earthquakes in
      Fennoscandia. Report No. 3–78, Seismological Institute,
      Uppsala, Sweden.

2.   Amos, D.E. and Kopmans, L.H.: 1963, Tables of the distribu-
      tion of the coefficient of coherence for stationary bi-
      variate gaussian processes, Sandia Corporation Monograph
      SCR–483.

3.   Capon, J.: 1969, Proc. IEEE, __57__, 1408–1418.

4.   Capon, J.: 1974, Bull. Seism. Soc. Am., <u>64</u>, 235-266.

5.   Landers, T. and Fitch, T.: 1978, Phase velocity of Lg in
      North America, Lincoln Laboratory Semiannual Technical
      Summary, Seismic Discrimination, September 1978.

6.   Panza, G.F. and Calcagnile, G.: 1975, Geophys. J.R. astr.
      Soc., <u>40</u>, 475-487.

7.   Mrazek, C.P., Der, Z.A., Barker, B.W. and O'Donnell, A.:
      1980, Seismic Array Design for Regional Phases, In:
      <u>Studies</u> <u>of</u> <u>Seismic</u> <u>Waves</u> <u>at</u> <u>Regional</u> <u>Distances</u>, AL-80-1,
      Teledyne-Geotech.

P AND S-VELOCITY JUMP AT THE INNER-CORE BOUNDARY FROM
PKP AMPLITUDES

Heinz Häge

Universitätsinstitut für Meteorologie und
Geophysik, 6000 Frankfurt/M., F.R. Germany

ABSTRACT

Amplitudes of long-period PKP phases from 16 large
earthquakes are investigated in the distance ranges
110°-134° and 142°-170°. A parameterized amplitude
distribution in the distance range 110°-134° is compared with theoretically computed amplitudes using
different velocity and density modifications of model
1066B in the inner core. An attempt is made to estimate the velocity jumps of P and S waves as well as
the density jump across the inner-core boundary (ICB).
The results of a preliminary analysis are:

a) The P-velocity jump can be restricted between 0.6
   and 0.7 km/sec with a preferred value of 0.64
   km/sec,

b) The shear wave velocity at the top of the inner
   core lies between 2.5 and 3.0 km/sec,

c) The density jump at the ICB is practically not restricted by the data because the amplitudes are not
   so sensitive to this jump and because of its inherent trade-off with the jump in shear wave velocity.

INTRODUCTION

Recent investigations concerning the driving mechanism

*E. S. Husebye and S. Mykkeltveit (eds.), Identification of Seismic Sources - Earthquake or Underground
Explosion, 483–496.*
Copyright © 1981 by D. Reidel Publishing Company.

of the geodynamo clearly show the need for a reliable
knowledge of the density distribution in the Earth's
core. Whereas there is wide agreement among various
earth models in the mantle and in the outer core,
there remains considerable disagreement on the jump of
the elastic parameters and density across the ICB. It
can be assumed that there exists no transition zone
above the ICB (Müller, 1975), and therefore we deal
with a discontinuous change of density and velocities
at the ICB alone whose radius we take as $r = 1216$ km.

The observational data used to infer the velocity
and density jumps at the ICB comprise traveltimes,
body wave amplitudes and eigenfrequencies of free
oscillations of the Earth. Moreover, valuable conclusions about the state of the matter have been drawn
from shock wave experiments (Al'tshuler et al., 1968;
Ahrens, 1980). For the jump in P velocity at the ICB
Buchbinder (1971) derived a value of 0.58 km/sec, and
Müller (1973) restricted this jump to 0.6-0.7 km/sec.
Earth model 1066B (Gilbert and Dziewonski, 1975) has
a jump of 0.67 km/sec and the PEM models (Dziewonski
et al., 1975) show velocity jumps of 0.83 km/sec. The
mean shear wave velocity in the inner core seems to
be restricted between 2.95 km/sec (Julian et al.,
1972) and an upper limit of 4.0 km/sec (Müller, 1973).
The former value is in doubt because of the questionable observability of the PKJKP phase (Doornbos, 1974).
The density contrast across the ICB seems to be the
least constrained parameter in all earth models.
Values for it range from 0.2 g/cm^3 in model C2 (Anderson
and Hart, 1976) to 1.8 g/cm^3 (Bolt, 1972). Free oscillation data can be well satisfied by 0.56 g/cm^3 in
1066B. Masters (1979) obtained a density jump of
0.87$\pm$0.32 g/cm^3 using eigenfrequencies of normal modes.
Shock wave experiments on pure Fe point to a density
jump of 1.0 g/cm^3 (Al'tshuler, Simakov and Trunin,
1968) under the assumption of an inner core consisting
mainly of pure Fe and maybe some Ni.

It is the aim of this paper to estimate the jumps
in seismic velocities and density at the ICB using
long-period body waves which travel through the upper
part of the inner core (PKIKP or PKP_{DF}) or are reflected from it (PKiKP or PKP_{CD}). Both phases are sensitive
to the jumps in velocities and density across the ICB.
For this purpose observed amplitudes from these core
phases are compared with theoretically calculated
amplitudes in the distance range from 110° to 134°. An
appropriate normalization procedure is applied to the

observed amplitudes of each event such that a cumulative plot of amplitudes can be produced. It can be shown from theoretical amplitude distributions that slope and intercept of a regression line through the logarithmic normalized amplitudes are directly related to the velocity and density jump at the ICB in the earth model.

THE EARTHQUAKE DATASET

Seismograms of the vertical components of long-period PKP phases from 16 earthquakes in the southwest Pacific Ocean and in South America have been investigated (see Table 1). The events were chosen such that their fault-plane solutions show nearly an optimal P-wave radiation vertically downwards. Fig. 1 shows an example of a digitized seismogram section. Besides PP and P_{diff} the PKIKP phase can be clearly traced from 119° up to 175°. Amplitudes in the lower distance range (110°-134°) are influenced not only by PKIKP but also by the reflection PKiKP from the ICB. From the original seismograms, peak-to-peak amplitudes of the first-arriving

Table 1: Selected earthquakes used in this study (data from ISC and USCGS)

Event No.	Date	Origin Time	Epicenter	Depth in km	Magnitude
1	3 NOV 1965	01:39:03.1	9.1S/71.4W	593	6.2
2	25 JUL 1968	07:23:07.8	30.8S/178.4W	60	6.4
3	12 SEP 1964	22:07:03.2	49.1S/164.2E	33	6.9
4	12 JAN 1972	09:59:10.3	6.8S/71.8W	575	5.7
5	20 MAY 1968	20:05:48.0	30.8S/178.2W	33	6.0
6	15 FEB 1967	16:11:11.8	9.0S/71.3W	598	6.1
7	14 SEP 1966	23:18:41.9	60.3S/27.2W	27	5.9
8	21 JUL 1977	11:53:22.3	53.8S/158.8E	33	6.2
9	23 MAY 1968	17:24:16.8	41.7S/172.0E	21	6.1
10	2 APR 1977	07:15:22.9	16.8S/172.0W	33	6.4
11	10 AUG 1970	15:15:20.7	13.9S/166.6E	46	5.9
12	11 AUG 1970	10:22:20.0	14.1S/166.6E	20	6.1
13	22 MAY 1972	20:45:55.1	17.7S/175.2W	227	6.1
14	17 JAN 1975	09:30:36.6	17.9S/174.6W	99	5.8
15	8 JAN 1970	17:12:40.6	34.8S/178.8E	189	6.1
16	26 SEP 1968	18:02:47.0	30.5S/178.0W	12	5.8

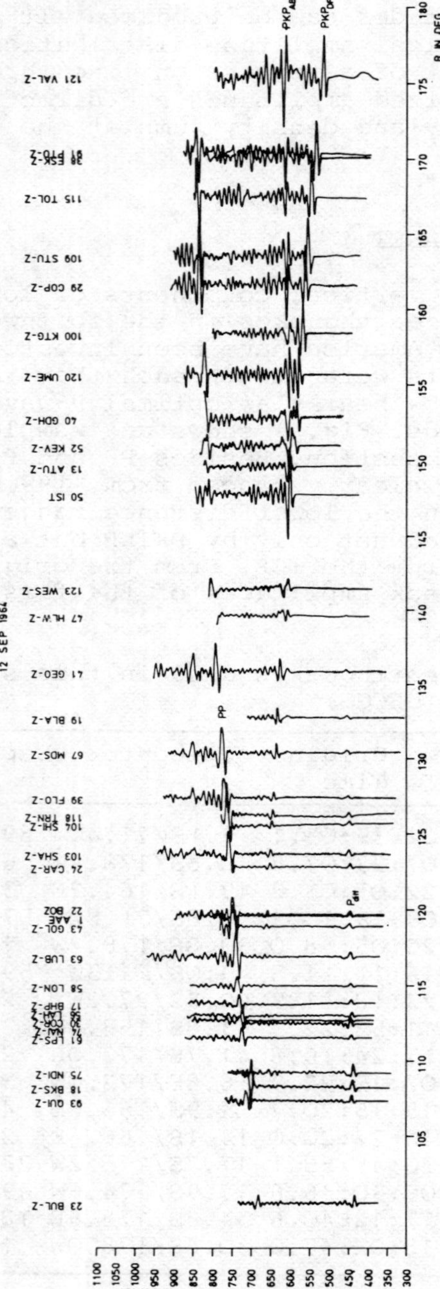

Fig. 1: Digitized seismogram section (vertical component at long-period WWNSS stations) for event No. 3 including Pdiff, PKP and PP. All records are on the same amplitude scale.

PKP phase have been read for stations between 110°
and 170° and normalized by use of their station magnifications. Thus, between 110° and 134° one reads the
amplitudes of PKIKP+PKiKP. Amplitudes between 134°
and 142° have been left out from the analysis because
they are influenced by the diffraction from the B
caustic. Around 146° the amplitude contribution mainly comes from the ABC branch. Fig. 2 shows a plot of
the logarithm of observed amplitudes versus distance
for event No. 3. Such amplitude-distance plots were
produced for each individual earthquake. For the deep
events No. 1, 4, 6, 13 and 15 distance corrections
have been applied according to Buchbinder (1969). This
is necessary in order to compare amplitudes from earthquakes at different depths consistently. The maximum
distance correction is   2° for the PKP_{AB} branch.

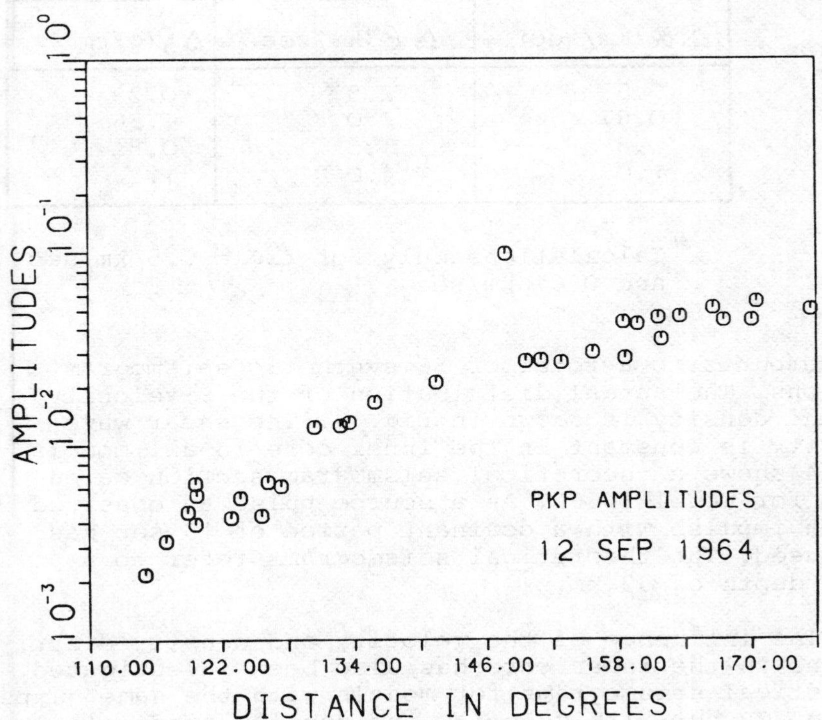

Fig. 2: Logarithmic plot of first-arriving PKP amplitudes for event No. 3. The maximum amplitude near
148° is mainly due to PKP_{ABC}.

ICB MODELS INVESTIGATED

A series of theoretical seismograms have been computed
using the reflectivity method. The underlying earth
model was 1066B whose velocity and density distribution outside the inner core was held constant throughout all computations. From model to model one of the
three jumps, $\Delta\alpha$, $\Delta\beta$ or $\Delta\rho$ ($\Delta\alpha$ is P-velocity
jump, $\Delta\beta$ is S-velocity jump and $\Delta\rho$ is density jump),
was changed and the corresponding influence on the
amplitude distribution was investigated. Table 2 shows
the jumps which were used in our calculations.

Table 2: Jumps in P-wave velocity ($\Delta\alpha$), S-wave velocity ($\Delta\beta$) and density ($\Delta\rho$) across the ICB for
which theoretical seismograms were calculated.

$\Delta\alpha$ (km/sec)	$\Delta\beta$ (km/sec)	$\Delta\rho$ (g/cm^3)
0.5	2.5 *	0.2
0.67	3.0	0.56
0.8	3.5	0.8
1.0	4.0	1.2

* Calculations only for $\Delta\alpha$ = 0.5 km/sec
and 0.67 km/sec

This amounts to a total of 56 synthetic seismogram
sections. The actual distribution of the P-velocity
and the density is shown in Fig. 3. The shear wave
velocity is constant in the inner core in all models.
Fig. 4 shows a theoretical seismogram section calculated for model 1066B. As a source pulse an observed
P-wave impulse with a dominant period of 15 sec has
been used. The theoretical seismograms refer to a
focal depth of 32 km.

The influence of the velocity and density distributions in the inner core has also been investigated.
Theoretical seismograms for models with the same jumps
at the ICB but with velocity and density gradients in
the inner core different from those in Fig. 3 practically show the same amplitudes. This justifies the restriction to parameter jumps at the ICB as variations
of the earth model.

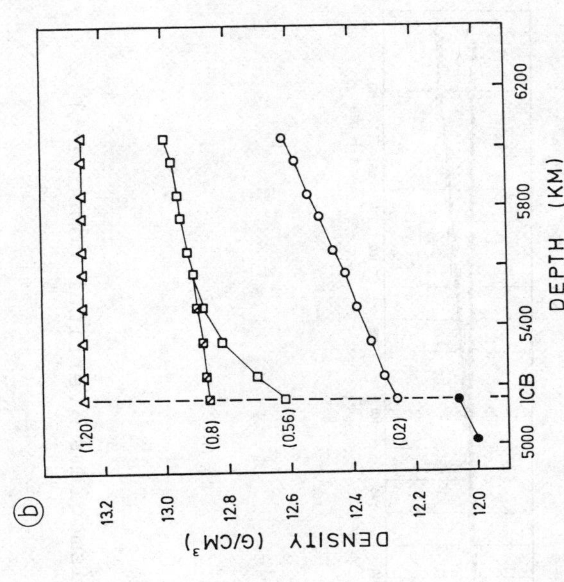

b) Density distribution in the inner core used in calculating theoretical seismograms. Values in brackets denote the density jump at the ICB. Open squares represent model 1066B.

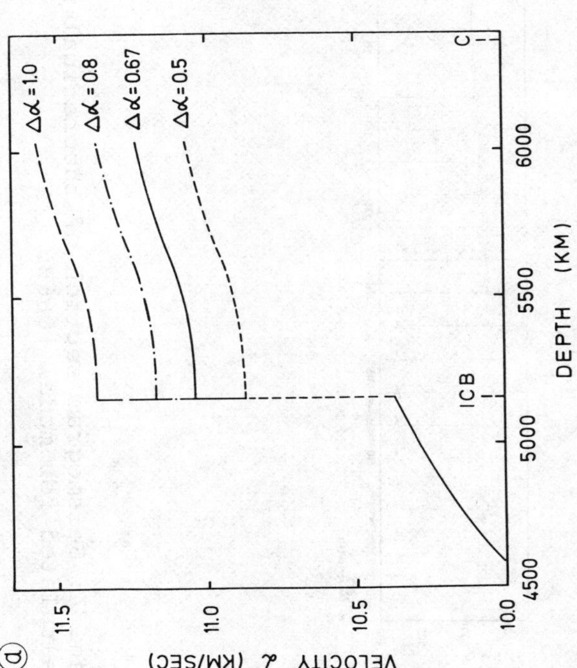

Fig. 3: a) P-velocity distributions in the inner core used in calculating theoretical seismograms. $\Delta\alpha$ is the P-velocity jump at the ICB. $\Delta\alpha = 0.67$ km/sec corresponds to model 1066B.

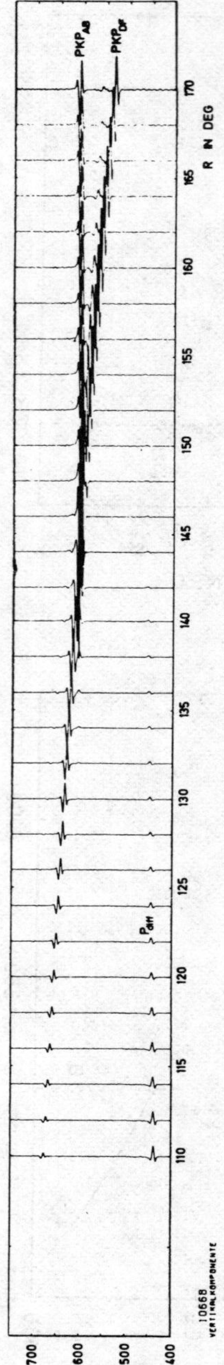

Fig. 4: Seismogram section of theoretical seismograms (vertical component) calculated for model 1066B.

INTERPRETATION METHOD

From the theoretical seismograms, peak-to-peak amplitudes were read in the same distance ranges and from
the same phases (DF+CD, DF+AB and DF) as from observed
seismograms. One of the synthetic amplitude distributions is depicted in Fig. 5. A striking feature is
the linear distribution of logarithmic amplitudes
between 110° and 134° which appear in all theoretical
calculations. Furthermore it turned out that changes
in $\Delta\alpha$, $\Delta\beta$ and $\Delta\rho$ influence only the amplitudes
between 110° and 134° leaving the amplitudes in the
distance range 142°-170° unchanged. For each model the
amplitudes in the lower distance range were fitted by

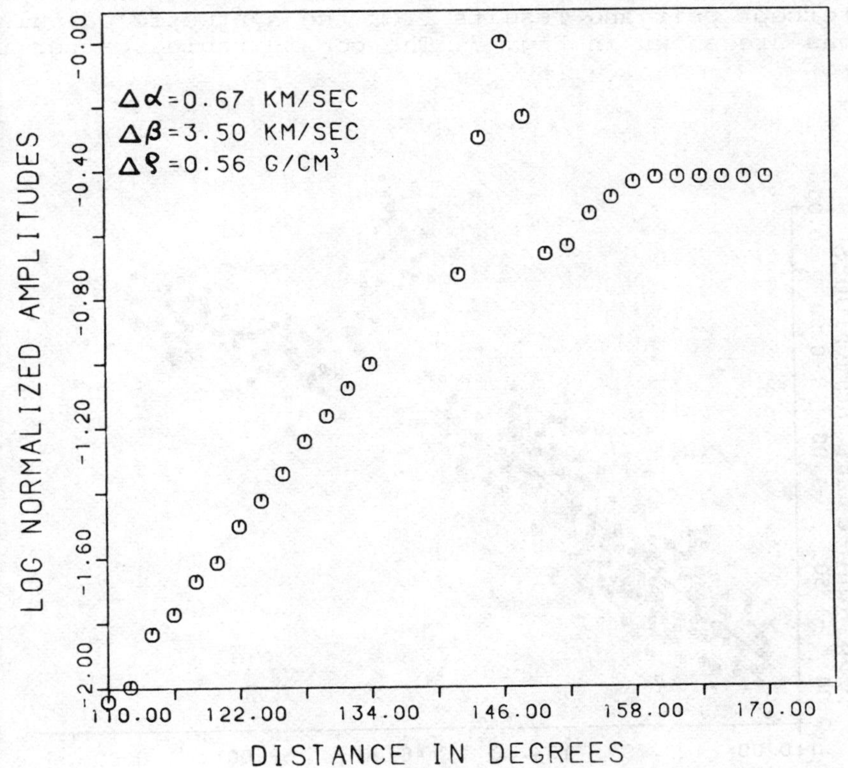

Fig. 5: Theoretical logarithmic amplitude distribution
for model 1066B. Amplitudes are normalized to maximum
amplitude at 146°. Note linear relationship between
110° and 134°.

a straight line whose slope and intercept were used as
parameters to distinguish between different models of th
ICB. A high-order polynomial was fitted to the theoretical amplitudes between 142º and 170º. Hereafter,
this polynomial was fitted to each earthquake amplitude distribution in the least-square-sense and in the
same distance range. This was done to find the usually
not observed maximum amplitude at 146º for each event
upon which the observed amplitudes were normalized.
With this normalization procedure all amplitudes from
the 16 events were put into one amplitude-distance
plot (Fig. 6). This plot contains a total of 429 observations of which 210 lie between 110º and 134º.
Again, a straight line was fitted to this cumulative
amplitude distribution and its slope and intercept
were determined. The comparison of the observed slope-
intercept pair and results from the synthetic calculations are shown in Fig. 7. The considerable scatter in

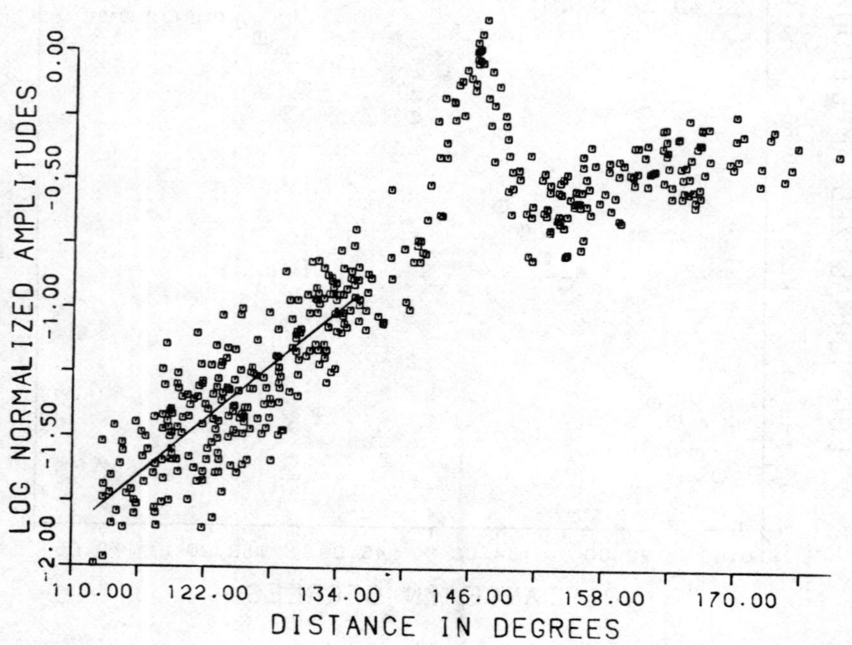

Fig. 6: Cumulative logarithmic amplitude plot containing the normalized PKP-amplitudes from all events.

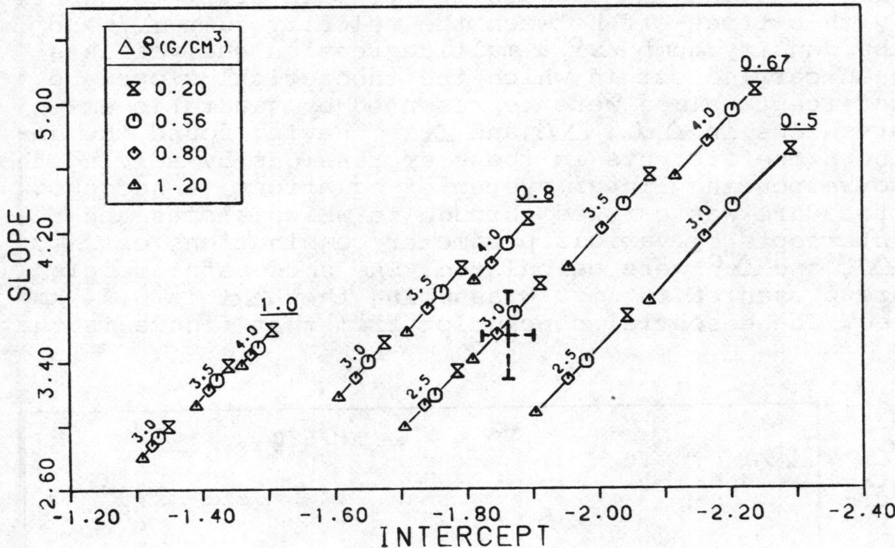

Fig. 7: Slopes and intercepts of regression lines for
synthetic amplitude distributions in arbitrary units.
Underlined large numbers denote the P-velocity jump
(in km/sec) at the ICB, the small numbers indicate
the shear wave velocities in the inner core. The 90%
confidence limits of observed regression parameters
are indicated by the dashed cross.

the amplitude data causes the relative large error
bars which represent the 90% confidence intervals of
slope and intercept.

DISCUSSION

From an inspection of Fig. 7 it can be concluded that
the jump in P velocity at the ICB lies roughly between
0.6-0.7 km/sec supporting the result obtained by
Müller (1973). Model with $\Delta\alpha$ outside this range would
require unreasonable values for the shear wave velocity and/or density jump. Successful models seem to lie
in the neighbourhood of $\Delta\alpha$ = 0.64 km/sec. There is
also a clear trade-off between $\Delta\beta$ and $\Delta\rho$ in the
sense that keeping $\Delta\beta$ constant and increasing $\Delta\rho$

leads to the same slope-intercept values as keeping
$\Delta\varrho$ constant and decreasing $\Delta\beta$. In order to quanti-
fy this trade-off between the velocity jump $\Delta\beta$ and
the density jump $\Delta\varrho$ a multiregression analysis has
been carried out in which the theoretical slope and
intercept values were represented by quadratic ex-
pressions in $\Delta\alpha$, $\Delta\beta$ and $\Delta\varrho$. Having found the unknown coefficients in these expressions by solving the
corresponding linear system of equations, a hedgehog
procedure was carried through in which slopes and
intercepts of various parameter combinations of $\Delta\alpha$,
$\Delta\beta$ and $\Delta\varrho$ were calculated. The successful models
are presented in Fig. 8 assuming that $\Delta\alpha$ is 0.64 km/
sec. The essential conclusion from that figure is that

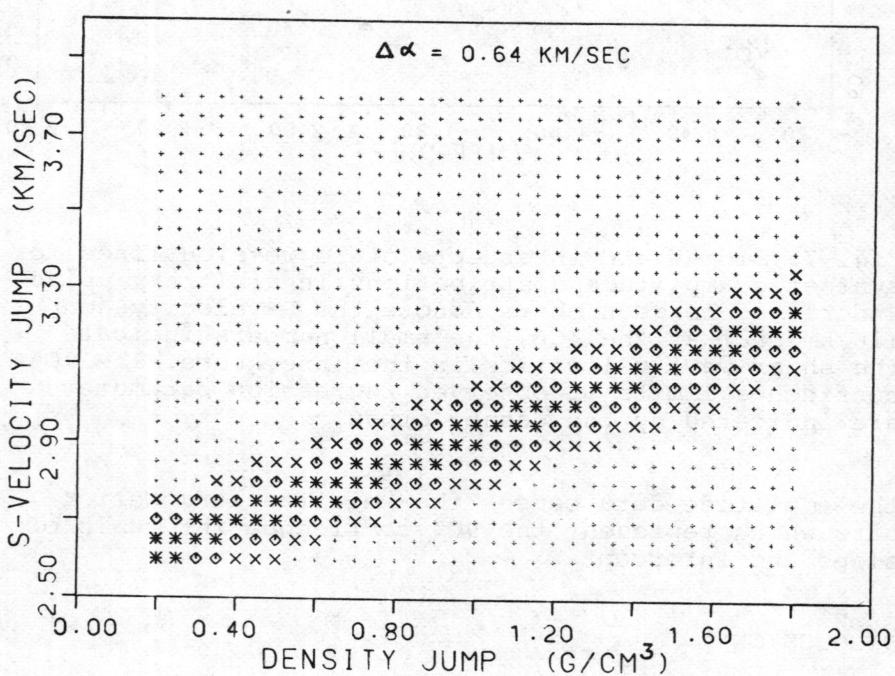

Fig. 8: Trade-off between $\Delta\beta$ and $\Delta\varrho$ for $\Delta\alpha$ = 0.64
km/sec. Large symbols indicate those models which
agree with the observed slope-intercept pair (including the confidence interval which has been subdivided
into 3 concentric rectangular areas indicated by ✳,
◇ and ✕).

the mean shear velocity in the inner core should be at the lower boundary of the 3.0-4.0 km/sec range. Values for $\Delta\beta$ which are considerably larger than 3.0 km/sec would lead to very high density jumps at the ICB for which there is not enough other evidence. This behaviour does not change very much if one assumes a somewhat higher value for $\Delta\alpha$ (e.g. $\Delta\alpha$ = 0.7 km/sec). The value of $\Delta\alpha$ = 0.83 km/sec which Cormier and Richards (1977) found comparing observed and theoretical amplitudes separately on different events cannot be matched in the present analysis without taking quite unreasonable values for $\Delta\beta$ and $\Delta\rho$ at the ICB.

From this preliminary analysis it can be concluded that there is high certainty for $\Delta\alpha$ at the ICB lying between 0.6-0.7 km/sec with a preferred value of 0.64 km/sec. This implies a shear wave velocity, representative of the top of the inner core, which lies between 2.5 km/sec and approximately 3.0 km/sec if one assumes that the density jump at the ICB has an approximate upper limit of 1.2 g/cm^3 which is indicated by shock wave experiments. Considering the mean shear wave velocity in the inner core of 3.5 km/sec which follows from eigenoscillation data the relatively low S velocity below the ICB which we found from PKP amplitudes suggests that there may exist an increase in the S velocity in the inner core with depth. The original intention to determine the density jump at the ICB from long-period PKP amplitudes could not be realized due to the data scatter and the inherent trade-off between $\Delta\beta$ and $\Delta\rho$.

ACKNOWLEDGMENTS

This work was supported by a grant from the Deutsche Forschungsgemeinschaft. The computations were made at the computing centres of the University of Karlsruhe and the University of Frankfurt/M. I am especially grateful to G. Müller for critically reading the manuscript and preceeding discussions. I would also like to thank Ingrid Hörnchen for typing it. I am also indebted to the anonymous referee for his constructive criticism.

REFERENCES

Ahrens, T.J., 1980: Dynamic compression of earth materials, Science, Vol. 207, No. 4435, pp.1035-1041.

Al'tshuler, L.V., Simakov, G.V., and Trunin, R.F.,
    1968: On the composition of the Earth's core, Isv.
    Earth Phys., 1, p. 3.
Anderson, D.L., and Hart, R.S., 1976: An earth model
    based on free oscillations and body waves, J. Geo-
    phys. Res., 81, pp. 1461-1475.
Bolt, B.A., 1972: The density distribution near the
    base of the mantle and near the earth's center,
    Phys. Earth Planet. Int., 5, pp. 301-311.
Buchbinder, G.G.R., 1969: Distance corrections for deep
    focus earthquakes, Geophys. J.R. astr. Soc., 17,
    pp. 195-202.
Buchbinder, G.G.R., 1971: A velocity structure of the
    earth's core, BSSA, 61, pp. 429-456.
Cormier, V.F., and Richards, P.G., 1977: Full wave
    theory applied to a discontinuous velocity increase:
    the inner core boundary, J. Geophys., 43, pp. 3-31.
Doornbos, D.J., 1974: The unelasticity of the inner
    core, Geophys. J.R. astr. Soc., 38, pp. 397-415.
Dziewonski, A.M., Hales, A.L., and Lapwood, E.R., 1975:
    Parametrically simple earth models consistent with
    geophysical data, Phys. Earth Planet. Int., 10,
    p. 12.
Gilbert, F., and Dziewonski, A.M., 1975: An application
    of normal mode theory to the retrieval of structural
    parameters and source mechanisms from seismic spectra
    Phil. Trans. R. Soc. Lond., Ser. A 278, pp. 187-269.
Julian, B.R., Davies, D., and Sheppard, R.M., 1972:
    PKJKP, Nature, 235, pp. 317-318.
Masters, G., 1979: Observational constraints on the
    chemical and thermal structure of the Earth's deep
    interior, Geophys. J.R. astr. Soc., 57, pp. 507-
    534.
Müller, G., 1973: Amplitude studies of core phases,
    J. Geophys. Res., 78, pp. 3469-3490.
Müller, G., 1975: Further evidence against disconti-
    nuities in the outer core, Phys. Earth Planet.
    Int., 10, pp. 70-73.

CONVERSION PHASES FROM MANTLE TRANSITION ZONES

Sonja Faber

Institut für Meteorologie und Geophysik,
Feldbergstr. 47, 6000 Frankfurt/M. Germany

Conversion phases from the transition zones between
the upper and lower mantle beneath North America have
been observed at North American stations and at the
GRF-array in Germany. Synthetic seismograms have been
calculated in order to determine the velocity contrast
and thickness of the transition zones at which conversion took place. The observations can be explained by
models having two well developed transition zones at
depths of 400 and 670 km.

Sp PHASES

Precursors to S and SKS were observed in long-period
SRO and WWSSN seismograms of the Romanian earthquake
(March 4, 1977) and of 2 Tonga-Fiji earthquakes (March
30, 1972 and February 22, 1975), recorded in the United
States at distances from 68° to 96°. These precursors
appeared to be Sp phases generated by conversion from
S to P at the transition zone between the upper and
lower mantle below the receivers.

Theoretical seismograms have been calculated for
earth models which differed in the velocity contrast
and thickness of the interfaces or transition zones at
which conversion took place. The observations are
compatible with velocity models having pronounced
transition zones at depths of 400 and 670 km as they
have been proposed for the western United States
(Helmberger and Engen, 1974; Burdick and Helmberger,

E. S. Husebye and S. Mykkeltveit (eds.), Identification of Seismic Sources - Earthquake or Underground
Explosion, 497–504.

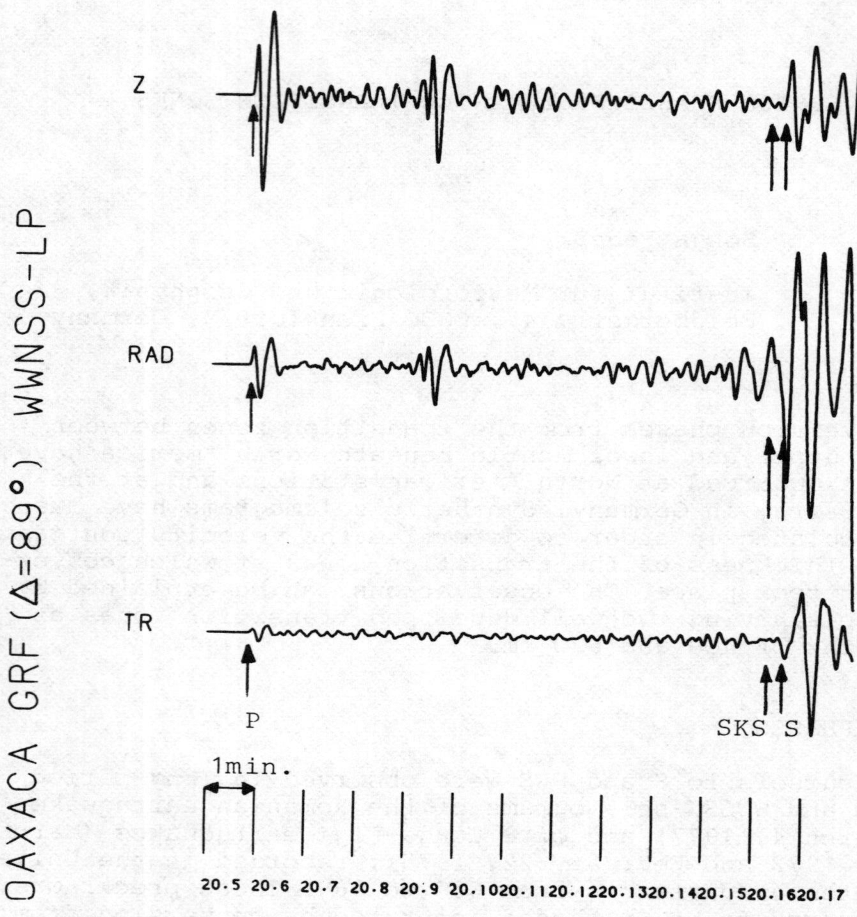

Fig. 1: Oaxaca earthquake of November 29, 1978: long-period WWSSN-simulation of the digital broadband data at the station GRF.

1978) and they exclude much smoother structures. Details of the velocity structure could only be determined within the resolution limits of long-period data. This investigation has been published (Faber and Müller, 1980).

pS PHASES

The record of the November 29, 1978 Oaxaca earthquake shown in Figure 1 is a long-period WWSSN simulation of the digital broad-band data observed at the station GRF (Δ = 89°). Only the long-period simulation is presented here and dealt with in the calculation of synthetic seismograms. An analysis of the frequency content of the conversion phases in the broad-band data, aiming at a better resolution of the thickness of the transition zones, is worked on. The arrival times of S and SKS were determined from travel-time differences to P, taken from the Jeffreys-Bullen tables. SKS cannot be identified on the vertical component; S energy is expected to be rather weak at European stations, since the radiation in this direction is close to the T-axis. In addition SKS might in part interfere destructively with the precursor phase which arrives about 50 sec before SKS and is very strong on the radial component. The explanation of this precursor with prevailing horizontal polarisation is that it originates mainly from a P to SV conversion (called pS in the following) at the transition zones between the upper and lower mantle beneath North America.

Synthetic seismograms for P-waves (Figure 2) as well as for SV-waves and pS below the focus and Sp below the receivers (Figure 3) have been calculated with the reflectivity method (Kind and Müller, 1975), using model SP-1 which explained the Sp phases observed at North American stations within the limits of resolution (Faber and Müller, 1980). This model has two first order discontinuities at depths of 395 km and 670 km with S-velocity changes from 4.72 to 5.07 km/sec and from 5.62 to 6.04 km/sec and P-velocity changes from 8.82 to 9.24 km/sec and from 10.4 to 10.8 km/sec. The double couple point source at a depth of 18 km was orientated according to the fault-plane solution of the Oaxaca earthquake. Surface reflections at the source were included in the calculation. The input signal for far-field displacement was one sine-oscillation with a dominant period of 20 sec corresponding to the period of the P impulse in the WWSSN simulation.

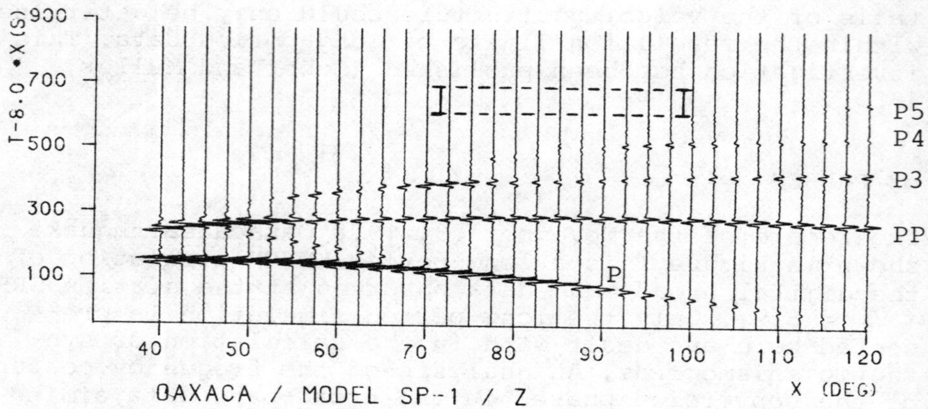

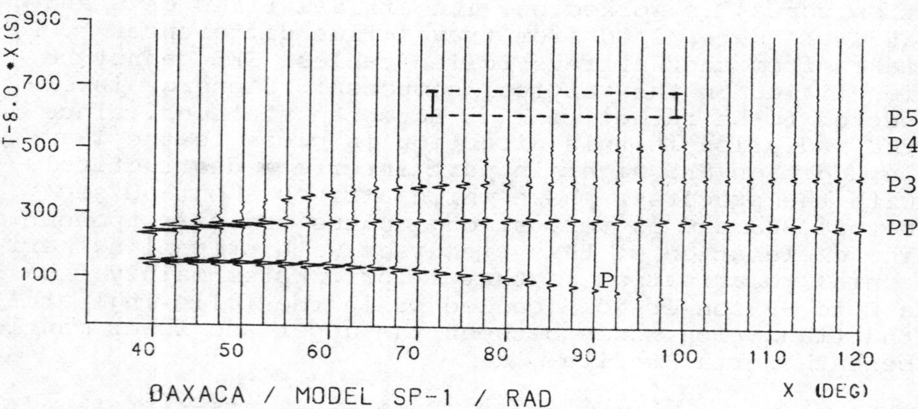

Fig. 2: Synthetic seismograms of P-waves calculated for the Oaxaca earthquake and model SP-1 (Faber and Müller, 1980). Surface multiples of P, sP, pP were included in the calculations. Arrival times of the conversion phases are marked (I-----I).

The arrival times of the conversion phases are marked in the record sections of Figure 2; no strong P multiples are expected in this time interval. The conversions from P to SV generated at the boundaries between the upper and lower mantle below the focus give rise to mantle and core SV waves, and the conversions from SV to P below the station which are of minor influence in the case of the radiation pattern of the Oaxaca earthquake originate from mantle and core SV-waves. These converted waves form through interference a complex wave field preceeding S and SKS in the distance range between 70° and 100° (Figure 3).

Figure 4 shows the observed records of GRF and the corresponding theoretical seismograms plotted

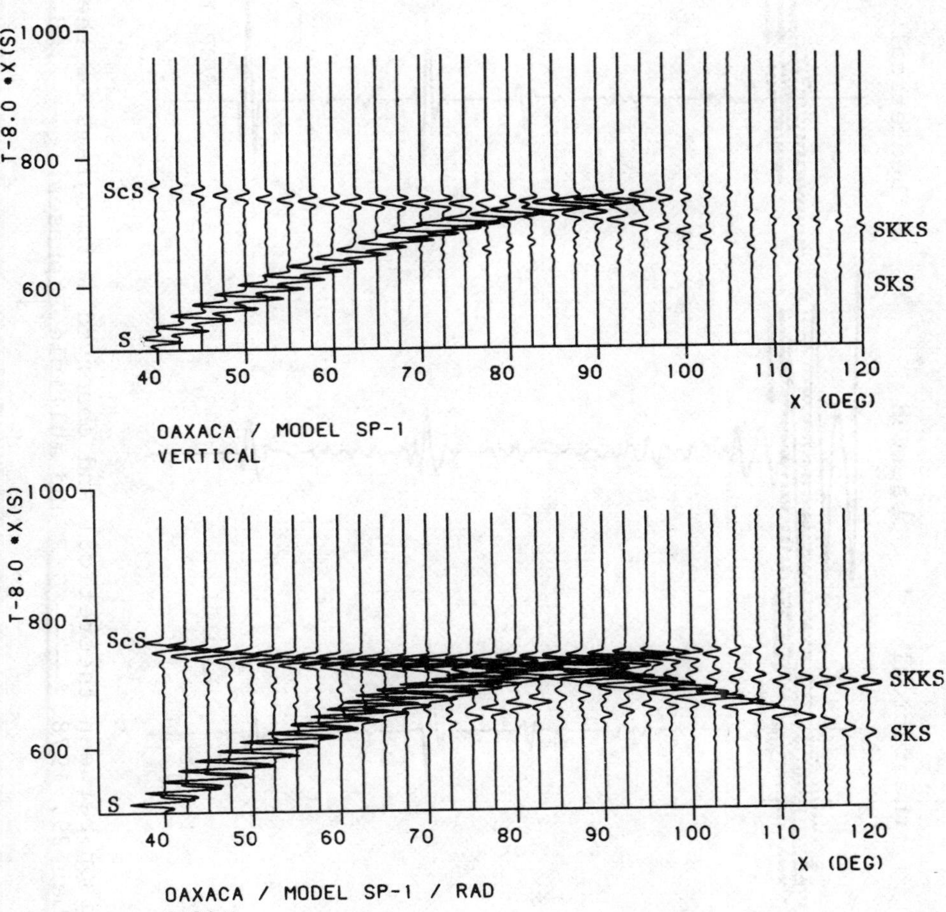

Fig. 3: Synthetic seismograms for model SP-1, including mantle and core S-waves, P to S conversions between the upper and lower mantle beneath the source and S to P conversions below the receivers. Surface reflections at the source are included. Anelastic attenuation was disregarded.

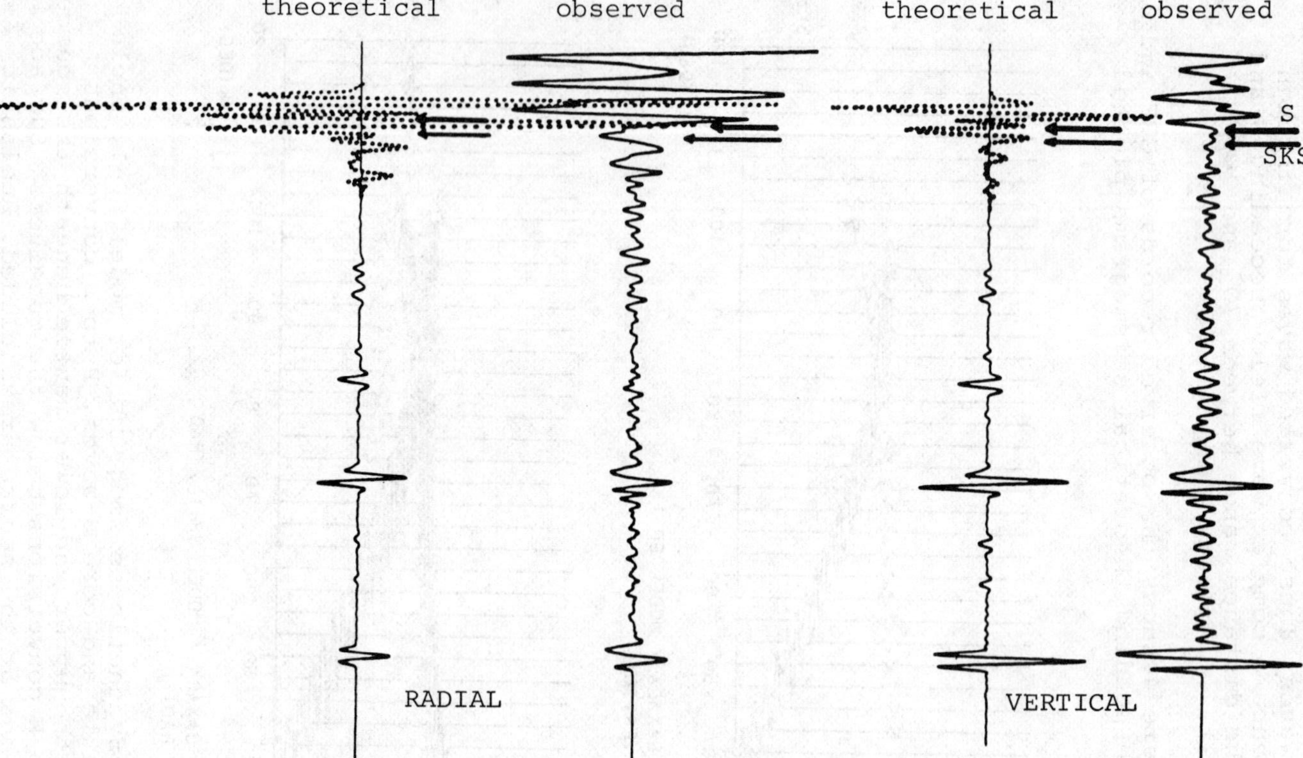

Fig. 4: Comparison between theoretical and observed seismograms of the Oaxaca event on November 29, 1978. ———— , P and multiple P phases; ········· , S and pS phases.

together at the same time-scale. The amplitudes are
scaled in a way to match amplitudes of the P impulse
on the vertical records to equal size. There is a
larger P-coda energy in the observations than predicted
from theoretical seismograms which interferes with the
pS phases. This observed irregular P-coda is in part
due to reverberations within the structures of the
uppermost mantle and lithosphere which were not included as reflecting layers in the calculations.

The observed amplitudes of the converted waves on
the radial component compared with the system SKS+S
are much larger than those in the theoretical seismograms. Before drawing from this any conclusions on
model changes concerning the velocity contrast at the
discontinuities between the upper and lower mantle
more data of other events has to be analysed to find
out if the existence of such energetic converted waves
from the same region is confirmed.

One has also to consider that the SV radiation in
the vicinity of the T-axis varies strongly and small
uncertainties in the direction of the fault planes may
result in rather large changes in S energy and hence
in large discrepancies between observed and theoretical amplitudes of S-waves and between observed and
theoretical amplitude ratios of converted waves and S-
waves. In the case of the Oaxaca earthquake it is
probably more reliable to take P as a reference phase.
As the converted phases travel most of their path as
S-waves and hence are more effected by anelastic damping
as P, the amplitudes have to be corrected for this attenuation using recently published Q models (Anderson
and Hart, 1978; Burdick, 1978). This means that the
amplitudes of the converted phases in the theoretical
seismograms have to be reduced by about 45% and the P-
amplitudes by about 15%, resulting in an amplitude ratio
of converted waves and P-waves of about 0.65. There
still remains a discrepancy of about 35% - the amplitudes of the converted phases being too low in the
theoretical seismograms - compared with the amplitude
ratio of nearly 1 in the observations.

DISCUSSION

Amplitude investigations of converted phases from
earthquakes with high SV- or P-radiation in the direction of the stations offer a powerful means to investigate the velocity structure of the transition

zones between the upper and lower mantle. In order to
explain conversion phases originating from those depths
beneath North America two well developed transition
zones had to be included in the models. The resolution
of the thicknesses of those transition zones is of the
order of half to one S-wavelength which leads to an
upper limit of 100 to 150 km according to the S-wave-
periods in long-period WWSSN records. From a systematic
investigation of converted phases observed at European
stations with special emphasis on the broad-band data
of the GRF array a higher resolution of the velocity
contrast and thickness of these transition zones should
result.

ACKNOWLEDGMENTS

This work was supported by a grant from the Deutsche
Forschungsgemeinschaft. The author wishes to thank
Gerhard Müller for discussions and reading the manuscript, Karen McNally and Eric Chael for providing the
information about the fault-plane solution of the
November 29, 1978 Oaxaca earthquake. Ingrid Hörnchen
typed the manuscript.

REFERENCES

Anderson, D.L., and Hart, R.S., 1978: Q of the Earth,
    J. Geophys. Res. 83, pp. 5869-5882.
Burdick, L.J., 1978: t* for S waves with a continental
    ray path, BSSA 68, pp. 1013-1030.
Burdick, L.J., and Helmberger, D.V., 1978: The upper
    mantle P velocity structure of the western United
    States, J. Geophys. Res. 83, pp. 1699-1712.
Faber, S., and Müller, G., 1980: Sp phases from the
    transition zone between the upper and lower mantle,
    BSSA 70, pp. 487-508.
Helmberger, D.V., and Engen, G.R., 1974: Upper mantle
    shear structure, J. Geophys. Res. 79, pp.4017-4028.
Kind, R., and Müller, G., 1975: Computations of SV
    waves in realistic earth models, J. Geophys. 41,
    pp. 149-172.

THE EXCITATION AND ATTENUATION OF SEISMIC CRUSTAL PHASES IN TURKEY

B. Maddison, A. Necioglu and N. Turkelli

Dept. of Geological Engineering, The Middle East
Technical University, Ankara, Turkey

ABSTRACT

The dominant crustal phases of Turkish earthquakes as recorded
on the short period vertical component of the ANTO station,
Ankara, are Pn, Pg and Lg. Pn usually has very low amplitude and
is prominent only due to its being the first arrival. Pg and Lg
appear as trains of waves which attenuate as dispersed surface
waves. For 1 second period Pg and Lg the coefficients of anelastic
attenuation are approximately 0.0032 km^{-1} and 0.0036 km^{-1}, respectively. A formula is given to compute body wave magnitudes
of Turkish earthquakes from the amplitudes of 1 sec period,
vertical component Lg waves.

INTRODUCTION

Closely following the approach used by Nuttli (1) in his study
of Iranian earthquakes at regional distances, the excitation
and attenuation of the Pg and Lg phases were estimated for
Turkey. Seismograms of the Ankara Seismic Research Observatory,
ANTO, were used for this study. All of the earthquakes selected
occurred within, or very close to, the boundaries of Turkey.
The ANTO station commenced operations in August 1978, and earthquakes from that time until September 1979 provided 55 usable
seismograms. Magnitudes, focal depths and epicentral coordinates
for 50% of the data were taken from Preliminary Determination of
Epicenters bulletins of the USGS. For the remaining 50% these
parameters were calculated using a local earthquake hypocenter
and magnitude determination computer program (2) using the first
arrival time readings from ANTO and from the Istanbul Kandilli

E. S. Husebye and S. Mykkeltveit (eds.), Identification of Seismic Sources – Earthquake or Underground Explosion, 505–511.

seismic network. For the earthquakes selected the body wave magnitude m_b varied from 3.0 to 5.7, with a median of 4.5, and the focal depth varied from 0 to 33 km, with a median of 10 km. For this study the seismograms of the short period vertical component, operating at a 1-sec period magnification of 50,000, were used exclusively.

The crustal phases chosen for analysis were Pg and Lg. Other phases such as Pn and Sn were not used because of the extremely weak amplitudes of the Pn phase for distances up to about 1000 km and because of the difficulty of identifying the Sn phase. The Pg and Lg phases appear as groups of waves whose onsets have group velocities of 5.89 ∓ 0.3 km/s and 3.55 ∓ 0.3 km/s, respectively, and whose dominant period is 1 sec. For both the Pg and Lg phases the measured amplitude A was the zero-to-peak of the sustained maximum motion of the 1 sec period wave. Sustained maximum motion is used in the sense originally defined by Nuttli (3).

All amplitude data were equalized to those of an m_b 4.5 earthquake, assuming $\Delta(\log A) = \Delta m_b$, where Δm_b is equal to 4.5 minus m_b. Because the m_b values of most of the earthquakes studied did not depart by more than 0.75 units from 4.5 this equalization process is not expected to introduce large errors.

The theoretical amplitude versus distance curves for surface waves were calculated using the equation given by Ewing et al (4)

$$A \sim (\sin \Delta)^{-\frac{1}{2}} \Delta^{-1/3} \mathrm{Exp}(-\gamma\Delta) \qquad\qquad (1)$$

where Δ is the epicentral distance in degrees and γ is the coefficient of anelastic attenuation. These curves can also be applied to the Pg phase (3).

Equation (1) can be rewritten as:

$$A^* = 56.306 \; A \; (\sin\Delta)^{\frac{1}{2}} \; \Delta^{1/3} \sim \exp(-\gamma\Delta) \qquad\qquad (2)$$

where 56.306 is a normalization factor to make the product $56.306 \; (\sin\Delta)^{\frac{1}{2}} \; \Delta^{1/3}$ equal to unity at 10 km (0.09 deg). The plot of these A^* data on semi-log paper will be spread out in such a way that the value of γ can be adequately estimated.

THE Lg WAVE

The A^* values for Lg and theoretical curves according to eq. (2) for various values of γ are plotted in Figure 1. The curves

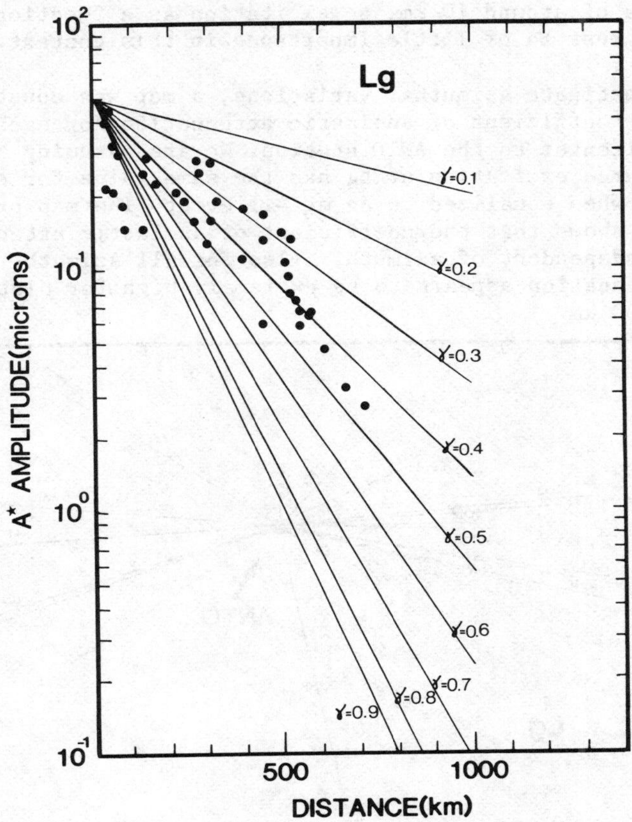

Figure 1. Plot of A* values (see text) vs epicentral distance
for the vertical component 1 second period Lg wave. Amplitudes
have been equalized to an m_b 4.5 earthquake. The theoretical
attenuation curves are fitted by eye.

are fitted by eye. With the exception of the low amplitude
values up to about 100 km epicentral distance, it can be seen
that most of the data points lie between the curves $\gamma = 0.5$
deg^{-1} and $\gamma = 0.2$ deg^{-1}, the mean being approximately $\gamma = 0.4$
deg^{-1}. The 10 km intercept of the theoretical curves corresponds
to an A* value of 50 microns.

As suggested by Nuttli (3), the obvious scatter in the data
in Figure 1 can be explained as follows: variation of excitation
with focal depth, azimuthal variation in the source radiation,
incorrect estimates of m_b, and variation in the absorption coef-

ficient over the area of Turkey. Most earthquakes studied had
focal depths of around 10 km, so excitation as a function of
focal depth must be of little importance in this context.

To investigate azimuthal variations, a map was constructed
showing the coefficient of anelastic attenuation for each path
from the epicenter to the ANTO station. We are assuming here
that the source excitation of Lg has the same value for each
earthquake, when equalized to an m_b 4.5 event. The map presented
in Figure 2 shows that the coefficient of anelastic attenuation
is nearly independent of azimuth. Also for all azimuths the anelastic attenuation appears to be extremely high for distances
less than 100 km.

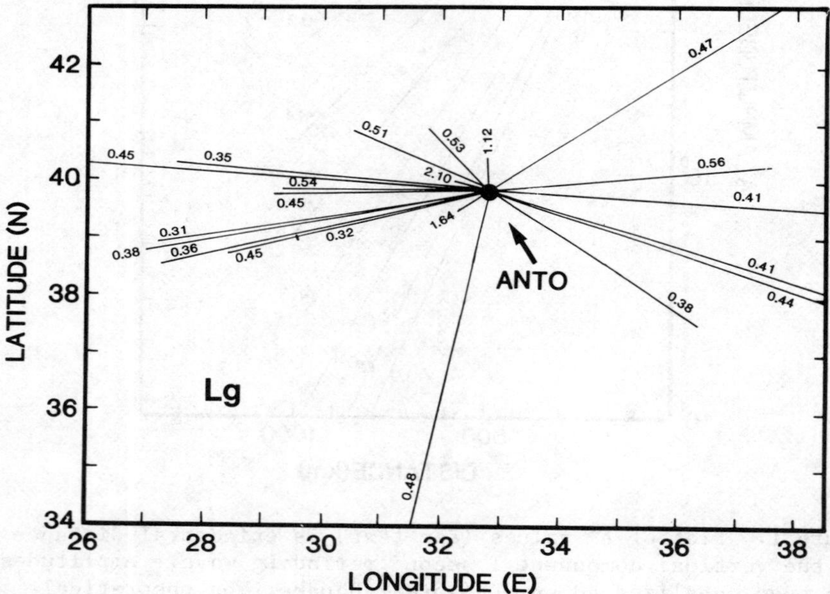

Figure 2.   Map showing individual γ values for Lg for each
earthquake path. Only a sample of all earthquakes studied have
been included in the plot.

The data in Figure 1 can be approximated by a series of
straight line segments and these in turn can be used in the
derivation of formulae relating the amplitude of the vertical
component, 1 sec period Lg waves to m_b. The formula obtained
for epicentral distances in the range 200 to 1000 km is

$$m_b = \log A + 3.10 \log\Delta - 3.09$$

where A is the zero-to-peak sustained maximum amplitude of 1 sec period Lg waves in microns, and Δ is the epicentral distance expressed in kilometers. In the range 200 to 400 km this formula is somewhat uncertain but may still be used for a rough m_b-estimation. Data scatter is too great at distances less than 200 km to obtain a reliable magnitude formula.

THE Pg WAVE

The A* plot for Pg is given in Figure 3. It is noticeable that the data scatter is even greater than for Lg. However, most of

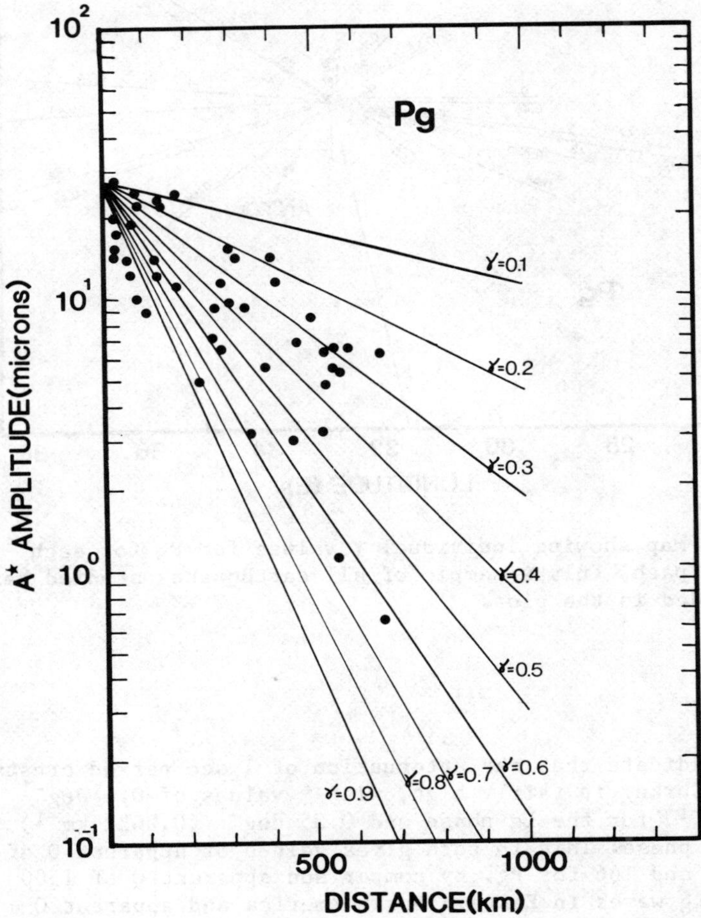

Figure 3. Plot of A* values (see text) vs epicentral distance for the vertical component 1 second period Pg wave. Amplitudes have been equalized to an m_b 4.5 earthquake. The theoretical attenuation curves are fitted by eye.

the data points can be seen to lie between the γ values 0.6 deg^{-1} and 0.2 deg^{-1} with a mean of approximately 0.35 deg^{-1} and a 10 km intercept of 25 microns. Attempts at fitting magnitude formulae to the data were not made due to the large scatter of the data. A map showing the individual γ values for each earthquake path is presented in Figure 4. The results of this map are similar to those for the Lg wave.

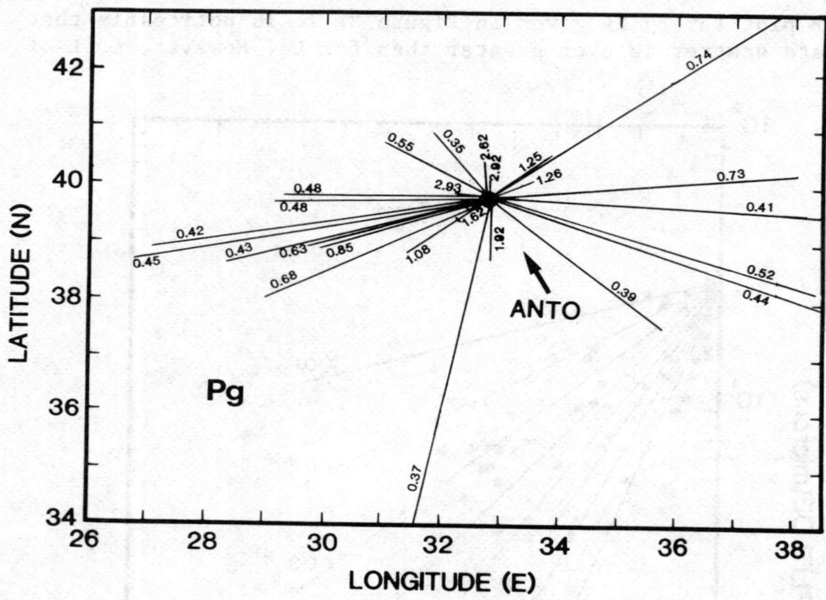

Figure 4. Map showing individual γ values for Pg for each earthquake path. Only a sample of all earthquakes studied has been included in the plot.

CONCLUSIONS

The data indicate that the attenuation of 1 sec period crustal phases in Turkey is fairly high, with γ values of 0.4 deg^{-1} (0.0036 km^{-1}) for the Lg phase and 0.35 deg^{-1} (0.0032 km^{-1}) for the Pg phase. This in turn gives values of apparent Q of 245 for Lg and 166 for Pg. By comparison apparent Q of 1500 for 1 sec Lg waves in Eastern North America and apparent Q of 200 for Lg and 125 for Pg in Iran have been reported by Nuttli (1,3).

The low amplitudes observed for both Pg and Lg at near distances (below 100 km) are not easily explained. However the two maps presented in Figures 2 and 4 indicate that the low amplitudes are not azimuthally dependent and suggest that they could be due to an apparently high γ value around the area of the ANTO station. Otherwise, there is no great variation in the value of the coefficient of anelastic attenuation over Turkey.

ACKNOWLEDGEMENTS

We should like to express our thanks to Professor O.W. Nuttli of Saint Louis University, who suggested the idea for this study.

REFERENCES

1.   Nuttli, O.W.:1980, Bull. Seism. Soc. Am., 70, 469–485.
2.   Lee, W.H.K., and Lahr, J.C.: 1972, Hypo 71; A computer program for determining hypocenters, magnitude and first motion pattern of local earthquakes, U.S.G.S. Open File Report.
3.   Nuttli, O.W.: 1973, J. Geophys. Res., 78, 876–885.
4.   Ewing, M., Jardetzky, W., and Press, F.: 1957, Elastic Waves in Layered Media, McGraw-Hill, New York.

COMPARISON OF WAVEFORM INVERSION SCHEMES FOR HORIZONTALLY
STRATIFIED MEDIA

K.-A. Berteussen[1] and B. Ursin[2]

1) Dept. of Geophysics, University of Oslo, Norway
2) SINTEF, Division of Petroleum Technology, Trondheim,
   Norway

Five different methods for inversion of seismic reflection
data are considered for a model with vertically travelling waves
in a medium with homogeneous, isotropic and lossless horizontal
layers. The first three of these methods involve application of
a detection scheme, that is, having removed the effect of, say
the first k layers, either on the surface seismogram or by downward
continuation, the reflection coefficient and travel time for layer
k+1 is found by detecting the primary reflection from this layer
and estimating its amplitude and time of arrival. These three
methods are:

A)   Downward continuation in the time domain (1). The method
     implies that the wavefield is split into up- and down-
     going waves which are both estimated.

B)   Layer removal in the frequency domain. This implies that the
     effect of a known layer may be removed by applying a non-
     linear formula in the frequency domain, whereupon the next
     layer may be estimated.

C)   Surface calculations in the time domain. Given that the k
     first layers are known, the effect of these are removed
     from the registered seismogram by calculating a synthetic
     seismogram. From the residual seismogram the parameters of
     layer k+1 are then estimated.

The two last methods are what we may call direct inversion
schemes, that is, the parameters of the medium are estimated
directly from the reflection seismogram. These methods are:

513

*E. S. Husebye and S. Mykkeltveit (eds.), Identification of Seismic Sources - Earthquake or Underground
Explosion, 513–514.*

D)   The classical method (2, 3).

E)   The iterative method in the frequency domain (4).

The classical method is applied to a layered medium with
layers of equal one-way travel time. The iterative method is
adapted to a model with continuously varying parameters. In practice a discretization has to be made, and a model with layers of
equal travel time is used.

The performance of the five inversion schemes has been compared by applying them to a synthetic seismogram for a seven-
layered model. The seismogram was convolved with a standard source
function to which Gaussian noise was added. For very low noise
levels the first three methods all recover the acoustic impedance
function. The two last methods have degraded performance because
it is not possible to deconvolve the source function perfectly,
i.e., one cannot make a perfect spike seismogram.

For higher noise levels, method A works better than C, which
again is better than B. Methods D and E have practically the same
performance and they are also in this case clearly inferior to
the first three methods. For the first three methods a standard
pulse detection procedure was used for a signal model with an
unknown number of pulses of known shape but unknown arrival time
and amplitude. The three methods have different performances
because i) they have different effects on the noise, and ii)
the effects of uncertainties (previous mistakes) in estimates of
reflection coefficients and travel times propagate differently.

There is still much work to be done before inversion schemes
of the types briefly described will find practical applications
(for further details, see (5)). Although it would be feasible to
retrieve only reasonably reliable impedance contrasts for the
strongest reflectors, we remark that this kind of information
is of most interest to geologists.

REFERENCES

1.   Habibi-Ashrafi, F. and Mendel, J.M.: 1980, Estimation of
      parameters in lossless layered media systems. Manuscript
      in press.
2.   Kunetz, G.: 1963, Geophysical Prospecting 11, 409-422.
3.   Claerbout, J.F.: 1968, Geophysics 33, 264-269
4.   Gjevik, B., Nilsen, A. and Høyen, J.: 1976, Geophysical
      Prospecting 24, 492-505.
5.   Ursin, B. and Berteussen, K.-A.: 1980, Inverse methods
      for wave propagation in layered media. Manuscript in
      preparation.

ATTENUATION AND SCATTERING OF SHORT-PERIOD SEISMIC WAVES IN THE
LITHOSPHERE

Keiiti Aki

Department of Earth and Planetary Sciences, Massachusetts
Institute of Technology, Cambridge, Massachusetts 02139

Abstract

The irregularities in seismic wave fronts observed at
arrays such as LASA can be ascribed to heterogeneities in the
lithosphere beneath the array. These heterogeneities are
important in detection seismology because of their effect on
seismic waveforms to be used for source identification.

The present author has studied the following effects of
heterogeneities in the lithosphere on short period seismic waves:
(1) the fluctuation of the amplitude and phase of teleseismic P
waves observed at a seismic array, (2) the attenuation of S
waves through the lithosphere, and (3) the backscattering effect
using coda waves of local earthquakes.

We found that the backscattered energy required to explain
coda amplitude agrees with the loss of energy from S waves,
suggesting that attenuation of S waves in the lithosphere may be
primarily due to scattering. The frequency dependence of Q
and its systematic change with the degree of tectonic activity
suggests a trend that the scale length of inhomogeneity is
shorter and the magnitude of velocity fluctuation is greater for
younger and more active regions.

Introduction

The goal of discrimination seismology in the next decade is
to develop a reliable discriminant for small events to which the
M_S-m_b discriminant become difficult to apply. Since small

*E. S. Husebye and S. Mykkeltveit (eds.), Identification of Seismic Sources - Earthquake or Underground
Explosion, 515–541.*

explosions do not generate long-period surface waves, the new discriminant must rely on the wave-forms or spectral contents of short-period seismic waves. In order to develop such a discriminant, we must first understand the complex effect of earth's structure on the attenuation and scattering of these short-period waves.

In the present paper I shall discuss the attenuation and scattering of short-period seismic waves in the lithosphere around the world. The lithosphere appears to be the most heterogeneous region of the earth, judging from my general experience with the regional array data that most of the fluctuations and irregularities of arrival times (and sometimes amplitudes) can be explained by heterogeneities in the lithosphere. The asthenosphere may be the most absorptive region (except near-surface soil), but probably not very heterogeneous because the ductility there may not sustain lateral heterogeneity. In this paper, we are concerned only with wave propagation in the lithosphere and not in the asthenosphere.

The frequency range to be considered is from about 0.5 Hz to about 25 Hz. We shall cover various regions of the world, including stable areas such as Baltic shield and active areas such as California, Hawaii and Japan.

The following three methods of investigation will be used to study the wave propagation in the lithosphere.

(1) Interpretation of the fluctuation of amplitude and phase of teleseismic P waves observed at a seismic array by characterizing the lithosphere as a random medium.

(2) Study of the attenuation of short-period S waves using a single-station method in which coda wave spectra are used to eliminate the source effect.

(3) Study of the backscattering effect of the heterogeneous lithosphere using coda waves of local earthquakes.

Synthesizing the results from the above three approaches, we are led to a rather surprising conclusion that the attenuation of short-period seismic waves in the lithosphere may be primarily due to scattering. The corresponding Q^{-1} for a region and its frequency dependence change systematically with the degree of tectonic activity of the region.

Scattering of P waves

Let us first review the results obtained by Aki (1973) from observed fluctuation of amplitude and arrival time of tele-

seismic P waves recorded at a Large Aperture Seismic Array (LASA)
located in the eastern Montana. Fig. 1 shows typical variation
of P waveforms across the array. The records demonstrate that
a large variation of signal amplitude exists and is not simply
related to the local effect of seismograph site such as the
sediment thickness but depends in a complicated way on the
direction of wave approach (Mack, 1969). The pattern of
variation, however, repeats exactly for the same direction of
wave approach. Moreover, waves having greatly different paths
such as P, PKP, PKKP show the same pattern of variation so long
as they share nearly the same direction of approach at LASA.
On this basis, Engdahl and Felix (1971) concluded that the cause
of the fluctuation of arrival time must lie almost entirely in
the crust and upper mantle beneath the array.

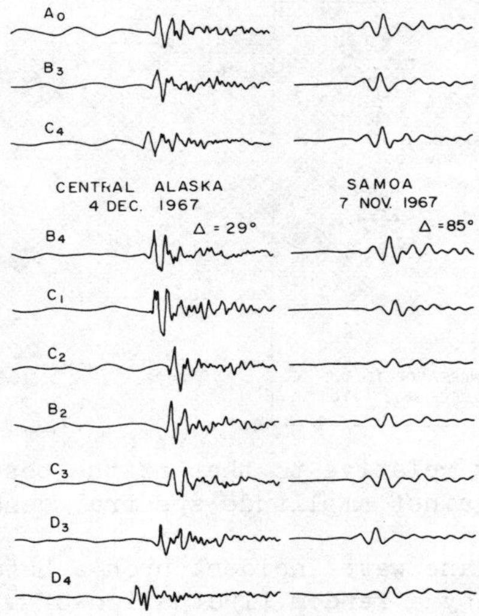

Fig. 1. P waves recorded at sub-array centers of the Montana
LASA for two events.

The Fourier amplitude and phase delay are determined for
P waves recorded at 66 seismographs spread over a 100 x 100 km
square in the following manner (Aki, 1973). First, the arrival
time of first motion of each event is read for 21 subarray
centers using a computer-plotted record. The azimuth and phase
velocity of the best-fitting plane waves are determined from
them by the least squares method. Then, the P-waves of 66
seismographs are Fourier-transformed using the window that
propagates across LASA with the velocity and direction of the
best-fitting plane waves. Thus, the computed phase delay was

relative to the time of window opening at each station. The
window length was determined by visual inspection of computer-
plotted records, usually about 5 sec long. 17 events were
Fourier-analyzed in this manner.

Fig. 2 shows how the amplitude and phase of P-waves
fluctuate from station to station within LASA for some typical
examples. The fluctuations are very large; a factor of 5 to 10
for amplitude and $\pi/2$ to π for phase delay. Fig. 2 shows a
clear positive correlation between the phase and amplitude
fluctuation. The variances of phase and amplitude fluctuation
as well as the correlation coefficient between them are known
in closed forms for a particular case of random medium by
Chernov (1960).

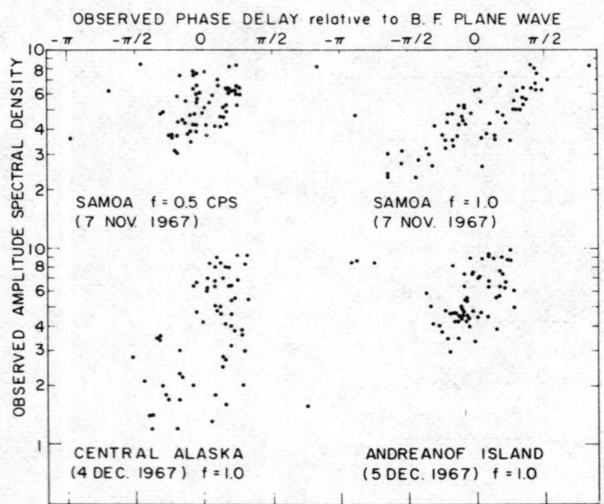

Fig. 2. Observed phase delay relative to that of the best-
fitting plane wave plotted against amplitude spectral density.

Consider a sinusoidal plane wave incident upon a heterogeneous medium characterized by a random fluctuation of wave
velocity α. The fluctuation of refractive index $\bar{\alpha}/\alpha$, where $\bar{\alpha}$
is the average velocity, is described by its rms $\langle\mu^2\rangle^{1/2}$ and
its spatial autocorrelation $N(r)$. For the sake of simplicity,
we shall assume that $N(r) = e^{-r^2/a^2}$, where the correlation
distance a represents the scale length of inhomogeneity.

When the incident plane wave travels through the random
medium over a distance L, the logarithmic amplitude and phase
fluctuations of the following variance will be produced:

$$\sigma_A^2 = [(\pi)^{1/2}/2]\langle\mu^2\rangle k^2(aL)(1-(1/D)\tan^{-1}D)$$

$$\sigma_\phi^2 = [(\pi)^{1/2}/2]\langle\mu^2\rangle k^2(aL)(1+(1/D)\tan^{-1}D)$$

(1)

where k is the wave number, and D is called "wave parameter"
defined by

$$D = 4L/ka^2 \tag{2}$$

which is square of the ratio of the radius of first Fresnel
zone to the correlation distance.

The wave propagation is in the region of ray-theory or
diffraction-theory, depending whether D is smaller or greater
than 1. The correlation coefficient between the logarithmic
amplitude and phase fluctuation is a function of D alone, and
is given by

$$\rho = \frac{1}{2} \frac{\log(1+D^2)}{[D^2-(\tan^{-1}D)^2]^{1/2}} \tag{3}$$

The above formulas were obtained by two main approximations,
namely, the first Born and Fresnel approximations. The
applicability of these approximations to the data from LASA was
discussed in detail and justified by Aki (1973), together with
the applicability of scaler wave solution to the vector wave
problem using the result of Knopoff and Hudson (1967).

In addition to the constraints on model parameters provided
by Equations (1), (2), and (3), the spatial autocorrelation
functions were computed for amplitude and phase fluctuation as
shown in Fig. 3, which constrained the correlation distance a
to be about 10 km. Other parameters were, then, obtained from
observed $_A$, and . It was found that the rms fluctuation of
velocity is about 4% and the thickness of heterogeneous layer is
60 km.

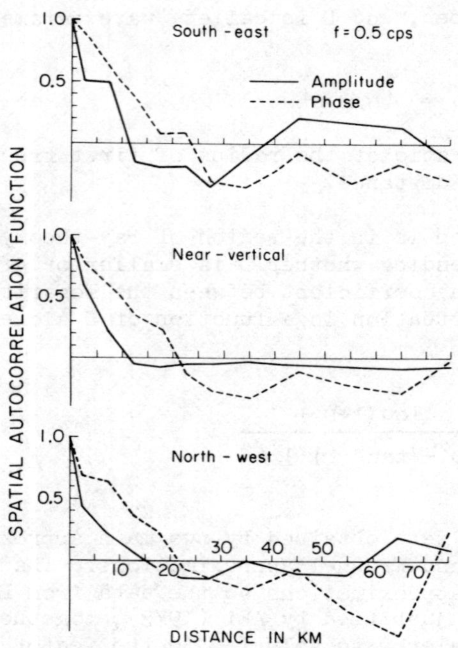

Fig. 3. Spatial autocorrelation functions for logarithmic
amplitude and phase fluctuations. Earthquakes are grouped into
three directions of approach to LASA, and autocorrelation
functions are computed separately for each direction.

A similar characterization of the crust and upper mantle
under LASA by a random media was made independently by Capon
(1974) who gave the estimates of a, L and $<\mu>^{1/2}$ to be 12 km,
120 km and 2% respectively.

More recently, Aki et al. (1976) applied the 3-D inversion
method of Aki, Christoffersson and Husebye (1977) to 3026
readings of P-time at 17 inner subarray centers of LASA for
178 teleseismic events tabulated by Chiburis and Ahner (1973).
In addition to the deterministic 3-D seismic image of the
lithosphere under LASA, they found that the lower limit of the
rms of the true velocity fluctuation is 3.2%, which is intermediate between the estimate by Aki (1973) and Capon (1974).
The true depth range of the heterogeneous region may be also
intermediate between 60 and 120 km. We may summarize the above
results by stating that the rms velocity fluctuation of about
3% with correlation distance about 10 km exists to the depth
of about 100 km within the lithosphere under LASA.

The lower limit of the rms of the true velocity fluctuation
has been determined also for two other arrays from the tele-

seismic P time data. Husebye et al. (1976) gave the estimate of
3.1% for the USGS central California array, and Aki et al. (1977)
gave 3.4% for NORSAR.

Berteussen et al. (1975), Dahle et al. (1975), and
Berteussen et al. (1977) have used an estimation method due to
Christoffersson (1975) based on the Chernov theory. They
preferred deeper extent of heterogeneous region and obtained
somewhat low values for velocity fluctuation. For example,
putting the thickness of heterogeneous region to be 250 km, the
rms velocity fluctuation was obtained as 1.0% for LASA, 3.0%
for NORSAR and 0.3% for the Gauribidanur array in southern India.
Their relatively low estimates of velocity fluctuation are
partly due to their method of estimation in which only the part
of amplitude and phase variance that behaves according to the
Chernov theory is attributed to the scattering effect.

Coda waves of local earthquakes

Coda waves of local earthquakes recorded at a station and
passed through a band-path filter with center frequency f have a
remarkable property that the envelop $A(f|t)$ of filter output
plotted as a function of lapse time t measured from the
earthquake-origin time can be expressed as

$$A(f|t) = S(f)C(f|t) \qquad\qquad (4)$$

for t greater than about twice the shear wave travel time
(Rautian and Khalturin, 1978), where $S(f)$ depends on source
alone, and $C(f|t)$ is a function of f and t alone and is
independent of locations of the source and receiver.

The above property of coda waves expressed by Equation (4)
has been confirmed for local earthquakes in many parts of the
world. It was first recognized by Aki (1969) for aftershocks of
the Parkfield, California earthquake of 1966. Later, many
investigators demonstrated the validity of Equation (4) for
earthquakes in various areas such as California, Hawaii, Japan,
Mexico, Central Asia, and other areas in the USSR, Norway, etc.
(Aki and Chouet (1975), Chouet (1976), Rautian and Khalturin,
(1978), Tsujiura (1978), and others). An example of such
demonstration due to Tsujiura (1978) is given in Fig. 4, which
shows the envelope of coda amplitude measured on the output of
a band-pass filter with center frequency at 6 Hz for local
earthquakes observed at Tsukuba, Japan. The curves for
individual earthquakes do not intersect each other and suggest
the existence of a common decay curve for all earthquakes as
implied in Equation (4). The dashed curve in Fig. 4 is such a
common curve which is determined by a least-squares fitting of
the form shown in the figure. The factor C_i represents the
strength of source at 6 Hz for the ith earthquake.

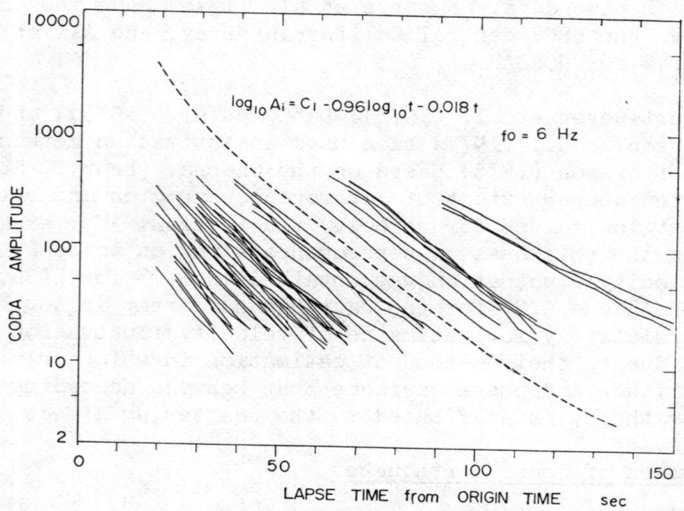

Fig. 4. The envelope of coda amplitude measured on the output
of a band-pass filter with center frequency at 6 Hz for local
earthquakes observed at Tsukuba, Japan. The curves for
individual earthquakes do not intersect each other and suggest
the existence of a common decay curve for all earthquakes as
implied in Equation (4). The dashed curve is such a common
curve which is determined by a least-squares fitting of the
form shown in the figure. The factor C_i represents the
strength of source at 6 Hz for the ith earthquake. This figure
is reproduced from Tsujiura (1978).

The above remarkable property of coda waves can be explained
if they are backscattered waves from heterogeneities distributed
more or less uniformly in the Earth. Consider for simplicity,
an unbounded medium containing a seismic source and a receiver.
The primary waves spreading from the source and passing the
receiver will meet heterogeneities and send the scattered waves
back to the receiver. Let t be the time measured from the time
when the waves left the source. Then, the heterogeneities that
send the backscattered waves to the receiver within the range
(t, t+Δt) lie within a spheroidal shell having the receiver and
source as their foci.

Consider a second source at a location different from the
first. Then, for a given (t, t+Δt), the spheroidal shell for
the first source and that for the second have the same size
partially overlapping with each other. As t increases, the
spheroid becomes closer to a sphere, and the overlapped part
will increase. Even for the non-overlapping part, the
properties of scatterers within the two shells may be quite
similar statistically, and the sum (average) of contributions

from numerous scatterers will produce similar coda waves.
Since the contributions from different scatterers are
incoherent, the similarity is not in deterministic wave form
(wriggle to wriggle correspondence), but in statistical
property (power spectrum and envelope shape).

Evidences supporting the backscattering hypothesis for
coda waves were summarized in Aki and Chouet (1975), Rautian
and Khalturin (1978) and Tsujiura (1978). In particular, the
station site effects on coda and S waves studied by Tsujiura
(1978) strongly supports the idea that coda waves are
primarily back-scattered S waves. Fig. 5 shows the locations
of seismographs at Mt. Dodaira used in his study. Some of
them are located on crystalline rocks marked as S, and others
are on sedimentary rocks marked as K_S. The spacing between
the seismographs are only several hundred meters. The signals
from local earthquakes are passed through band-pass filters
with center frequencies at 0.75, 1.5, 3, 6, 12 and 24 Hz.
Fig. 6 shows the amplitude of filter output at three sites
(H1, H3 and H7) relative to the station H5 located on the
crystalline rock for both S and coda waves. For S waves,
we find a surprisingly large variation of the amplitude ratio
for different earthquakes. The amplitude ratio can vary up
to a factor of 5 even though the stations are separated only
by a few hundred meters.

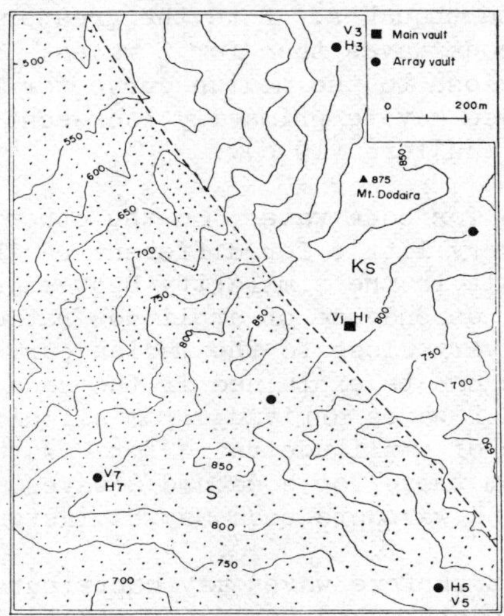

Fig. 5. Seismograph locations at Mt. Dodaira used by Tsujiura
(1978). Dotted area corresponds to crystalline rocks, and K_S
indicates Kasayama group including slate, chert and schalstein
rocks. V and H indicates vertical and horizontal seismographs
respectively. Reproduced from Aki (1980).

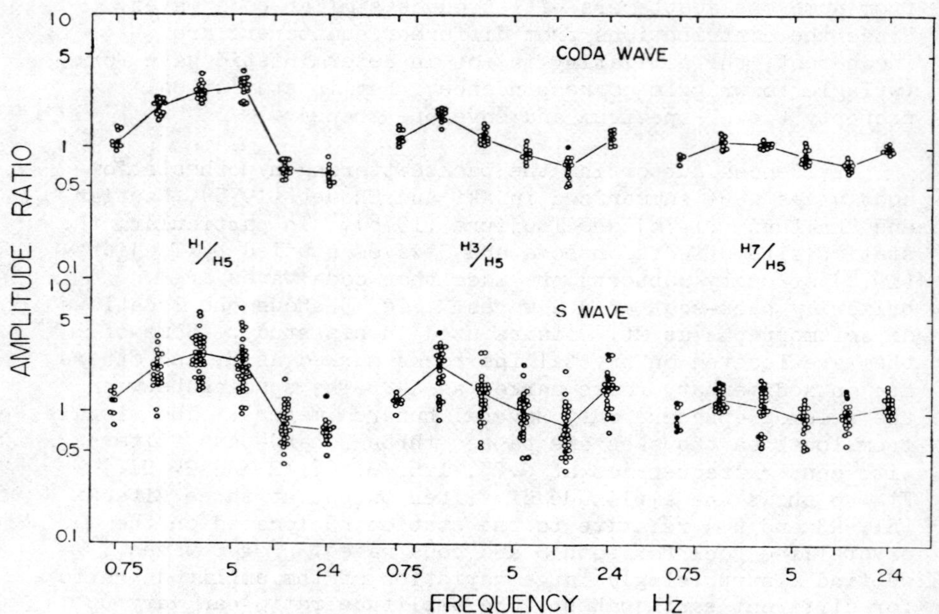

Fig. 6. The amplitude of band-pass filtered signals, at three
sites relative to the crystalline rock site H5 (Fig. 5) for coda
and S waves from local earthquakes. Note the greater stability
of amplitude ratio for coda waves than for S waves. The curve
for coda waves is very close to the median curve for S waves,
suggesting that coda waves may be primarily back-scattered S
waves. Reproduced from Tsujiura (1978).

The amplitude ratio for coda waves, on the other hand, is
very stable and varies very little for different earthquakes.
The most remarkable result is the similarity between coda and
S waves in the frequency dependence of amplitude ratio. The
curve for coda waves is very close to the median curve for S
waves. This observation can be explained if the coda waves are
primarily back-scattered S waves arriving from all directions.
The frequency dependence of amplitude ratio for P waves are
distinctly different from those for coda and S waves, suggesting
that the contribution of P waves to coda may be insignificant.

The possibility that surface waves may contribute significantly to coda of local earthquakes was discussed by Aki and
Chouet (1975) in order to explain the strong frequency
dependence of apparent Q estimated from the coda wave data.
This possibility was conclusively excluded by H. Sato (personal
communication) at least for central Japan, by showing that the
nature of coda waves observed by a bore-hole seismograph placed
at a depth of about 3 km was nearly identical to that observed

at the surface. For frequencies higher than 1 Hz, surface waves
would not penetrate to the depth of 3 km.

With these supporting observations, it may be justified to
construct a more detailed quantitative model of coda waves
assuming that they are back-scattered S waves.

Theory of coda waves as back-scattered S waves

Since the observed coda wave amplitude is independent of
distance from the surface to receiver, we can simplify our
problem by placing the source and receiver at the same point.
Let us write the Fourier transform of displacement due to
primary S waves as $\phi_0(\omega|r)$, where r is the distance from the
source point. · Assume that the scatterers responsible for coda
waves are distributed randomly in space, and let $N(r)$ be the
number of scatterers within r radius of the station. The number
of scatterers in the zone bounded by $(r, r+\Delta r)$ will be
$(dN/dr)\Delta r$. We shall assume that both primary and scattered
waves are S waves with velocity β. The back-scattering waves
from scatterers in $(r, r+\Delta r)$ will arrive at the receiver in the
time interval $(t, t+\Delta t)$, where $t = 2r/\beta$ and $\Delta t = 2\Delta r/\beta$. For a
Δr long enough so that the corresponding Δt is greater than the
duration of an individual backscattering wavelet, and for a
random distribution of scatterers, the total energy of coda
contained in the interval $(t, t+\Delta t)$ will be approximately equal
to the sum of energy carried by the individual wavelets
arriving in the same interval. On the other hand, for an
arbitrary stationary time series $f(t)$, the autocorrelation
function estimated from a finite sample of $f(t)$ for a time
interval T may be written as

$$\Phi(\tau) = <f(t)f(t+\tau)>$$

$$\sim \frac{1}{T} \int_0^T f(t)f(t+\tau)dt \tag{5}$$

Putting the power spectra of $f(t)$ as $P(\omega)$, we have

$$P(\omega) = \int_{-\infty}^{\infty} \Phi(\tau)e^{i\omega\tau}d\tau$$

$$\sim \frac{1}{T} |F(\omega)|^2 \tag{6}$$

where $F(\omega)$ is the Fourier transform of the finite sample of $f(t)$.
Applying the above relation to our case, the power spectral

density $P(\omega|t)$ of the coda waves can be written as

$$P(\omega|t) = \frac{1}{\Delta t} \sum_{r < r_n < r+\Delta r} |\phi_n(\omega)|^2 \tag{7}$$

where $\phi_n(\omega)$ is the Fourier transform of backscattering wavelet from the nth scatterer located at r_n. Assuming that a scatterer at distance r generates a backscattering wavelet with the Fourier transform $\phi(\omega|r)$, we obtain

$$P(\omega|t) = \frac{\Delta r}{\Delta t} \frac{dN}{dr} |\phi(\omega|r)|^2 \tag{8}$$

Let us now separate the effect of seismic source and that of scattering in $\phi(\omega|r)$. In general, consider a plane wave $A_0\,e^{i\omega(t-x/\beta)}$ propagating in the x-direction and incident upon a heterogeneous region of volume V. At a long distance R from V, we measure the scattered waves which may be expressed as $A_1\,e^{i\omega(t-R/\beta)}$. We introduce the directional scattering coefficient $g(\theta)$ which is 4π times the fractional loss of energy by scattering per unit travel distance of primary waves and per unit solid angle at the radiation direction θ measured from the x-direction. Then, we can write

$$|A_1|^2 = |A_0|^2 \frac{Vg(\theta)}{4\pi R^2} \tag{9}$$

In our case, putting the Fourier transform of primary wave as $\phi_0(\omega|r)$, we consider that they are locally plane waves for large r. Then, we can write $A_0 = \phi_0(\omega|r)$. The volume of the heterogeneous region is $4\pi r^2 \Delta r$ and the number of scatterers is $\frac{dN}{dr}\Delta r$. Assuming random contribution from each scatterer, we have $|A_1|^2 = (dN/dr)\Delta r|\phi_0(\omega|r)|^2$. Putting these quantities into equation (9), we obtain

$$\frac{dN}{dr} |\phi(\omega|r)|^2 = g(\pi)|\phi_0(\omega|r)|^2 \tag{10}$$

where we put $\theta=\pi$, assuming a literal backscattering. Putting (10) into (8), we obtain the formula for coda power spectrum in terms of backscattering coefficient $g(\pi)$ and primary wave spectrum $\phi_0(\omega|r)$.

$$P(\omega|t) = \frac{\beta}{2} g(\pi)|\phi_0(\omega|\beta t/2)|^2 \tag{11}$$

This equation was first derived by Aki and Chouet (1975).

Equation (11) was derived under the assumption of single scattering or the first Born approximation. Multiple scatterings are neglected. Under this assumption, the loss of energy from primary waves is of the second order and should be neglected. In fact, it was neglected in deriving equation (11). On the other hand, the primary waves before entering the heterogeneous volume under consideration have lost some energy by absorption (change to heat) and by scattering. We shall write down $|\phi_0(\omega|r)|$ explicitly as $|\phi_0(\omega|r)| = |S(\omega)|r^{-1}e^{-\omega r/(2Q_c\beta)}$. Including also the attenuation for propagation from the scatterers to the receiver, we obtain the equation (11)

$$P(\omega|t) = \frac{\beta}{2} g(\pi)| S(\omega)|^2 (\frac{\beta t}{2})^{-2} e^{-\omega t/Q_c} \tag{12}$$

We denote the effective quality factor as Q_c, which can be obtained by fitting equation (12) to the observed envelope of coda waves. If our S to S single-scattering theory is correct, Q_c should be equal to Q_β determined directly from S waves.

In an earlier paper (Aki and Chouet, 1975), we interpreted Q_c as only due to absorption. This interpretation is correct for Q used in the diffusion theory (Dainty et al., 1974, Aki and Chouet, 1975), but not for Q_c in equation (12) in the context of the single scattering theory, in which Q_c should be considered as an effective Q including both absorption and scattering effects.

Q_c as a function of frequency has been determined for various regions of the earth by Aki and Chouet (1975), Chouet (1976), Rautian and Khalturin (1978), Tsujiura (1978), Herrmann (1980), and others. It was found that Q_c varies considerably from region to region at about 1 Hz, showing low values around 100 for tectonically active regions and values as high as 1000 for stable regions. But they show values higher than 1000 at about 20 Hz for most areas.

As mentioned earlier, we now discard our earlier interpretation (Aki and Chouet, 1975) that the frequency dependence of Q_c is due to an increased contribution of surface waves sampling the shallow low Q region. Therefore, a crucial test on our coda model based on the S to S scattering would be to find if Q_β of shear waves and Q_c of coda waves agree and have the same frequency dependence. For this, we need a reliable method of measuring Q_β for frequencies higher than 1 Hz.

Determination of Q_β for high-frequency S waves

Reliable measurements of Q for high-frequency seismic waves are difficult for several reasons.

The use of artificial sources near the surface does not provide much help in measuring Q because the wave path to the surface receiver always has a turning point, and the amplitude and spectral contents of such waves are very sensitive to the velocity gradient at and near the turning point. As we have seen an example earlier, the local site effect at a receiver or a source is another factor which causes great scatter and bias in Q measurements. If the site effects vary greatly, the addition of a new data point may necessitate the introduction of an additional unknown factor, rendering the experiment practically useless.

The two major difficulties mentioned above can be avoided by a single-station method using deep sources, thereby minimizing the number of site effects and using direct body waves instead of waves which went through a turning point. A drawback of earlier single-station methods, (Asada and Takano, 1963, and Fedotov and Boldyrev, 1969) was over-simplifying assumptions made for eliminating the source effect. This difficulty can now be overcome by the use of coda waves which have the remarkable property of source effect separation as expressed by equation (4). The amplitude of coda waves observed at a fixed time t_o is directly proportional to the source spectra averaged over all directions. Therefore, by taking the amplitude ratio of S to coda waves, one can eliminate the average source effect. The remaining effect of radiation pattern is smoothed out by taking an appropriate average over numerous events located in a wide range of direction from the station.

In order to show how we determine Q_β of S waves, let us write the absolute values of Fourier transform of S waves as

$$A_S(f) = S(f,\theta)R^{-1}\exp(-\frac{\pi f R}{\beta Q_\beta}), \tag{13}$$

where R is the distance from the source to receiver, and $S(f,\theta)$ represents the source factor including the radiation pattern effect. The above distance dependence may be justified for ray paths from deep sources. Rewriting the coda amplitude measured at time t_o as $A_c(f|t_o)$, we have the following relation from equations (4) and (13),

$$\ln \frac{R.A_S(f)}{A_C(f|t)} = \ln \frac{S(f,\theta)}{S(f)} - \ln C(f|t_o) - \frac{\pi f R}{\beta Q_\beta} . \tag{14}$$

We average the left hand side of equation (14) over many events which lie in a distance range $(R-\Delta R, R+\Delta R)$. Assuming that any systematic variation of focal mechanism with distance may be smoothed out by the above average, we obtain

$$\left< \ln \frac{R.A_S(f)}{A_C(f|t)} \right>_{R\pm\Delta R} = a - bR, \tag{15}$$

where a is independent of R and

$$b = \frac{\pi f}{\beta Q_\beta} . \tag{16}$$

Thus, the slope b of the averaged logarithmic ratio plotted against distance R gives Q_β directly.

The above method was applied by Aki (1980) to the data obtained by Tsujiura (1978) using a real time band-pass filtering seismograph operated at Tsukuba (TSK) and Dodaira (DDR) Japan. Fig. 7 shows a typical example of the outputs of the filters with center frequencies at 1.5, 3, 6, 12, and 24 Hz and an octave band-width. The coda amplitude was measured at the lapse time of 50 sec, and the Fourier transform of S waves is estimated from the peak amplitude.

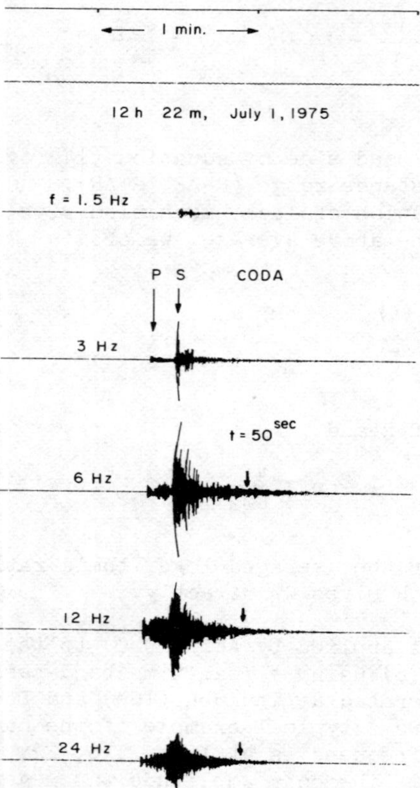

Fig. 7. An example of Tsujiura's band-pass filtered seismogram, for frequency bands centered at 1.5, 3, 6, 12 and 24 Hz. Coda amplitudes were measured at the lapse time 50 seconds.

In order to test the validity of our method, we applied it to the data from TSK and DDR independently. More than 450 earthquakes recorded at TSK and more than 400 earthquakes at DDR were used in the analysis. Their epicenters (crosses for focal depth greater than 35 km, circles for shallower) are shown in Figs. 8 and 9. As shown in Fig. 10, the focal depth distribution has a peak at 0-10 km and another at about 70 km. Most of the earthquakes are in the magnitude range 3-4. The epicenters are grouped into three areas A, B, and C. Area A represents the southern end of the zone of high Bouguer gravity along the coast of northeastern Honshu, where the Pacific plate is subducting under the Eurasian plate. The eastward extension of Line 2 (separating area A from B) meets the Japan trench at a corner where the trench axis is bent southward. Under area B, the Philippine-Sea plate and the Pacific plate are subducting from south and east respectively.

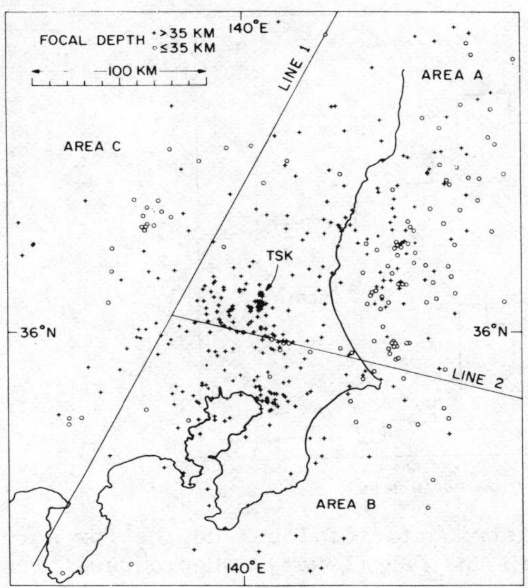

Fig. 8. Epicenters of earthquakes used in the determination of Q_β from TSK records. Crosses and circles represent, respectively, focal depths greater and less than 35 km. Reproduced by Aki (1980).

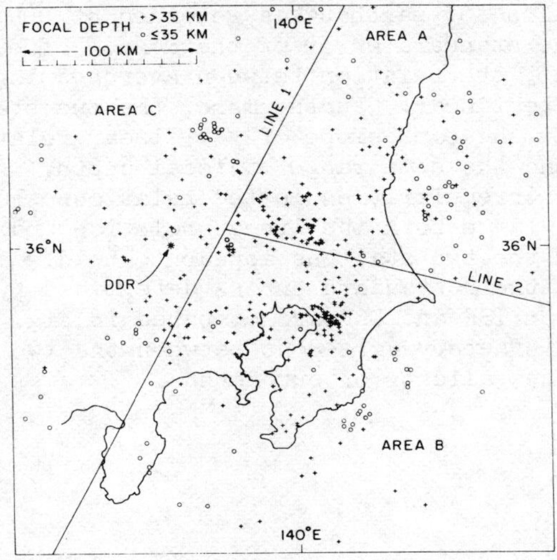

Fig. 9. Epicenters of earthquakes used in the determination of Q_β from DDR records. Crosses and circles represent, respectively, focal depths greater and less than 35 km. Reproduced from Tsujiura (1978).

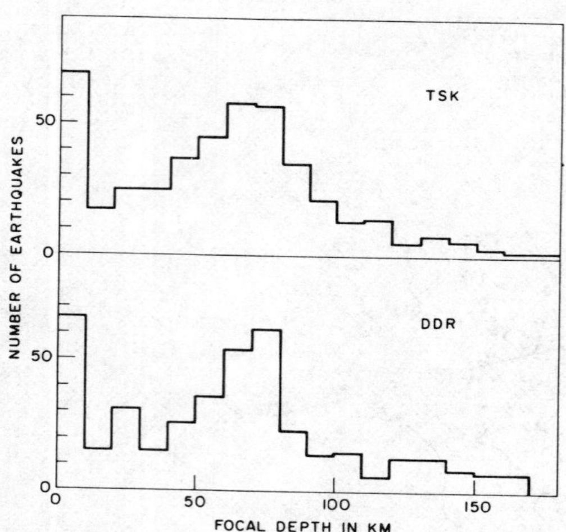

Fig. 10.   Frequency distribution of focal depths for earthquakes
recorded at TSK (upper) and DDR (lower).   Reproduced from Aki
(1980).

Fig. 11 shows a plot of averaged logarithm (left-hand-side
of Equation (15)) against distance R for earthquakes recorded
at TSK.   The results for all earthquakes with focal depth less
than 35 km are shown separately from those with deeper foci.
Fig. 12 shows the results for earthquakes recorded at DDR.   The
error bar indicates the standard error of the mean.   If our
assumptions are correct, the relation between averaged logarithm
and distance R should be linear.   Furthermore, the two stations
DDR and TSK should give the same slope of the linear relation
for the same channel and the same range of focal depth.   Although
there are considerable irregularities in the relations shown in
Fig. 11 and 12, by and large both the linear relation and the
equality of slopes for the two stations appear to hold.   The
values of slope b with their standard errors determined by the
least-squares method for TSK and DDR are compared in Fig. 13.
The results show a satisfactory agreement between the two
stations, confirming the validity of our method.

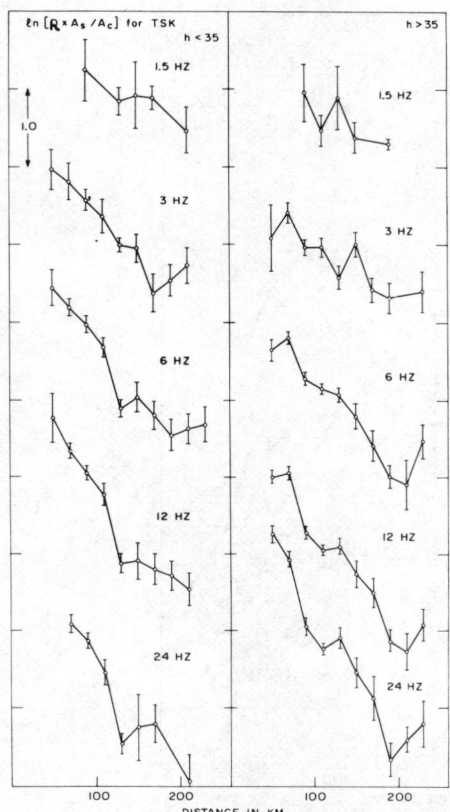

Fig. 11. The average of natural logarithm of S to coda amplitude ratio multiplied by the source-receiver distance R plotted against R for TSK. According to Equations (15) and (16), the slope is proportional to Q_β^{-1}. Reproduced by Aki (1980).

Then, we divided the earthquakes into four groups according to focal depth and area, and the resultant values of b for TSK and DDR are averaged for each group, as shown in Fig. 14. For both area A and B+C, the attenuation coefficient is slightly (10∿20%) higher for the group with focal depths shallower than 35 km than for the deeper group. This difference, however, disappears in the corresponding Q_β values because of the difference in β between the shallower and deeper part.

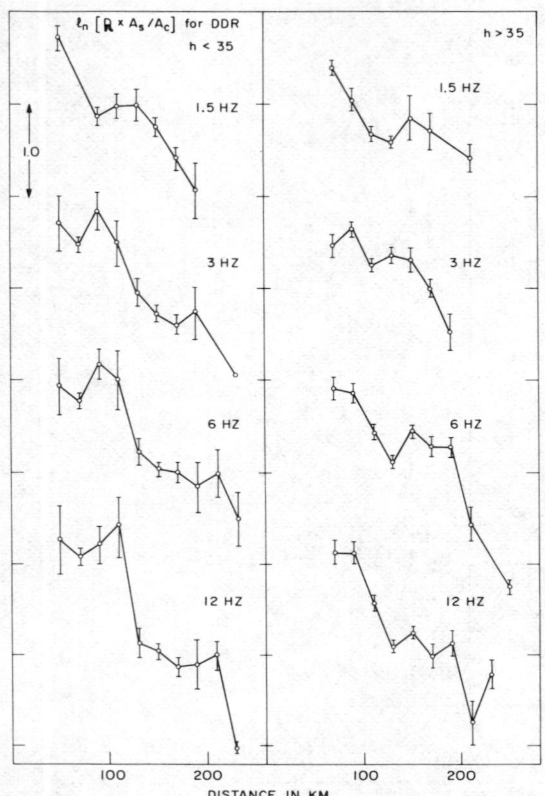

Fig. 12.   The average of natural logarithm of S to coda amplitude ratio multiplied by the source-receiver distance R plotted against R for DDR. According to Equations (15) and (16), the slope is proportional to Q_β^{-1}. Reproduced by Aki (1980).

The value of Q_β^{-1} is plotted as a function of frequency in Fig. 15. For both areas, we find a strong frequency dependence of Q_β quite similar to the frequency dependence of Q_c of coda waves discussed earlier. In Fig. 15, we also show identical Q_β and Q_c for the Garm area in central Asia determined by Rautian and Khalturin (1978) using the method similar to ours. Herrmann (1980) studied Q of Lg and Q_c for various regions in the U.S., and found that they agree well. Fig. 15 also shows Q determined for Lg waves at periods around 1 sec for California (Herrmann, 1980), Iran (Nuttli, 1980), and the central and eastern U.S. (Street, 1976, Bollinger, 1978, Nuttli, 1973). What emerges from these results is the strong variation of Q^{-1} from place to place for periods around 1 Hz apparently correlated with the

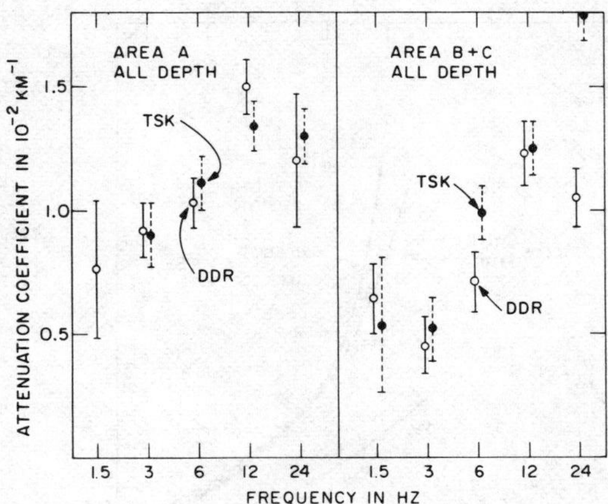

Fig. 13. Comparison of attenuation coefficients (b in Equation
15) determined from the TSK records and the DDR records.
Agreement between the two confirms the validity of our method.
Reproduced from Tsujiura (1978).

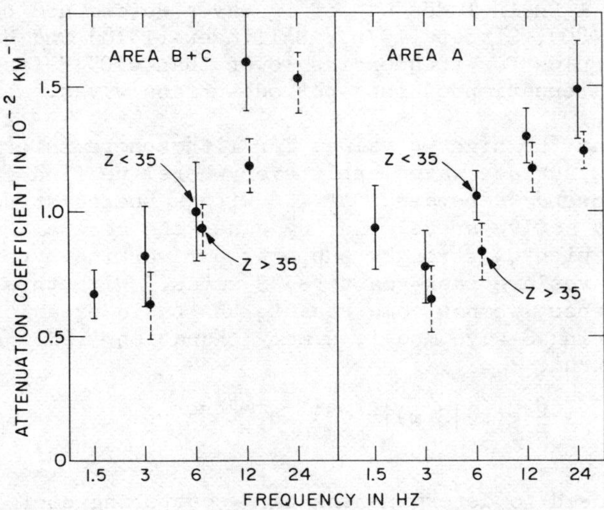

Fig. 14. Attenuation coefficients (b in equation 15) are
slightly higher for earthquakes with focal depths (z) less than
35 km than for those with z greater than 35 km.

degree of tectonic activity, while Q^{-1} converges to a very low
value at high frequencies irrespective of the tectonic activity.

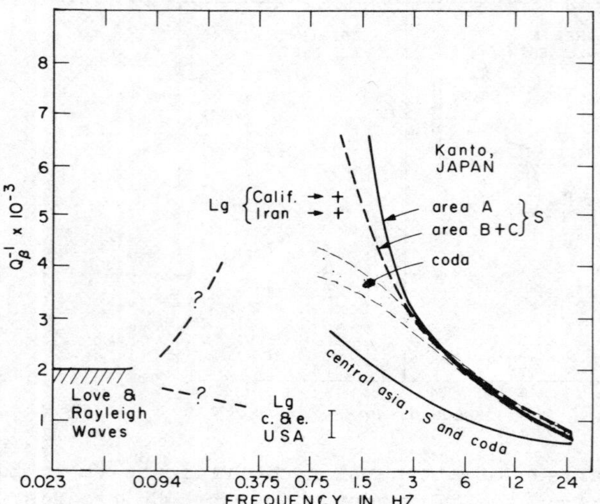

Fig. 15. Q_β^{-1} of S waves for the lithosphere under various
regions. The results for frequencies higher than 1 Hz are from
Rautian and Khalturin (1978), Tsujiura (1978) and Aki (1980).
The results for 1 Hz includes Q_β^{-1} of Lg waves determined by
Nuttli (1973, 1980), Street (1976), Bollinger (1978) and Herrmann
(1980). The results for frequencies lower than 0.05 Hz are
inferred from attenuation of long-period surface waves.

If we assume that the high Q_β values for lithosphere inferred
from long period surface waves apply everywhere, we find an
interesting frequency dependence of Q_β^{-1} with a peaked structure
for tectonically active areas. In any case, the general
agreement of Q_β with Q_c strongly supports our working hypothesis
that the coda waves are back-scattered S waves. With this
supporting evidence, we now come back to the basic equation for
the back-scattering S-wave model, namely, Equation (12) for the
coda power spectrum,

$$P(\omega|t) = \frac{\beta}{2} g(\pi) |S(\omega)|^2 (\frac{\beta t}{2})^{-2} e^{-\omega t/Q_c} \qquad (16)$$

Let us proceed to determine the back-scattering coefficient
$g(\pi)$ which will tell us something about the inhomogeneity. We
can find $g(\pi)$ using Equation (12) if we can determine
simultaneously the coda power spectrum $P(\omega|t)$ and the S wave
source spectrum $|S(\omega)|$.

Estimation of backscattering coefficient

The estimation of coda power spectrum $P(\omega|t)$ from the peak-
to-peak amplitude A_c of coda can be made by the following formula
(Aki and Chouet, 1975),

$$A_c = 2(2\Delta f \ P(\omega|t))^{1/2} \tag{17}$$

where we considered that manually smoothed peak-to-peak amplitude
corresponds roughly to twice the rms value of the signal
amplitude. On the other hand, the peak amplitude A_S of S waves
measured on the band-pass filter trace may be related to the
Fourier transform by

$$A_S = 2\Delta f \ |F(\omega)| \tag{18}$$

The above equation, however, is applicable only to an impulsive
waveform as pointed out by Rautian and Khalturin (1978). Care
has been taken in our earlier studies (Aki and Chouet, 1975) to
choose the smallest and closest events with impulsive S waveforms. Unfortunately, as shown in Fig. 7, the typical output
of band-pass filters shows non-impulsive waveforms especially
for higher frequency bands. The condition for the impulsiveness
appears to be met only for 3 Hz or lower frequencies. Further
inspection of the records indicated that for the area B+C the
condition is barely met even for 3 Hz.

Thus, we shall attempt to obtain here the back-scattering
coefficient only for frequencies 1.5 and 3 Hz and for area A.
To extend the study to higher frequencies and to area B + C, we
need to develop a recording and processing scheme which will give
the Fourier transform for appropriately windowed S-waves.

The value of a in Equation (15) (intercept of the curves in
Fig. 11 and 12 with the vertical coordinate) obtained from the TSK
and DDR records of 900 earthquakes can be used to determine
$g(\pi)$. Writing the amplitude spectrum of S waves as

$$|F(\omega)| = |S(\omega)|r^{-1}\exp(-\frac{\omega r}{2\beta Q_\beta}) \tag{19}$$

and putting Equations (17), (18), (19) into Equation (12),
we obtain

$$\frac{RA_S}{A_c} = \frac{(\Delta f)^{1/2}(\beta t_0)e^{-\omega R/(2\beta Q_\beta)}}{2(\beta g(\pi))^{1/2}e^{-\omega t_0/(2Q_c)}} \tag{20}$$

Comparing the above equation with Equation (15), we find

$$e^a = \frac{(\Delta f)^{1/2}\beta t_0}{2(\beta g(\pi))^{1/2}e^{-\omega t_0/(2Q_c)}} \tag{21}$$

Using the values of a and Q_c obtained earlier, $g(\pi)$ was
determined to be about 0.02 km^{-1} at both 1.5 Hz, and 3 Hz with

the standard error of a factor of 1.5. This result means that
S waves with frequencies 1.5 to 3 Hz lose energy by 2% at
every kilometer of propagation through backscattering. The
attenuation of S waves measured from the S to coda ratio in the
preceding section gives also 2% loss of energy per 1 kilometer
propagation. Thus, we come to a conclusion that the attenuation
of S waves for frequencies 1.5 to 3 Hz can be explained by
backscattering due to inhomogeneities. In other words, provided
that the single scattering approach used in deriving equation
(12) is valid for coda waves, the back-scattering energy loss
required to explain the observed coda wave amplitude totally
accounts for the observed attenuation of S waves.

Our estimates of $g(\pi)$ are consistent with earlier ones by
Sato (1978) who studied earthquakes in the same area as ours
using a different approach based on the total energy estimates
by the Gutenberg-Richter magnitude-energy formula and gave the
value of $g(\pi)$ in the range of 0.4 to 4% per km for the
frequency range 1 to 30 Hz.

Our conclusion is supported also by Herrmann (1980) who
studied the coda waves of local earthquakes in various parts of
the United States, and observed that the coda excitation for a
given earthquake size was greater in the area where Q^{-1} is
greater.

Frequency dependence of Q_β^{-1}

Furthermore, the scattering hypothesis for S wave
attenuation may adequately explain the frequency dependence of
Q_β^{-1} presented in Fig. 15. Consider a random medium characterized by the auto-correlation function $N(r)$ for refractive index
fluctuation such as discussed in the section on scattering of
P waves. In the case of acoustic waves studied by Chernov
(1960), the detail of correlation function does not affect the
frequency dependence of scattered energy for small ka, where k
is the wave number and a is the correlation distance defined,
for example, inIthe forms $N(r) = e^{-r/a}$ or $N(r) = e^{-r^2/a^2}$. For
small ka, the scattering is so called "Rayleigh scattering", and
Q^{-1} corresponding to energy loss by scattering is proportional
to k^3 or f^3.

On the other hand, for large ka, the scattered energy is
mostly directly forward within an angle of about 1/(ka) radian.
Then, depending on the length of time window to be assigned for
the primary wave, some of the scattered energy will be included
in the primary wave and should not be considered lost. In
other words, the scattered energy should be integrated for the
range of angle excluding the forward bundle of rays which arrive

within the primary wave time window. If we take this into
account, exclude the loss within a fixed solid angle about the
forward direction, the corresponding Q^{-1} will decrease with
frequency. For example, for $N(r) = e^{-r/a}$, Q^{-1} will be
proportional to k^{-1} or f^{-1}. According to Wu (1980), if the one-
dimensional power spectrum of refractive index fluctuation
$P(k)$ (Fourier transform of $N(r)$ with respect to r) is proportional to k^{-p}, Q^{-1} will be proportional to k^{1-p} or f^{1-p}.
Observed frequency dependence for Q_β and Q_c shows that p is in
the range from 1 to 2.

The peak of Q^{-1} will appear at a frequency for which ka is
a certain constant of the order of unity. The peak value of
Q^{-1} will be proportional to the variance $\langle \mu^2 \rangle$ of refractive
index fluctuation. Observed Q^{-1} summarized in Fig. 15 suggests
that the scale length of inhomogeneity is greater but the
variance is less for older and more stable areas.

For a quantitative estimate of parameters of the inhomogeneity, we need more complete observations of Q^{-1} filling the
gap in the frequency range between 0.05 and 1 Hz, and extending
the measurements to many areas. We also need to improve the
scattering theory including conversion from S to P waves, and
vice versa.

Conclusion

From the observations of amplitude and phase fluctuation of
teleseismic P waves across the Montana LASA, we found that a few
percent of velocity fluctuation with correlation distance about
10 km exists throughout the lithosphere. The small scale
inhomogeneities in the lithosphere generates back-scattered
waves observed as coda waves of local earthquakes. We conclude
that coda waves are primarily back-scattered S waves at least
for earthquakes in central Japan. Supporting evidences are
the similarities between S and coda in the apparent attenuation
as well as in the site effect. The backscattered energy required
to explain coda amplitude agrees with the loss of energy from
S waves, suggesting that the attenuation of S waves in the
lithosphere may be primarily due to scattering. The frequency
dependence of apparent Q^{-1} and its systematic change with the
degree of tectonic activity suggests a trend that the scale
length of inhomogeneity is shorter and the magnitude of velocity
fluctuation is greater for younger and more active regions.

Such information on inhomogeneities is essential for understanding the teleseismic P waveform, on which the future
discriminant for smaller events must rely.

Acknowledgement

The author is grateful to Masaru Tsujiura of the
Earthquake Research Institute, Tokyo University for the data
on which the present work is partially based. The author also
thanks Dr. Haruo Sato and Ru-shan Wu for their helpful discussions. This work was supported by the National Science
Foundation under Grant No. PFR8005720. The research was also
supported in part by the Advanced Research Project Agency of
the Department of Defense and was monitored by the Air Force
Office of Scientific Research under contract No. F44620-75-C-0064.

References

Aki, K.: "Analysis of the seismic coda of local earthquakes as
scattered waves", 1969, J. Geophys. Res., 74, pp. 615-631.
Aki, K.: "Attenuation of shear-waves in the lithosphere for
frequencies from 0.05 to 25 Hz", 1980, Phys. Earth Planet. Int.,
21, pp. 50-60.
Aki, K., A. Christoffersson, and E.S. Husebye: "Three-dimensional
seismic structure of the lithosphere under Montana LASA", 1976,
Bull. Seism. Soc. Am., 66, pp. 501-524.
Aki, K., A. Christoffersson, and E.S. Husebye: "Determination of
the three-dimensional seismic structure of the lithosphere",
1977, J. Geophys. Res., 82, pp. 277-296.
Aki, K., and B. Chouet: "Origin of coda waves: source, attenuation and scattering effects", 1975, J. Geophys. Res., 80, pp.
3322-3342.
Asada, T. and K. Takano: "Attenuation of short-period P waves in
the mantle", 1963, J. Phys. Earth., 11, pp. 25-34.
Berteussen, K.A., A. Christoffersson, E.S. Husebye, and A. Dahle:
1975, "P-wave anomalies at NORSAR and LASA", Geophys. J., 42,
pp. 403-417.
Berteussen, K.A., E.S. Husebye, R.F. Mereu, and A. Ram:
"Quantitative assessment of the crust-upper mantle heterogeneities
beneath the Gauribidanur seismic array in southern India", 1977,
Earth Planet. Sci. Lett., 37, pp. 326-332.
Bollinger, G.A.: "Attenuation of the Lg phase and the determina-
tion of m_b in the southeastern United States", 1979, Bull. Seis.
Soc. Am., 69, pp. 45-63.
Capon, J.: "Characterization of crust and upper mantle structure
under LASA as a random medium", 1974, Bull. Seis. Soc. Am., 64,
pp. 235-266.
Chernov, L.A.: "Wave propagation in a random medium", 1960,
McGraw-Hill, New York.
Chouet, B.: "Source, scattering and attenuation effects on high
frequency seismic waves", 1976, Ph.D. Thesis, M.I.T.
Christoffersson, A.: "Estimation of parameters characterizing
random medium-theoretical", in Exploitation of Seismograph
Networks, 1975, K.G. Beauchamp, ed., Nordhoff, Leiden.

Dahle, A., E.S. Husebye, K.A. Berteussen, and A. Christoffersson: "Wave scattering effects and seismic velocity measurements", in Exploitation of Seismograph Networks, 1975, K.G. Beauchamp, ed. Nordhoff, Leiden.

Engdahl, E.R. and C.P. Felix: "Nature of travel-time anomalies at LASA", 19 , J. Geophys. Res., 76, pp. 2706-2715.

Fedotov, S.A. and S.A. Boldyrev: "Frequency dependence of the body-wave absorption in the crust and the upper mantle of the Kuril-Island chain", 1969, Izv. Akad. Sci. USSR, Phys. Solid Earth, No. 9, pp. 17-33.

Herrmann, R.B.: "Q estimates using the coda of local earthquakes", 1980, Bull. Seism. Soc. Am., 70, pp. 447-468.

Husebye, E.S., A. Christoffersson, K. Aki, and C. Powell: "Preliminary results on the 3-dimensional seismic structure of the lithosphere under the USGS central California seismic array", 1976, Geophys. J.R. astr. Soc., 46, pp. 319-340.

Mack, H.: "Nature of short-period P-wave signal variations at LASA", 1969, J. Geophys. Res., 74, pp. 3161-3170.

Nuttli, O.W.: "Seismic wave attenuation and magnitude relations for eastern North America", 1973, J. Geophys. Res., 78, pp. 876-885.

Nuttli, O.W.: "The excitation and attenuation of seismic crustal phases in Iran", 1980, Bull. Seism. Soc. Am., 70, pp. 469-486.

Rautian, T.G. and V.I. Khalturin: "The use of coda for determination of the earthquake source spectrum", 1978, Bull. Seism. Soc. Am., 68, pp. 923-948.

Sato, H.: "Mean free path of S waves under the Kanto district of Japan", 1978, J. Phys. Earth, 26, pp. 185-198.

Street, R.L.: Scaling northeastern United States/southeastern Canadian earthquakes by their Lg waves", 1976, Bull. Seism. Soc. Am., 66, 1525-1537.

Tsujiura, M.: "Spectral analysis of the coda waves from local earthquakes", 1978, Bull. Earthquake Res. Inst., Tokyo Univ., 53, 1-48.

Wu, R.S.: "Attenuation of seismic waves due to scattering in the Earth", 1980, in preparation.

SOURCE LOCATION IN LATERALLY VARYING MEDIA

David Gubbins

Bullard Laboratories, Department of Earth Sciences,
Madingley Rise, Madingley Road, Cambridge CB3 OEZ

ABSTRACT

This paper deals with the interpretation of travel times of seismic waves from earthquakes and explosions. The location of seismic sources comes ultimately from travel time data but the same data are used in evaluating the wave speeds within the Earth itself, so the two problems of source location and velocity determination cannot be separated. In the early work of Jeffreys and Bullen a large set of travel times from earthquakes was used to refine the locations as well as to calculate the travel times of waves as a function of distance. Since that time instruments have improved and more data has become available but the J-B model is still in date. The principal problem with improving travel time models has been that the Earth is not spherically symmetric and it is simply not possible to fit the data with a spherically symmetric model. For example Herrin *et al.*[22] give an improved set of travel time tables that are considered to be a better fit to the oceanic regions of the world. This reflects improved station coverage rather than any new understanding about Earth. It is clear that further developments must take account of lateral variations within the Earth.

Considerable efforts have been made recently towards finding lateral variations immediately beneath arrays of seismometers (e.g. Aki *et al.*[1]). These studies are often restricted to regions where there happens to be an array of seismometers, such as at NORSAR or LASA. The velocity models derived from the data are in many instances rather ambiguous. A more serious problem is that *all* the travel time anomalies are assumed to arise from lateral variations beneath the array whereas we know that the most inhomogeneous parts of the Earth are near sources and so a possible bias in the results will come from source effects. While the idea of finding lateral variations without relocating the sources is an attractive one, there is no sure alternative to finding locations and velocity models simultaneously, as Jeffreys & Bullen did for spherically symmetric models.

E. S. Husebye and S. Mykkeltveit (eds.), Identification of Seismic Sources - Earthquake or Underground Explosion, 543–573.

Simultaneous source and structure determination is a nonlinear inverse problem which can be solved iteratively. The formalism for this is presented here, together with an algorithm for its solution which is so fast that the process is competitive with conventional location techniques. This general formalism also contains other methods of accounting for lateral variations, mainly that group known as "master event" methods. One particularly well located event (e.g. an explosion) is chosen as the master event and other, close, events are located relative to it. The relative locations are well determined and insensitive to lateral variations because the ray paths for close events are all affected in a similar manner. The joint inversion method allows for the assessment of errors in this technique as well as generalizing it. There are two approaches to this inverse problem. One can either regard the wave speed as a continuous function of position and apply Backus-Gilbert theory to it, or the velocity model can be parameterized and the parameters estimated by least squares. Both methods are being developed but neither has yet been applied to a sufficient number of real situations to be assessed.

1. INTRODUCTION

1.1 General Formulation of the Problem

The infinite frequency geometrical ray theory approximation is assumed to hold everywhere within the Earth. The *data* are the travel times (T_{ij}) from the ith source to the jth receiver, where i runs from 1 to n_e and j from 1 to n_s. The *model* we wish to determine from the data is the wave speed function $v(\mathbf{r})$, the origin times of the events ($T_i^{(0)}$) and the hypocentral coordinates ($\mathbf{X}_i$). The data are related to the model by:

$$T_{ij} = T_i^{(0)} + \int_{\Gamma_{ij}} \frac{dl}{v} \tag{1}$$

where the integral is taken over the ray path Γ_{ij} from source to receiver. This ray path depends on the hypocentral coordinates and the velocity function v and must be found from ray theory (for example Fermat's principle). For a discussion of the calculation of ray paths in the seismological context see for example Julian[34]. The dependence of Γ_{ij} on the unknown model parameters means that (1) is a complicated nonlinear relation between velocity and hypocenter parameters and the travel times. The velocity may be parameterized or left as a continuous function, or represented in a particularly simple way such as by a time-distance function $T(\Delta)$. Once $T(\Delta)$ is determined from the data it is a separate problem to find v from $T(\Delta)$ which has been the subject of much recent study. In all but one particular special case, to be discussed in §2, (1) remains nonlinear and the inverse problem is solved by iteration.

1.2 Linearization

The dependence of T_{ij} on the origin time is already linear. The change in time due to small changes in the hypocentral coordinates will be given by the partial derivatives $\partial T/\partial \mathbf{X}_i$:

$$\delta \mathbf{X}_i \cdot \nabla_{x_i} T$$

Suppose the ray path at the source makes an angle i with the line through the center of the Earth and leaves with azimuth z. Then geometrical considerations give:

$$
\nabla_{x_i} T = \begin{pmatrix} -\dfrac{\cos i}{v} \\[2mm] \dfrac{\sin i \, \cos z}{v} \\[2mm] \dfrac{\sin i \, \sin z}{v} \cos X_{i_2} \end{pmatrix} \tag{2}
$$

where X_{i_1}, X_{i_2}, X_{i_3} represent the depth, latitude and longitude of the ith event (Fig. 1).

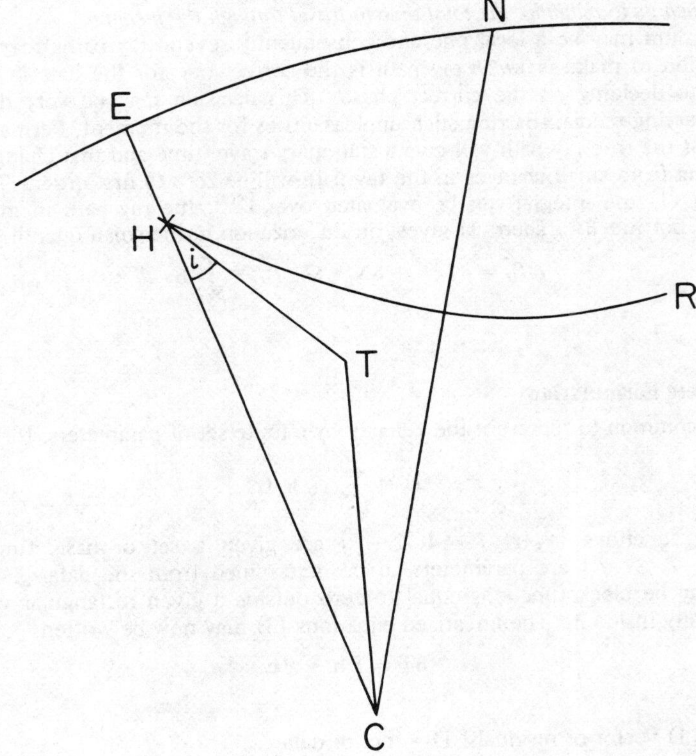

FIGURE 1: Geometry for calculating partial derivatives of the travel time with respect to the hypocentral coordinates. E is the epicenter, HR the ray path, H the hypocenter, C the center of the Earth, N the north pole and HT the tangent to the ray at H. i is the angle CHT between HT and HC. z, the azimuth, is the angle between the planes CEN and CHT, measured positive clockwise from north.

For a spherically symmetric Earth the ray path is completely and conveniently described by the ray parameter $p = \dfrac{r \, \sin i}{v}$ and azimuth z. (2) becomes:[10]

$$r \nabla_{x_i} T = \begin{pmatrix} -[(r^2/v^2) - p^2]^{1/2} \\ p \cos z \\ p \sin z \cos X_{i_2} \end{pmatrix}$$

Note here that z is the azimuth of the receiver from the source. For a spherically symmetric Earth p is constant and equal to $dT/d\Delta$. For laterally varying media i and z have to be calculated by tracing the ray path.

The dependence of T_{ij} on velocity may be linearized using Fermat's principle of least time which states that:

*"the ray path through a medium in which the wave speed varies with position will
be such as to minimize the total time to travel through the medium."*

The minimum may be a local one and consequently several ray paths may exist. The first decision to make is *which* ray path is the correct one for the data in hand. This amounts to deciding on the correct phase. This decision may be very difficult with laterally varying media. Barring such unpleasantries for the moment, Fermat's principle asserts that the true ray path will give a stationary travel time and that changes in travel time arising from small changes in the ray path will be zero to first order. Therefore in linearizing (1) the integral can be evaluated over $\Gamma_{ij}^{(0)}$, the ray path in medium with velocity v, not $v + \delta v$. Then (1) gives, on linearization in the small quantities,

$$\delta T_{ij} = \delta T_{ij}^{(0)} + \delta \mathbf{X}_i \cdot \nabla_{x_i} T_{ij} + \int_{\Gamma_{ij}^{(0)}} \delta s \, dl \tag{3}$$

where $s = v^{-1}$.

1.3 Discrete Formulation

It is common to represent the velocity by a finite set of parameters, for example

$$s(\mathbf{r}) = \sum_{k=1}^{P} c_k \, w_k(\mathbf{r}) \tag{4}$$

where the functions $\{w_k(\mathbf{r}); k = 1, 2... P\}$ are given a set of basis functions and $\{c_k; k = 1, 2 \cdots P\}$ are parameters to be determined from the data. For example $\{w_k(\mathbf{r})\}$ may be block functions equal to zero outside a given rectangular volume and equal to unity inside it. The linearized equations (3) may now be written:

$$\delta \mathbf{T} = B\mathbf{h} + A\mathbf{c} \tag{5}$$

where

$\delta \mathbf{T}$ is the D-vector of residuals, D $=$ no. of data

$\mathbf{h}$ is the $4n_e$ vector of hypocentral parameters

$\mathbf{c}$ is the P-vector of velocity parameters

B is the $D \times 4n_e$ matrix of partial derivatives given by (2)

A is a $D \times P$ matrix of partial derivatives given by:

$$A_{ij, p} = \int_{\Gamma_{ij}^{(0)}} w_p(\mathbf{r}) \, dl$$

Although we have assumed the specific form (4) for representing the velocity, the equations of condition (5) would have the same form if we had parameterized the $T(\Delta)$ function instead of v or used station corrections to allow for the effects of lateral variations. All my remarks on solving (5) apply equally well provided the model parameters can be separated into $\mathbf{h}$ and $\mathbf{c}$. Examples of solutions of (5) for this problem are where $\mathbf{c}$ is a vector of corrections to a travel time curve $T(\Delta)$,[10] for travel time curve *and* station corrections[40] and for velocity parameters with the $\{w_p(\mathbf{r})\}$ giving blocks for a three-dimensional structure and homogeneous layers respectively.[3, 13] All these methods use some form of least squares solution of (5), mostly with damping. (5) are combined to the standard form

$$\delta\mathbf{T} = (B \mid A)\left(\frac{\mathbf{h}}{\mathbf{c}}\right) = C\mathbf{x} \tag{6}$$

(e.g. Crosson[13]) which are solved by a stochastic inversion or some generalized inverse, e.g.

$$\mathbf{x} = (C^T C + \theta^2 I)^{-1} C^T \delta\mathbf{T} \tag{7}$$

Thus all the theory of least squares is available from standard methods[28, 51, 39] and the only judgement required is the choice of smoothing parameter, θ.

1.4 Continuous Formulation

Alternatively we may treat the velocity as a continuous function of position and seek to estimate spatial averages from the data. The dangers of prejudicing the result by choice of parameterization is therefore avoided but at the expense of considerably more arithmetical work. Referring back to (3) we may write

$$\delta\mathbf{T} = B\mathbf{h} + \int_{\Gamma^{(0)}} \delta s \, dl \tag{8}$$

for the linearized expression. This is in the standard form for a Backus & Gilbert[5] approach except for the slight complication of the extra discrete parameters $\mathbf{h}$. This arises in other applications, for example the density jumps in the Earth when free oscillation frequencies are inverted for the density gradient[41].

Johnson & Gilbert[33] have applied this approach to finding velocities from a travel time curve for a spherically symmetrical model, but only recently have Chou & Booker[11] and Pavlis & Booker[46] included lateral variations and studied simultaneous event location. The inversion is not perfectly straightforward because geometrical rays sample velocity along a line rather than over a volume in space. In the spherically symmetric case the data kernels have square root singularities at the turning points of the rays. Johnson & Gilbert[33] remove these singularities by "quelling" by integration by parts. In three-dimensional media Chou & Booker[11] find a straightforward application of inverse theory to be unstable and introduce another form of quelling by a volume average of the model.

1.5 Description of the Rest of this Paper

Most of the work done to date has dealt with spherically symmetric Earth models and this topic is reviewed in §2, with particular emphasis on event location and recent methods for finding velocity functions from travel time curves. The very considerable recent effort on finding structure beneath arrays is reviewed in §3 and the assumptions underlying these methods are discussed. In §4 simultaneous event location and structure determination is described for both continuous and discrete formulations. The formalism developed is applied to joint epicenter and master event methods. Section 4 ends with an elementary application to location of events in a straight line which serves

datasets this error is much too large. Evernden[15] demonstrates this using data from
to illustrate the methods.

2. THE SPHERICALLY SYMMETRIC EARTH

2.1 Basic Results for Travel Times

The properties of ray paths in spherically symmetric media enable the problem to
be simplified. A complete treatment is given by Bullen, Ch. 7.[10] A few necessary properties are listed here.

(i) $p = \dfrac{r \sin i}{v}$ is a constant along the ray path because of Snell's law and is called
the *ray parameter*.

(ii) T depends on Δ, the angular distance between event and receiver, and to a lesser
extent on the depth.

(iii) $p = dT/d\Delta$ so that take-off angles can be estimated from the travel time table.

(iv) $T = 2 \displaystyle\int_{r(p)}^{r_0} \dfrac{r}{v^2(r^2/v^2 - p^2)^{\frac{1}{2}}}\, dr$ is a useful formula for calculating the travel time
for a ray with a parameter p and velocity of medium $v(r)$. Here $r(p)$ is the
deepest point of the ray path and r_0 the position of the receiver. The event is
assumed to be at the surface.

(v)   Another useful function is $\tau = T - p\Delta$ because it is a single valued function of Δ
whereas T may not be. Note that: $d\tau/dp = -\Delta$.

2.2 Construction of a Travel Time Table

It is clumsy to represent the earth structure by a velocity function and usually a
function $T(\Delta)$ is preferred. T is also allowed to vary with depth of the event. Bullen,
Ch. 10,[10] describes the standard method of finding a travel time curve for a set of
events in a particular area. The curve $T(\Delta)$ is described by a set of parameters, $\{\xi_p\}$
say, giving the equations of condition:

$$\delta\mathbf{T} = B\mathbf{h} + D\boldsymbol{\xi}$$

which are again in the standard form for least squares. Bullen describes two
modifications to the standard procedures:

(i)   The normal equations can be solved by a simple method of successive approximation. This method only works because the matrix of coefficients is diagonally
dominant.

(ii)   The errors are believed not to be normally distributed, because of systematic
errors, and this is corrected for by the *method of uniform reduction*.[30] In its simplest form the method involves making a histogram of the errors and removing n
observations from each group, with n sufficiently large to isolate a central group of
normally distributed errors. A more sophisticated alternative is to weight observations according to their error.

2.3 The ISC Approach

The International Seismological Center locates earthquakes routinely using data from a global network of stations and the Jeffreys-Bullen[31] travel time tables, with a correction for the Earth's ellipticity. Strictly speaking each new event provides additional data for improving the travel time tables but lateral variations in the Earth provide the main source of error and further refinement would only bias the tables towards different regions. The ISC procedure has evolved as a sensible one for locating earthquakes from a global network of stations assuming the structure is known. It is described in the ISC Bulletin[26] and outlined below.

(i)   The raw data are arrival times reported by individual observations. The problems of accurate timing and the fact that different seismograms are read by different observers make the errors in this dataset large.

(ii)   A phase is *associated* with an event with the following order of priority:

    (a) P_g, P^*, P_n if $\Delta < 15°$ and Residual $< 70s$

    (b) P, PKP, P_{diff} if Residual $< 70s$

    (c) S_g, S^*, Sn if $\Delta < 15°$ and Residual $< 70s$

    (d) S, SKS if Residual $< 70s$

When a phase is associated with more than one event the one with lower residual and higher priority is taken. The starting location of the event is typically an estimate from some other agency.

(iii)   The iterative method described in §1.2 is used to improve the location using the method of uniform reduction to normalize the errors. The *associated* phases are revised according to the new residuals and phases are rejected if they duplicate each other or have residuals greater than 20s. Each event is examined individually before another refinement of the focal parameters is carried out.

(iv)   The raw data is searched for other phases and these are associated with the event if $\Delta > 25°$ and residual $< 7.5s$, but they are not used to relocate the event.

(v)   If the depth is not adequately constrained by P data it is set arbitrarily at 33km. Three reports of $pP-P$ phases are the minimum acceptable for good depth determination.

The problem of depth determination is discussed further in §2.5.

2.4 Small Networks

Procedures appropriate to large, global networks need to be modified when dealing with small amounts of data. Buland[9] points out that Jeffreys' remarks about errors do not apply to local array data because:

(a)   The arrivals are high frequency and easy to pick.

(b)   They are closely spaced in time so there are no clock problems.

(c)   All seismograms are read by the same observer.

The random errors should therefore be Gaussian, unlike ISC data, although there are still systematic errors due to similarities of waveform across the array and errors in Earth structure. Buland also reinforces a point made by Evernden[15] about the method for calculating confidence ellipses for locations. With large datasets it is satisfactory to find the usual unbiased estimator for the RMS error in the travel times but with small nuclear explosions with known source locations. An independent *a priori* estimate of the travel time errors must be used. The confidence ellipses for the locations can then be found by a χ^2 test.

Buland[9] also comments on the numerical methods used to refine the focal parameters, pointing out that the convergence of the iteration scheme can be assisted by "step-length damping" - i.e. damping the normal equations at each linear step. Solving the normal equations can also be facilitated by QR decomposition which is a stable numerical method for solving poorly conditioned linear systems. He examines the non-linearity of the problem of relocation in a uniform half space. The errors in location were found to lie in the linear range, i.e. they could be predicted from changes in source location with the linearized formula, which simplifies calculation of the error ellipsoids for the locations. He found that step length damping improved convergence as did S-wave data. Surprisingly S-wave data gives much better convergence and accuracy than a similar additional amount of P-data. Buland[9] also applied his method to two real datasets of microearthquake studies.

2.5 Use of Different Phases

Referring to (3) we see that differencing travel times for two phases will eliminate the dependence of the data on some of the model parameters. For example times for $pP - P$ will not depend on the origin time of the event at all, and for distant stations the velocity of the medium will have little effect except near the source because the ray paths for the two phases are so close together (Fig. 2).

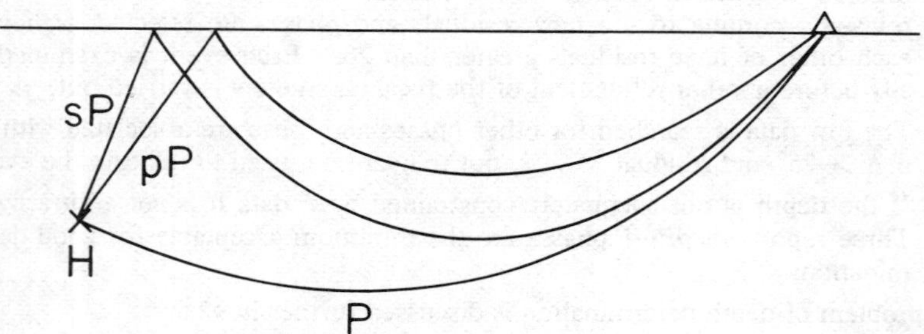

FIGURE 2: The effect of changing the velocity of the medium will be similar for both pP and P phases, because the ray paths are similar. Changing the depth of H will have markedly different effects on the two phases because of the differences in take-off angles between the two rays.

The difference in take-off angle for the two rays means that the differential travel times will be very sensitive to changes in depth of the hypocenter. In terms of inverse theory the data kernels for the differential travel times $pP - P$ are small except for the depth kernel. This is the reason why pP phases are so desirable in finding depths of events. Similar remarks apply to the phase sP.

Jackson[29] has studied the problem of finding depths using P-data, with applications to events in the Zagros mountains of Iran. A common problem that arises when the station coverage is poor is that errors in depth have the same effect on the data as errors in origin time. Jackson shows that this trade off can be broken by including core phases which give a shorter travel time when the depth is increased.

S-wave data provide another source of useful differential travel times. We have already seen that it is very useful in improving convergence and accuracy of locations. The S-wave ray path lies very close to the corresponding P-wave path but the disadvantage is that the velocity of the S-wave may not be known, and the S-data therefore add extra uncertainties.

2.6 Finding the Velocity Function

The subject has been reviewed by Kennett[37] although his paper does not include the work of Garmany[17,18] or Orcutt[44]. The usual problem is to find a velocity function v from a travel time curve. For exact data the Herglotz-Wiechert method can be used provided that v/r is everywhere an increasing function of depth (no low velocity zones). Gerver & Markushevich[19,20] proved that travel time data without amplitude data gives no information about v within a low velocity zone. Bessonova *et al.*[7,8] make use of the variable

$$\tau = T - p\Delta$$

and the relationship

$$r(q) = \frac{1}{\pi} \int_0^{\tau(q)} [p^2(\tau) - q^2]^{-\frac{1}{2}} \, d\tau \qquad (9)$$

to find $r(q)$, the bottom point of the ray path for a given ray parameter. This approach has been generalized to include low velocity zones so that v can be found (within bounds) in normal regions beneath low velocity zones where v/r increases again. Moreover inexact data is treated by placing bounds on the observationally determined function $\tau(p)$ which map into uncertainties in velocity. Model calculations of Kennett[36] show that the inherent nonlinearity of this approach is severe and this makes resolution difficult to estimate.

Garmany[17] has changed our whole outlook on $\tau - p$ inversions. He shows for a plane half space (which can be simply transformed to the spherical case) the function $\tau(p)$ may be written as

$$\tau(p) = 2 \int_p^{U_{max}} \frac{r(u) \, u \, du}{\sqrt{u^2 - p^2}} \qquad (10)$$

where U_{max} is the inverse velocity at the surface and there are no low velocity zones. This relationship is a *linear* one between τ and $r(p)$ and hence we can apply standard linear inverse theory. For an even simpler relation set $t^2 = u^2 - p^2$ so that (10) becomes

$$\tau(p) = \int_{-\infty}^{\infty} z \, dt \qquad (11)$$

where z is depth. These simple relations provide elegant proofs of Bessonova *et al's*[7,8] results. Garmany[17] applies his method using (10) with a Backus & Gilbert approach and the second Dirichlet condition. Garmany *et al.*[18] use linear programming methods and compare their results with those of Bessonova *et al.*[8] and Kennett.[36]

Low velocity zones can be treated in the same way by separating the half space into layers in which dv/dz is of one sign. This procedure is model dependent and the depths of the layers would change with a change in v. This makes the problem inherently *nonlinear*. Garmany[17] also points out that the three-dimensional inverse problem is bound to contain "low velocity zones," in some sense, and will therefore be nonlinear.

Since our main interest lies in inverting for lateral variations this section closes with a more detailed description of the Johnson & Gilbert[33] method. It is the only one to have been generalized to three dimensions.

Johnson & Gilbert[33] begin with the relation:

$$\tau(p) = 2 \int_{r(p)}^{r_0} r^{-1} \sqrt{\frac{r^2}{v^2} - p^2} \, dr$$

from which for small changes δv:

$$\delta\tau(p) = -2 \int_{r(p)}^{r_0} \frac{r}{v^3} \left[\frac{r^2}{v^2} - p^2 \right]^{-\frac{1}{2}} \delta v \, dr \tag{12}$$

Equation (12) is in the standard form for a linearized inverse problem with data kernel

$$K_\tau(r) = \frac{-2r}{v^3 \sqrt{\frac{r^2}{v^2} - p^2}} \qquad r(r) \leqslant r \leqslant r_0$$

$$= 0 \qquad\qquad r < r(p)$$

$$\delta\tau(p) = \int_0^{r_0} K_\tau(r) \, \delta v(r) \, dr$$

The data kernel has a square root singularity at the bottom of the ray path where $r/v = p$ which must be removed by integration by parts. The model becomes changed to the *velocity gradient* while the new data kernels are integrals of $K_\tau(r)$. Johnson & Gilbert apply their method to a dataset of travel times and direct observations of $dT/d\Delta$ using arrays to obtain a new Earth model.

3. FINDING LATERAL VARIATIONS

3.1 The Method

The "ACH" method[1] uses teleseismic data from a closely spaced network or array. Residuals are collected for a selection of events, chosen so as to give a good distribution on the focal sphere. The distance between the events and the network must be much larger than the aperture of the array itself so that mislocation of the source or lateral variations away from the array will affect all travel times from a single event equally. Source effects are then removed from the data by subtracting the average residual from all recordings of each event. The resulting travel time residuals are assumed to be caused by lateral variations immediately beneath the network. The only available check on this assumption is one of internal consistency. Provided the zone beneath the receivers is well crossed with ray paths it will be well sampled and any reduction in residuals will probably reflect significant lateral variations. The level of significance can be determined by an F test in the usual way. Sampling becomes poorer with increasing

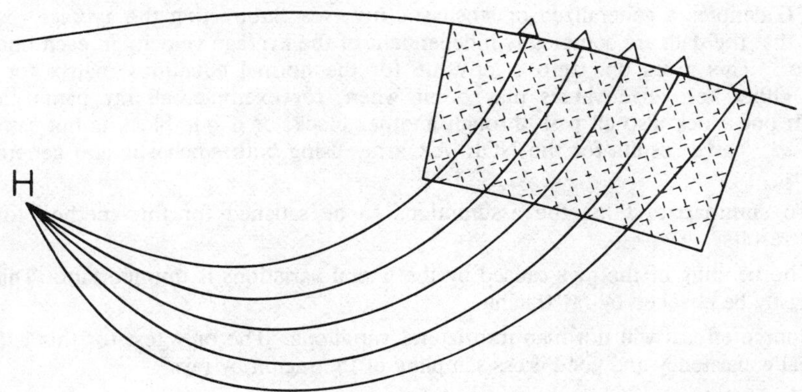

FIGURE 3: Geometry of seismic rays for ACH method. Rays from a distant earthquake to a network of stations are shown. The source is assumed to be sufficiently distant, and the network sufficiently small, that rays travel along similar paths except near the receivers and hence lateral variations outside the zone immediately beneath the array will affect all ray paths equally. The source is further assumed to be accurately located (for example by additional teleseismic data). All travel time residuals except for the average are then due to lateral variations near the network.

depth and it is impossible to resolve anomalies at depths greater than about the diameter of the array. Taking the anomalous zone to be too deep gives two possible effects. If core phases are not included in the dataset then very few ray paths pass through the deeper regions, because they are curved out away from directly beneath the sensors. As a result, no anomalies are detected in this region. Core phases will sample this deep region but the ray paths are all nearly vertical and parallel. The inversion procedure will find anomalies but it is impossible to distinguish between small anomalies of deep extent and larger anomalies in a correspondingly thinner layer. The variations so found may also affect time residuals caused by source effects or other systematic errors because there is very little cross sampling of the ray paths.

There are several variants of this method which differ mainly in the parameterization of the anomalies. The original method involved no ray tracing - the ray paths are simply straight lines with the appropriate azimuth and slowness, and the velocity is represented in blocks. In each horizontal layer the ray path is assumed to travel entirely within the block in which it spends the most time. The partial derivatives matrix, A, can be calculated very straightforwardly and the equations of condition follow from (5) with

$$\delta \mathbf{T} = A \mathbf{c} \qquad (13)$$

Subtracting the average residual for each event requires modification of the A matrix:

$$\delta \mathbf{T} - \overline{\delta \mathbf{T}} = (A - \overline{A})\mathbf{c} = G\mathbf{c}$$

The least squares solution is

$$\mathbf{c} = [(A - \overline{A})^T (A - \overline{A})]^G (A - \overline{A})^T (\delta\mathbf{T} - \overline{\delta\mathbf{T}})$$

where G denotes a generalized or stochastic inverse. Subtracting the average residual means that the data are completely independent of the average velocity in each horizontal layer. This gives one ·zero eigenvalue for the normal equations matrix for each layer. Other zero eigenvalues may occur when, for example, all ray paths passing through one block also all pass through another block, or if one block is not sampled. Aki *et al.*[2] give results for the NORSAR array using both stochastic and generalized inverses.

To summarize, I list the assumptions to be satisfied for this method to give correct results:

(1)   The bending of the rays caused by the lateral variations is unimportant. This can easily be checked by ray tracing.

(2)   Source effects will not map into lateral variations. The only test for this is internal consistency and good cross sampling of the region by rays.

(3)   The major anomalies must lie within the zone and not be so deep that they cannot be resolved by the network of receivers.

(4)   The variations must not be so rapidly varying that non-geometrical effects become significant.

3.2 Applications and Extensions of the Method

NORSAR array data provided a test of the method.[2] Timings are very accurate and it is easy to correlate traces from different sensors to pick the onset accurately. Errors in the picks are probably around 0.05s. The global distribution of earthquakes gives quite a good coverage of sources on the focal sphere, particularly for core phases. Travel time residuals give an RMS of 0.25s after the averages have been subtracted.

Aki *et al.*[2] restrict the anomalies beneath NORSAR to the uppermost 125km and give the parameterization five layers with each layer having 9 x 9 blocks horizontally. The upper layer is poorly resolved. This is a common occurrence and is caused by poor cross sampling. The anomalies reflect the local geology to some extent. The largest velocity anomalies, about 7%, are in the bottom layer, where resolution is quite good. The RMS residual is reduced by 60-70% leaving about 0.1s of unexplained error which is probably due to errors in picking, and source effects. The stochastic inverse solution gives more smoothing than the generalized inverse and is consequently preferred. That the largest anomalies are in the deepest layer suggests that the true anomalies lie even further down in the mantle. Further evidence supports this view. Haddon & Husebye[21] have to place the anomalies at 200km depth in order to satisfy amplitude data and Husebye *et al.*[25] report some precursors to $P'P'$ scattered from depths too great to be resolved by travel time data alone because they exceed the aperture of the array.

Christoffersson & Husebye[12] have developed some modifications to the basic method and applied it to an improved NORSAR dataset. The time consuming part of the calculation lies in evaluating the partial derivatives rather than in solving the normal equations and so it is worthwhile finding ways to compute a variety of models from the same normal equations. Christoffersson & Husebye[12] do this by introducing a linear transformation from the basic model, $\mathbf{c}$, to some other model, $\boldsymbol{\theta}$:

$$\mathbf{c} = D\boldsymbol{\theta}$$

where the matrix D is known, and need not necessarily be square. For example suppose our basic model c has five elements and we wish to set $c_1 = 0$ and $c_2 = c_3 = c_4$. Then $\theta_1 = c_2$ and $\theta_2 = c_5$ and the matrix D has the form:

$$D = \begin{pmatrix} 0 & 0 \\ 1 & 0 \\ 1 & 0 \\ 1 & 0 \\ 0 & 1 \end{pmatrix}$$

Thus we may set all of the anomalies in a particular layer to zero, or set those in several of the blocks equal to each other in order to fit some known geological feature. Christoffersson & Husebye find a satisfactory fit to the NORSAR data with just a single layer at 170km depth, while the fit is improved signficantly by a crustal layer constrained to align with the Oslo graben. Substituting for c in the solution gives:

$$D^T (G^T G) D \, \theta = D^T G^T \delta T$$

to be solved for θ. Once the normal equations matrix, $G^T G$, and the right hand sides, $G^T \delta T$, have been found and stored, the equations can be solved to give any model θ corresponding to a given choice of D.

Aki et al.[1] have inverted a similar dataset from the Montana LASA array. The velocity anomalies are smaller than those beneath NORSAR but they do exhibit low velocity under the central and northeast part of the array persisting down to depths of 100km. Berteussen et al.[6] have looked at travel time data from the Gauribidanur array in Southern India and interpreted the anomalies in terms of a random medium. The residuals here are so small that the underlying mantle must be relatively homogeneous.

There are very few permanent arrays of seismometers with good central timing and an even spatial distribution, and those that do exist are sited in rather uninteresting areas. To gain more information about other areas various authors have used datasets of lower quality. Husebye et al.[24] used the USGS Central Californian network of stations, Menke[42] the Tarbela network in Pakistan, Mitchel et al.[43] the St. Louis seismic array, and Iyer[27] a network in the Yellowstone area.

Ellsworth[14] has done a very complete study using the Hawaiian network, using teleseismic data for the most part. The inversion shows a low velocity region in the upper mantle beneath the array. He also gives several improvements to the basic ACH method:

(1) The partial derivatives are computed taking account of the exact length of each ray path in each block. This removes the smoothing effect of the original "quantized" method in which each path was assumed to pass entirely through one block in each layer.

(2) A 2D continuous velocity is represented by Hanning basis functions:

$$w_{ij}(x) = (1 + \cos p_i)(1 + \cos q_j)$$

where

$$p_i = \frac{\pi}{\Delta x}(x - x_i)$$

$$q_j = \frac{\pi}{\Delta y}(y - y_j)$$

Rays were traced through the three dimensional medium to check the linearizing assumption. This question will be discussed in detail in §3.3.

(3) Data from local earthquakes were included and relocated at the same time as determining the velocity, using Crosson's method.[13] The method will be discussed in §4. It gives a model with broadly similar features to that derived from teleseismic data alone.

Ellsworth finds that he can explain 70% of the residuals with a low velocity upper mantle under the center of the array and high velocity around the perimeter, suggestive of a pipe of low velocity material associated with the vulcanism.

A high reduction in residuals - 60 or 70% - suggests that the random errors in these datasets is very small (about 0.1s). Bulletin data such as from the ISC is of very much poorer quality with reading errors up to 0.5s and one would not expect so much success at reducing residuals with a laterally varying model. Romanowitz[47] has looked at Bulletin data from US stations and also from Western Europe.[48, 23] Reduction in residuals is only 10-20% but provided the anomalies are still well resolved the inversions can produce interesting results. It is relatively quick to invert bulletin data that is already in machine readable form and there are likely to be many more such studies in the near future. Time will tell whether or not the derived models are meaningful.

3.3 Ray Tracing

The validity of the linear assumption can be checked by tracing rays through the final, laterally varying model. This will test whether the true travel time can be estimated by a linear perturbation from a spherically symmetric Earth, but may not answer the more difficult question of whether this ray path is indeed the first arrival. There may be other completely different paths, particularly when reflections occur, that arrive earlier and which can only be found by a thorough search of all possible paths. A scheme has been developed by Smith et al.[49] to incorporate ray tracing into travel time inversions in an efficient manner. The velocity of the medium is represented by cubic spline functions which have continuous first and second derivatives. This facilitates ray tracing. Beginning with a spherically symmetric model the rays are traced and a first inversion done to find a laterally varying model. Rays are re-traced through this model and a second inversion is performed. The process is repeated until there are no further changes in the model. The technique used for ray tracing is the boundary value method described by Julian & Gubbins[35] ("bending" method). One computes the ray from a fixed point at the receiver to a given direction at the base of the anomalous zone. The method is appropriate because it relies on iteratively improving a ray path to give minimum time. Thus ray paths can be computed quickly for a given velocity by starting from the ray path for the previous iteration.

Assessing the nonlinearity of the inversion is not completely straightforward. The first question is whether the linear formula correctly predicts the travel times. This will depend on the precise nature of the anomaly as well as its magnitude but some insight may be gained from a simple example. Consider a low slowness slab with sinusoidal variation across it and exponential fall-off with depth:

$$S = S_0 (1 - F)$$

$$F = A \, H \, (k^2 - b^2 y^2) \exp (cz) \, [1 + \cos (by)]/2$$

where y is measured across the slab, z down and H is the Heaviside function. Dip and strike are appropriate for the North Island of New Zealand and the synthetic calculations model real event-station pairs. The example is taken from Spencer & Gubbins.[50] The only parameter of interest here is A, the amplitude of the anomaly, which can be expressed as a percentage of the spherically symmetric velocity. Table 1 gives results for travel times for three values of A evaluated both by ray tracing and by a linear perturbation from a simple spherically symmetric velocity. In most cases the error due to nonlinearity is below 0.1s in a residual of 1s. This is probably adequate for most purposes. 791-MNG for $A = 10\%$ is an exception. The error is 1s in a residual of 4s. The nonlinearity may become important rather quickly. The error is systematic, the linear prediction always giving a longer travel time because of Fermat's principle. The partial derivatives used in the linear calculation are also shown, together with those based on ray tracing through the 3D velocity medium.

Linearity in the inversion procedure itself also depends on the partial derivative matrix A. The elements of A are computed from the ray paths and will therefore be slightly in error under the linear assumption. Moreover Fermat's principle of stationarity applies only to the travel time and not the ray paths and so A will probably be more in error than the travel times themselves. This expectation is borne out by the data in Table 1. When linear estimates of the residuals are calculated from these partial derivatives and a given model the errors tend to cancel as might be expected from Fermat. The actual result of an inversion depends less on the partial derivatives than on the misfit and so we would hope that these errors are relatively unimportant.

Now consider the iteration towards a nonlinear solution. With perfectly accurate data and a good model we would iterate until the residuals became insignificant. When errors are present the decision of when to stop is not an easy one to make. Suppose a stochastic inverse has been used for the first iteration. The associated damping parameter must be chosen in the usual way as a compromise between reducing the residuals and producing anomalies of reasonable size. The procedure will produce a reasonable looking model but, because our only test that the model is realistic is one of internal consistency, it may be wrong as a result of a false linearizing assumption. Further iteration will be useless because we have already generated substantial anomalies and we would judge the smoothing parameter to be very large to damp out further anomalies. The first hint of nonlinearity comes when the rays are re-traced through the laterally varying medium. If the new (time) residuals are the same as the linear predictions there is clearly no point in continuing with the iteration. However, calculating the linear predictions is very tedious and most authors are content to quote an RMS residual for the whole dataset, which is easy to compute from the normal equations, and compare this one number with the RMS after ray tracing. We have found this simple comparison to be rather misleading. Table 2 shows residuals at 22 NORSAR subarrays from a single event. The data residuals are typically 0.25s. A velocity model similar to that published by Aki et al.[2] reduces the RMS to about 0.1s. However the errors due to linearizing are, for this event, about 0.1s, unacceptably large. The largest errors occur for rays passing through both high and low velocity regions which have only a small overall residual. For this particular calculation the RMS residual did not change after ray tracing showing that this criterion is perhaps oversimplified. The procedure adopted by Smith et al.[49] was to subtract the average residual at each iteration, even though the strict justification for it requires that the linear assumption holds. This reduces the error involved because the travel times after ray tracing are all smaller.

The only thorough check on linearity would seem to be the following. Perform one iteration to give a satisfactory model for the slowness anomaly. Now adjust the smoothing parameter to yield a model with anomalies about half as big and iterate again from this model. If the problem is truly linear there will be some value of the smoothing parameter that gives a model on second iteration that is very close to the first one-

TABLE 1: Comparison of residuals calculated for rays passing through a slab of high velocity material modelling New Zealand using both the linear prediction and three-dimensional ray tracing.

Event No.	Station	A%	Symmetric Time	Residual (Linear)	Residual (Ray Trace)	Partial Derivative (sec/100%)	Partial Derivative (Ray Trace)
303	ECZ	2		-.1948	-.1841		-10.790
		5	36.4578	-.4870	-.5062	-9.5817	-12.217
		10		-.9740	-1.1285		-13.641
791	MNG	2		-.6650	-.7680		-39.583
		5	60.6770	-1.6625	-1.7279	-35.975	-41.142
		10		-3.2490	-4.0762		-42.038
872	GNZ	2		-.4450	-.4498		-22.557
		5	47.1753	-1.1126	-1.1336	-22.2139	-22.961
		10		-2.2252	-2.2933		-23.458
87	CRZ	2		-.0422	-.0500		-2.092
		5	82.206	-.1055	-.1249	-2.1099	-2.100
		10		-.2110	-.2509		-2.115
200	GPZ	2		-.2312	-.2765		-14.473
		5	70.9857	-.5781	-.7670	-12.7800	-17.786
		10		-1.1561	-1.8500		-24.597
536	TUA	2		-.1925	-.1955		-9.855
		5	38.304	-.4812	-.4989	-9.6232	-10.265
		10		-.9623	-1.0337	-9.6232	-11.017

TABLE 2: Travel time residuals for single event (18) for the NORSAR array. Typical residuals from a spherically symmetric model are 0.2s after subtracting the average residual. With a velocity model similar to one which explains 70% of the residuals,[2] the misfit has an RMS of 0.07s. The error due to not ray tracing but making the linear assumption is .07s RMS which reduces to 0.05s if the average is subtracted. The errors due to not ray tracing are therefore only marginally significant according to the RMS. Close inspection of individual errors, however, show that for some ray paths such as at station 6, 7, 17, 19 the error is quite large. The last two columns give the error due to nonlinearity from the starting model and also from the final 3D model, i.e. T_{1D} minus T_{PRED} from the partial derivatives for the 3D model. These values are quite different for some stations (e.g. 20) giving further indication of nonlinearity.

Station	Observed Residual	T_{3D}-T_{PRED} with Average Subtracted	Departure from Linearity for 1D Model	Departure from Linearity for 3D Model
1	-.19	-.01	.05386	.04774
2	.01	-.06	.00412	.00458
3	-.08	-.01	.05192	.04980
4	-.43	-.04	.01511	.01972
5	-.40	-.05	.01360	.00718
6	-.22	-.10	.17537	.10406
7	.10	-.08	.14257	.11484
8	.14	-.04	.02515	.02129
9	.19	-.03	.03091	.03084
10	.18	-.02	.04693	.02438
11	.08	-.05	.00958	.00654
12	-.03	.00	.05900	.05473
13	-.04	.01	.07643	.06476
14	-.45	-.05	.01331	.01505
15	-.32	-.03	.03083	.01768
16	-.33	.05	.11879	.07536
17	-.03	.07	.11487	.15728
18	.42	.00	.02731	.16393
19	.44	.06	.08111	.25275
20	.49	.01	.05161	.12241
21	.30	.00	.07135	.04529
22	.09	-.06	.00391	.00369
RMS	.275	.047		

Other RMS values:

$$T_{OBS} - T_{1D} = .275; \quad T_{OBS} - T_{PRED} = .071;$$
$$T_{OBS} - T_{3D} = .076; \quad T_{3D} - T_{PRED} = .070 \text{ (before subtracting averages).}$$

step result. This calculation is being done for NORSAR data at the moment. Ellsworth[14] reports on an iterative calculation in which he achieved a further reduction in residuals on the second iteration, holding the damping parameter constant. He did not discuss whether the same model could be obtained by reducing the damping parameter and using a single iteration, and without such a comparison we cannot accept the calculation as an estimate of the importance of nonlinearity.

3.5 Use of Backus-Gilbert Theory

Wu[52] and Chou & Booker[11] have adopted a continuous velocity model and used Backus & Gilbert's theory of linear inference. The method is similar to a three-dimensional analogue of Johnson & Gilbert's[33] approach for a spherically symmetric velocity. They assume throughout that the model is linearly close to a one-dimensional starting velocity. The data samples the velocity along lines and do not adequately cross sample at a point for ordinary methods to work. Chou & Booker[11] elect to try to estimate averages of the velocity over rectangular regions. The linearized relationship between the travel time residual and velocity model is written as:

$$\delta T_{ij} = - \int_{\Gamma_{ij}} \frac{dl}{v_0} \frac{\delta v}{v_0} = - \int_{\Gamma_{ij}} \delta m \, \frac{dl}{v_0} \tag{14}$$

where v_0 is the one-dimensional starting model and dl represents an element of length, and $m = log(v)$. Define the functions:

$$J_{ij}(r) \quad = \quad 0 \qquad \qquad \text{off } ij^{\text{th}} \text{ ray path}$$

$$= \quad 1 \qquad \qquad \text{on it}$$

$$f_{ij}(r)dV \quad = \quad - J_{ij}(r) \, \frac{dl}{v_0}$$

then (14) becomes:

$$\delta T_{ij} = \int_v f_{ij} (\mathbf{r}) \, \delta m \, dV \tag{15}$$

Now make an estimate of the average of δm, $\delta \tilde{m}$, over a rectangular volume with dimensions w_x, w_y, w_z centered on $\mathbf{r}_0$.

$$\delta \tilde{m} = \int_v H(\mathbf{r}, \, \mathbf{r}_0, \, \mathbf{w}) \, \delta m \, (\mathbf{r}) \, dV \tag{16}$$

where H is the appropriate three-dimensional box car function. This expression is integrated by parts (quelled) to give:

$$\delta \tilde{m} = \int_v R (\mathbf{r}, \, \mathbf{r}_0, \, \mathbf{w}) \, \frac{\partial^3 m}{\partial x \partial y \partial z} \, dV \tag{17}$$

where R is the three-dimensional function formed by integrating the box car. Now the problem is set up to be treated by Backus' method of linear inference, as modified by Parker[45]. The integrals (15) are first quelled and then linear combinations of the data are taken so that the integrand is close in form to the ramp function R:

$$\sum a_{ij} \, \delta T_{ij} = \int_v \{ \sum a_{ij} \, F_{ij} \, (\mathbf{r}) \} \, \frac{\partial^3 \delta m}{\partial x \partial y \partial z} \, dV \tag{18}$$

$$= \int_v \tilde{R} \, \frac{\partial^3 \delta m}{\partial x \partial y \partial z} \, dV$$

The shape of $\tilde{R}$ can be assessed simply by evaluating its three dimensions. The volume integrals are transformed again to line integrals which makes them easy to do.

Chou & Booker[11] apply their method to accurate synthetic data from an ellipsoidal shaped low velocity bowl in a uniform half space and also a fault model with low velocity on one side of the fault. The low velocity zones are recovered but there are gaps in detection where there are no ray paths. For inaccurate data a trade off curve is found in the usual way.

4. SIMULTANEOUS SOURCE LOCATION AND VELOCITY DETERMINATION

4.1 A Fast Algorithm for solving the Normal Equations

I present an algorithm for solving (5) efficiently. The algorithm is due to Spencer & Gubbins[50] and relies on the separation of hypocenter and velocity parameters. Define the error, ϵ to be:

$$\epsilon = \delta T - Bh - Ac \tag{19}$$

Minimizing $|\epsilon|^2$ with respect to first c, then h gives the conditions

$$A^T \delta T = A^T A\, \tilde{c} + A^T B\, \tilde{h}$$

$$B^T \delta T = B^T A\, \tilde{c} + B^T B\, \tilde{h}$$

where $\tilde{c}$ and $\tilde{h}$ are the least squares estimate for c and h. Rearranging gives two equations for $\tilde{c}$ and $\tilde{h}$:

$$\tilde{c} = (OA)^{-1}\, O\delta T \tag{20}$$

$$\tilde{h} = (B^T B)^{-1}\, [B^T \delta T - B^T A \tilde{c}] \tag{21}$$

where

$$O = A^T - A^T B\, (B^T B)^{-1}\, B^T$$

Now if there are P velocity parameters then (20) is a set of P simultaneous equations, the same as for velocity determination by the ACH method. $B^T B$ is a square, $4n_e \times 4n_e$ matrix which must be inverted. It is block diagonal with all elements zero except for 4 x 4 block matrices down the diagonal. The inverse of such a matrix is another block diagonal matrix with each small block matrix the inverse of the corresponding 4 x 4 matrix of $B^T B$. Thus solution of (21) is very simple and involves only inverting one 4 x 4 matrix per event. There is little more effort involved in relocating the events and finding lateral variations than in the ACH method where events are not relocated.

Spencer & Gubbins[50] show that the solutions (20) and (21) are identical with the solution (6) which requires a great deal more work. Moreover the generalized and stochastic inverses of (20) and (21) have analogues in the full matrix solution (6). What is more interesting is that the individual 4 x 4 matrices in $B^T B$ can be decomposed in terms of their eigenvectors. Each of these 4-length eigenvectors give some information about the resolving power of the data for each individual event. Any information about trade offs between velocity parameters and locations will be locked up in the matrices OA and $B^T A$. Spencer & Gubbins[50] apply their method to data from North Island, New Zealand, with a simple high velocity slab with five free parameters.

The solution involves inverting or decomposing 5 x 5 and 4 x 4 matrices only. The eigenvectors of the 4 x 4 submatrices are remarkably close to being orthogonal and each can be identified with one of the hypocentral parameters. The corresponding eigenvalue gives an indication of how well each eigenvector is resolved. For the data from New Zealand the depths are the most poorly determined parameters.

An interesting aspect of this inversion noted by Spencer & Gubbins is that the relation between residuals and model is more nearly linear for the velocity parameters than for the hypocenter coordinates. The inversion can then be carried out iteratively without ray tracing, provided the linear assumption continues to hold. The partial derivatives for the hypocenter coordinates will be in error without ray tracing but provided the procedure converges the solution will be nearly correct. This approximation will have to be tested for both synthetic and real data.

The algorithm applies to *any* description of the velocity model, whether **c** represents a velocity function, station corrections or points on a curve. For example it could have been used by Lillwall & Douglas[40] to drastically reduce the size of their matrices, as well as by Aki & Lee[3] and Crosson.[13] So far there have been very few published results for real data. Experience will no doubt lead to improvements in the method. One important practical technique will be to control how much of the residual is taken up with the velocity model compared with the locations. It is easy to use the Spencer algorithm to selectively damp out either the location or velocity variations. More results are needed with real data.

4.2 A Fast Algorithm for the Continuous Case

Pavlis & Booker[46] have analyzed the relocation problem when the velocity function is an unknown, continuous function of position. The linear assumption is made:

$$\delta T = B\mathbf{h} + \int_{\Gamma^{(0)}} \delta s \; dl = B\mathbf{h} + \int \mathbf{f} \; m \; dV \tag{22}$$

where the integral represents a column vector of integrals and the line integrals have been transformed to volume integrals.[11] This is an example of a linear inverse problem of mixed type in which the model is expressed partly as a continuous function (δs) and partly by discrete parameters ($\mathbf{h}$). The procedure is first to find a linear transformation which, when applied to the equations, removes the dependence on $\mathbf{h}$. This leaves a straightforward continuous inverse problem to be solved for δs.

Any matrix B can be decomposed into:[38]

$$B = U \Lambda V^T$$

where if B is a $D \times P$ matrix, U is $D \times D$ orthogonal, Λ is diagonal $D \times P$ and V is a $P \times P$ orthogonal matrix. The usual Moore-Penrose inverse of A is

$$B^+ = V \Lambda^{-1} U^T$$

U can be partitioned into:

$$U = [U_2 \quad U_1 \quad U_0]$$

where U_2 contains the first r columns of U where $r \leqslant P$ is the number of non-zero singular values of B (i.e. its rank), U_1 the next $P - r$ columns and U_0 the last $D-P$ columns. It follows therefore that

$$U_1^T B = 0, \quad U_0^T B = 0$$

The Moore-Penrose inverse can also be written as:

$$B^+ = V \, \Lambda_r^{-1} \, U_2^T \tag{23}$$

where Λ_r is the $P \times r$ diagonal matrix of non-zero singular values of A. Now when U^T is applied to (22) we get:

$$\left[\frac{U_1^T}{U_0^T} \right] \delta \mathbf{T} = \int \left[\frac{U_1^T}{U_0^T} \right] \mathbf{f} \, m \, dV \tag{24}$$

and

$$U_2^T \, \delta \mathbf{T} = U_2^T \, B \mathbf{h} + \int U_2^T \, \mathbf{f} \, m \, dV \tag{25}$$

Equation (24) is a standard inverse problem for m which can be solved in the usual way. Equation (25) is used to find $\tilde{\mathbf{h}}$ once m, the velocity model, is known. Using (23) and rearranging (25) gives:

$$\tilde{\mathbf{h}} = V \, \Lambda_r^{-1} \, U_2^T \, \delta \mathbf{T} + \int V \, \Lambda_r^{-1} \, U_2^T \, \mathbf{f} \, m \, dV \tag{26}$$

$$= B^+ \, \delta \mathbf{T} + \int \mathbf{N} \, m \, dV$$

So far the analysis is applicable to any mixed continuous-discrete inverse problem. For the hypocenter problem B has a block diagonal form in which each submatrix is of order $n_s \times 4$ where n_s is the number of reports for each event. It is not necessary to decompose the whole of the B matrix, only each submatrix - a much easier task. The matrix U has a similar block diagonal form and can be treated in the same way. Each submatrix is square and $n_s \times n_s$. The computations required for the solution are thus reduced to decomposing $n_s \times n_s$ matrices to effect the separation of hypocenters and velocity function. Solution of (24) still involves matrices of order at least (D-P) as with all Backus-Gilbert schemes. For the NORSAR dataset D-P is many thousands making the method very costly if not impractical. The benefits arising from a full treatment of the velocity as a continuous function may justify the method for smaller datasets.

Pavlis & Booker[46] apply their method to synthetic data for a one-dimensional velocity structure. The only source of error is round-off in the machine. They also find that to include data from badly constrained events is a mistake because the errors in their locations can map into errors in the velocity model.

4.3 Master Event Techniques

The joint inversion procedure described in §4.1 and 4.2 can be seen as a generalization of master event techniques. I take this opportunity to review some joint techniques and put them into the general formalism. The idea behind a master event method is that an event can be located *relative* to another event more accurately than in absolute terms. This is particularly true when the events are clustered together. The results can be very interesting when the earthquakes are clustered in a single tectonic region. There are several variants of the method. Usually a single event is chosen as the "master" and the residuals for this event are subtracted from the corresponding residuals for other events. Hypocenters are found relative to the location of the master event, which remains undetermined. The master event may have been selected as being particularly well located by other data such as teleseismic information, surface breaks, or it may be an explosion.

Ansell & Smith[4] call their version of the master event method the *homogeneous station method*. They subtract the mean *station* residual for groups of events from similar localities. This has the effect of eliminating systematic errors due to lateral variations in the velocity structure or errors in the travel time tables. They apply their method to earthquakes from the seismic zone beneath North Island, New Zealand.

Fitch[16] uses a master event technique with the additional feature of allowing the velocity near the hypocenters to be a variable parameter. The data are ISC residuals from deep events from the Fiji-Tonga and Peru-Brazil seismic zones. The relocation procedure is iterative and very similar to the joint inversion, except here there is just one velocity parameter.

Equation (5) relates residuals to the hypocenter coordinates and velocity:

$$\delta T_{ij} = (B\mathbf{h})_{ij} + (A\mathbf{c})_{ij}$$

The master event technique relies on the ray paths being similar. Subtracting the residual from the master event $i = 0$ gives:

$$\delta T_{ij} - \delta T_{0j} = \sum_{k=1}^{4} (B_{ij,k} - B_{0j,k})\, h_k + \sum_{r=1}^{P} (A_{ij,r} - A_{0j,r})\, c_r \tag{27}$$

If all the ray paths are similar then the partial derivatives $(A_{ij,r} - A_{0j,r})$ will be very small. The master event method assumes it to be zero. The joint inversion method can be used to improve on the annihilation of the A matrix and remove the uncertainty in the velocity model still further. It will also give an independent estimate of the uncertainty in the velocity, and how this maps into the locations. No authors have yet used this sort of formalism to estimate errors in the master event technique.

Figure 4 shows ray paths from two close events. Differencing the residuals for each station will eliminate the dependence on velocity along the ray paths, except near the events. The differential residuals therefore depend on the relative positions of A and B and the velocity of the medium there. Thus, differential residuals could be used to estimate the velocity near the sources.[16] This dependence of differential times on locations would be reflected in the partial derivative matrices in (27), the A matrix being zero except for those elements which relate to the velocity near the sources.

4.4 A Simple Example - Location on a Line

The methods of joint inversion described in §4.1 and 4.2 are illustrated here by a simple example. The data are travel "times" from "events" on a straight line segment $0 \leqslant x \leqslant 1$ to "stations" on the same line. The time is computed as

$$T_{AB} = \left| \int_{A}^{B} s(x)\, dx \right|$$

where $s(x)$ is the slowness function along the line. Note that the dependence of time on slowness is fully linear because there can be no bending of the ray paths - rays are constrained to be along the straight line. The dependence on the event locations is still nonlinear so the problem must in general be solved iteratively. Let $s(x)$ be the starting slowness function and h_i the initial location of event i. Then for the jth station at X_j:

$$\delta T_{ij} = \left| \int_{h_i}^{X_j} \delta s\, dx \right| + s(h_i)\, \delta h_i\, \text{sgn}\, (h_i - X_j) + O(\delta h^2) \tag{28}$$

This is the mixed continuous-discrete formulation of the problem. To make it fully

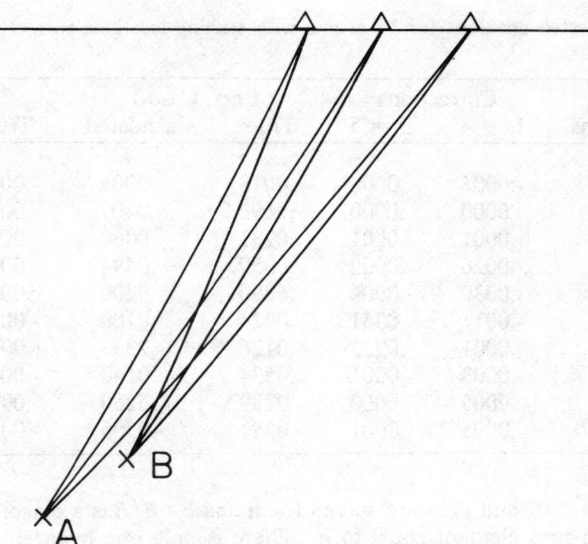

FIGURE 4: Geometry of ray paths for the master event technique.

discrete we represent the slowness by block functions

$$\delta s = \sum_{k=1}^{P} c_k \ w_k \ (x)$$

where

$$w_k(x) \ = \ 1 \quad \left[\frac{k-1}{P}\right] \leqslant x \leqslant \left[\frac{k}{P}\right]$$

$$= \ 0 \quad \text{otherwise}$$

and (28) becomes

$$\delta T_{ij} = \sum_{k=1}^{P} c_k \ | \int_{h_i}^{x_j} w_k(x) \ dx \ | + s(h_i) \ sgn \ (h_i - x_j) \ \delta h_i \qquad (29)$$

This equation has our standard form:

$$\delta \mathbf{T} = A\mathbf{c} + B\mathbf{h}$$

where B is $D \times n_e$ and A is $D \times P$.

The data are travel time residuals for ten randomly positioned events to five randomly placed stations. The locations are shown in Table 3. The travel times were computed for a slowness function

$$s(x) = 1 + A \ \sin \ \pi \ x$$

with $A = 0.1$. The starting velocity model had $s(x) = 1$.

TABLE 3: Relocated positions for 10 events with starting locations correct and in error, with calculated results.

Event Locations	Corrections P = 4	P = 5	Corr. Case 1 True	Calculated	Corr. Case 2 True	Calculated
1	.6289	-.0001	.0000	-.0018	.0005	-.0002	-.0000
2	.5666	.0000	.0000	-.0692	-.0407	-.0069	-.0041
3	.0001	.0001	.0001	.0231	-.0060	.0023	-.0019
4	.0290	-.0026	-.0002	.0059	+.0494	.0006	.0026
5	.0194	-.0030	-.0004	.6890	.9200	.0069	.0090
6	.9525	-.0011	.0351	-.0224	.0100	-.0022	-.0018
7	.8964	.0001	.0223	.0126	-.2333	.0013	-.0032
8	.2588	-.0003	-.0001	.0534	.0140	-.0053	-.0008
9	.3865	.0005	.0000	.0789	.5260	.0079	.0039
10	.3998	-.0005	-.0001	-.0281	-.0305	-.0028	-.0025

Equations (20) and (21) are solved for $\tilde{\mathbf{h}}$ and $\tilde{\mathbf{c}}$. $B^T B$ is a diagonal $D \times D$ matrix with every non-zero element equal to n_s. There is only one hypocentral parameter for each earthquake, rather than four, which is why $B^T B$ is truly diagonal rather than having 4 x 4 non-zero matrices down the diagonal as in the real case. For two events and three stations B has the form:

$$B = \begin{pmatrix} 1 & 0 \\ 1 & 0 \\ 1 & 0 \\ 0 & 1 \\ 0 & 1 \\ 0 & 1 \end{pmatrix}$$

while the elements of A are simply the length of the ij^{th} ray path in the kth block. If we wanted to make an analogue of the "quantized" procedure[1] we could make these elements simply P^{-1} or 0 depending on whether or not the interval (h_i, X_j) included any of the kth block. To solve (21) we evaluate the matrices $A^T A$, $A^T B$ and hence OA and the vector $O\delta T$ and solve the $P \times P$ linear system.

$$OA\mathbf{c} = O\delta T$$

The inverse of $B^T B$ is trivial because it is diagonal. The first calculations were for $P = 4, 5, 6$ with the correct starting locations. For $P = 4$ the velocity parameters were:

$$\mathbf{c} = (.0500, .0940, .0935, .0441)$$

in good agreement with the true velocity (see Figure 5). The corrections to the hypocenters are found by substituting $\mathbf{c}$ into (21) which only requires evaluating 4-length vectors and the trivial inverse of $B^T B$. The solutions are all small (Table 3) as they should be. The solution for $P = 5$ is

$$\mathbf{c} = (.0284, .0855, .0993, .0879, -.2042)$$

which is reasonable except for the fifth value. The corresponding location corrections are small except for 6 and 7. There is only one station *just* inside this last block and so it is very poorly resolved. The velocity inside this block can be compensated for by changing the locations of the two events, 6 and 7, that also lie in the block. In this case the procedure is unstable and errors due to the inability of a block model to completely fit a continuous function is giving errors which map into the fifth element of **c** and the locations 6 and 7.

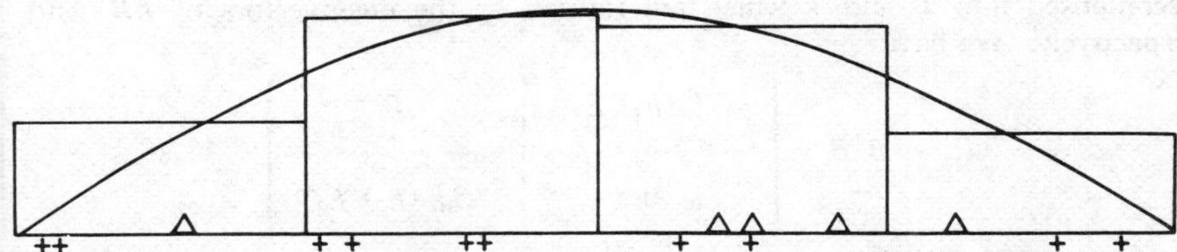

FIGURE 5: Locating earthquakes on a straight line. The data are travel times computed through a slowness model with sinusoidal variation as shown. Stations and events are also shown, together with the solution for the velocity model parameterized with four block functions. The starting solution for the inversion was a slowness equal everywhere to unity and the correct event locations.

When $P = 6$ the last block is not sampled at all. The last row of the matrix is zero. This means that the vector $\mathbf{v}^{(6)} = (0, 0, 0, 0, 0, 1)$ is an eigenvector of OA with zero eigenvalue and this component of the model is completely unresolved by the data. This uncertainty will map into the hypocenter corrections as $-(B^TB)^{-1} B^TA \, \mathbf{v}^{(6)}$ giving some indication of the uncertainty there.

Further calculations were done with the starting locations of the events in error. The errors were chosen from a normally distributed set with mean zero and standard deviation 0.05 for case 1, .005 for case 2. Here the errors in location trade off against velocity parameters. For $P = 4$ the velocity models are:

$$\text{Case 1} \quad \mathbf{c} = (-.3829, .3234, .0305, .1443)$$
$$\text{Case 2} \quad \mathbf{c} = (.0006, .1295, .0789, .0643)$$

The corresponding locations are in Table 3. The results for **h** reflect the biased coverage of this particular set of stations. Clearly blind use of the inversion procedure in the three-dimensional case will be fraught with similar difficulties, and every calculation will have to be treated on its own merits. This inversion was not iterated.

The same problem can be attacked by the Pavlis & Booker method.[46] Consider, for simplicity the three-station and two-event case. Here

$$\delta T_{ij} = (B\ \delta \mathbf{h})_{ij} + \int_0^1 R_{ij}\ \delta s\ dx$$

(c.f. Eqn. 22). The B matrix has the same form as before:

$$B = \begin{pmatrix} S_0(h_1) & 0 \\ S_0(h_1) & 0 \\ S_0(h_1) & 0 \\ 0 & S_0(h_2) \\ 0 & S_0(h_2) \\ 0 & S_0(h_2) \end{pmatrix}$$

where $S_0(x)$ is the initial slowness function. R_{ij} is the appropriate step function. B is decomposed into U and V which are formed by the eigenvectors of BB^T and B^TB respectively. We have:

$$B^TB = \left(\begin{array}{c|c} S_0^2\ (h_1)\ E_3 & 0 \\ \hline 0 & S_0^2\ (h_2)\ E_3 \end{array} \right)$$

where

$$E_3 = \begin{pmatrix} 1 & 1 & 1 \\ 1 & 1 & 1 \\ 1 & 1 & 1 \end{pmatrix}$$

and

$$B^TB = \begin{pmatrix} 3\ S_0^2\ (h_1) & 0 \\ 0 & 3\ S_0^2\ (h_1) \end{pmatrix}$$

The eigenvalues of E_3 are $\lambda = 3$, 0 and 0 with corresponding orthogonal eigenvectors $(1,\ 1,\ 1)$, $(\frac{1}{2},\ -1,\ \frac{1}{2})$, $(1,\ 0,\ -1)$. BB^T has two non-zero eigenvalues $3\ S_0^2\ (h_1)$ and $3\ S_0^2\ (h_2)$. For simplicity take $S_0\ (x) = 1$. Then U_0^T has the form:

$$U_0^T = \begin{pmatrix} 2^{-\frac{1}{2}} & 0 & -2^{-\frac{1}{2}} & 0 & 0 & 0 \\ 6^{-\frac{1}{2}} & (2/3)^{\frac{1}{2}} & 6^{-\frac{1}{2}} & 0 & 0 & 0 \\ 0 & 0 & 0 & 2^{-\frac{1}{2}} & 0 & -2^{-\frac{1}{2}} \\ 0 & 0 & 0 & 6^{-\frac{1}{2}} & (2/3)^{\frac{1}{2}} & 6^{-\frac{1}{2}} \end{pmatrix}$$

which has the property $U_0^TB = 0$. Multiplying by U_0^T gives the equations relating δ_s to the "annulled" dataset $U_0\ \delta \mathbf{T}$ which has four members:

$$U_0\ \delta \mathbf{T} = \begin{pmatrix} 2^{-\frac{1}{2}}\ (\delta T_{11} - \delta T_{13}) \\ 6^{-\frac{1}{2}}\ (\delta T_{11} - 2\delta T_{12} + \delta T_{13}) \\ 2^{-\frac{1}{2}}\ (\delta T_{21} - \delta T_{23}) \\ 6^{-\frac{1}{2}}\ (\delta T_{21} - 2\delta T_{22} + \delta T_{23}) \end{pmatrix}$$

The transformed Frechet derivatives are:

$$N_1 \quad = \quad 2^{-\frac{1}{2}} (R_{11} - R_{13}) \quad = \quad -1 \quad X_1 \leqslant x \leqslant X_3$$

$$= \quad 0 \quad \text{otherwise}$$

$$N_2 \quad = \quad 6^{-\frac{1}{2}} (R_{11} - 2R_{12} + R_{13}) \quad = \quad -1 \quad X_1 \leqslant x \leqslant X_2$$

$$= \quad 1 \quad X_2 \leqslant k \leqslant X_3$$

$$= \quad 0 \quad \text{otherwise}$$

with similar expressions for N_3 and N_4.

Solution of this inverse problem proceeds in the usual way. We take a linear combination of the data, $\gamma = U_0 \, \delta T$,

$$\sum_{i=1}^{4} a_i \, \gamma_i = \int_0^1 \sum_{i=1}^{4} a_i \, N_i \, (x) \, \delta s \, dx = \delta \bar{s} \, (x_0)$$

where $\{a_i\}$ are chosen to make the *averaging function*

$$A \, (x, \, x_0) = \sum_{i=1}^{4} a_i \, N_i \, (x)$$

peak near $x = x_0$. A popular criterion for choosing the $\{a_i\}$ is to minimize some scalar measure of the *spread* or *width* of A. For simplicity here we choose the First Dirichlet Criterion which can be heuristically thought of as minimizing:

$$\int_0^1 [A - \delta(x - x_0)]^2 \, dx$$

which gives

$$\sum_{k=1}^{4} \int_0^1 N_k \, N_e \, dx \, a_k = N_e \, (x_0)$$

The *spread matrix* $S_{kl} = \int_0^1 N_k \, N_e \, dx$ has elements

$$S_{ii} = S_{13} = S_{24} = X_3 - X_1$$

$$S_{12} = S_{14} = S_{23} = -(X_1 - 2X_2 + X_3)$$

$$S_{ij} = S_{ji}$$

Without proceeding further with the details of this solution we can see that the averaging function must always be a combination of the $\{N_i\}$ which are square functions that change sign only at station locations. The averaging function for the slowness will at best have a width equal to the distance between the closest stations. It is interesting that the averaging function does not depend on the event locations in any way, which is a rather surprising result.

To complete the solution of the two-event, three-station problem, the matrix U_1^T is:

$$U_1^T = 3^{-\frac{1}{2}} \begin{pmatrix} 1 & 1 & 1 & 0 & 0 & 0 \\ 0 & 0 & 0 & 1 & 1 & 1 \end{pmatrix}$$

and operating on (22) gives

$$U_1^T \, \delta T = U_1^T \, B \, \delta h + \int_0^1 U_1^T \, \mathbf{R} \, \delta s \, dx$$

Substituting for δs, (26) can be solved for the hypocenter coordinates $\delta \mathbf{h}$.

The extension of the solution to five stations and ten events is straightforward. BB^T will be a block diagonal, 50 x 50 square matrix. The submatrices will be multiples of E_5. E_5 has four zero and one non-zero eigenvalue. U_0^T will be 40 x 50 with 4 x 5 non-zero matrices down the diagonal. The spread matrix will be 40 x 40 and herein lies the extra computational effort for the continuous formulation.

ACKNOWLEDGMENTS

This work was partially supported by Natural Environment Research Council Grant GR3/3905. Carl Spencer and Colin Thomson helped with some of the calculations.

REFERENCES

(1)   Aki, K., Christoffersson, A. and Husebye, E.S.: 1976. Three dimensional structure of the lithosphere under Montana LASA. *Bull. Seism. Soc. Am. 66*, pp. 501-524.

(2)   Aki, K., Christoffersson, A. and Husebye, E.S: 1977. Determination of the three dimensional seismic structure of the lithosphere. *J. Geophys. Res. 82*, pp. 277-296.

(3)   Aki, K. and Lee, W.H.K.: 1976. Determination of three-dimensional velocity anomalies under a seismic array using first P arrival times from local earthquakes.
1. A homogeneous initial model. *J. Geophys. Res. 81*, pp. 4381-4399.

(4)   Ansell, J.H. and Smith, E.G.C.: 1975. Detailed structure of a mantle seismic zone using the homogeneous station method. *Nature 253*, pp. 518-520.

(5)   Backus, G.E. and Gilbert, F.: 1967. Numerical applications of a formalism for geophysical inverse problems. *Geophys. J. R. Astr. Soc. 13*, pp. 247-276.

(6)   Berteussen, K.A., Husebye, E.S., Mereu, R.F. and Ram, A.: 1977. Quantitative assessment of the crust-upper mantle heterogeneities beneath the Gauribidanur seismic array in Southern India. *Earth Planet. Sci. Lett. 37*, pp. 326-332.

(7)   Bessonova, E.N., Fishman, V.M., Ryaboyi, V.Z. and Sitnikova, G.A.: 1974. The tau method for the inversion of travel times - I. Deep seismic sounding data. *Geophys. J. R. Astr. Soc. 36*, pp. 377-398.

(8)   Bessonova, E.N., Fishman, V.M., Johnson, L.R., Shnirman, M.G. and Sitnikova, G.A.: 1976. The tau method for the inversion of travel times - II. Earthquake data. *Geophys. J. R. Astr. Soc., 46*, pp. 87-108.

(9)   Buland, R.: 1976. The mechanics of locating earthquakes. *Bull. Seism. Soc. Am. 66*, pp. 173-187.

(10)   Bullen, K.E.: 1965. *An Introduction to the Theory of Seismology*. Cambridge Univ. Press, London, 381 pp.

(11)   Chou, C.W. and Booker, J.R.: 1979. A Backus-Gilbert approach to inversion of travel time data for three-dimensional velocity structure. *Geophys. J. R. Astr. Soc. 59,* pp. 325-344.

(12)   Christoffersson, A. and Husebye, E.S.: 1979. On three dimensional inversion of P wave time residuals: Option for geological modelling. *J. Geophys. Res. 84,* pp.6168-6176.

(13)   Crosson, R.S.: 1976. Crustal structure modeling of earthquake data. 1. Simultaneous least squares estimation of hypocenter and velocity parameters. *J. Geophys. Res. 81,* pp. 3036-3046.

(14)   Ellsworth, W.L.: 1977. Three dimensional structure of the crust and mantle beneath the Island of Hawaii. Ph.D. thesis, Massachusetts Institute of Technology.

(15)   Evernden, J.F.: 1969. Precision of epicenters obtained by small numbers of world-wide stations. *Bull. Seism. Soc. Am. 59,* pp. 1365-1398.

(16)   Fitch, T.J.: 1975. Compressional velocity in source regions of deep earthquakes: An application of the master earthquake technique. *Earth Planet. Sci. Lett. 26,* pp. 156-166.

(17)   Garmany, J.: 1979. On the inversion of travel times. *Geophys. Res. Lett. 6,* pp. 277-279.

(18)   Garmany, J., Orcutt, J.A. and Parker, R.L.: 1979. Travel time inversion: a geometrical approach. *J. Geophys. Res. 84,* pp. 3615-3622.

(19)   Gerver, M.L. and Markushevich, V.M.: 1966. Determination of seismic wave velocity from the travel-time curve. *Geophys. J. R. Astr. Soc. 11,* pp. 165-173.

(20)   Gerver, M.L. and Markushevich, V.M.: 1967. On the characteristic properties of travel time curves. *Geophys. J. R. Astr. Soc. 13,* pp. 241-246.

(21)   Haddon, R.A.W. and Husebye, E.S.: 1978. Joint interpretation of P-wave time and amplitude anomalies in terms of lithospheric heterogeneities. *Geophys. J. R. Astr. Soc. 55,* pp. 19-44.

(22)   Herrin, E., Arnold, E.P., Bolt, B.A., Clawson, G.E., Engdahl, E.R., Freedman, H.W., Gordon, D.W., Hales, A.L., Lobdell, J.L., Nuttli, O, Romney, C., Taggart, J. and Tucker, W.: 1978. Seismological tables for P phases. *Bull. Seism. Soc. Am. 58,* pp. 1193-1241.

(23)   Hovland, J., Gubbins, D. and Husebye, E.S.: 1980. Upper mantle heterogeneities beneath Central Europe. *Geophys. J. R. Astr. Soc.,* in press.

(24)   Husebye, E.S., Christoffersson, A., Aki, K. and Powell, C.: 1976. Preliminary results on the 3-dimensional seismic structure of the lithosphere under the USGS central California seismic array. *Geophys. J. R. Astr. Soc. 46,* pp. 319-340.

(25)   Husebye, E.S., Haddon, R.A.W. and King, D.W.: 1977. Precursors to $P'P'$ and upper mantle discontinuities. *J. Geophys. 43,* pp. 535-543.

(26)   ISC Bulletin of the International Seismological Centre. July 1977. ISC, Newbury Berkshire, UK.

(27)   Iyer, H.M.: 1974. Teleseismic evidence for the existence of low velocity material deep into the upper mantle under the Yellowstone Caldera. (abstract) *EOS Trans. A. Geophys. Union 56,* pp. 1190.

(28)   Jackson, D.D.: 1972. Interpretation of inaccurate, insufficient and inconsistent data. *Geophys. J. R. Astr. Soc. 28,* pp. 97-109.

(29)   Jackson, J.: 1980 Errors in focal depth determination and the depth of seismicity in Iran and Turkey. *Geophys. J. R. Astr. Soc. 61,* pp. 285-301.

(30)   Jeffreys, H.: 1961. *Theory of Probability.* Oxford University Press, Oxford, England, 447 pp.

(31)   Jeffreys, H.A. and Bullen, K.E.: 1970. Seismological Tables. British Association for the Advancement of Science, Gray Miln Trust, London.

(32)   Johnson, L.E. and Gilbert, F.: 1972a. A new datum for use in the body wave travel time inverse problem *Geophys. J. R. Astr. Soc. 30,* pp. 373-380.

(33)   Johnson, L.E. and Gilbert, F.: 1972b. Inversion and inference for teleseismic ray data. *Methods in Comp. Phys. 12,* pp. 231-266.

(34)   Julian, B.R.: 1970. Ray tracing in arbitrarily heterogeneous media. *Lincoln Lab. Tech. Note, 1970-45.*

(35)   Julian, B.R. and Gubbins, D.: 1977. Three dimensional seismic ray tracing. *J. Geophys. 43,* pp. 95-113.

(36)   Kennett, B.L.N.: 1976. A comparison of travel time inversions. *Geophys. J. Astr. Soc. 44,* pp. 517-536.

(37)   Kennett, B.L.N.: 1978. Ray theoretical inverse methods in geophysics. *In* Applied Inverse Problems, ed. P.C. Sabatier, Springer-Verlag, Berlin.

(38)   Lanczos, C.: 1961. *Linear Differential Operators.* Van Nostrand, London, 564 pp.

(39)   Lawson, C.L. and Hanson, R.J.: 1973. *Solving Least Squares Problems.* Prentice-Hall Inc., Englewood Cliffs, New Jersey, 340 pp.

(40)   Lillwall, R.C. and Douglas, A.: 1970. Estimation of P-wave travel times using the joint epicentre method. *Geophys. J. R. Astr. Soc. 19,* pp. 165-181.

(41)   Masters, T.G.: 1979. Observational constraints on the chemical and thermal structure of the Earth's deep interior. *Geophys. J. R. Astr. Soc. 57,* pp. 507-534.

(42)   Menke, W.H.: 1977. Lateral inhomogeneities in P velocity under the Tarbela array of the Lesser Himalayas of Pakistan. *Bull. Seism. Soc. Am. 67,* pp. 725-734.

(43)   Mitchell, B.J., Cheng, C.C. and Stauder, W.: 1977. A three dimensional velocity model of the lithosphere beneath the new Madrid seismic zone. *Bull. Seism. Soc. Am. 62,* pp. 1061-1074.

(44)   Orcutt, J.A.: 1980. Joint linear, extremal inversion of seismic kinematic data. *J. Geophys. Res. 85,* pp. 2649-2660.

(45)   Parker, R.L.: 1977. Linear inference and underparameterized models. *Rev. Geophys. Space Phys. 15,* pp. 446-456.

(46)   Pavlis, G.L. and Booker, J.R.: 1980. The mixed discrete-continuous inverse problem: Application to the simultaneous determination of earthquake hypocenters and velocity structure. *J. Geophys. Res.,* in press.

(47)   Romanowicz, B.: 1979. Seismic structure of the upper mantle beneath the United States by three dimensional inversion of body wave arrival times. *Geophys. J. R. Astr. Soc. 57,* pp. 479-506.

(48)   Romanowitz, B.A.: 1980. Large scale lateral variations of P velocity in the upper mantle beneath Western Europe. *Geophys. J. R. Astr. Soc.*, in press.

(49)   Smith, M.L., Julian, B.R., Engdahl, E.R., Gubbins, D. and Gross, R.: 1979. Linearized inversion of travel times for three dimensional Earth structure (Abstract). *EOS Trans. Amer. Geophys. Union 59*, p.12.

(50)   Spencer, C. and Gubbins, D.: 1980. Travel time inversion for simultaneous earthquake location and velocity structure determination in laterally varying media. *Geophys. J. R. Astr. Soc. 62.*, in press.

(51)   Wiggins, R.: 1972. The general linear inverse problem: Implications of surface waves and free oscillations on Earth structure. *Rev. Geophys. Space Phys. 10*, pp. 251-258.

(52)   Wu, J.C.: 1977. Inversion of travel time data for seismic velocity structure in three dimensions. Ph.D. Thesis, University of Washington.

OPTIMUM APPROACHES TO MAGNITUDE MEASUREMENTS

Anders Christoffersson and Frode Ringdal

Dept. of Statistics   NTNF/NORSAR
University of Uppsala   Kjeller, Norway
Uppsala, Sweden

ABSTRACT

Conventional magnitude estimates from a seismic network
are based on measurements from only those stations which actually
detect a given event, thus ignoring the data from stations where
the signal amplitude is below the detection threshold. The topic
of this paper is to review some recent developments in magnitude
estimation methods, using information from both detecting and
non-detecting stations, and it is shown that this leads to significant improvments in magnitude estimates for small events.
The method is applied to study the linearity of the m_b:M_s relation
of earthquakes, and it is found that the apparent curvature of
observed m_b:M_s relationships can be explained through detectability considerations alone. Thus, from the available data,
there is no need to assume a change of the m_b:M_s slope at low
magnitudes. Finally, the method is used to obtain a separation
curve between earthquakes and explosions on the m_b:M_s diagram
which represents an improvement compared to conventional
approaches.

1.   INTRODUCTION

The basic problem addressed in this paper is related to the
fact that no seismic station has a 'perfect' detection capability.
Traditionally, seismologists have been working only with seismic
phases that can be seen on any given seismogram, while ignoring
stations where the phases are not detectable. As has been amply
demonstrated in the literature, this procedure may easily lead to
biased estimates of event magnitudes and amplitude ratios of
different phases, e.g., P:PcP, P:LR (the m_b:M_s criterion), etc.

575

*E. S. Husebye and S. Mykkeltveit (eds.), Identification of Seismic Sources - Earthquake or Underground
Explosion, 575–588.*
Copyright © 1981 by D. Reidel Publishing Company.

A typical example of the importance of this problem in earthquake-
explosion discrimination is the use of so-called 'negative evidence' in the m_b:M_s criterion, i.e., classification of small
events for which P-waves are detected, whereas surface waves
cannot be seen because of low M_s magnitude.

A proper statistical treatment of the above problem requires the introduction of advanced estimation methods, such
as maximum likehood estimation, where the information from detecting as well as non-detecting stations is included in the model.
Maximum likelihood estimation relating to these problems has been
applied by Vinnik and Dashkov (1) to determine the P:PcP amplitude ratio, by Kelly and Lacoss (2) to estimate seismicity and
detection potential, by Ringdal (3,4) for estimating magnitude
and detection thresholds, by Christoffersson (5,6) to obtain a
unified approach to model the entire seismic field thereby synthesizing and expanding a number of the above approaches and by
Elvers (7) to develop methods for reducing false associations of
phases from a global network.

Related problems have also been addressed by Elvers (8) regarding the use of 'negative evidence', by von Seggern and
Blandford (9) in establishing models for event detectability, and
by Pirhonen et al (10) and Ringdal et al (11) in estimating the
detection potential of existing seismological stations.

Seismic magnitude has long been accepted as the most reasonable measure of the 'size' of an earthquake. However, there are
large variations up to one magnitude unit in measurements at different stations for the same event. These variations are caused
by various factors: focal radiation patterns, variation in attenuation, focusing, etc. To improve the magnitude measurements,
single stations are usually combined into networks and the
'network magnitude' is estimated as the average magnitude of the
individual stations. This procedure works well for larger events,
but for events around and smaller than the 50% detection threshold
it tends to overestimate the magnitude as shown by, for example,
Herrin and Tucker (12). The reason for this bias is that information from non-detecting stations is ignored. To improve the magnitude measurements, it is necessary to use information on detecting
properties (thresholds, etc.) and region-station bias for both
detecting and non-detecting stations in the network.

In this paper we will mainly review some method-development
aimed at improving the accuracy of network magnitude determinations
and also discuss some recent developments concerning the classical
m_b:M_s relation for discrimination between earthquakes and explosions.

The work presented here is for simplicity expressed in terms of magnitude, whereas from an optimization point of view log A/T is better. It should also be stressed that the methods put forward here are not restricted to m_b and M_s measurements. They can easily be applied to measurements of the amplitudes of any phase.

2. BASIC MODEL

The basic model is of the following simple form:

$$m_b = m + B + \varepsilon \tag{1}$$

where

m_b is the magnitude of the arriving signal at a given station generated by a seismic event in a specific region

m is the 'true' magnitude of the event, and can be thought of as the average magnitude of a hypothetical uniformly spaced, dense global network.

ε is the scattering effects due to source radiation patterns and inhomogeneities in the earth.

If we regard ε as the sum of many independent scattering sources, it can be shown that the distribution of m_b for given m is Gaussian with expectation $m + B$ and standard deviation σ (= standard deviation of the scattering).

It is further assumed that the station has a detection curve of the form

$$\Phi\left(\frac{m_b - G}{\gamma}\right) = \int_{-\infty}^{(m_b - G)/\gamma} \frac{1}{\sqrt{2\pi}} \exp\{-t^2/2\}\ dt \tag{2}$$

i.e., the Gaussian distribution function where G is the mean (50% detection threshold) and γ the corresponding standard deviation. The justification for a detection curve of this form has been discussed by Ringdal (3). Other types of detection curves might be considered, but this one is regarded as flexible enough and convenient to work with. The slope of the detection curve is inversely proportional to γ at $m_b = G$.

From this we can deduce the distribution of observed m_b at a station for a given 'true' m. The distribution is

$$H(m_b/m) = f(m_b/m)\Phi\left(\frac{m_b - G}{\gamma}\right) \bigg/ \Phi\left(\frac{B + m - G}{\sqrt{\sigma^2 + \gamma^2}}\right) \tag{3}$$

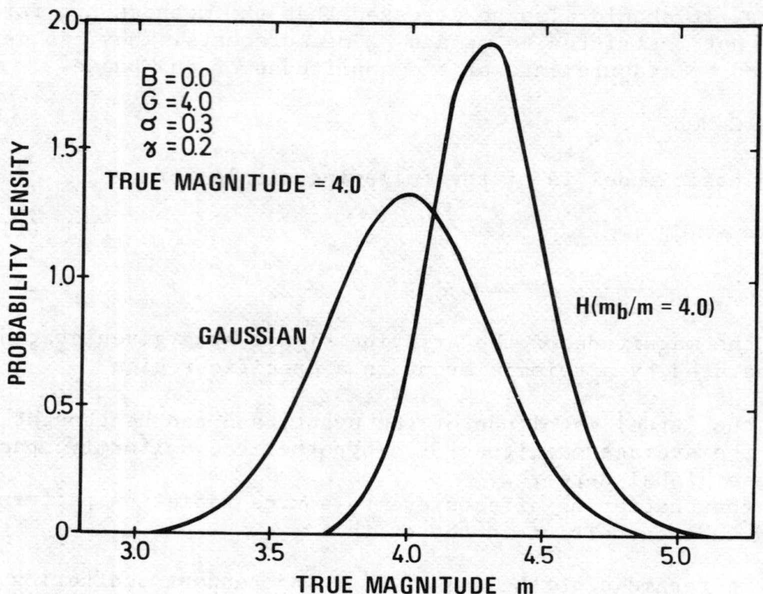

Figure 1. Frequency distribution $H(m_b/m=4.0)$ of observed magnitude at a station with detection threshold 4.0 for an event of true magnitude m=4.0. The Gaussian distribution corresponding to observed magnitude at a hypothetical station with 'perfect' detectability is shown for comparison.

where

$$f(m_b/m) = \frac{1}{\sqrt{2\pi}\,\sigma} \exp\left\{-(m_b-B-m)^2 \,/2\sigma^2\right\}$$

and $\Phi(\ )$ as in 2.

Figure 1 shows $H(m_b/m)$ for G=4.0, B=0.0, σ=0.3, γ=0.2 and the 'true' magnitude m=4.0. As a comparison the Gaussian distribution corresponding to the case where the station records each and every event are shown. From the figure we see that the bias in observed m_b is not negligible.

It is also possible to compute the expected observed magnitudes for a given 'true' magnitude.

$$E(m_b/m) = B + m + (\sigma^2/\sqrt{\sigma^2+\gamma^2})\,\Phi'(x)/\Phi(x) \qquad (4)$$

where

$$x = (B + m + G)/\sqrt{\sigma^2+\gamma^2}$$

Figure 2 shows $E(m_b/m)$ for $G=4.0$, $B=0.0$, $\sigma=0.3$ and $\gamma=0.2$.

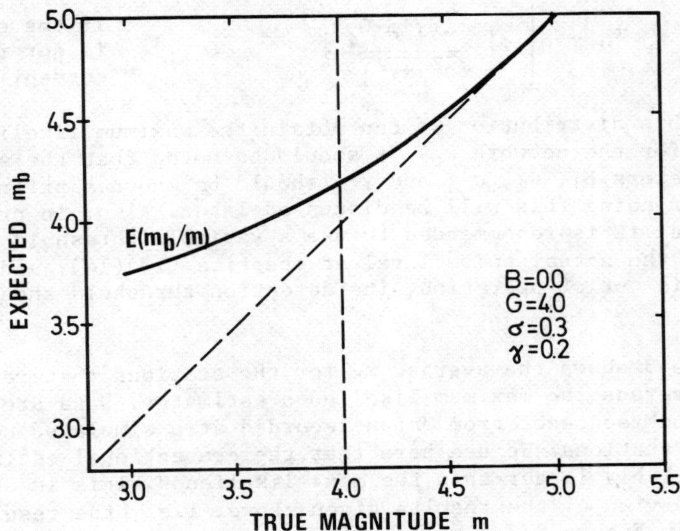

Figure 2. Expected m_b at a station as a function of true magnitude. The station detection threshold is assumed to be $m=4.0$, and the significant positive bias at low magnitudes is clearly indicated.

If we consider a network of stations we can obtain the joint distribution of observed (and not observed) m_b over the network given that at least one station observed the event. In this case the distribution is

$$H(m_{b1}, m_{b2} \cdots m_{bM}/m)\, dm_{b1} \cdots dm_{bM} = \qquad\qquad (5)$$

$$= \prod_{i=1}^{M} h_i(m_{bi}/m)dm_{bi} \bigg/ \left(1 - \prod_{i=1}^{M} \Phi\left\{\frac{(B_i+m-G_i)}{\sqrt{\sigma^2_i+\gamma^2_i}}\right\}\right)$$

where

$$h_i(m_{bi}/m)dm_{bi} = \begin{cases} f(m_{bi}/m)\ \Phi\left\{\dfrac{m_{bi}-G_i}{\gamma_i}\right\}\ dm_{bi} & \text{if the event is recorded at station } i \\[3em] \Phi\left\{\dfrac{-(B_i+m-G_i)}{\sqrt{\sigma^2_i+\gamma^2_i}}\right\} & \text{if the event is not recorded.} \end{cases}$$

Based on this distribution we can obtain the maximum likelihood
estimator for the network m_b. It should be noted that the station parameters B_i, G_i, σ^2_i and γ^2_i should be known a priori.
Methods for doing this will be discussed later. Also, in practical
applications it is recommended to use a variable threshold closely
related to the actual noise level at the time (13)(14), and if
a station is out of operation, its detection threshold should be
set to $+\infty$.

Figure 3 shows the average m_b for the stations that record
the event versus the maximum likelihood estimator. Data are from
an aftershock sequence from Japan recorded at a simulated network
of 15 U.S. stations. We see here that the conventional estimates
are considerably larger than the max. likelihood. This is as
expected in view of the results given above, i.e., the results
displayed in Figure 2 for the single station case carry over
directly to networks. There are some outliers in Figure 3. The
first (event 2 seen by only five of the stations) illustrates
the importance of having a variable threshold because the size
of the event is so large that it should have been recorded by
all the stations. Those that did not were probably out of operation or having extremely high noise levels due to coda from other
events in the sequence. The other outliers on the other hand mainly
reflect the bias in the conventional averaging process (small
events detected by few stations (1 to 3)). The maximum likelihood
estimator based on the distribution in eq. (5) is clearly much
more efficient than the conventional averaging process but somewhat more complicated as we need to know the station parameters.

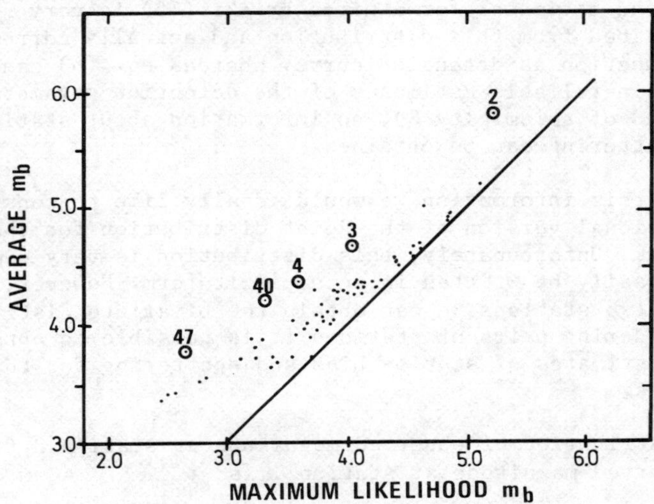

Figure 3. Average m_b and maximum likelihood m_b for a simulated
network of 15 stations. The data, marked as points, are based on
real observations from an aftershock sequence of 72 earthquakes
from Japan. Five 'outliers' in the data set are marked with
circles.

3.   UNCONDITIONAL DISTRIBUTION OF EARTHQUAKE MAGNITUDE

     In order to estimate the station parameters we have to
leave the realm of conditional distributions and turn to the
unconditional. By assuming the usual linear relation between
magnitude and logarithmic frequency of earthquake occurrence,
we find the unconditional distribution of observed magnitude at
a station

$$G(m_b) = \exp\{\beta G - \gamma^2 \beta^2\} \exp\{-\beta m_b\} \; \Phi(\frac{m_b - G}{\gamma}) \tag{6}$$

It would be observed that this distribution is independent of
scattering (σ) and region-station bias (B). Although not explicitly

stated, the likelihood equation corresponding to this distribution
was considered by Kelly and Lacoss (2) who in addition were estimating the total number of earthquakes in a given time period.
The likelihood estimator for β given by Aki (15) is very similar
to that obtained from this distribution and actually corresponds
to a step function as detection curve. Whereas eq. (6) can be
used to obtain reliable estimates of the detection parameters
G and γ (and of seismicity β), no information about station
bias and scattering can be obtained.

To get this information we would ideally like to consider
the unconditional version of the joint distribution for the complete network. Unfortunately, this distribution is very complex
and cannot easily be written in an explicit form. However, for
the case of two stations we can obtain the bivariate distribution.
And by considering pairs of stations it is possible to obtain
consistent estimates of station bias and scattering for the complete network.

The distribution of observed magnitude at station 2 for
a given observed magnitude at station 1 is

$$g(m_{b2}/m_{b1}) = \Phi\left(\frac{m_{b2}-G_2}{\gamma_2}\right) \frac{1}{\sqrt{2\pi}\sigma} \exp\left\{-(m_{b2}-m_{b1}-B)^2/2\sigma^2\right\}$$

$$\Phi\left\{\frac{m_{b1}+B-G_2}{\sqrt{\sigma^2+\gamma^2_2}}\right\}$$

where $B = B_2 - B_1 - \beta\sigma^2_1$

$\sigma^2 = \sigma^2_1 + \sigma^2_2$

Identified parameters are B, G_2, σ^2 and γ^2_2.

We can also derive the regression curve $E(m_{b2}/m_{b1})$ which turns
out to be

$$E(m_{b2}/m_{b1}) = B + m_{b1} + (\sigma^2/\sqrt{\sigma^2+\gamma^2_2})\ \Phi'(x)/\Phi(x) \qquad (8)$$

$$\text{with} \quad x = (m_{b1} + B - G_2)/\sqrt{\sigma^2+\gamma^2_2}$$

Again we note the similarity with the single station case. Figure
4 shows an example of this.

Another approach is to compute just the probability of detecting an event at station 2 for given observed magnitude at
station 1 as done by Ringdal (3).

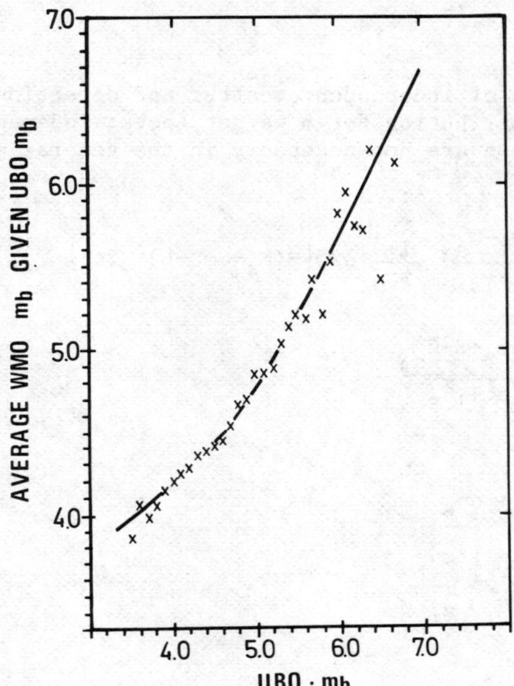

Figure 4.   Average observed magnitude at WMO for given observed
magnitude at UBO for 1958 events in the distance range 30°–90°.
The data are well fitted by the curve which is derived from the
model in the text. For details, it is referred to (6).

4.   APPLICATION OF THE m_b:M_s RELATIONSHIP OF EARTHQUAKES

Of special interest for discrimination is the generalization
of the above to the m_b:M_s situation. To do this we first rewrite
the basic model (1) as

$$m_b = B_b + \alpha_b m + \varepsilon_b \qquad\qquad (9)$$

$$M_s = B_s + \alpha_s m + \varepsilon_s$$

with detection curves

$$\Phi(\frac{m_b - G_b}{\gamma_b}) \quad \text{and} \quad \Phi(\frac{M_s - G_s}{\gamma_s}) \quad \text{respectively.}$$

In the special case of independent scatter and detection and the usual loglinear distribution for m we get (note: independent scatter and detection are not necessary in the general model)

$$K(M_s/m_b) = \Phi(\frac{M_s - G_s}{\gamma_s}) \frac{1}{\sqrt{2\pi}\sigma} \exp\{-(M_s - \alpha m_b - B)^2/2\sigma^2\} \;/ \qquad (10)$$

$$\Phi \; (\frac{\alpha m_b + B - G_s}{\sqrt{\sigma^2 + \gamma^2_s}})$$

with

$$\sigma^2 = \sigma^2_b \cdot \alpha^2 + \sigma^2_s$$

$$B = B_s - \alpha B_b - \alpha\beta \frac{\sigma^2_b}{\alpha_b}$$

$$\alpha = \frac{\alpha_s}{\alpha_b}$$

Identified parameters are α, B, σ^2, G_s and γ_s and

$$E(M_s/m_b) = B + \alpha m_b + \sigma^2/\sqrt{\sigma^2 + \gamma^2_s} \; \Phi'(x)/\Phi(x)$$

$$\text{with } x = (\alpha m_b + B - G_s)/\sqrt{\sigma^2 + \gamma^2_s} \qquad (11)$$

Note the similarity with eq. (7), i.e., the distribution of m_b at one station for given m_b at another.

It has been a common practice when looking at the $m_b:M_s$ relation to assume different slopes for large and small events. For large events the slope is typically ~ 1.6 and for small events some investigators, e.g., Evernden (16) argue that the slope is 1.0. In our model, the apparent decrease in slope for smaller events is just a consequence of the detection properties of the recording station.

Figure 5 shows a plot of Uppsala M_s versus NORSAR m_b for a sample of just over 400 events. It should be noted that the estimate of the slope is rather uncertain because there are very few

large events in the data and the slope is determined mainly by
large events. (A model with α fixed to 1.59 (as in Gutenberg and
Richter (17)) would fit the data almost as well as α =2.19.)

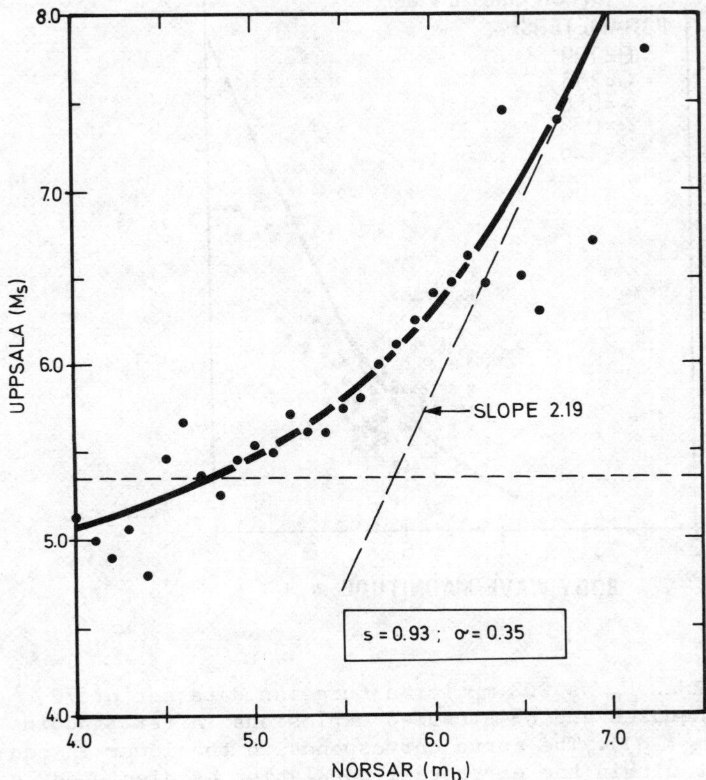

Figure 5. M_s:m_b plot for M_s observed at Uppsala and m_b observed
at NORSAR. The dots are the average observed M_s for given m_b. The
larger scatter at very small and very large m_b is due to the small
number of observations. The figure shows that the apparent decline
in the M_s:m_b slope at low magnitudes can be explained through
detectability considerations alone, as indicated by the curve fit
based upon the method in the text.

5.   APPLICATION TO SEISMIC DISCRIMINATION

The model presented in the preceding section can be applied
to the seismic discrimination problem. Ideally we would like to
have a classification rule where the probabilities of misclassification can be calculated, i.e., the probability of classifying

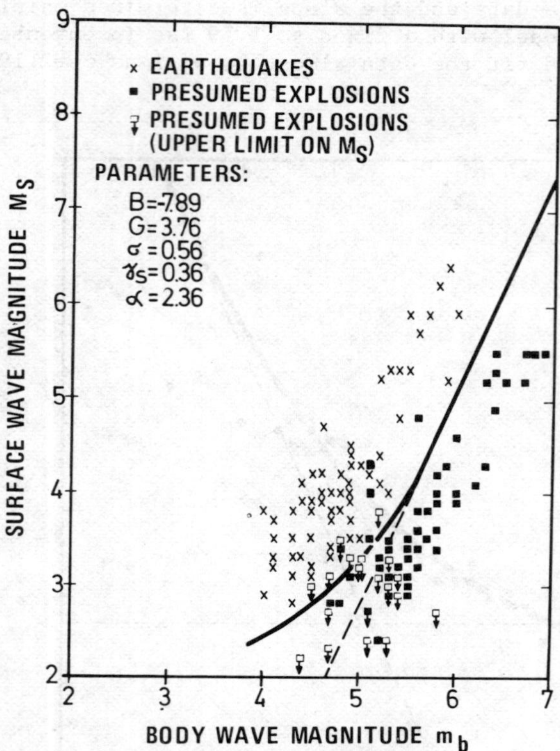

Figure 6. NORSAR M_S vs PDE m_b for a Eurasian data set of 72 shallow earthquakes and 83 presumed explosions (after Sandvin and Tjøstheim (18)). The curve corresponds to the lower 1% confidence limit within the earthquake population as discussed in the text. At low magnitudes, this curve gives an improved separation between the two populations compared to the straight (stippled) line.

an earthquake as an explosion and the probability of classifying
an explosion as an earthquake.

Is it then possible to compute this latter probability? In
order to do that we must know the joint distribution of observed
$m_b:M_s$ for explosions, implying some sort of stochastic process
generating the explosions. And the idea of people setting off
nuclear explosions in a random manner is by no means appealing.
We might of course use subjective probabilities, but for classification purposes this does not seem to be appropriate. So for
the time being there seems to be no reasonable way of calculating
this latter probability. For earthquakes, on the other hand, the
distribution of $m_b:M_s$ can be used to compute the probability
of classifying an earthquake as an explosion. To illustrate this
we consider the following data set used by Sandvin and Tjøstheim
(18). This data set consists of 72 shallow earthquakes from the
Eurasian continent and 83 presumed explosions from Eurasia.
Figure 6 shows a plot of NORSAR M_s/U.S.PDE m_b for the data set.
Fitting the above model to the earthquake data gives the following
results: B=-7.89, G=3.76, γ=0.36, σ=0.56 and α=2.36. The curve
on Figure 6 is the 1% lower confidence limit of expected M_s for
a given m_b; that is, the probability of an earthquake falling
below the curve is 1%. Those presumed explosions marked with
arrows in the figure are events where no Rayleigh waves were
detected and the M_s reading is actually an upper limit based on
the max. noise amplitude preceding the signal (in a 1 minute
time window covering the expected arrival time of the 20 s Rayleigh
component). The difference compared to the classification by
conventional discriminant analysis is apparent in the lower part
of the magnitude scale where the conventional method assumes
linear relation between observed m_b and M_s. We see that by including detectability considerations in the model, we achieve
a significantly improved separation curve between the earthquake and explosion populations at low magnitudes.

In conclusion, this model seems to be able to explain the
observed $m_b:M_s$ relation for earthquakes with just one slope parameter. For the explosions on the other hand, more research is
necessary in order to at least approximately describe the joint
observed $m_b:M_s$ distribution.

REFERENCES

1.   Vinnik, L.P. and Dashkov, G.G.: 1970, Phys. Sol. Earth
      (Eng. trans.) 4-9, Jan 1970.

2.   Kelly, E.J. and Lacoss, R.T.: 1969, Tech. Note 1969-41
      Lincoln Laboratories, MIT, Cambridge, Mass.

3.   Ringdal, F.: 1975, Bull. Seism. Soc. Am. 65, 1631-1642.

4.   Ringdal, F.: 1976, Bull. Seism. Soc. Am. 66, 789-802.

5.   Christoffersson, A.: 1978, NTNF/NORSAR Sci. Rep. No. 1-77/
      78, Kjeller, Norway.

6.   Christoffersson, A.: 1980, Phys. Earth Planet. Int. 21,
      237-260.

7.   Elvers, E.: 1980, FOA Report C, Stockholm, Sweden.

8.   Elvers, E.: 1974, Bull. Seism. Soc. Am. 64, 1671-1683.

9.   von Seggern, D. and Blandford, R.R.: 1976, Bull. Seism.
      Soc. Am. 66, 753-788.

10.   Pirhonen, S.E., Ringdal, F. and Berteussen, K.-A.: 1976,
      Phys. Earth Planet. Int. 12, 329-342.

11.   Ringdal, F., Husebye, E.S. and Fyen, J.: 1977, Phys. Earth
      Planet. Int. 15, P24-P32.

12.   Herrin, E. and Tucker, W.: 1972, Tech. Rep. to AFOSR, SMU,
      Dallas, Texas.

13.   von Seggern, D. and Rivers, D.W.: 1978, Bull. Seism. Soc.
      Am. 68, 1543-1546.

14.   Ringdal, F.: 1978, Bull. Seism. Soc. Am. 68, 1547-1548.

15.   Aki, K.: 1965, Bull. Earthq. Res. Inst. Tokyo Univ. 43,
      237-239.

16.   Evernden, J.F.: 1975, Bull. Seism. Soc. Am. 65, 359-391.

17.   Gutenberg, B. and Richter, C.F.: 1956, Ann. Geofis. 9, 1-15.

18.   Sandvin, O.A. and Tjøstheim, D.: 1978, Bull Seism. Soc. Am.
      68, 735-756.

THREE-DIMENSIONAL SEISMIC VELOCITY IMAGE OF THE UPPER MANTLE
BENEATH SOUTHEASTERN EUROPE

Jan Hovland[1] and Eystein S. Husebye[1,2]

[1] Dept. of Geology, Oslo University, Norway
[2] NTNF/NORSAR, Post Box 51, N-2007 Kjeller, Norway

ABSTRACT

P-wave travel time residuals for a network of stations in
southeastern Europe were used in an inversion experiment for
mapping upper mantle heterogeneities in the network region.
For Level 1 (depth range 0-100 km), the estimated velocity anomalies correlated reasonably well with heat flow observations
and other geophysical and tectonic features. Level 2 anomalies
(depth range 100-300 km) imply that the Pannonian Basin has an
asthenospheric root, while a velocity low over the northern
Aegean Sea supports McKenzie's (9) hypothesis of asthenospheric
upwelling as part of his model for the extensional tectonics of
the Aegean Sea. Also in the deeper part of the upper mantle
(levels 3 & 4, depth ranges 300-500 km and 500-600 km) pronounced
velocity anomalies are observed, although their interpretation
in terms of specific thermal or compositional anomalies is not
yet feasible.

1.   INTRODUCTION

To advance our knowledge of plate tectonic processes and
associated convection hypothesis there is an increasing demand
for structural details bearing on the lithosphere and the
asthenosphere. In this context recent methodological developments like 3-dimensional (3-D) inversion of 2-D seismic travel
time and amplitude data (1,2,3,4,5) as well as crustal reflection profiling of the COCORP type (6) are very suitable for
producing detailed seismic images of the lithosphere/astheno-
sphere. There is, however, a problem that is not easy to ignore

*E. S. Husebye and S. Mykkeltveit (eds.), Identification of Seismic Sources - Earthquake or Underground
Explosion, 589–605.*

associated with these modern analyzing techniques, namely, that
good quality seismic data are a prerequisite for high resolution
mapping. Such data are rather scarce as, for example, the number
of dedicated seismograph networks like NORSAR and that in Southern
California are rather few (1,7). The prospects for the future
are good, in view of the rapidly increasing number of digital
seismic recording systems becoming operational. For the time
being, however, it may be worthwhile to exploit existing, easily
available seismic data bases like the P-travel time residual
listings by the International Seismological Centre (ISC) for
globally distributed stations. Although the data quality and
associated network configurations are not optimum, Romanowicz
(8) and Hovland et al (4) among others have demonstrated that
the ISC observations permit lithosphere/asthenosphere seismic
imaging which is not matched by more conventional studies.

In this contribution we attempt a 3-D seismic mapping of
the upper mantle beneath the Aegean microplate and adjacent
areas using ISC listed P wave time residuals for the local seismograph network shown in Figure 1. This region is particularly
interesting in view of its complex tectonics still subject
to debate (9,10).

2.   DATA AND METHOD OF ANALYSIS

As mentioned above, the P-wave travel time residuals used in
analysis were taken from ISC files and cover the period 1964-
76. The number of events occurring in this interval was very
large, and in consequence the following selection criteria were
adopted; the minimum number of stations reporting an event was
set to 50 and at the same time we required that 30% of the network stations also had detected the event in question. Still,
with these selection criteria, a formidable number of events
remained which in turn were subdivided in distance/azimuth
branches (reckoned from the network center) of 10 deg, respectively. Within each bracket, the consistency of individual
station reportins were compared and in the end one or two
events were retained for further analysis. Due to relatively
modest earthquake activities in the Indian and Atlantic Oceans,
the event sampling as a function of distance/azimuth become
somewhat uneven. In the end a total of 132 events remained
the geographical distribution of which is shown in Figure 2.

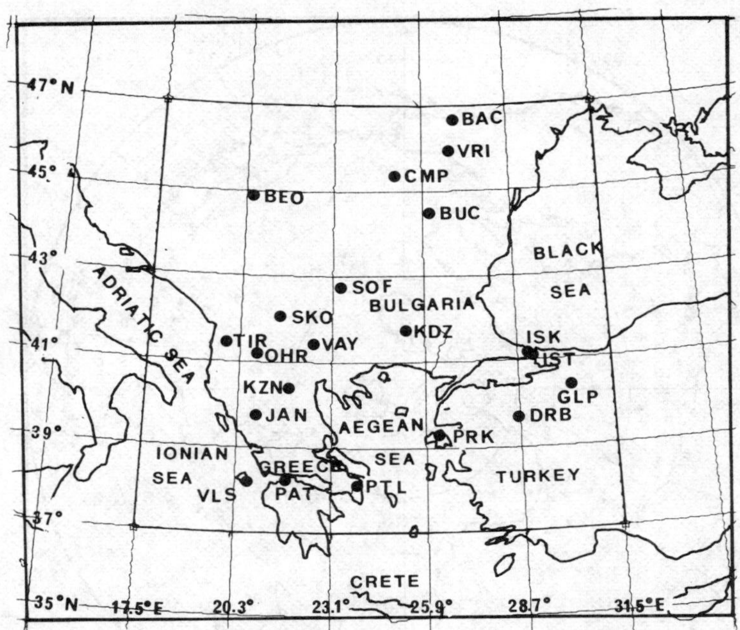

Figure 1. The southeastern Europe seismograph station network used in analysis. The intersections of the latitude/longitude grid system correspond to the knots in the time residual inversion described in section 2.

For each event, the average network time residual was estimated and then subtracted from the individual station reportings. This constituted the data base which in turn was subjected to inversion analysis following the procedure detailed by Hovland et al (4). A prerequisite here is that the residuals (relative to J-B tables) are assumed to be caused by lateral variations in P-wave velocity within a confined volume immediately beneath the receivers. The surface expression of this volume is marked in Figure 1 and extends to a depth of 600 km. Also, the velocity structure is represented by a smooth cubic interpolation between slowness values on a three-dimensional grid of 4x6x6 knots. This means that the upper mantle beneath the seismograph network is subdivided into 4 levels (0-100 km, 100-300 km, 300-500 km, 500-600 km) with slowness estimates at individual grids of 6x6 knots.

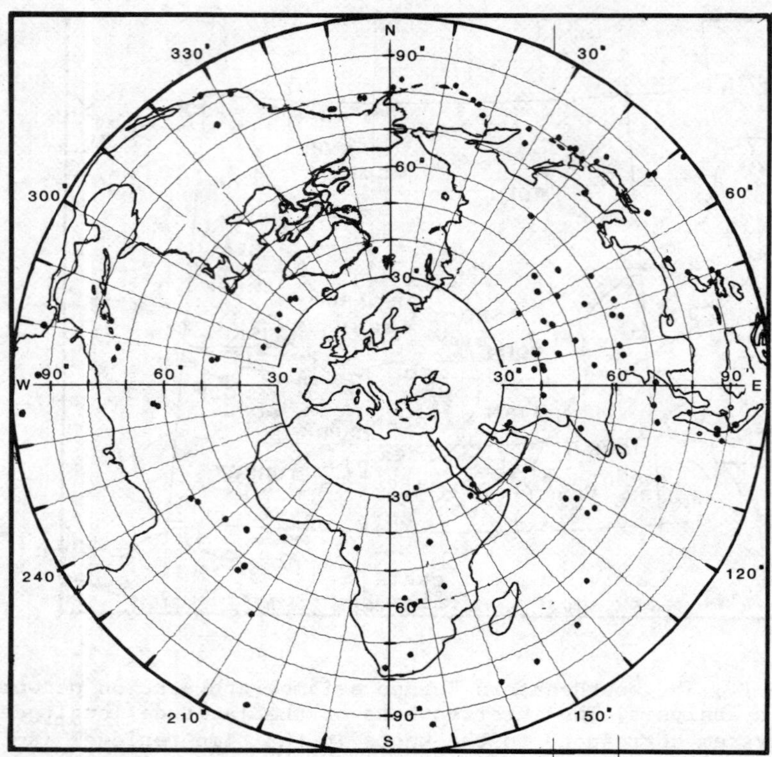

Figure 2.   Epicenter map of the 132 events used in analysis.
The associated P wave travel time residuals as reported by
stations in the southeastern Europe station network were
extracted from the International Seismological Centre (ISC)
bulletin tape files covering the time interval 1964-1976.

The relative slowness/velocity perturbations at the individual knots are estimated by minimizing the following
quantity:

$$(\delta\underline{T}-A\underline{m})*(\delta\underline{T}-A\underline{m}) + \theta^2\underline{m}*\underline{m} \tag{1}$$

where $\delta\underline{T}$ is observed travel time residuals, A is a travel path
matrix, θ^2 a stochastic inverse smoothing parameter while the
$\underline{m}$ vector contains the grid elements of the unknown slowness (s)/
velocity (v) perturbations. Interestingly, the smoothing has a
different influence on the solution depending on whether we
choose δs, δv or δv/v for unknowns. The reason for this is that

depth. This in turn affects the size of the diagonal elements of
the inverse matrix needed for solving eq. (1), as the stochastic
inverse smoothing parameter θ^2 is usually expressed in fractions
of the largest diagonal element. We decided to use $\delta v/v$ for
model parameterization because i) the deeper part of the model
is less well cross-sampled by the incoming rays and ii) outside
the model box the earth is presumed homogeneous. For computational
details on 3-D inversion of seismic travel time residuals, including estimation of standard errors and resolution for each
knot, reference is made to Hovland et al (4) and Gubbins (3).

3.   RESULTS

The network of stations in Figure 1 covers a region from
37.0° to 47.0°N and from 17.5° to 31.5°E, a square of side
approximately 1100 km. The thickness of the anomalous zone
was set at 600 km which conveniently includes the whole of the
upper mantle, implicitly indicating that the network data are
sensitive to anomalies down to depths comparable with the radius
of its aperture. The data will not discriminate too sharply
between anomalies of different depths, and so only a relatively
coarse representation is needed in this dimension. Horizontally,
the 6x6 grid of knots represents a compromise between adequate
resolution for small-scale features and having a reasonable
number of unknowns. In general, the velocity at any point will
be affected most by the value of the nearest knot. Thus the
slowness in the depth range 0-100 km is dominated by the values
at the top level of knots, for 100-300 km the second level,
300-500 km the third level and 500-600 km by the bottom level.

     The choice of the smoothing parameter in equation (1) is
somewhat subjective; if θ^2 is too small, the derived model
will exhibit large anomalies that fluctuate rapidly between
adjacent knots. Increasing θ^2 reduces the RMS of the velocity
perturbations but impairs the fit of the model to the data.
A trade-off curve between RMS velocity perturbations versus
the reduction in RMS travel time residual is shown in Figure 3,
from which a θ^2 value of 654.30 was chosen. Other parameters
of importance for judging the quality of the velocity anomaly
solution is the resolution and standard errors whose estimates
are displayed in Table 1. This information proved invaluable in
contouring the velocity anomaly maps shown in Figures 4a-4d and
should be kept in mind during the discussion of these results.
Level 1, shown in Figure 4a, is representative of the uppermost
100 km and corresponds roughly to the lithosphere. The resolution
of the knots along the edges is mostly poor (see Table 1) so the
corresponding values were not included in the anomaly contouring.
Outstanding features are velocity lows over southern Greece and
southwest Turkey. Further to the north there is also an area of

                                             J. HOVLAND AND E. S. HUSEBYE

Table 1. Resolution and standard error estimates for all
4x6x6 knots used in the seismic inversion experiment of the
upper mantle in Southeastern Europe. The velocity model itself
gives a reduction in the travel time residual observations
of 12.5 per cent while the corresponding variance reduction
is 23 per cent for a smoothing parameter value of θ^2 = 654.3.
The velocity anomalies for the four levels are shown in Figure 4.

RESOLUTION						STD. ERROR					
LEVEL 1											
0.0	0.0	0.0	0.5	0.3	0.0	0.1	0.3	0.4	1.1	1.1	0.2
0.0	0.5	0.4	0.6	0.3	0.0	0.3	0.9	1.1	1.0	1.1	0.2
0.0	0.6	0.7	0.4	0.1	0.0	0.4	1.3	1.0	1.1	0.5	0.3
0.2	0.7	0.7	0.6	0.7	0.6	1.1	1.0	1.0	1.1	0.9	1.1
0.0	0.7	0.5	0.7	0.7	0.2	0.3	1.0	1.1	0.9	1.1	0.8
0.0	0.5	0.3	0.1	0.0	0.0	0.2	1.2	1.1	0.5	0.2	0.1
LEVEL 2											
0.0	0.4	0.5	0.8	0.8	0.1	0.5	1.4	1.4	1.1	1.0	0.8
0.2	0.6	0.7	0.7	0.8	0.1	1.1	0.8	0.9	0.8	0.9	0.7
0.5	0.8	0.8	0.8	0.8	0.5	1.4	0.8	0.8	0.8	1.0	1.4
0.6	0.8	0.8	0.8	0.7	0.8	1.3	0.8	0.8	0.8	0.9	1.0
0.4	0.8	0.8	0.7	0.7	0.6	1.3	0.9	0.8	0.8	0.9	1.3
0.1	0.6	0.7	0.5	0.1	0.0	0.9	1.1	1.2	1.3	0.9	0.3
LEVEL 3											
0.1	0.5	0.7	0.7	0.8	0.4	0.7	1.4	1.2	1.3	1.1	1.5
0.6	0.7	0.8	0.8	0.8	0.6	1.4	1.0	0.9	0.9	0.8	1.4
0.6	0.8	0.8	0.8	0.8	0.7	1.3	0.8	0.8	0.7	0.8	1.1
0.7	0.8	0.8	0.8	0.8	0.8	1.3	0.9	0.8	0.7	0.8	1.1
0.6	0.8	0.8	0.8	0.8	0.7	1.2	0.9	0.8	0.8	0.8	1.2
0.3	0.5	0.7	0.6	0.6	0.1	1.4	1.4	1.2	1.2	1.3	1.0
LEVEL 4											
0.1	0.2	0.4	0.3	0.4	0.2	1.0	1.1	1.4	1.3	1.4	1.1
0.2	0.5	0.6	0.5	0.5	0.3	1.2	1.4	1.3	1.3	1.3	1.4
0.3	0.5	0.6	0.7	0.6	0.5	1.3	1.3	1.2	1.2	1.3	1.4
0.3	0.4	0.6	0.7	0.6	0.4	1.3	1.3	1.2	1.2	1.2	1.5
0.3	0.4	0.5	0.5	0.5	0.3	1.3	1.4	1.3	1.2	1.3	1.3
0.1	0.2	0.3	0.3	0.3	0.1	0.7	1.1	1.3	1.2	1.2	0.8

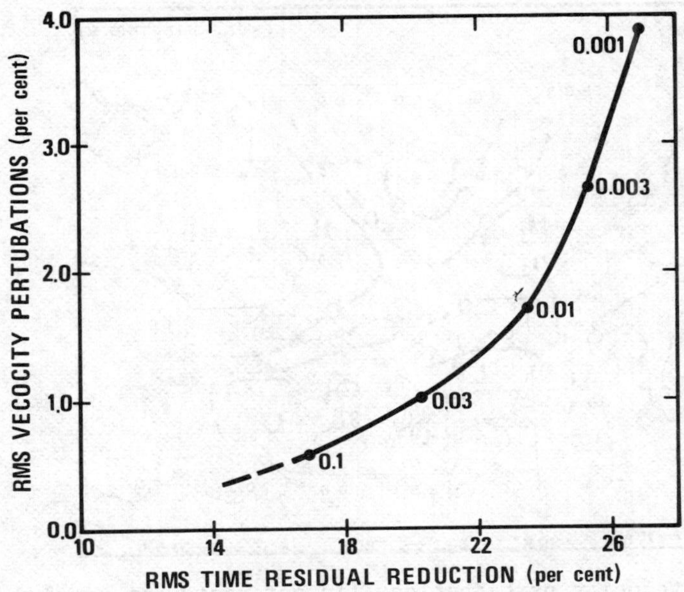

Figure 3. RMS slowness (velocity) perturbation versus RMS
time residual reduction for various values of the effective
smoothing parameter F. The relation between F and the stochastic
damping parameter θ^2 =F · max diagonal element of $A^T A$). A value
of F = 0.01 which corresponds to θ^2 = 654.30 was used in this
study. For computational strategy and details, reference is
made to Gubbins (3) and Hovland et al (4).

velocity low which straddles the southeastern part of the
part of the Pannonian basin. The anomaly picture changes dramat-
ically from Level 1 to Level 2 (Figure 4b), which is not unexpected in view of the plate tectonic axiom of a general decoupling between the lithosphere and asthenosphere. The areas of low
velocities are now found in the Aegean Sea and the coastal areas
of Bulgaria, while southern Greece and northwards are characterized by velocity highs. The velocity anomalies of level 3
(Figure 4c) are quite different from those of level 2, the main
dissimilarity being that the velocity highs now are in the Aegean
Sea-western Turkey while a velocity low is found over the Black
Sea. At level 4 (Figure 4d) the velocity anomaly pattern is
broadly similar to that of level 3, the main difference being
that the Black Sea anomaly has migrated somewhat westward. A
comparison of the velocity anomalies at individual levels gives
that the strongest ones are found at level 1 and level 4, while
those at level 2 and level 3 hardly exceed 2 per cent. The same

J. HOVLAND AND E. S. HUSEBYE

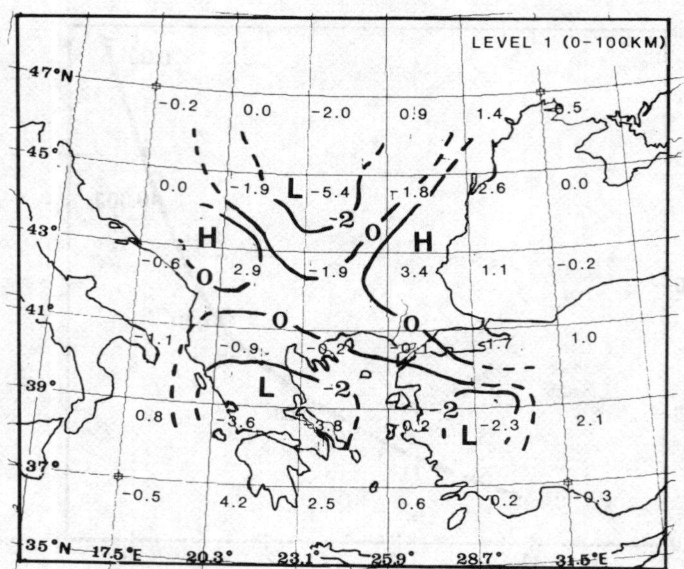

Figure 4a. Velocity perturbations (in per cent) for Level 1. Areas of high and low velocities are indicated by captial letters H and L. Resolution and standard errors for all knots are listed in Table 1.

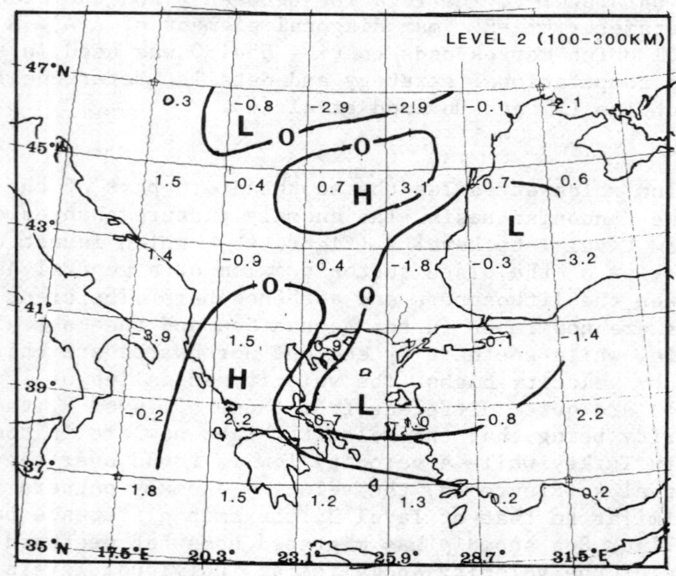

Figure 4b. Velocity perturbations (in per cent) for Level 2. Otherwise caption as for Figure 4a.

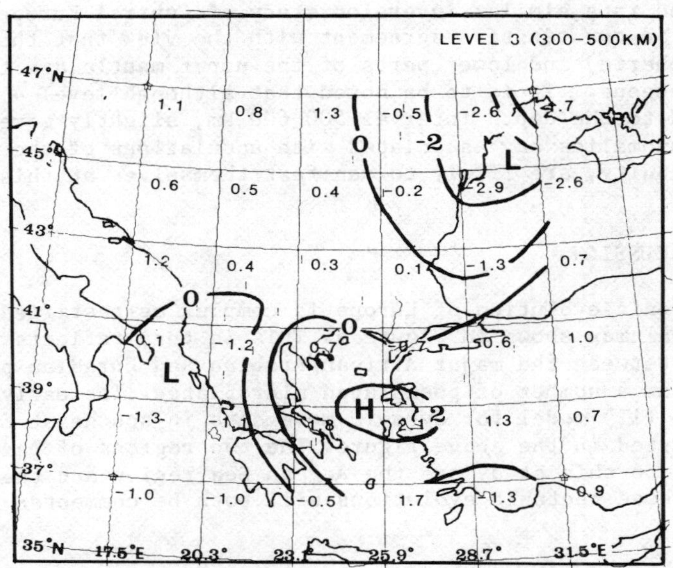

Figure 4c. Velocity perturbations (in per cent) for Level 3.
Otherwise caption as for Figure 4a.

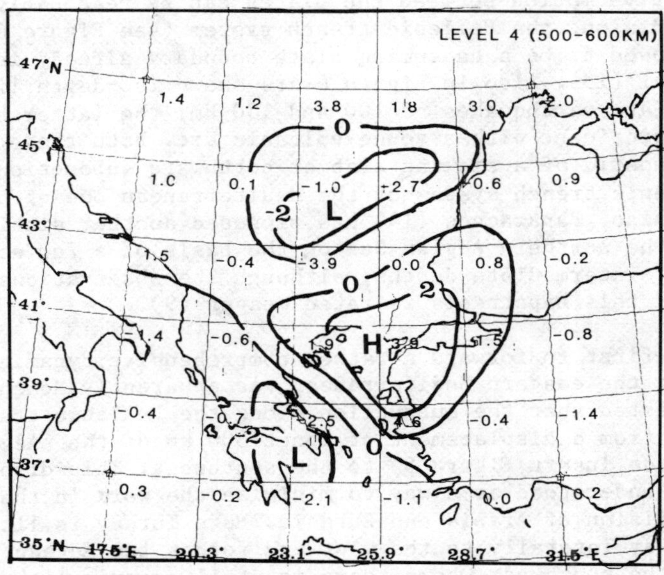

Figure 4d. Velocity perturbations (in per cent) for Level 4.
Otherwise caption as for Figure 4a.

was found in a similar inversion study of Central Europe (4).
This is in quantitative agreement with the view that the upper
(lithospheric) and lower parts of the upper mantle are the most
heterogeneous. It is to be noted that although level 4 is restricted to the depth interval 500-600 km, slightly more deep-
seated anomalies say associated with undulations of the 650 km
discontinuity, are likely to manifest themselves at this level.

4.   DISCUSSION

The tectonic evolution of Europe is complex as testified by the
geological map shown in Figure 5. This in turn reflects interactions between the major African/Arabian and Eurasian plates
as well as a number of postulated microplates. The early
McKenzie (11) model for dynamic movements in southeast Europe
are inserted in the above figure. The two regions of particular
interest to this study are the Aegean Sea region and the Pannonian
basin, whose tectonic evolutions will both be commented upon
below.

The relative pattern of plate motion in the Eastern Mediterranean region has been dominated during the last 70 m.y. by
the collision between the mentioned major plates, resulting in
a north-south shortening of several hundred kilometers (10).
The relative motion between the plates can be reasonably well
estimated along the Hellenic trench system (see Figure 6), which
was proposed to be a consuming plate boundary already in 1970 by
Ryan et al (13). Also in Figure 6 are shown iso-depth lines for
intermediate earthquakes of 100 and 150 km, the latter being
nearly coincident with a young volcanic arc. Both these features
are diagnostic of a sinking slab or northward subduction along
the Hellenic trench system of the Mediterranean oceanic lithosphere. Also, Papazachos (14) has proposed another sinking slab
beneath the northern Aegean Sea on the basis of a few earthquakes of intermediate depths, although the observational evidence for this hypothesis is rated meager (9).

The first to forward a rather comprehensive dynamic plate
model for the eastern Mediterranean was apparently McKenzie (11)
who suggested that the subduction along the Hellenic trenches
resulted from a displacement of about 300 km of the 'Aegean
plate' (see insert Figure 5) to the southwest. This displacement
could be understood as a way to minimize the work in the northward collision of Arabia and Eurasia. Here Turkey is literally
chased away laterally to the west and forces the Aegean plate to
move to the southwest where there is still oceanic lithosphere
to consume. However, independent geological and geophysical observations are not in direct conformity with this somewhat sim-
plistic plate tectonic model, the main objection being that

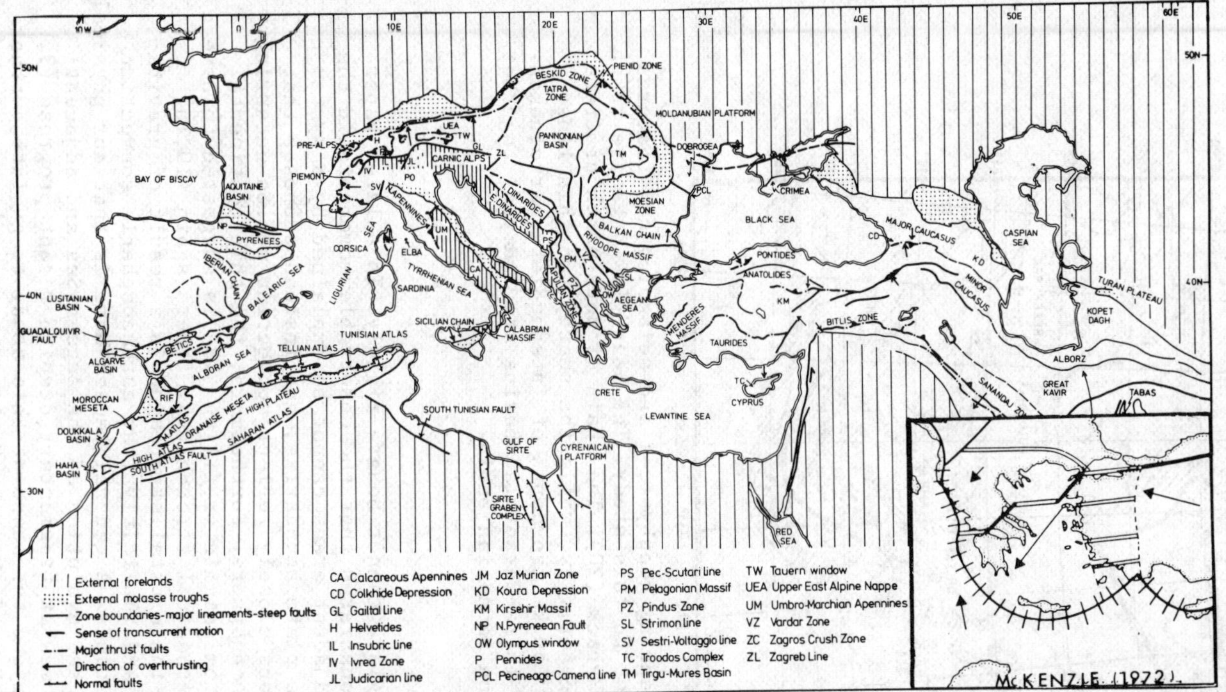

Figure 5. Schematic tectonic outline of the Alpine system of Europe, North Africa and the Middle East (after Dewey et al (12)). Inserted to the right is McKenzie's (11) sketch of the plate boundaries and motions in the Aegean Sea area. Extensional plate boundaries are shown by a double line, transform faults by a single heavy line and boundaries where shortening is occurring by a solid line cross by short lines at right angles. The fine dashed lines in western Turkey mark the E and W boundaries of the region of active E-W normal faults. The arrows show the direction of motion relative to Eurasia and their lengths are proportional to the magnitude of the velocity. The thick line in northwest Turkey is the North Anatolian Fault which terminates in the Marmara Sea area.

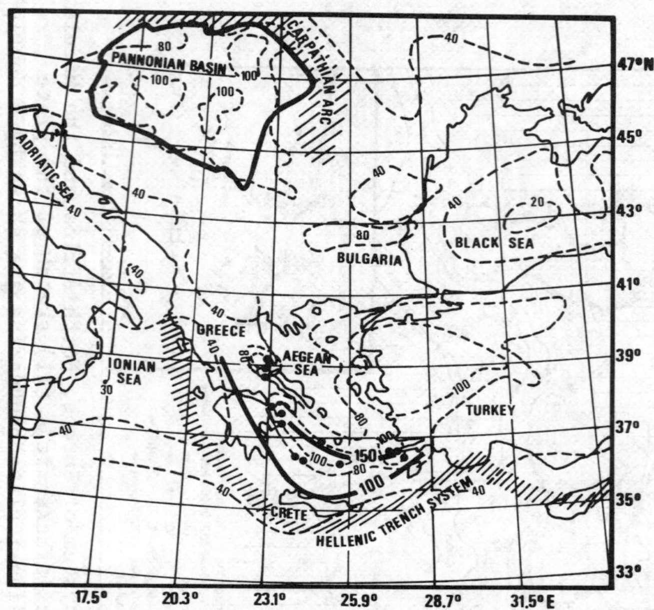

Figure 6.   Outstanding geophysical and seismological features
of direct interest to this study. The heat flow data are taken
from Cermak and Ryback (20). The chain of dots north of Crete
indicate volcanic activity, while the nearby two thin lines
mark the 100 and 150 km iso-depth of intermediate earthquakes.
To the north the intra-Carpathian basins are 'lumped' together
under the notation Pannonian Basin.

the Aegean microplate cannot behave rigidly. In a later paper
McKenzie (9) demonstrates that the observable movements in the
Aegean Sea and adjacent areas cannot be explained by a system of
forces applied to plate boundaries, and instead proposes that
the surface motions closely reflect convection motion occurring
below in the asthenosphere. Furthermore, the thin continental
crust reported from seismic profiling (15) is suggested to have
been halved by mechanical stretching. Others again (10,16)
hypothesize that the crustal stretching is a result of gravitational spreading and implicitly that asthenospheric convection
is of minor importance. A general remark here is that our physical
understanding of the evolution of the Aegean Sea and adjacent
areas still is somewhat fragmentary despite recent progress (9,
10), and it is beyond the scope of this study to go into further
details here in view of the relatively coarse resolution of the
seismic data at hand.

The mechanics of sedimentary basin formation have recently been subject to rather comprehensive studies (17,18). Of interest to us is the formation study by Sclater et al (19) of the intra-Carpathian basins; the peripheral ones like Vienna, W. Danube, Transcarpathian and Transylvanian and the central intra-Carpathian ones like E. Danube, Little Hungarian and Great Hungarian (Pannonian). In Figure 6 the notation Pannonian comprises most of these basins. Anyway, the basins are believed to be thermal in origin and are the direct result of the continental collision which formed the Carpathian arc. Dominant feature of the formation of these basins is stretching/attenuation of the whole lithosphere or parts thereof, which creates a very thin lithosphere, gives a rapid thermal subsidence and high heat flow. The latter feature is quite obvious from Figure 6.

Returning to our seismic images of the upper mantle displayed in Figure 4, the following comments apply. The estimated velocity anomalies for level 1 (Figure 4a) are the most interesting because the depth range is roughly equivalent to the lithospheric thickness, and thus good correlation with other geophysical and geological observations is feasible. A striking feature here is the correlation between heat flow (20) and estimated velocity anomalies as quantified in Figure 7. Pronounced velocity lows are found for S Greece and SW Turkey in agreement with heat flow observations and the extensional tectonics of the Aegean microplate. However, the lack of negative velocity anomalies in the northern Aegean Sea are contrary to expectations (10). The underthrusting slab is weakly manifested in the Peloponnesus area although nearly all peripheral knots are poorly resolved (see Table 1). High velocities are found over Bulgaria and the thermally cold Black Sea. The southern termination of this velocity high is the Marmara Sea and the North Anatolian Fault (see insert in Figure 5). The velocity low to the north straddles the Pannonian Basin which is characterized by high heat flow.

The level 2 results in Figure 4b reflect seismic sampling of a substantial part of the asthenosphere and thus would have a direct bearing on implied deep-seated features of the Aegean Sea extensional tectonics, the Pannonian basin formation and the slab subduction associated with the Hellenic Trench consuming boundary. In the latter respect, the high velocity zone north of Crete coincides with the intermediate earthquake isodepth of 150 km and thus clearly outlines the northern extension of the relatively cool, high velocity subducting slab. Another, somewhat coarser, approach for mapping of this slab has been outlined by Gregersen (21). This velocity high also extends far below southwest Greece but these anomalies are unlikely to

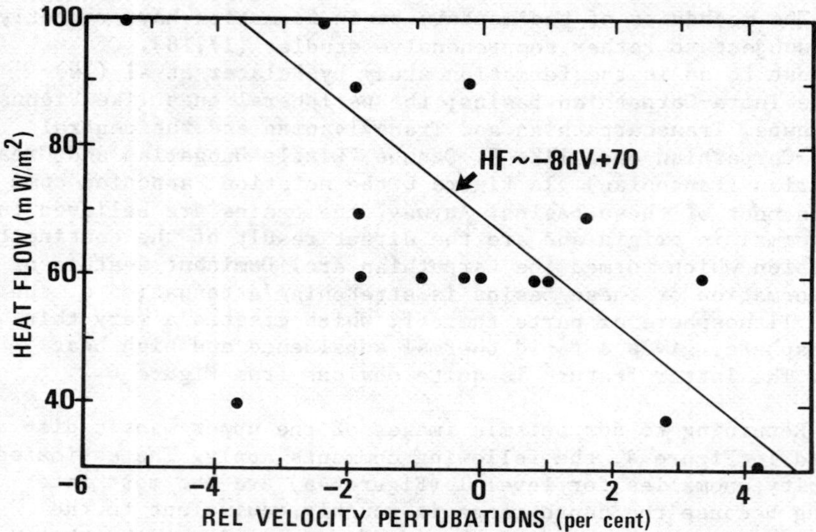

Figure 7. Estimated seismic velocity anomalies versus heat flow
in all level 1 knots with resolution better than 0.4. The re-
gressive coefficient of line drawn correspond to a velocity
anomaly of 1.2 per cent for each 10 mW/m² change in heat flow
for a 100 km thick layer (for details see Hovland et al (4).
The correlation between observed heat flow and estimated velocity
anomalies is deemed remarkably good. Furthermore, the heat flow
range of 30–100 mW/s⁻¹ matches that of velocity anomalies of
−5.4 to 3.4 per cent. An exceptional observation is the velocity
low of −3.6 in the Ionian Sea but we consider the corresponding
low heat flow value not particularly representative.

be associated with present-day subduction, although the hypothesis of a remnant slab from subduction prior to say 15 m.y. ago
cannot be ruled out. The velocity low beneath the northern
Aegean Sea and Bulgaria – the velocity anomalies are mostly
just larger than their associated std. errors – is considered
significant, and moreover interpreted as supporting McKenzie's
(9) hypothesis of convective upwelling in the asthenosphere
as part of his extensional tectonics model for the Aegean Sea.
The low velocities to the north are taken to indicate that the
'thermal origin' of Pannonian basin is rooted deeply in the
asthenosphere. Sandwiched between the Pannonian Basin and the
Black Sea there is a velocity high whose significance, if any,
is unclear.

The levels 3 and 4 are broadly similar to each other, albeit the most pronounced anomalies are slightly displaced relative to each other. In view of this similarity our hypothesis is that these velocity undulations in the deeper parts of the upper mantle cannot be explained in terms of known thermal or compositional anomalies but are plausibly accounted for by variations in the depth to the phase change at about 400 km. As demonstrated by Hovland et al (4) a velocity anomaly of 4% would lead to a depth change of this discontinuity of 80 km. Similar depth variations of the 650 km discontinuity may take place as well, and in this particular case are correlated with that at 400 km, given that this explanation in principle is correct. In the Central Europe study (4), the Level 2 and 3 anomaly patterns were broadly similar, but on the other hand quite different from that of Level 4.

Finally, we should like to remark that the inversion technique of Aki, Christoffersson and Husebye (1) or variants hereof have been widely used in earth structural studies, using P wave travel time reportings from local, regional and global seismograph networks. On the other hand, the level of sophistication in interpreting the estimated 3-D seimsic images of earth structures is somewhat superficial, essentially amounting to a comparison with independent tectonic and geophysical information. In this context there is an obvious need for an improved understanding of the relationship between seismic anomalies and physical properties of earth structures as well as their implications for mantle convection hypothesis. It is gratifying to see that this problem complex is emerging as a major seismological research area.

5.   CONCLUSION

3-D inversion of P wave travel time residuals as reported by ISC during 1964-76 for seismograph stations in southeastern Europe gave the following results:

-    Estimated lithospheric velocity anomalies correlate reasonably well with observed heat flow and other geophysical and tectonic features. In particular, pronounced low velocities are found over the Pannonian Basin, southern Greece and southwest Turkey.

-    In the asthenosphere, level 2 anomalies in the depth range 100-300 km, pronounced low velocity areas found under the Pannonian Basin and norther Aegean Sea. The latter result is in quantitative agreement with McKenzie's (9) hypothesis of asthenosphere convection for explaining the Aegean Sea extensional tectonics. The Hellenic trench subducting slab is 'seen' north of Crete.

-   The level 3 and 4 (depth range 300–500 km and 500–600 km)
    velocity anomalies are broadly similar but cannot be
    explained in terms of known thermal or compositional
    anomalies. It is hypothesized that these anomalies are
    associated with the 400 and 650 km phase transition zones.

ACKNOWLEDGEMENT

This research was supported by the Advanaced Research Projects
Agency of the Department of Defense and monitored by AFTAC,
Patrick AFB FL 32925, under contract F-08606-79-0001.

One of us, JH, is supported by a postdoctoral fellowship from
the Norwegian Council for Science and Humanities.

REFERENCES

1.   Aki, K., Christoffersson, A. and Husebye, E.S.: 1977,
     J. Geophys. Res. 82, 277–296.

2.   Christoffersson, A. and Husebye, E.S.: 1979, J. Geophys.
     Res., 84, 6168–6176.

3.   Gubbins, D.: 1981, Source location in laterally varying
     media (ibid).

4.   Hovland, J., Gubbins, D. and Husebye, E.S.: 1981, Geophys.
     J.R. astr. Soc., in press.

5.   Haddon, R.A.W. and Husebye, E.S.: 1978, Geophys. J.R. astr.
     Soc., 55, 19–44.

6.   Cook, F.A., Albaugh, D.S., Brown, L.D., Kaufman, S. and
     Oliver, J.E.: 1979, Geology, 7, 563–567.

7.   Raikes, S.A. and Hadley, D.M.: 1979, Tectonophysics, 56,
     89–96.

8.   Romanowicz, B.A.: 1980, Geophys. J.R. astr. Soc., 63,
     217–232.

9.   McKenzie, D.P.: Geophys. J.R. astr. Soc. 55, 217–254.

10. LePichon, X. and Angelier, J.: 1979, Tectonophysics, 60,
     1–42.

11. McKenzie, D.P.: 1972, Geophys. J.R. astr. Soc., 30, 109-185.

12. Dewey, J.F., Pitman, W.C., Ryan, W.B.F. and Bonnin, J.: 1973,
     Geol. Soc. Am. Bull., 84, 3137-3180.

13. Ryan, W.B.F., Stanley, D.J., Hersey, J.B., Fahlquist, D.A.
     and Allan, T.D.: 1970, In: A. Maxwell (ed.), The Sea,
     Vol. 4, II. Wiley-Interscience, New York, N.Y., 387-492.

14. Papazachos, B.C.: 1976, Tectonophysics, 33, 199-209.

15. Makris, J.: 1977, Geophysical Investigations of the Hel-
     lenides, Hamburger Geophys. Einzerlschriften, 34, Hamburg
     Univ., FRG.

16. Tapponier, P.: 1977, Bull. Soc. Geol. Fr., 7, XIX, 3:
     437-460.

17. Jarvis, G.T. and McKenzie, D.P.: 1980, Earth Planet.
     Sci. Lett., 48, 42-52.

18. Sclater, J.G. and Christie, P.A.F.: 1980, J. Geophys.
     Res., 85, 3711-3739.

19. Sclater, J.G., Royden, L., Horvath, F., Burchfield, B.C.,
     Seemben, S. and Stegena, L.: 1980, Earth Planet. Sci.
     Lett., 51, 139-162.

20. Cermak, V. and Ryback, L. (eds.): 1979, Terrestrial Heat
     Flow in Europe, Springer Verlag, Berlin, 328 p.

21. Gregersen, S.: 1977, Tectonophysics, 37, 83-93.

TAU INVERSION OF UPPER MANTLE ARRAY DATA WITH APPLICATION TO THE
PROBLEM OF THE SCATTERING OF SEISMIC WAVES IN THE LITHOSPHERE

R.F. Mereu

Dept. of Geophysics. University of Western Ontario,
London, Ontario, Canada N6A 5B7

An analysis of the records from several hundred earthquakes
in the 12–35° distance range recorded at the WRA array in Australia
(1, 2), the GBA array in India (3), the YKA array in Canada (4),
and the EKA array in Scotland (5) shows that the upper mantle
structure in different regions of the earth is extremely complex.
The analysis method used in all of the above-mentioned studies
was an adapative processing technique originally developed by
Gangi and Fairborne (6). This method was modified by King, Mereu
and Muirhead (7) and Mereu and Ram (8) so that arrival times,
apparent velocities, and azimuths of all portions of the signal
wave trains crossing the UKAEA type arrays could be measured
automatically along the whole seismic trace. Efforts were then
made to identify and locate the upper mantle travel-time branches
associated with the 450 and 650 km discontinuities.

An examination of the results of these experiments showed
that in general different values of $dT/d\Delta$ and azimuth were detected for various portions of the signals along the individual
seismic traces thus showing that multipathing had taken place.
However, when all the results from any one region were put together, the observed pattern was very complex but indicated or
suggested that some of the later arrivals could be associated
with the forward branches. The most surprising feature of the
results was the complete lack of almost any evidence for the
retrograde branches when in fact these branches should theoretically have the largest amplitudes and be well recorded. Because
of the complexity of the results and the difficulty in position-
ing the later branches accurately, no simple earth model was

*E. S. Husebye and S. Mykkeltveit (eds.), Identification of Seismic Sources - Earthquake or Underground
Explosion, 607–609.*

possible which could satisfy the results in any one region
satisfactorily.

A simplified linearized Tau inversion analysis based on the
theory of Garmany and Orcutt (9) was applied to the observed
dt/dΔ, t and Δ observations of both the first arrival and later
arrivals to yield extremal values of velocity of ±0.5 km/sec
deviation from the Herrin model at any particular depth. The
results obtained suggested that the upper mantle can best be described as a non-layered region of the earth's interior with
high average velocity gradients on which is superimposed a set
of random vertical and lateral inhomogeneities. These random
variations have an associated correlation distance of about 15 km
and can deviate the velocities laterally by several per cent at
any particular depth. A set of numerical experiments was done in
which rays were traced step by step through the numerical models
such as that described above by Mereu and Ojo (10). The patterns
of scattering obtained was very similar to that observed with
real data. The continuous upper mantle travel-time curve associated with the average model was in general broken up into numerous
overlapping segments yielding multiple arrivals of dt/dΔ with
much scatter. The scattered points all lie later than the average
model curve as would be expected due to the path lengthening effects of the random medium. Of particular note was the fact that
both the results from the observed data and from the random
model data show no distinct retrogrades. In other words it may
be said that in order to explain upper mantle short-period
seismic array results, it is not necessary to invoke the upper
mantle discontinuities into the various models if instead lateral
and vertical velocity perturbations are allowed in the models.

REFERENCES

1.   Mereu, R.F., Simpson, D. and King, D.W.: 1977, Earth and
      Planet. Sci. Lett. 21, 439-447.

2.   Simpson, D., Mereu, R.F. and King, D.W.: 1974, Bull. Seism.
      Soc. Am. 64, 1757-1788.

3.   Ram, A. and Mereu, R.F.: 1977, Geophys. J.R. astr. Soc. 49,
      84-114.

4.   Ram, A., Weichert, D. and Mereu, R.F.: 1978, Can. J. Earth
      Sci. 15, 227-236.

5.   England, P.C., Worthington, M. and King, D.W.: 1977, Geophys.
      J.R.astr. Soc. 48, 71-79.

6.   Gangi, A.F. and Fairborne, J.W.: 1968, Suppl. Nuovo Cim. 1,
     6, 105-115.

7.   King, D.W., Mereu, R.F. and Muirhead, J.: 1973, Geophys. J.R.
     astr. Soc. 35, 137-167.

8.   Mereu, R.F. and Ram, A.: 1975, in "Exploitation of Seismo-
     graph Networks", Advanced Studies Institute Series E, No. 11,
     pp. 327-339 (ed. Beauchamp, K.G.).

9.   Garmany, J. and Orcutt, J.: 1979, J. Geophys. Res. 84, 3615-
     3622.

10. Mereu, R.F. and Ojo, S.B.: 1981, Phys. Earth Planet. Sci.
     (in press).

OBSERVATION OF SCATTERING IN THE LITHOSPHERE

Anton M. Dainty

School of Geophysical Sciences, Georgia Institute of Technology, Atlanta, Georgia U.S.A.

Three types of observations of scattering in the lithosphere may be made. Observations of forward scattering have been carried out using the fluctuations of amplitude and phase of teleseismic P arrivals at arrays. Back scattering of shear waves may be observed in the coda of local earthquakes following direct S. Total scattering may be observed in studies of seismic attenuation (Q), where loss of energy due to scattering is described by the turbidity $g = w/(vQ_s)$, where w = angular frequency, v = seismic velocity, and Q_s = Q due to scattering. Two single-scattering theories have been used in the past to describe scattering. Random medium theory, in which the seismic velocity is considered to have random fluctuations with a specified autocorrelation, has been used to interpret fluctuations and Q. Discrete scatterer theory, in which the scattering has been considered as due to scatterers of a finite size randomly distributed, has been used in backscattering observations. A significant difference between the theories is seen in backscattering and total scattering situations. The individual scatterers in discrete scattering theories in general backscatter and total scatter efficiently for ka $\geq$ 1, where k is wavenumber of the seismic wave and a is the linear dimension of the individual scatterers. In random medium theory, however, efficient backscattering only occurs near kA = 1, where A is the scale length of the fluctuations. The amount of total scattering in random medium theory is uncertain because it is not clear how much forward scattered energy should be included.

To distinguish between these theories, an integrated attack on all aspects of scattering is needed. As a first step, the variation of Q for shear waves between 1 and 25 Hz has been studied using observations reported in the literature and the

611

discrete scatterer theory. It is found that a simple model may explain the observations. If the total Q obeys the relation $1/Q = 1/Q_i + 1/Q_s$, where Q_i = intrinsic (anelastic) Q, the observations in any one region may be adequately explained using a constant Q_i and a constant turbidity g as defined above. Since $g = n\sigma$, where σ is the total scattering cross-section of each scatterer and n is the number of scatterers per unit volume, this means that σ must be constant with frequency for the frequency range 1-25 Hz. This is possible provided $ka \geq 1$ for all scatterers of importance in this frequency range. Aki, however, has noted that observations of the Q of Rayleigh and Love waves show that Q is high for periods of 20 sec and greater, implying that $ka < 1$ (low scattering efficiency) for these periods. This places bounds of 1-10 km on a, the linear dimension of the scatterers. Q_i is found to be high, Q_i = 2000 for the regions studied (Japan and Central Asia); g is different for the two regions.

LATERAL VARIATIONS IN THE EARTH'S CRUST AND THEIR EFFECT ON
SEISMIC WAVE PROPAGATION

R.F. Mereu and S. Ojo

Dept. of Geophysics, University of Western Ontario,
London, Ontario, Canada

ABSTRACT

A review of the results of a number of crustal seismic experiments which were carried out over the Canadian Shield over the
past fifteen years in general shows that precise values of
crustal velocities are not well determined and are very sensitive to how the data set was sampled. When the seismic data is
combined with gravity data, the only reasonable solutions to
the inverse problem are ones which have large lateral velocity
variations in structure (1). There is also very strong evidence
from both the seismic and gravity data of many experiments
that indicates that large velocity gradients exist with depth
in the crust such that the velocity contrast at the Moho is not
large (2). In recent years considerable progress has been made
in using random media theory to explain precursors to various
earthquake phases and the problem of amplitude and travel-time
fluctuations across seismic arrays (3, 4, 5, 6).

In this paper the results of a number of numerical experiments performed by Mereu and Ojo (7) are presented in which
seismic waves are propagated through a crust modelled as a random medium. The starting model is a single-layered crust with a
vertical velocity gradient. A two-dimensional set of small smooth
random velocity deviations is then superimposed on the vertical
gradient model. The resulting models show short reflectors at
various depths in agreement with many deep seismic reflection
experiments (8, 9). Ray-tracing experiments using the method of
Gebrande (10) showed that the effect of the lateral and vertical
anomalies is to scatter the energy and break up the continuous
travel-time curve from a vertical gradient model into segments

E. S. Husebye and S. Mykkeltveit (eds.), Identification of Seismic Sources – Earthquake or Underground
Explosion, 613–614.

of different slope similar to those observed in many long-range
refraction experiments. Many of the numerical experiments produced
a Pg segment and a P* segment with an 'apparent' Conrad discontinuity at a depth of 10 to 20 km. As the correlation distance
was increased, the apparent depth tended to increase. For small
correlation distances there was a tendency for the travel-time
curve to break up into three segments indicating that the PnP
branch should be observable above the noise but the position of
the associated cusp is very sensitive to the effects of scattering.

REFERENCES

1.   Mereu, R.F. and Jobidon, G.: 1971, Can. J. Earth Sci. 8,
     1553-1583.

2.   Mereu, R.F., Majumdar, S.B. and White, R.E.: 1977, Can. J.
     Earth Sci. 14, 196-208.

3.   Aki, K.: 1973, J. Geophys. Res. 78, 1334-1346.

4.   Cleary, J.R., King, D.W. and Haddon, R.A.W.: 1975, Geophys.
     J.R. astr. Soc. 43, 861-872.

5.   Berteussen, K.-A., Christoffersson, A., Husebye, E.S. and
     Dahle, A.: 1975, Geophys. J.R. astr. Soc. 42, 403-417.

6.   Aki, K., Christoffersson, A. and Husebye, E.S.: 1977, J.
     Geophys. Res. 82, 277-296.

7.   Mereu, R.F. and Ojo, S.B.: 1981, Phys. Earth Planet. Inter.,
     in press.

8.   Cumming, J.R. and Chandra, N.N.: 1975, Can. J. Earth Sci. 12,
     539-557.

9.   Smithson, S.B., Brewer, J.A., Kaufman, S., Oliver, J.E. and
     Hurich, C.A.: 1979, J. Geophys. Res. 84, 5955-5972.

10. Gebrande, H.: 1976, in "Explosion Seismology in Central
     Europe" pp. 162-167, (ed. Giese, P., Prodehl, C. and Stein,
     A.).

FUNDAMENTALS OF MULTIDIMENSIONAL TIME-SERIES ANALYSIS

M. P. Ekstrom and T. L. Marzetta

Schlumberger-Doll Research
Old Quarry Road
Ridgefield, Connecticut 06877

This paper presents a survey of recent research in multi-dimensional signal processing. As a means of illustrating the fundamental principals, the presentation focuses on two-dimensional (2-D) results. A tutorial section in the paper addresses the basic subjects of 2-D systems representations (convolution and difference equations), 2-D Z-transforms, stability theorems/tests, and modeling. Both deterministic and random formalisms are described. To give some flavor for on-going research in this field, two new results are also described. The first deals with the extension of Wiener's realizable filtering technique to 2-D and the second with the concept of autoregressive models of 2-D random fields.

1. INTRODUCTION

Time-series methodologies have long found application in detection seismology, usually in processing signals from the large-aperture seismic arrays used to detect weak seismic events. Because the arrays are made-up of spatially distributed receivers, their response to a source excitation is a multidimensional (M-D) space-time signal. While these signals are M-D, they have conventionally been processed, for detection/estimation purposes using one-dimensional (1-D) techniques. Beamforming is an obvious example of this.

More recent research has focused on both characterizing and processing array signals in their natural M-D setting (for example, see Lacoss et al, 1969 and Woods, 1976). Not surprisingly, many classical array processing problems involving

615

E. S. Husebye and S. Mykkeltveit (eds.), *Identification of Seismic Sources - Earthquake or Underground Explosion, 615–647.*
Copyright © 1981 by D. Reidel Publishing Company.

modeling, estimation, and detection are quite naturally cast in
the M-D case. Consider the following seismic detection problem.
The seismic event to be detected may arise from a wide-variety
of sources/locations. Thus, its array signature may be best
characterized as a sample function from a M-D random field. In
the presence of additive environmental noise, a reasonable
detection strategy (under appropriate conditions, it is optimal)
involves i) forming an optimal estimate of the M-D signal, ii)
correlating the estimate with the array measurement, and iii)
performing a subsequent likelihood ratio test. Construction of
this "estimator-correlator" type receiver (Van Trees, 1971)
therefore requires addressing both the modeling and estimation
of M-D fields.

The purpose of this paper is to present in a unified framework, some recent extensions of modern time-series analysis to
multidimensional problems of this nature. This work is collected
from diverse sources (almost exclusively outside the seismology
community), and generally categorized under the heading of multidimensional signal processing. In order to simplify the
description, the presentation will focus on 2-D problems/results.
This will involve an improvement in clarity and no loss of
information. For the most part, the steps required in going
from the 2-D to M-D cases will be apparent.

The paper is divided into two main parts. The first is a
tutorial, providing a comprehensive overview of some basic 2-D
signal processing results. Both deterministic and random
formalisms are described. The second section contains examples
of recent research, seemingly relevant to applications in
detection seismology.

2. FUNDAMENTALS OF 2-D SIGNAL PROCESSING

We begin this section with a review of classical 1-D time
series results. This review will be presented in such a way
that the 2-D results can be developed analogously. In this
manner, we hope to draw on the reader's familiarity with time
series, in evolving some physical insight and understanding into
the 2-D formalisms.

2.1 Review of Time Series (1-D)

Consider the two-sided ℓ^1 sequence x(k) k=0, $\pm$1, $\pm$2...
This sequence has a corresponding Z-transform

$$X(z) = \sum_{k=-\infty}^{+\infty} x(k)\, z^{-k} \;, \tag{1}$$

where z is a complex variable, and X(z) is defined over a region
of convergence (ROC):

$$\{(z): \sum_{k=-\infty}^{\infty} \left| x(k) z^{-k} \right| < \infty \} \ . \tag{2}$$

The original sequence can be recovered from X(z) via the
inverse transform

$$x(k) = \frac{1}{2\pi j} \int_c X(z) \ z^{k-1} dz \tag{3}$$

where c is a CCW closed contour, encircling the origin. A
transform pair is indicated by the relation

$$x(k) \longleftrightarrow X(z) \ . \tag{4}$$

For some interesting cases, there exists a fundamental
relationship between the support of x(k) and the ROC of X(z).
Specifically, for x(k) a right-sided sequence, i.e. x(k)=0 k<0,
its transform must have a ROC for which $R_- < \left| z \right|$. Conversely, for
x(k) a left-sided sequence, X(z) has a ROC for which $\left| z \right| < R_+$. As
might be expected, transforms of two-sided sequences have ROC
which are annular and described by $R_- < \left| z \right| < R_+$. These relations
along with that of a finite sequence are summarized in Fig. 1.

<u>x(k)--Support</u>   <u>X(z)--ROC</u>

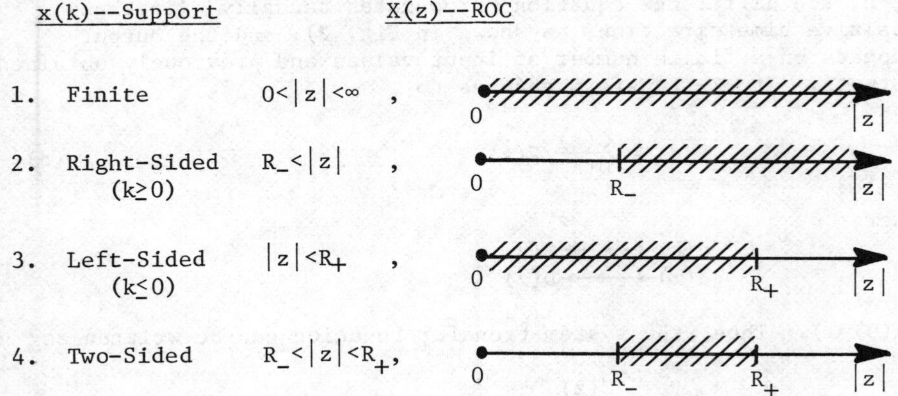

Figure 1. Support of sequences and corresponding ROC for
          transform. (Cross-hatching indicates ROC.)

These transforms are commonly used for their ease of handling input/output descriptions of linear, shift-invariant systems. Letting h(k) denote the unit sample response of such a system, it follows from the system properties that its input x(k) and output y(k) are related by the discrete convolution

$$y(k) = \sum_{n=-\infty}^{+\infty} h(k-n)x(n), \quad \forall\, k. \tag{5}$$

Correspondingly, from the convolution property of the Z-transforms, we have

$$Y(z) = H(z)X(z). \tag{6}$$

The convolution summation in (5) is sometimes called a <u>nonrecursive</u> form because the output y(k) at any point only depends on values of the input sequence, x(k). In some cases, the input/output can be related by a <u>recursive</u> form, i.e. a difference equation

$$y(k) = \sum_{n=0}^{C} c(n)x(k-n)$$

$$- \sum_{n=1}^{D} d(n)y(k-n) \quad . \tag{7}$$

Here, the difference equation "propagates causally" (in the positive time direction, as shown in Fig. 2), and the output depends on a finite number of input values and previously-obtained outputs. Its Z-transform reduces to

$$Y(z) = \frac{C(z)}{D(z)} X(z) \tag{8}$$

where

$$c(k) \longleftrightarrow C(z)$$
$$d(k) \longleftrightarrow D(z)$$

(d(0)=1). Thus, the system transfer function can be written as

$$H(z) = \frac{C(z)}{D(z)} \quad . \tag{9}$$

The system is said to be <u>causal</u> or <u>realizable</u> if h(k)=0, k<0 (i.e. the unit sample response is one-sided or unilateral). If $h(n) \epsilon \ell^1$, the system is said to be <u>stable</u>. It follows that if the system is both realizable and stable, then H(z) has a ROC

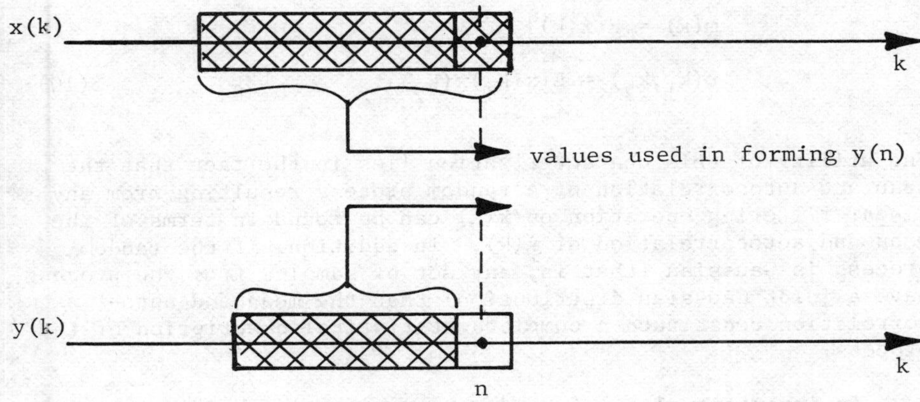

Figure 2.   Components of the input and output sequences which
            contribute to y(n) via the difference equation.
            (Causally propagated.)

which includes the $|z|=1$ (unit) circle. For the recursive form,
this implies $D(z)\neq 0$ $\{z:|z|\geq 1\}$. System stability is, of course,
an extremely important property, and has received a great deal
of attention, particularly in the controls literature. There
exists a multitude of algorithms for testing system stability
(Freeman, 1965); they include direct root location, algebraic,
and Nyquist-based approaches.

    Clearly, the above relations are valid for both deterministic
and random sequences. In the case of random sequences, however,
it is of interest to characterize additional properties of the
sequences.

    In our following discussion, we take x(k) to be a sample
function from a 1-D, discrete-time random process. A complete
statistical description of a random process involves the
specification of the joint probability density of any finite set
of samples from the process. In practice, such complete statistical knowledge if often not available, and it is necessary
either to assume a simplified model for the random process
which <u>does</u> have a complete statistical description, or to work
with <u>only</u> a partial statistical specification for the random
process.

    The most common partial statistical characterization is a

specification of its first and second moments, the mean and the autocorrelation, denoted respectively by

$$m(k) = E\{x(k)\}$$

$$r(k_1,k_2) = E\{x(k_1)x(k_2)\}. \tag{10}$$

The utility of this characterization lies in the fact that the mean and autocorrelation of a random process resulting from any _linear_ filtering operation on $x(k)$ can be found in terms of the mean and autocorrelation of $x(k)$. In addition, if the random process is Gaussian (that is, any set of samples from the process have a joint Gaussian distribution) then the mean and auto-correlation constitute a complete statistical description of the process.

An important class of random processes are those which are _wide-sense_ stationary (wss), where the mean is a constant (with regard to the time index), and the autocorrelation depends only on the difference of the two arguments:

$$m(k_1) = m(k_2)$$

$$r(k_1,k_2) = r(k_1-k_2,0) \ . \tag{11}$$

For the case of a wss random process, we denote the autocorrelation function by $r(k)$ where

$$r(k) = E\{x(\tau+k)x(\tau)\} \ . \tag{12}$$

Again, for a wss random process, we define the power density spectrum, $S(z)$, to be the Z-transform of the autocorrelation function:

$$r(k) \longleftrightarrow S(z) \tag{13}$$

For all $|z|=1$, $S(z)$ is real-valued and non-negative; furthermore the symmetry of the autocorrelation function implies that $S(z)=S(1/z)$.

For the remainder of this section we assume that $x(k)$ is zero-mean and wss.

In estimation and hypothesis-testing problems, it is often useful, at least conceptually, to replace a wss random process having _correlated_ samples, by an equivalent wss process whose samples are _uncorrelated_ (i.e. a white noise process). Such problems typically require the inversion of a covariance matrix, so if they can be formulated in terms of a white-noise process, the inversion operation is trivial.

Therefore, given a zero-mean wss random process $x(k)$, we would like to find a causal and causally invertible linear shift invariant filter which whitens $x(k)$. If this minimum-phase whitening filter exists (minimum-phase implies that all poles and zeros of the filter are located strictly within the unit circle) then $x(k)$ is said to have an innovations representation, and the white-noise process is called the innovations process. The present and past values of the innovations process contain all of the information in the present and past values of the original process, since the two processes are related by a minimum phase operator. The concept of the innovations representation is closely related to some results from linear algebra: the Gram-Schmidt procedure, and the Cholesky factorization.

Assuming that $x(k)$ has an innovations representation, we denote the minimum-phase whitening filter by $A(z)$, its inverse by $B(z)$, and the innovations process by $w(k)$. The following can be proved:

1. $A(z)$ is unique to within a multiplicative constant;

2. The power density spectrum of $x(k)$ is related to $A(z)$ and $B(z)$ by the following formulas:

$$S(z) = P \cdot B(z)B(1/z) = \frac{P}{A(z)A(1/z)} \qquad (14)$$

   where P is the variance of the innovations process, $w(k)$;

3. $w(k)$ is uncorrelated with all past values of $x(k)$:

$$E\{w(k)x(k-\tau)\} = P \cdot \delta_\tau, \quad \tau \geq 0 ; \qquad (15)$$

4. If

$$A(z) = \sum_{n=0}^{\infty} a(n)z^{-n}, \text{ then the quantity}$$

$$- \frac{1}{a(0)} \sum_{n=1}^{\infty} a(n)x(k-n) \qquad (16)$$

   is the minimum mean-square linear prediction estimate for $x(k)$ given all past samples; the corresponding prediction error variance, σ^2, is given by the formula:

$$\sigma^2 = \exp\left\{\frac{1}{2\pi j}\int_{|z|=1} z^{-1}\ell n S(z)dz\right\}, \tag{17}$$

where the quantity in the exponent is called the "entropy" of x(k).

In what follows, we assume that x(k) has an innovations representation, and furthermore that the minimum-phase whitening filter is rational (that is, a ratio of two finite minimum-phase polynomials):

$$A(z) = \frac{C(z)}{D(z)}, \tag{18}$$

where M_C and M_D are the orders of the numerator and denominator polynomials. Then x(k) is said to have an autoregressive moving-average (ARMA) representation, and it satisfies a difference equation of the form:

$$x(k) = \frac{1}{c(0)}\left[-\sum_{n=1}^{M_C} c(n)x(k-n) + \sum_{n=0}^{M_D} d(n)w(k-n)\right]. \tag{19}$$

Since c(z) is minimum-phase, it follows that the difference equation is stable when propagated causally. Given such a model, the minimum-phase whitening filter can be obtained by inspection, and the innovations process is the solution to a stable recursive difference equation driven by x(k):

$$w(k) = \frac{1}{d(0)}\left[-\sum_{n=1}^{M_D} d(n)w(k-n) + \sum_{n=0}^{M_C} c(n)x(k-n)\right]. \tag{20}$$

Two special cases of the ARMA representation are the moving average (MA) representation, where $M_C=0$, and the autoregressive (AR) representation, where $M_D=0$.

We now state some results concerning the existence of a minimum-phase whitening filter for a particular random process. In general, the problem of determining the minimum-phase whitening filter given the power density spectrum is called the spectral factorization problem.

The two general cases to consider are when the spectrum is rational, and when it is nonrational.

If $S(z)$ is rational, and finite and strictly positive for all $|z|=1$, then it can be shown, using the fundamental theorem of algebra, that there is always a rational minimum-phase whitening filter.

If $S(z)$ is nonrational, but analytic for all z in some neighborhood of the unit circle, and strictly positive for all z on the unit circle, then there is always a nonrational minimum phase whitening filter which is analytic for all z in some neighborhood of the unit circle.

2.2 Extensions to Two-Dimensions

Consider the two-sided, ℓ^1 bi-sequence $x(k,\ell)$ $k,\ell=0, \pm1, \pm2,...$ It is a function of two integer variables, and correspondingly has a 2-D Z-transform

$$X(z_1,z_2) = \sum_{k,\ell=-\infty}^{+\infty} x(k,\ell) z_1^{-k} z_2^{-\ell} \; . \tag{21}$$

Both z_1 and z_2 are complex variables, and the ROC of $X(z_1,z_2)$ is

$$\{(z_1,z_2): \sum_{k,\ell=-\infty}^{+\infty} |x(k,\ell) z_1^{-k} z_2^{-\ell}| < \infty\} \; . \tag{22}$$

When discussing 1-D Z-transforms, we dealt with analytic functions having (generally) annular ROC. Here, we deal with holomorphic functions (analytic functions of multidimensions) and Reinhardt domains (if $(z_1,z_2) \; \epsilon$ ROC, then $(e^{ju}z_1, e^{jv}z_2) \; \epsilon$ ROC), in direct analogy.

The inverse transform is given by

$$x(k,\ell) = \frac{1}{(2\pi j)^2} \int_{c_1} \int_{c_2} X(z_1,z_2) z_1^{k-1} z_2^{\ell-1} dz_1 dz_2 \tag{23}$$

where c_1,c_2 are CCW, closed contours encircling the origin and within the ROC of $X(z_1,z_2)$. Because the kernel of the transform is separable (i.e. a Cartesian product of two 1-D functions), the inversion can be viewed conceptually as two-each, sequential 1-D inverse transforms. As a practical matter, however, the 2-D inversion can seldom be performed using the standard 1-D techniques (based on the method of residues), as the singularities of $X(z_1,z_2)$ are not generally isolated poles.

While the fundamental relationship between sequence support and the ROC of its transform carries over to the 2-D setting, it is somewhat more complicated, if for no other reason than the diversity of support cases increases. The most apparent 2-D generalizations of support are so-called quarter-planes and half-planes. A first-quadrant, quarter-plane bi-sequence is one which only takes values for $k,\ell \geq 0$ (i.e. unilateral with respect to both variables). Its transform has a ROC for which $R_- < |z_1|$, $S_- < |z_2|$. As an example of half-plane bi-sequences, an upper half-plane is one which only takes values $\forall k$, and $\ell \geq 0$ (i.e. unilateral with respect to only one variable). Its ROC is the region for which $|z_1|=1$, $S_- < |z_2|$. The relations for these and other bi-sequences which are commonly encountered are in Fig. 3. We will return to these results shortly, when dealing with 2-D stability issues.

By appropriately defining a 2-D unit sample function, the input/output mapping of a 2-D, linear shift-invariant system can be described by

$$y(k,\ell) = \sum_{m,n=-\infty}^{+\infty} h(k-m,\ \ell-n)x(m,n) \quad \forall k,\ell \tag{24}$$

where $h(k,\ell)$ is the system's unit sample response. Transforming, (24) becomes

$$Y(z_1,z_2) = H(z_1,z_2)\ X(z_1,z_2)\ . \tag{25}$$

The 2-D discrete convolution in (24) is of <u>nonrecursive</u> form in that only values from the input bi-sequence enter into the sum. The 2-D <u>recursive</u> form, in which both input and "previously obtained" output values are used is the 2-D difference equation

$$y(k,\ell) = \sum_{m,n\epsilon\ M} c(m,n)x(k-m,\ell-n)$$

$$- \sum_{m,n\epsilon\ M'} d(m,n)y(k-m,\ell-n) \tag{26}$$

The sets M and M' define the input and output values incorporated in $y(k,\ell)$.

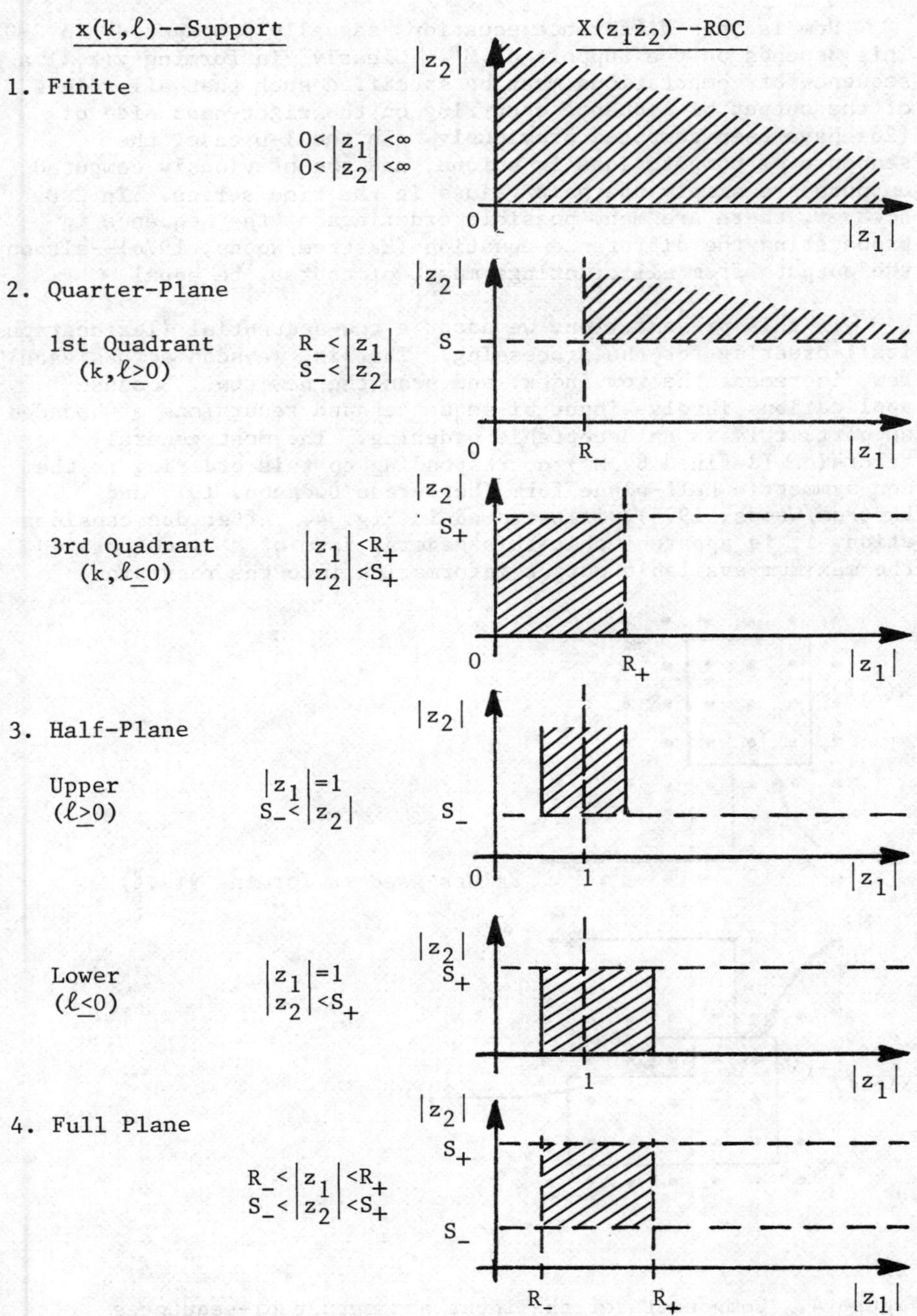

Figure 3. Selected examples of bi-sequences and corresponding
           ROC for 2-D transform. (Cross-hatching indicates ROC)

How is this difference equation "casually propagated" in 2-D?
This depends on the support of M'. Clearly, in forming y(k,ℓ) a
sequence of computations must be specified such that all values
of the output bi-sequence occurring on the right-hand side of
(26) have been computed previously. In the 1-D case, the
sequence of computations is unique, and the previously computed
outputs are simply the past values in the time series. In 2-D,
however, there are many possible orderings of the sequence in
propagating the difference equation (Ekstrom/Woods, 1976)--although
the outputs from all orderings must, of course, be equal. •

In this presentation, we adopt a row-sequential (lexicograph-
ical) ordering for the processing. That is, we scan across each
row, increment the row index, and scan the new row. Because
applications involve input bi-sequences and recursions of bounded
support, this is an acceptable ordering. The most general
recursion (defined by M') corresponding to this ordering is the
non-symmetric half-plane form (Mersereau/Dudgeon, 1975 and
Ekstrom/Woods, 1976), illustrated in Fig. 4. After due consideration, it is apparent that this general form of M' incorporates
the maximum available output information into the recursion.

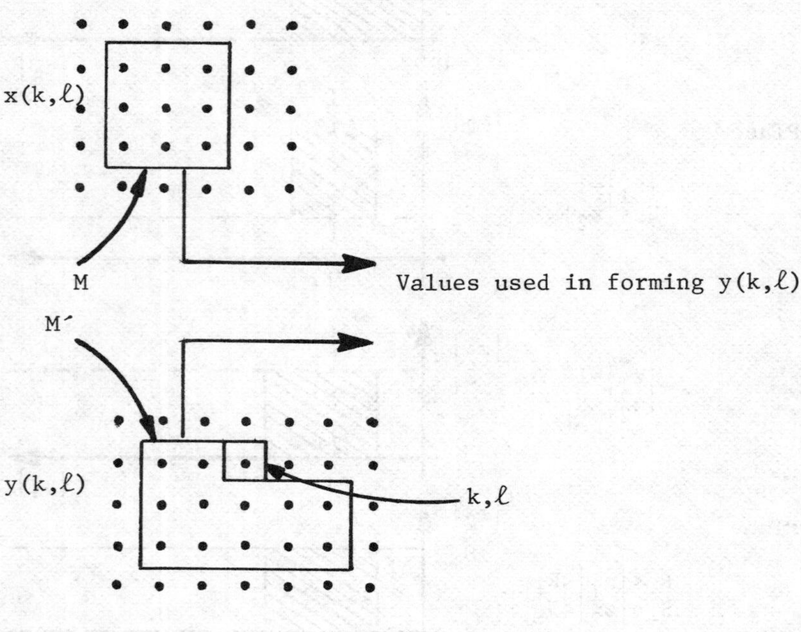

Using the convolution property of the 2-D Z-transform, (26) becomes

$$Y(z) = \frac{C(z_1, z_2)}{D(z_1, z_2)} X(z) \, , \tag{27}$$

hence

$$H(z_1, z_2) = \frac{C(z_1, z_2)}{D(z_1, z_2)} \, . \tag{28}$$

This system is said to be realizable if $h(k, \ell)$ takes support on the non-symmetric half-plane U^+, defined by

$$U^+ \overset{\Delta}{=} \{(k, \ell): (k \geq 0, \ell \geq 0) \cup (k < 0, \ell > 0)\} \, , \tag{29}$$

and illustrated in Fig. 5. Notice that U^+ is the upper half-plane with the negative k-axis removed. It is in this sense that $h(k, \ell)$ is considered to be unilateral. If $h(k, \ell) \ \varepsilon \ \ell^1$, the system is said to be stable. If it is both realizable and stable, the ROC of $H(z_1, z_2)$ must include the unit bi-circle ($|z_1| = |z_2| = 1$).

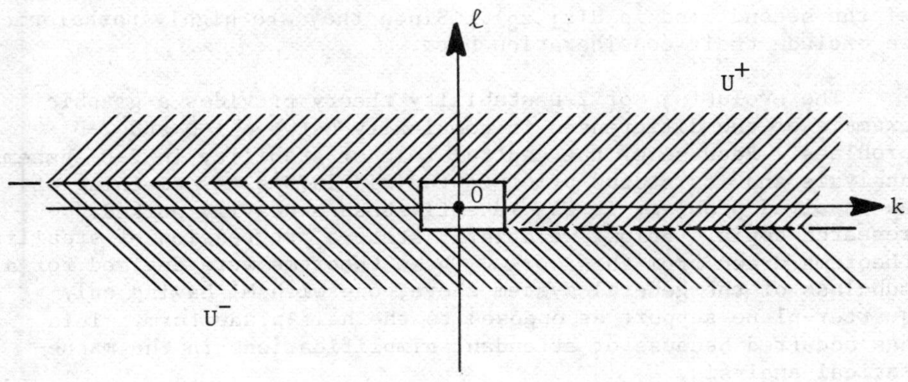

Figure 5.   Support for the non-symmetric half-planes U^+ and U^-.
            Overlapping occurs at the origin.

Using this result and the relations in Fig. 3, we can state the following:

<u>Theorem 1.</u>   (Ekstrom/Woods, 1976)

$H(z_1,z_2)$ in (28) is stable iff

   i) $D(z_1,z_2)\neq 0$ $\{(z_1,z_2) : |z_1|=1, 1\leq|z_2|\}$

     and   (30)

   ii) $D(z_1,\infty)\neq 0$ $\{z_1 : 1\leq|z_1|\}$.

Condition i) ensures an upper half-plane support for $h(k,\ell)$, and
condition ii) ensures that $h(k,0)$ will be a right-sided sequence.

   In the quoted reference, a stability test was also presented
which was based on the above theorem. The test involves a
computation and partitioning of the 2-D cepstrum, and is computationally quite attractive.

   At this point, a qualification is in order. While the above
conditions are sufficient, they are not necessary for stability
(Goodman, 1977). This is due to an interesting peculiarity in
the M-D case. It is theoretically possible to have pole-zero
cancellations at isolated points on the unit bi-circle, with
numerator/denominator polynomials which are mutually prime (Rudin,
1969). These are due to the presence of non-essential singularities
of the second kind in $H(z_1,z_2)$. Since they are highly pathological,
we exclude their consideration here.

   The evolution of 2-D stability theory provides a graphic
example of the differences in complexity between 1-D and 2-D
problems. Because of the central role of stability in 2-D systems-
analysis and its status prior to the mid-1970's as something of
an unsolved problem, it has recently been a heavily worked
research topic. Below, we briefly state a progression of stability
theorems drawn from this work. These theorems were devised for a
subclass of the general system above, one with M' having only
quarter-plane support as opposed to the half-plane form. This
has occurred because of attendant simplifications in the mathematical analysis.

   Fortunately, a simple relation exists which allows the
quarter-plane theorems to be applied to the half-plane case. It
is stated in the following lemma:

<u>Lemma 1</u> (O'Connor/Huang, 1979)

The system $h(k,\ell)$ is stable iff $h(c_1k+c_2\ell, c_3k+c_4\ell)$ is stable,
for c_i integer and $c_1c_4-c_2c_3\neq 0$. Thus, any half-plane form can be
shifted into a quarter-plane form (via an appropriate coordinate

transformation), and the quarter-plane form tested for stability.

The following theorems use as their premise that $H(z_1,z_2)$ has a denominator bi-sequence $d(k,\ell)$ taking support on a quarter-plane.

<u>Theorem 2.</u>   (Shanks/Treitel/Justice, 1972)

$H(z_1,z_2)$ is stable iff

$$D(z_1,z_2) \neq 0 \quad \{(z_1,z_2) : 1 \leq |z_1|, \ 1 \leq |z_2|\} \tag{31}$$

The stability test based on and presented with this theorem, essentially involved direct function evaluation of $D(z_1,z_2)$ to detect its zeros. This is not a very satisfactory test.

<u>Theorem 3.</u>   (Huang, 1972)

$H(z_1,z_2)$ is stable iff

$$\text{i)} \quad D(z_1,z_2) \neq 0 \quad \{(z_1,z_2) : |z_1|=1, \ 1 \leq |z_2|\}$$

$$\text{ii)} \quad D(z_1,\infty) \neq 0 \quad \{z_1 : 1 \leq |z_1|\} \tag{32}$$

Curiously, these conditions are identical to those given for half-plane forms in Theorem 1 above. That they are "sufficient" is apparent, since quarter-planes are subsets of half-planes. The "necessary" part is something of a surprise. Algebraic stability algorithms based on this theorem were presented by Maria/Fahmy (1973), Anderson/Jury (1973), and Siljak (1975). Although they all involve a "finite" number of computational steps in their implementation, their complexity is such that they have never been used for system orders of any practical interest.

<u>Theorem 4.</u>   (Strintzis, 1977)

$H(z_1,z_2)$ is stable iff

$$\text{i)} \quad D(z_1,z_2) \neq 0 \quad \{(z_1,z_2) : |z_1|=|z_2|=1\}$$

$$\text{ii)} \quad D(z_1,\infty) \neq 0 \quad \{z_1 : 1 \leq |z_1|\} \tag{33}$$

$$\text{iii)} \quad D(\infty,z_2) \neq 0 \quad \{z_2 : 1 \leq |z_2|\}$$

<u>Theorem 5.</u>   (Decarlo/Murray/Saeks, 1977)

$H(z_1,z_2)$ is stable iff

i) $D(z_1,z_2) \neq 0$ $\{(z_1,z_2) : |z_1|=|z_2|=1\}$

$$(34)$$

ii) $D(z,z) \neq 0$ $\{z : 1 \leq |z|\}$

The simplifications available with these last two theorems appear
to be substantial. Both involve a condition for zeros on the unit
bi-circle, and 1-D minimum-phase conditions. The 1-D conditions
are easily tested using classical algorithms, and Shaw (1978) and
O'Connor/Huang (1979) have devised cepstral-based approaches for
testing condition i). Their methods, however, have yet to receive
universal acceptance.

The theory of 2-D discrete parameter random processes has
undergone considerable development during the past five years, in
much the same manner as 2-D stability theory. Previously, there
was considerable confusion among researchers in this field
because of the apparent impossibility of generalizing some well-
known results in 1-D random process theory to the 2-D case. Now
it is safe to say that there are 2-D versions of virtually all
of the important results in 1-D random process theory.

We denote a 2-D discrete-parameter random process by $x(k,\ell)$
where k and ℓ are integers. In analogy with our 1-D development,
we focus our attention on the class of 2-D random processes
which are zero-mean, and wide-sense stationary with autocorrelation function $r(k,\ell)$, where

$$r(k,\ell) = E \{x(k+m,\ell+n) \, x(m,n)\} .$$

$$(35)$$

We denote the power density spectrum by $S(z_1,z_2)$ where

$$r(k,\ell) \longleftrightarrow S(z_1,z_2)$$

$$(36)$$

The idea of the 2-D innovations representation is similar to
that of the 1-D innovations representation: to find a "causal"
and "causally invertible" whitening filter for a particular 2-D
random process. Conceptual problems have occurred in the 2-D
case, because unlike the 1-D case, there is usually no physical
motivation for a definition of causality. But it turns out that
even in the 1-D case, the <u>physical</u> notion of causality is of
little importance in filter and random process theory. For
instance, any 1-D random process with a causal innovations representation, also has an innovations representation in terms of an
anticausal whitening filter, and considering first and second
moments only, there is no way to determine which way the process

was actually generated. In general, except where "real time"
filtering is required (as in a closed-loop control system) the
notion of a 1-D causal filter is useful, not because it corresponds to a physical notion of causality, but rather because it
contains the class of recursive filters, which have attractive
computational properties.

Therefore, consistent with our earlier discussion, we define
a 2-D linear shift-invariant filter $H(z_1,z_2)$ to be causal (or
realizable) if it has a half-plane support (see Fig. 5). In this
case,

$$H(z_1,z_2) = \sum_{k=0}^{\infty} h(k,0)z_1^{-k} + \sum_{k=-\infty}^{\infty}\sum_{\ell=1}^{\infty} h(k,\ell)z_1^{-k}z_2^{-\ell}. \qquad (37)$$

This definition is equivalent to defining past, present, and
future, and then restricting the support of the filter to include
the origin and all points in the future of the origin. For any
point (m,n) the past is defined to be the set $\{(k,\ell) \mid (k<m,\ell=n);$
$(-\infty \le k \le \infty, \ell < n)\}$ and the future is defined to be the set
$\{(k,\ell) \mid (k>m,\ell=n); (-\infty \le k \le \infty, \ell>n)\}$ (illustrated in Fig. 6).

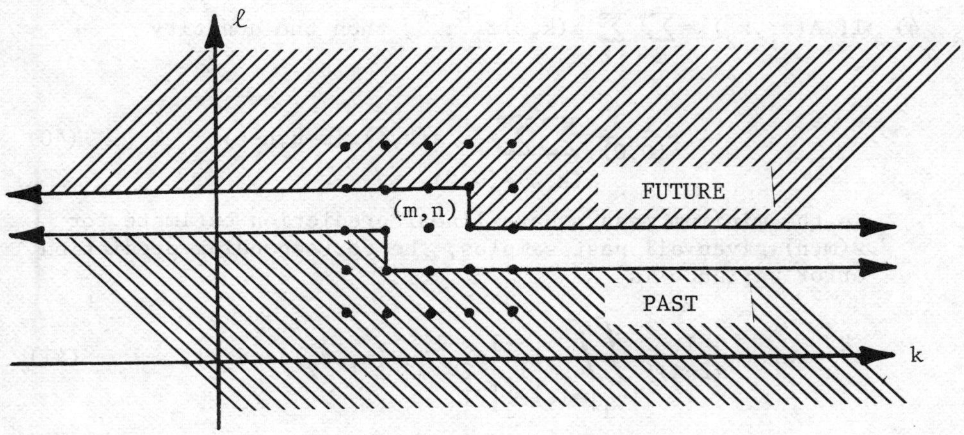

Figure 6. Definition of past, present, and future.

We say that $H(z_1,z_2)$ is minimum-phase if it is causal and
stable, with a causal and stable inverse.

We say that the zero-mean wss random process $x(k,\ell)$ has an
innovations representation if it has a minimum-phase whitening

filter. Assuming that $x(k,\ell)$ has an innovations representation, and denoting the minimum-phase whitening filter by $A(z_1,z_2)$ and its inverse by $B(z_1,z_2)$, the following can be proved: (Marzetta, 1978)

1) $A(z_1,z_2)$ is unique to within a multiplicative constant;

2) The power spectrum of $x(k,\ell)$ is related to $A(z_1,z_2)$ and $B(z_1,z_2)$ by the formulas:

$$S(z_1,z_2) = P\, B(z_1,z_2)\, B(1/z_1,1/z_2)$$

$$\qquad\qquad (38)$$

$$= \frac{P}{A(z_1,z_2)\, A(1/z_1,1/z_2)}$$

where P is the variance of the innovations process, $w(k,\ell)$;

3) $w(k,\ell)$ is uncorrelated with all past values of $x(k,\ell)$:

$$E\{w(k,\ell)\ x(k-m,\ell-n)\} = P\,\delta_m\delta_n\ ,$$

$$\qquad\qquad (39)$$

$$(m\geq 0,\ n=0);\ (-\infty\leq m\leq\infty,\ n>0).$$

4) If $A(z_1,z_2) = \displaystyle\sum_k \sum_\ell a(k,\ell)z_1^{-k}z_2^{-\ell}$, then the quantity

$$-\frac{1}{a(0,0)} \sum\sum_{(k,\ell)\neq(0,0)} a(k,\ell)x(m-k,n-\ell) \qquad\qquad (40)$$

is the minimum mean square linear prediction estimate for $x(m,n)$ given all past samples; the corresponding prediction error variance is

$$\sigma^2 = \exp\left\{ \frac{1}{(2\pi j)^2} \iint_{|z_1|=1,\,|z_2|=1} z_1^{-1}z_2^{-1}\ln S(z_1,z_2)dz_1dz_2 \right\}, \qquad (41)$$

where the term in the exponent is called the "entropy" of $x(k,\ell)$.

If we assume that $x(k,\ell)$ has an innovations representation, and furthermore that the minimum-phase whitening filter is rational (that is, a ratio of finite-order minimum-phase polymonials):

$$A(z_1,z_2) = \frac{C(z_1,z_2)}{D(z_1,z_2)} \tag{42}$$

then $x(k,\ell)$ is said to have an ARMA representation, and it satisfies a stable, recursive difference equation driven by the innovations process:

$$x(k,\ell) = \frac{1}{c(0,0)}\left[-\sum_{(m,n)\neq(0,0)}\sum c(m,n)x(k-m,\ell-n) \right.$$
$$\left. + \sum_{(m,n)}\sum d(m,n)w(k-m,\ell-n) \right] \tag{43}$$

Likewise, since $A(z_1,z_2)$ is minimum-phase, the innovations process satisfies a stable, recursive difference equation driven by $x(k,\ell)$:

$$w(k,\ell) = \frac{1}{d(0,0)}\left[-\sum_{(m,n)\neq(0,0)}\sum d(m,n)w(k-m,\ell-n) \right.$$
$$\left. + \sum_{(m,n)}\sum c(m,n)x(k-m,\ell-n) \right]. \tag{44}$$

Two special cases of the ARMA representation are the MA representation, where $C(z_1,z_2)$ is a constant, and the AR representation, where $D(z_1,z_2)$ is a constant.

Up until now, all of the 2-D random process theory discussed has been virtually identical to the 1-D theory. Some differences occur in the 2-D spectral factorization problem; specifically, rational 2-D spectra are as difficult to factor as nonrational 2-D spectra.

Consider first the case where $S(z_1,z_2)$ is nonrational, analytic for all z_1 and z_2 in some neighborhood of the unit circles, and strictly positive for all z_1 and z_2 simultaneously on the unit circles. Then there is always a nonrational minimum-phase whitening filter, analytic in some neighborhood of the unit circles (Whittle, 1954; Ekstrom/Woods, 1976). In this respect, the 2-D spectral factorization problem is identical to the 1-D problem.

Now suppose that $S(z_1,z_2)$ is rational (i.e. a ratio of finite-

order 2-D polynomials) and that the numerator and denominator
polynomials are each strictly positive for all z_1 and z_2 simultaneously on the unit circles. Then it can be shown that there
is <u>almost never</u> a rational minimum-phase whitening filter (Whittle,
1954; Woods, 1972). Of course, there is always a <u>nonrational</u>
factor in this case; it has been shown that the minimum-phase
whitening filter is the ratio of two minimum-phase filters, each
of which has <u>infinite</u> support in z_1, and <u>finite</u> support in z_2
(Marzetta, 1978).

The problem of designing a 2-D recursive digital filter with
a prescribed magnitude frequency response has motivated the
development of a number of approximate 2-D spectral factorization
algorithms. Some representative approaches that have been taken
are those of Ekstrom and Woods (1976), Chang and Aggarwal (1978),
Marzetta (1978, 1979), and Ekstrom et al (1980).

3. ADVANCED TOPICS IN 2-D SIGNAL PROCESSING

In this section, we present the 2-D extension of two classical
problems in time series analysis: autoregressive spectral estimation and Wiener's minimum, mean-square error filtering. Both
extensions are quite recent results, and further illustrate the
fact that standard 1-D methodologies/theory can be broadened to a
2-D setting, if the underlying concepts are appropriately
generalized.

3.1 2-D Autoregressive (AR) Spectral Estimation

In this section we discuss the problem of 2-D autoregressive
spectral estimation, which involves estimating the parameters of
an AR model, given a finite set of samples from a 2-D wss random
process.   One-dimensional AR spectral estimation has been used
extensively in a number of applications because of its "high
resolution" when compared with classical methods (Lacoss, 1971).
In addition it provides a spectral estimate which is already in
factored form, an advantage when the minimum-phase whitening
filter is required for further data processing. For these reasons,
there has been considerable interest in developing 2-D AR spectral
estimation algorithms. This has been an active area of research
for a number of years (Whittle, 1954; Ong, 1971; Woods, 1976),
but it was only recently that the theoretical aspects of the
problem were explained, and that will be the focus of this section.

As in the 1-D case, there are two basic approaches to 2-D
AR spectral estimation: 1) To use the available data to estimate
first the autocorrelation function for a certain set of arguments
and then to use the estimated autocorrelation samples to determine

the AR model; 2) To estimate the AR model parameters directly
from the data. In the 1-D case the first technique is usually
called the "autocorrelation method" or the "maximum entropy"
method; there are several techniques for the second method,
including the "covariance method" and the "Burg algorithm."

A problem which we will not address is that of deciding the
order of the AR model. Even in the 1-D case there is no
completely satisfactory solution to this problem, and in practice
it is usually a matter of trading-off resolution for statistical
stability.

Considering first the 2-D maximum entropy method, suppose
that estimates for the following samples of the 2-D autocorrelation
function are available:

$$\{r(k,\ell); \quad -K \leq k \leq K, \quad -L \leq \ell \leq L\} \tag{45}$$

and that the corresponding covariance matrix is positive-definite.
The object is to find a positive-definite extension of the autocorrelation estimate over the entire plane, or equivalently to
find a power spectrum which is consistent with (45). If any
single extension exists, then there are an infinite number of
valid extensions, so it is necessary to do the extension according
to some criterion. A common criterion in the 1-D case (i.e. L=0)
is the maximum entropy criterion, which leads to a K-order AR
model. Applying this criterion to the 2-D case, we want to find
a spectrum, consistent with the estimated autocorrelation samples,
with the maximum entropy, where the entropy is defined by (41).
If such a solution exists, then it can be shown (Woods, 1976),
using calculus of variations, that the maximum entropy spectrum is
of the form

$$S(z_1, z_2) = \frac{1}{\displaystyle\sum_{k=-K}^{K} \sum_{\ell=-L}^{L} \lambda(k,\ell) z_1^{-k} z_2^{-\ell}} \tag{46}$$

The remaining problem is to choose the $\{\lambda(k,\ell)\}$ so that (46) is
consistent with (45). In the 1-D case, this is done by finding
a one-step, minimum mean-square linear prediction estimator:

$$\hat{x}(k) | \{x(k-1), \ldots, x(k-K)\} = -\sum_{m=1}^{K} a(m) x(k-m), \tag{47}$$

where the predictor coefficients, and the prediction-error

variance, P, are the solution to the "normal" (Yule-Walker) equations (Edward and Fitelson, 1973),

$$r(k) + \sum_{m=1}^{K} a(m)r(k-m) = P\,\delta_k, \quad 0 \le k \le K. \tag{48}$$

It turns out that P is always positive, and the prediction error filter (PEF), $A_K(z)$,

$$A_K(z) = 1 + \sum_{k=1}^{K} a(k)z^{-k}, \tag{49}$$

is always minimum-phase (Markel and Gray, 1973), so it is a valid whitening filter for an AR model. Another important property of the PEF is the so-called correlation-matching property (Makhoul, 1975; Dubroff, 1975), according to which given a positive P, and a minimum-phase $A_K(z)$, there is a unique positive-definite auto-correlation sequence, $\{r(0),\ldots, r(K)\}$ such that (48) is satisfied. Therefore, we have a one-to-one relation

$$\{r(0),r(1),\ldots,r(K)\} \xleftrightarrow{\ (48)\ } \{P;\, A_K(z)\}\ , \tag{50}$$

and the maximum entropy spectrum (46) becomes

$$S(z) = \frac{P}{A_K(z)A_K(1/z)}\ . \tag{51}$$

The same solution technique does not work in the 2-D case. For example, the following 2-D prediction estimator can be found by solving a 2-D version of the normal equations,

$$x(k,\ell)\,|\,\{x(k-m,\ell-n);\ (1 \le m \le K,\ n=0),$$
$$(0 \le m \le K,\ 1 \le n \le L)\} \tag{52}$$

to obtain a 2-D PEF, $A(z_1,z_2)$, and a prediction error variance, P. Contrary to the 1-D case, $A(z_1,z_2)$ is not always minimum-phase, as can be shown by counterexample (Genin and Kamp, 1975). More-over, even when $A(z_1,z_2)$ happens to be minimum-phase, the corre-lation-matching property almost never holds (Marzetta, 1978, 1980), so the AR spectrum is inconsistent with the given autocorrelation array. (The autocorrelation array contains approximately twice as many parameters as are contained in $A(z_1,z_2)$ and P, so an infinite number of different autocorrelation arrays will yield the same $A(z_1,z_2)$ and P.)

The fundamental difficulty with the 2-D problem is that, given our definition of past, present, and future, there are an infinite number of points simultaneously in the future of the origin, and in the past of (K,L): $\{(k,0),\ k>0;\ (k,\ell),\ -\infty<k<\infty,\ 1\leq\ell\leq(L-1);\ (k,L),\ k<K\}$, and the autocorrelation function is given at only a finite number of these points. Consequently, the 2-D autocorrelation extension problem is actually a combined problem of <u>extrapolation</u> and <u>interpolation</u>. In contrast, the usual 1-D problem is entirely an extrapolation problem. Precisely the same difficulties encountered in the 2-D problem occur in the 1-D problem when the autocorrelation sequence is missing one or more points. Suppose, for example, that we want to find the maximum entropy spectrum based on the following 1-D autocorrelation sequence having discontinuous support (missing the point $r(2)$):

$$\{r(0),\ r(1),\ r(3),\ r(4)\} = \{1,\ 0,\ -.2,\ -.9\}. \tag{53}$$

(This autocorrelation sequence could correspond to that of a linear array of sensors with relative positions of 0, 3, and 4 units.) It can be shown that the maximum entropy spectrum (if it exists) is of the form

$$S(z) = \frac{1}{\lambda(0)+\lambda(1)(z^{-1}+z)+\lambda(3)(z^{-3}+z^3)+\lambda(4)(z^{-4}+z^4)}. \tag{54}$$

If we use (53) to find the prediction estimator,

$$\hat{x}(k)\,|\,\{x(k-3),\ x(k-4)\}\ , \tag{55}$$

we obtain $P=3/20$ and $A(z)=(1+.2z^{-3}+.9z^{-4})$, which is not minimum-phase. Moreover, even if $A(z)$ happened to be minimum-phase, we would not have solved the problem, since the autocorrelation sequence contains 4 parameters, while P and $A(z)$ contain only 3 parameters. Applying the rational spectral factorization theorem to (54), we find that the maximum-entropy spectrum is of the form

$$S(z) = \frac{P}{A(z)A(1/z)}\ , \tag{56}$$

where

$$A(z) = 1 + \sum_{k=1}^{4} a(k)z^{-k} \tag{57}$$

is minimum-phase, and the $\{a(k)\}$, are in general, all non-zero. $A(z)$ and P satisfy the normal equations,

$$r(k) + \sum_{m=1}^{4} a(m)r(k-m) = P\delta_k, \quad 0 \leq k \leq 4, \tag{58}$$

but this <u>cannot</u> be used to obtain $A(z)$ and P, since its solution requires that we know $\{r(k), 0 \leq k \leq 4\}$, and $r(2)$ is missing.

Returning to the original 2-D problem, if we apply the 2-D rational spectral factorization theorem (Marzetta, 1978) to (46), we find that the maximum-entropy spectrum is of the form

$$S(z_1,z_2) = \frac{P}{A(z_1,z_2)\, A(1/z_1,1/z_2)} , \tag{59}$$

and $A(z_1,z_2)$ is minimum-phase with

$$A(z_1,z_2) = \left[1 + \sum_{k=1}^{\infty} a(k,0)z_1^{-k} + \sum_{k=-\infty}^{\infty} \sum_{\ell=1}^{L-1} a(k,\ell)z_1^{-k}z_2^{-\ell} \right. \tag{60}$$
$$\left. + \sum_{k=-\infty}^{K} a(k,L)z_1^{-k}z_2^{-L} \right]$$

where the $\{a(k,\ell)\}$ are, in general, non-zero. $A(z_1,z_2)$ is said to have <u>continuous</u> <u>support</u>, since it includes $(0,0)$, (K,L), and <u>all</u> intermediate points. $A(z_1,z_2)$ and P satisfy the 2-D normal equations,

$$\left[r(k,\ell) + \sum_{m,n} \sum a(m,n)r(k-m,\ell-n) \right] = P\delta_m\delta_n \tag{61}$$

$$\{(k>0,\ell=0),(-\infty \leq k \leq \infty,\ 1 \leq \ell \leq L-1),(k \leq K,\ell=L)\} ,$$

but this <u>cannot</u> be used (even conceptually) to find $A(z_1,z_2)$ and P, since, in order to solve it, we need to know the auto-correlation function at $(0,0)$, (K,L), and <u>all</u> intermediate points, and <u>all</u> but a finite number of these autocorrelation points are missing. (If the entire continuous-support autocorrelation array <u>is</u> available, and if (61) is solved, then $A(z_1,z_2)$ is guaranteed to be minimum-phase, and the correlation-matching property always holds (Marzetta, 1978, 1980).)

Consequently, there is no closed-form solution to the 2-D maximum-entropy extension problem, just as there is no closed-form solution to the 1-D maximum-entropy extension problem for a 1-D autocorrelation sequence with missing points. Iterative solutions for the $\{\lambda(k,\ell)\}$ have been proposed (Ong, 1971; Woods, 1976), but even they have some theoretical difficulties (apart from purely numerical problems). Specifically, in our discussion, we tacitly assumed the existence of a positive-definite extension of the auto-correlation array (45) over the entire plane. In fact, it was recently shown that this extension does not always exist (Dickinson, 1980).

We have seen that the chief difficulty with the 2-D auto-correlation method is that it requires the interpolation of the autocorrelation array. This problem is avoided in AR modeling methods where the AR parameters are obtained directly from the data, with no intermediate step of estimating samples of the autocorre-lation function. For some reason, little or no attention has been paid to these methods in the 2-D literature. In the 1-D case the so-called covariance method fits an AR model directly to the data by choosing the coefficients of a PEF to minimize the sum of the squares of prediction errors (Box & Jenkins, 1976; Makhoul, 1975). The method can be easily extended to the 2-D case.

Assume that we have a finite set of samples from the 2-D random process, $\{x(k,\ell), \ (k,\ell)\epsilon T\}$, and that we want to estimate the parameters of an AR model of some specified order. The 2-D covar-iance method can be derived formally by assuming the random process to be Gaussian, and using maximum-likelihood parameter esimation. Ignoring edge effects, the joint probability density for the data samples given the AR model, $\{A(z_1,z_2); P\}$, is approximately

$$Pr \cong \frac{1}{(2\pi)^{N_{T'}/2}} \cdot \frac{1}{P^{N_{T'}/2}}$$

$$\cdot \exp\left[-\frac{1}{2P}\sum_{(k,\ell)\epsilon T'}\sum\left[x(k,\ell)+\sum_{m,n}\sum a(m,n)x(k-m,\ell-n)\right]^2\right], \quad (62)$$

where T' is some subset of T (generally chosen, in the 1-D case, so that the PEF is not run off the edge of the data), and $N_{T'}$ is the number of samples contained in T'. Maximizing the likelihood over P, we have that

$$P = \frac{1}{N_{T'}}\sum_{(k,\ell)\epsilon T'}\sum\left[x(k,\ell)+\sum_{m,n}\sum a(m,n)x(k-m,\ell-n)\right]^2 . \quad (63)$$

Substituting (63) into (62), we see that $A(z_1,z_2)$ is obtained by minimizing the quadratic expression,

$$\sum_{(k,\ell)\in T'}\left[x(k,\ell)+\sum_{m,n}a(m,n)x(k-m,\ell-n)\right]^2 , \tag{64}$$

which can be done by solving a set of linear equations. As in the
1-D case, the covariance method is not guaranteed to provide a
minimum-phase estimate for $A(z_1,z_2)$.

An alternative algorithm can be obtained by writing the likelihood function in terms of the "backward" PEF, $A(1/z_1,1/z_2)$ as
well as the "forward" PEF, $A(z_1,z_2)$. $A(z_1,z_2)$ is then obtained by
minimizing a sum of <u>forward</u> and <u>backward</u> squared errors. In the
1-D case, this method has sometimes been found to be superior to
the ordinary covariance method (Ulrych & Clayton, 1976) (though it
still does not guarantee a minimum-phase estimate).

A 1-D variation on the forward-backward covariance method is
the Burg algorithm (Burg, 1975), which represents the PEF in terms
of a "reflection" ("partial correlation") coefficient sequence,
which is found by successively fitting a sequence of PEFs of increasing order to the data. Computationally, the Burg algorithm
is very convenient; moreover it guarantees a minimum-phase PEF.
The reflection coefficient representation for 2-D minimum-phase
filters was discovered recently (Marzetta, 1978, 1980), and a 2-D
Burg algorithm was proposed, but no numerical experiments have
been performed.

A possible modification of the 2-D covariance methods, which
<u>would</u> guarantee the minimum-phase condition, would involve applying
the penalty function-type method of Ekstrom et al (1980) (originally developed for 2-D recursive filter design).

In conclusion, the theoretical aspects of 2-D AR spectral
estimation are now well-understood, but further development of
practical algorithms is needed.

3.2 2-D, "Causal" Wiener Filtering

Wiener's minimum mean-square error (MMSE) filtering is a
classical approach to the estimation of wss time series. Here, we
address a 2-D version of the estimation problem considered by
Wiener in his canonical work (Wiener, 1949): an arbitrary bi-
sequence $x(k,\ell)$ has been observed or measured, and it is desired
to estimate a signal $s(k,\ell)$ from these observations. Both $x(k,\ell)$
and $s(k,\ell)$ are taken to be sample functions from 2-D, wss processes
with zero means and autocorrelations, $r_x(k,\ell)$ and $r_s(k,\ell)$, respectively. Their cross-correlation is given by

$$r_{sx}(k,\ell) = E\{s(k+m,\ell+n)x(m,n)\} \tag{65}$$

The estimate $\hat{s}(k,\ell)$ is chosen to be a weighted, linear combination of "present and past" observations (in the sense of Fig. 5 and 6), hence takes the form of the filtering operation

$$\hat{s}(k,\ell) = \sum_{m,n\epsilon U^+} x(k-m,\ell-n)h(m,n) \quad . \tag{66}$$

The weights $h(m,n)$ make-up the filter point spread response and completely describe the estimator. Because the filter uses only "present and past" input values, it is "casual" (or realizable) and $h(m,n)$ takes support on U^+.

According to the MMSE approach, we choose to define the optimal estimator, $h_o(m,n)$, as the filter which minimizes the error

$$\xi = E\{\left[s(k,\ell)-\hat{s}(k,\ell)\right]^2 \tag{67}$$
$$\text{for all } k,\ell$$

where $\hat{s}(k,\ell)$ is constrained to be of the form (66). It can be easily shown that the minimizing bi-sequence is such that each residual is orthogonal to all other data used in the estimate, that is

$$E\{\left[s(k,\ell)-\hat{s}(k,\ell)\right]x(k-m,\ell-n)\} = 0 \tag{68}$$
$$\text{for } m,n \;\epsilon\; U^+$$

Substituting (66) into (68) and using the defining correlation relations, the optimal filter can be shown to be the solution of the system of equations

$$r_{sx}(k,\ell) = \sum_{m,n\epsilon U^+}\sum r_x(k-m,\ell-n)h_o(m,n) \tag{69}$$
$$\text{for } k,\ell \;\epsilon\; U^+ \quad .$$

This is a Wiener-Hopf type equation of the first kind. Its solution is greatly complicated by virtue of the fact that the system is not defined for all k,ℓ (rather only for $k,\ell \;\epsilon\; U^+$), and that $h_o(m,n)$ is constrained to be causal. However, for the moment let us ignore these conditions, assuming that (69) is in fact defined for all k,ℓ and that $h_o(k,\ell)$ is unconstrained. We will correct for these assumptions below. Now, we have the following system

642   **M. P. EKSTROM AND T. L. MARZETTA**

$$r_{sx}(k,\ell) = \sum_{m,n=-\infty}^{+\infty} r_x(k-m,\ell-n)h_o(m,n) \tag{70}$$

$$\text{for all } k,\ell.$$

Using the convolution property of 2-D z-transforms (see (24) and (25)), we transform (70) to obtain

$$S_{sx}(z_1,z_2) = S_x(z_1,z_2)\, H_o(z_1,z_2) \quad, \tag{71}$$

where

$$h_o(m,n) \longleftrightarrow H_o(z_1,z_2) \quad. \tag{72}$$

The power spectral density $S_x(z_1,z_2)$ can be spectrally factored into minimum and maximum-phase terms as in (38) (with $P=1$) :

$$S_x(z_1,z_2) = B_x(z_1,z_2)B_x(1/z_1,1/z_2) \,, \tag{73}$$

where $B_x(z_1,z_2)$ is the inverse of a minimum-phase, whitening filter. Incorporating this factorization into (71), we have

$$S_{sx}(z_1,z_2) = B_x(z_1,z_2)B_x(1/z_1,1/z_2)H_o(z_1,z_2) \tag{74}$$

At this point, we introduce the corrections into our formalisms for the assumptions made above. First, because (69) was only defined for $(k,\ell) \in U^+$, we project (74) onto U^+ to obtain

$$\left[S_{sx}(z_1,z_2)\right]_+ = \left[B_x(z_1,z_2)B_x(1/z_1,1/z_2)H_o(z_1,z_2)\right]_+ \tag{75}$$

where the notation $[\cdot]_+$ indicates the component of the bracketed term whose inverse 2-D z-transform takes support on U^+. Because $B_x(1/z_1,1/z_2)$ is maximum-phase ("anti-causal"), its inverse transform takes support on U^-. Hence, (75) can be written as

$$\left[\frac{S_{sx}(z_1,z_2)}{B_x(1/z_1,1/z_2)}\right]_+ = \left[B_x(z_1,z_2)H_o(z_1,z_2)\right]_+ \,, \tag{76}$$

that is, values on U^+ for the convolution product of an anti-causal bi-sequence and an arbitrary bi-sequence, only depend on values of the arbitrary bi-sequence in U^+.

The second correction involves constraining $h_o(k,\ell)$ to be causal. As $B_x(z_1,z_2)$ is minimum-phase, its inverse transform takes support on U^+, i.e. causal. It is straightforward to show that the convolution of two causal bi-sequences is itself causal. Therefore,

$$\left[B_x(z_1,z_2)H_o(z_1,z_2)\right]_+ = B_x(z_1,z_2)H_o(z_1,z_2) \quad . \tag{77}$$

Consequently, (76) can be written as

$$B_x(z_1,z_2)H_o(z_1,z_2) = \left[\frac{S_{sx}(z_1,z_2)}{B_x(1/z_1,1/z_2)}\right]_+ \quad . \tag{78}$$

Solving for the optimal filter, we have

$$H_o(z_1,z_2) = \frac{1}{B_x(z_1,z_2)}\left[\frac{S_{sx}(z_1,z_2)}{B_x(1/z_1,1/z_2)}\right]_+ \quad . \tag{79}$$

This is the general form for the 2-D, causal Wiener filter. A comparison of (79) with the 1-D filter form indicates they are functionally identical (with appropriate generalization of the operations to 2-D, of course). There is a substantial difference in implementation, because the operations required in deriving the exact filter form, the 2-D spectral factorization and half-plane projection, cannot generally be performed using polynomial algebra, even if the spectra are rational. However, both can be calculated with algorithms based on sectioning cepstral bi-sequences (Ekstrom & Woods, 1976). Whether dealing with rational or nonrational 2-D spectra, these decompositions must be done in approximation (Ekstrom et al, 1980), but this does not appear to be a serious practical limitation. Using the optimal filter (79), expressions for the causal MMSE can be directly derived, thereby characterizing the estimator performance.

4.   CONCLUDING COMMENTS

As indicated in the Introduction (Section 1), we have sought to demonstrate the extension of many, classical time-series re-sults to multidimensions, by describing their extension to 2-D. In so doing, we have attempted to give some articulation of the character of the extensions without the attendant complexity of the full M-D case. For a sampling of the M-D literature (and its complexity), see Helson/Lowdenslager (1958 and 1961), Justice/ Shanks (1973), Chan (1980), and Goodman/Ekstrom (1980).

The fundamental point to be derived from our presentation is that almost all 1-D approaches can be adapted to a M-D setting with a careful generalization of the underlying concepts. Algo-

rithms for implementing these approaches are more complicated in
the higher dimensional problems (sometimes substantially) due in
large measure to the absence of a M-D equivalent to the Fundamental Theorem of Algebra. Consequently, most operations must
be performed in numerical approximation; this is clearly a subject which will receive much future attention in the literature.

REFERENCES

Anderson, B. and Jury, E., "Stability Test for Two-Dimensional
    Recursive Filters," *IEEE Trans. Audio Electroacoust.*, Vol.
    AU-21, pp. 366-372, 1973.

Box, G., and Jenkins, G., *Time Series Analysis: forecasting and
    control*, Holden-Day Inc., 1976.

Burg, J., *Maximum Entropy Spectral Analysis*, Ph.D. Thesis,
    Stanford University, Dept. of Geophysics, Stanford, Cal.,
    May 1975.

Chan, D., "The Structure of Recursible Multidimensional Discrete
    Systems," *IEEE Trans. Automat. Control*, Vol. AC-25, 1980
    (in press).

Chang, H. and Aggarwal, J., "Design of Two-Dimensional Semicausal
    Recursive Filters," *IEEE Trans. Circuits Systems*, Vol. CAS-25,
    pp. 1051-1059, December 1978.

Decarlo, R., Murray, J., and Saeks, R., "Multivariable Nyquist
    Theory," *Int. J. Control*, Vol. 25, pp. 657-675, 1977.

Dickinson, B., "Two-Dimensional Markov Sprectrum Estimates Need
    Not Exist," *IEEE Trans. Inf. Th.*, Vol. IT-26, No. 1, pp. 120-
    121, January 1980.

Dubroff, R., "The Effective Autocorrelation Function of Maximum
    Entropy Spectra," *Proc. IEEE*, pp. 1622-1623, November 1975.

Edward, J. and Fitelson, M., "Notes on Maximum Entropy Processing,"
    IEEE Trans. Inf. Th., Vol. IT-19, pp. 232-234, March 1973.

Ekstrom, M., Twogood, R., Woods, J., "Two-Dimensional Recursive
    Filter Design - A Spectral Factorization Approach," *IEEE
    Trans. ASSP*, Vol. ASSP-28, No. 1, pp. 16-26, February 1980.

Ekstrom, M. and Woods, J., "Two-Dimensional Spectral Factorization
    with Applications in Recursive Digital Filtering," *IEEE Trans.
    ASSP*, Vol. ASSP-24, No. 2, pp. 115-127, April 1976.

Freeman, H., _Discrete-Time Systems_, Wiley, 1965.

Goodman, D., "Some Stability Properties of Two-Dimensional Linear
     Shift-Invariant Digital Filters," _IEEE Trans. Circuits Syst._,
     Vol. CAS-24, pp. 201-208, 1977.

Goodman, D. and Ekstrom, M., "Multidimensional Spectral Factor-
     izations and Unilateral Autoregressive Models," _IEEE Trans._.
     Automat. Contrl., Vol. AC-25, pp. 258-262, 1980.

Genin, Y. and Kamp, Y., "Counterexample in the Least Square
     Inverse Stabilization of 2D-Recursive Filters," _Electronics_
     Letters, Vol. 11, pp. 330-331, 1975.

Helson, H. and Lowdenslager, "Prediction Theory and Fourier Series
     In Several Variables," I and II, _Acta Math_, Vol. 99, pp. 165-
     202, 1958 and Vol. 106, pp. 175-213, 1961.

Huang, T., "Stability of Two-Dimensional Recursive Filters,"
     IEEE Trans. Audio Electroacoust., Vol. AU-20, pp. 158-163,
     1972.

Justice, J. and Shanks, J., "Stability Criterion for N-Dimensional
     Digital Filters," _IEEE Trans. Automat. Contr._, Vol. AC-18,
     pp. 284-286, 1973.

Lacoss, R., "Data Adaptive Spectral Analysis Methods," _Geophysics_,
     Vol. 36, pp. 661-675, 1971.

Lacoss, R., Kelly, E. and Toksoz, M., "Estimation of Seismic
     Noise Structure Using Arrays," _Geophysics_, Vol. 34, pp. 21-
     38, 1969.

Makhoul, J., "Linear Prediction: A Tutorial Review," _Proc. IEEE_,
     Vol. 63, pp. 561-580, April 1975.

Maria, G. and Fahmy, M., "On the Stability of Two-Dimensional
     Digital Filters," _IEEE Trans. Audio Electroacoust._, Vol. AU-21,
     pp. 470-472, 1973.

Markel, J. and Gray, A., "On Autocorrelation Equations as Applied
     to Speech Analysis," _IEEE Trans. Audio Electroacoust._,
     Vol. AU-21, No. 2, pp. 69-79, April 1973.

Marzetta, T., _A Linear Prediction Approach to Two-Dimensional_
     Spectral Factorization and Spectral Estimation, Ph.D Thesis,
     Mass. Inst. of Tech., Dept. of Elec. Eng. & Comp. Sci.,
     Cambridge, Mass., February 1978.

Marzetta, T., "The Design of 2-D Recursive Filters in the 2-D Reflection Coefficient Domain," Proc. 1979 Int. Conf. ASSP, 79CH1379-7 ASSP, pp. 32-35, 1979.

Marzetta, T., "Two Dimensional Linear Prediction: Autocorrelation Arrays, Minimum-Phase Prediction Error Filters, and Reflection Coefficient Arrays," Accepted by IEEE Trans. ASSP (1980).

Mersereau, R. and Dudgeon, D., "Two-Dimensional Digital Filtering," Proc. IEEE, Vol. 63, pp. 610-623, 1975.

O'Connor, B. and Huang, T., "Stability of General Two-Dimensional Recursive Digital Filters," IEEE Trans. Acoust., Speech, and Signal Proc., Vol. ASSP-26, pp. 550-560, 1978.

Ong, C., "An Investigation of Two New High-Resolution Two-Dimensional Spectral Estimate Techniques," Long Period Array Processing Report #1, Texas Instruments, Inc., April 1971.

Rudin, W., Function Theory in Polydiscs, Benjamin, 1969.

Shanks, J., Treitel, S., and Justice, J., "Stability and Synthesis of Two-Dimensional Recursive Filters," IEEE Trans. Audio Electroacoust., Vol. AU-20, pp. 115-128, 1972.

Shaw, G. A., "An Algorithm for Testing Stability of Two-Dimensional Digital Recursive Filters," Proc. 1978 IEEE Internat'l Conf. on Acoustics, Speech, and Signal Processing, pp. 769-772, 1978.

Siljak, D., "Stability Criteria for Two-Variable Polynomials," IEEE Trans. Circuits Syst., Vol. CAS-22, pp. 185-189, 1975.

Strintzis, M., "Tests of Stability of Multidimensional Filters," IEEE Trans. Circuits Syst., Vol. CAS-24, pp. 432-437, 1977.

Ulrych, T. and Clayton, R., "Time Series Modelling and Maximum Entropy," Phys. Earth, Planet. Int., Vol. 12, 1976.

Van Trees, H., Detection, Estimation and Modulation Theory. Part 1, Wiley, 1968.

Whittle, P., "On Stationary Processes in the Plane," Biometrika, Vol. 41, pp. 434-449, December 1954.

Wiener, N., Extrapolation, Interpolation, and Smoothing of Stationary Time Series, Wiley, 1949.

Woods, J., "Two Dimensional Discrete Markov Fields," *IEEE Trans. Inf. Th.*, Vol. IT-18, pp. 232-240, March 1972.

Woods, J., "Two Dimensional Markov Spectral Estimation," *IEEE Trans. Inf. Th.*, Vol. IT-22, No. 5, pp. 552-559, September 1976.

THE INSTANTANEOUS AMPLITUDE, PHASE AND FREQUENCY IN SEISMIC EVENT DETECTION, TIMING AND IDENTIFICATION

Rudolf Unger

Facultad de Ciencias Exactas y Tecnología,
Universidad Nacional de Tucumán, Argentina

The essential time-domain information of seismic waveforms
is quantized in terms of the instantaneous amplitude, phase and
frequency. The quantized information is used in the automatic
detection, timing and identification of seismic event signals.
Phase detection is in principle 6 dB better than amplitude (envelope) detection, and can be applied successfully to automatically
detect and time well-dispersed long-period surface waves. Interference by early multiple signals precludes phase detection of
short-period signals. For the latter, it is feasible to design
an efficient, automatic envelope detector and timer with a controllable false alarm rate, based on a Gaussian noise model. For
Eurasian events, short-period phase and frequency related identification parameters, produced automatically by the envelope
detector, are believed to show an inverse relation between the
size and the amount of tectonic energy release ruptures; the
ruptures triggered by explosions then appear to be smaller than
the spontaneous ruptures in earthquakes.

INTRODUCTION

We summarize recent research to quantize the essential time-
domain information of seismic waveforms in terms of the instantaneous amplitude, phase and frequency, and to use this quantized
information in the automatic detection, timing and identification
of seismic event signals (1,2,3,4,5). The use of these parameters
in seismic signal analysis was also suggested by Farnbach (6).

Following parameter definition and resolution, vector diagram
representation of signal and noise interaction leads us to algo-

649

*E. S. Husebye and S. Mykkeltveit (eds.), Identification of Seismic Sources - Earthquake or Underground
Explosion, 649–662.*
Copyright © 1981 by D. Reidel Publishing Company.

rithms for the automatic detection and timing of both long-period
and short-period signals, by means of phase detection and amplitude (envelope) detection, respectively. We then show that envelope detectors with a stable false alarm rate control can be designed, based on a Gaussian noise model. Finally, we discuss the
discrimination potential exhibited by quantities derived from the
instantaneous phase and frequency. We believe the relationship
between these quantities to be indicative of the character of
rupture propagation in earthquakes and underground explosions.

PARAMETER DEFINITION AND RESOLUTION

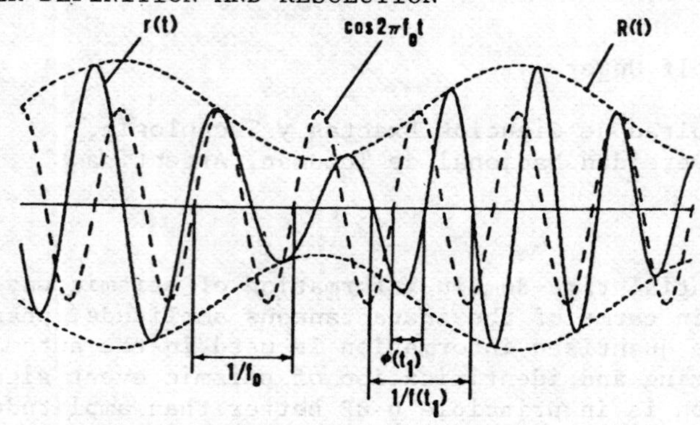

Figure 1. Waveform representation in terms of the in-
          stantaneous amplitude or envelope R(t), the
          instantaneous phase $\phi(t)$, and the instanta-
          neous frequency f(t).

A waveform r(t) can be expressed in terms of its instantaneous amplitude or envelope R(t) and its instantaneous frequency
f(t), or, equivalently, in terms of its instantaneous amplitude
and its instantaneous phase $\phi(t)$ with respect to a monochromatic
waveform of frequency f_o and zero phase (Figure 1):

$$r(t) = R(t) \cos\{2\pi \int f(t)dt\} \tag{1}$$

or

$$r(t) = R(t) \cos \{2\pi f_o t + \phi(t)\}. \tag{2}$$

These parameters can be viewed as the amplitude modulation, and
the frequency or phase modulation, of the monochromatic "carrier"
wave. Although the frequency f_o of the latter may be chosen arbitrarily, it is convenient to choose f_o as the center of the frequency band of interest.

$R(t)$ and the angular argument can be resolved through the
Hilbert transform, defined as the convolution with $1/\pi t$, and also
known as a quadrature filter or a 90° phase shift operator (7,8),
resulting in

$$\tilde{r}(t) = R(t) \, \sin\{2\pi \int f(t)dt\} \tag{3}$$

or

$$\tilde{r}(t) = R(t) \, \sin\{2\pi f_o t + \phi(t)\}. \tag{4}$$

We then obtain

$$R(t) = \{r^2(t) + \tilde{r}^2(t)\}^{\frac{1}{2}} \tag{5}$$

and

$$\phi(t) = \arctan \frac{\tilde{r}(t)}{r(t)} \pm k.2\pi - 2\pi f_o t. \tag{6}$$

The $\pm k.2\pi$ ambiguity is resolved by incorporating the term $-2\pi f_o t$
in the arctangent argument, and accumulatively adding and subtracting 2π radians at discontinuities less than $-\pi$ and greater than
$+\pi$ radians, respectively. The instantaneous frequency is then obtained by differentiation:

$$f(t) = \frac{1}{2\pi} \frac{d\phi(t)}{dt} + f_o. \tag{7}$$

AUTOMATIC SIGNAL DETECTION AND TIMING

With the above parameters, we can represent the waveform $r(t)$
at any instant as a vector or phasor $\vec{r}(t)$ of length or modulus
$|\vec{r}(t)|=R(t)$, and angular argument $2\pi \int f(t)dt=2\pi f_o t+\phi(t)$. For our
study of phase variation, we may omit the term of constant rotation
velocity $2\pi f_o t$ in the vector diagram representation.

We may think this vector to be the result of the interaction
between a signal, $s(t)$, with zero phase angle, and noise, $n(t)$,
with phase angle $\phi_n(t)$, as pictured in Figure 2. In signal envelope
detection, we expect the waveform's envelope to be greater when a
signal is present, than when the waveform consists only of noise.
For a given signal-envelope-to-noise-envelope-ratio (snr), this
occurrence depends on the noise phase angle, and its probability

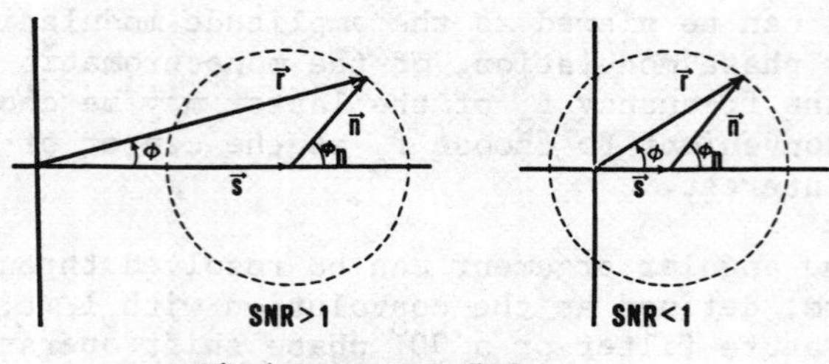

Figure 2. Waveform representation of signal and noise
interaction

equals the thickened arc in Figure 3a, divided by 2π. Also, when
a signal is present, the waveform's phase angle will be statistically biased. For instance, for a given snr, the probability
that the phase is between $-\pi/2$ and $+\pi/2$, depends similarly on the
noise phase angle, and equals the thickened arc in Figure 3b,
divided by 2π. We observe that, for a given snr, this phase bias
probability is higher than the envelope detection probability.

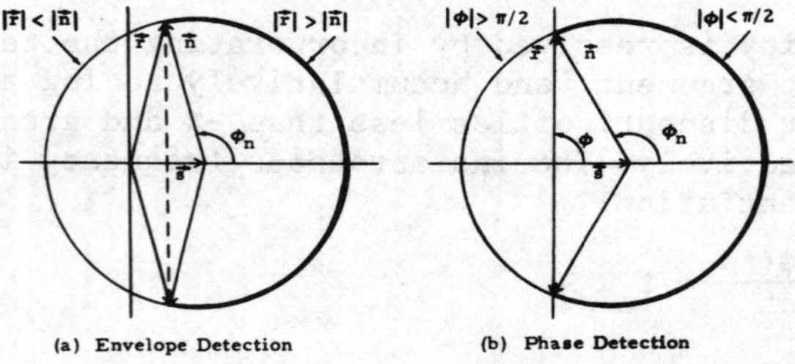

Figure 3. Vector diagram geometry for envelope and
phase detection

From the vector diagram geometry of Figure 3 follow explicit
expressions of these probabilities as a function of snr. These
functions are given in Figure 4, which shows that, in principle,
signal detection by phase bias observation is 6 dB better than the
suggested signal envelope detection procedure.

Automatic signal detection and timing by phase bias observation can be established by performing a second order, moving-
window regression on the time series of the instantaneous phase,
and counting, within each window, the number of times that the
phase deviation from the regressed phase polynomial is within $\pi/2$

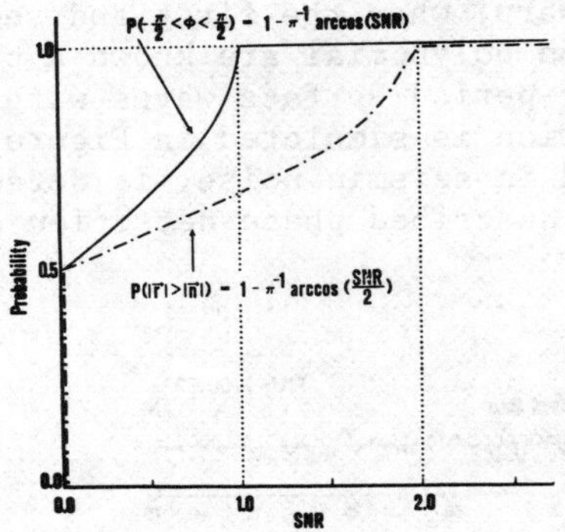

Figure 4. Envelope detection and phase bias probability
distributions as functions of snr.

radians. This phase bias probability becomes maximum, and the
regression residu, or phase standard deviation minimum, when the
window start time coincides with a signal onset. The maxima and

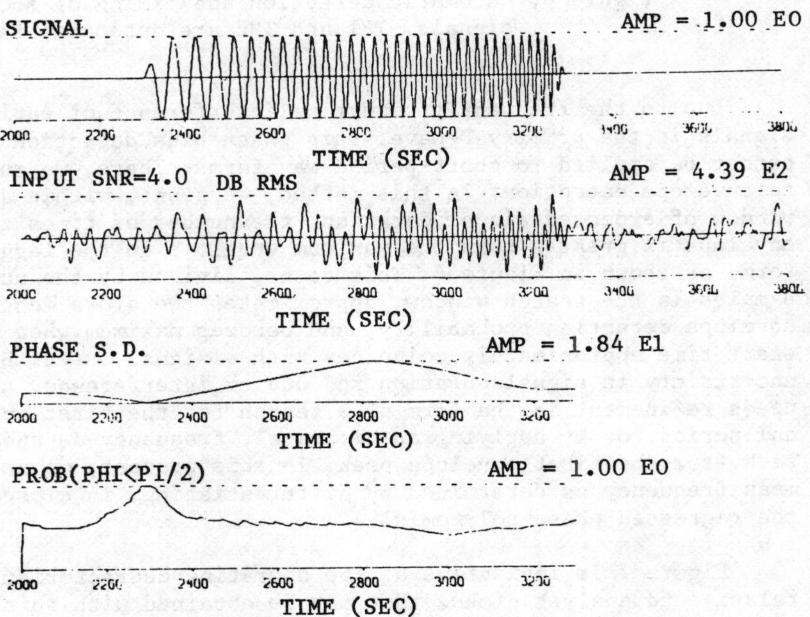

Figure 5. Automatic detection and timing of simulated
long-period surface wave signal by moving-
window phase regression algorithm.

minima become particularly sharp, when the first and second order
coefficients of the regression polynomial are known a priori, for
instance, in the case of long-period surface waves with a known
dispersion curve. This situation is simulated in Figure 5, where
a linear chirp signal, buried in seismic noise, is detected and
timed correctly by the above described phase detection and timing
algorithm.

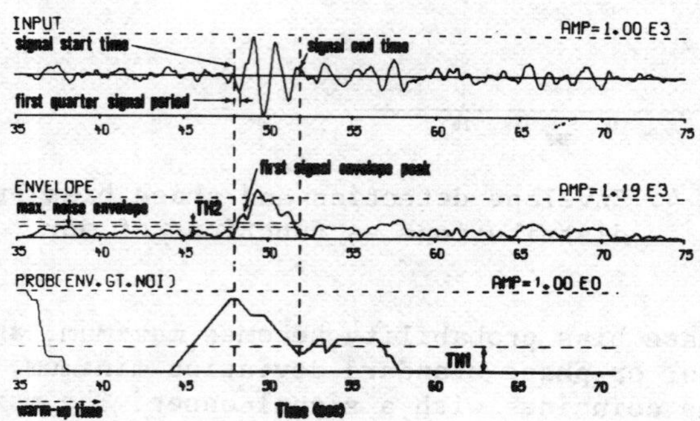

Figure 6. Automatic detection and timing of short-period
signals. TH1 and TH2 are detection thresholds.

Due to the frequently occurring interference of early multiple
signals in the primary P-wave, this phase bias detection method
cannot be applied to short-period waveforms. There, we must resort
to envelope detection. In this method, we count, within a moving
window of expected signal duration, the number of times that the
envelope is greater than the maximum envelope in the lagging waveform, as shown in Figure 6. This count, divided by the number of
samples in the search window, approximates the afore mentioned
envelope detection probability, and becomes maximum when the window
start time approximately coincides with a signal onset. Because of
uncertainty in signal duration and due to interference, the timing
needs refinement in the form of a search for the first quarter signal period, or by applying an empirical, frequency-dependent step-
back from the first envelope peak. In this process, we use the
mean frequency as determined by differentiating, in closed form,
the regressed phase polynomial.

Figure 7 is indicative of the operating characteristics,
relative to analyst picks, that may be obtained with this type
of detector-timer algorithm. It shows its performance on a small
set of Norwegian Seismic Array (NORSAR) single-site data. The
detection ratio and the false alarm rate vary with detection threshold settings as expected. The RMS timing error is 0.23 seconds.
Note that the timing error does not seem to depend on the snr.

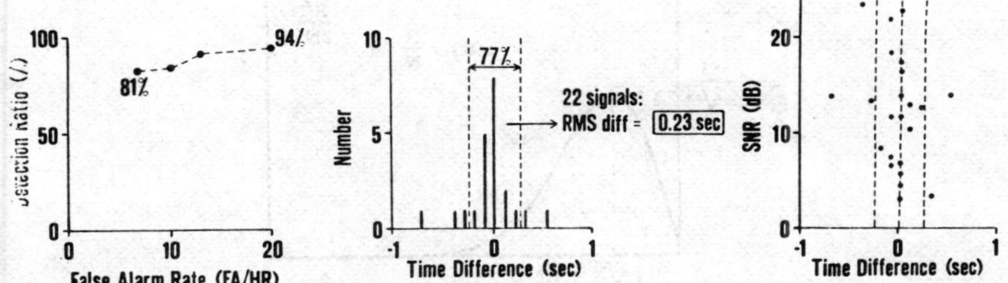

Figure 7. Tentative operating characteristics of automatic short-period signal detection and timing
algorithm for NORSAR single-site data, for snr
settings of 2 to 3 dB.

To our knowledge, no comparable operating characteristics
have been published for other automatic detectors. Detector performance comparison would require a unique definition of "false
alarm" and "detection", and evaluating the operating characteristics of different detectors in an objective (preferably automatic)
manner on one particular set of data. Candidates for such an evaluation are the NORSAR and Seismic Research Observatory (SRO) short-
term-average-over-long-term-average (STA/LTA) detectors, an STA/
LTA detector with controllable false alarm rate designed by
Swindell and Snell (9), a detector designed by Allen (10), a Walsh
transform detector (11), and a multiple-narrowband detector (12).
Of these, only the Allen detector has been designed for accurate
timing, however, with emphasis on local event signals.

In principle, the algorithms have been designed for on-line
application, but have not yet been tested under such conditions.
They can probably be implemented on micro-computers for portable
field instrument application.

FALSE ALARM RATE CONTROL OF SHORT-PERIOD ENVELOPE DETECTORS

In the context of a worldwide seismic surveillance system it
is important to be able to control the false alarm rate of automatic detectors. If the detection statistic is stationary, the
probability of false alarm can be controlled by selecting a suitable detection threshold. Below, we analyze in this sense the
statistical behavior of short-period noise envelopes.

If noise stems from a large number of different sources, the
noise process tends to be Gaussian. The envelopes of a zero-mean
Gaussian process are Rayleigh-distributed (8), as shown in the
upper part of Figure 8. Since the most likely value, R_0, equals
the waveform's RMS value, this distribution is not stationary,

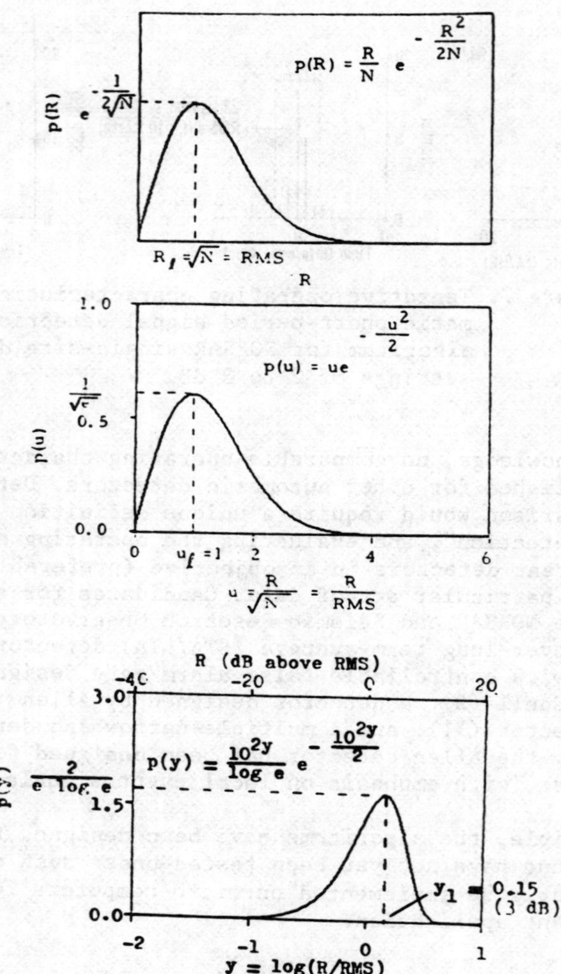

Figure 8. Theoretical distribution of the envelope, the
normalized envelope, and the base-ten logarithm
of the normalized envelope, for a zero-mean
Gaussian process.

but will change shape according to the variations in noise power.
A stationary detection statistic is obtained when we divide the
envelope by the RMS value. We will call R/RMS the normalized envelope; its distribution is drawn in the center part of Figure 8.
Since we usually set detection thresholds in terms of dB snr, we
study the distribution of the base-ten logarithm of the normalized
envelope, log(R/RMS), obtained by transformation of the R/RMS distribution, and shown in the lower part of Figure 8. For instance,

for a single detection trial in a short-period waveform, an R/RMS
threshold of 12 dB would result in a false alarm rate of approximately 13 false alarms per hour, assuming the interval between
independent samples to be 0.4 seconds. We also see why, typically,
the ratio of maximum-amplitude-over-RMS-value in a noise seismogram is taken as 10 dB, and why the false alarm rate increases
steeply for R/RMS thresholds of less than 10 dB, i.e., approximately 0 dB snr. The false alarm rate can be reduced significantly by using multiple detection trials, for instance, as in
the above described envelope detection scheme.

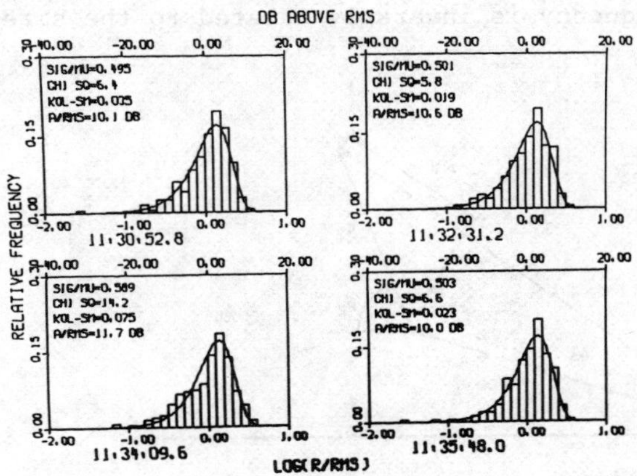

Figure 9. Goodness-of-fit of KSRS single-site, short-
period noise (histogram bars) to the Gaussian
noise model (solid curve), measured with the
normalized envelope logarithm.

Figure 9 shows that empirical histograms of normalized envelope logarithms of seismic noise, recorded at a short-period single
site of the Korean Seismic Research Station (KSRS), fit the solid
curve of the theoretical normalized envelope logarithm distribution of a Gaussian process to an extent that stable false alarm
rate control can be established, by appropriate threshold selection in the righthand tail of the distribution. The figure, and
the annotated goodness-of-fit parameters (σ/μ, ideally 0.525 for
the Rayleigh distribution; chi-square; and Kolmogorov-Smirnov),
show that this holds true also in cases of a relatively poor fit,
down to the 5% significance level of model retention. Thus, the
Gaussian noise model seems an adequate basis for control of the
false alarm rate.

EVENT IDENTIFICATION

We now turn to the discrimination potential of the parameters
mentioned. Figure 10 shows a vector diagram representation of mul-
tiple-signal interaction. The onset of a secondary signal causes
a sudden change in both amplitude and phase of the waveform vector.
By differentiation, sudden phase changes show up as approximate
delta functions in the time series of the instantaneous frequency.
Moving-window regression on the time series of the instantaneous
phase would show a relatively high residu or phase standard deviation for a window containing multiple signals. The phase standard
deviation thus is a measure of pulse complexity. As is generally
accepted, the frequency is inversely related to the size of the
event.

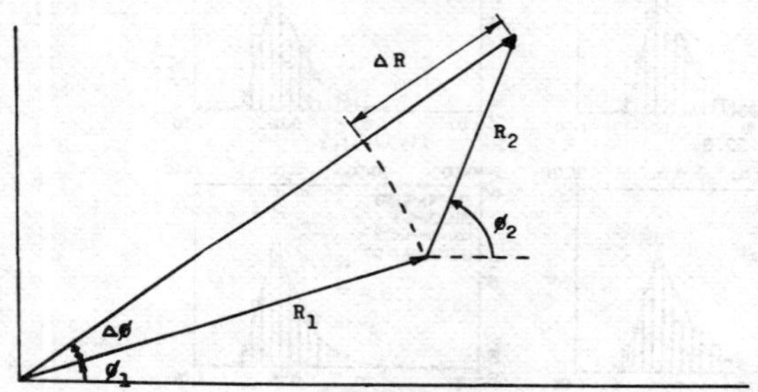

Figure 10. Vector diagram representation of secondary
signal interaction, resulting in sudden
changes in amplitude (ΔR) and in phase (Δϕ).

Figure 11 shows the waveforms, and the time series of instantaneous amplitude, phase, frequency and phase standard deviation,
of a shallow Eurasian earthquake, an Eastern Kazakh presumed underground nuclear explosion, and a Nevada Test Site presumed underground nuclear explosion. The smooth trace in the time series of
the instantaneous frequency is the mean instantaneous frequency,
computed by closed form differentiation of the time-variant phase
regression polynomial. The vertical dotted lines indicate the signal onset times automatically determined by the envelope detector.
Note the higher frequency, and the high amount of multiple signals
in the EKZ event relative to the Eurasian earthquake, resulting in
a relatively high phase standard deviation. In contrast, the NTS
event is nearly monochromatic, with a low pulse complexity and of
low frequency. The difference in characteristics exhibited by the
time-domain parameters for the three types of events suggest their
usefulness in seismic event discrimination.

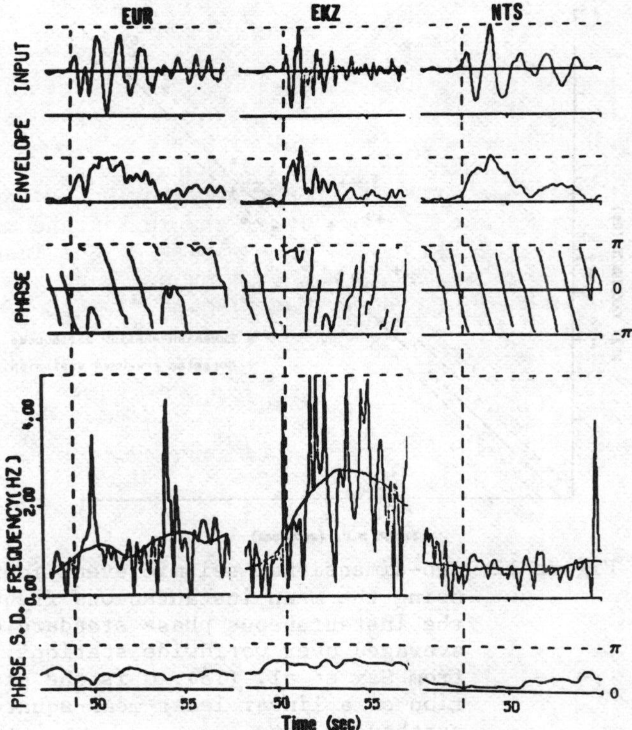

Figure 11. Potential of time-domain parameters to discriminate between a Eurasian shallow earthquake (EUR), an Eastern Kazakh presumed underground nuclear explosion (EKZ), and a
Nevada Test Site presumed underground nuclear
explosion (NTS), recorded at a NORSAR short-
period single site.

In a recent, multi-dimensional event identification experiment (13), the mean instantaneous frequency and the instantaneous
phase standard deviation were routinely reported for Eurasian
event signals automatically detected with the envelope detector
algorithm described previously. These values were then averaged
over a number of worldwide. detecting stations. The worldwide averages of mean instantaneous frequency and instantaneous phase
standard deviation are plotted against each other in Figure 12.
The Eurasian shallow earthquake data show a linear relation between these quantities. The Eurasian presumed explosions show a
similar trend, but with a larger spread and of higher frequency
for a given phase standard deviation. One presumed explosion falls
within the earthquake population; this classification was corrected
by the other discriminants of the event identification experiment.

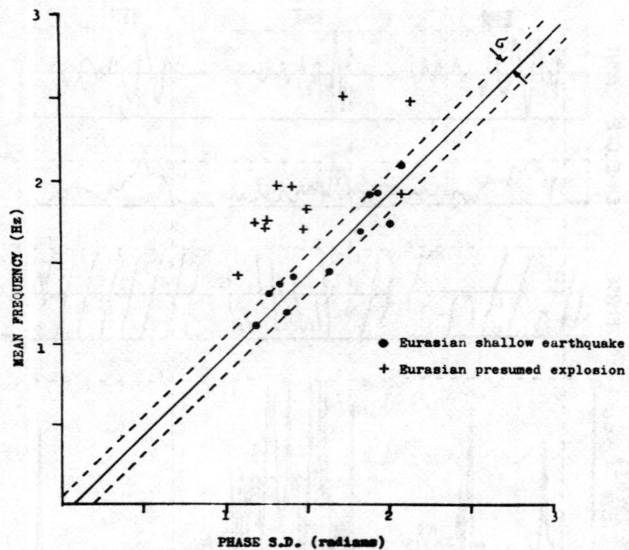

Figure 12. Two-dimensional seismic event identification
using the mean instantaneous frequency and
the instantaneous phase standard deviation
averaged over worldwide stations. After data
from Sax et al. (13). σ is the standard deviation of a linear least-mean-square fit to the
earthquake data.

These phenomena may reflect that in material of low critical
stress, rupture probably occurs discretely within a few seconds
at several places, each rupture being relatively small, resulting
in the relatively high frequency of signals with a high pulse
complexity. This hypothesis has some similarity with a dynamic
rupture model suggested by Day (14). Our observations seem to
agree with some of his resulting synthetic signal characteristics.
In material of high critical stress, failure may be more coherent
and large, resulting in a relatively simple signal of low frequency.
Tectonic energy release ruptures (either multiple or simple), triggered by underground explosions then would be smaller than the
spontaneous earthquake ruptures, resulting in the relatively high
frequency of the presumed explosion signals.

CONCLUSIONS

The instantaneous amplitude, phase and frequency are useful
parameters in the automatic detection, timing and identification
of seismic event signals. It seems feasible to design efficient,
automatic detection-and-timing algorithms, based on phase detection

for long-period surface waves, and on envelope detection for short-period body waves. The latter can be equipped with a stable false alarm rate control based on a Gaussian noise model.

For Eurasian shallow earthquake signals, the frequency and the pulse complexity (reflecting the amount of multiple signals within the first few seconds after the P-wave onset) exhibit a linear relation which seems indicative of certain rupture propagating characteristics. Eurasian presumed underground nuclear explosions show a similar relation, but at a higher frequency. These findings yet must be confirmed with a larger data base containing earthquake and presumed explosion signals from various regions worldwide.

ACKNOWLEDGMENTS

Invaluable suggestions by Dr. R.L. Sax, in particular on event identification data interpretation, contributed greatly to this study. Presentation of this paper was made possible by a grant from NORSAR. This research was supported by the Advanced Research Projects Agency of the U.S. Department of Defense, and was monitored by the U.S. Air Force Technical Application Center under Contract No. F08606-77-C-0004, and by the U.S. Air Force Office of Scientific Research under Contract No. F49620-79-C-0098.

REFERENCES

1. Unger, R.: 1976, *Seismic Event Discrimination Using the Instantaneous Envelope, Phase and Frequency*, Abstract, EOS 57, p. 286.

2. Unger, R.: 1976, *The Instantaneous Phase of Long-Period Waveforms*, Abstract, EOS 57, p. 759.

3. Unger, R.: 1978, *Automatic Detection, Timing and Preliminary Discrimination of Seismic Signals with the Instantaneous Amplitude, Phase and Frequency*, Technical Report No. 4, Texas Instruments Report No. ALEX(01)-TR-77-04, AFTAC Contract Number F08606-77-C-0004, Texas Instruments Incorporated, Dallas, TX.

4. Unger, R.: 1978, *Short-Period Envelope Statistics: A Basis for Envelope Detector Design*, Technical Report No. 17, Texas Instruments Report No. ALEX(01)-TR-78-05, AFTAC Contract Number F08606-77-C-0004, Texas Instruments Incorporated, Dallas, TX.

5. Sax, R.L. and Unger, R.: 1980, *Source Dimension versus Rupture Propagation: A Potential Discriminant*, Abstract, EOS 61, p. 295.

6. Farnbach, J.S.: 1975, *The Complex Envelope in Seismic Signal Analysis*, Bull. Seismol. Soc. Am. 65, pp. 951-962.

7. Papoulis, A.: 1965, *Probability, Random Variables and Stochastic Processes*, McGraw Hill, Inc., New York.

8. Schwartz, M., Bennett, W.R., and Stein, S.: 1966, *Communication Systems and Techniques*, McGraw Hill, Inc., New York.

9. Swindell, W.H., and Snell, N.S.: 1977, *Station Processor Automatic Detection Sysytem, Phase I*, Final Report, Station Processor Software Development, Texas Instruments Report No. ALEX(01)-FR-77-01, AFTAC Contract Number F08606-76-C-0025, Texas Instruments Incorporated, Dallas, TX.

10. Allen, R.V.: 1976, *Automatic Earthquake Recognition for a Single Trace*, Abstract, EOS 57, p. 960.

11. Goforth, T.T., and Herrin, E.: 1980, *An Automatic Seismic Signal Detection Algorithm Based on the Walsh Transform*, Abstract, EOS 61, p. 303.

12. Bache, T.C., Cherry, J.T., Lambert, D.G., Masso, J.F., and Savino, J.M.: 1976, *A Deterministic Methodology for Discriminating Between Earthquakes and Underground Nuclear Explosions*, Systems, Science and Software Final Technical Report Number SSS-R-76-2925.

13. Sax, R.L. and Technical Staff: 1978, *Event Identification – Application to Area of Interest Events*, Technical Report No. 20, Texas Instruments Report No. ALEX(01)-TR-78-08, AFTAC Contract Number F08606-77-C-0004, Texas Instruments Incorporated, Dallas, TX.

14 Day, S.M.: 1981, *Three-Dimensional Finite Difference Simulation of Fault Dynamics*, Proc. NATO ASI Identification of Seismic Sources - Earthquake or Underground Explosion, Oslo, Norway.

15. Rodean, H.C.: 1981, *Inelastic Processes in Seismic Wave Generation by Underground Explosions*, Proc. NATO ASI Identification of Seismic Sources - Earthquake or Underground Explosion, Oslo, Norway, this volume, p. 97.

MULTIDIMENSIONAL DISCRIMINATION TECHNIQUES - THEORY
AND APPLICATION.

Dag Tjøstheim

Norwegian School of Economics and
Business Administration
Hellevn. 30, 5000 Bergen, Norway

ABSTRACT. The problem of discriminating between earthquakes and
underground nuclear explosions is formulated as a problem in pattern recognition. As such it may be separated into two stages,
feature extraction and classification. Various ways of doing
feature extraction will be discussed. Among the techniques mentioned will be univariate and multivariate autoregressive representation, Karhunen-Loève expansions, geophysical parameters and
spectral parameters. The ordinary multivariate Gaussian classification algorithm will be reviewed, but some more recent methods
will also be mentioned and practical problems in designing a classifier will be discussed. The theory described will be illustrated on a relatively large data base of Eurasian earthquakes and
explosions.

1. INTRODUCTION.

The problem of discriminating between earthquakes and underground
nuclear explosions is of a very general nature: On the basis of
the observational raw data vector $\underline{X} = [X(1),\ldots,X(N)]^T$, which
may represent the digitized short-period (SP) and long-period
(LP) wave traces from one or more seismic stations, as well as
geophysical parameters such as m_b , M_S and depth and location
estimates, the task is to recognize the vector and to decide which
of two populations it belongs to. With this formulation the seismic discrimination problem essentially is a problem in pattern
recognition or more specifically multidimensional waveform discrimination. As such it can be separated into two stages, feature extraction and classification. Usually the dimension N of
the data vector is so high that an implementation of any of the

*E. S. Husebye and S. Mykkeltveit (eds.), Identification of Seismic Sources - Earthquake or Underground
Explosion, 663–694.*

standard multivariate statistical classification algorithms directly on the raw data is impossible or highly impractical. The
feature extraction stage consists in reducing the original data
vector $\underline{X}$ to a feature vector $\underline{Z} = [Z(1),...,Z(M)]^T$, where it
is desirable that M is small compared to N while $\underline{Z}$ is still
preserving as much information as possible from the original data
vector $\underline{X}$. The classification then proceeds on the vector $\underline{Z}$.
The transformation f mapping $\underline{X}$ into $\underline{Z}$ is called a feature
extractor and is said to be linear if f is linear.

   A number of techniques for doing feature extraction and classification exist in the statistical pattern recognition literature.
A survey of some of the most important techniques and their properties is given in Sections 2 and 3. Only a few of these techniques have been used in seismic discrimination and some illustrations on observed seismic data are given in Section 4.

2. FEATURE EXTRACTION.

Roughly speaking pattern recognition techniques may be grouped into statistical techniques and structural (or syntactic) techniques. In statistical pattern recognition one tries to extract
characteristic statistical features from the observed data, and
statistical criteria are used when evaluating patterns. The techniques are entirely general and very flexible and have been applied on such diverse areas of application as seismic discrimination, electrocardiography, speech recognition, X-ray photographs
and natural resource estimation (cf [1, Sec. 9]). A disadvantage
of this method is that little consideration is given to the context in which the data patterns originate. In structural pattern
recognition a main point is to take this contextual framework into account. Typically, in seismic discrimination one could try
to use contextual information of a geophysical nature such as information based on source theory and source related parameters
(m_b and M_S are examples).

   The statistical techniques are much better developed both
theoretically and when it comes to applications. At the present
time the existing theories of general structural pattern recognition must be said to be rather formal. They are associated
with the structure of grammars of abstract languages (e.g. see
[2]). It is not easy to apply this formal theory to concrete problems in discrimination. Instead, in practice one often chooses
an ad hoc procedure exemplified by the $m_b:M_S$ procedure of seismic discrimination. In the absence of general applicable structural methods we shall be concerned in this section almost exclusively with statistical feature extraction, but we realize the
importance of an integrated statistical-structural pattern recognition approach to seismic discrimination. We now discuss some
of the most used feature extraction techniques.

2.1. Transform techniques.

Given digitized SP or LP data in time domain it may be adven-
tageous to transform these to obtain a representation better suited for classification. The most commonly used transformation
is probably the discrete Fourier transform, and indeed some of the
earliest seismic SP discriminants were based on features in frequency domain. Thus the spectral ratio discriminant was proposed
by Kelly [3] in 1968, while the third moment of frequency was
suggested by Weichert [4] in 1971. These and other SP discriminants have subsequently been combined and tested by a number of
scientists and we refer to [5], [6] for reviews.

It has not been entirely clear how much of the discriminating
information contained in the raw data these discriminants represent, or in what sense, if any, they can be considered to be optimal. An interesting recent contribution in this connection,
however, is [7], which gives a justification for spectral ratio
discriminants using the Kullback-Leibler information measure.

The discrete Fourier transform is not the only which has been
considered. Other transforms include the Walsh-Hadamard transform,
the Haar transform and the Chebyshev transform named after the
orthogonal set of functions used in representing the signal.
Applications of these transforms to seismic discrimination has
been considered in e.g. [8] and [9]. Chen [8] concludes that the
Fourier transform appears to be the most appropriate one since
teleseismic data is highly sinussoidal in nature. Chen [8] also
tries the complex cepstrum which is a digital filtering technique
which preserves both the amplitude and phase information of the
Fourier spectrum. For the data set studied by him the complex
cepstrum gives better results than the spectral ratio and third
moment of frequency.

2.2. Feature extraction based on covariance properties.

Probably the most commonly employed feature extraction method in
statistical pattern recognition is the principal component method
or the Karhunen-Loève expansion as it is often called in the pattern recognition literature. Principal component analysis has
been applied before in detection seismology, e.g. [10], but it
has found surprisingly little use in seismic discrimination (see
[11], [12] and [13], however). We now give a brief description.
For a full account we refer to [14] and [15].

Let $\underline{Y} = [Y(1),...,Y(M)]^T$ be an M-dimensional column vector
characterizing one of the elements (seismic events) of our data
base. We assume that $\underline{Y}$ is a random sample vector distributed
according to one of two multivariate probability distributions,
characterizing for example the earthquake and explosion popula-
tion of Y-vectors. Let $\underline{h}_1,...,\underline{h}_M$ be an orthonormal basis, which
is yet to be determined. Decomposing $\underline{Y}$ along $\underline{h}_1,...,\underline{h}_M$ we have

$$\underline{Y} = \sum_{i=1}^{M} Z(i)\underline{h}_i \tag{2.1}$$

where

$$Z(i) = \underline{Y}^T \cdot \underline{h}_i = \sum_{k=1}^{M} Y(k)h_i(k) \tag{2.2}$$

The idea in a principal component analysis of $\underline{Y}$ is to pick $\underline{h}_1,\ldots,\underline{h}_M$ in such a way that the main part of the information carried by $\underline{Y}$ is decomposed along a few of the vectors $\underline{h}_i$, and the problem is to find $\underline{h}_1,\ldots,\underline{h}_m$ where m is small compared to M such that

$$E\{[\underline{Y} - \sum_{i=1}^{m} Z(i)\underline{h}_i]^T \cdot [\underline{Y} - \sum_{i=1}^{m} Z(i)\underline{h}_i]\} \tag{2.3}$$

is minimized under the constraint that the $\underline{h}_i$'s are orthonormal. The letter E is here used to denote the expectation operator. Using Lagrange multipliers it is not difficult to show that the vectors $\underline{h}_i$, $i=1,\ldots,m$ must satisfy the eigenvalue problem

$$R\underline{h}_i = \gamma_i\underline{h}_i \tag{2.4}$$

where R is the symmetric positive semi-definite matrix given by $R = E(\underline{Y}\,\underline{Y}^T)$. Using the relation $Z(i) = \underline{Y}^T \cdot \underline{h}_i$ it follows that

$$E\{[\underline{Y} - \sum_{i=1}^{m} Z(i)\underline{h}_i]^T \cdot [\underline{Y} - \sum_{i=1}^{m} Z(i)\underline{h}_i]\}$$

$$= E\{\underline{Y}^T \cdot \underline{Y}\} - \sum_{i=1}^{m} E|Z(i)|^2 \tag{2.5}$$

But from (2.4) we have $E|Z(i)|^2 = \gamma_i$ and the first m principal components are obtained by choosing the m largest eigenvalues. In practice R, $\underline{h}_i$ and γ_i have to be estimated. This is done by estimating $R = R_j$, $j = 1,2$, by

$$\hat{R}_j = \frac{1}{N_j} \sum_{k=1}^{N_j} \underline{Y}_{k,j}\,\underline{Y}_{k,j}^T \tag{2.6}$$

for the two classes $j = 1,2$ (e.g. earthquakes and explosions). Here N_j is the number of available observations in a learning data set for class j . Estimates $\hat{\underline{h}}_{i,j}$ and $\hat{\gamma}_{i,j}$ are subsequently obtained by solving

$$\hat{R}_{j} \hat{\underline{h}}_{i,j} = \hat{\gamma}_{i,j} \, \hat{\underline{h}}_{i,j} \tag{2.7}$$

for $j = 1,2$. It should be noted that a pooled covariance matrix for the two classes is often used in the pattern recognition literature [16].

The method of principal components has proved to be very efficient in pattern recognition applications. The method is general in nature and is not limited to time series data. Its theoretical properties are well known, and it is optimal in the above linear least square sense. However, there are several disadvantages associated with the method both for statistical pattern recognition in general and for seismic discrimination in particular.

1) The method becomes very impractical when the dimension of the input data vector $\underline{Y}$ is high. This is exactly what happens in seismic discrimination, so if it should be applied here it will usually be necessary to do some preprocessing or primary feature extraction to reduce the dimension of the raw data vector [11].

2) The method is linear, while the structure of the data could be highly nonlinear such that a nonlinear representation will be more effective. Gnanadesikan [17] has considered a nonlinear principal component method. In this connection it should be mentioned that there exists a number of other techniques for doing nonlinear feature extraction [1].

3) It was pointed out by Fukunaga and Koontz [18] that one of the big disadvantages of the principal component technique as far as pattern recognition is concerned is that the technique does not necessarily select important features for separating pattern classes. When two pattern classes share similar important features, the corresponding eigenvalues are large and dominant. But the corresponding eigenvectors may not be important for separating pattern classes it is claimed by Fukunaga and Koontz, and they therefore propose in [18] a kind of normalization procedure in order to extract the important features separating the two pattern classes. The idea is to obtain a representation where both classes cannot share common important features. We have applied this method to earthquake-explosion discrimination in [11], [12] and [13]. The results did not indicate that the modified method works better than the ordinary one. On the contrary the opposite seemed to be the case. This is consistent with the criticism of the modified method in [19]. Foley and Sammon [20] suggest an alternative method based on so-called optimal discri-

minant vectors. To our knowledge the method has not been applied
to seismic discrimination.

2.3. Features based on parametric time series models.

Digitized time series from seismic events are often very long.
Thus, at the seismic research observatory at NORSAR a single SP
data trace typically comprises 1200 data points. This requires
some form of preprocessing before applying a principal component
technique, especially when several time series are available for
each seismic event. One possibility is to use orthogonal transforms as in Section 2.1, but a more promising approach seems to
be to use parametric time series models. This has in fact been
done in [11], [12] and [13] with good results. It should be noted
that underline univariate parametric models have been used in other fields
for classification purposes: mainly speech analysis and synthesis
[21], [22] and speech signal pattern recognition for speaker verification [23]. We will now present the main ideas in a multivariate
setting. It should be noted that unlike the principal component
technique, the features do not depend on a learning set and thus
can be obtained for each seismic event separately.

Assume that an observed multiple (seismic) time series can be
described by a q-dimensional weakly stationary random process $\underline{X}(t)$.
It can be shown that such a process can be approximated by a so-
called autoregressive AR(p) process if the order p is chosen high
enough. An AR(p) vector time series is defined by a difference
equation

$$\underline{X}(t) - A_1\underline{X}(t-1) - \ldots - A_p\underline{X}(t-p) = \underline{W}(t) \qquad (2.8)$$

where $A_1, \ldots, A_p$ are qxq matrices and $W(t)$ is a multichannel
white noise series; i.e., $E[\underline{W}(t), \underline{W}^T(s)] = V\delta_{ts}$. Here δ_{ts} is the
Kronecker delta and V is the covariance matrix of the white noise
series $\underline{W}(t)$. We assume that $\underline{X}(t)$ has zero mean. Let I be the
identity matrix. It is assumed that $X(t)$ satisfies the usual
stability condition, namely, that $\det[I-A_1z - \ldots - A_pz^p]$ has its
zeros outside the unit circle in the complex z-plane. A q-dimensional vector series as described by (2.8) has $q^2p+q(q+1)/2$ parameters, namely, the q^2p elements of the coefficient matrices
$A_1, \ldots, A_p$ and the $q(q+1)/2$ distinct elements of the symmetric
covariance matrix V. The important point as far as feature extraction is concerned is that these parameters [24, p. 29] completely
describe the second order (covariance) properties of $\underline{X}(t)$.

Estimates $\hat{A}_1, \ldots, \hat{A}_p$ of the coefficient matrices $A_1, \ldots, A_p$
are obtained by solving the matrix Yule-Walker equation

$$[\hat{A}_1, \ldots, \hat{A}_p] \begin{bmatrix} \hat{C}(0) & \hat{C}(1) & \ldots & \hat{C}(p-1) \\ \hat{C}^T(1) & \hat{C}(0) & & \hat{C}(p-2) \\ \vdots & & & \vdots \\ \hat{C}^T(p-1) & \hat{C}^T(p-2) & & \hat{C}(0) \end{bmatrix} = [\hat{C}(1), \ldots, \hat{C}(p)]$$

$$(2.9)$$

where

$$\hat{C}(k) = N^{-1} \sum_{t=1}^{N-|k|} \underline{X}(t+k)\underline{X}^T(t) \qquad (2.10)$$

An estimate $\hat{V} = \hat{V}(p)$ of the covariance matrix V of the residual process $\underline{W}(t)$ is now obtained from

$$\hat{V}(p) = \frac{N}{N-pq}[\hat{C}(0)-\hat{A}_1\hat{C}^T(1) - \ldots - \hat{A}_p\hat{C}^T(p)] \qquad (2.11)$$

More details concerning multivariate time series estimation algorithms are given for example in [25]. When it comes to the difficult problem of estimating the order p of the approximating AR(p) process, one possibility [25] is to use the multivariate version of the Akaike FPE criterion. Another possibility is to try to generalize some of the other univariate criteria. See [26] for a review of these.

Some properties of univariate autoregressive feature extraction is considered by Kashyap [27]. The features are sufficient in the sense that once the AR parameters are given, the likelihood function for the observed sample is uniquely determined if the measurements are assumed to be normally distributed. Kashyap [27] also gives an interesting extension. He suggests that each pattern class C_i could be represented by a probability density in the AR parameter space. This may be a reasonable way of looking at things for the earthquake-explosion AR modelling, where it seems natural to allow the AR parameters to vary within each class depending for example on parameters such as seismic region and depth. Kashyap's model can be extended to the multivariate case and could be of value in practical discrimination work.

An objection to AR models in seismic discrimination has been that they are not "structural", i.e., they are merely ad hoc descriptions of the data without due consideration to the seismic mechanisms generating the data. In this connection, however, it should be noted that in a recent paper Dargahi-Noubary et al [28] show that univariate AR(p) models of the form

$$(1-e^{-k}B)^p X(t) = W(t) \qquad (2.12)$$

where B is the shift operator $B:X(t) \rightarrow X(t-1)$, can be related to
deterministic seismic source models proposed by Haskell [29], [30],
Von Seggern and Blandford [31] and Aki [32]. Thus [28] seems to
provide some geophysical justification for AR models in addition
to the purely data reduction arguments in favour of such models.

AR models are just special cases of the more general ARMA
(autoregressive-moving average) models. A vector ARMA (p,q)
model $\underline{X}(t)$ is defined by a difference equation

$$\underline{X}(t)-A_1\underline{X}(t-1)-\ldots-A_p\underline{X}(t-p) = \underline{W}(t)-C_1\underline{W}(t-1)-\ldots-C_q\underline{W}(t-q) \qquad (2.13)$$

where $A_1,\ldots, A_p; C_1,\ldots, C_q$ are coefficient matrices and $\underline{W}(t)$ is
a multiple white noise series. An ARMA model may sometimes yield
a representation with fewer parameters than a pure AR model for a
given time series and Chen [33, Appendix] argues for the use of
ARMA models for the purpose of seismic event representation.
However, although these models are very good for feature extraction
in the sense that few parameters are required, in our view they are
inferior for discrimination purposes because of the instability of
ARMA representations. ARMA models with entirely different sets of
parameters may give rise to virtually the same covariance function.
It is clear that this may have very serious effects if a classification is undertaken based on these parameters. Pure AR (or MA)
models do not seem to suffer from this instability problem.

2.4. Selecting a subset of feature extractors. Some cautionary
      remarks.

Having arrived at a set of say M feature extractors for
example using one of the techniques described above, it may be
tempting to search further among these features for a subset of
them which is optimal with respect to classification error rate.
In this connection it is an annoying fact that the set of K individually best discriminating features is not necessarily the best
discriminating feature set of size K, even for the case of independent features [34]. The only way to ensure that the best subset
of features from a set M is chosen is to examine all $\binom{M}{L}$ possible
combinations for $L = 1, \ldots, M$. For M large this is clearly impractical and various suboptimal search procedures have been
suggested [35][36].

It should be noted, however, that such a search for an optimal
subset may produce somewhat misleading results if, as is sometimes
the case, the performance of each subset is tested in terms of a
fixed learning set, where it is known apriori which elements
(events) belong to which class. Typically, and maybe contrary to
intuition, the best classification error rates do not decrease monotonically with increasing size of the subset. The error rates de-

crease initially, but then increase, and it is sometimes possible to find a relatively small subset which gives much better discrimination for the learning data base than the entire set of M features. Murray [37], however, shows that the error rate estimates as obtained from the learning set may be very substantially biased downwards, and typically the bias is largest for a subset of size K approximately equal to M/2. This is because it is for these values of K that the largest number of possible combinations $\binom{M}{L}$ exists, and this increases the possibility of choosing a subset of features which is especially tailored to the specific learning set under consideration with the resulting dangers of seriously underestimating the error rates. This effect is clearly illustrated in [37] by simulation experiments where $M = 10$ and $K = 4$, and where the true error rate is approximately 30% while the "best" error rate estimate based on misclassified samples of the learning set is below 15%. This clearly shows the dangers inherent in manipulating features using just a learning set. We will return to a related set of problems in Section 3.4.

3.   CLASSIFICATION.

The classification may be termed purely statistical, structural or mixed according to the choice of feature extractors. Structural features such as m_b and M_S could of course also be discussed using statistical arguments (see e.g. [38] and [39]). The main emphasis in this section will be on discussing various techniques used in multivariate statistical analysis.

3.1. Basic principles.

Once the feature extraction has been performed and each observed data vector $\underline{X}$ has been reduced to an appropriately dimensioned vector $\underline{Z}$, the problem of classification consists in deciding which of several classes $\underline{Z}$ belongs to. We will assume that we have two classes 1 and 2 only, and that the feature vector $\underline{Z}$ is distributed according to multivariate probability density functions $P_1(\underline{z})$ or $P_2(\underline{z})$ if it comes from class 1 or class 2, respectively. For a given $\underline{Z}$ we compute $P_1(\underline{Z})$ and $P_2(\underline{Z})$ and assign $\underline{Z}$ to the class i having the largest likelihood $P_i(\underline{Z})$ of having $\underline{Z}$ occurring. This is a common procedure [14, Ch. 6], [15] in problems of this sort, and, as is well known, it can be generalized by assigning prior probabilities π_i and a loss function r_{ij} representing the loss in identifying a member of the ith class as a member of the jth class, leading to a so-called Bayesian decision

rule which is optimal in the sense that it minimizes the average
risk of misclassification with respect to the given a priori
probabilities and loss functions [14, Ch. 6]. In our concrete
application to the earthquake/explosion problem we have found it
convenient to omit π_i and r_{ij} since they are difficult to specify,
and the loss function will to a certain extent have to be based
on political decisions.

3.2. Gaussian discrimination. The Fisher discriminant.

Generally the density functions P_1 and P_2 are unknown and
will have to be estimated. This is complicated in the general
multivariate situation, but if the functional form is assumed
known, the estimation will be reduced to the estimation of a
finite number of parameters. Usually P_1 and P_2 are assumed to be
multivariate normal in which case the estimation reduces to estimation of mean vectors and covariance matrices.

Let $\underline{\mu}_i$ and K_i, $i=1,2$, denote the mean vector and covariance
matrix for the class 1 and class 2 $\underline{Z}$-vectors respectively. In
the Gaussian case the probability densities $P_i(\underline{z})$ are given by

$$P_i(\underline{z}) = (2\pi)^{-M/2}|K_i|^{-\frac{1}{2}}\exp[-\tfrac{1}{2}(\underline{z}-\underline{\mu}_i)^T K_i^{-1}(\underline{z}-\underline{\mu}_i)] \tag{3.1}$$

for $i=1,2$, where $M = \dim \underline{Z}$ and $|K_i|$ is the determinant of K_i.
Taking logarithms and omitting the common factor $(2\pi)^{-M/2}$ an
equivalent discriminant score is given by

$$S_i(\underline{z}) = -\tfrac{1}{2}\log|K_i| - \tfrac{1}{2}(\underline{z}-\underline{\mu}_i)^T K_i^{-1}(\underline{z}-\underline{\mu}_i) \tag{3.2}$$

and the classification rule is to assign $\underline{Z}$ to the population i
having the largest value $S_i(\underline{Z})$. The discrimination algorithm
(3.2) is nonlinear in $\underline{Z}$. However, if $K_1 = K_2 = K$, then the term
$-\tfrac{1}{2}\log|K_i| - \tfrac{1}{2}\underline{z}^T K_i^{-1}\underline{z}$ is common to S_i, $i=1,2,$ and can be eliminated,
and the discriminant (3.2) is equivalent to

$$S_i(\underline{z}) = -\tfrac{1}{2}\underline{\mu}_i^T K^{-1}\underline{\mu}_i + \underline{\mu}_i^T K^{-1}\underline{z} \tag{3.3}$$

the so-called Fisher discriminant, which is linear in $\underline{z}$.

In general neither $\underline{\mu}_i$ nor K_i are known, but have to be
estimated by averaging the feature vectors $\underline{Z}$ in a learning data
base. Thus

$$\hat{\underline{\mu}}_i = N_i^{-1} \sum_{i=1}^{N_i} \underline{Z}_{j,i} \tag{3.4}$$

and

$$\hat{K}_i = N_i^{-1} \sum_{j=1}^{N_i} (\underline{Z}_{j,i} - \hat{\underline{\mu}}_i)(\underline{Z}_{j,i} - \hat{\underline{\mu}}_i)^T \tag{3.5}$$

where $i=1,2$ and N_1 and N_2 are the number of class 1 and class 2 feature vectors in the learning data set, and where $\underline{Z}_{j,1}$ is the M-dimensional feature vector for the jth event of the learning set of class 1 and similarly for $\underline{Z}_{j,2}$. We can now discriminate between class 1 and class 2 using (3.2) or (3.3), but with theoretical values replaced by estimated ones. The resulting discriminants can be justified from considerations of minimizing expected risk of misclassification but only in an asymptotic sense when N_1 and N_2 tend to infinity in which case $\hat{K}_i$ and $\hat{\underline{\mu}}_i$ tend to K_i and $\underline{\mu}_i$, respectively. In this case asymptotic expansions for the errors of misclassification can also be obtained as indicated in [14, p. 135] but no closed form expression is available in the finite sample case.

The discriminants (3.2) and (3.3) have been reported [40] to relatively robust to deviations from normality. An interesting alternative to the linear Fisher discriminant (3.3) is to do discrimination [41], [42] using logistic regression which is more robust to distributional assumptions. Efron [41] has shown that if $K_1 = K_2$ and $\underline{Z}$ is normally distributed, then asymptotically logistic regression typically is between one half and two third as effective as the Fisher discriminant. In an application to two empirical data sets which were known to be non-normal (in fact a mixture of continuous and discrete variables were used) Wilson and Press [42] in both cases found that the Fisher discriminant was outperformed by the logistic regression approach.

3.3. Nonparametric classification.

Nonparametric classification methods may roughly be divided in two main groups; nonparametric density estimation and nearest neighbour classification rules. In nonparametric density estimation the functional form of $P_i(\underline{z})$ is assumed unknown and the multivariate density $P_i(\underline{z})$ is sought estimated directly from the observed feature vectors $\underline{Z}$. Various ways of doing this have been suggested in the literature. For an examination and comparison of five commonly used probability density function estimators we refer to Cover [43] who gives 87 references on nonparametric density estimation. An interesting recent contribution is [44].

The nearest neighbour classification method is a non-parametric method where the functional form of $P_i(\underline{z})$ is unknown and where one does not attempt to estimate $P_i(\underline{z})$. The idea of the method is as follows: Let $A = [\underline{Z}_1, \ldots, \underline{Z}_n]$ be a set of n feature vectors, where the classification of each vector Z_i is known. When a new pattern vector X with a corresponding feature vector Z is observed, we calculate the distances in feature space between Z and each vector Z_i of A. The distance may be the usual Euclidian distance function given by

$$d(Z, Z_i) = (Z_i - Z)^T \cdot (Z_i - Z) \qquad (3.6)$$

or a weighted distance function. Let $Z_j^* \in A$ be the feature vector for which $d(Z, Z_i)$ is minimized. The vector Z_j^* is then called the nearest neighbour of Z in feature space, and Z is classified according to the class of Z_j^*. This is an intuitive and simple classification rule. It avoids the difficult problem of density estimation altogether. One may think that this amounts to neglecting too much information. However, it can be shown [45] that asymptotically as $n \to \infty$ the error rate using nearest neighbour classification is always bounded by twice the error rate when $P_i(z)$, $i=1,2$, is completely known. This is a rather remarkable result since it holds independent of functional form of $P_i(z)$. There are indications [1, p. 709], however, that the small sample behaviour of nearest neighbour classification rules may be quite bad for special classes of density functions. The nearest neighbour classification rule can be generalized to the so-called k-nearest neighbour classification rule. In this case Z is assigned to the class to which a majority of its k nearest neighbours belong.

3.4. Designing a classifier in practice.

In practical discrimination work it is customary to divide the available data set consisting of say L_1 and L_2 data vectors X from classes 1 and 2 into a learning set of N_1 and N_2 data vectors and an independent data set of L_1-N_1 and L_2-N_2 vectors. The learning set may be used both for the purpose of feature extraction and for the construction of classifiers. Thus, typically the learning data set will be used in estimating the principal components in (2.6) and (2.7) as well as in estimating the density functions $P_1(z)$ and $P_2(z)$ independent of whether parametric or non-parametric methods are being used. Leaving part of the data set out has the advantage that there will be a data base of independent events available for testing constructed discriminants. If the entire data set of available samples L_1 and L_2 is used to design a classifier it is well known [46], [37] that optimistic bias will result if future performance is predicted to be that obtained on the design set itself.

Unfortunately it is not clear what is the best way of splitting the data set if a realistic error rate estimate shall

be obtained for the classifier under consideration. The simplest
method is to draw at random N_1 and N_2 events from the original
class 1 and class 2 data sets. In most applications, except for
exceedingly large L_i's it seems natural to require $N_i > L_i/2$,
$i=1,2$, but it is not clear which values of N_1 and N_2 will be
optimal. Based on experimental comparisons reported by Lachenbruch
and Mickey [47] and others the "leave one out" method appears to
be a good approach. For this method, given $L = L_1 + L_2$ samples,
a classifier is designed using $L-1$ samples, tested on the remaining
sample, and then the error rates resulting from <u>all</u> such partitions
of size $L-1$ for the learning set and 1 for the independent set are
averaged. A disadvantage of this method is that it is very time
consuming for L large or moderately large.

A generalization of the "leave one out" method is considered
by Toussaint [48] who proposes using an independent test set of
size k, $k \leq L/2$ and L/k an integer, and a learning set of size $L-k$
and letting the test set rotate over the L/k possible disjoint
test sets and again using the average error rate over these test
sets as an estimate of the true error rate. In [48] typically
$L/k = 10$, and we refer to this work for more details.

Another important problem in designing a classifier is the
problem of choosing an optimal dimension M of the feature vector
in relation to L_1 and L_2 or rather N_1 and N_2. Clearly this is a
very important problem. It has been attacked mostly in some
special cases [49], [50]. For Gaussian features with a common
unknown covariance matrix (cf. (3.3)) a rule of the thumb is that
N_i/M, $i=1,2$, ought to be larger than five. Furthermore for any M,
there is a lower bound on the number of samples N_i per class required to obtain a low variance estimate.

When it comes to the error estimate itself essentially this
can be done either by simply counting misclassified samples or by
estimating the Bayes error probability using density estimation
techniques. For further references we refer to the review paper
[1, Section 7].

4.   APPLICATIONS TO SEISMIC DISCRIMINATION.

In this section we illustrate some of the methods discussed
in this paper by considering discrimination between earthquakes
and underground nuclear explosions. This material is mainly taken
from [11], [12] and [13] to which we refer for more details and
background information. ·

4.1. Data sets.

The main data set to be used consists of 83 presumed explosions (hereafter called explosions) and 72 presumed earthquakes (hereafter called earthquakes) which occurred in Eurasia in the time period between March 1971 and September 1976. Relevant data for the events are tabulated in Tables 1 and 2 of [12]. The geographical location of the corresponding epicenters is depicted in Fig. 1. An m_b:M_S diagram for the data set is shown in Fig. 2. This could be considered as an especially simple form of structural feature extraction since the features m_b and M_S are selected on the basis of seismic theory which predicts that explosions should produce less surface waves than earthquakes for a given body wave magnitude m_b. It is seen that the m_b:M_S diagram contains considerable overlap between explosions and earthquakes.

One of the major difficulties of the m_b:M_S criterion is the problem of detecting surface waves from weak explosions. Thus for 20 of the explosions on Figs. 1 and 2 vertical component Rayleigh waves were not detected. These events have been provided with an upper limit for a surface wave magnitude M_S by measuring the amplitude of the largest noise cycle of the 20s Rayleigh wave component.

Motivated by these difficulties, considerable efforts have been made to construct discriminants based exclusively on body wave or SP information. We will therefore treat pure SP and mixed SP-LP discrimination seperately.

4.2. SP discrimination. Feature extraction.

Some simple SP features (including third moment of frequency and complexity) commonly used in seismic discrimination are discussed in [5]. To get an impression of the quality of these features we have plotted in Fig. 3 the two high order autoregressive coefficients $a_{31}(C)$ and $a_{32}(C)$ of a third order autoregressive model fitted to the coda section of a wave trace from a single NORSAR subarray for each event. According to [5, pp. 248-249] where the $a_{31}(C)$-$a_{32}(C)$ discriminant is included in a comparison of 6 SP discriminants, this discriminant is fairly typical for the quality of SP discriminants. It is seen from Fig. 3 that the overlap between explosions and earthquakes is very substantial, certainly much worse than for the m_b:M_S discriminant of Fig. 2.

Using the theory of Section 2.3 on multiple AR feature extraction a natural extension if array data are available, is to try to fit a multivariate AR model to several subarray beams. In [12] we used data from three NORSAR subarrays. The onset of the signal was determined separately for each subarray beam, and the traces were subsequently aligned so that the signal onset corresponds to the first samples of each of the component series

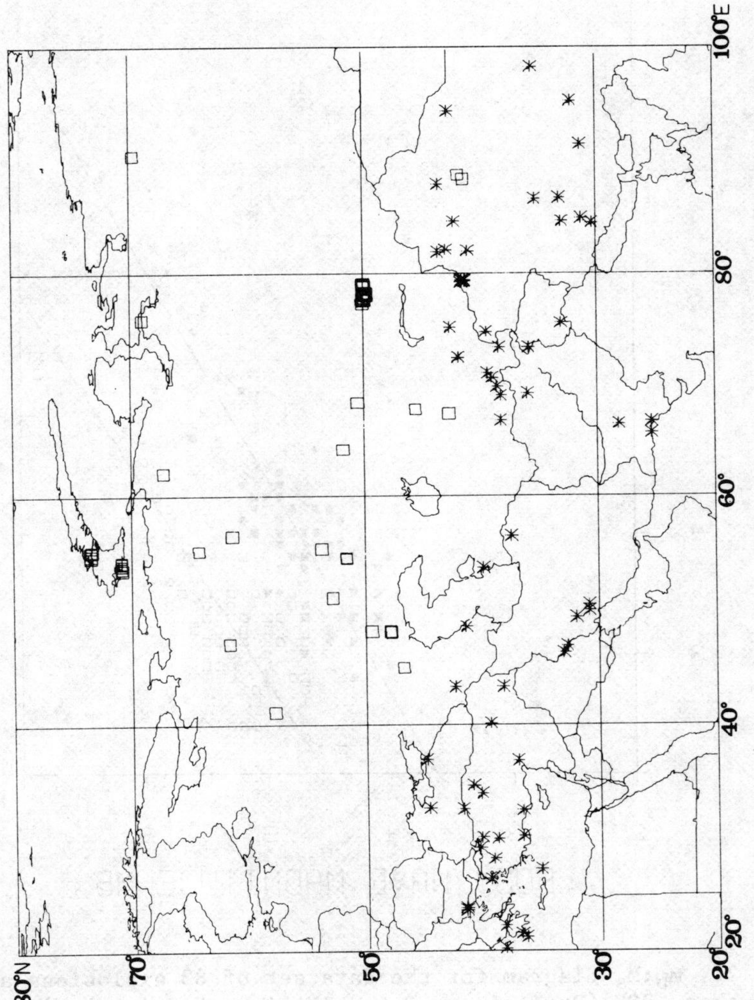

Figure 1. The geographic distribution of the 83 explosions and 72 earthquakes used in the SP discrimination. Explosions are depicted by squares and earthquakes by stars in this figure as well as in Figures 2, 3, 4, 5 and 6.

                                                                          D. TJØSTHEIM

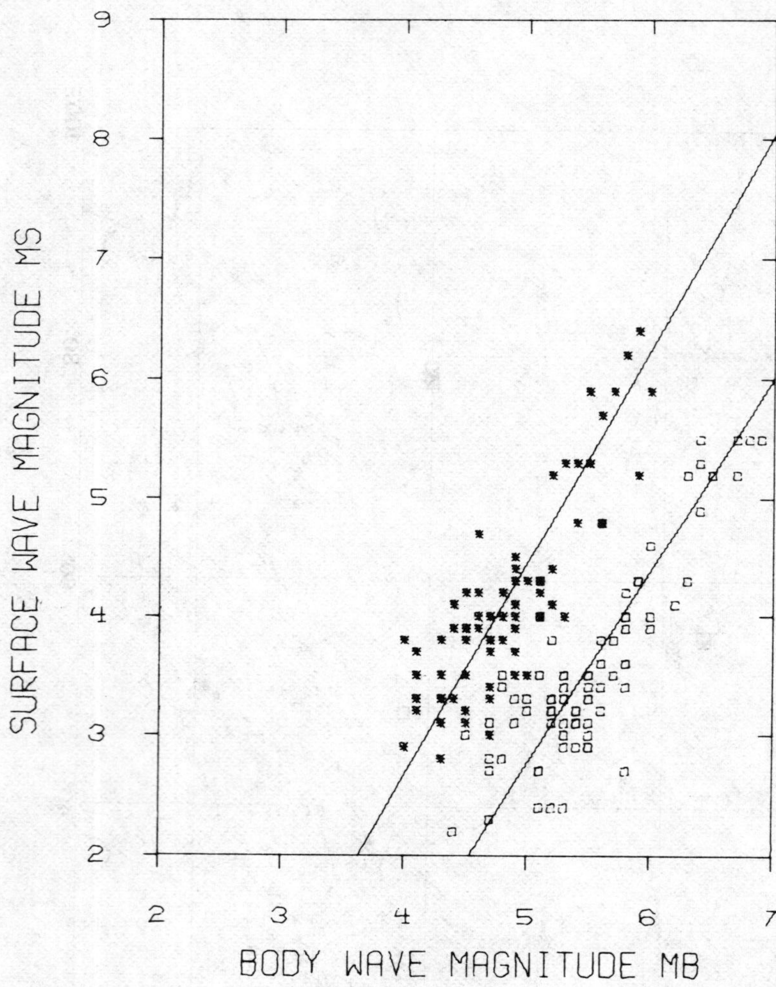

Figure 2. m_b:M_S diagram for the data set of 83 explosions and 72 earthquakes. PDE (Preliminary Determination of Epicenter) m_b and NORSAR M_S values have been used.

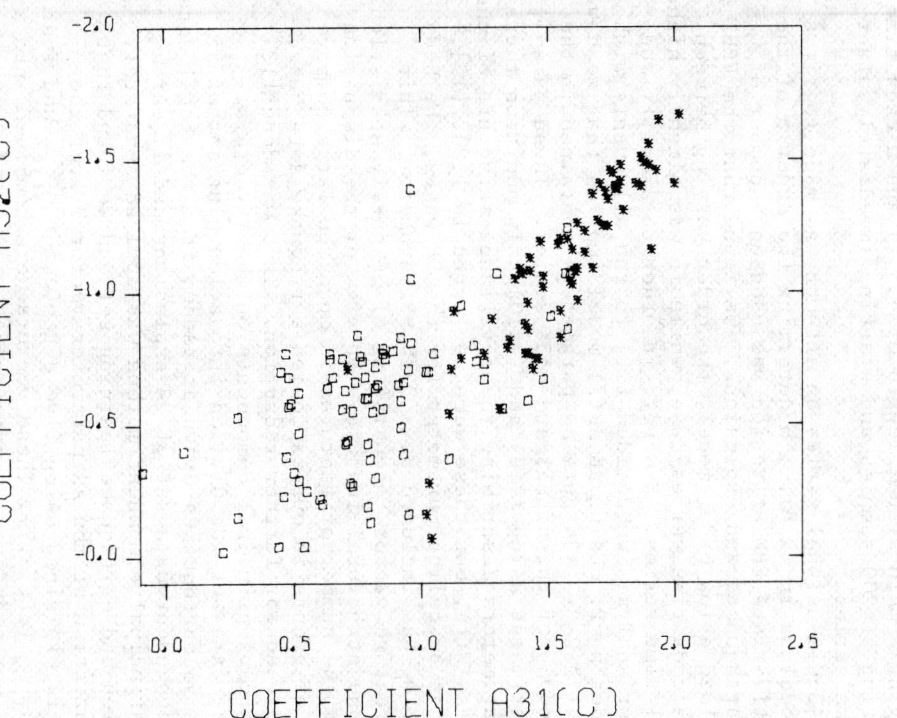

Figure 3. First and second order autoregressive $\hat{a}_{31}(C)$ and $\hat{a}_{32}(C)$ for the coda time series for the complete data base of 83 explosions and 72 earthquakes. A third-order univariate model has been assumed for each event, and the parameters have been calculated from a subarray beam.

$X_i(t)$, $i=1,2,3$. A sampling rate of 10 Hz was used and the vector signal $\underline{X}(t)$ of (2.8) will here represent the nonfiltered aligned seismic time series data as measured for these subarrays. Furthermore, each signal-wave train was divided in two parts, a main signal series extending from the onset of the signal to sample 65 (giving a main signal length of 6.5 s) and a coda time series consisting of 200 samples and starting immediately after the signal time series.

To each 3-dimensional series an attempt was made to fit a multiple AR model as in (2.8). To determine the order p of the model the multivariate FPE criterion was used on a number of events both earthquakes and explosions. We found that the FPE criterion yielded a low order model, the order varying between 2 and 5 in the coda time series case, while it was slightly higher for the signal time series. However, in order to simplify and standardize our analysis we decided [12] to run all events with a fixed AR order $p = 2$. Using an AR(2) model implies that we have at our disposal $2 \cdot 2 \cdot 3 \cdot 3 = 36$ pure AR parameters contained in the matrices A_1 and A_2 of the signal and coda series. These parameters were estimated using (2.9) and (2.10). In addition there are $2 \cdot 6 = 12$ parameters describing the covariance structure of the residual process $\underline{W}(t)$, but these were neglected in [12], [13], and, instead, as a scaling parameter for the SP waves, the body wave magnitude m_b was included as a 37th SP parameter. This results in a feature vector $\underline{Y}$ consisting of 36 purely statistical features and one structural feature. In view of the discussion of Section 3.4 the number of features is much too large compared to the number of events in our explosion/earthquake data base, and there is a need for further reduction. This was done [12] using the method of principal components on the $\underline{Y}$ vectors. To this end the data set was divided into a learning set of 55 explosions and 55 earthquakes drawn at random from the total data set and an independent data set of 28 explosions and 17 earthquakes. The principal component method was applied separately to the explosion and earthquake learning set and we found that most of the variation of the AR parameter vector $\underline{Y}$ over the learning sets is explained using five eigenvector $\hat{\underline{h}}_{i,EX}$ and four eigenvectors $\hat{\underline{h}}_{i,EQ}$ (cf. 2.7). These vectors were combined into a 9-dimensional (not necessarily orthonormal) basis, and for each event with a primary feature vector $\underline{Y}$ of characteristic parameters, a secondary vector $\underline{Z}$ is constructed as

$$\underline{Z} = [Z_{EX}(1), \ldots, Z_{EX}(5); Z_{EQ}(1), \ldots, Z_{EQ}(4)] \qquad (4.1)$$

where $Z_{EX}(i) = \underline{Y}^T \cdot \hat{\underline{h}}_{i,EX}$ and similarly for $Z_{EQ}(i)$. We thus obtain a ratio 55/9 between the number of events of each learning set and the dimension of the feature vector $\underline{Z}$, which should be quite acceptable according to Section 3.4.

4.3. SP discrimination. Classification.

The secondary feature vectors $\underline{Z}$ were assumed to be distributed according to a multivariate normal distribution. This was done mostly out of convenience so that the classification algorithm (3.2) could be used. The multivariate normal assumption was checked by plotting marginal sampling distributions for the vectors $\underline{Z}$, and the fit to a normal distribution was found to be moderately good only.

The results of the classification are depicted in Fig. 4 and summarized in Table 1, where a marginal event is defined as a correctly classified event having absolute value less than 2.5 for $\hat{S}_1(\underline{Z})-\hat{S}_2(\underline{Z})$ (cf. formula (3.2)). In Fig. 4 we have also shown the results of using the algorithm (3.2) on $\underline{Z} = (m_b, M_S)$ and $\underline{Z} = (\hat{a}_{31}, \hat{a}_{32})$. It is seen that as far as the learning set is concerned the pure SP discriminant based on m_b and multivariate autoregressive features is far superior in quality to the $a_{31}:a_{32}$ discriminant.

We also examined the effect of introducing the surface-wave parameter M_S as a 38th structural feature of the vector $\underline{Y}$ and subsequently reducing to a 9-dimensional $\underline{Z}$-vector as before, but now resulting in a mixed SP-LP feature vector. From Fig. 4 and Table 1 it is seen that this leads to significant improvement in discrimination potential for the learning set. Finally, we did a classification based on 9-dimensional $\underline{Z}$-vectors obtained by using the modified principal component method of [18] on the 37- and 38-dimensional primary vectors $\underline{Y}$. As is seen from Fig. 4 d) for our data set the modified method performed slightly worse than the ordinary one on the 38-dimensional vector $\underline{Y}$ with M_S included. For the 37-dimensional pure SP vector $\underline{Y}$ the results pointed in the same direction with an even larger discrepancy between the two methods. These results support the statements made at the end of Section 3.2 concerning the two methods.

To get a check on the error rate estimates produced by the learning data set the events of the independent data set were next classified by decomposing the vectors $\underline{Y}$ along the basis vectors $\hat{h}_{i,EX}$ and $\hat{h}_{i,EQ}$ of the learning data set, thus obtaining feature vectors $\underline{Z}$ and subsequently inserting the $\underline{Z}$ vectors in the classification algorithm (3.2) using estimated covariance matrices $\hat{K}_i$ and mean vectors $\hat{\mu}_i$ from the learning set. The results are given in Fig. 5 and in the lower part of Table 1 and tend to support our statements concerning the relative quality of the discriminants discussed, although it is somewhat disturbing that the error rate of the pure SP discriminant based on m_b and multivariate AR features increases from 4/110 to 3/45. Clearly a more thorough investigation along the lines suggested in Section 3.4 would have been desirable here.

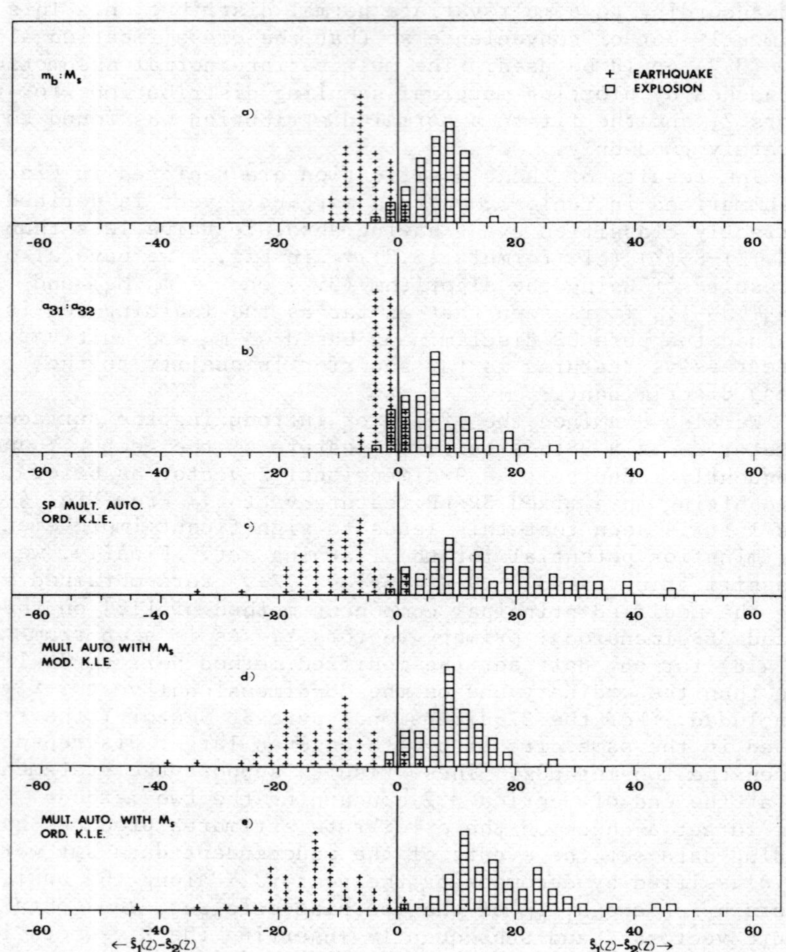

Figure 4. Histogram of discrimination scores $\hat{S}_1(Z)-\hat{S}_2(Z)$ as determined by (3.2) and using the learning set of 55 explosions and 55 earthquakes. The figure has been divided in 5 parts as follows: (a) $Z = (m_b, M_s)$; (b) $Z = (a_{31}, a_{32})$; (c) 9-dimensional Z-vector obtained by using the modified Karhunen-Loeve expansion on the SP 37-dimensional primary feature vector Y; (d) 9-dimensional Z-vector obtained by using the modified Karhunen-Loeve expansion on the 38-dimensional primary feature vector Y with M_s included; (e) same as (d) but using the ordinary Karhunen-Loeve expansion.

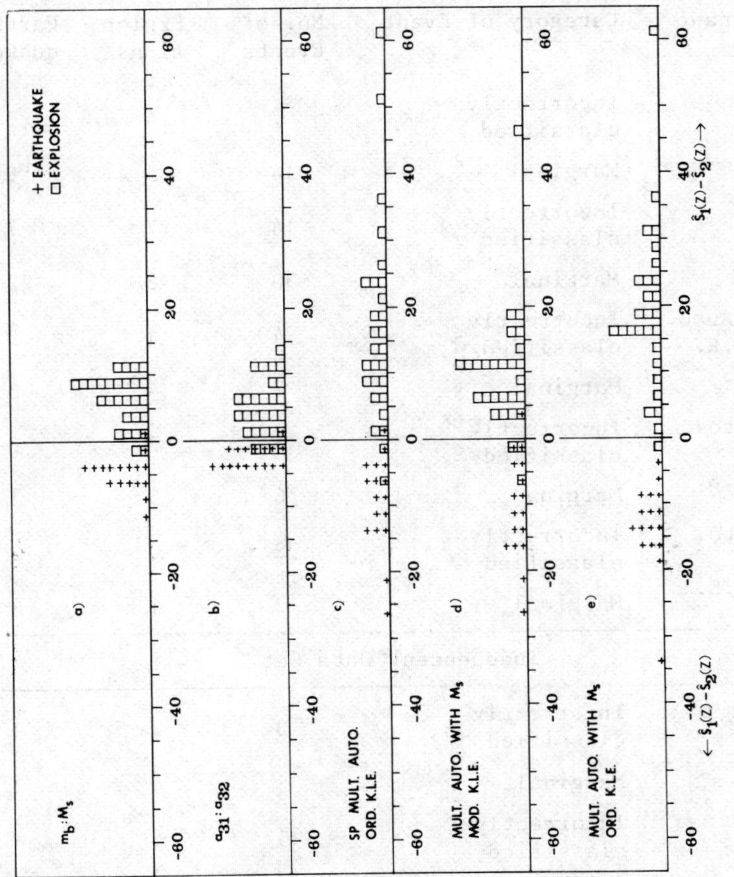

Figure 5. Same as Figure 4 for the independent data set of 28 explosions and 17 earthquakes.

Table 1

Discriminant	Category of Event	No. of Events	Explosions	Earthquakes
Learning Data Set				
$m_b:M_s$	Incorrectly classified	7	4	3
	Marginal	14	5	9
$a_{31}:a_{32}$	Incorrectly classified	14	7	7
	Marginal	30	8	22
SP Mult.Auto. Ord. K.L.E.	Incorrectly classified	4	1	3
	Marginal	7	4	3
Mult. Auto. with M_s Mod. K.L.E.	Incorrectly classified	5	2	3
	Marginal	6	5	1
Mult. Auto. with M_s Ord. K.L.E.	Incorrectly classified	2	0	2
	Marginal	3	1	2
Independent Data Set				
$m_b:M_s$	Incorrectly classified	3	2	1
	Marginal	5	4	1
$a_{31}:a_{32}$	Incorrectly classified	5	4	1
	Marginal	12	5	7
SP Mult.Auto. Ord. K.L.E.	Incorrectly classified	3	2	1
	Marginal	3	2	1
Mult. Auto. with M_s Mod. K.L.E.	Incorrectly classified	4	3	1
	Marginal	2	0	2
Mult. Auto. with M_s Ord. K.L.E.	Incorrectly classified	1	1	0
	Marginal	1	1	0

4.4. Mixed SP-LP discrimination.

The marked improvement obtained by introducing the surface wave parameters M_S (or the equivalent noise measurement in cases where surface waves were not detected) suggests that even further improvements may be obtained by a more accurate description of surface waves. A preliminary investigation of this sort was done in [51], where M_S was replaced by a special feature $\hat{P}_{20}$ for the incoming surface waves. To be more exact $\hat{P}_{20}$ is the estimated power spectral density when using a 5th order AR model to describe the incoming LP time series. $\hat{P}_{20}$ is measured over a suitable group velocity window and at a frequency of 0.05 Hz. It should be noted that, unlike M_S, $\hat{P}_{20}$ does not depend on visual inspection of seismograms, and it can be computed irrespective of the signal-to-noise ratio for the events in question. Of course for weak signals the noise contributes significantly to the estimate $\hat{P}_{20}$, but the results in [11] indicate that it still carries valuable information for discrimination purposes. When using $\hat{P}_{20}$ as an estimate of surface wave energy it should of course be compensated for epicentral distance. This has been done [11] by using the distance corrections in existing M_S formulas and results in energy estimates E_{20} with values in the familiar surface wave magnitude range.

The quantities $\hat{P}_{20}$ and E_{20} were computed not only for vertical Rayleigh waves but also for horizontal Rayleigh waves and Love waves and altogether 9 different group velocity windows were used. The resulting 27 LP features could be characterized as a mixture of statistical and structural features. These LP features were supplemented in [11] by 10 SP features, namely 9 univariate autoregressive features obtained by fitting 3rd order AR models to the signal, coda and the noise time series preceeding the signal and one structural SP feature, namely m_b. Thus we have a primary feature vector $\underline{Y}$ of dimension 37. Feature vectors $\underline{Y}$ were computed in [11] for a data set of 67 explosions (among these 15 for which surface waves could not be detected) and 73 earthquakes. With the exception of 3 events the data set is a subset of the main data set used in Figs. 1 and 2. In [13] these data were analysed further by using the principal component method as a means of secondary feature extraction on a randomly selected data set of 45 explosions and 45 earthquakes and with an independent data set constituted by the remaining events. The feature vector $\underline{Z}$ was required to have dimension 8 to guarantee (cf. Section 3.4) a proper ratio between the number of events and the dimension of $\underline{Z}$.

The results are depicted in Fig. 6 and Table 2. It is seen that the Z-discriminant based on the combined SP and LP features works far better that the $m_b:M_S$ discriminant both for the learning and independent data set. It is difficult to get a good estimate of the error rate since none of the events is misclassified for the learning set and only one for the independent data set.

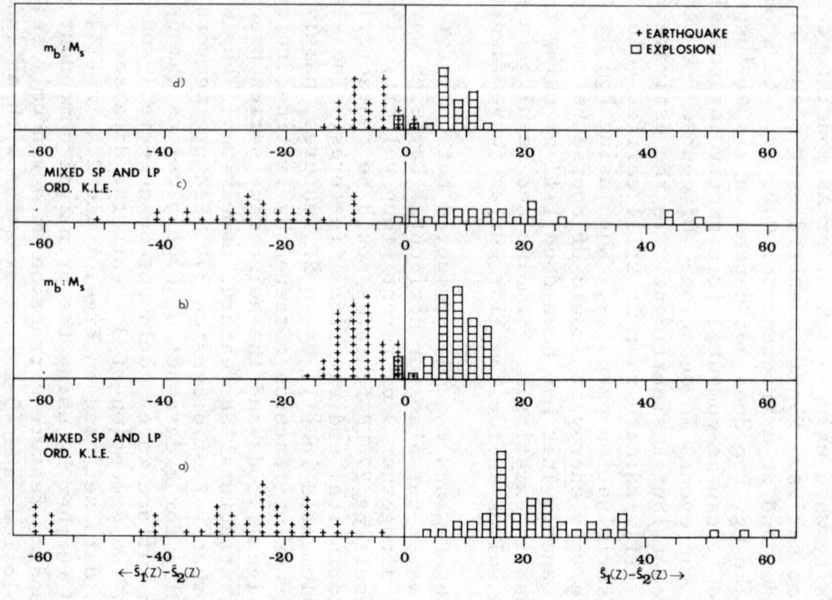

Figure 6.   Histogram of discrimination scores $\hat{S}_1(Z)-\hat{S}_2(A)$ as determined by (3.2) and using a data set of 67 explosions and 73 earthquakes. The figure is divided in 4 parts as follows: (a) 8-dimensional Z-vector obtained from a 37-dimensional primary feature vector Y consisting of 10 univariate SP parameters and 27 univariate LP parameters. A randomly selected learning data set of 45 explosions and 45 earthquakes has been used; (b) $Z = (m_b, M_s)$ with the same data set as in (a); (c) same as (a) for the independent data set of 22 explosions and 28 earthquakes; (d) same as (b) for independent data set.

Table 2

Learning Data Set				
Discriminant	Category of Event	No. of Events	Explosions	Earthquakes
$m_b : M_s$	Incorrectly classified	4	3	1
	Marginal	6	1	5
Mixed SP & LP Ord. K.L.E.	Incorrectly classified	0	0	0
	Marginal	0	0	0
Independent Data Set				
$m_b : M_s$	Incorrectly classified	4	2	2
	Marginal	4	1	3
Mixed SP & LP Ord. K.L.E.	Incorrectly classified	1	1	0
	Marginal	2	2	0

Again it probably would have been advantageous if a rotating
independent data sample scheme had been used and possibly with
a probability density error rate estimate.

The mixed SP-LP discriminant also seems to be superior to
any of the multivariate AR related discriminants on Figs. 4
and 5 although an exact comparison cannot be made due to a few
nonoverlapping events for the two data sets. The results in
Fig. 6 all refer to the ordinary principal component method.
Again the modified expansion gave clearly inferior results.
In fact the discrepancy was considerably larger than the one
illustrated on Figs. 4 and 5.

5.   SUMMARY REMARKS.

The pattern recognition approach sketched in this paper is
very flexible. It can easily be extended to include multistation
data and new features. All that is needed is to extend the primary
vector $\underline{Y}$ to allow for the desired number of features from each
station. A reduction to an appropriately dimensioned classifier
can then be achieved by secondary feature extraction using, for
example, the principal component method.

A multistation scheme should be particularly well suited
to the data from the network of SRO (Seismic Research Observatories) stations which employ modern digital data acquisition
and a recording system especially designed for seismic discrimination.

To make effective use of such data networks it is imperative
that good methods of feature extraction are found. Some of the
existing methods were described in Section 2. Several of these
have not been tried in seismic discrimination, among them the
nonlinear principal component method [17] and the optimal discriminant vector approach of [20]. A natural supplement to the
E_{20} features discussed in Section 4.3 would be (possibly multivariate) AR features for LP waves. Other possible time series
features are the parameters of the covariance matrix of the residual process $\underline{W}(t)$ of a multiple AR model. It may be natural
to adjust these for epicenter location as was done for the energy
estimates E_{20}. These features could be considered as a mixture
of structural and statistical features. It is clearly a major
task to find more and better structural features, i.e. to extract
meaningful geophysical parameters which could be combined with
statistical features such as AR coefficients.

In this connection an interesting possibility is to try to
use some form of regionalization to characterize the local conditions in a seismic region. A relevant statistical technique
in this context is the technique of cluster analysis [52] which
possibly could be used to organize seismic events in clusters

and subclusters according to the properties of the seismic region in which they occur. Here it should be noted that in this paper - as is the case for virtually all previous attempts of seismic discrimination - the explosions are generally in stable areas while the earthquakes are in tectonic areas. This fact alone may account (cf. [6]) for some of the sprectral differences. The importance of testing the methods proposed on explosions and earthquakes from a region which is geophysically uniform should therefore be realized.

When it comes to classification most of the work done in seismic discrimination has been limited to rather elementary Gaussian discriminants such as (3.2) and (3.3) or simply to graphical displays such as the m_b:M_S diagram. It should not be ruled out that more advanced techniques, some of which are mentioned in Sections 3.2 and 3.3 (e.g. logistic regression or nonparametric classification) could lead to improvements. An indication that this may indeed be so is our result that the fit to a normal distribution for the $\underline{Z}$-vectors chosen in [12] is moderately good only.

Finally the question of error estimates is very important if these methods are to be used in a test ban treaty context. Up to now this problem has received little attention in the seismic literature. Some exceptions are the error studies made in connection with the m_b:M_S discriminant [39]. Section 3.4 indicates that the general question of error estimation is a difficult one, but the methods and references mentioned there could form a convenient point of departure for further analysis. It is clear that such studies will have to require larger data sets than those considered in the present paper.

REFERENCES

1. Kanal, L. "Patterns in pattern recognition: 1968-1974",
    IEEE Trans. on Information Theory, IT-20, pp. 698-722, 1974.

2. Fu, K.S. "Syntactic Methods in Pattern Recognition,"
    Academic Press 1974.

3. Kelly, E.J. "A study of two short period discriminants",
    M.I.T., Lincoln Lab. Tech. Note 1968-8-1968.

4. Weichert, D.H. "Short period sprectral discriminant for
    earthquake and explosion differentiation", Z. Geophys.,
    37, pp. 147-152, 1971.

5. Dahlman, O. and H. Israelson, "Monitoring Underground
    Nuclear Explosions", Elsevier, Amsterdam, 1977.

6. Douglas, A. "Seismic source identification: "A review of
    past and present research efforts", Proceedings of NATO
    advanced study institute on Identification of Seismic
    Sources-Earthquake and Underground Explosions, E.S. Husebye,
    editor pp. 1-48, 1981.

7. Dargahi-Noubary, G.R. and P.J. Laycock, "Spectral ratio
    discriminants and information theory," Tech. Report No. 116,
    Dept. of Mathematics, Univ. of Manchester, 1979.

8. Chen, C.H. "Seismic pattern recognition", Geoexploration,
    16, pp. 133-146, 1978.

9. Brolley, J.E. "Preprocessing of seismic signals for pattern
    recognition", Proceeding of NATO advanced study instute on
    Pattern Recognition and Signal Processing, C.H. Chen,
    editor, pp. 367-386, Sijthoff & Noordhoff 1978.

10. Christofferson, A. and E.S. Husebye, "Least squares signal
    estimation techniques in analysis of seismic array recorded
    P-waves", Geophys. J.R. Astron, Soc., 38, pp. 525-552, 1974.

11. Tjøstheim, D. "Improved seismic discrimination using pattern
    recognition", Phys. Earth. Planet. Inter., 16, pp. 85-108,
    1978.

12. Sandvin, O. and D. Tjøstheim, "Multivariate autoregressive
    representation of seismic P-wave signals with application to
    short-period discrimination," Bull. Seism. Soc. Am., 68,
    pp. 735-756, 1978.

13. Tjøstheim, D. and O. Sandvin, "Multivariate autoregressive
     feature extraction and the recognition of multichannel
     waveforms", IEEE Trans. on Pattern Analysis and Machine
     Intelligence, PAMI-1, pp. 80-86, 1979.

14. Anderson, T.W., "Introduction to Multivariate Statistical
     Analysis", Wiley, New York, 1958.

15. Rao, C.R. "Linear Statistical Inference and Its Applica-
     tions", Wiley, New York, 2nd. ed., 1973.

16. Young, T.Y. and T.W. Calvert, "Classification, Estimation
     and Pattern Recognition", American Elsevier, New York, 1974.

17. Gnanadesikan, R. "Methods for Statistical Data Analysis of
     Multivariate Observations," Wiley, New York, 1977.

18. Fukunaga, K. and W.L.G. Koontz, "Application of the
     Karhunen-Loeve expansion to feature selection and ordering",
     IEEE Trans. on Computers, C-19, pp. 311-318, 1970.

19. Foley, D.H. "Orthonormal expansion study for waveform pro-
     cessing system," Rome Air Develop. Center, AF Systems
     Command, Griffiss AFB, New York, Tech. Rep. RADC-TR-73-168,
     1973.

20.   Foley, D.H. and J.W. Sammon Jr., "An optimal set of discri-
     minant vectors", IEEE Trans. on Computer, C-24, pp. 281-289,
     1975.

21. Atal, B.S. and S.L. Hanauer, "Speech analysis and synthesis
     by linear prediction of the speech wave," J. Acoust. Soc.
     Amer., 50, pp. 637-655, 1971.

22. Itakura, F. and S. Saito, "A statistical method for estima-
     tion of speech spectral density and formant frequencies,"
     Electron Commun. Japan, 53-A, pp. 36-43, 1970.

23. Rosenberg, A.E. and M.R. Sambur, "New techniques for auto-
     matic speaker verification," IEEE Trans. Acoust. Speech
     Signal Processing, ASSP-23, pp. 169-176, 1975.

24. Whittle, P. "Prediction and Regulation," Van Nostrand,
     Princeton, 1963.

25. Jones, R.H. "Multivariate autoregression estimation using
     residuals", Applied Time Series Analysis, Proceeding from
     Symposium in Tulsa, Oklahoma, D.F. Findley, Editor,
     pp. 139-162, Academic Press, 1978.

26.	Sandvin, O. and D. Tjøstheim,	"A numerical comparison of two criteria for determining the order of AR processes",	Journal Matematische Operationsforschung und Statistik, Series Statistics, to appear 1980.

27.	Kashyap, R.	"Optimal feature selection and decision rules in classification problems with time series",	IEEE Trans. Inform. Theory, IT-24, pp. 281-288, 1978.

28.	Dargahi-Noubary, G.R., Laycock, P.J. and T. Subba Rao, "Non-linear stochastic models for seismic events with applications in event identification",	Geophys. J.R. Astron. Soc., 55, pp. 655-668, 1978.

29.	Haskell, N.A.	"Total energy and energy spectral density of elastic wave radiation from propagating faults, 2, A Statistical source model,"	Bull. Seism. Soc. Amer., 56, pp. 125-140, 1966.

30.	Haskell, N.A.	"Analytic approximation from elastic radiation from a contained underground explosion".	J. Geophys. Res., 72, pp. 2583-2587, 1967.

31.	Von Seggern, D. and R. Blandford,	"Source time functions and spectra of underground nuclear explosions",	Geophys. J.R. Astron. Soc. 31, pp. 83-97, 1972.

32.	Aki, K.	"Scaling law of seismic spectrum."	J. Geophys. Res., 72, pp. 1217-1232, 1967.

33.	Chen, C.H.	"A review of statistical pattern recognition", Proceedings of NATO advanced study institute on Pattern Recognition and Signal Processing, C.H. Chen, editor, pp. 117-132, 1978.

34.	Cover, T.	"The best two independent measurements are not the two best.", IEEE Trans. Syst., Man., Cybern. (Corresp.), SMC-4, pp. 116-117, 1974.

35.	Mucciardi, A.N. and E.E. Gose,	"A comparison of seven techniques for choosing subsets of pattern recognition properties",	IEEE Trans. Comput., C-20, pp. 1023-1031, 1971.

36.	Fu, K.S.	"Sequential Methods in Pattern Recognition and Machine Learning, Academic Press, New York, 1968.

37.	Murray, G.	"A cautionary note on selection of variables in discriminant analysis", Appl. Statist., 26, pp. 246-250, 1977.

38. Elvers, E. "Seismic event identification by negative
     evidence", Bull. Seism. Soc. Am., 64, pp. 1671-1683, 1974.

39. Weichert, D.H. and P.W. Basham, "Deterrence and false alarms
     in seismic discrimination", Bull Seism. Soc. Am.,
     pp. 1119-1132, 1973.

40. Azen, S.P., Breiman, L. and W.S. Meisel, "Modern approaches
     to Data Analysis". Course Notes. Technology Service
     Corporation, Santa Monica, Calif., 1975.

41. Efron, B. "The efficiency of logistic regression compared to
     normal discriminant analysis", Journal of the American
     Statistical Association, 70, pp. 892-898, 1975.

42. Press, S.J. and S. Wilson, "Choosing between logistic
     regression and discriminant analysis", Journal of the
     American Statistical Association, 73, pp. 699-705, 1978.

43. Cover, T.M. "A hierarchy of probability density function
     estimates", in Frontiers of Pattern Recognition, S. Watanabe,
     editor, Academic Press, New York, pp. 83-98, 1972.

44. Breiman, L., Meisel, W. and E. Purcell, "Variable kernel
     estimates of multivariate densities and their calibration",
     Technology Service Corporation, Santa Monica, Calif., 1975.

45. Cover, T.M. and P.F. Hart, "Nearest neighbor pattern classi-
     fication", IEEE Trans. Inform, Theory, IT-13, pp. 21-27, 1967.

46. Hills, M. "Allocation rules and their error rates," J. Roy.
     Stat. Soc., Ser. B., Vol. 28, pp. 1-31, 1968.

47. Lachenbruch, P.A. and R.M. Mickey, "Estimation of error
     rates in discriminant analysis", Technometrics, 10,
     pp. 1-11, 1968.

48. Toussaint, G.T., "Bibliography on estimation and misclassifi-
     cation", IEEE Trans. Inform. Theory, IT-20, pp. 472-479, 1974.

49. Kanal, L. and Chandrasekaran, "On dimensionality and sample
     size in statistical pattern classification", Pattern
     Recognition, 3, pp. 225-234, 1971.

50. Foley, D.H. "Considerations of sample and feature size",
     IEEE Trans. Inform. Theory, IT-18, pp. 618-626, 1972.

51. Tjøstheim, D. and E.S. Husebye, "An improved discriminant
     for test ban verification using short and long period spectral
     parameters," Geophys. Res. Lett., 3, pp. 499-502, 1976.

52. Anderberg, M.R. "Cluster Analysis for Applications",
     Academic Press, New York, 1973.

SEISMIC DISCRIMINATION PROBLEMS AT REGIONAL DISTANCES

Robert R. Blandford

Teledyne Geotech, Alexandria, Virginia

ABSTRACT

The amplitude ratio of the maximum motion before S_n (P_{max}) to the maximum after S_n (L_g) is a good discriminant between earthquakes and explosions at regional distances. Experimental data shows that the ratio is not seriously affected by site or source geology or by event depth. However propagation effects can be severe, especially from the USSR to the South, and can lead to the requirement for regionalization. The amplitude-distance relation for P_{max} and L_g is $A \sim r^{-3}$ in the Western United States (WUS), and $r^{-2.5}$ and $r^{-2.0}$ for P_{max} and L_g respectively in the EUS and within the USSR. The radial to transverse ratio for L_g does not appear to be a discriminant, being controlled by the local geology. Evidence is presented that there are no useful spectral discriminants in the 1 to 10 Hz band. It appears that spectra of L_g are contaminated by compressional wave coda which can lead to an overestimate of the high frequency energy in the direct L_g wave.

INTRODUCTION

In 1958 the Geneva Conference of Experts suggested that a worldwide network of 180 seismic stations might constitute a feasible seismic monitoring system to enforce a comprehensive nuclear test ban (CTB). Such a large number of stations naturally included several inside the United States and the USSR, and there was, therefore, considerable research carried out in succeeding years on distance-amplitude relations and discrimination capabilities at "regional" distances of less than say, 20°.

E. S. Husebye and S. Mykkeltveit (eds.), Identification of Seismic Sources – Earthquake or Underground Explosion, 695–740.

In time it became less clear that internal stations would be
allowed, national means of verification were emphasized, and
furthermore, seismologists realized that teleseismic signals were
not so dominated by effects of complicated crustal structures
as were the regional signals. Since the teleseismic signals were
easier to understand, they could perhaps be relied upon more for
discrimination. Thus, around 1963, emphasis shifted from studies
at regional distances to those at teleseismic distances.

In 1977 it began again to seem possible that the USSR would
allow stations within its borders, and so interest has been
revived in regional discrimination. In retrospect, it seems clear
that research was on the verge of major discoveries on the subject of regional discrimination in 1963, when emphasis was
shifted to the teleseismic distances. In the interim, due to work
related to earthquake risk, plate tectonics, and the ARPA discrimination program, substantial improvement in our understanding of
the propagation of crustal phases has nonetheless occurred.

After a brief sketch of the literature we shall begin this
paper with a qualitative review of the characteristics of
regional phases, using as a prime example the well-recorded
signals of the event SALMON and a companion Alabama earthquake
in the Eastern United States (EUS). Among the topics to be
examined for the different phases are the frequency, clarity of
onset, travel times, and the relative amplitudes of vertical,
radial and transverse components. Differences between earthquakes and explosions will be discussed, and we will examine
similar data as recorded at NORSAR from Eurasian events. Some
qualitative remarks will be made about the detectability of
frequencies above and below the standard 1 to 4 Hz short-period
band, and array data will be used to delineate the differences
between phase and group velocity of regional phases. The
qualitative review will conclude with a survey of the types of
geology which block regional phases such as the blockage of
S_n by mid-ocean ridges, and the blockage of L_g by ocean basins
and by the mountain range on the southern border of the USSR.

The next section of the study is on array design for regional
phases and on the use of arrays for location, together with a comparison of the azimuthal precision obtainable by arrays as compared to 3-component processing.

The next section is on amplitude-distance relationships;
some of the substantial literature on the subject dating back to
the early 1960's is briefly reviewed and then some more recent
results are presented. It is essential to have reliable distance-
amplitude results if network-averaged phase ratios are to be
combined for discrimination in a simple and systematic way. If

such results are unobtainable, then a more complicated, detailed regionalization is the only alternative.

The causes of variability in phase ratios is the subject of the next section. The influences of station geology and of event depth are discussed. There is a lack of experimental data bearing on the questions of variations due to fault-plane orientation.

Finally, a compressional-to-shear ratio discriminant is outlined and applied to events in the United States and the USSR. Some authors have found that a compressional-to-shear ratio has not worked as a discriminant. Some possible causes for this are outlined and various remedies are discussed.

CHARACTER OF REGIONAL PHASES

Early work on the qualitative properties of L_g was published by Press and Ewing(1), Båth(2), and Lehman(3). Press and Gutenberg(4) first discussed P_g. Among the early important papers discussing the phase S_n is that of Molnar and Oliver(5). Later work on these phases touched on such questions as the theoretical explanation for the phases(6–20), travel-times(21–28), amplitude-distance relationships(27–35), and observations of geological characteristics which block the propagation of phases(5,12,20,23, 25,32,36,37).

A comparison of the events SALMON of October 22, 1964, and the Alabama earthquake of February 18, 1964 offers a good way to introduce the qualitative characteristics of the phases P_n, P_g, P_{max}, S_n, L_g, and R_g(38).

Figure 1(39) shows the relationship of recording stations to SALMON and the earthquake. Figures 2 and 3(38) are of SALMON and the Alabama earthquake at two stations of similar geology(40) selected to be on the same path and at nearly equal distances from their respective events. The character of the different phases can be seen in Figures 2 and 3 because the appropriate time windows have been extracted, starts aligned on the phase velocity, and scales enlarged in both the time and amplitude dimensions. The requirement for such a complex figure illustrates the difficulties in proper graphic display of regional phases. Most of the conclusions which we will deduce later on in this study by means of spectral analysis can be deduced by a careful exam- ination of Figures 2 and 3. The most important single fact that is clear from comparison of Figures 2 and 3 is that on the verti- cal component $L_g(_{max})/P_{max}$ is greater for the earthquake than for the explosion by about a factor of 5. We note also that the trans- verse to radial ratio is about 2 for both earthquake and explosion, suggesting that this ratio cannot serve as a discriminant. Finally, the P_n for the earthquake is far more emergent.

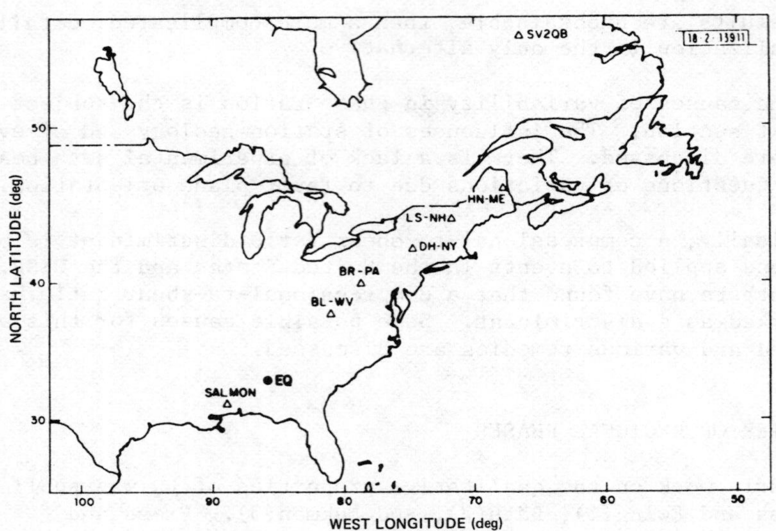

Figure 1.   East coast LRSM stations for SALMON.   Note how signals
from SALMON must pass near the epicenter for the 18 February 1964
earthquake on the way to stations in the profile.

A recent book(41) has reviewed much of the discrimination
work based on short-period discriminants.   Most of this work which
has been <u>apparently</u> successful suffers from the severe defect that
the explosion and earthquake sources are in different geological
regions.   The events in Figure 1 do not suffer from this defect,
and Figure 4(38) shows that they have nearly identical P spectra.
(Unless otherwise noted the spectra in this paper are of a window
approximately 20 seconds long.)   It has been pointed out(42) that
much of the short-period discrimination could be due to the cancellation of low frequencies by the surface reflection pP and that
any discriminant based on such cancellation could be spoofed by
varying the depth of burial.   However, such major effects of the
SALMON pP occur below 0.5 Hz and are obscured by the noise level
apparent in Figure 4.

Further analysis of the P signal is provided in Figure 5
where we see that the transverse component within 20 seconds of P
is of slightly higher frequency suggesting that it is scattered(38)
It cannot, of course, have been scattered anywhere except near the
receiver since if it were scattered near the source, it would have
to arrive predominantly on the radial component and could not have
a transverse component of nearly equal amplitude.   If it were
scattered at large distances from the direct path it would arrive
later than 20 seconds after P.

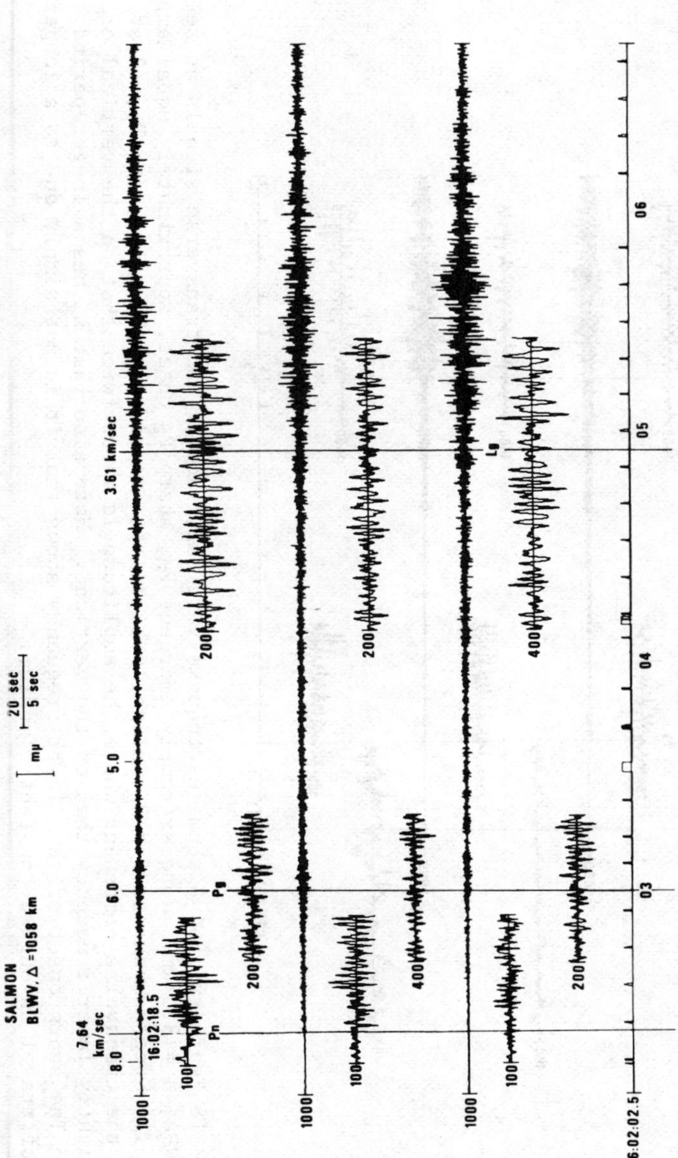

Figure 2. SALMON vertical, radial and transverse signals as seen at station BLWV, Δ = 1058 km.
Note clear Pn, small Pn transverse, no clear onset for Pg, apparently similar frequency content
for Pn and Pg, approximately equal amplitude Pg on vertical, radial and transverse, lack of clear
onset for Lg, but with some indication of lengthening period, transverse Lg approximately twice the
amplitude of the vertical and three times the radial. Note also that Lg has a longer period than
Pn and Pg.

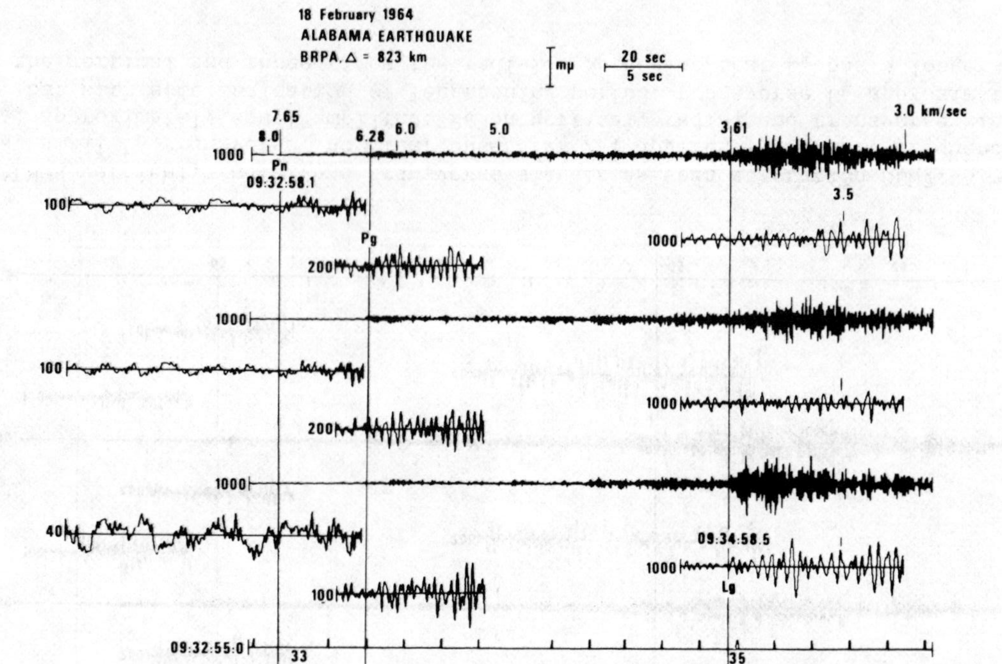

Figure 3. 18 February 1964 Alabama earthquake vertical, radial, and transverse signals as seen at station BRPA, Δ = 823 km. Note extremely emergent P_n, sharp P_g onset, approximately equal amplitudes on all three components for P_g; slightly higher frequencies on the transverse P_g, clear onsets for L_g on the transverse component where the amplitude is about twice that on the vertical or radial and of slightly lower frequency than on the vertical. Note also that L_g has a longer period than P_n or P_g. The sharp onset of this lower frequency shows that it is not simply due to a longer duration of travel through a constant Q medium.

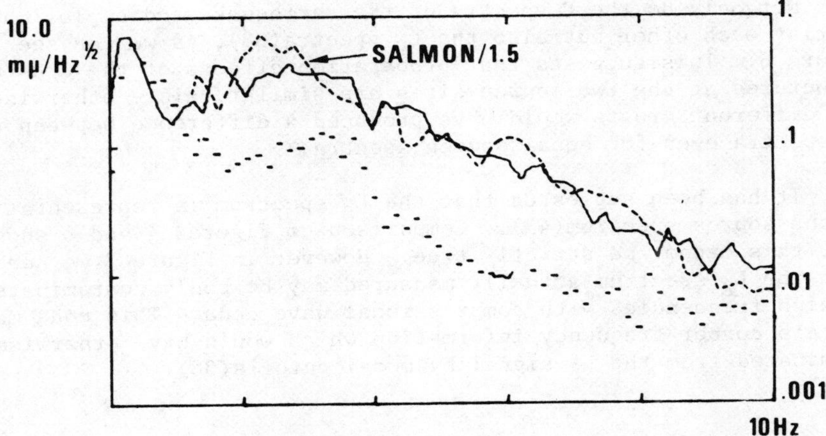

Figure 4. P_n vertical amplitude signal and noise spectrum from
the 18 February 1964 earthquake at BRPA with an overlay of the
SALMON P_n signal spectrum (divided by 1.5) from BLWV for frequencies at which S/N > 1. We see that between 1 and 10 Hz the spectra are almost identical in shape thus making spectral discrimination in this band probably impossible. Except as specifically
noted all spectra in this study are corrected for instrument
response only at 1 Hz.

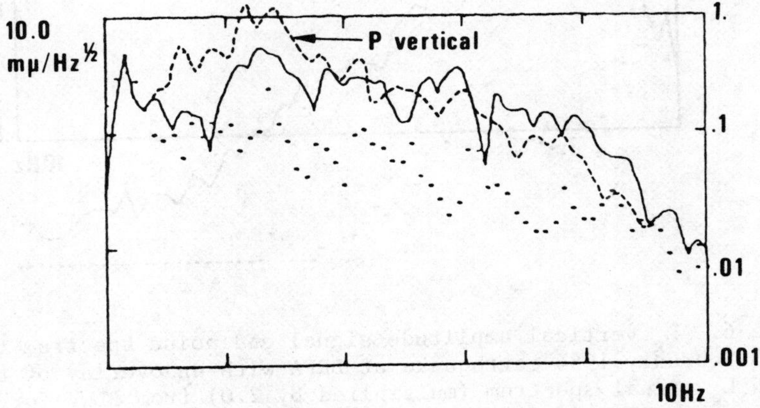

Figure 5. P_n transverse amplitude signal and noise spectrum from
SALMON at BLWV with an overlay of the SALMON vertical signal spectrum for frequencies at which S/N > 1. We see that the transverse
component is proportionally rich in <u>high</u> frequencies; a result
consistent with the hypothesis that higher frequencies are preferentially scattered and that the scattering occurs near the
station. (Note that the scattering could not have occurred near
the source and the energy propagated as a transverse component to
arrive with the first arrival.)

Not only do the P spectra of the earthquake and explosion overlay each other but also the L_g spectra(38), as we can see in Figure 6. This suggests that propagation effects of the crustal structures at the two source sites are similar, since otherwise the different crusts would have produced a difference between the L_g spectra even for equal source spectra.

It has been suggested that the L_g spectrum is representative of the source spectrum(43). Comparison of Figures 4 and 6 show that this cannot be strictly true. However in Figure 7 we see that the L_g spectrum actually measured may be badly contaminated at high frequencies with compressional wave coda. This coda may contain corner frequency information which would have otherwise been eliminated from the L_g signal by modal cutoffs(38).

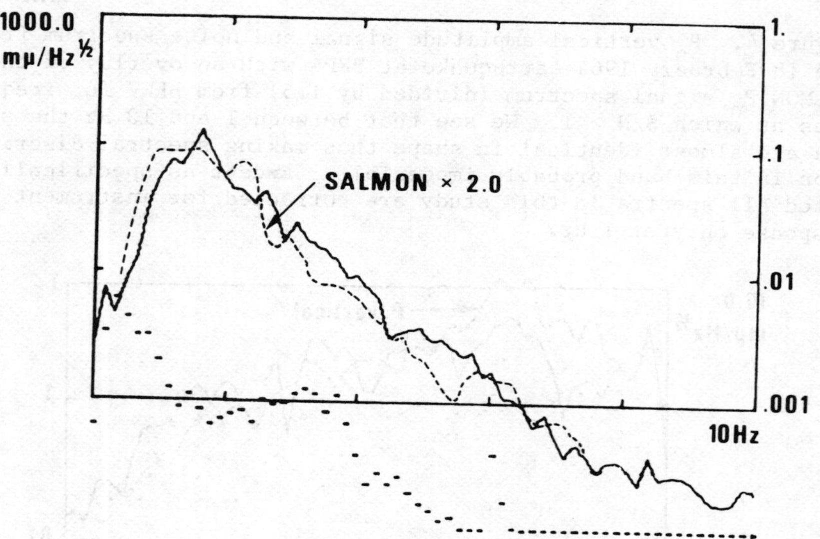

Figure 6. L_g vertical amplitude signal and noise spectrum from the 18 February 1964 earthquake at BRPA with an overlay of the SALMON L_g signal spectrum (multiplied by 2.0) from BLWV for frequencies at which S/N > 1. We see that between 0.5 and 7 Hz the spectra have identical shapes: a much lower frequency shape than the P_n spectrum of Figure 4. Der and McElfresh(69) have shown that the P_n spectrum is the same as that of the SALMON reduced displacement potential. Thus the L_g spectrum is different from the source spectrum for both earthquakes and explosions. The difference between multiplying by 2.0 in this figure, and dividing by 1.5 in Figure 4 is a reflection of the discrimination capability in the P_{max}/L_g ratio.

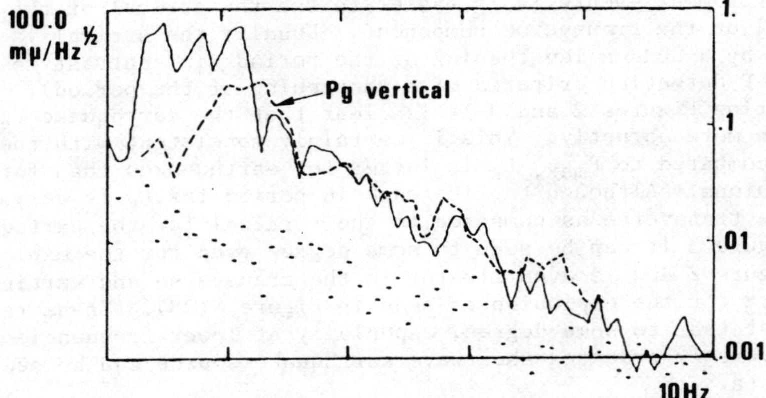

Figure 7. L$_g$ vertical amplitude signal and noise spectrum from
SALMON at BLWV with an overlay of the SALMON P$_g$ signal spectrum
for frequencies at which S/N > 1. We see that near 2 Hz the L$_g$
spectrum merges into the P$_g$ spectrum. This suggests that apparent
L$_g$ energy above 2 Hz is really P$_g$ coda. This throws doubt on
uncritical comparisons of observed and theoretical Lg spectra.

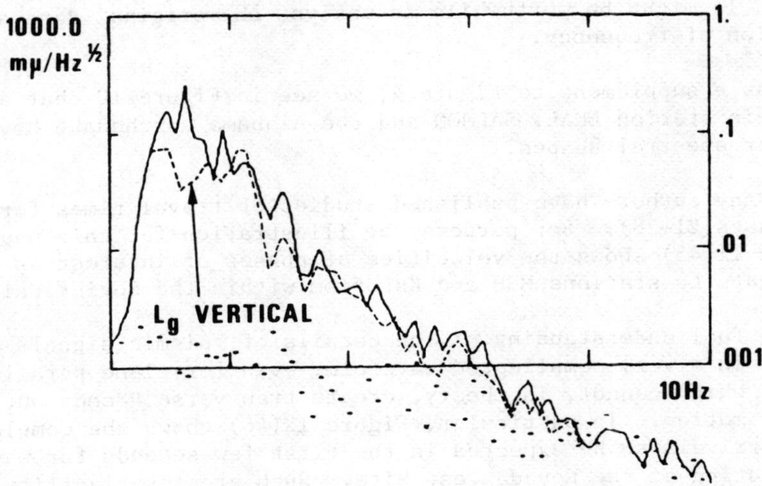

Figure 8. L$_g$ transverse amplitude signal and noise spectrum from
SALMON at BLWV with an overlay of the SALMON vertical signal
spectrum for frequencies at which S/N > 1. We see that the transverse component is proportionally rich in <u>low</u> frequencies. This
is not consistent with the hypothesis that the transverse component is composed of preferentially scattered high frequenices.
It is consistent with the hypothesis that the transverse L$_g$ is
propagating as an independent wave.

For most events it is easier to see the arrival of the L_g signal on the transverse component. Usually the arrival is indicated by a sudden lengthening of the period (in contrast to the usual P detection criteria of a shortening of the period). In comparing Figures 2 and 3 it is clear that the earthquake L_g starts more abruptly. This is certainly consistent with the idea that compared to P_{max}, L_g is larger for earthquakes than for explosions. Although the increase in period for L_g is very clear on the transverse as compared to the vertical for the earthquake in Figure 3 it can be seen to some degree even for the explosion in Figure 2 and is even clearer in the transverse and vertical <u>spectra</u> for the explosion as seen in Figure 8(38). These results suggest that to some degree, especially at lower frequencies, the transversely oriented shearwave earthquake source can be seen in the data.

It is certainly true, however, that the source is greatly obscured. A clear example of the obscuring mechanisms at work can be seen in Figure 9(44) where the ratio of the maximum transverse velocity to the maximum radial increases from 0.1 to nearly 1.0 as the distance changes from 3 to 30 km. The frequency of measurement is approximately 5 Hz, high, but still in the range of interest and certainly suggests that at perhaps 300 km one might have a thoroughly contaminated transverse component even at 1 Hz. It might be worthwhile to analyze the original data as a function of frequency.

As a supplement to Figure 4, we see in Figure 10 that at the close-in station EUAL, SALMON and the Alabama earthquake have similar spectral shapes.

Many authors have published studies of travel times for region al phases(21-28). For purposes of illustration for this paper, Figure 11(45) shows the velocities of phases of interest as they propagate to stations MSH and KBL from within the Soviet Union.

A full understanding of the details of seismic signals can result in a very complicated waveform, even for plane parallel layers which cannot, in theory, create transverse P coda out of radial motion. In particular, Figure 12(46) shows the complex set of P arrivals to be expected in the first few seconds for waves originating at the Nevada Test Site. Such great variability even in theoretical models certainly lends credence to Nersesov's(25) observations that the maximum motion before the arrival of S_n is to be preferred for magnitude estimates to a precisely defined amplitude of any particular phase.

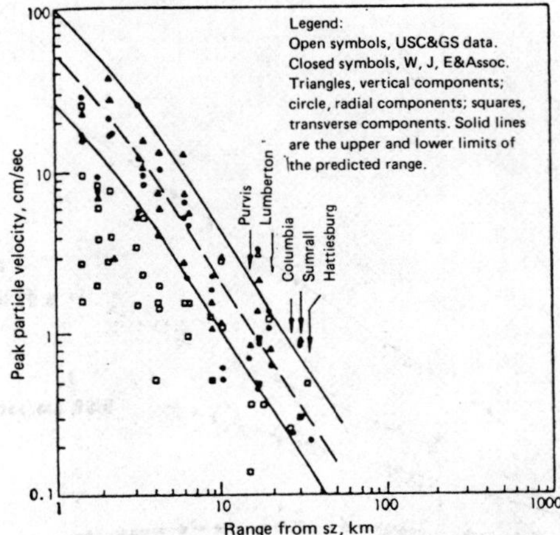

Figure 9. Predicted and observed peak particle velocities at the surface versus range from surface zero (SZ) for the SALMON event. Note that the transverse to radial ratio changes from approximately 1:10 at 2 km to approximately 1:1 at 30 km. Thus the transverse energy observed at regional distances for SALMON almost surely represents converted radial energy which has settled into an "equilibrium." One must conclude also that transverse L_g from earthquakes must be highly contaminated by radial source energy for $f \gtrsim 1$ Hz.

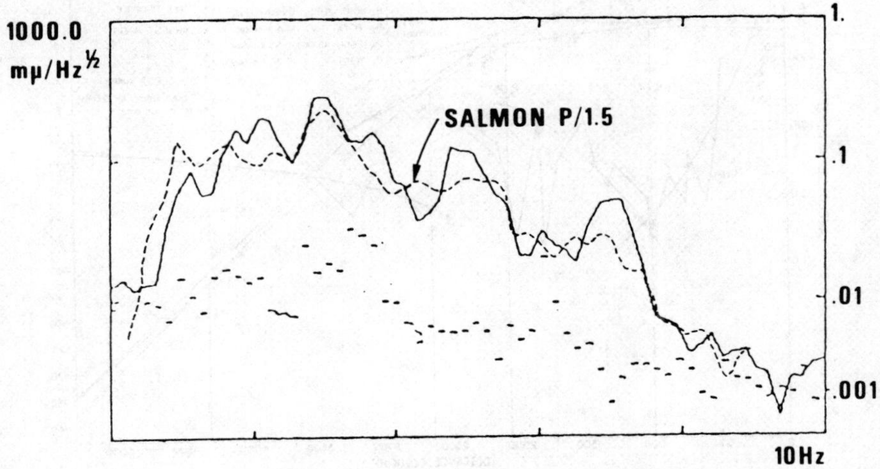

Figure 10. P_n vertical amplitude signal and noise spectrum from the 18 February 1964 earthquake at EUAL, a distance of 311 km, with an overlay of the SALMON P_n spectrum, also at EUAL at a distance of 242 km, for frequencies at which S/N > 1. We see that between 0.5 and 9 Hz the spectra are nearly identical, a result similar to that of Figures 4 and 6.

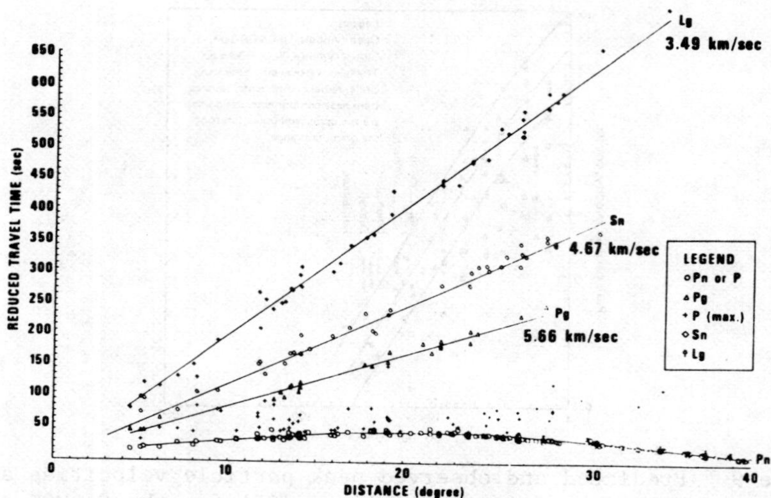

Figure 11. Travel-time curves of regional phases based on data
from 19 shallow earthquakes in Western Russia as seen at stations
outside. The travel times are reduced by $\Delta/10.0$ where Δ is
in km. Note that for these signals P_{max} generally arrives well
before P_g. P_n first motion velocities range from 8.26 to 12.61
km/sec.

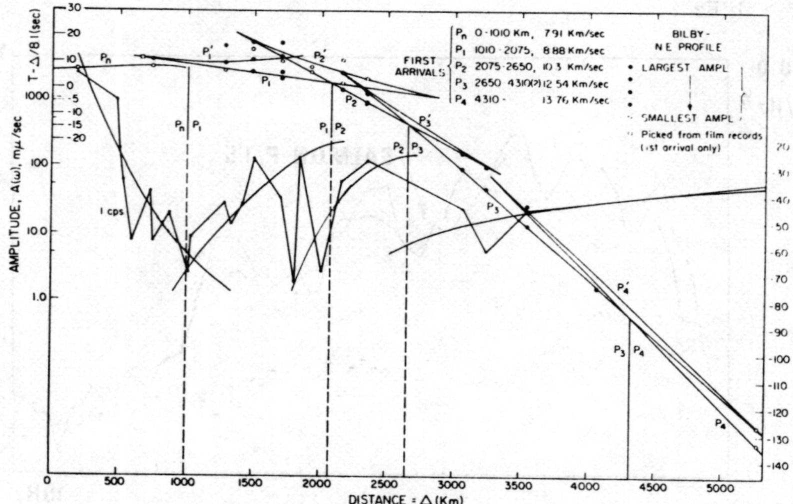

Figure 12. Example of the complexity of detailed P motion in the
first few seconds of the signal. There are multiple arrivals
deriving from velocity discontinuities in the upper mantle. The
shape and relative amplitudes of the signal change rapidly with
distance. Results such as this suggest that P_{max} might be a
more stable estimate of magnitude than P_n.

The features discussed above for SALMON and the Alabama earthquake as seen at BLWV and BRPA are in general valid for all other stations in the EUS that I have examined; this includes the stations seen in Figure 1 and other LRSM stations along a northern profile. Similar results can be seen on NORSAR recordings of events shown in Figure 13(47). The event of September 4, 1972 is the closest nuclear explosion to NORSAR. Figure 14(38) shows that the P and L_g spectra and relative shape are very similar to those seen in Figures 4 and 6. Figure 15(47) for an explosion at slightly greater distance is very similar except that the dominant frequencies are slightly smaller and the Lg amplitude is equal to that of the explosion instead of being a factor of 2 higher.

Figure 16, spectra of an Austrian earthquake(47), shows a dramatic difference with respect to Figure 15; the Lg is much larger than the P showing the same discrimination between earthquakes and explosions as we saw for SALMON and the Alabama earthquake. However, we can see from Figure 13 that the events are not near one another as SALMON and the Alabama earthquake were seen to be in Figure 1. The difference in propagation paths is then a candidate for the cause of the difference in peak spectral frequency as seen by comparison of Figures 15 and 16. The earthquake has lower frequencies for both P and L_g. Now, if the spectral shapes are different (and perhaps caused by propagation),

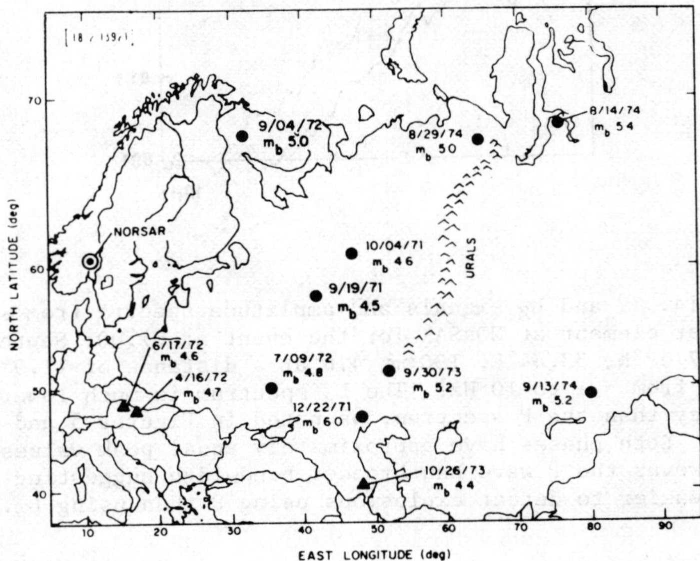

Figure 13. Map showing epicenters of selected presumed explosions in Western Russia and three earthquakes recorded at NORSAR.

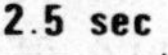

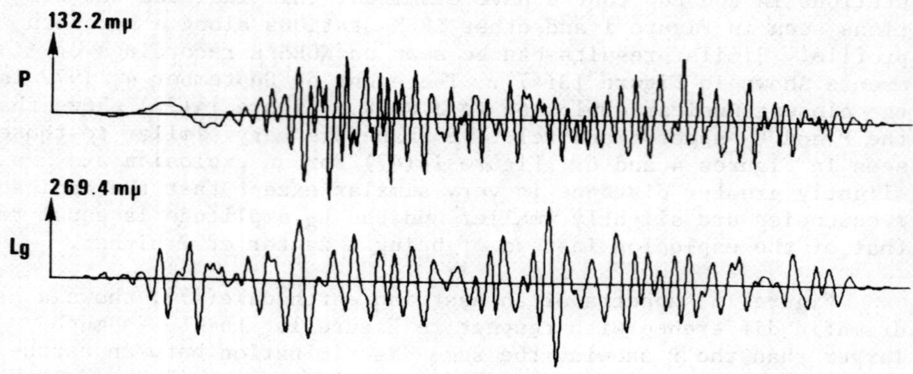

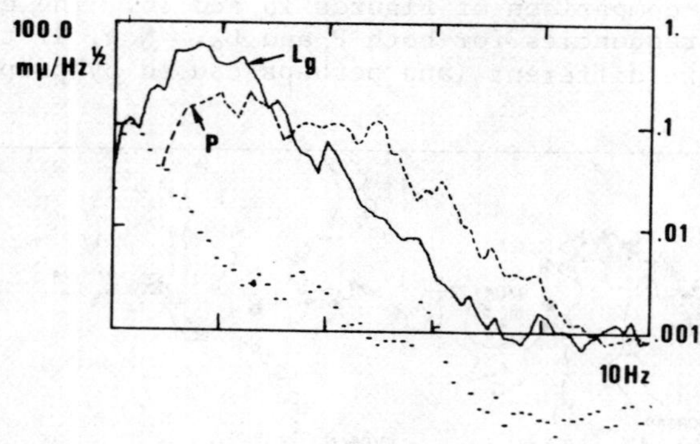

Figure 14. P and Lg signals and amplitude spectra from subarray B2 center element at NORSAR for the event at 07:00, September 4, 1972, 67.69°N, 33.44°E, ISC m_b 4.6 at a distance of 11.9°. Note S/N > 1 from ~ 1 to 10 Hz. The Lg spectrum is much lower in frequency than the P spectrum, as noted in Figures 5 and 6 for SALMON. Both phases have approximately equal peak values of S/N; however the P wave has broader bandwidth suggesting that it may be easier to detect explosions using P than using L_g.

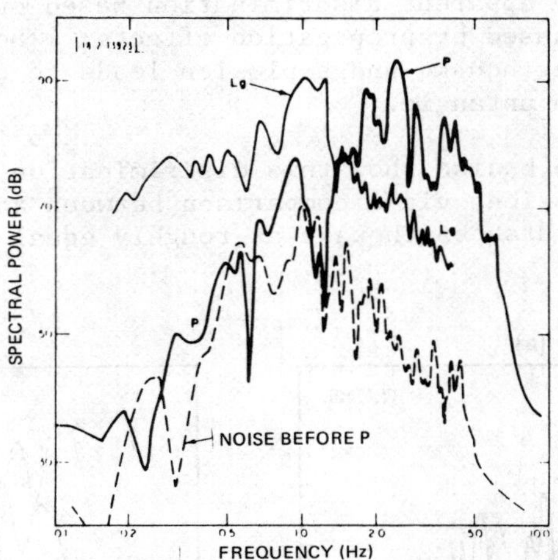

Figure 15. Single-element NORSAR spectra of 20-sec windows of P,
L_g and noise before P of the PNE of 9 July 1972. Note that the
P wave has better S/N than L_g and is of higher frequency. Note
also that the peak spectral amplitudes through the instrument
response are about equal, in contrast to the 20 dB difference
seen in Figure 16 for the Austrian earthquake.

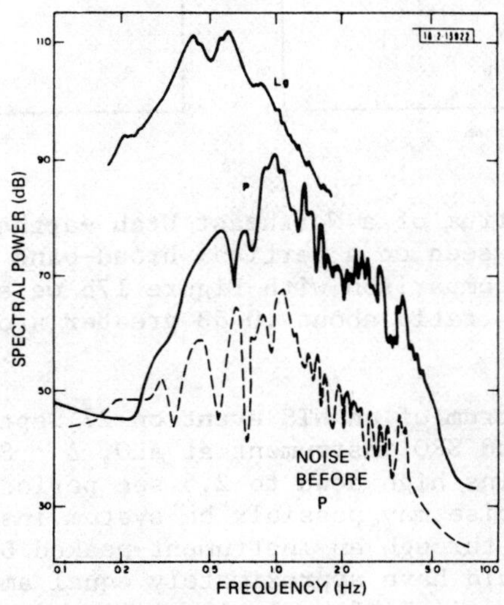

Figure 16. Single-element NORSAR spectra of 20-sec windows of P,
L_g and noise before P of Austrian earthquake of 16 April 1972.
Note that signal/noise ratio of P phase is nearly 20 dB from 1 to
4 Hz, that L_g is of lower frequency than P and that L_g is approximately 20 dB larger than P, in contrast to Figures 14 and 15
which are both explosion spectra. m_b = 4.7, Δ = 13.5°.

then perhaps the apparent discrimination based on maximum amplitudes is also caused by propagation effects. The geographical
separation of earthquake and explosion leads to questions which
are difficult to untangle.

Figures 17a,b also show this discrimination between earthquakes and explosions via a comparison between an NTS explosion
and a northeast Utah earthquake at roughly equal distances from
ALQ(48).

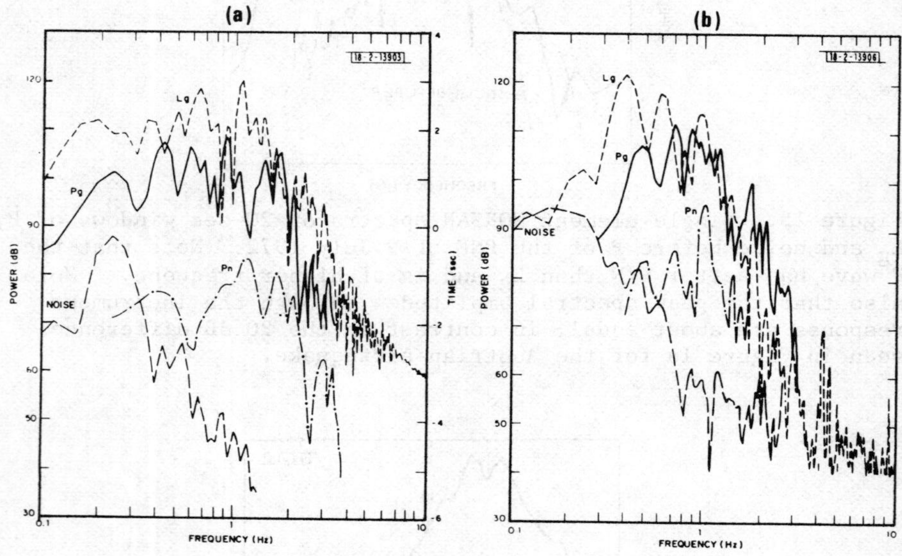

Figure 17a. Spectrum of a Northeast Utah earthquake on 30
September 1977 as seen on a vertical broad-band SRO instrument at
ALQ, $\Delta \sim 6°$. By comparison with Figure 17b we see that the earthquake has an L_g/P_g ratio about 10 dB greater around 1 Hz than does
the explosion.

Figure 17b. Spectrum of an NTS event on 27 September as seen on a
vertical broad-band SRO instrument at ALQ, $\Delta \sim 8°$. Here we see
that the S/N remains high down to 2.5 sec period. Some of the
lower frequency noise may possibly be system instead of earth
noise. Note that through an instrument peaked between 1 and 5 Hz
that L_g and P_g would have approximately equal amplitude as seen
previously to be typical of explosions.

Because of the possibility that decoupled explosions may
be more easily detected at high frequencies than fully coupled
events(42), there is great interest in the propagation of high
frequency signals especially in shield and platform regions
similar to those surrounding salt beds in the Soviet Union. The
most promising results in the literature to date are shown in
Figure 18(49). The question of the nature of the noise around
30 Hz is crucial. Is it earth or system noise and if the latter,
can the noise be significantly reduced by placing the system down-
hole? The ways to answer these questions include careful system
design and monitoring of the recorded noise for day-night variations. It is conceivable that in certain regions, if system and
ambient noise can be adequately reduced, that the peak in the S/N
ratio may be pushed to frequencies higher than the 1 to 4 Hz band
seen in spectra thus far in this paper.

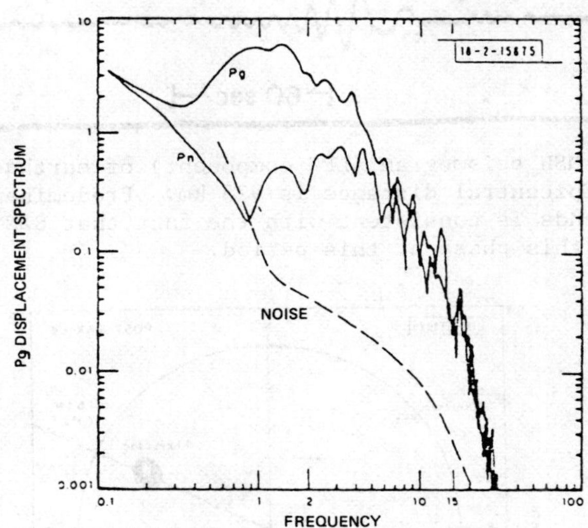

Figure 18. Relative displacement spectra for P_n and P_g from an
18 April 1979 Maine earthquake at GNT, a station in Canada at a
distance of 346 km. The data are sampled at 60 sps. Note S/N >
1 at 30 Hz.

It has been suggested, see for example Figure 19(50), that for
L_g there may be a second maximum in S/N near 4 second period.
More spectra from systems limited only by earth noise need to be
computed to explore this possibility, and advanced processing
of arrays and three-components may offer additional S/N enhancement in this frequency band.

The theoretical discussions of the nature of regional phases(6-20) show that their phase and group velocity should be different. This fact has been shown experimentally by several authors(51-55) and is illustrated by Figures 20 to 22. Note that not only L_g but also P_g and even P_n have phase velocities larger than their group velocities. It has been shown(53) that the phase velocity is higher for deep events and that it varies from event to event reflecting the modal composition of the signal.

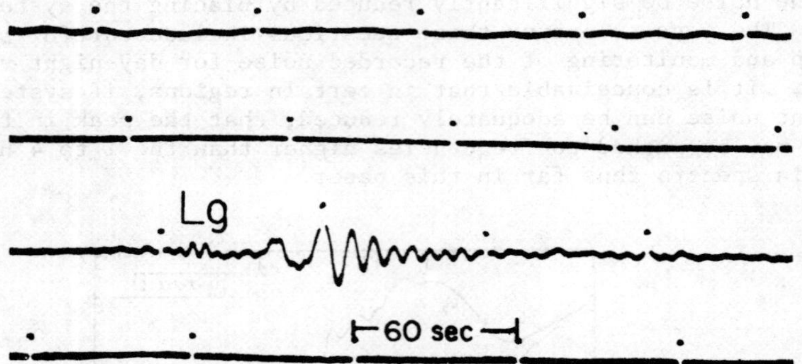

Figure 19.   MSH seismogram (LPZ component) of earthquake of May 24, 1973.   Epicentral distance is 925 km.   Predominant period of L_g at 4 seconds is consistent with the fact that S/N is often maximum for this phase at this period.

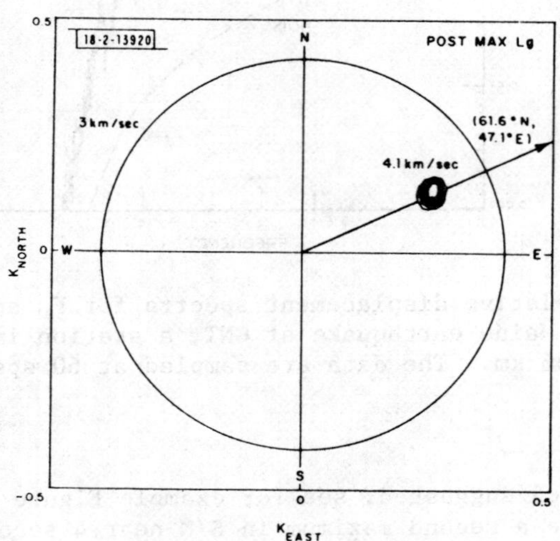

Figure 20.   Wavenumber spectrum at 1.2 Hz of time window immediately following maximum L_g amplitude (K = frequency/phase velocity).   The observed phase velocity of 4.1 km/sec is greater than the group velocity of 3.5 km/sec.

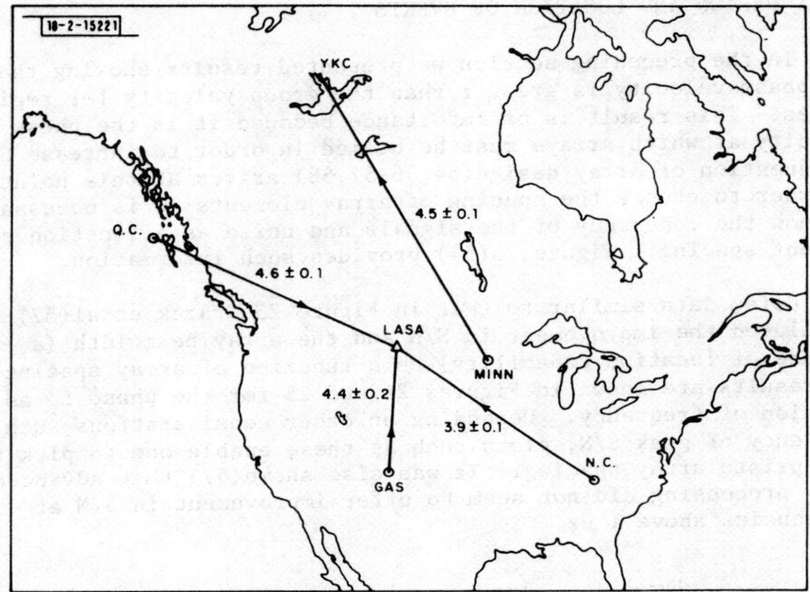

Figure 21. Phase velocity of vertical component of L$_g$ at group velocity 3.5 km/sec at 1 Hz.

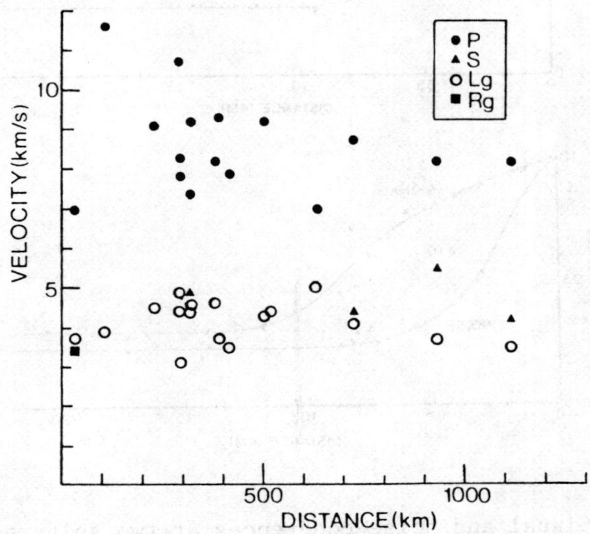

Figure 22. Phase velocities as determined by processing of NORESS data from a set of regional events.

ARRAY DESIGN AND LOCATION OF EVENTS

In the preceding section we presented results showing that the phase velocity is greater than the group velocity for regional phases. This result is of importance because it is the phase velocity at which arrays must be beamed in order to increase S/N. The question of array design(54,56,57,58) arises at this point. In order to choose the spacing of array elements it is necessary to know the coherency of the signals and noise as a function of element spacing. Figure 23(54) provides such information.

Using data similar to that in Figure 23 Mrazek et al(57) calculated the improvement in S/N and the array beamwidth (a measure of location capability) as a function of array spacing. The results are shown in Figures 24 and 25 for the phase L_g as a function of frequency. Depending on other considerations such as frequency of peak S/N, plots such as these enable one to pick the appropriate array spacing. It was also shown(57) that advanced array processing did not seem to offer improvement in S/N at frequencies above 1 Hz.

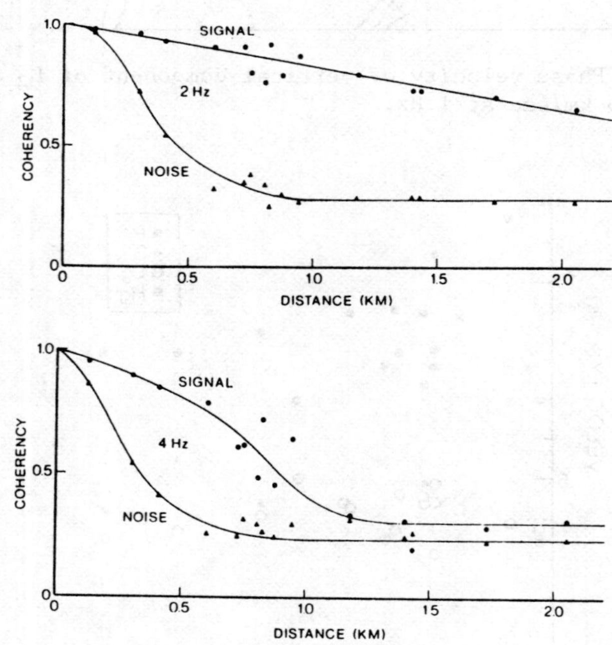

Figure 23. Signal and noise coherences at two selected frequencies for L_g phases as a function of sensor distance within NORESS. Frequencies shown are 2 Hz (top) and 4 Hz (bottom).

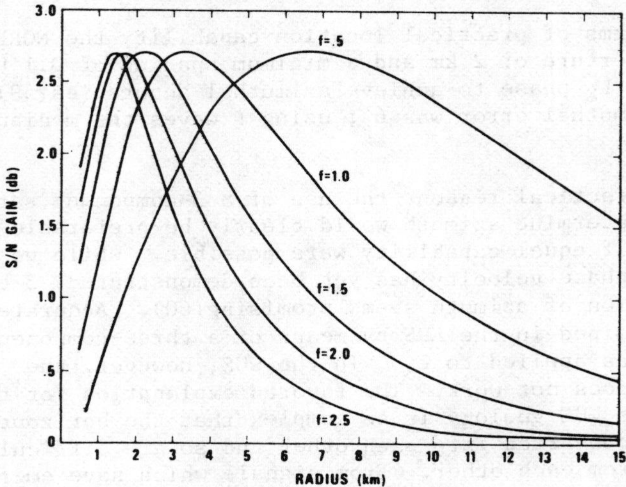

Figure 24. Theoretical S/N gain as a function of frequency and radius for a seven element array, (phase L_g). Gain is calculated as a function of measured coherency of signals and noise at LASA, CPO, and NORSAR. The maxima represent a trade-off between signal and noise coherency. Note that the maxima are well below the 8.5 dB to be expected from coherent signals and incoherent noise.

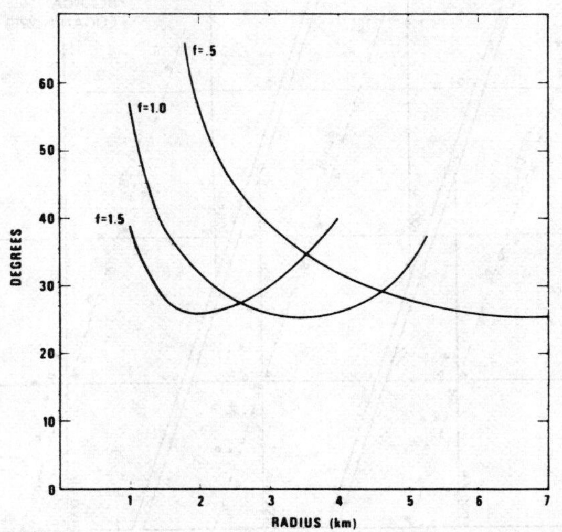

Figure 25. Half beamwidth as a function of radius for a seven element array, (phase L_g). Half beamwidth is calculated from the co- and cross-spectra using the measured coherency of signals and noise at LASA, CPO, and NORSAR. The minima represent a trade-off between aperture and signal coherency.

In terms of practical location capability the NORESS array
with an aperture of 2 km and a minimum spacing of 0.1 km was able
to use the L_g phase to achieve azimuthal accuracies(59). The
median azimuthal error was 6°; using P waves the median error was
3°.

For practical reasons the use of a 3-component set of instruments to determine azimuth would clearly be preferable to use of
an array, if equal capability were possible. While no method to
determine phase velocity has yet been demonstrated, 3-component
determination of azimuth seems promising(60). Accurate azimuths
were determined in the EUS by means of a three-component
processor as applied to L_g. In the WUS, however, the
technique does not work. The favored explanation for the failure
is that the WUS geology is so complex that the horizontal components are converted into each other and so are difficult to distinguish from each other. From signals which have emerged from
the WUS into the platform and shield areas good back azimuths can
be obtained. Accurate back azimuths can also be obtained from P
waves but this technique has not yet been tested in the WUS.

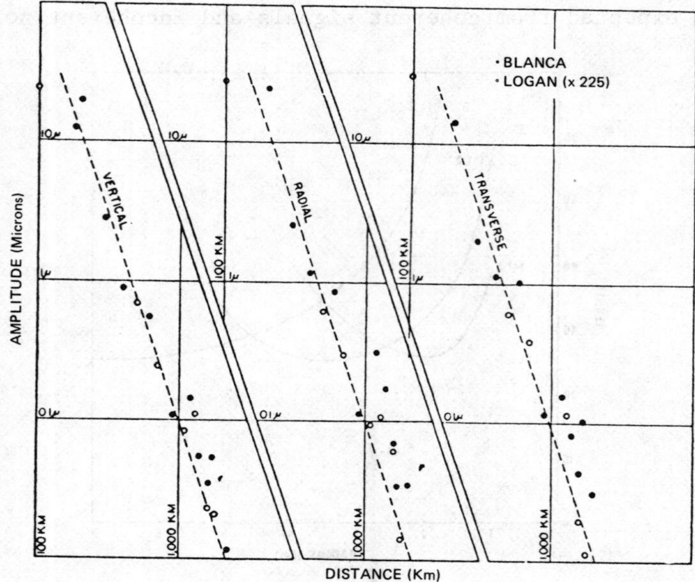

Figure 26. Amplitudes of 3.5 km/sec shear waves from a 19 kt
underground explosion. Curves are proportional to the inverse
cube of the distance, and can be seen to be slightly too steep.
Note that the amplitudes seem to be equal on all three components.

AMPLITUDE DISTANCE AND RELATIVE PHASE AMPLITUDES IN THE US AND USSR

   Among the earliest amplitude-distance studies are those by
Romney(35) and Romney et al(27) from which we have taken Figures
26 and 27. One of the interesting points to be seen in these

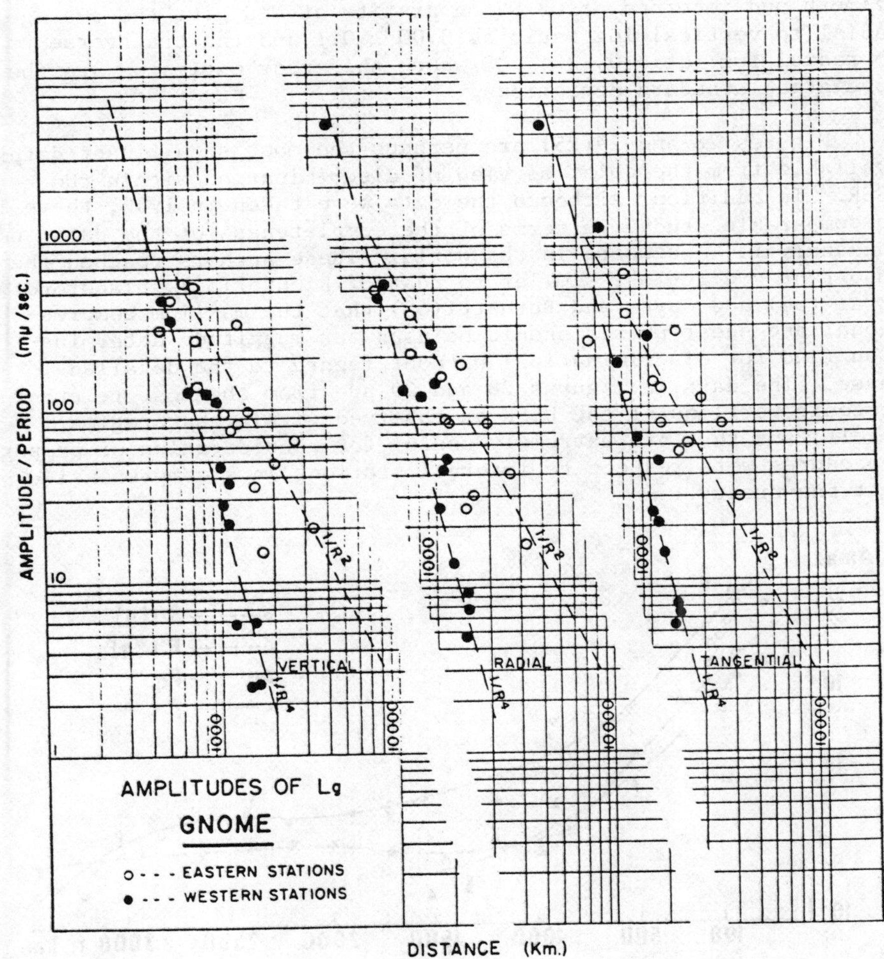

Figure 27. Amplitude graphs for three components of Lg versus
distance for eastern and western stations. The decay rate is
clearly different for the two directions. Close station by
station inspection will show that on the eastern profile the
transverse component is larger than the radial and vertical
component, while on the western profile the amplitudes are equal,
as they are for BLANCA and LOGAN in Figure 26.

three-component data is that in the WUS the three components of
L_g motion are more nearly equal to each other in amplitude than in
the EUS where the transverse is typically 1.7 to 2.0 times the
radial, as seen for example in Figures 2 and 3. Quantitative
rotated waveform data for the WUS is available in Barker et al(61).
Analysis of the plots of 27 good S/N events well distributed in
azimuth and recorded at OB2NV on granite at NTS give the average
radial to vertical log ratio as 0.00 ± 0.1 and the transverse
to radial log ratio 0.12 ± .08 where the error estimates are the
population standard deviations.

Figures 28 and 29(25) are perhaps the most significant data
available from the point of view of discrimination within the
USSR. In addition, although the data were taken in 1962, there is
no comparable study, in terms of the completeness of the data and of
the analysis, elsewhere in the world. These authors reached the
important conclusion, similar to that of Richter(62), Blandford
et al (38) and Gupta and Burnetti(63) that the maximum compressional and shear phases should be used for magnitude determination (and for discrimination) without regard to the detailed
phase. The data in Figures 28 and 29 are from the most relevant
tectonic portions of the USSR as observed within that region
itself, and thus are extremely useful for consideration of proper
procedures with respect to internal stations in a comprehensive
test ban treaty.

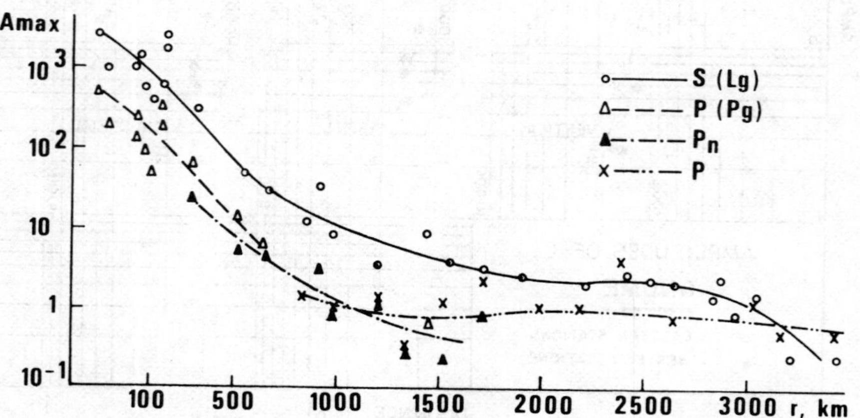

Figure 28. Amplitude distance relations for $S(L_g)$, $P(P_g)$, P_n
and P for the 31 January 1962 Garm earthquake along a profile to
Lake Baikal. Maximum horizontal peak-to-peak/2 amplitudes are
picked within 30 seconds after first arrival for L_g, and on the
vertical 10 seconds after P_g, and 5 seconds after P_n and P.
CKB IIIM instruments are flat to displacement between 1 and 10 Hz.

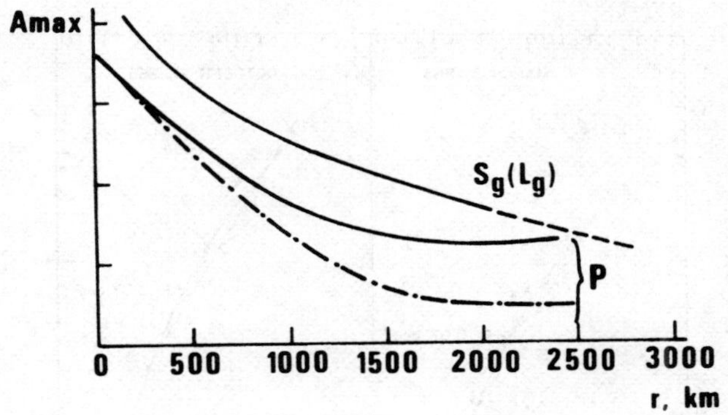

Figure 29. Average amplitude graph for $S_g(L_g)$ and the maximum in the P wavetrain. Average is over several events in each of five Soviet regions: Central Asia, Dzhungaria, Altai-Sayan, Eastern Sayans and, for L_g, Pribaikal. Only the Pribaikal P curve is significantly different from the others. From Nersesov and Rautian (25) who say "For energy determination of earthquakes from longitudinal waves the most consistent amplitude dependence is obtained neither for individual waves nor their first arrivals, but rather for the maximum amplitudes of longitudinal waves for the entire (velocity range) in which they exist. From such an examination of the amplitude curves we obtained a single type dependence for almost all the regions."

Note the large ratio of L_g/P_{max} for these data, these are similar to the earthquake ratio we have seen in Figure 3 for the EUS and suggest that experience in the EUS may be relevant to the USSR for stations within the USSR. Of course, for stations outside the USSR the $S_n(13)$ and $L_g(36,64)$ phases are usually badly distorted making discrimination difficult at such stations.

Early work in L_g distance amplitude relations was carried out by Baker(29) which was followed by the classic paper of Nuttli(33) from which Figure 30 was taken. Other amplitude-distance data(38) for L_g and P_{max} are seen in Figures 31 and 32. A critical survey of the literature(27-35,62,65-67) on amplitude distance relations combined with analysis of approximately 20 new events using high dynamic range LRSM data led to the result that from 300 to 2000 km the amplitude distance relation for the WUS for L_g and P_{max} was r^{-3} while for the EUS the relations were r^{-2} and $r^{-2.5}$ respectively(38). The EUS results are consistent with the curves in Figure 29 from Nersesov and Rautian(25) as can be easily seen in Figures 31 and 32 from the dashed lines which have been adjusted in the vertical for best fit to the data of the Hebgen Lake event. Thus we may conclude that within the USSR the amplitude-distance relations are very similar to those of the EUS.

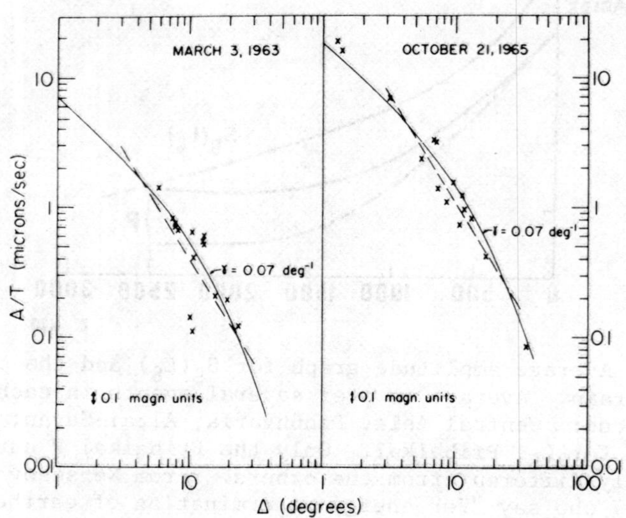

Figure 30. Observed A/T values for the Z component of 1-sec
period L_g waves for the earthquakes of March 3, 1963 and October
21, 1965 at stations in eastern North America. The solid line is
a theoretical attenuation curve for an Airy phase with Δ = 0.07
deg⁻¹. The dashed line is a 1.66 log distance straight line
approximation to the theoretical curve over the distance interval
3° < Δ < 30°; over these distances it does not differ from the
theoretical curve by more than about 0.1 magnitude unit. A slope
of 2.0 log distance could not be rejected using these data.

Attempts have been made to determine the amplitude-distance
relations <u>within</u> the USSR by analysis of data recorded <u>outside</u> the
USSR. Such approaches must plot several events at different
distances as recorded at a single external station, with each
amplitude suitably normalized for event magnitude. Figures 33
and 34 are examples of such approaches(47,48). In general
terms they verify the validity of EUS formulas for L_g propagation from the USSR into Scandinavia, and suggest that the Urals
and the Baltic Sea do not form serious obstacles to L_g propagation. There is some indication in Figure 33 that the amplitude-
distance relation in the distance range 10° to 20° falls off much
more rapidly than $r^{-1.66}$ or even r^{-2}. This may, however, be an
artifact of the normalization procedure; one must look into possible biases in the m_b values of these magnitude 4.5 to 5.0 events
much more carefully before concluding that the amplitude-distance
relations of Nersesov and Rautian are in error.

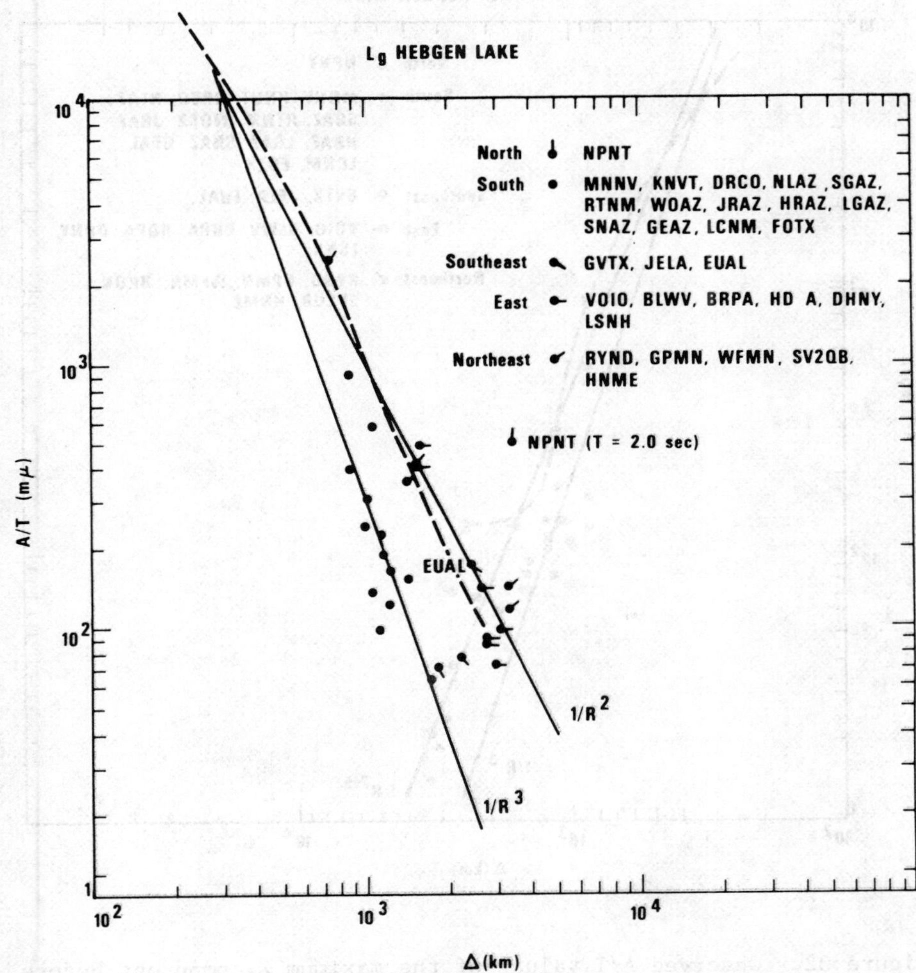

Figure 31.   Observed A/T values of the maximum Z component of L_g
waves for the Hebgen Lake earthquake of 21 October 1964 which is on
the boundary between WUS and EUS according to Der et al. (1975).
Note the more rapid attenuation to the South (within the western
United States) than toward other directions.   Note that the more
northeasterly the azimuth the higher the amplitude.   The data are
consistent with a decay rate of r^{-3} in the WUS and r^{-2} in the EUS.
The dashed line is taken from Nersesov and Rautian(25) (see Figure
29, this paper) for the maximum L_g averaged over five seismic
regions of the USSR.   We see that it is consistent with r^{-2} in the
EUS.

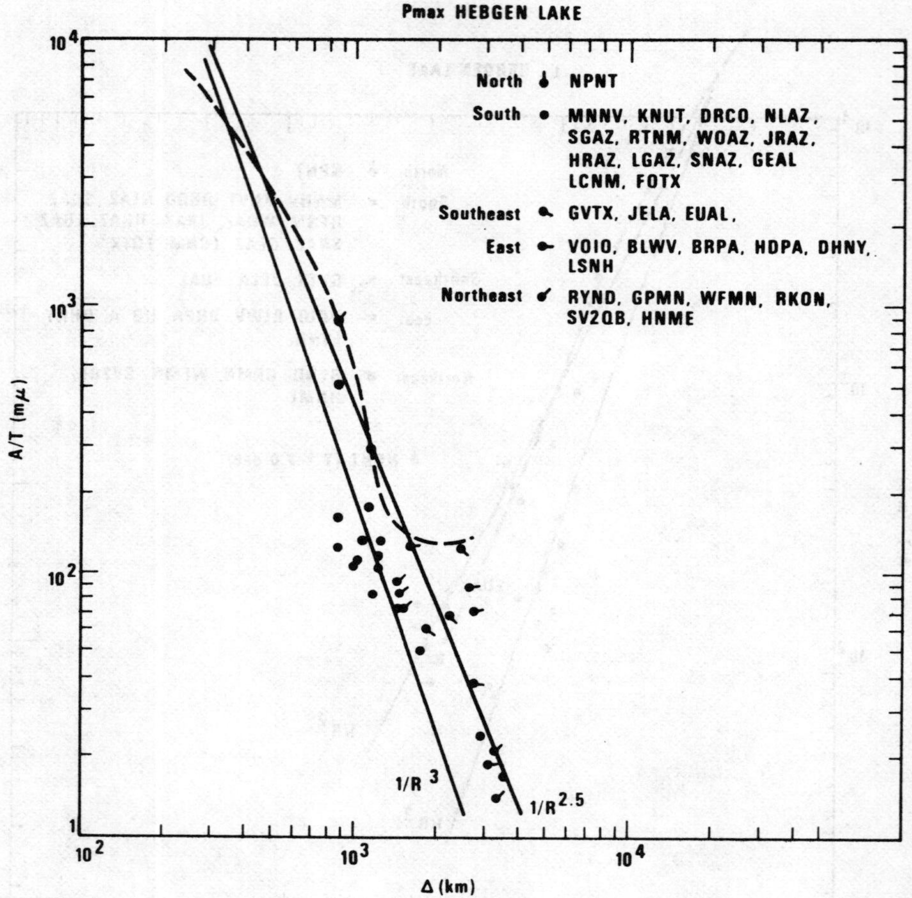

Figure 32. Observed A/T values of the maximum Z component before the arrival of S_n for the Hebgen Lake earthquake of 21 October 1964. Results are consistent with a decay rate of r^{-3} in the WUS and $r^{-2.5}$ in the EUS. The dashed line is taken from Nersesov and Rautian(25)(see Figure 29, this paper) for P_{max} averaged over four regions of the USSR. The P_{max} curve for Pribaikal would fit the Hebgen Lake EUS data better for $\Delta > 2000$ km.

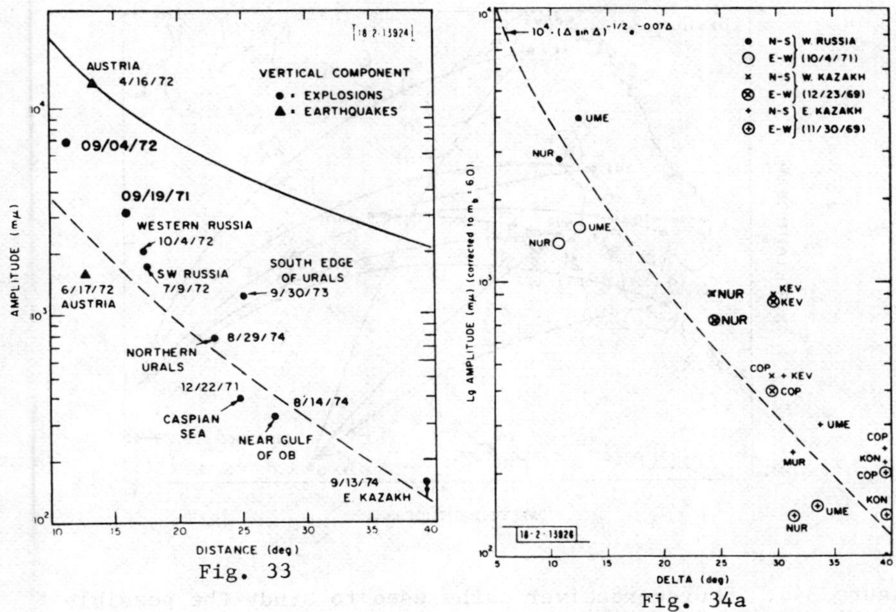

Fig. 33

Fig. 34a

Figure 33.   NORSAR measurements of L_g amplitudes normalized to m_b 6.0.   See map of epicenters in Figure 13.   This was done for each event by scaling L_g amplitude by factor 10^{6-m_b}.   Dashed line is Nuttli's theoretical trend of L_g amplitude versus distance for earthquakes in EUS shifted here to pass through explosion data. Solid curve is Nuttli's empirical variation of L_g with distance for $m_b = 6.0$ earthquakes.   Separation of the two lines shows the discrimination between earthquakes and explosions assuming that most European earthquakes will fall near the solid line.   The Austrian earthquake of 6/17/72 fails to discriminate and deserves further study.   The two closest explosion points have been added to the original figure.

Figure 34a. L_g amplitudes, separated into measurements from short-period north-south and east-west component seismograms. Also shown is Nuttli's relation for amplitude of L_g with $k = 10^4$. Since the event East of the Urals falls in with other events one may conclude that the Urals have little effect on the amplitude.   Since amplitudes to East and West of the Baltic are about the same, one may conclude that the Baltic has little effect.   Finally, note that the north-south component, which is nearly transverse, is always larger than the east-west component.

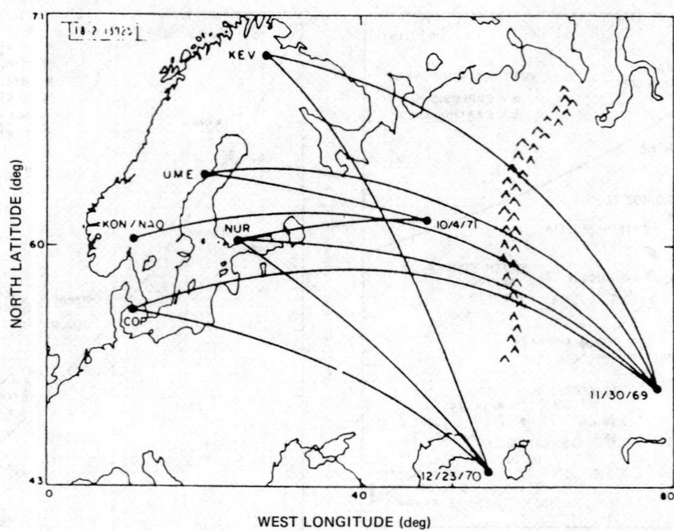

Figure 34b. Source-receiver paths used to study the possible blocking of L_g by the Urals and by the Baltic Sea. Urals are indicated by shaded region.

CAUSES OF VARIATION IN THE P_{max}/L_g DISCRIMINANT

If the ratio P_{max}/L_g is to be used as a discriminant it is important to know what, other than the nature of the source, can change it. In general terms, as we shall see, the ratio seems very stable, the major exception being propagation blockages which strongly decrease the amplitudes of L_g and P_g.

Figures 35 and 36 show that although a soft sedimentary site can increase the amplitude of regional phases by nearly a full unit of magnitude(61), (much more than teleseismic magnitudes are increased(40)), the ratio P_{max}/L_g remains unchanged. These facts show how important it is that the compressional amplitudes be measured at the same stations as the shear amplitudes; and that the two readings make equal contributions to their respective network magnitudes. In particular, troubles may arise if teleseismic m_b or regional P_n is used to substitute for P_{max}.

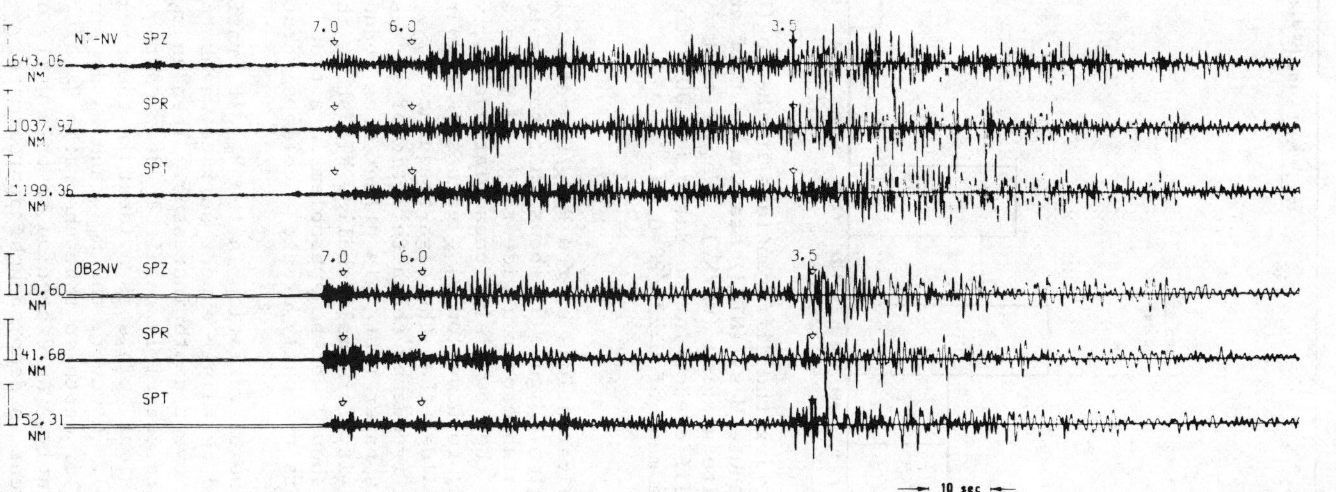

Figure 35. Vertical, radial and transverse motion at Nevada Test Site for the event 18 October 1976; 17:26:52, 32.7N, 117.9W California-Mexico Border Region, NEIS m_b 4.6, distance to OB2NV 529 km. Event recorded at OB2NV on Climax Stock granite and at NTNV on Pahute Mesa tuff. At NTNV all amplitudes are about 6 times larger and the L_g coda is extended in time. Note, however, that on the vertical traces the P_{max}/L_g ratio seems to be the same.

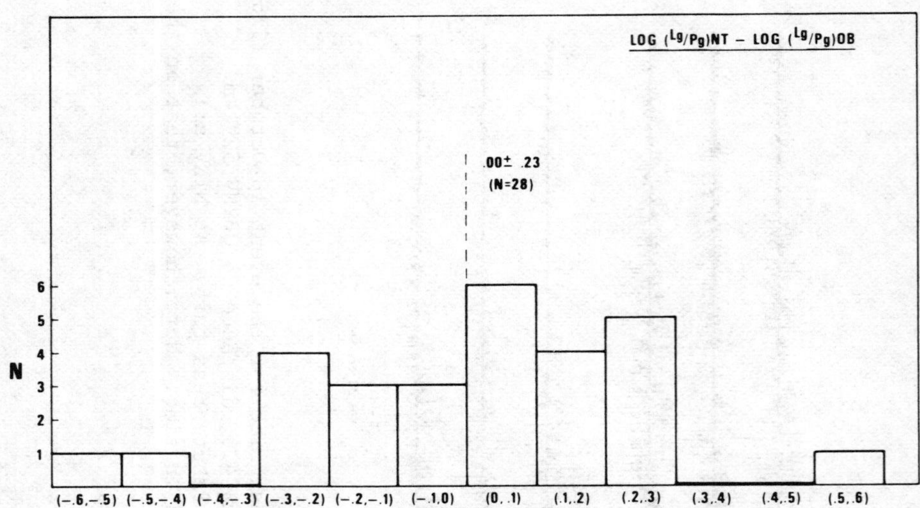

Figure 36. Histogram showing the differentials in the $\log(L_g/P_g)$ amplitudes between the Pahute Mesa (NTNV) station on tuff and the Climax Stock on granite station (OB2NV). We see that averaged over 28 events the difference in the ratio is .00 with a standard deviation of the mean of $0.23/\sqrt{28}=0.04$.

In terms of possible variations of the P_{max}/L_g ratio with respect to the source region, Figure 37 shows that for explosions at NTS the ratio is very stable over a wide range in yield and medium. The tendency for the low yield Yucca flat events, which are detonated in alluvium, to have low P_{max} values as compared to L_g is consistent with the idea that the high-frequency P_{max} is reduced by the low corner frequency characteristic of such shots. We shall see later in this paper that this effect does reduce our discrimination capability for small shallow NTS shots in dry alluvium. However, this should not be a problem in a true test ban treaty since such shots typically create a collapse crater.

Another possible cause of variation in the P_{max}/L_g ratio is the event depth. We would certainly expect surface waves to decline in amplitude as the event depth increases. Then an m_b/L_g or P_n/L_g discriminant would certainly result in deep earthquakes being classified as explosions. This is evident in the results displayed in Figure 38(66). However, the P_{max} and L_g decline in parallel as a function of depth so that the ratio P_{max}/L_g does not change significantly. An exception to this will, of course, arise when the event is deep enough, approximately 100 km, that the initial P is invariably the largest compressional

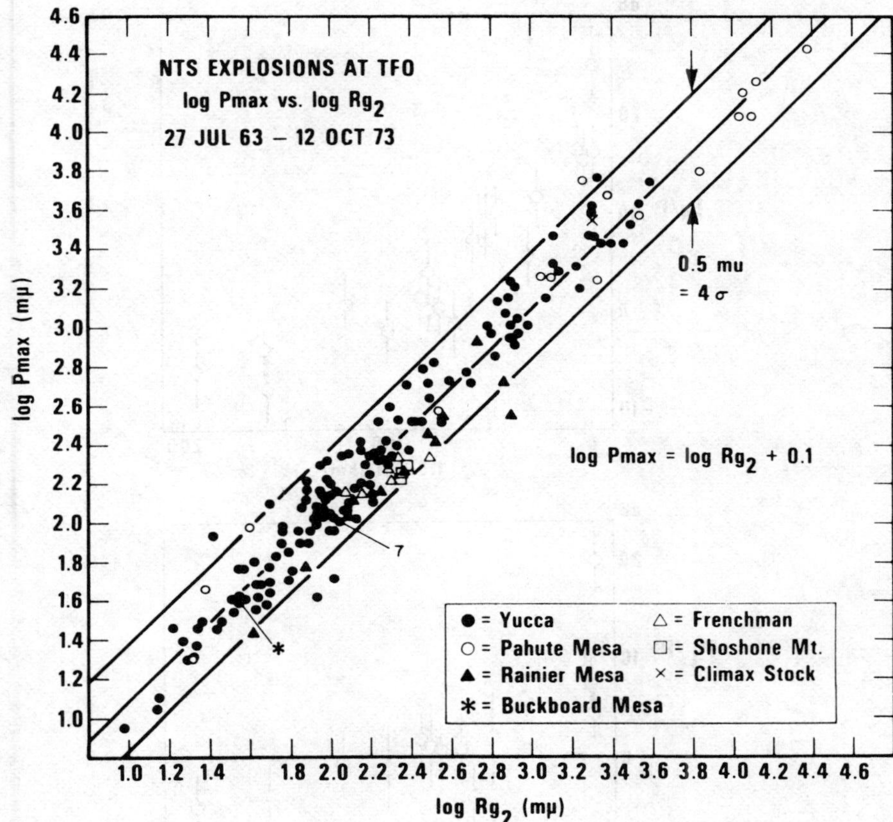

Figure 37. P_{max} versus L_g as measured on the vertical component at Tonto Forest Observatory (TFO) for NTS events. Every measurable event is included ranging from events in granite to those in alluvium and from megatons to less than 1 kiloton. The standard deviation is 0.15 magnitude units showing great stability for this discriminant ratio.

signal. If events of such depth are thought to be a problem, then perhaps P_{max} should be redefined to exclude the first 3 to 5 seconds of the signal at regional distances.

Thus we see that only propagation blockages, which can be mapped and allowed for, seem to significantly alter the P_{max}/L_g ratio, suggesting that it can function as a stable discriminant. No experimental study has yet appeared on the change in P_{max}/L_g as a function of azimuth and fault-plane mechanism. However, what has been seen thus far in this paper would suggest that component conversion must mix the radial and transverse components, thus smoothing out any nodes.

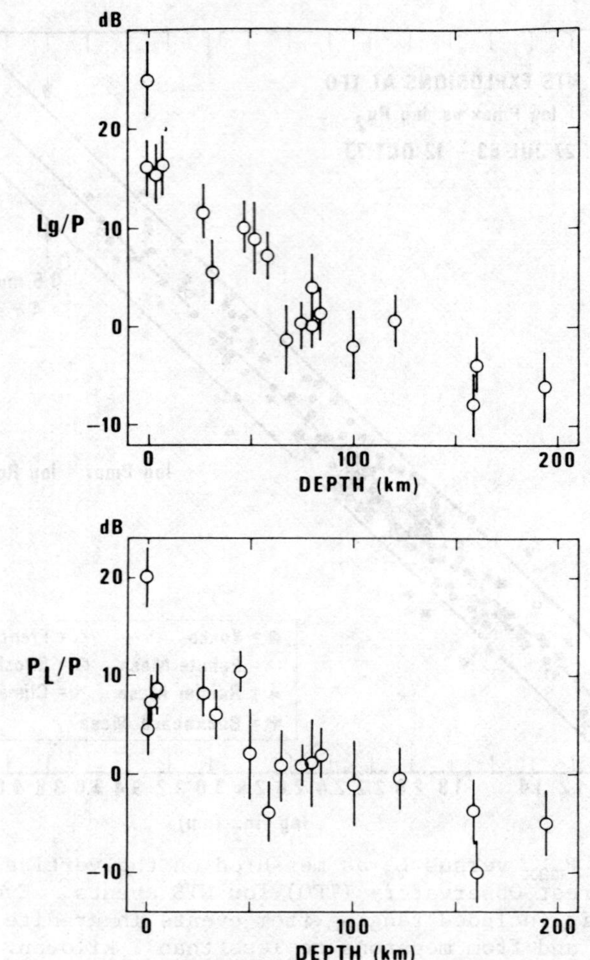

Figure 38. Average L_g/P and P_L/P amplitude ratio as averaged over a network of five Alaskan stations at distance of 5° to 10° for events in Alaska as a function of event depth. P is defined as the maximum in the first 3 seconds and P_L as the maximum after the first 3 seconds until the predicted arrival of S_n. For earthquakes alone the variation offers a good estimate of depth. The figures also show that an L_g/P_{max} discriminant will not be severely affected by event depth, especially if P_{max} at regional distances is taken to exclude the first 3 seconds.

NETWORK DISCRIMINATION USING P_{max}/L_g

Early work on regional discrimination(70-72) generally
showed that a P_g/L_g discriminant had promise but work along these
lines was generally halted when emphasis shifted to teleseismic
discrimination. Figure 39 shows the P_{max} and L_g amplitudes as a
function of distance for SALMON and the Alabama earthquake(38).
A general idea can be gained from this figure as to the stability
of the P_{max}/L_g discriminant. A similar number of data points was
gathered for the events plotted in Figure 40 and the appropriate,
EUS or WUS, distance-amplitude formula was used to calculate a
mean amplitude at 1000 km(38). We see that the populations
separate in both EUS and WUS and that the mean earthquake in the
WUS is shifted by about 0.2 magnitude units to the right of the
EUS line reflecting the large P_g in the WUS. The poor discrimination of the small, shallow explosions in alluvium has already
been discussed in terms of the low corner frequencies for such
shots. A similar explanation is possible for BILBY which cube-
root-scaling ensures will have a low corner frequency. The explosion perhaps most typical of a test-ban treaty environment is BUTEO,
a small overburied shot, and we see that it discriminates extremely
well.

Discrimination studies of events within the USSR are hampered
by the absence of data from within the USSR and by the geographical separation of the explosions and earthquakes. These factors
interact with the severe propagation effects discussed in previous
sections to make analysis very difficult. We have seen in the
discussion of Figure 33(47) that there is some suggestion of good
discrimination between earthquakes and explosions using Scandina-
vian data but that not enough data have been analyzed.

Nuttli(73) has analyzed a number of USSR earthquakes and explosions at WWSSN stations in Scandinavia and to the south of the
USSR. He concluded that the P/L_g discriminant showed an average
separation between earthquakes and explosions but that the scatter
was too large for the ratio to be a useful discriminant. However,
the analysis differed in some respects from what might be optimum.
The P_{max} was picked only in the first 5 to 10 seconds (Nuttli,
personal communication) and while this is often adequate, for
selected crucial seismograms in his paper the maximum occurs much
later. Secondly, the Veith-Clawson distance amplitude relation
was used for normalization. This is certainly the correct normalization for the initial P wave which Nuttli measured, but is
not in agreement with the amplitude-distance results of Nersesov(25)
for P_{max} which we have seen can be approximated as $A \sim r^{-2.5}$.
Finally, the L_g amplitudes were corrected for distance using a
formula which varied according to an absorption coefficient determined from the waveform coda. While this is a very interesting

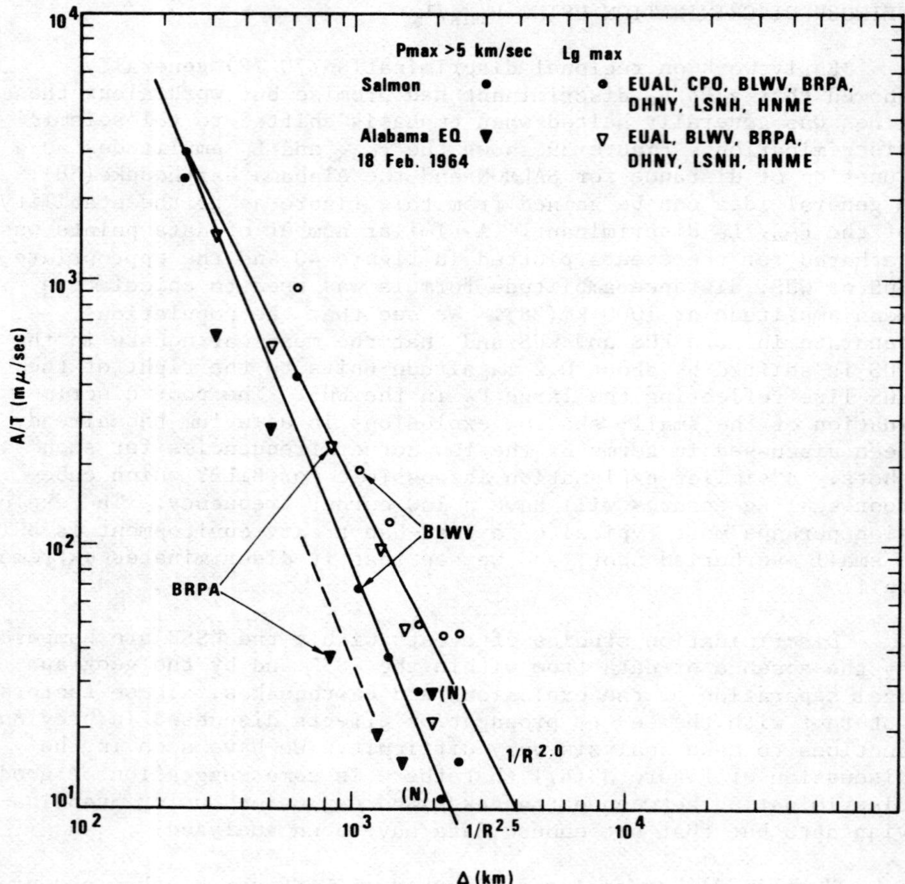

Figure 39. P_{max} and L_g amplitude on the vertical component as a
function of distance on the Northeast profile out of SALMON.   The
L_g amplitudes of the Alabama earthquake and of SALMON are equal
at equal distances but the P_{max} values for SALMON are well above
those for the earthquake.   The appropriate decay rates for L_g and
P_{max}; $r^{-2.0}$ and $r^{-2.5}$ respectively are drawn in on the figure, and
the dashed line shows the separation of the two events in terms
of P_{max}.   The greater separation between L_g and P_{max} for the
earthquake at BRPA than for SALMON at BLWV, as seen in Figures 2
and 3, can be seen in this figure to be in keeping with the
overall pattern.   (N) denotes an upper limit to the signal
as taken from a noise measurement.

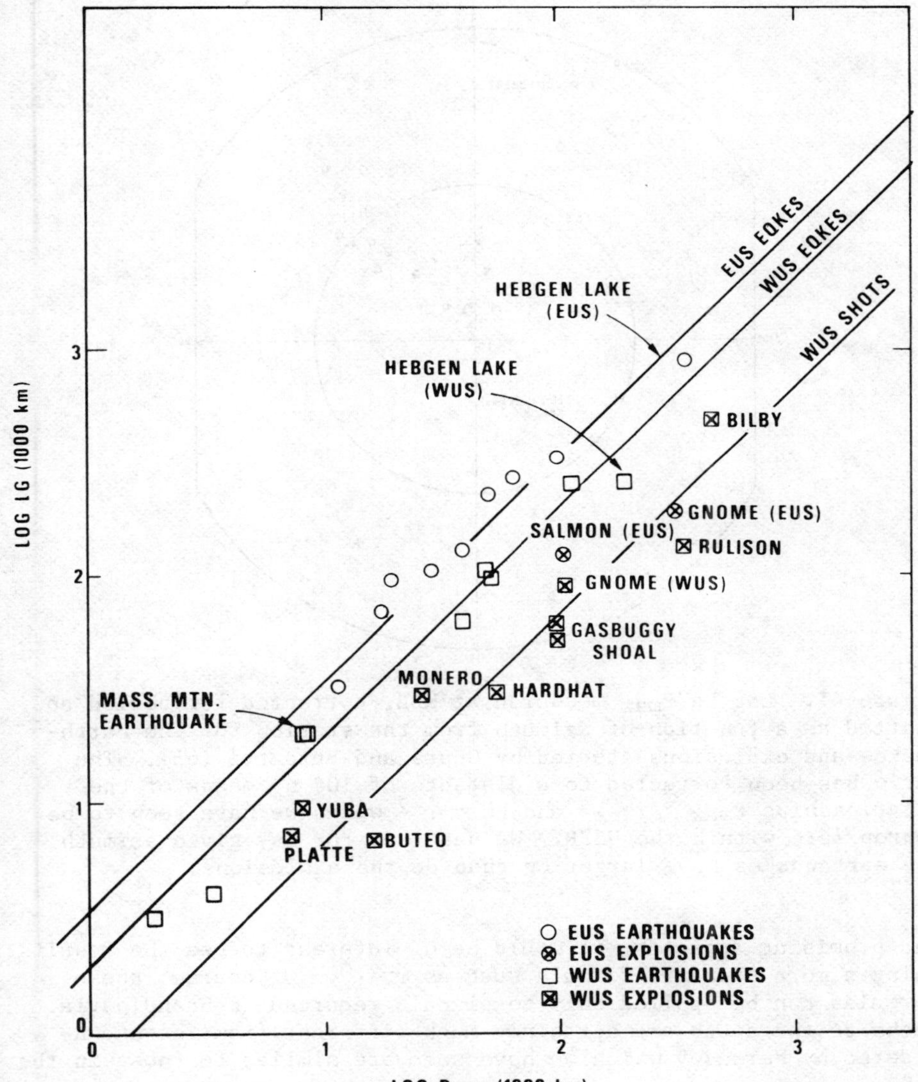

Figure 40. Network L_g plotted as a function of network P_{max} for EUS and WUS events. Amplitudes have been corrected to 1000 km before averaging by use of EUS or WUS formulas as appropriate. Separation between explosions and earthquakes is seen in both EUS and WUS. Poor discrimination of YUBA, PLATTE, MONERO, and BILBY is probably due to low corner frequency which should not be a problem in a comprehensive test ban treaty. Note reduced amplitude of Hebgen Lake and GNOME in WUS as compared to EUS. Note distinct separation of the Massachusetts Mountain earthquake, located at NTS.

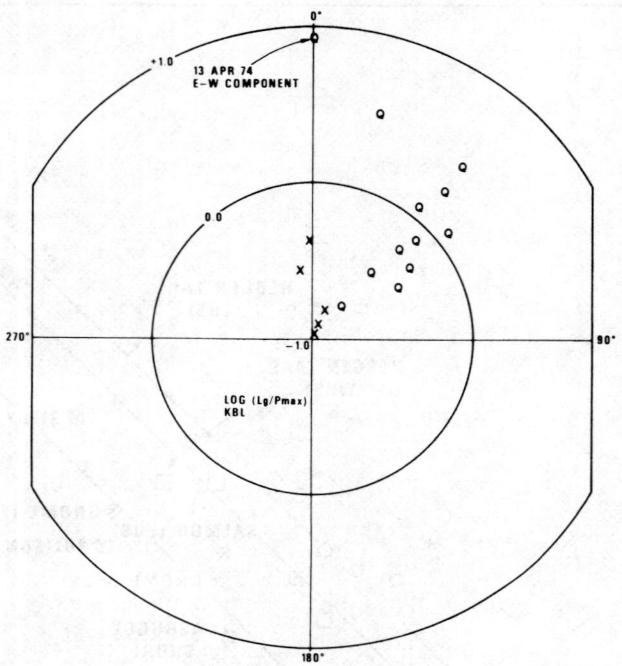

Figure 41. Log Lg/P_{max} measured at KBL, corrected for period and plotted as a function of azimuth from the station for the earthquakes and explosions studied by Gupta and Burnetti (63). The ratio has been corrected to a distance of 10° by means of the relationships $P_{max} \sim r^{-2.5}$ and $Lg \sim r^{-2}$ which we have seen to be appropriate <u>within</u> the USSR. We see that for any given azimuth the earthquakes have larger Lg than do the explosions.

and promising approach, it would be of interest to see the results using a more standard formula such as $r^{-2.0}$. Of course, these formulas can be applied only to signals recorded in Scandinavia. Paths to the south are certainly much different from those considered by Nersesov which we have seen are similar to those in the EUS.

Finally, recent revisions of earlier work (63,64) have resulted in Figure 41. Even within the azimuthal range of efficient propagation of Lg there are significant variations in Lg propagation from within the USSR to station KBL. Thus it is advantageous to plot the P_{max}/Lg discriminant as a function of azimuth in order to minimize these effects. The fair discrimination seen in Figure 41 suggests that if we must rely on external measurements for regional discrimination it will be very important to regionalize the P_{max}/Lg ratio.

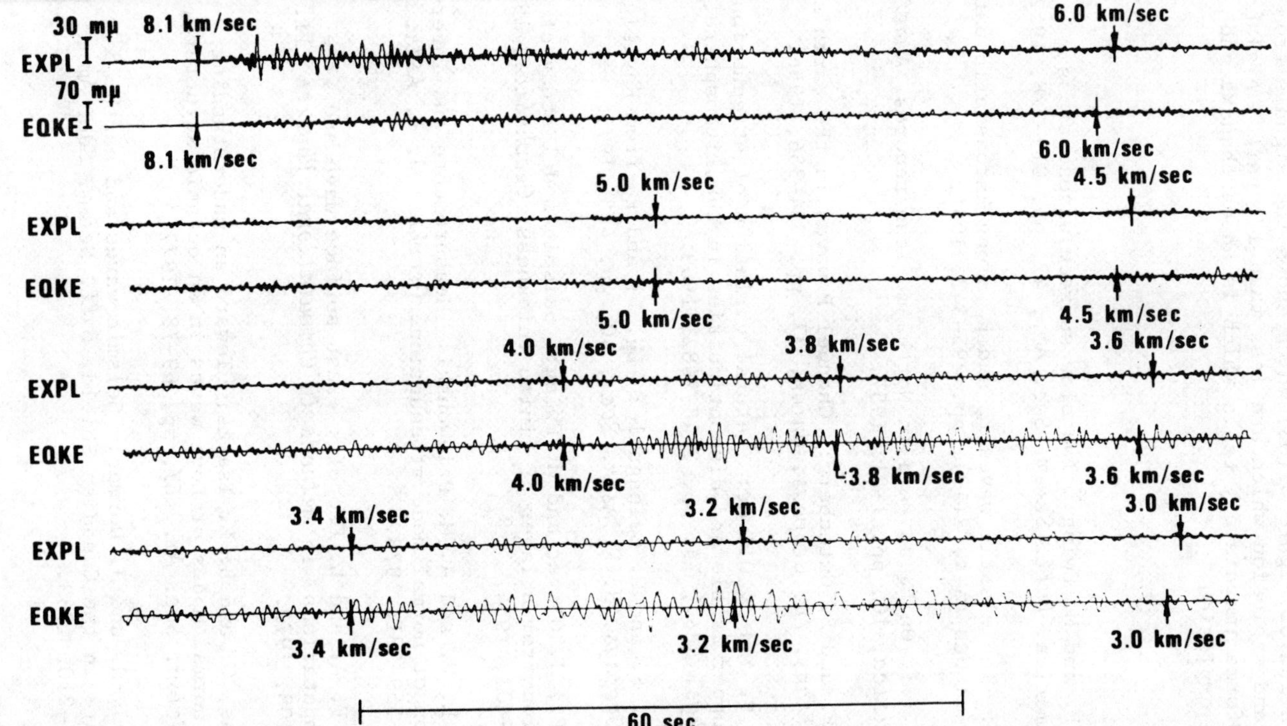

Figure 42. A comparison of the MSH records of an explosion and an earthquake occurring near Semipalatinsk. Boundaries of ten group velocity windows on each record are indicated by arrows. Note the strong P wavetrain on the explosion record, the more emergent earthquake P-wave, and large-amplitude earthquake L_g. Explosion: January 15, 1976, 49.9°N, 78.2°E, m_b 5.2, $\Delta = 19.2°$; Earthquake: March 20, 1976, 50.1°N, 77.3°E, m_b 5.1, $\Delta = 18.8°$.

The gains possible with this approach may be best appreciated by consideration of Figure 42(63) which shows the signals from an earthquake and explosion which are less than 1° apart. A glance at the waveforms immediately tells which is an earthquake and which is an explosion.

REFERENCES

1. Press, F. and M. Ewing: Two slow surface waves across North America, *Bull. Seism. Soc. Am.*, 42, pp. 219-228, 1952.

2. Båth, M: The elastic waves L_g and R_g along Euroasiatic paths, *Arkiv. für Geofysik*, 2(13), pp. 295-342, 1954.

3. Lehmann I.: On L_g as read in North American records, *Anneli di Geofisica*, 10, pp. 1-21, 1957.

4. Press, F. and B. Gutenberg: Channel P waves in the earth's crust, *Trans. Am. Geophys. Union*, 37, pp. 354-356, 1956.

5. Molnar, P. and J. Oliver: Lateral variation of attenuation in the upper mantle and discontinuities in the lithosphere, *J. Geophys. Res.*, 74, pp. 2648-2682, 1969.

6. Panza, G. F. and G. Calcagnile: L_g, L_i and R_g from Rayleigh modes, *Geophys. J. R. Astr. Soc.*, 40, pp. 475, 1975.

7. Press, F., J. Oliver and M. Ewing: Seismic model study of refraction from a layer of finite thickness, *Geophysics*, 19, pp. 388-401, 1954.

8. Sailor, R. V. and A. M. Dziewonski: Measurements and interpretation of normal mode attenuation, *Geophys. J. R. Astr. Soc.*, 53, pp. 559-581, 1978.

9. Schwab, F. A. and L. Knopoff: Fast surface wave and free mode computations, in *Methods in Computational Physics*, 11, pp. 87-108, 1972.

10. Stephens, C. and B. L. Isacks: Toward an understanding of S_n, normal modes of Love waves in an oceanic structure, *Bull. Seism. Soc. Am.*, 67, pp. 69-78, 1977.

11. Brune, J. N. and J. Dorman: Seismic waves and earth structure in the Canadian Shield, *Bull. Seism. Soc. Am.*, pp. 167-210, 1963.

12. Knopoff, L., F. Schwab, K. Nakanishi, and F. Chang: Evaluation of Lg as a discriminant among different continental crustal structures, *Geophys. J. R. Astr. Soc.*, 39, pp. 41-70, 1974.

13. Molnar, P. and J. Oliver: Lateral variations of attenuation in the upper mantle and discontinuities in the lithosphere, *J. Geophys. Res.*, 74, pp. 2648-2682, 1969.

14. Oliver, J. and M Ewing: Higher modes of continental Rayleigh waves, *Bull. Seism. Soc. Am.*, 47, pp. 187-204, 1957.

15. Oliver, J. and M. Ewing: Normal modes of continental surface waves, *Bull. Seism. Soc. Am.*, 48, pp. 33-49, 1958.

16. Oliver, J. and M. Ewing: The effect of surficial sedimentary layers on continental surface waves, *Bull. Seism. Soc. Am.*, 48, pp. 339-354, 1958.

17. Saha, B. P.: The seismic Lg waves and their propagation along the granitic layer of the crust of the Indian subcontinent, *J. Geophys. Res.*, 12, pp. 609-618, 1961.

18. Shurbet, D. H.: The P phase transmitted by crustal rock to intermediate distance, *J. Geophys. Res.*, 65, pp. 1809-1814, 1960.

19. Mantovani, E., F. Schwab, H. Liao and L. Knopoff: Teleseismic S_n, a guided wave in the mantle, *Geophys. J. R. Astr. Soc.*, 51, pp. 709-726, 1977.

20. Haskell, N. A.: The leakage attenuation of continental crustal P waves, *J. Geophys. Res.*, 71, pp. 3955-3967, 1966.

21. Bollinger, G.: Travel time study of central Appalachian earthquakes, *Bull. Seism. Soc. Am.*, 60, pp. 629-637, 1970.

22. Gumper, F. and P. W. Pomeroy: Seismic wave velocities and earth structure on the African continent, *Bull. Seism. Soc. Am.*, 60, pp. 651-668, 1970.

23. Jordan, J. N., W. V. Mickey, W. Helterbran, and D. M. Clark: Travel times and amplitudes from the SALMON explosion, *J. Geophys. Res.*, 71, pp. 3469-3482, 1966.

24. Massé, R. P. and S. S. Alexander: Compressional velocity distribution beneath Scandinavia and Western Russia, *Geophys. J. R. Astr. Soc.*, 39, pp. 587-602, 1964.

25.   Nersesov, I. L. and T. G. Rautian: Kinematics and dynamics
      of seismic waves to distance of 3500 km from the epicenter,
      Akad Nauk SSSR, Trudy Inst. Fiziki Zemli, 32, pp. 63-87.
      (Translated by A. Ryall, DARPA, 12/8/77).

26.   Pasechnik, I. P.: Characteristics of seismic waves from
      nuclear explosions and earthquakes, Nauka, Moscow. Trans-
      lated by C. Shishkevish, Geosciences Bulletin, Series A,
      Volume 1, Rand Corporation, Washington, D.C., 1970.

27.   Romney, C., B. G. Brooks, R. H. Mansfield, D. S. Carder,
      J. N. Jordan and D. W. Gordon: Travel times and amplitudes
      of principal body phases recorded from GNOME, *Bull. Seism.
      Soc. Am.*, 52, pp. 1057-1074, 1962.

28.   Ryall, A. and D. J. Stuart: Travel times and amplitudes
      from nuclear explosions, Nevada Test Site to Ordway,
      Colorado, *J. Geophys. Res.*, 68, pp. 5821-5835, 1963.

29.   Baker, R. G.: Determining magnitude from L_g, *Bull. Seism.
      Soc. Am.*, 60, pp. 1907-1919, 1970.

30.   Everden, J. F.: Magnitude determination at regional and
      near-regional distances in the United States, *Bull. Seism.
      Soc. Am.*, 66, pp. 1609-1922, 1976.

31.   Gibowicz, S. J.: The relationship between teleseismic body-
      wave magnitude M and local magnitude M_L from New Zealand
      earthquakes, *Bull. Seism. Soc. Am.*, 62, pp. 01-11, 1972.

32.   Jones, F. B., L. T. Long, and J. H. McKee: Study of
      the attenuation and azimuthal dependence of seismic-wave
      propagation in the Southeastern United States, *Bull. Seism.
      Soc. Am.*, 67, pp. 1503-1513, 1977.

33.   Nuttli, O. W.: Seismic wave attenuation and magnitude rela-
      tions for Eastern North America, *J. Geophys. Res.*, 78,
      pp. 876-885, 1973.

34.   Press, F.: Seismic wave attenuation in the crust, *J. Geophys.
      Res.*, 69, pp. 4417-4418, 1964.

35.   Romney, C.: Amplitudes of seismic body waves from under-
      ground nuclear explosions, *J. Geophys. Res.*, 64, pp. 1489-
      1498, 1959.

36.   Ruzaikin, A. I., I. L. Nersesov, V. I. Khalturin, and
      P. Molnar: Propagation of L_g and lateral variations in
      crustal structure in Asia, *J. Geophys. Res.*, 82, pp. 307-
      316, 1977.

37. Isacks, B. L. and C. Stephens: Conversion of S_n and L_g at
     a continental margin, *Bull. Seism. Soc. Am.*, 65, pp. 235–
     244, 1975.

38. Blandford, R. R., R. A. Hartenberger and R. Naylor: Regional
     discrimination between earthquakes and explosions, *EOS, Trans.
     Am. Geophys. Union*, 59, pp. 1140, 1978.

39. McCowan, D. W.: Sonograms for an Eastern U. S. LRSM profile
     from the nuclear explosion SALMON, *Lincoln Laboratory Semi-
     annual Technical Summary*, Seismic Discrimination, June 1978.

40. Der, Z. A, T. W. McElfresh and C. P. Mrazek: Interpretation
     of short-period P-wave magnitude anomalies at selected LRSM
     stations, *Bull. Seism. Soc. Am.*, 69(4), pp. 1149–1160, 1979.

41. Dahlman, O. and H. Israelson: Monitoring underground nuclear
     explosions, *Elsevier*, pp. 440, 1977.

42. Blandford, R.: Discrimination between earthquakes and under-
     ground explosions, *Ann. Rev. Earth Planet. Sci.*, 5, pp. 111–
     122, 1977.

43. Street, R. L.: Scaling Northeastern United States/Canadian
     earthquakes by their L_g waves, *Bull. Seism. Soc. Am.*, 66,
     pp. 1525–1538, 1976.

44. Werth, G. and P. Randolph: The SALMON seismic experiment,
     J. Geophys. Res., 71, pp. 3405–3414, 1966.

45. Gupta, I. N., B. W. Barker, J. A. Burnetti and Z. A. Der:
     A study of regional phases from earthquakes and explosions
     in Western Russia, *Bull. Seism. Soc. Am.*, 70, pp. 851–872,
     1980.

46. Archambeau, C. B., E. A. Flinn and D. G. Lambert: Fine
     structure of the upper mantle, *J. Geophys. Res.*, 74, pp. 5825–
     5864, 1969.

47. Frazier, C. W. and R. M Sheppard: L_g measurements at NORSAR
     from Soviet explosions and two earthquakes, *Lincoln Laboratory
     Semiannual Technical Summary*, Seismic Discrimination, 9 June
     1978.

48. Fitch, T. J. and R. M. Sheppard: Crustal phases recorded on
     short-period and broad-band systems, *Lincoln Laboratory
     Semiannual Technical Summary*, Seismic Discrimination, June
     1978.

49. Fitch, T. J. and M. W. Shields: Spectra of crustal phases
     recorded digitally in eastern Canada, *Lincoln Laboratory
     Semiannual Technical Summary*, Seismic Discrimination, 8
     February 1980.

50. Nuttli, O. W.: The excitation and attenuation of seismic
     crustal phases in Iran, *Bull. Seism. Soc. Am.*, 70, pp. 469-
     485, 1980.

51. Landers, T.: The mechanisms of L_g propagation, *Lincoln
     Laboratory Semiannual Technical Summary*, Seismic Discrimina-
     tion, June 1978.

52. Landers, T. and T. Fitch: Phase velocity of L_g in North
     America, *Lincoln Laboratory Semiannual Technical Summary*,
     Seismic Discrimination, September 1978.

53. B. W. Barker, Z. A. Der and R. H. Shumway: Phase velocities
     of regional phases P_n, S_n, P_g, and L_g observed at the
     Cumberland Plateau Observatory and LASA, AL-80-1, Teledyne
     Geotech, Alexandria, Virginia, 1980.

54. Mykkeltveit, S., F. Ringdal and H. Bungum: An experimental
     small subarray within the NORSAR array: Crustal phase velo-
     cities and azimuths from local and regional events, Project
     Report VT/0702/B/PMP, NTNF/NORSAR, 1980.

55. Pomeroy, P. W. and T. A. Nowak: An investigation of seismic
     wave propagation in eastern North America, Rondout Associates,
     Inc., Stone Ridge, New York.

56. Blandford, R. and D. Clark: A note on seismic array designs,
     Bull. Seism. Soc. Am., 65, pp. 787-791, 1975.

57. Mrazek, C. P., Z. A. Der, B. W. Barker and A. O'Donnell:
     Seismic array design for regional phases, In Studies of
     Seismic Waves at Regional Distances, AL-80-1, Teledyne
     Geotech, Alexandria, Virginia.

58. Barley, B. J.: On the use of seismometers arrays to deter-
     mine azimuth of higher mode Rayleigh waves, AWRE Report,
     Blacknest, England.

59. Ringdal, F. and S. Mykkeltveit: An experimental small sub-
     array within the NORSAR array: Location of local and region-
     al events, Project Report VT/0702/B/PMP, NTNF/NORSAR, 1980.

60. Smart, Eugene: Observations of apparent seismic surface wave
     scattering in the Western United States, AL-80-1, Teledyne
     Geotech, Alexandria, Virginia, 1980.

61. Barker, B. W., Z. A. Der and C. P. Mrazek: The effect of crustal structure on the regional phases P_g and L_g at NTS, AL-80-1, Teledyne Geotech, Alexandria, Virginia, 1980, in press *J. Geophys. Res.*

62. Richter, C. F.: Elementary Seismology, W. H. Freeman, San Francisco, 1958.

63. Gupta, I. N. and J. A. Burnetti: An investigation of discriminants for events in Western USSR based on regional phases at a single station, AL-80-1, Teledyne Geotech, Alexandria, Virginia, 1980, in press *Bull. Seism. Soc. Am.*

64. Gupta, I. N., B. W. Barker, J. A. Burnetti and Z. A. Der: A study of regional phases from earthquakes and explosions in Western Russia, AL-80-1, Teledyne Geotech, Alexandria, Virginia, 1980.

65. Shields, M. W., T. J. Fitch and R. E. Needham: P_n, S_n, and L_g amplitude-distance relationships for eastern North America, *Lincoln Laboratory Semiannual Technical Summary,* Seismic Discrimination, June 1978.

66. Noponen, I. and J. Burnetti: Alaskan regional data analysis, AL-80-1, Teledyne Geotech, Alexandria, Virginia, 1980.

67. Pomeroy, P. W.: Regional seismic wave propagation, Rondout Associates, Inc., Stone Ridge, New York, Report No. 4, 1979.

68. North, R. G.: Propagation of the L_g phase across the Urals and the Baltic, *Lincoln Laboratory Semiannual Technical Summary.*, Seismic Discrimination, June 1978.

69. Der, Z. A. and T. W. McElfresh: Short-period P-wave attenuation along various paths in North America as determined from P-wave spectra of the SALMON nuclear explosion, *Bull. Seism. Soc. Am.*, 66, pp. 1609-1622, 1976.

70. Willis, E. E., J. DeNoyer and J. T. Wilson: Differentiation of earthquakes and underground nuclear explosions on the basis of amplitude characteristics, *Bull. Seism. Soc. Am.*, 53, pp. 979-987, 1963.

71. Willis, E. E.: Comparison of seismic waves generated by different types of sources, *Bull. Seism. Soc. Am.*, pp. 965-986, 1963.

72. Booker, A. and W. Mitronovas: An application of statistical discrimination to classify seismic events, *Bull. Seism. Soc. Am.*, 54, pp. 961-971, 1964.

73. Nuttli, O. W.: On the attenuation of L_g waves in western and central Asia and their use as a discriminant between earthquakes and explosions. Submitted to *Bull. Seism. Soc. Am.*, 1980.

LINEAR DISCRIMINATION FOR SAMPLES OF LIMITED SIZE

V.F. Pisarenko, A.F. Kushnir, F.M. Pruchinka and
S.L. Zvang

Institute of Physics of the Earth, Moscow, USSR

ABSTRACT

The problem of linear discrimination of r-dimensional vectors is considered. Learning samples have sizes n_1 and n_2,
whereas r/n_1, r/n_2 are not very small. Three questions concerning this problem are considered: error probability estimation,
the use of qualitative components in linear discriminator, and
the choice of the most informative components. Results are illustrated by an artificial example and by some earthquake discrimination problems.

1. ERROR PROBABILITY ESTIMATION

Let us suppose we have sample of r-dimensional vectors, corresponding to two classes:

$$\bar{x}_1 = (x_1^{(1)},\ldots,x_1^{(r)}); \qquad \bar{y}_1 = (y_1^{(1)},\ldots,y_1^{(r)});$$
$$\vdots \qquad\qquad \vdots \qquad\qquad \vdots \qquad\qquad \vdots$$
$$\bar{x}_{n1} = (x_{n1}^{(1)},\ldots,x_{n1}^{(r)}); \qquad \bar{y}_{n2} = (y_{n2}^{(1)},\ldots,y_{n2}^{(r)})$$

The number of vectors are n_1 and n_2. In the most interesting
cases for practical application the values r/n_1, r/n_2 are not too
small. We consider linear discriminators of the type:

E. S. Husebye and S. Mykkeltveit (eds.), Identification of Seismic Sources - Earthquake or Underground
Explosion, 741–745.

$$\ell(\overline{z}) = \sum_{j=1}^{r} \alpha_j \, z^{(j)} + jr$$

where $\overline{z}$ is an observed r-dimensional vector and where jr, α_j are discriminator coefficients. If

$$\ell(\overline{z}) \geqslant C$$

we decide that $\overline{z}$ belongs to the first class; otherwise we decide that it is from the second class. Two types of errors are possible: to attribute $\overline{z}$ to the first class, when it belongs in fact to the second class and vice versa, to attribute $\overline{z}$ to the second class, when it belongs in fact to the first one. The3 probabilities of these errors are denoted by p_1 and p_2 respectively. The efficiency of a discriminator can be characterized by the mean error probability p:

$$p = \frac{p_1 + p_2}{2}$$

It is known that for multidimensional Gaussian classes with mean vectors μ_1, μ_2 and a common covariance matrix B the minimal error probability $-p$ discriminator is linear

$$\overline{\alpha} = B^{-1}(\overline{\mu}_1 - \overline{\mu}_2) \; ; \; \alpha_j = \sum_{k=1}^{r} \beta_{jk}(\mu_1^{(k)} - \mu_2^{(k)})$$

where $\{\beta_{jk}\}$ denotes the elements of the inverse matrix B^{-1}. The error probability p depends in this case on one parameter, namely, the D^2 Mahalonobis statistic:

$$D^2 = (\overline{\mu}_1 - \overline{\mu}_2)^T \cdot B^{-1} \cdot (\overline{\mu}_1 - \overline{\mu}_2)$$

This D^2 statistic is a generalization of the squared signal/noise ratio. Large values of D^2 correspond to small values of p, in fact:

$$p = \tfrac{1}{2}\left[1 + \phi(C/D - D/2) - \phi(C/D + D/2)\right]$$

where $\phi(x)$ is the standard Gaussian distribution function.

Since μ_1, μ_2 and B are unknown, they are estimated by

$$\overline{m}_1 = \frac{1}{n_1} \sum_{i=1}^{n_1} x_1 \; ; \; \overline{m}_2 = \frac{1}{n_2} \sum_{i=1}^{n_2} y_i$$

$$\hat{B} = \frac{1}{n_1+n_2} \left[\sum_{i=1}^{n_1} (\bar{x}_i-\bar{m}_1)(\bar{x}^i-\bar{m}_1)^T + \sum_{i=1}^{n_2} (\bar{y}_i-\bar{m}_2)(\bar{y}_i-\bar{m}_2)^T \right]$$

If these estimates are inserted into $\ell(\bar{z})$ instead of the true values μ_1, μ_2, B, we get a randomly distributed discriminator $\ell(\bar{z})$. It is very difficult to find the exact distribution of $\ell(\bar{z})$. An asymptotic expansion for this distribution is derived under the conditions:

$$r/n_1 \rightarrow \lambda_1 \; ; \quad r/n_2 \rightarrow \lambda_2 \; ; \quad n_1,n^2 \rightarrow \infty$$

where λ_1, λ_2 are some non-zero constants. The first approximation to the limit distribution is the Gaussian distribution with parameters:

$$M_1 \ell(\bar{z}) = \frac{n_1+n_2-2}{n_1+n_2-r-1} \cdot \left[D^2/2 + \frac{(n_1+n_2)(r-1)}{2n_1n_2} \right]$$

when $\bar{z}$ belongs to the first class and

$$M_2 \ell(\bar{z}) = \frac{n_1+n_2-2}{n_1+n_2-r-1} \cdot \left[-D^2/2 + \frac{(n_1+n_2)(r-1)}{2n_1n_2} \right]$$

when $\bar{z}$ belongs to the second class.

$$\text{Var } \ell(\bar{z}) = \frac{(n_1+n_2-2)^2}{(n_1+n_2-r-1)^2} \cdot \frac{(n_1+n_2-1)(n_1+n_2+1)}{(n_1+n_2-r)(n_1+n_2)}$$

$$\left[D^2 + \frac{(n_1+n_2)(r-1)}{n_1n_2} \right]$$

These expressions depend on the unknown parameter D^2. For practical purposes one has to derive an estimator of D^2. An estimator

$$d_1^2 = (\bar{m}_1-\bar{m}_2)^T \cdot \hat{B}^{-1} \cdot (\bar{m}_1-\bar{m}_2)$$

has a bias, since

$$Md_1^2 = \frac{n_1+n_2-2}{n_1+n_2-r-3} \cdot (D^2 + r/n_1 + r/n_2)$$

It is better to use one of the two following estimators:

$$d_2^2 = \frac{n_1 + n_2 - r - 3}{n_1 + n_2 - 2} \cdot d_1^2 - r/n_1 - r/n_2 \; ;$$

$$d_3^2 = \frac{n_1 + n_2 - r - 3}{n_1 + n_2 - 2} \cdot d_1^2$$

The estimator d_3^2 is preferable, when n_1, n_2 are not too large. In this case the estimator d_2^2 can sometimes be negative.

Inserting d_3^2 or d_2^2 into $M_1\ell$, $M_2\ell$ and $\mathrm{Var}\,\ell$ we get estimates for the parameters of the distribution of $\ell(\bar{z})$. In order to obtain estimated error probabilities p, p_1, p_2 one has to use the standard Gaussian distribution:

$$\hat{p}_1 = 1 - \phi\left(\frac{C - M_1\ell}{\sqrt{\mathrm{Var}\,\ell}}\right) \; ; \quad \hat{p}_2 = \phi\left(\frac{C - M_2\ell}{\sqrt{\mathrm{Var}\,\ell}}\right) \; ; \quad \hat{p} = \frac{\hat{p}_1 + \hat{p}_2}{2}$$

2. THE USE OF QUALITATIVE COMPONENTS IN LINEAR DISCRIMINATION

Some components of vectors $\bar{x}_i$, $\bar{y}_i$ are often of qualitative nature. Suppose, for example, that x_1 is a qualitative component and can be in one of k_1 possible states $V_1, \ldots, V_{k1}$. For example x_1 can be one of two possible answers (yes or no) to some question. In this case one cannot use directly the linear discriminator $\ell(\bar{z})$. However, if we assign to the qualitative states $V_1, \ldots, V_{k1}$ some numerical values $t_1, \ldots, t_{k1}$, we are able to use $\ell(\bar{z})$. Usually values $t_1, \ldots, t_{k1}$ are chosen arbitrarily. For example, one takes:

$$t_1 = 1 \; ; \; t_2 = 2 \; ; \; \ldots \; ; \; t_{k1} = k_1$$

Such a choice, however, is not efficient. We suggest another choice that would be optimal for multinormally distributed and independent components. Using this choice we get

$$t_1 = \lg\left(\frac{n_1^{(1)} + \tfrac{1}{2}}{n_1 + \tfrac{1}{2}} \Big/ \frac{n_1^{(2)}}{n_2 + \tfrac{1}{2}}\right) \; ; \quad t_2 = \lg\left(\frac{n_2^{(1)} + \tfrac{1}{2}}{n_1 + \tfrac{1}{2}} \Big/ \frac{n_2^{(2)} + \tfrac{1}{2}}{n_2 + \tfrac{1}{2}}\right)$$

where $n_1^{(1)}$ is the number of occurrences of state V_1 among vectors $\bar{x}_1, \ldots, \bar{x}_{n1}$; $n_1^{(2)}$ is the number of occurrences of state V_1 among vectors $\bar{y}_1, \ldots, \bar{y}_{n2}$ and so on. The terms $1/2$ are added into nominators and denominators of t_1, t_2 in order to avoid undesirable situations $n_1^{(1)} = 0$, or $n_1^{(2)} = 0$ and so on.

3. THE CHOICE OF THE MOST INFORMATIVE COMPONENTS

The choice of the most informative components in discrimination analysis is one of the most important problems. Usually, it is not desirable to use too many components since a large number of components needs very large sample sizes n_1, n_2 to obtain good estimates.

In order to choose dimensionality r in practice one needs to know the dependence of the d_1^2 statistics on r. For this purpose we suggest starting with one component ($r = 1$) and to choose the component with the largest value of $d_1^2(1)$. Next, trying one by one of the remaining components, we choose the second component in such a way that it would give, in combination with the first component, the largest value of $d_1^2(2)$ among all pairs, and so on.

Inserting values $d_1^2(r)$ into $p(r)$ we obtain the corresponding $\hat{p}(r)$. Now we choose the value r corresponding to a minimum of $\hat{p}(r)$. It should be noted that if the minimum of $\hat{p}(r)$ is not very pronounced, we suggest taking a smaller value r. Sometimes $\hat{p}(r)$ decreases monotonically as r increases, but the rate of decrease of $\hat{p}(r)$ changes more or less abruptly in the vicinity of some $r = r_0$. In such situations we recommend taking r_0 components for discrimination. In any case of doubt a smaller number of components is preferable, since this makes the estimation and interpretation easier.

THE GLOBAL DIGITAL SEISMOGRAPH NETWORK: A STATUS REPORT

Jon Peterson

U.S. Geological Survey

ABSTRACT

A global, digital seismograph network is being established
to provide high-quality digital data for seismological research.
It comprises several different types of seismograph systems plus
data management and distribution facilities. Most of the Seismic
Research Observatories, which are major elements of the digital
network, have been operational for several years. Design objectives have been met for these observatories, but there are limits
to the application of the data for analysis of large signals. The
capabilities of the Seismic Research Observatories are not fully
utilized at present and the potential exists to record additional
data including broadband and very-long-period signals. Digital
recorders are being installed at selected WWSSN stations to augment the geographical coverage of the existing digital network.
They will also extend the capacity of the network to record large-
amplitude signals. Data collected from the digital network are
organized onto network-day tapes. These data files contain station
and calibration information as well as waveform data.

INTRODUCTION

The Global Digital Seismograph Network (GDSN) is the term
used by the U.S. Geological Survey (USGS) to describe the worldwide network of digital-recording observatories shown in Figure 1.
It comprises several different types of instrumentation -- the
Seismic Research Observatories (SRO), modified high-gain, long-
period (ASRO) stations, and the digital World-Wide Standardized
Seismograph Network (DWWSSN). There are 12 SRO and 5 ASRO stations

*E. S. Husebye and S. Mykkeltveit (eds.), Identification of Seismic Sources - Earthquake or Underground
Explosion, 747–761.*

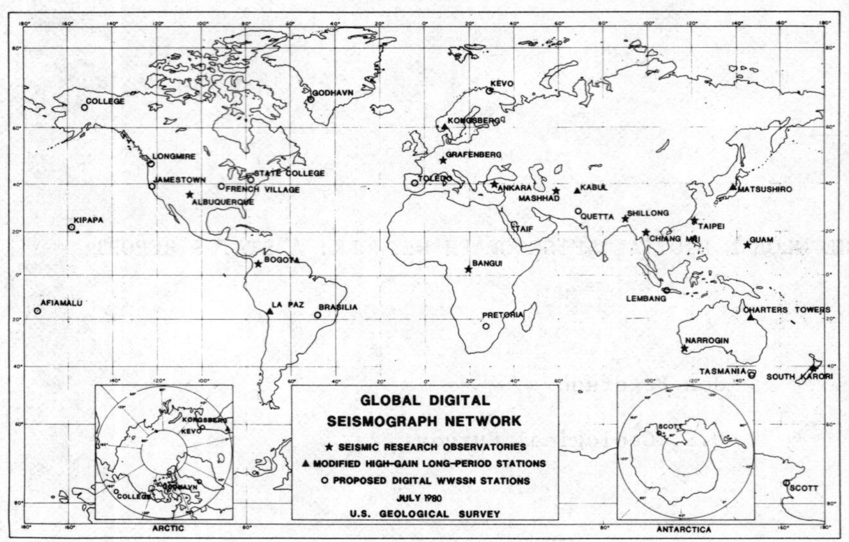

Figure 1. Planned distribution of digital-recording seismographs.

in operation and plans for 1 additional SRO station and 17 DWWSSN
stations. An important element of the network not shown on the map
in Figure 1 is the data processing facility used to merge the digital data into user-compatible network-day tapes.

The fundamental purpose of the GDSN has been to provide high-
quality digital data for seismological research from earthquake
and explosion sources. The purpose did not change when the USGS
assumed management responsibility for the network several years
ago from the Defense Advanced Research Projects Agency. The USGS
does place a high priority on making the network data available
and useful for a wide range of research objectives. With this in
mind, ways to modify or augment the existing data systems to improve the general applicability of the GDSN data base will be
continually reviewed.

In this paper each element of the Global Digital Seismograph
Network will be described briefly, without dwelling on technical
details, but rather with the intention of characterizing the design objectives, the potential, and the limitations of the data
systems. The methods used to collect, process, and distribute the
network data will also be described.

SEISMIC RESEARCH OBSERVATORIES

The SRO network has been operational for several years. In fact, one station has been operating for more than 6 years. Both the concept and a technical description of the Seismic Research Observatory have been fully documented (1,2). Nevertheless, it will be useful in this forum to review the major features of the SRO systems because they are the centerpiece of the digital network and they have the potential for producing additional data over an extended bandwidth.

The principal design objective in developing the Seismic Research Observatory was to lower the surface-wave detection threshold in the 20- to 40-second-period band. This was achieved using a newly developed borehole seismometer (Figure 2) that could be installed deeply enough in rock to avoid most of the surface noise

Figure 2. SRO borehole seismometer being prepared for installation.

generated by wind. The borehole seismometer is broadband in the
sense that the fundamental signals are generated in the feedback
loop of force-balance accelerometers and they are proportional to
mass position (or earth acceleration) from 0 to about 1 Hz. The
mass-position signals from each component are filtered and ampli-
fied to produce the "data outputs." The amplitude response of a
data output signal is approximately flat relative to earth acceleration from 1 to 50 second periods (see Figure 3). The three data-

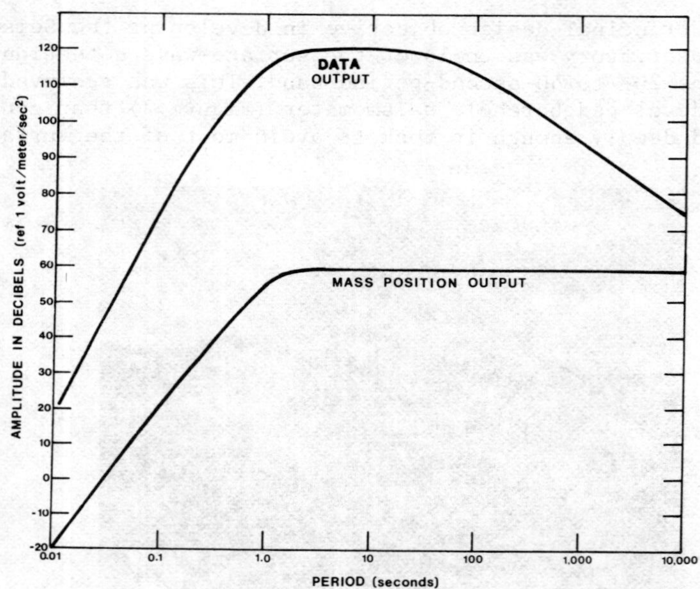

Figure 3. Relative amplitude response of data-
output channel and mass-position channel to an
input of earth acceleration.

output signals and one mass-position signal, which may be manually
switched between the three components, are cabled to a wellhead
terminal at the surface. In the wellhead terminal the data-output
signals are further filtered to produce two separate signals for
recording: a short-period signal (vertical-component only) and
long-period signals. Displacement response curves for these two
bands are shown in Figure 4 and typically recorded signals are
shown in Figure 5. An example of a recording made from the mass-
position channel is shown in Figure 6. This signal is not currently
being recorded but the example does illustrate the potential for

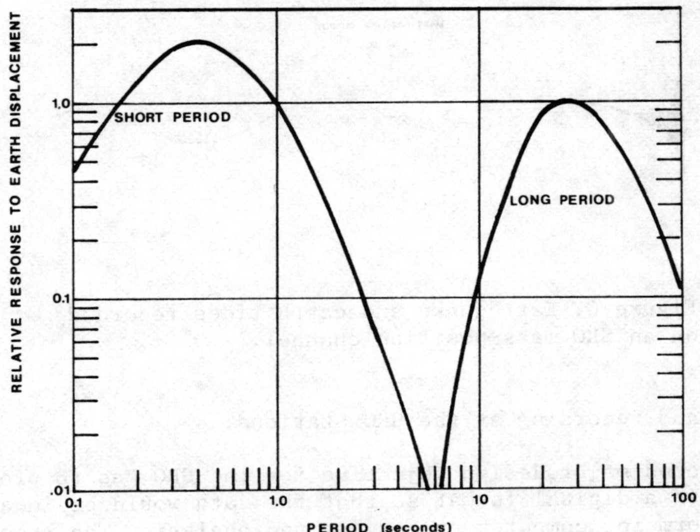

Figure 4. Relative amplitude response of SRO short-period and long-period channels to inputs of earth displacement.

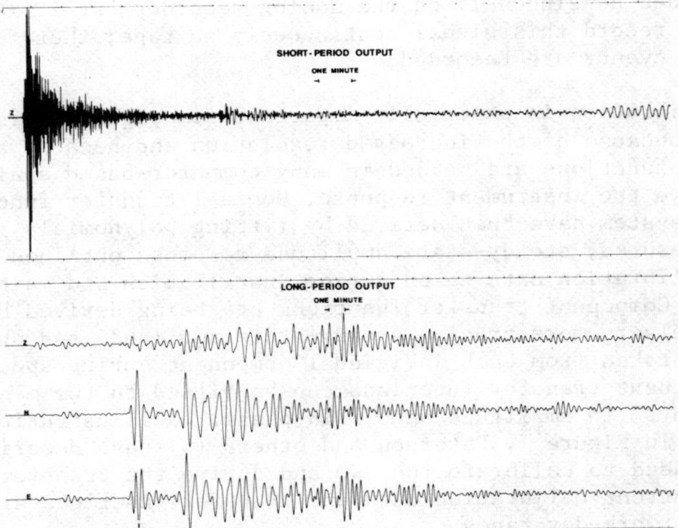

Figure 5. Typical short- and long-period signals recorded at an SRO station.

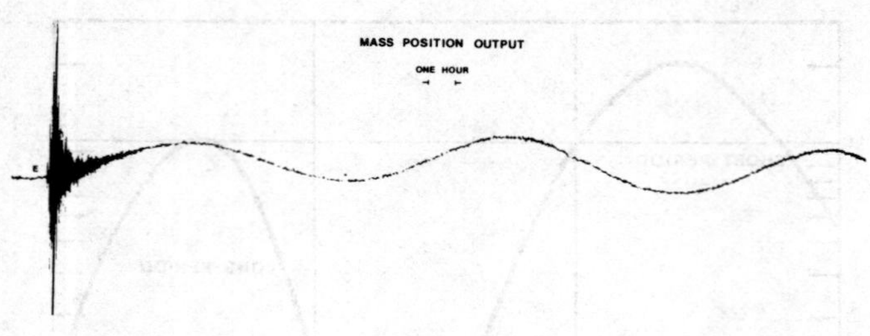

Figure 6. Earthquake and earth tides recorded
on an SRO mass-position channel.

broader-band recording at the SRO stations.

A second major design objective for the SRO was to provide
the data in a digital format so that the data would be in a convenient form for computer processing and analysis. The recording
system developed for the Seismic Research Observatory provides for
both analog and digital recording. The digital data are recorded
in 2,000 byte records on magnetic tape drives. The 16-bit gain-
ranged digital data word provides a signal recording range of 126
dB. Long-period signals are recorded continuously on the analog
recorders and on tape. The vertical-component, short-period signal
is recorded continuously on the analog recorder but it is impractical to record this signal continuously on tape; therefore, only
detected events are recorded.

Calibration is especially important for digital seismograph
systems because of the increased resolution and because accurate
transfer functions are needed in many computer-based analyses to
deconvolve the instrument response. Nominal transfer functions for
the SRO system have been derived by fitting polynomials to the
mean, measured, steady-state amplitude response obtained by averaging calibration data taken during installation and maintenance
testing. Component transfer functions are being derived by fitting
polynomials to more precise steady-state amplitude and phase measurements taken from each individual component during special tests.
The component transfer functions can be fitted to the measured data
to within 2% in amplitude and 2° in phase. An illustration of this
is shown in Figure 7. Peterson and others (3) have described the
methods used to calibrate the SRO and derive the transfer functions. Current calibration data and transfer functions are listed
on the network-day tapes.

Linearity is another subject that receives close attention in
digital system analysis. The increased signal amplitudes that can
be accommodated in a digital system places severe demands on sys-

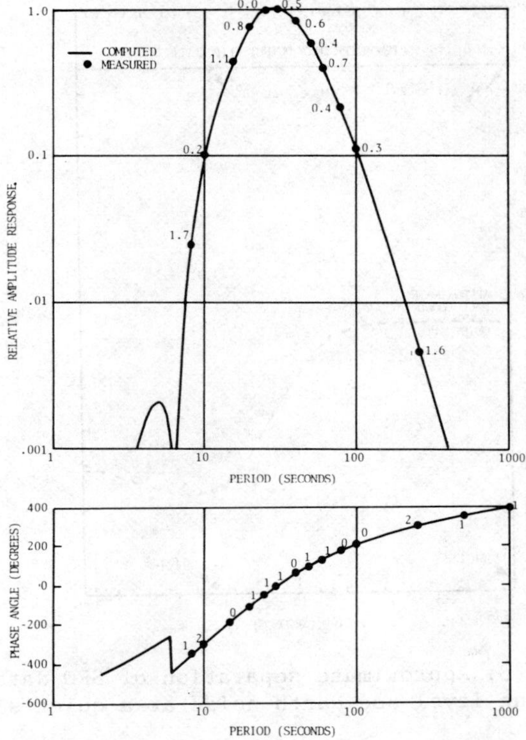

Figure 7. Computed and measured amplitude and phase
responses of an SRO long-period channel for an input
of earth displacement. Differences between computed
and measured points are shown as percent of deviation
for amplitude and degrees of deviation for phase.

tem linearity. All measuring systems become nonlinear at some
point and the SRO system is no exception. Nonlinearities in an
instrument cause distorted signals, sometimes obvious as in the
case of truncated signals when the clipping threshold has been
exceeded, and sometimes less obvious as a result of the spurious
signals generated by intermodulation distortion. Tests have shown
that intermodulation distortion is not a significant problem within the normal operating range of the SRO system (3). A more serious problem that confronts SRO data users is the relatively low
clipping threshold. Unfortunately, the clipping threshold cannot
be raised without affecting the ability to resolve small signals
in the long-period band. This is illustrated in Figure 8. The SRO
system was optimized for recording long-period signals and the
full recording range is needed in this band. Because of this

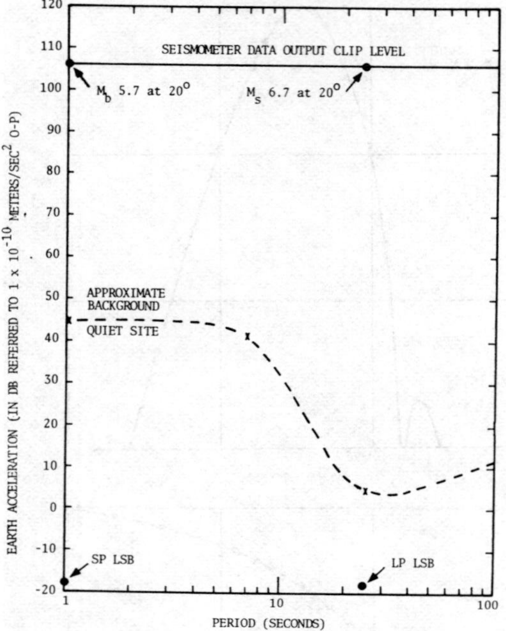

Figure 8. Approximate separation of SRO data-output
clipping level and earth noise at a quiet site.

requirement, the range between background and clipping levels is
compressed at shorter periods. Clipping generally occurs in the
data-output line drivers, which are ahead of the bandpass filters.
An unfortunate effect of this is that clipped signals are smoothed
by the bandpass filters and may be mistaken for legitimate data.
In order to alert users to this possibility, the SRO systems will
be modified so that data are flagged when they are out of range.

Although the Seismic Research Observatory is not well suited
for recording strong motion, it is unexcelled in resolving small
surface-wave signals, which is its major intended purpose. Over
the years the problem in improving long-period signal detection
has been to reduce instrumental noise and noise from environmental
sources (temperature, pressure, wind) sufficiently so that both
vertical and horizontal components of earth background noise could
be resolved. Figure 9 illustrates the relative levels of noise
measured during evaluation tests of the SRO system during a quiet
period when earthquake signals were not observed. The components
of noise (earth, instrumental, environmental) were separated from
total noise by measuring the coherence between signals derived
from parallel instruments (3). An earthquake signal is shown for
reference. Earth noise in the long-period band does not vary sig-

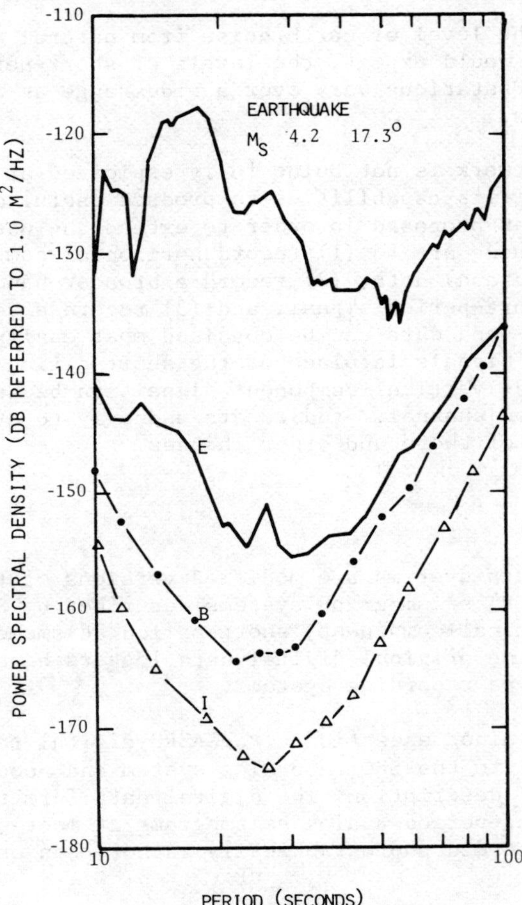

Figure 9. Approximate levels of noise power for an
SRO long-period channel. The curve labeled E represents earth noise, the curve labeled B represents
environmental noise, and the curve labeled I represents instrumental noise.

nificantly from site to site. This was a conclusion of Murphy and
others (4) based on their observations of high-gain, long-period
station data. Observations of noise spectra obtained from the SRO
stations supports their conclusion as levels do not vary in power
by more than 10 dB in the 20- to 40-second-period band (5).

There are no significant differences between the SRO short-
period data and short-period data obtained from conventional seismometers. Signal resolution in the short-period band depends

entirely upon the level of earth noise from natural and cultural sources. As one would expect, the levels of short-period earth noise at the SRO stations vary over a wide range as a function of station location.

The SRO network is not being fully exploited at the present time in terms of its capabilities to produce useful data. Three changes have been proposed in order to extend the usefulness of the SRO data. These are to (1) record horizontal-component, short-period (or broadband) data, (2) record a broader-band signal in place of the short-period signal, and (3) record a very-long-period signal. Broader-band data can be obtained most easily by recording the data-output signals in place of the short-period signals. A very-long-period, vertical-component signal can be derived from the mass-position channel. Studies are underway to evaluate the cost and impact of these and other changes.

ASRO SYSTEMS

The ASRO data systems are modified versions of the high-gain long-period (HGLP) seismograph systems described by Savino and others (6). Vertical-component, short-period seismometers have been added and the original digital data loggers have been replaced by SRO-type recording systems.

With a few minor exceptions, the ASRO digital recording system is identical to the SRO recording system and needs no additional technical description. The digital data format is identical as well. The long-period analog seismograms at most of the ASRO stations are recorded photographically rather than on visual recorders.

The major difference between the ASRO and SRO systems is that conventional seismometers are used in the ASRO system rather than the borehole seismometer. Therefore, the operating environment is not as well controlled, although the long-period seismometers are installed in airtight tanks and most of the systems are located underground in well-constructed vaults. Long-period noise spectra obtained from the ASRO stations show more variation between components as a result of environmental noise sources, but in general noise levels during quiet periods are not significantly different from noise levels measured at the SRO stations (5). It is during windy periods that the near-surface ASRO stations will show an appreciable rise in noise level. There are no discernible differences between ASRO and SRO short-period data. The ASRO stations are producing excellent digital data and they are a very useful supplement to the SRO network.

DWWSSN SYSTEMS

The original concept leading to the digital WWSSN was to use
microprocessor-based technology to develop and install relatively
simple, low-cost digital recorders at a number of WWSSN stations
as an economical way of augmenting the digital data base. The
technical requirements became more demanding as development pro-
ceeded, and the digital recorders have become more complex and
expensive as a result. However, the potential value of the data
has been enhanced as well. Installation of the digital recorders
has just begun.

A WWSSN digital recorder (shown in Figure 10) consists of a

Figure 10. A DWWSSN recording console.

16-bit, fixed-point analog-to-digital converter, microprocessor
controls, and a single magnetic tape drive. The analog recorder
is furnished so that any of the digital channels may be monitored

during operation. A time code is obtained from a digital time en-
coder that is installed in place of the electronic programmer in
the WWSSN timing console. Amplifiers and filters are furnished to
boost and condition seismic signals derived from the existing
WWSSN seismometers.

Three bands are recorded digitally -- short, long, and intermediate period (see Figure 11). The three long-period signals are

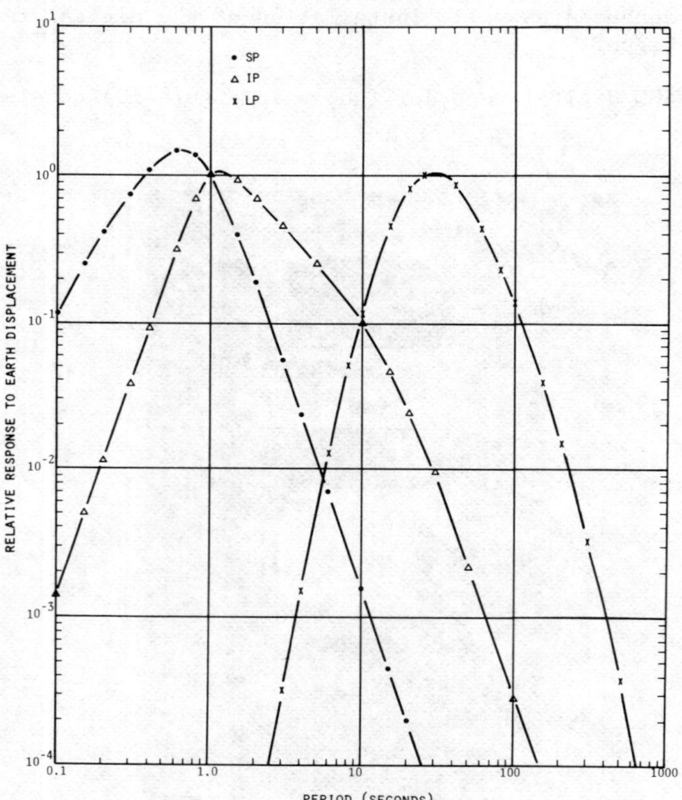

Figure 11. Relative amplitude response curves for
the DWWSSN short-, intermediate- and long-period
recording channels.

recorded continuously. A vertical-component, short-period signal
is recorded on an event-only basis. Three intermediate-period signals are derived from the long-period seismometers and also recorded on an event-only basis. The sensitivity of the intermediate-

period channel is set low enough so that signals from large earthquakes (M_S 8 earthquakes at 20°) do not overdrive the recorder.
An example of body waves from a large earthquake recorded on an
intermediate-period channel is shown in Figure 12. The signal
amplitude is about one fourth of the clipping level.

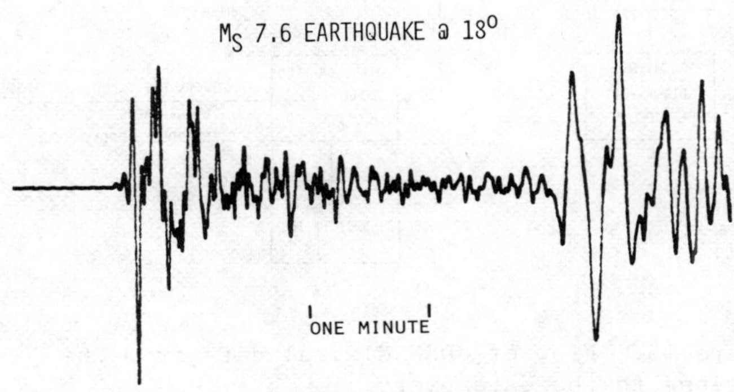

Figure 12. Body waves from a large earthquake
recorded on a DWWSSN intermediate-period channel.

GDSN DATA MANAGEMENT

The Global Digital Seismograph Network is operated to provide standardized seismic data to the entire scientific community.
Hence, the procedures used to process and distribute the data are
a very important part of the program. The flow of digital data
from the stations to the external data users is shown in Figure 13.

The station tapes generally contain about 2 weeks of data.
They are mailed, together with the analog seismograms, to the USGS
Albuquerque Seismological Laboratory. At Albuquerque, the tapes
are first processed by a review and edit program, then the data
are placed in a large disk file that holds about 15 days of network data. Automated data review procedures include the reduction
of calibration data, detection and correction of errors in timing,
parity, and format, listing of dropouts and outages, and other
quality-control operations. When data from all stations are in the
disk for overlapping periods of time, the data for successive days
are read from the disk and organized onto network-day tapes. The
length of time between data recording and the compilation of the
day tape is about 8 weeks.

Copies of the network-day tapes are furnished to the USGS in
Golden, Colorado, to the National Oceanic and Atmospheric Admin-

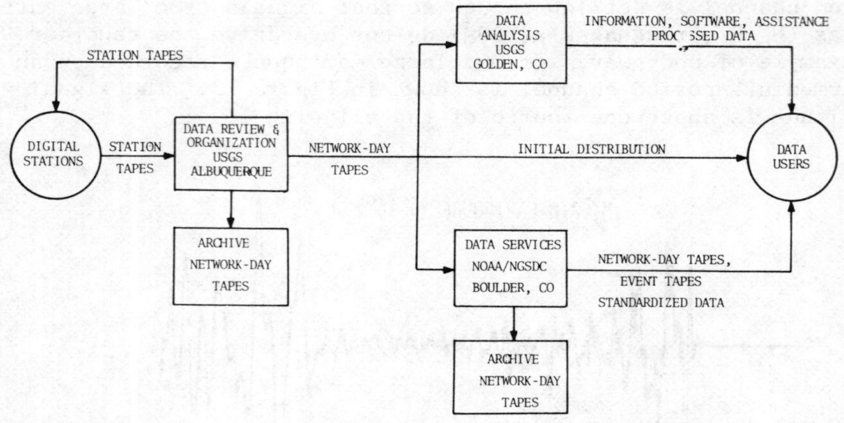

Figure 13. Flow of GDSN digital data from the
stations to the data users.

istration (NOAA) in Boulder, Colorado, and to several other organizations. The USGS group in Golden uses the digital data for in-
house research, but it also has the responsibility for developing
transportable software to read and process the day tapes and this
software package, when completed, will be made available on request. Data users normally obtain copies of the network-day tapes
through the NOAA data center in Boulder.

The network-day tape is the principal end product of the
GDSN. Its format has been carefully designed to provide all of
the information needed by an analyst to use and process the digital data. Logs on the tape list station parameters, calibration
data, including transfer functions, timing corrections, periods
of outages, and other useful information. The format used to compile the network-day tapes is described by Hoffman (7).

REFERENCES

1. Peterson, J. and Orsini, N.A.: 1976, *Seismic research observatories: upgrading the worldwide seismic data network*, EOS Trans.
   AGU, 57, pp.548-556.
2. Peterson, J., Butler, H.M., Holcomb, L.G., and Hutt, C.R.: 1976,
   The seismic research observatory, Bull. Seism. Soc. Am., 62,
   pp. 2049-2068.
3. Peterson, J., Hutt, C.R., and Holcomb, L.G.: 1980, *Test and
   calibration of the seismic research observatory*, USGS Open-File
   Report 80-187.

4. Murphy, A.J., Savino, J., Rynn, J.M.W., Choy, G.L., and
   McCamy, K.: 1972, *Observations of long-period (10-100 sec)
   seismic noise at several worldwide locations*, J. Geophys. Res.,
   77, pp. 5042-5049.
5. Peterson, J.: 1980, *Preliminary observations of noise spectra
   at the SRO and ASRO stations*, USGS Open-File Report
6. Savino, J., Murphy, A.J., Rynn, J.M.W., Latham, R., Sykes, L.R.,
   Choy, G.L., and McCamy, K.: 1972, *Results from the high-gain
   long-period seismograph experiment*, Geophys. J., 31, pp. 179-203
7. Hoffman, J.: 1980, *The global digital seismograph network-day
   tape*, USGS Open-File Report 80-289.

BROAD-BAND SEISMOMETRY - A UNIFIED APPROACH TOWARDS A KINEMATIC AND DYNAMIC INTERPRETATION OF SEISMOGRAMS

H.-P. Harjes

Institute of Geophysics
Ruhr University, D-4630 Bochum

An important aspect of a global seismic system for identification of seismic sources concerns the standardisation of seismographs on a modern level. The development of digital broad-band seismographs is a successful technical approach. A broad-band seismogram is an overall picture of the seismic wave field at the recording station. The high dynamic range of digital data acquisition allows to focus upon special features by digital filters. These filters can simulate arbitrary seismographs. Thereby an old dispute among seismologists on the best instrument may after all be solved. There is no longer a need for restriction to special instruments which would make standardization more difficult because different instruments have been found optimal by different seismological groups in the world. The lecture describes the seismological contraints and the technical realization of digital broad-band seismographs at the Central Seismological Observatory Graefenberg (GRF). Using data from this instrumentation the filter effect of conventional narrow-band seismographs will be demonstrated. These seismographs filter the original broad-band seismic signal and reduce its information content. The loss of information is irreversible and depends on bandwidth, resolution and instrumental noise.

1.   INTRODUCTION

An important technical aspect of a global seismic data exchange system for discrimination of seismic events concerns the standardization of seismographs on a high-quality level. When discussing seismographs it is important to consider the frequency and amplitude characteristics of seismic signals.

E. S. Husebye and S. Mykkeltveit (eds.), Identification of Seismic Sources - Earthquake or Underground Explosion, 763–786.

The bandwidth of teleseismic signals covers a range from
0.01 Hz to 5 Hz, the actual width depending to some extent on the
size of the event and on the transmission properties of the earth
between the source and the recording station. The dynamic range
of teleseismic signals ranges from around 1 nm for P-wave signals
from weak events to more than 10 mm for long-period surface waves
from strong events. Because of this large amplitude range conventional seismic records are usually split into separate short-
period and long-period frequency bands obtained at different sensitivities.

Short-period instruments have their maximum sensitivity at
frequencies of about 1 Hz. Long-period seismographs have their
maximum sensitivity at frequencies of about 0.05 Hz and show surface waves most clearly which is valuable for the identification
of earthquakes and explosions.

In the sixties a world-wide seismograph system using these
short-period and long-period instruments was installed by the
United States. Until now this network produced the basic data
for seismological research in many countries. Whereas these band-
limited seismograms solve best the detection problem (on-sets of
signals) and the location problem (travel times), they suppress
most of the information in the spectral range between the short-
period (around 1 s) and long-period (20 s) window. This information, however, is highly desirable for the identification of seismic signals. Therefore the concept of broad-band seismometry,
i.e., one instrument covering the whole frequency band from 5 Hz
to 0.01 Hz, has repeatedly been discussed. Broad-band instruments
have so far been used almost entirely within the USSR and the East
European Countries, but interest in such instruments for the purpose of the seismic monitoring system has also been shown by the
United Kingdom for several years (CCD/401, 1973[1]).The difference
in instrumentation is not only a technical problem, it also resulted in significant discrepancies for important seismic parameters. It is, e.g., well known that the magnitude of seismic
events determined from broad-band recordings is somewhat higher
than from short-period instruments.

The main disadvantage of broad-band recordings in the past
was their insufficient dynamic range. Due to the large seismic
background noise (around 6s - 8s) analog recording techniques
offer no possibility to escape the trade-off between low detection
level and broad-band recording. In order to overcome this disadvantage and to use the benefits of broad-band characteristics for
seismological research, digital data acquisition is the appropriate concept.

A digital broad-band array has been established at the Central
Seismological Observatory Graefenberg (GRF) in Southern Germany.
The full 13-element-array (Fig. 1) became operational in April
1980, although continuous recordings of the first subarray are
available since 1976 (Harjes, et al. 1980[2]). Using data from
this array it will be shown that such instruments offer a raw

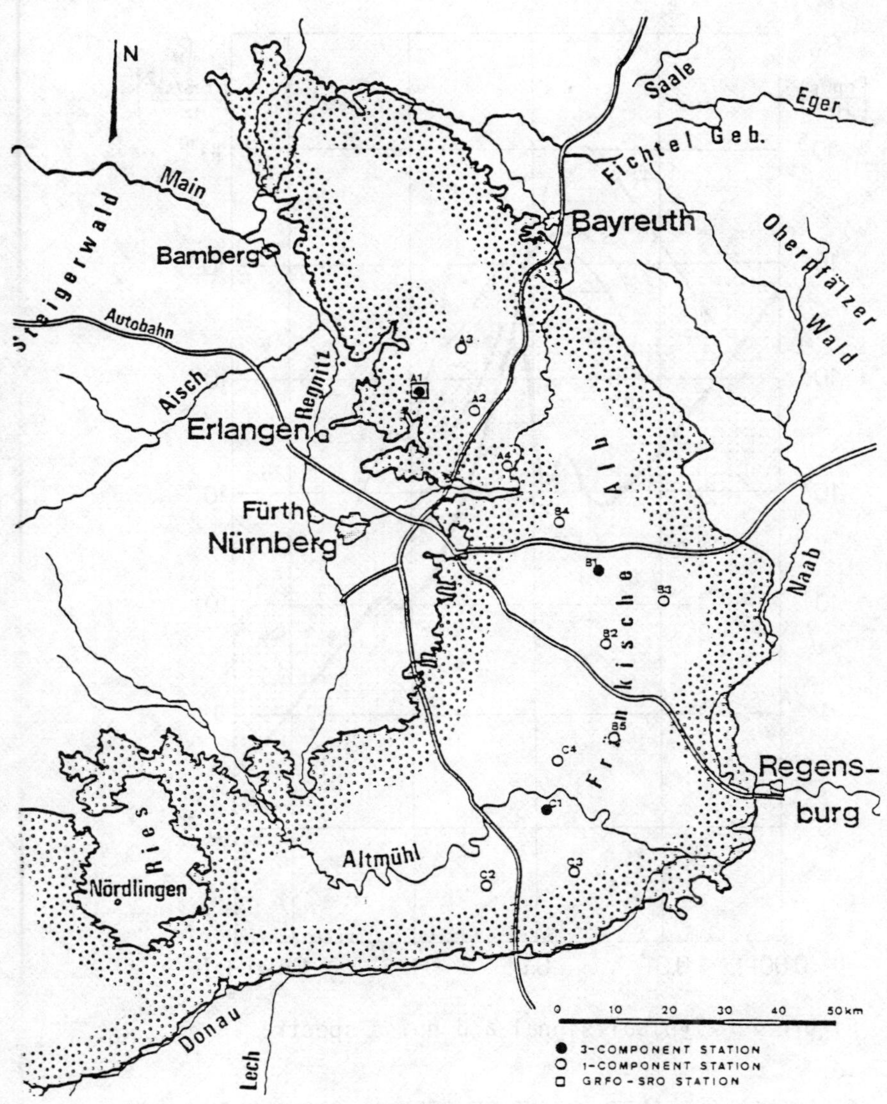

Figure 1. Location of GRF-array stations

data base from which different special instrumental characteristics, optimal for detection and/or identification of seismic
sources, can be simulated by numerical methods.

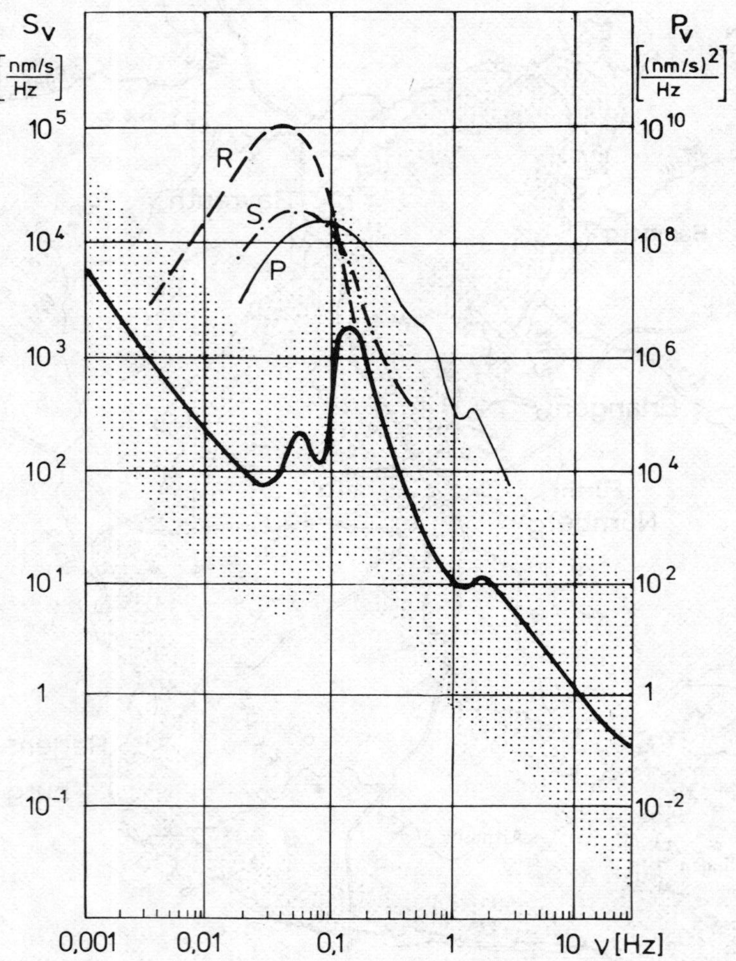

Figure 2. Typical signal and noise spectra

2.   FREQUENCY CHARACTERISTICS OF SEISMIC SIGNALS AND NOISE

The spectral content of seismic waves from earthquakes, like
P-, S-, or surface waves is generally well known (Fig. 2). Velocity energy spectra of body and surface waves increase with decreasing frequency above the so called corner frequency where they
reach their maximum. At lower frequencies the spectral curves fall
down again.
For shallow earthquakes the spectral amplitudes of surface
waves are larger than those of P- and S-waves. Besides surface

wave spectra generally show the steepest slope. The position of
the spectral maximum mainly depends on the magnitude of the event,
less on the epicentral distance. With decreasing magnitude the
spectral maximum is shifted towards increasing frequency.

The maximum of P-waves for a magnitude 7 event is located
between 10 sec and 20 sec (Fig. 2) compared to 1 sec to 2 sec for
a magnitude 4 event. Fig. 2 also shows the typical trend of the
S-wave spectrum and Rayleigh wave spectrum of a magnitude 7 event.

In contrast to the deterministic nature of seismic signals
the seismic noise can be described by a stochastic process. The
noise power spectrum shows a large spatial and temporal variation
which is indicated by the dotted area in Fig. 2. Similar to signal spectra increase the power spectra of seismic noise with decreasing frequency. The high frequency noise above 1 Hz originates from traffic and industry near to the seismometer site. This
noise level is strongly varying from site to site and it forms a
severe problem in a densely settled area like Western Europe.

This local noise is followed by the microseismic peak at
lower frequencies in the power spectrum with a maximum between
0.17 Hz and 0.12 Hz. A second maximum appears between 0.1 Hz and
0.05 Hz. The microseismic noise, which mainly contains Rayleigh
waves, is characterized by a large energy content and a strong
spatial coherency. Fig. 2 shows a relative spectral minimum at
frequencies between 0.025 Hz and 0.005 Hz, before the power spectrum increases for very low frequencies again. The low frequency
noise results from variations in atmospheric pressure and also
from instrumental effects. Different stationarity conditions are
given for high frequency industrial noise and microseisms. Whereas the first can be regarded as stationary in time intervals of the
order of minutes, the second remains stationary within hours. Because of meteorological conditions there can however appear large
variations of the microseismic noise within days. Fig. 3 shows the
power spectra of two noise samples of GRF-data which differ by
40 db at about 0.12 Hz to 0.17 Hz. These broad-band noise spectra
were computed by the periodogram method using 10 nonoverlapping
segments each of 400 seconds length (90 % confidence limit is 7 db).

From the shape of the spectral curves of earthquake signals
and earth noise there result two windows with optimal signal to
noise ratio around 1 Hz and 0.05 Hz, respectively. Conventional
analog seismographs record in these frequency bands. Because of
the limited dynamic range of about 60 db of analog seismographs
different systems are needed for the high frequency and low frequency window. These seismographs record with different sensitivities to prewhiten the noise. As an example Fig. 4 shows the velocity response curves of the WWSSN-system. The corresponding
maximum magnifications are 1 K for the low frequency and 200 K for
the high frequency system. The shaded area indicates the dynamic
range for direct recording. Low resolution for both systems in
the medium frequency range - due to the small overlapping area -
strongly limits the deconvolution of instrumental effects. The

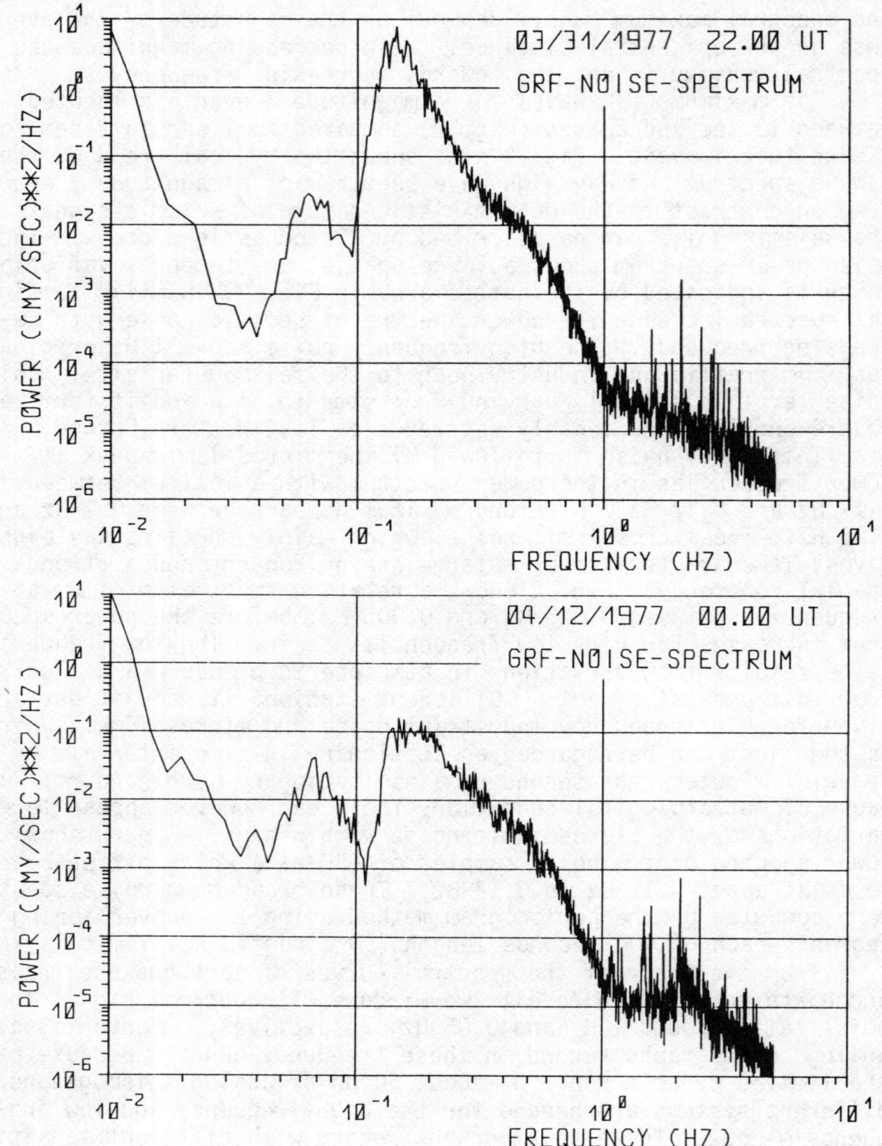

Figure 3. Temporal variations of noise power spectra at GRF

restitution problem becomes even more difficult when calibration
errors are taken into account. Fig. 5 shows the "true" noise

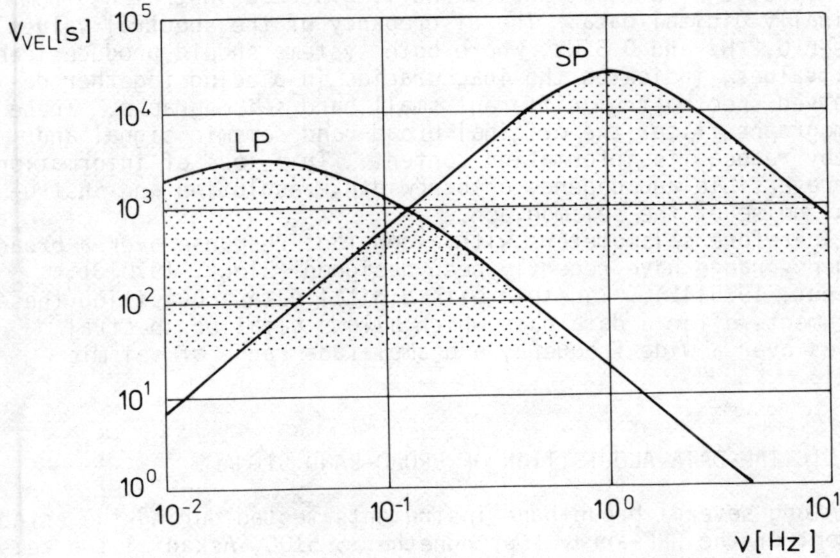

Figure 4. Velocity response characteristics and dynamic range of World Wide Standard Seismographs

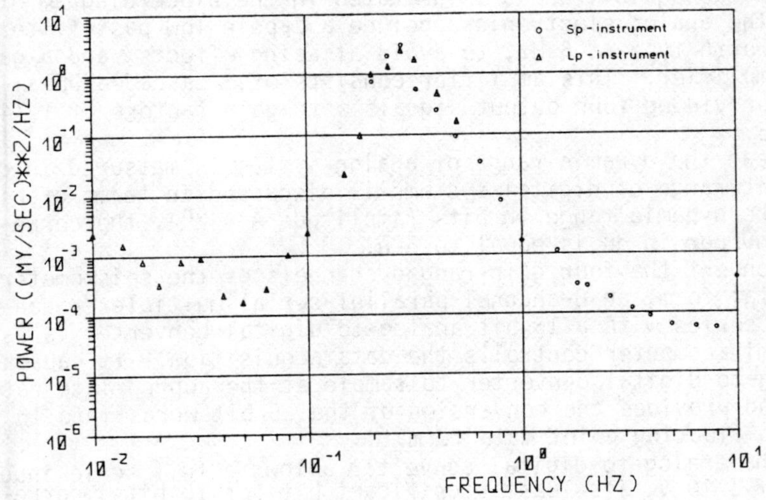

Figure 5. Microseismic noise spectrum from short-period and long-period seismograph

power spectra of a short period and long period instrument from
originally digital data. The discrepancy of the spectral values
between 0.2 Hz and 0.5 Hz, where both systems should produce identical values, indicates the inaccuracies in piecing together deconvolved spectra from different small band seismographs. These
seismographs filter the original broad-band seismic signal and
thereby reduce its information content. This loss of information
is irreversible and depends on bandwidth, resolution and instrumental noise of the seismograph.

Therefore seismometers with a constant response over a broad
frequency range have recently been designed (Block, 1970[3];
Wielandt, 1975[4]). Together with digital data acquisition these
instruments offer a data base which allows detailed spectral
studies over a wide frequency and amplitude range of seismic
signals.

3.   DIGITAL DATA ACQUISITION OF BROAD-BAND SIGNALS

Among several broad-band instruments tested during the establishment of the GRF-array (Sprengnether S-5100, Askania) the seismometer designed by Wielandt was finally installed. This seismometer is operated according to the force-balance principle. The
broad-band output produced by a feed-back circuit covers a dynamic
range of more than 140 dB. It is proportional to ground velocity
at frequencies above 0.05 Hz and falls down below the corner-fre-
quency with 12 dB/octave as shown in Fig. 6.
The data acquisition is illustrated in the block diagram of
Fig. 7. The analog electronics include a 7-pole low-pass filter
of Butterworth type at 5 Hz, to avoid aliasing effects, and a gain-
ranging amplifier. This amplifier consists of 3 cascaded DC-am-
plifiers providing four output signals with gain factors of 1, 4,
32 and 256.
Whereas the dynamic range of analog systems is measured in dB,
the dynamic range of digital systems is discussed in terms of bits.
If n is the dynamic range in bits (amplitude $A = 2^n$), the corresponding number in dB is equal to 6 n.
To convert the four gain-ranged channels of the seismometer
into digital form, an 8-channel parallel-serial multiplexer connected in series with a 15 bit analog-to-digital converter is used.
A 16-bit minicomputer controlls the data acquisition. It causes
the analog-to-digital converter to sample at the appropriate rates
(20 Hz) and provides the conversion of the 15 bit words into 16
bit pseudo-floating point data format.
As the analog-to-digital converter allows a full scale input
voltage of $\pm$ 10 V, its least significant bit (of 15 bits) corresponds to 0.61×10^{-3} V, which is in turn the smallest resolvable
voltage. For the highest amplification level the least significant bit equals a seismometer output voltage of 2.38×10^{-6} V.
So the ratio between the maximum and minimum voltage expressed in

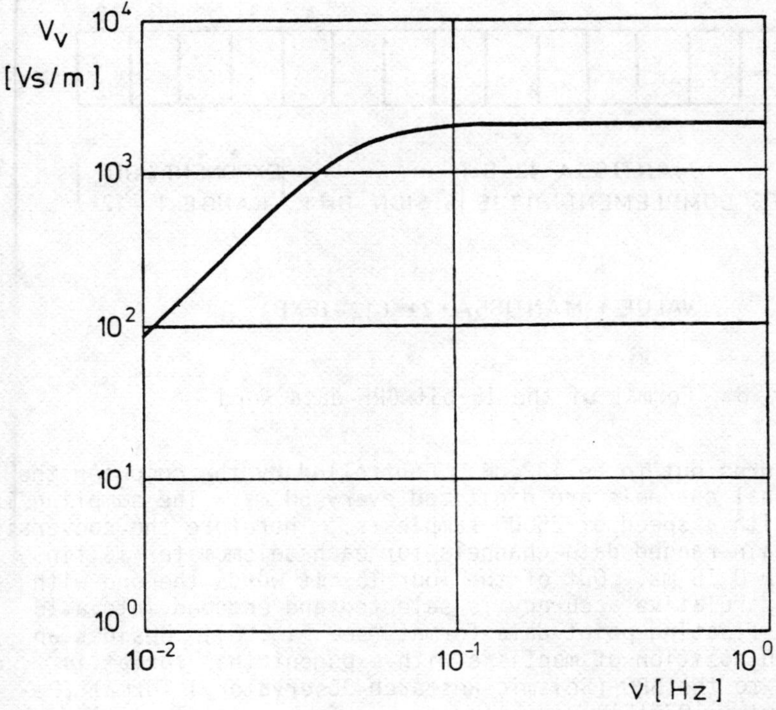

Figure 6. Transfer function of the Wielandt-broad-band seismometer

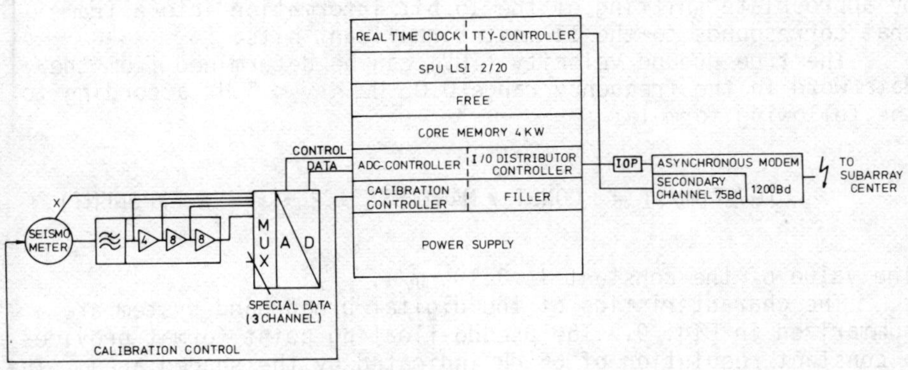

Figure 7. GRF data acquisition system

$$\text{VALUE} = \text{MANTISSA} * 2 ** (12 - \text{EXP})$$

Figure 8. Format of the 16-bit GRF data word

decibels turns out to be 132 dB. Controlled by the computer the
analog signal channels are digitized every 50 ms. The sampling is
executed with a speed of 25000 samples/s. Therefore the conversion
of the 4 gain-ranged data channels for each seismometer is fin-
ished after 0.16 ms. Out of the four 15 bit words the one with
the highest relative accuracy is selected and encoded into a 16
bit pseudo-floating point data format (see Fig. 8). Besides an
exchange in position of mantissa with exponent this format is
equivalent to the SRO-(Seismic Research Observatory) format (Peterson et al., 1976[5]).
     The most significant bit determines the sign followed by an
11 bit mantissa. The exponent, given by the four least significant bits, ranges between 1 and 12 in steps of 2. The representation of the exponent in powers of two instead of powers that
correspond to the amplification factors (see Fig. 7) is achieved
by appropriate shifting of the 15 bit information into a frame
that corresponds to the 11 most significant bits.
     The true ground velocity (TGV) can be determined from the
data word in the frequency range 0.05 Hz $< \nu <$ 5 Hz according to
the following formula:

$$\text{TGV [nm/s]} = \text{CONST} * \text{MANTISSA} * 2 ** (12-\text{EXPONENT})$$

The value of the constant is 1.19 nm/s.
     The characteristics of the digital broad-band system are
summarized in Fig. 9. The pseudo-floating point format provides
a constant resolution of 66 dB indicated by the shaded area. The
lowest and uppermost curves define the dynamic range (Harjes,
Seidl, 1978[6]). A simple comparison can explain the superiority
of the digital broad-band system to the conventional analog WWSSN-
system (Fig. 4) with photographic registration. Assuming that for

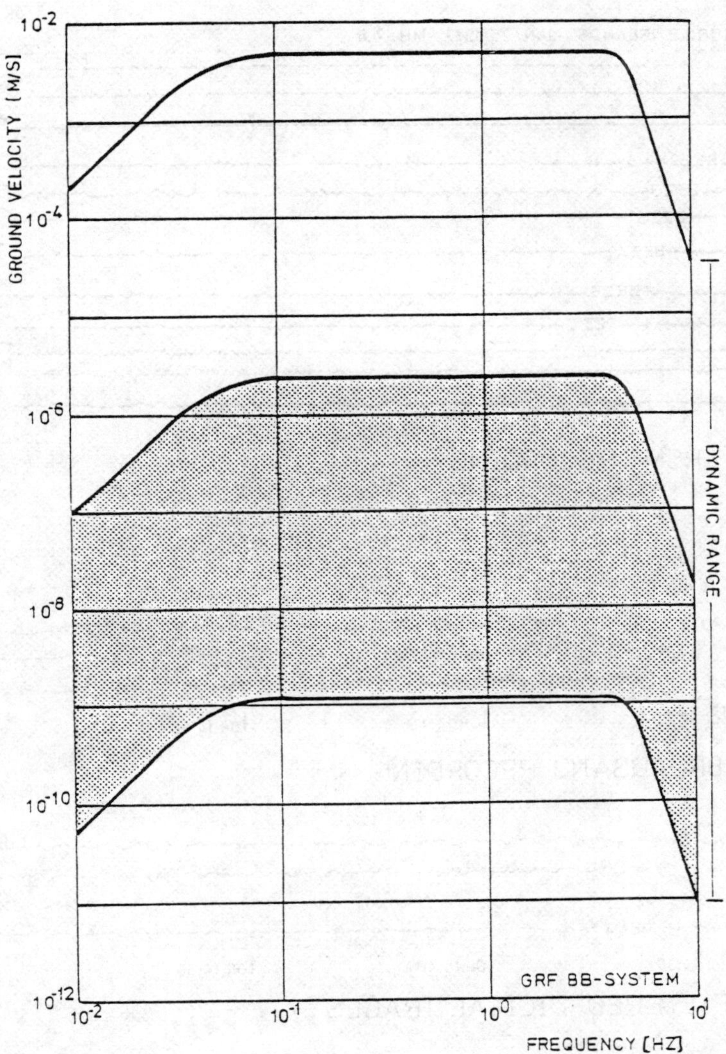

Figure 9. Velocity response characteristics, dynamic range,
and resolution (shaded area) of the GRF-broad-band system

the short-period instrument the smallest visible amplitude is
0.25 mm, this is equivalent to a ground velocity of about 9 nm/s
at a magnification of 200 K. In comparison the least significant
bit of the digital broad-band system represents about 1 nm/s which
means a higher sensitivity by about a factor of 10. Assuming

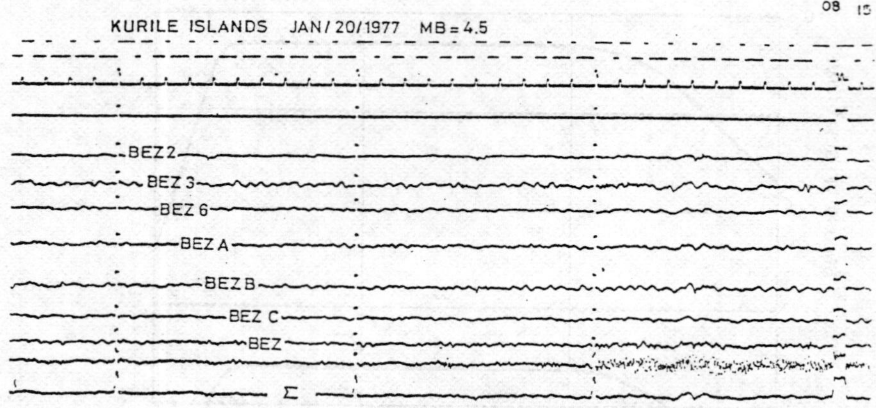

a) LRSM RECORDING

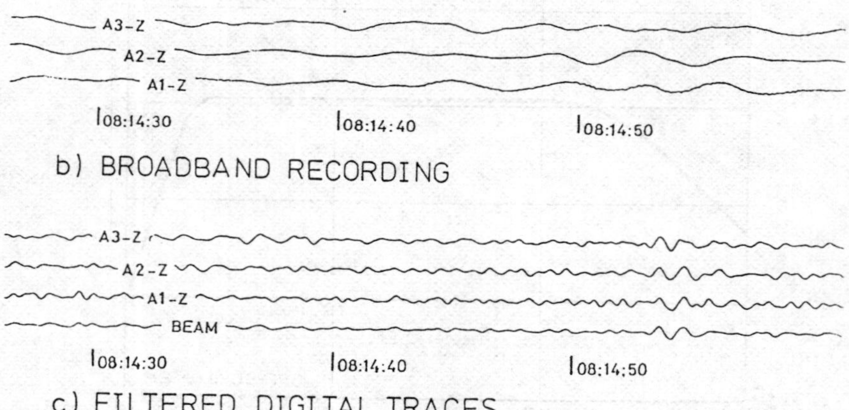

b) BROADBAND RECORDING

c) FILTERED DIGITAL TRACES

Figure 10 a.   LRSM-recording at GRF of a weak teleseismic
               event;
        b.   Digital broad-band recording;
        c.   Digital filtered traces and beam

similarly for the long-period film recording the largest amplitude
is 15 cm, this gives a saturation at 0.05 mm/s, which demonstrates
the well-known clipping problem of the long-period WWSSN-seismo-
graph. The digital broad-band system reaches its saturation at

about 1 cm/s (at 20 s), a factor of 200 above the long-period
WWSSN clipping point.

The superiority of a single digital broad-band instrument
compared to several small-band instruments does not only hold for
analog seismographs but also for digital instruments. One argument was already given at the end of the last paragraph with regards to the restitution problem. A more convincing reason will
be given in the next paragraph where different small-band seismograph recordings are simulated from the original broad-band output by recursive digital filters.
In this context the described gain-ranging data acquisition system,
using 11 bit for resolution, might appear rather limited, especially in a difficult detection environment where one might look
for weak explosions hidden in surface wave trains of large earthquakes. But also this difficulty can principally be solved by a
digital broad-band system using the full capability of a 15 (or
even more) bit analog to digital converter. The described GRF-
system was originally not designed for detection purposes but for
general seismological research, where 11 bit (or 66 dB) resolution
together with a dynamic range of 132 dB seemed to be sufficient.
These assumptions can be illustrated by examples of seismograms.

Fig. 10 shows a small Kuril-earthquake (m_b = 4.5) recorded
with a LRSM-array at GRF. (This instrumentation is no longer in
operation.) The event is near the detection level of this very
sensitive instrument (maximum magnification 200 K at 0.3 s, seismometer distance about 2 km). This small event is hidden in the
microseisms of the broad-band recording for the same time interval
(Fig. 10 b). By simple band-pass filtering this weak event (amplitudes about 4 nm/s) can easily be detected although it is
about 40 dB below the microseismic noise level.

On the opposite the largest amplitudes, recorded since the
installation of the broad-band instruments at GRF-observatory,
resulted from a seismogram of a Swabian Alb earthquake on Sept. 3,
1978 (Fig. 11). This largest instrumentally recorded earthquake
in the Federal Republic of Germany (M_L = 5.9, Δ = 220 km) reached
amplitudes of about 1 mm/s at the GRF observatory, which is more
than a factor of 10^5 above the recorded amplitudes of the Kuril
Islands event shown in the previous seismogram.

4.   SIMULATION OF SEISMOGRAPHS BY RECURSIVE FILTERS

A broad-band seismogram is an overall picture of the seismic
wave field at the recording station. The high dynamic range of
digital traces allows to focus upon special features by digital
filters. These filters can simulate arbitrary seismographs.
Therefore the installation of special instruments which optimize
the signal to noise ratio in small frequency bands is not necessary. The method was developed by Seidl[7], this paragraph
essentially refers to this work.

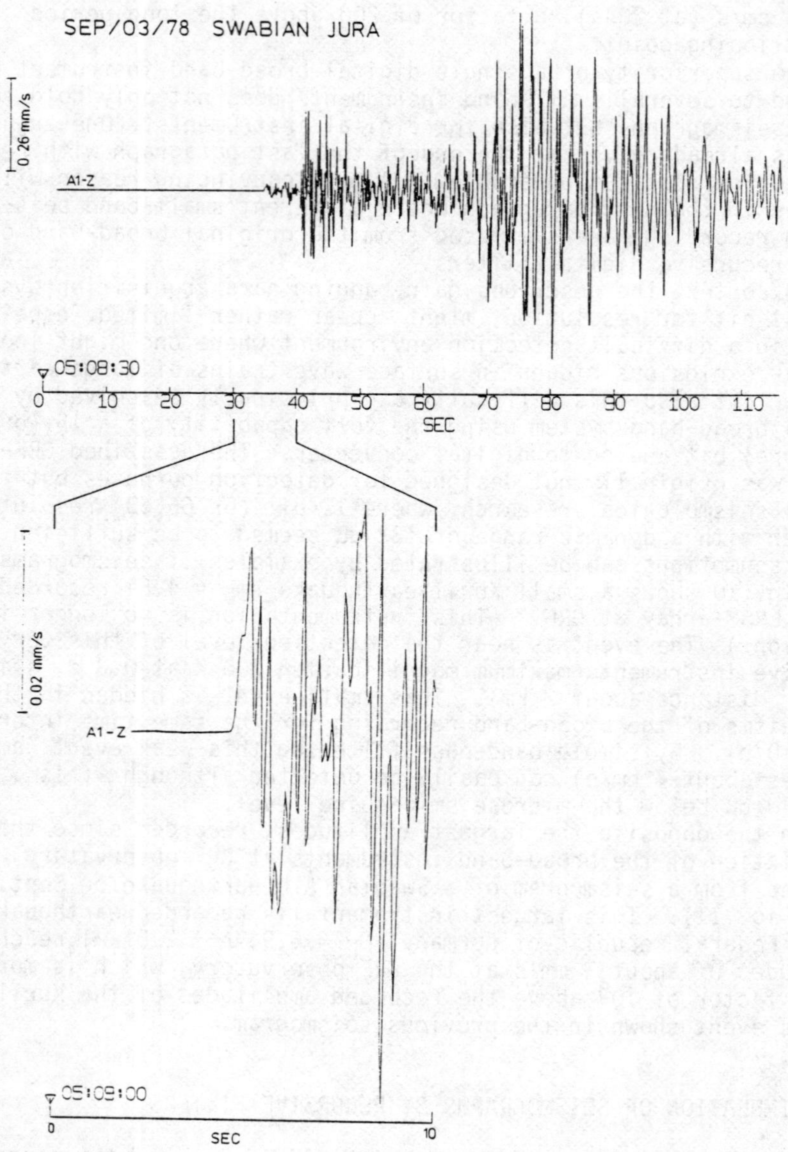

Figure 11. Dynamic range of GRF recording system demonstrated at a local earthquake of magnitude M_L = 5.9 (distance = 220 km)

Fig. 12 shows the flow diagram in input-output form of simulating special seismograms $U(\omega)$ for arbitrary seismometer-galvanometer combinations from a broad-band seismogram $X(\omega)$.

$B^{-1}(\omega)$ is the inverse transfer function of the broad-band seismometer, $S(\omega)$ and $G(\omega)$ are the transfer functions of the simulated seismometer and galvanometer, respectively. For the continuous system these transfer functions are found to be (Savarenski, Kirnos [8])

$$(1a) \qquad B^{-1}(\omega) = \frac{(i\omega)^2 + 2h_B\omega_B i\omega + \omega_B^2}{(i\omega)^3}$$

$$(1b) \qquad S(\omega) = \frac{(i\omega)^3}{(i\omega)^2 + 2h_S\omega_S i\omega + \omega_S^2}$$

$$(1c) \qquad G(\omega) = \frac{1}{(i\omega)^2 + 2h_G\omega_G i\omega + \omega_G^2}$$

where h_B, h_S, h_G and ω_B, ω_S, ω_G are the damping constants and eigen-frequencies of the broad-band seismometer, respectively the simulated seismometer and galvanometer.

From Fig. 12 an overall transfer function $Y(\omega)$ is derived

$$(2) \qquad Y(\omega) = \frac{S(\omega)}{B(\omega)} \cdot \frac{1}{G(\omega)}$$

A discrete description of the continuous transfer function is obtained by applying the z-transform (Oppenheim, Schafer [9])

$$(3) \qquad z = e^{-i\omega\Delta t}$$

where Δt is the sampling interval.
Substituting (3) for ω into (2)

$$(4) \qquad \omega = \frac{i}{\Delta t}\ln z$$

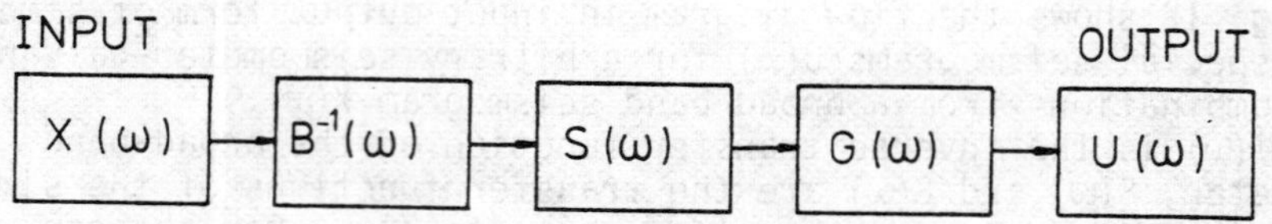

Figure 12. Digital simulation of a seismometer-galvanometer system

the transfer function Y is no quotient of polynomials in z and cannot easily be realized by digital filters. Therefore a first order approximation of the logarithm is used

$$(5) \qquad \widetilde{\omega} = \frac{2i}{\Delta t} \frac{z-1}{z+1}$$

Equation (5) defines the so-called bilinear z-transform which is ideally adapted to digital systems. Of course (5) is only an approximation of (4), therefore the bilinear z-transform produces a nonlinear distortion of the frequency scale. This distortion is small at low frequencies, but it strongly increases towards higher frequencies. The frequency distortion can be compensated for by a prewarping of the characteristic frequencies of the continuous system, i.e. a shifting of the zeros and poles of the overall transfer function Y.

The prewarped frequencies $\widetilde{\omega}_p$ can be found from the desired frequencies ω_d by substituting the exact expression (3) into the bilinear z-transform (5)

$$(6) \qquad \widetilde{\omega}_p = \frac{2i}{\Delta t} \frac{e^{-i\omega_d \Delta t} - 1}{e^{-i\omega_d \Delta t} + 1} = \frac{2}{\Delta t} \tan \frac{\Delta t}{2} \omega_d$$

The application of the bilinear z-transform (5) to the prewarped transfer function (2) finally gives [7]

$$(7a) \qquad S(z)(1+b_1 z+b_2 z^2) = X(z)(a_0+a_1 z+a_2 z^2)$$

$$(7b) \qquad U(z)(1+d_1 z+d_2 z^2) = S(z) \cdot c_0 (1+2z+z^2)$$

The coefficients a_i, b_i, d_i of the z-polynomials are easily derived from the instrumental parameters (eigenfrequencies, damping)
of the broad-band seismometer, simulated seismometer or galvanometer, respectively.
Using the shifting theorem of the z-transform

$$(8) \qquad z^k A(z) = a_{t-k}$$

equations (7a), (7b) read in the time domain

$$(9a) \qquad s_t = a_0 x_t + a_1 x_{t-1} + a_2 x_{t-2} - b_1 s_{t-1} - b_2 s_{t-2}$$

$$(9b) \qquad u_t = c_0 (s_t + 2s_{t-1} + s_{t-2}) - d_1 u_{t-1} - d_2 u_{t-2}$$

Thus the z-transform of the simulation of arbitrary seismometer-
galvanometer systems from broad-band inputs yield a cascade recursive filter.
The accuracy of the filter method is qualitatively demonstrated in Fig. 13. The seismogram on the bottom of this picture
shows the digital broad-band recording; the uppermost seismogram
is an analog record of a long-period seismometer-galvanometer
system (T_S = 20 s, T_g = 100 s, h_S = h_g = 1.0). This is found to
compare precisely with the digitally simulated seismogram in between both recordings.
The comparison of original broad-band seismograms with simulated narrow-band recordings clearly demonstrates the relationship
between bandwidth and information content (Seidl[10]). Fig. 14
shows the broad-band recording (uppermost trace) for the Bukarest
earthquake on March 4, 1977, as well as the velocity-impulse response function of the WWSSN-LP system (dotted). The pulse width
of this function is about 7 s. Fine structures with smaller time
constants are smoothed by convolution. Thus the multi-pulse structure of the P-wave group appears only as a broadening of the impulse with small shoulders in the WWSSN-LP simulation (bottom
trace). Using the broad-band recording from GRF observatory as a
reference seismogram, G. Müller et al.[11] were able to find similar
fine structures in other WWSSN-LP recordings. Therefore these
authors interpreted the Bukarest earthquake as a multi-shock event.
A SRO-instrumentation has been installed by the USGS at GRF-
observatory. The recordings of this borehole instrument can
directly be compared with broad-band seismograms of array station
A1 (see Fig. 1). The horizontal distance between the stations is
about 30 m, the vertical distance about 110 m. The bilinear

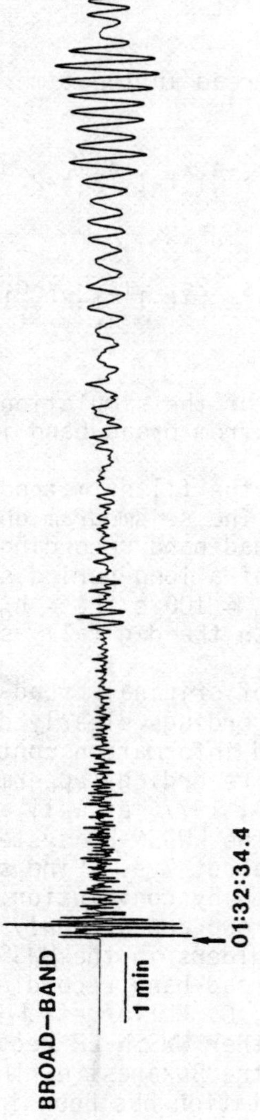

Figure 13. Analog seismogram recorded by a long-period seismometer–galvanometer system (T_s = 20 s, T_g = 100 s, h_s = h_g = 1.0)

BUKAREST 4.3.1977

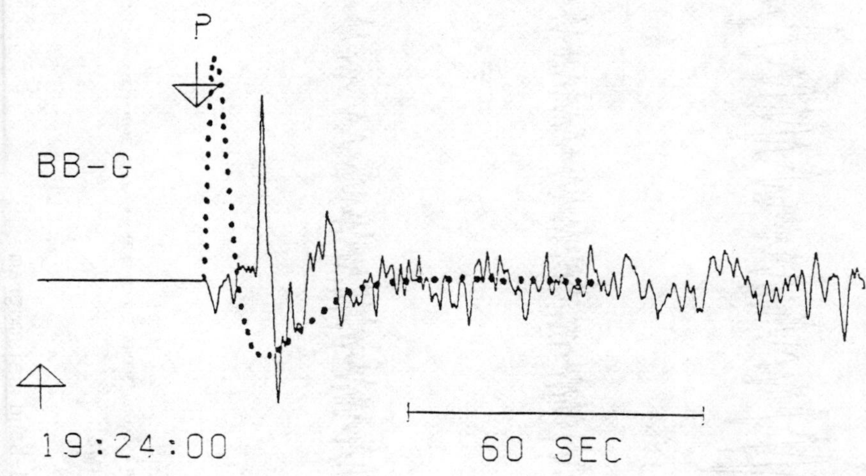

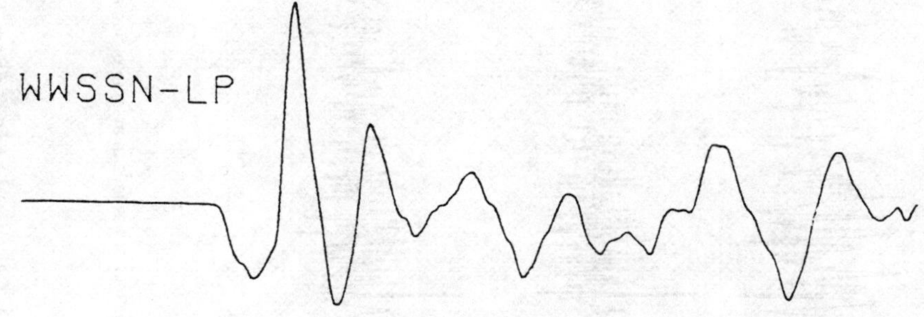

Figure 14. Broad-band recording (proportional velocity) and WWSSN-LP simulation of Romanian earthquake March 4, 1977

z-transform was used to simulate the SRO-output resulting in a recursive filter of 7th order for the LP-channel and a 5th order filter for the SP-channel. Fig. 15 shows a comparison of the original SRO-SP recording (bottom trace) with the simulated seismogram (middle trace) and the GRF-broad-band output on the top of the figure. This event is an earthquake of magnitude $MB_{NEIS} = 5.1$ from Szechwan Province, China. A slight reduction of the high-frequency noise on the seismogram of the borehole instrument can be seen.

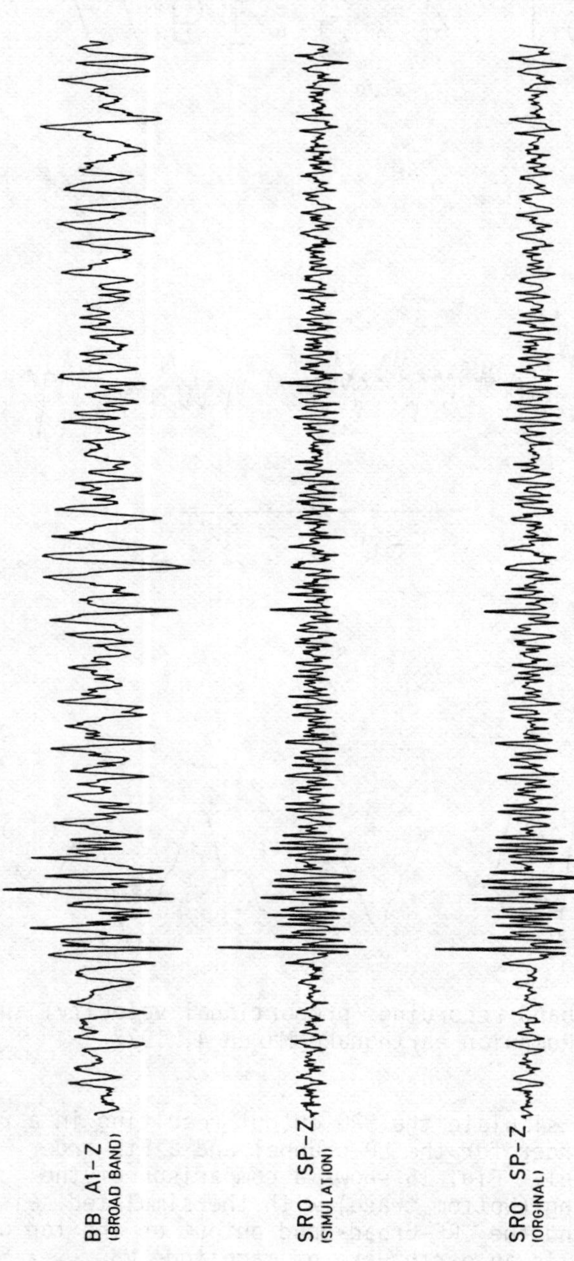

Figure 15. SRO-SP simulation from GRF-broad-band seismogram

Again the relationship between bandwidth and fine structure can be demonstrated using broad-band recordings proportional to ground velocity (BB-G) and ground displacement (BB-W) as well as simulated seismograms for the long-period WWSSN- and SRO-systems (Fig. 16). The fine structure is increasingly suppressed with decreasing bandwidth. Especially in the broad-band recording BB-G for a deep focus earthquake (focal depth 530 km) in the North Korea region on March 9, 1977, the P-wave group is a superposition of a lower frequency signal and a high frequency wave arriving shortly after the first onset. This information can only be found in the broad-band recording proportional to ground velocity.

Finally Fig. 17 qualitatively shows the benefits of broad-band seismograms for the identification of seismic events. Besides the original broad-band recording of an earthquake from China-Sea (upper half) and an E. Kazakh explosion (lower half) the P-wave group is plotted in extended time scale. To compare the surface waves a digital fourth-order Butterworth bandpass filter (0.025 Hz - 0.08 Hz) is applied to the broad band data. The LP-output gives an intuition of the detection capability of the broad-band system for surface waves. The comparison of the P-wave groups demonstrates the obvious differences of an earthquake and an explosion on broad-band seismograms. The enlarged spectral information can be used for further developing the conventional short-period discriminants (Dahlman, Israelson[12]). Recently Harjes[13] modelled the P-wave group by an autoregressive-moving average (ARMA) process. The ARMA-parameters are used for classification of seismic sources in a multidimensional variance analysis. The method was successfully applied to discriminate E. Kazakh-explosions and natural seismic events.

In the daily routine analysis at GRF-observatory the additional information in broad-band recordings allows the extraction of more sophisticated parameters like broad-band magnitude, energy, and seismic moment.

5.   CONCLUSION

Using data from the new digital broad-band array at GRF-observatory it was demonstrated that such instruments offer a raw data base from which different special instrumental characteristics, optimal for detection and identification of seismic events, can be simulated by numerical methods. The technical development of digital broad-band instruments also includes the SRO-stations, recently installed by the USGS, if these instruments are slightly modified for a continuous recording of the broad-band output. Because there will be modifications and supplements in evaluation schemes when (and after) a global seismic data exchange system comes into operation, it seems favourable to record as much information as possible with the standard instrumentation, which therefore should consist of digital broad-band seismographs.

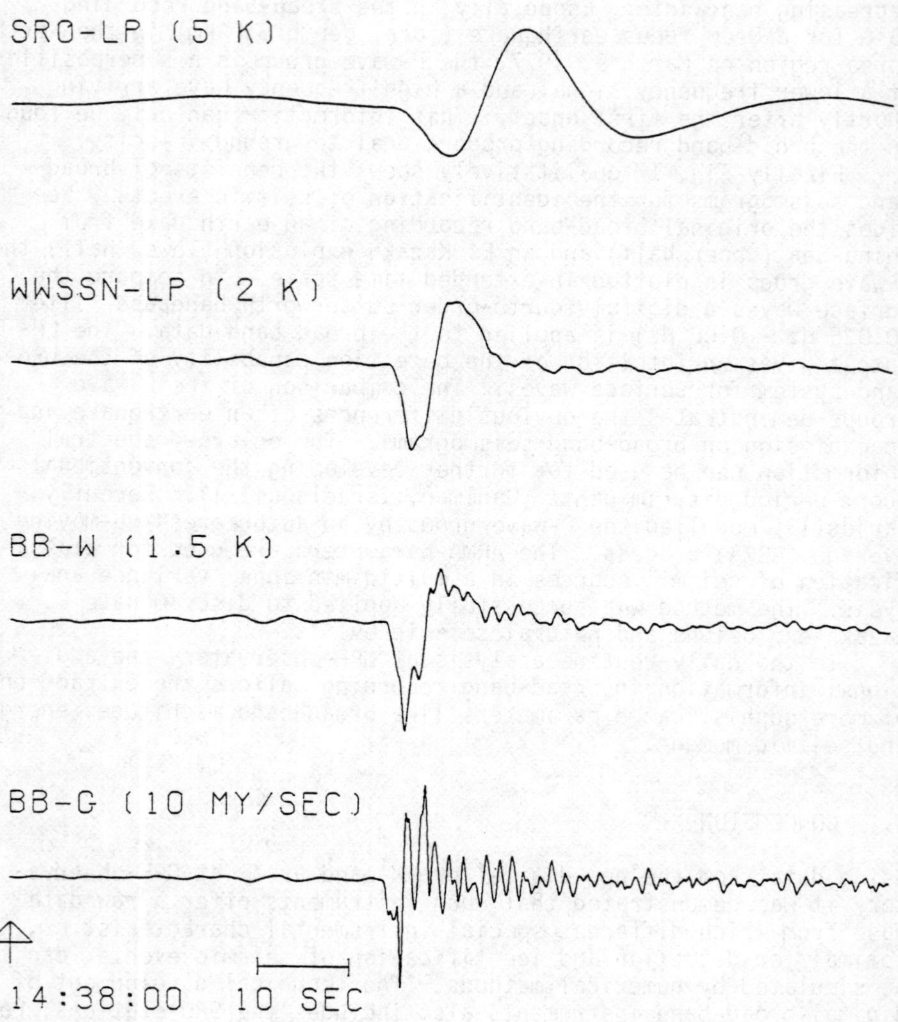

Figure 16. Comparison of resolution for different seismographs (BB-G: broad-band proportional velocity, BB-W: broad-band proportional displacement, WWSSN-LP: World Wide Standard Seismograph - long-period, SRO-LP: Seismic Research Observatory - long period channel)

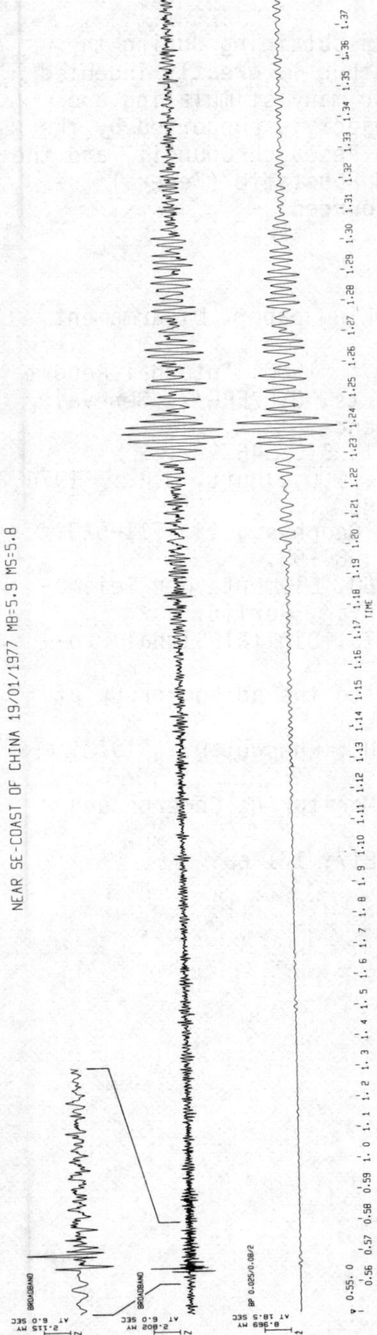
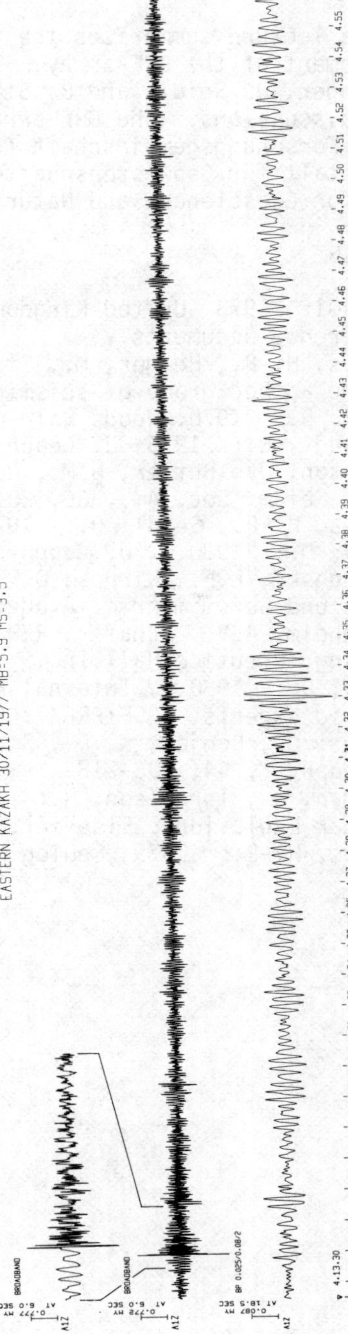

Figure 17. Broad-band recording and long-period filtered seismograms of an earthquake (upper traces) and an explosion (lower traces)

Acknowledgements

This lecture summarizes the results obtaining during the establishment of the GRF-array. The author is greatly indebted to M. Henger, D. Seidl, and B. Stork for many stimulating and helpful discussions. The GRF-array project is supported by the Deutsche Forschungsgemeinschaft (German Research Council) and the Bundesanstalt für Geowissenschaften und Rohstoffe (Federal In-stitute for Geosciences and Natural Resources).

REFERENCES

1. CCD/401:   1973, United Kingdom. Working paper. Disarmament Conference Documents.
2. Harjes, H.-P., Henger, M., Stork, B.:   1980, Internal Report to the ad hoc group of seismic experts. GSE/FRG/7. Geneva.
3. Block, B.:   1970, Woods Hole Conference.
4. Wielandt, E.:   1975, J. Geophys., 41, 545-546.
5. Peterson, J., Butler, H.M., Holcomb, L.G., Hutt, C.R.:   1976, Bull. Seism. Soc. Am., 66, 2049-2068.
6. Harjes, H.-P., Seidl, D.:   1978, J. Geophys., 44, 511-523.
7. Seidl, D.:   1980 a, J. Geophys., 48, 84-93.
8. Sawarenski, E.F., Kirnos, D.P.:   1960, Elemente der Seismo-logie und Seismometrie. Akademie-Verlag, Berlin.
9. Oppenheim, A.V., Schafer, R.W.:   1975, Digital Signal Pro-cessing. Prentice Hall Inc.
10. Seidl, D.:   1980 b, Internal Report to the ad hoc group of seismic experts. GSE/FRG/7. Geneva.
11. Müller, G., Bonjer, K.-P., Stöckl, H., Enescu, D.:   1978, J. Geophys., 44, 203-218.
12. Dahlman, O., Israelson, H.:   1977, Monitoring Underground Nuclear Explosions. Elsevier.
13. Harjes, H.-P.:   1979, Geolog. Jb., E17, 171 pp.

AUTOMATIC PROCESSING METHODS IN THE ANALYSIS OF DATA FROM A GLOBAL SEISMIC NETWORK

Frode Ringdal

NTNF/NORSAR, Post Box 51, N-2007 Kjeller, Norway

ABSTRACT

Automatic procesing of seismic data is today a key element
in the efforts to achieve an adequate seismic verification system
for a comprehensive test ban treaty. This paper reviews the current state-of-the-art in the field, with special reference to
the global system proposed by the seismic experts group established by the United Nations Committee on Disarmament. Current
automatic processing methods and research results obtained at
the Norwegian Seismic Array (NORSAR) form the basis for these
discussions. It is concluded that considerable progress has
been made in areas like automatic detection processing, while
seismic parameter measurements at individual stations and processing at global data centers still require substantial research
efforts before the goal of achieving reliable automatic processing
schemes can be attained.

1.   INTRODUCTION

With the development of digital seismograph recording
systems, procedures for automatic processing of seismic data
have become increasingly important. In particular, the ever-
decreasing cost and increased capacity of modern computers has
in the past few years made possible the implementation of very
sophisticated automatic processing algorithms. Today, the limitations in this regard are not on computer capacity, but rather on
the shortcomings of available automatic processing schemes.

*E. S. Husebye and S. Mykkeltveit (eds.), Identification of Seismic Sources – Earthquake or Underground
Explosion, 787–810.*

In March 1978, the seismic experts group established by the
Conference of the Committee on Disarmament of the United Nations
submitted a report with recommendations regarding an international
co-operative system to detect and identify seismic events. Such
a system would become a key part of the verification of a comprehensive test ban treaty(CTB), and therefore forms a suitable
framework for the discussion of automatic procedures in a global
network. Consequently, a brief review of the recommendations of
the experts group is given in Section 2. Noteworthy, the importance
of future research to improve current automatic processing algorithms was stressed in particular in the group's report.

Current state-of-the-art in automatic seismic signal processing is today reflected in the operation of the Norwegian
Seismic Array (NORSAR). A description of the NORSAR processing
system is given in Section 3, together with an evaluation of the
capabilities and limitations of the automatic algorithms used.

In Section 4, automatic processing methods are discussed
in general, with a special view to those techniques that have
proved effective in an operational environment. Areas in which
further research is needed are outlined in particular. Automatic processing at global data centers is addressed in Section
5, while Section 6 summarizes the conclusions as to the present
state of automatic seismic signal processing.

2.   THE GLOBAL SEISMIC NETWORK PROPOSED BY THE CD SEISMIC
      EXPERTS GROUP

In July 1976 the Conference of the Committee on Disarmament
(the CCD) of the United Nations established an _Ad Hoc_ group of
Government-appointed experts to consider and report on international co-operative measures to detect and identify seismic
events, so as to facilitate the monitoring of a comprehensive
test ban. The initial terms of reference of the Group were to
specify the characteristics of an international monitoring system
including _inter alia_

-    A global network of seismological stations, selected from
     existing and planned installations;

-    Data required from the stations to facilitate the analysis
     for detecting, locating and identifying seismic events;

-    Tranmission facilities for the timely exchange of data
     between seismological stations and data centers;

- Facilities, procedures and related financial implications
  with respect to contributing and receiving centers for
  detecting, locating and identifying seismic events throughout
  the world and facilitating the collation and dissemination
  of relevant documentation;

- The costs which would be incurred if an international monitoring system were established.

The Ad Hoc group presented its recommendations to the CCD in
March 1978 (Reference CCD/558 (1)). Subsequently, the group was
given a renewed mandate to continue its investigations, and this
mandate was later maintained by the Committee on Disarmament
(the CD) established in 1979. In the group's second report (CD/43
(2)) of July 1979, further elaboration of the technical aspects
of the proposed global system were presented. The Ad Hoc group
is currently pursuing its investigations further under a third
mandate, with the participation of scientific experts from about
30 countries.

I shall now briefly outline the structure of the global system
proposed by the CD experts group. This co-operative international
effort would have three main elements:

(i)   A systematic improvement in the observations reported from
      more than 50 seismological observatories around the globe.

(ii) An international exchange of these data over the Global
      Telecommunications System (GTS) of the World Meteorological
      Organization.

(iii) Processing of data at special international data centers for
      the use of participant states.

Selection of seismograph stations for the global network

The Ad Hoc group considered that a suitable global network
should comprise around fifty existing or planned seismic observatories, and considered several alternatives in this regard. An
example of such a network (Network III) is shown in Figure 1.
This network was judged by the group to be the best one based on
available information on existing and planned stations.

Data extraction at the stations

The group's recommendations are summarized as follows:

(i)   Data are to be reported from each station in standard form
      in two levels:

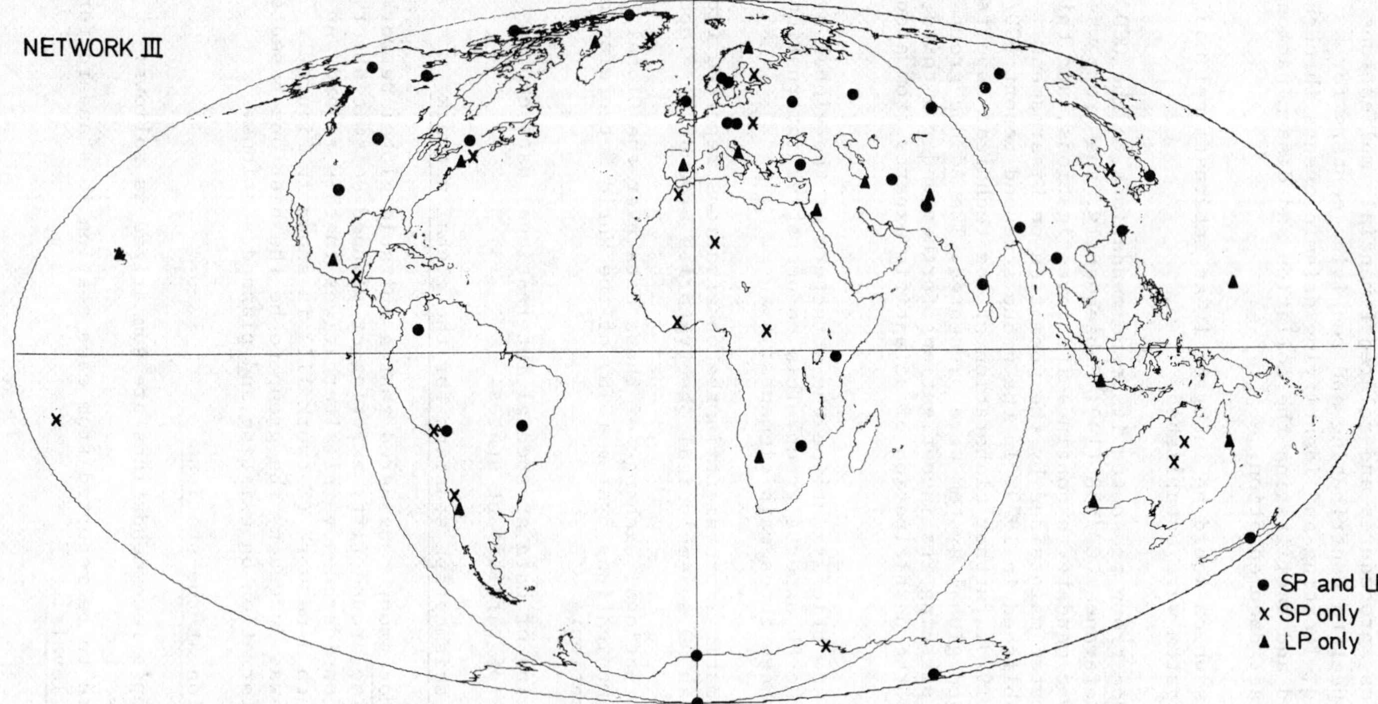

Figure 1. An example of a possible global network of existing and planned seismic observatories proposed by the CD Ad Hoc group of seismic experts (After CCD/558).

Level 1:   Routine reporting, with minimum delay, of basic
           parameters of detected seismic signals

Level 2:   Data transmitted as response to requests for
           additional information, mainly waveforms for
           events of particular interest.

(ii)   Compared to current seismological practice, increased
       emphasis is laid on parameters relevant to event identifica-
       tion.

(iii) Strict operational requirements are set forth as to scope,
       consistency, reliability and promptness in the reporting.

The procedures to be applied for detection, location and
evaluation of magnitude and depth of seismic events would follow
practice now standard at existing international seismological
centers.

International Data Centers

The Ad Hoc group considered that special international data
centers should be established for the global network. In order to
achieve a reliability acceptable to all, it was proposed that
more than one standardized international center be established,
each equipped with equivalent hardware and software and performing
equivalent processing functions.

The main tasks of the international centers will be:

(i)   to receive data of Levels 1 and 2 from the world network
       of seismic stations via the authorized Government facility
       of each State

(ii)   to apply agreed analysis procedures to available data for
       the estimation of the origin time, location, magnitude
       and depth of seismic events

(iii) to associate reported identification parameters with these
       events

(iv)   to distribute, in accordance with defined procedures and
       without interpretation of identification parameters, compila-
       tions of the complete results of these analyses

(v)   to act as an archive for reported data and results of
       analysis on those data.

A schematic illustration of the data flow between stations and
international data centers is shown in Figure 2.

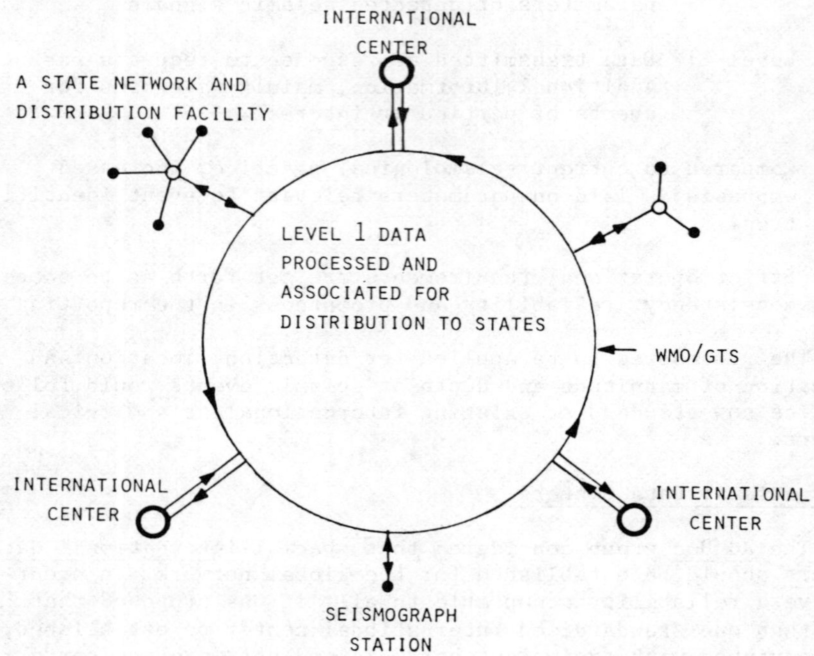

Figure 2. Schematic diagram of the main elements involved in the data exchange between stations of a global network and international data centers (After CCD/558).

Data exchange between stations and international data centers

After review and assessing several possible media for seismic data exchange at Levels 1 and 2 the following recommendations were made:

For Level 1 data (basic signal parameters) usage of the Global Telecommunication System (GTS) of the World Meteorological

Organization (WMO) is recommended because of its global availability, proven operation and low cost. The WMO/GTS network is schematically shown in Figure 3.

For Level 2 data (requested waveforms) which are generally more voluminous and not so critically dependent on fast communication, several solutions might be employed. Digital transmission (e.g., via WMO/GTS) should be used when possible; otherwise, telecopying of seismograms is recommended in preference to using mail services.

With respect to the time delays that would be involved in the data exchange, the group considered that expedient reporting and response to data requests was essential, and time delays both for Level 1 and Level 2 data should therefore be kept to a minimum, i.e., only a few days.

In the course of its work, the Ad Hoc group also identified several areas in which additional research is needed. One of the tasks considered most important was the further development of automatic processing techniques in the analysis of seismic data. This is needed both in the analysis of waveforms in digital form recorded at the seismograph stations as well as at the data centers, for phase association and computation of physical source parameters for the detected events. The main purpose of this paper is to review the research efforts that have been conducted in this regard, and I will start by describing the NORSAR system, which comprises one of the most advanced automatic seismic processing systems in operation today.

3.   THE NORSAR PROCESSING SYSTEM

One of the most amibitious research undertakings to date in connection with the detection and identification of underground nuclear explosions was the large array development program initiated by the Defense Advanced Research Projects Agency of the USA in the early 1960's. The major part of this program was the construction of the large aperture seismic arrays LASA in Montana, USA, and NORSAR in Norway and their associated automated data centers.

The Norwegian Seismic Array

NORSAR (Norwegian Seismic Array) was established as a joint undertaking by the governments of Norway and the United States. The field work and instrument installation started in 1968 and the array was fully operational in April 1971. Initially, NORSAR was configured with 22 subarrays (Figure 4), each containing one LP (3 component) and 6 SP (vertical) seismometers. In 1976, the array was reconfigured to 7 subarrays (especially marked in

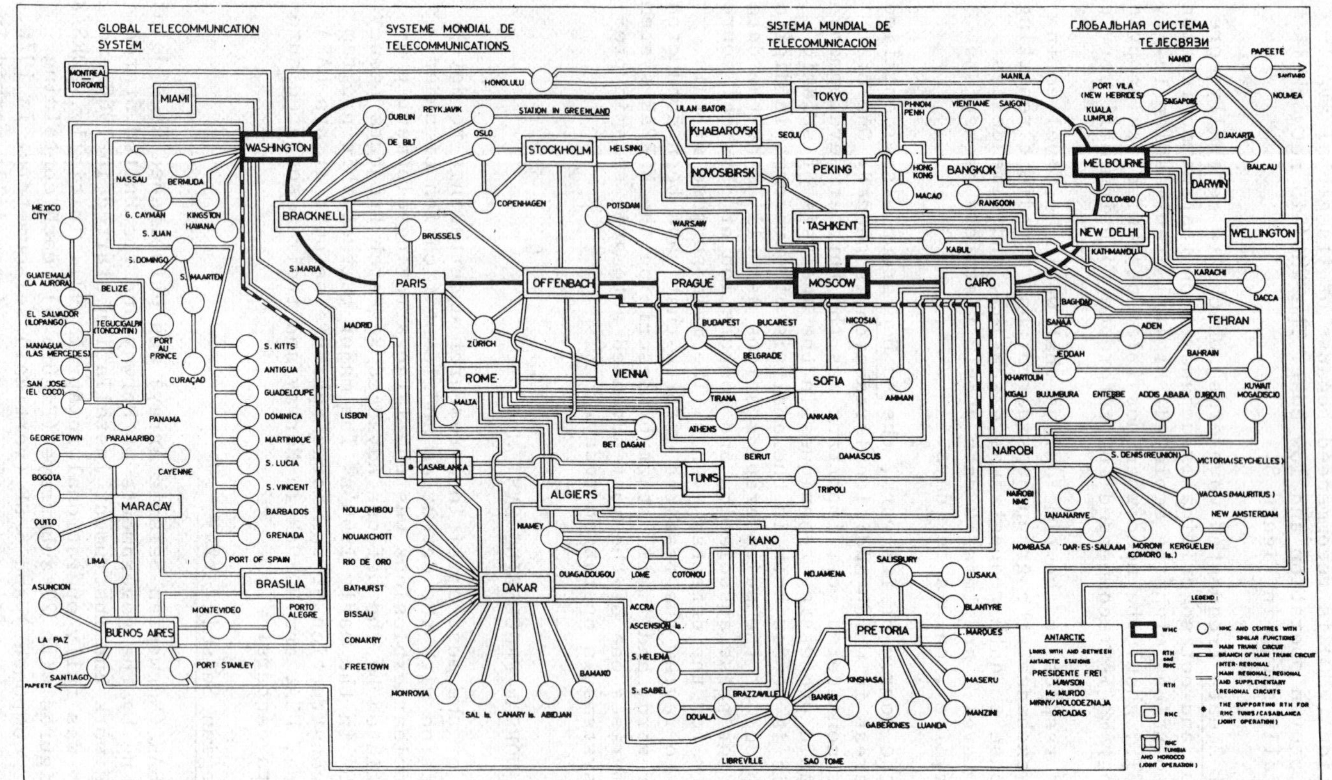

Figure 3. Schematic view of the WMO/GTS communications network. Level 1 data exchange in the global system proposed by the CD Ad Hoc group would take place via this network.

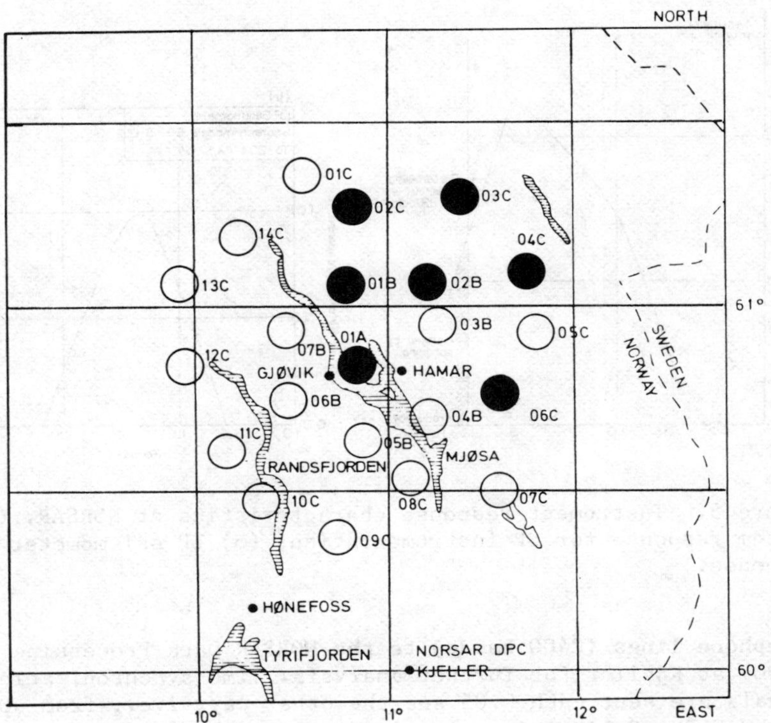

Figure 4. The Norwegian Seismic Array - NORSAR. The circles
mark the location of the original 22 subarrays, which each consists of 6 vertical SP and one three-component LP instruments.
The seven subarrays currently in operation are marked as filled
circles.

Figure 4), in order to make possible more nearly automatic operation of the array. Instrument response characteristics at NORSAR
are shown in Figure 5.

From the seismometers the recorded earth motions are transmitted via trenched cables to a Central Terminal Vault (CTV) at
the subarray center. The CTV is housing a so-called Short and
Long Period Electronic Module (SLEM) which multiplexes and digi-
tizes the 9 seismometer outputs into a single bit stream. The
sampling rate is 1 and 20 Hz for LP and SP seismometers, respectively. The data are then transmitted by means of ordinary

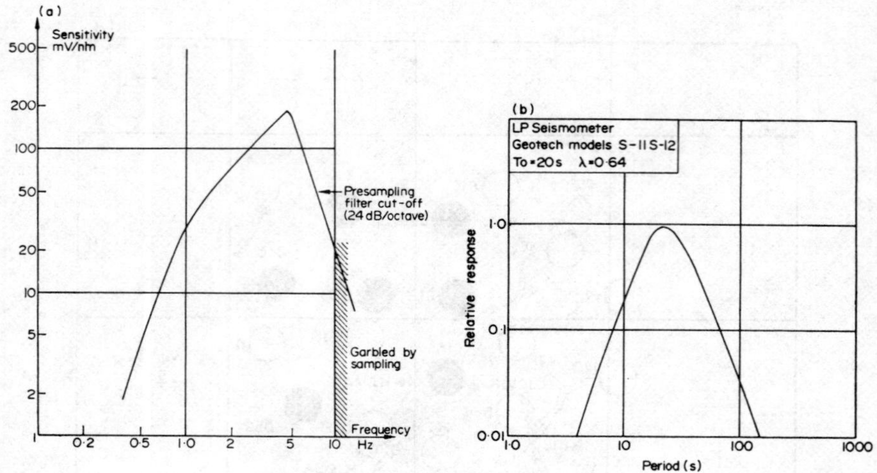

Figure 5.   Instrument response characteristics at NORSAR: (a) system response for SP instrumentation, (b) LP seismometer response.

telephone lines (2400 bauds) to the NORSAR Data Processing Center (NDPC) at Kjeller for further analysis. Time synchronization signals are sent each 0.05 sec the other way, i.e., from NDPC to the SLEM. Calibration of seismometers, SLEM and data transmission lines is performed by sending special commands from NDPC to signal generators (sine or pseudo random waves) which are part of the SLEM units. A trans-atlantic data link connects NORSAR to the Seismic Data Analysis Center in Alexandria, Virginia, USA. Data exchange between the two centers takes place in real time via this link, which is part of the computer-to-computer network ARPANET. The NORSAR computer system, which originally was based on IBM S/360 equipment, is now in the process of being upgraded, and the planned configuration is shown in Figure 6. For a more comprehensive description of the NORSAR array, it is referred to Bungum et al (3).

<u>NORSAR data analysis</u>

The data received at NDPC are processed and at the same time stored on magnetic tape for more permanent saving. The routine data analysis is performed in two steps, called detection and event processing.

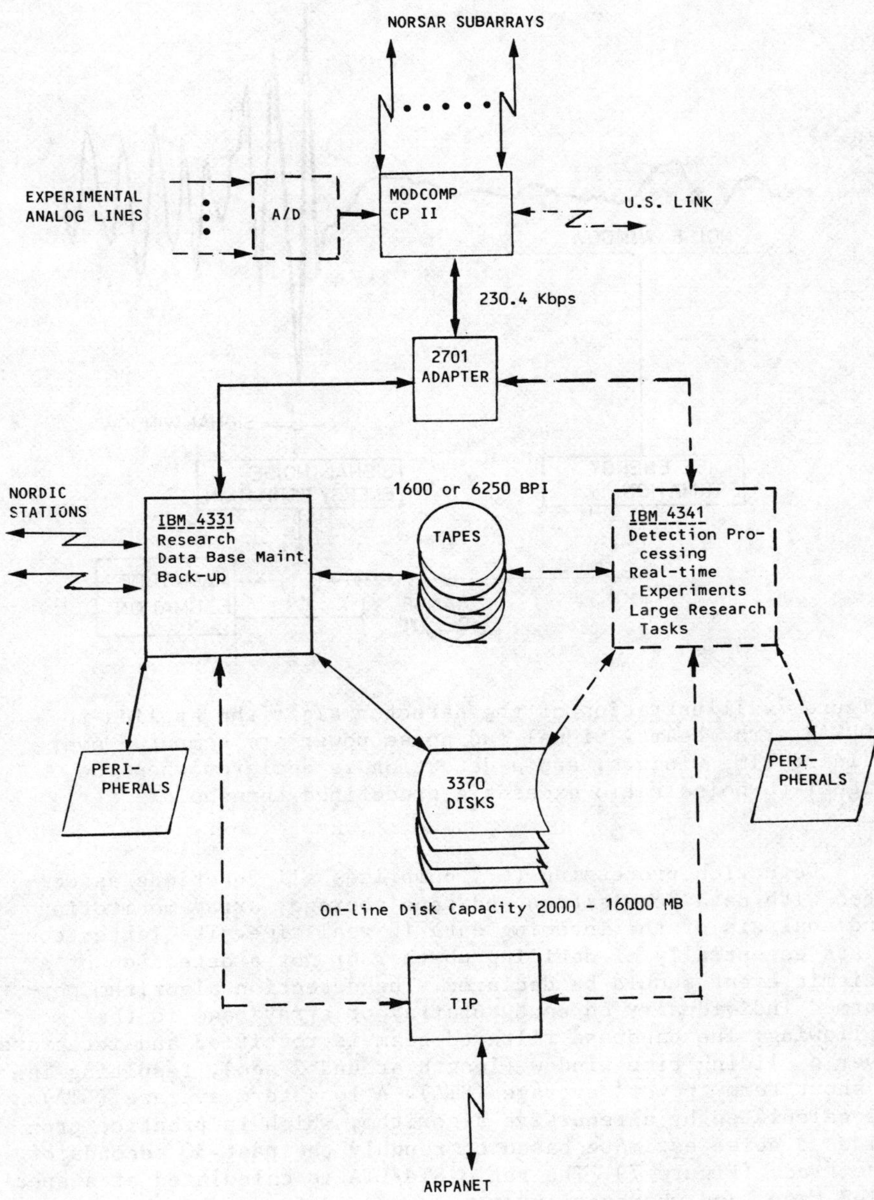

Figure 6. Planned computer configuration at the NORSAR data center.

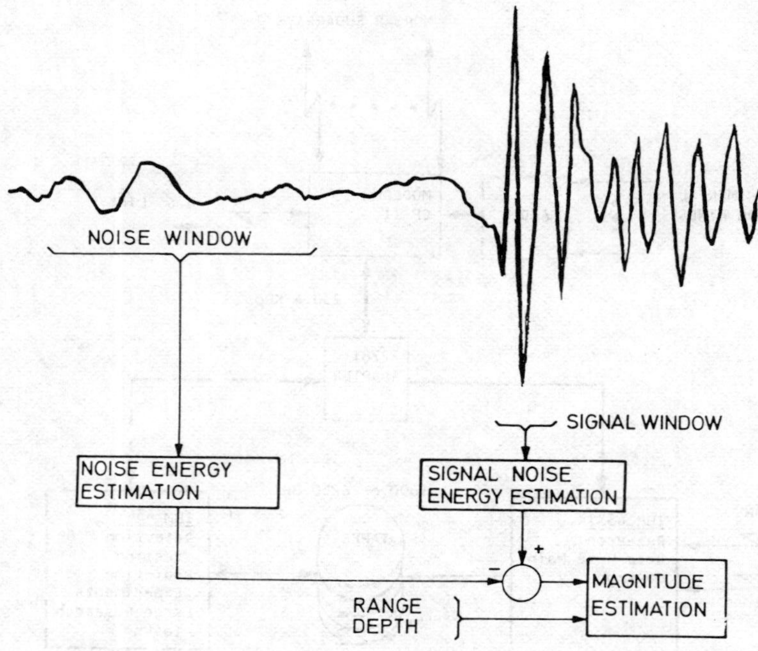

Figure 7. Illustration of the detector algorithm applied to NORSAR array beams. Signal and noise power are computed over sliding time windows, and a detection is declared when the signal-to-noise ratio exceeds a predefined threshold.

Detection processing (DP) comprises all functions associated with data acquisition and tape storage, array monitoring and analysis of the incoming data in real time. The latter consists essentially of deciding whether or not a detection of a seismic event should be declared. The detection algorithm performed individually on each subarray or array beam is the following: The bandpass filtered beam is rectified and integrated over a sliding time window (length around 2 sec), resulting in a short term 'power' average (STA). A long term average (LTA) is calculated by a recursive algorithm, which in practice provides a noise estimate based on roughly the past 30 seconds of the trace (Figure 7). The ratio STA/LTA is calculated at a specified rate, and whenever it exceeds a predefined threshold a number of successive times, a detection is declared.

During beamforming about 200 array beams (selected surveillance) are deployed over the most interesting seismic
regions. Each array beam is computed using a delay-and-sum
algorithm, thus adding together all seismometer inputs 'steered'
to a specific teleseismic area. In this way, a location estimate
is achieved simultaneously with obtaining improved signal-to-
noise ratio for detection purposes. Due to inhomogeneities in
the crust and upper mantle underneath the array, a 'plane wave'
fit of the incoming arrivals is inadequate, and introduction of
regionally dependent time delay corrections prior to beamforming
is therefore necessary (4). Such corrections are obtained from
'master events' via a cross-correlation procedure illustrated
in Figure 8. Each array beam is processed independently in the
subsequent detection processing.

Automatic Phase Delay Measurement System

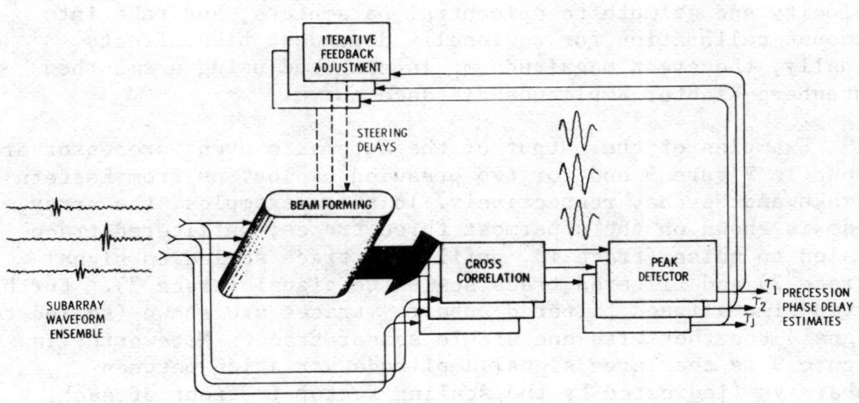

Figure 8. Illustration of the cross-correlation procedure used
at NORSAR to obtain accurate phase delays for incoming seismic
signals.

To ensure adequate coverage of both regional and teleseismic
areas, but at a lower detection capability, a general surveillance
is performed in parallel with the selected surveillance. This
second type of processing comprises so-called envelope beamforming of filtered subarray traces prior to calculating STA/LTA,
and is especially suited to achieve good detection performance
for signals with poor coherency across the array. Such signals
are usually seen from near events (distance less than 30 degrees),
and the envelope beam is in addition highly useful for detecting
the high frequency signals usually observed from underground
nuclear explosions (5).

Event processing (EP) satisfies two objectives, namely,
preparation of a daily bulletin of seismic events and support
of research through the formation of a seismic data base.
In the present NORSAR system, EP is a two-step process, the
first purely automatic and the second based on analyst interaction.

The automatic part of the event processing consists of
adopting the preliminary location estimate provided by the DP
array beam and computing P-wave onset time, amplitude and period
based upon the filtered array beam trace. This trace usually
has sufficient signal-to-noise ratio so that these parameters
may be obtained simply by scanning the trace for the maximum
zero-to-peak deflection to identify the cycle of highest amplitude, while the onset time can be identified by an envelope
fitting procedure. Subsequently, event origin time and epicenter
are estimated (using a restricted depth of 33 km) from specially
developed tables. These tables convert the computed phase
velocity and azimuth to epicentral parameters, and take into
account calibration for regionally dependent bias effects.
Finally, the event magnitude m_b is computed using a smoothed
Gutenberg-Richter amplitude-distance curve.

Examples of the output of the automatic event processor are
shown in Figure 9 and for two presumed explosions from Eastern
Kazakh and Nevada, respectively. In these examples, the array
beam is shown on the uppermost three traces; unfiltered trace
scaled to noise (trace 1), unfiltered trace scaled to signal
(trace 2) and filtered trace scaled to signal (trace 3). Further,
seven time-aligned filtered subarray traces are shown (scaled to
signal) together with one single sensor trace. Noteworthy in
Figure 9 is the large signal amplitude variation between
subarrays (indicated by the scaling factor in front of each
trace). These variations appear to be caused by focusing effects
in the upper mantle (6). The amplitude patterns have been found
to be stable within limited epicentral regions, whereas they
shift strongly from one epicentral region to another (7).

The interactive part of the event processing comprises
analyst review and in some cases reprocessing of the event
to improve upon the estimated source parameters. Currently, in
about thirty per cent of the cases such reprocessing is necessary. In conclusion, therefore, while a high degree of automation in the signal analysis at NORSAR has been achieved today,
a completely automatic EP process will need still further refinement of the processing algorithms now in use.

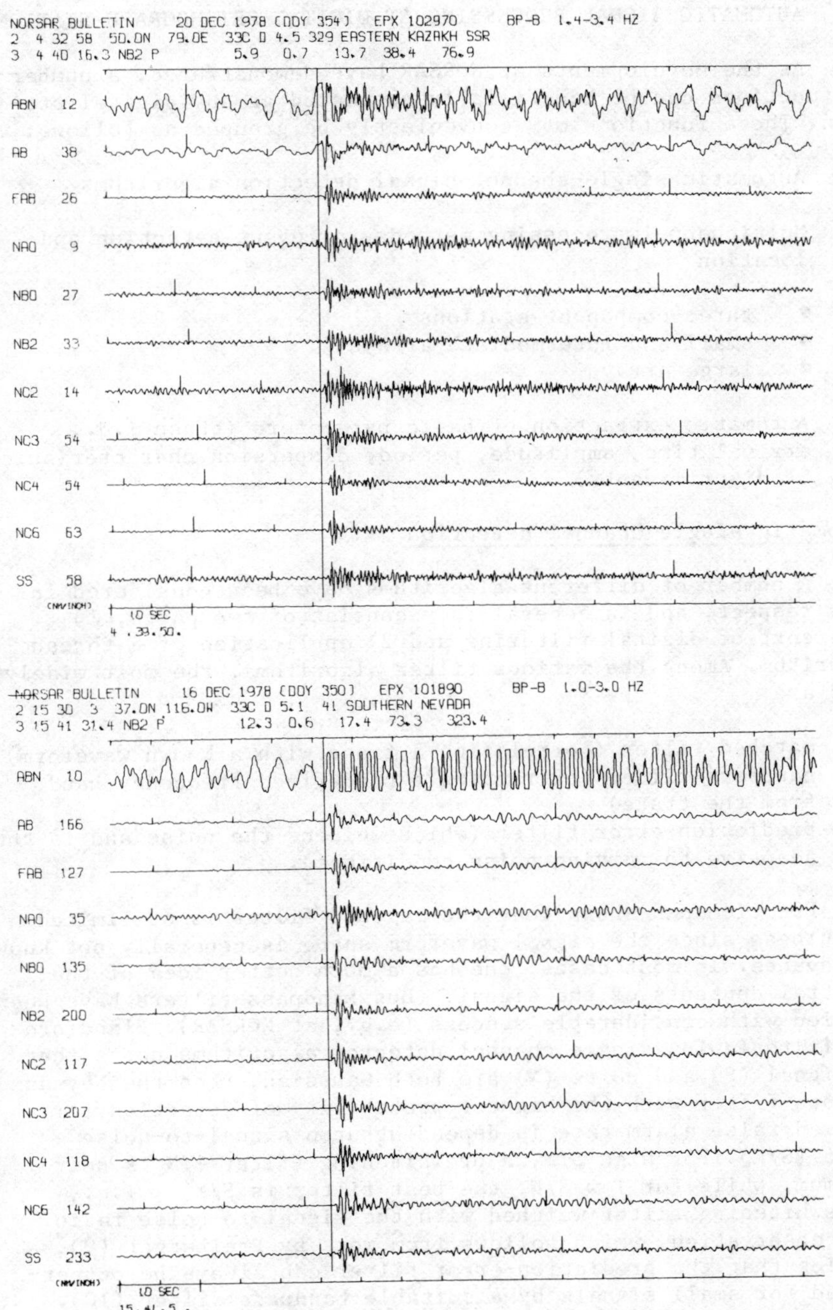

Figure 9. Two examples of automatically processed output from the NORSAR on-line system. Details are given in the text.

4.   AUTOMATIC SIGNAL PROCESSING AT DIGITAL SEISMOGRAPH STATIONS

As the developments at NORSAR have demonstrated, a number of functions can be beneficially automated at any digital station. These functions may conveniently be grouped as follows:

-    Automatic single-channel signal detection algorithms

-    Multichannel processing methods for event detection and location

  - three-component stations
  - small and intermediate arrays
  - large arrays

-    Automatic extraction of basic parameters (phase i.d., arrival time, amplitude, period, dispersion characteristics, SP discriminants).

Automatic single channel detection

A number of different algorithms have been considered in this respect, and in general they consist of two parts, 1) some sort of digital filtering and 2) application of a threshold algorithm. Among the various filter algorithms, the most widely used are:

-    matched filter (correlating a trace with a known waveform)
-    bandpass filter (extracting a predefined frequency band from the trace)
-    prediction-error filter (which whitens the noise and is thus adaptive to varying noise conditions).

Practice has shown that the matched filter is of limited usefulness since the signal waveform shape is generally not known in advance. In most cases, one has a much better idea of the spectral contents of the signal, thus bandpass filters have been applied with considerable success (e.g., at NORSAR). Blandford (8) in reviewing single channel detector algorithms notes that if signal (S) and noise (N) are both Gaussian, then the Neyman-Pearson filter with the highest probability of detection for a fixed false alarm rate is dependent upon signal-to-noise ratio (S/N). For high S/N, a prewhitening filter 1/N is near optimum, while for low S/N, the best filter is S/N^2 , i.e., a prewhitening filter weighed with the signal-to-noise ratio. This observation, which follows from work by Freiberger (9), implies that the prediction-error filter can always be outperformed for small signals by a suitable bandpass filter (10). This conclusion has been confirmed experimentally by Gjøystdal and Husebye (11).

The actual detector application to the filtered trace is most often achieved using squared or rectified moving average, where the short term power (typically over 1.5 seconds) is compared to the preceding long term average (typically over 30 seconds or more). Threshold setting is performed to achieve a tradeoff between good detection performance and low false alarm rates. This is illustrated in Figure 10, which is based on NORSAR data and shows that the number of detections (mostly false alarms) exhibit a sharp increase for SNR below 12 dB. The point of intersection between the two slopes on the figure corresponds roughly to the EP threshold at NORSAR. However, it has been found that using such a fixed SNR threshold generally leads to a non-constant false alarm rate, as the noise characteristics change. Even differences between day-time and night-time noise is significant in this respect, as shown in Figure 11 for NORSAR data. The figure also illustrates how a relatively constant false alarm rate may be achieved using a variable threshold algorithm developed by Steinert et al (13).

While the bandpass-and-threshold detector can be considered near optimum under the assumption of Gaussian noise and signal, little is known about what improvements may be obtained for 'real' seismic traces, which do not necessarily adhere to the Gaussian assumption. A number of algorithms have been proposed over the years, and some of these are discussed in (14), (15), (16) and (17).

Multichannel methods

Algorithms for detection and location using multiple input channels are even more diversified and numerous than single channel detectors. The beamforming algorithms described in the previous section are the most straightforward of these, and have been proved to be effective in operation at LASA, NORSAR and the United Kingdom arrays over the years.

Another relatively straightforward multichannel algorithm that is currently in widespread use is the so-called voting detector, in which a detection algorithm is applied independently to each channel, with a detection declared if a certain subset of the channels exceed a threshold. As shown by Wirth et al (18), such detectors are in practice not as sensitive as beamforming algorithms, but they have the advantage of reducing false alarms due to noise bursts and spikes at single channels.

Among other multichannel detectors, most of which have been tested in an operational environment, I mention

-    The Fisher-detector (19)
-    The FKCOMB-detector (20)

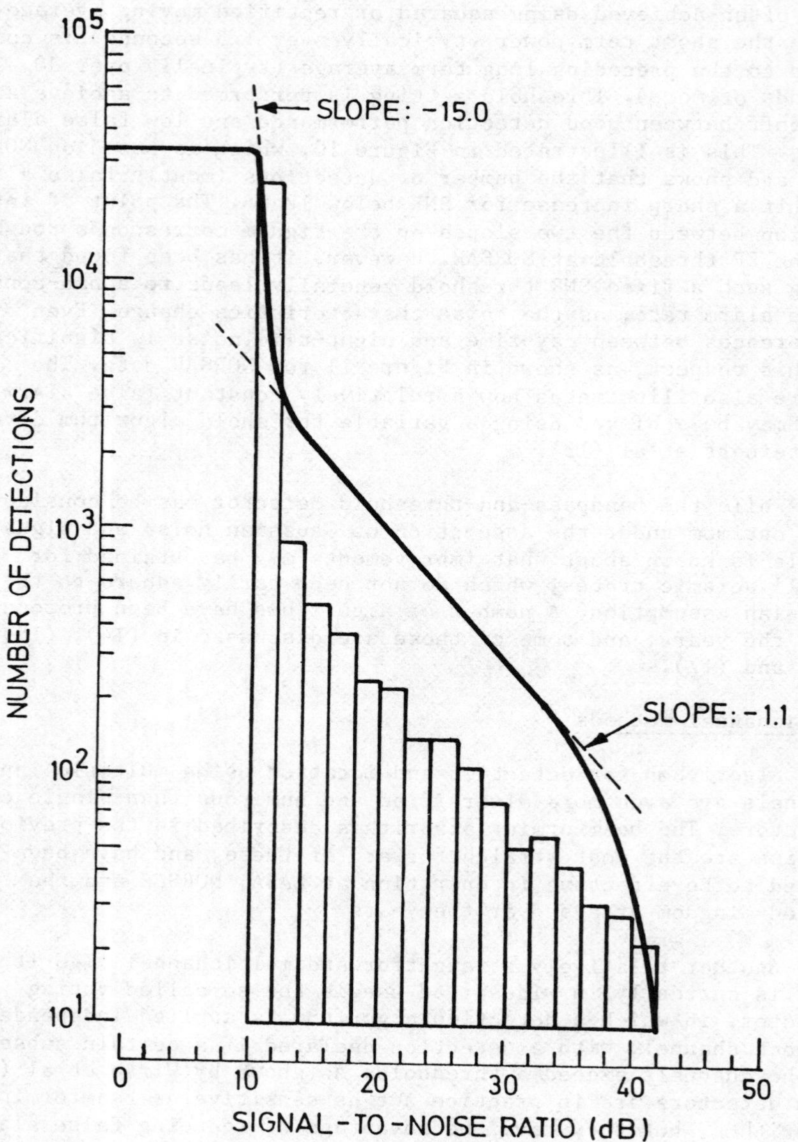

Figure 10. Number of detections by the automatic NORSAR detector as a function of STA/LTA (signal-to-noise) ratio. The number of detections increases sharply below an SNR of 12 dB, where false alarms generated by noise fluctuation dominate the detections triggered by real events.

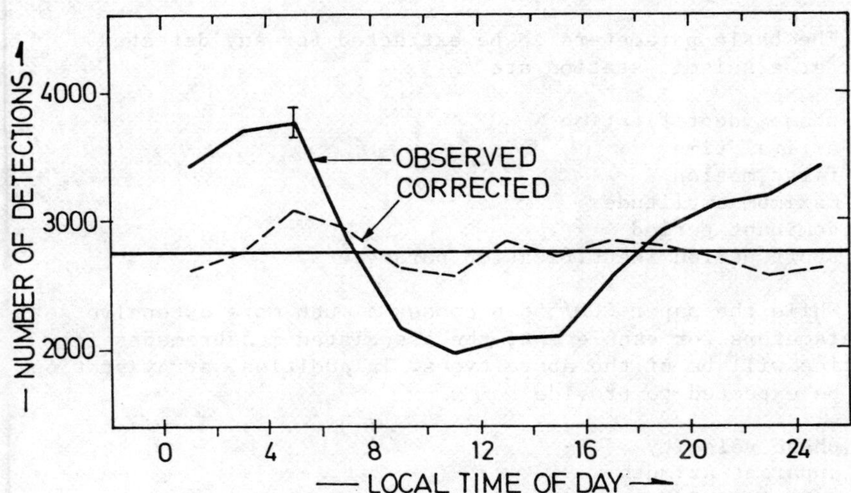

Figure 11.   Diurnal variation in the number of detections at NORSAR
over a 45-week interval in 1972. The large diurnal variation in
observed detection rates is due to statistical differences between
night-time and daytime noise characteristics. By introducing a
variable threshold, an approximately constant false alarm rate
may be achieved (stippled curve).

- Weighted beamforming detector (21)
- N'th root beamform detectors (22)
- Logarithmic beamform detectors (23)
- Adaptive beamforming dectectors (24)
- Three-component maximum likelihood detector (25)
- Three-component polarization filtering detector (26).

While all of these detectors seem to have advantages in
certain applications, it has, to my knowledge, not been documented that any such array detector would consistently provide
a significant improvement in comparison to the beamforming
algorithms currently in operation at NORSAR. Common to all
detectors is that the requirement for analyst verification of
the final output has not been completely eliminated. This objection notwithstanding, automatic detectors are today indispensable
at any digital station to relieve the burden on the analyst in
reading phases and obtaining preliminary event location.

Automatic signal parameter extraction

The basic parameters to be extracted for any detected
event at a seismic station are

- phase identification
- arrival time
- first motion
- maximum amplitude
- dominant period
- short period identification parameters.

While the paper CCD/558 proposes a much more extensive set
of parameters for each event, the associated measurements in
practice will be of the above types. In addition, array stations
will be expected to provide

- phase velocity
- apparent azimuth
- estimate of epicenter and origin time.

As previously described, these parameters are today extracted
in an automatic mode at the NORSAR array, although the automatic
process fails in a fairly high percentage of the cases. Attempts
to obtain improved algorithms for extracting the above information has been made by several investigators. I mention here in
particular the work by Unger (27) in using instantaneous phase,
amplitude and frequency in picking signal onset. Another promising
approach is the MARS process (28) which uses multiple narrow-band
filters to identify signal onset and measure amplitude and period.
Mykkeltveit and Ringdal (29) have obtained promising results in
regional phase identification using a small aperture array. Still,
much research work remains in this field, and some important
problems, such as the automatic extraction of depth phases, have
so far been met with little success.

5.   AUTOMATIC PROCESSING AT GLOBAL DATA CENTERS

In the report from the CD seismic experts group (CCD/558),
the importance of achieving automatic processing at the envisaged
international data centers was strongly stressed. The main reason
for this concern has been to eliminate subjective analyst judgements to the extent possible, thereby obtaining essentially
identical bulletins from these centers.

Briefly, the main processing functions for which automatic
algorithms would be desirable are:

A.   Association of detected phases to define seismic events, and
      to obtain criteria for false alarm reduction

B.   Event location and depth estimation

C.   Automatic association of SP and LP detections

D.   Magnitude determination

E.   Compilation of identification parameters to facilitate
      source classification.

The most critical part would be the phase association and
location algorithms to be employed at the center. Current data
centers apply Geiger's method, correlating the reported first
arrivals from a number of stations (typically four or more) to
obtain a location estimate. While this method works satisfactorily
for large and medium size events, the problem becomes much more
difficult for events for which only a small number of reportings
are available. Such low magnitude events are of course of prime
importancve in a CTB verification system, and improvements in
the standard procedures are clearly needed. Examples of possible
improvements are:

-     Use of array information and azimuth from 3-component
      stations to obtain improved initial estimates and reduce
      false alarms

-     Include amplitude information to remove spurious events
      which may result from random association of detected
      phases (30)

-     Use phase determination and analyst comments (e.g., local,
      regional) in the association process

-     Supplement the network data with observations from local
      stations or station networks.

At present it is difficult to see how a satisfactory
automatic procedure could be implemented for the above purpose.
Thus, automatic processing at global data centers must be considered an important research topic for the future.

6.   CONCLUSIONS

The ultimate goal of achieving a fully automated analysis
of digitally recorded seismic signals from the initial input at
the seismograph station level to the final collation and analysis
at global data centers is still far from being a reality. Nonetheless, promising advances have been made in several areas, and
recent research indicates further substantial progress in the
years to come.

The most successful attempts at automation have come in the field of detection of seismic signals. Automatic detection processing has proved its value over the past decade in an operational environment at stations ranging from single-sensor installations (such as the Seismic Research Observatories) to the large arrays LASA and NORSAR. By the nature of existing automated algorithms, a certain number of false detections will be generated, and analyst review is therefore still necessary to sort out the real seismic signals from spurious noise bursts causing the detector to trigger.

Automated parameter measurements are today in effect at only a few seismic stations (such as NORSAR), and experience obtained shows that analyst review and re-analysis is essential to obtain reliable results. Substantial progress in the methods in this field is still awaiting further research, although some promising new methods have been proposed.

At data centers, phase association and event location procedures are to a large extent automated today, but the elimination of spurious events and the extraction of small events depend to a considerable degree on the skills of the human analyst. This is probably the most difficult field in which to obtain a completely automated process, and completely new algorithms combined with substantially improved earth models would be necessary if this goal is to be attained.

REFERENCES

1.   Conference of the Committee on Disarmament paper CCD/558: 1978, Geneva.

2.   Committee on Disarmament paper CD/43: 1979, Geneva.

3.   Bungum, H., Husebye, E.S. and Ringdal, F.: 1971, Geophys. J.R. astr. Soc. 25, 115-126.

4.   Bungum, H. and Husebye, E.S.: 1971, Pure appl. Geophys. 91, 56-70.

5.   Ringdal, F., Husebye, E.S. and Dahle, A.: 1975, in NATO Adv. Study Inst. Series E Applied Science No. 11.

6.   Haddon, R.A.W. and Husebye, E.S.: 1978, Geophys. J.R. astr. Soc. 56, 97-118.

7.   Berteussen, K.-A. and Husebye, E.S.: 1974, NORSAR Sci. Rep. No. 1, 1974/75, Kjeller, Norway.

8.   Blandford, R.R.: 1980, VSC Tech. Note No. 38, Alexandria, Va., USA

9.   Freiberger, W.F.: 1963, Quarterly J. Appl. Math. 20, 373-378.

10. Vanderkulk, W., Rosen, F. and Lorenz, S: 1965, IBM Final Report, ARPA Contract SD-296, 15 July 1965.

11. Gjøystdal, H. and Husebye, E.S.: 1972, NTNF/NORSAR Tech. Rep. 48, Kjeller, Norway.

12. Ringdal, F. and Bungum, H.: 1977, Bull. Seism. Soc. Am. 67, 479-492.

13. Steinert, O., Husebye, E.S. and Gjøystdal, H.: 1975, J. Geophys. 41, 289-302.

14. Allen, R.V.: 1978, Bull. Seism. Soc. Amer. 68, 1521-1532.

15. Goforth, T. and Herrin, E.: 1980, Semiann. Tech. Rep. to AFOSR, SMU, Dallas, Texas, USA.

16. Shensa, M.J.: 1977, Report TR-77-03, Texas Instruments, Dallas, Texas, USA.

17. Masso, J.F., Archambeau, C.B. and Savino, J.M.: 1979, Systems Science and Software, Final Report SSS-R-79-3963.

18. Wirth, M.H., Blandford, R.R. and Shumway, R.H.: 1976, Bull. Seism. Soc. Amer. 66, 1375-1380.

19. Blandford, R.R.: 1974, Geophysics 39, 633-643.

20. Smart, E.: 1972, Tech. Rep. No. 9, Teledyne-Geotech, Alexandria, VA, USA.

21. Christoffersson, A. and Husebye, E.S.: 1974, Geophys. J.R. astr. Soc. 38, 525-552.

22. Kanasewich, E.R., Hemmings, C.D. and Alpaslan, T.: 1973, Geophysics 38, 327-338.

23. Weichert, D.H.: 1975, Geophys. Res. Lett 2, 121-123.

24. Shen, W.W.: 1979, Geophysics 44, 1088-1096.

25. Smart, E.: 1977, Report No. SDAC-TR-77-14, Teledyne-Geotech, Alexandria, VA, USA.

26. Shimshoni, M. and Smith, S.W.: 1964, Geophysics 24, 664-671.

27. Unger, R.: 1978, Report No. ALEX(01)-TR-77-4, Texas Instru-
     ments, Dallas, Texas, USA

28. Farrell, W.E., Goff, R.C. and Wang, J.: 1980, in AFTAC-TR-80
     -37, Alexandria, VA, USA.

29. Mykkeltveit, S. and Ringdal, F.: 1980, in Semiannual Techn.
     Summary, NORSAR Sci. Rep. No. 1/1980, Kjeller, Norway.

30. Elvers, E.: 1980, FOA Report C, Stockholm, Sweden.

INTERNATIONAL SEISMOLOGICAL DATA CENTER. DEMONSTRATION
FACILITIES IN SWEDEN.

Ola Dahlman

National Defense Research Institute
Stockholm, Sweden

SUMMARY

To facilitate the establishment of an international global seismological monitoring system for the verification of a Comprehensive Test Ban Treaty (CTBT), Sweden has offered to establish and
operate an International Seismological Data Center. To demonstrate one possible way to carry out the main tasks of such a
center, temporary data center facilities have been set up at the
Hagfors Observatory of the National Defense Research Institute
in Stockholm. These facilities were demonstrated on 12-14 July
1979 to representatives and experts from 26 countries and the
World Meteorological Organisation. This report contains a general description of the demonstration facilities and a summary
of the experience gained so far from these facilities.

*E. S. Husebye and S. Mykkeltveit (eds.), Identification of Seismic Sources - Earthquake or Underground
Explosion, 811–830.*

1. INTRODUCTION

It is generally accepted that seismological verification is
the key method to monitor compliance with a comprehensive
test ban treaty (CTBT). Thus, to verify an international
CTBT, a global seismological monitoring system is needed and
the three parties, UK, US and USSR which are negotiating a
CTBT, have agreed to provisions establishing such a system
(CD/130, 1980). The Committee on Disarmament, CD, Ad Hoc
Group of Seismic Experts has suggested that such a network
could consist of:

- Some 50 globally distributed seismological stations

- Data communication over the World Meteorological
  Organization Global Telecommunication Network

- Specially established international seismological
  datacenters.

International Datacenters are thus seen as an integral and
important part of a global verification system. The main
purpose with these Datacenters is to be service facilities
for all the States parties to a CTBT and participating in
the verification of the treaty. Datacenters should provide
compiled seismological data ready for national assessment.
The data provided by International Datacenters should be
obtained through standardized analyses of data reported from
a global network of seismological stations.

In order to demonstrate one possible way to carry out the
main tasks of International Seismological Datacenters and to
facilitate experimental studies in considerable technical
and scientific detail of the functions to be performed at
such centers, temporary international datacenter facilities
have been set up in Sweden. These facilities were demonstrated
for representatives and experts from 26 countries and WMO on
12-14 July 1979.

This paper gives a short description of the facilities that
were demonstrated and of the preliminary results so far
obtained. More comprehensive technical documentation of the
components of the demonstration facilities and of the results
obtained are under publication. A list of these publications
and the names of the scientists responsible for the reports
is given in an Appendix.

2. TASKS OF INTERNATIONAL DATACENTERS

The International Datacenters have a number of tasks (CCD/558, 1978 and CD/43, 1979). One is to receive all so-called Level I data which are routinely transmitted from the contributing stations through the World Meteorological Organisation Global Telecommunication Network (WMO/GTS) and to store these data in computer readable form. In the next step the Datacenters would put together those data which seem to come from one and the same seismic event. Seismic events are then established from these associated arrival times, and their locations on the earth are estimated. This procedure is quite similar to the one carried out today at existing seismological datacenters, established for purely scientific purposes. An International Datacenter for test ban verification must, however, in addition to these functions also compile the reported identification data. Such data will make it possible for indivudual countries to assess the nature of the event, that is to judge whether it is an earthquake or an explosion. No such assessment should be made at the International Datacenters.

Surface waves, which are seismic waves that propagate along the surface of the earth, are generally recognized as important identifiers. The Datacenters should associate reported surface waves with located events and compile and report the results. Earthquakes and explosions can also be identified using the differences in the recorded so-called short period signals propagating through the interior of the earth (Dahlman and Israelson, 1979). Such short period identification data are not reported today but are most important data for test ban verification. International Datacenters should also compile and report such identification data.

International Datacenters should carry out their analyses promptly and report the result to the participating countries, using the WMO telecommunication network, within a week of the occurrence of the event. The result of the analyses should also be stored at the Datacenters for possible future requests. The International Datacenters should also play a role in the exchange of complete record sections, known as Level II data. Request for such data from individual countries and the data received at Datacenters as the result of such requests, should be retransmitted and also stored at the Datacenters. The exchange of Level II data is expected to be an important part of the international data exchange and thus of the work at Datacenters.

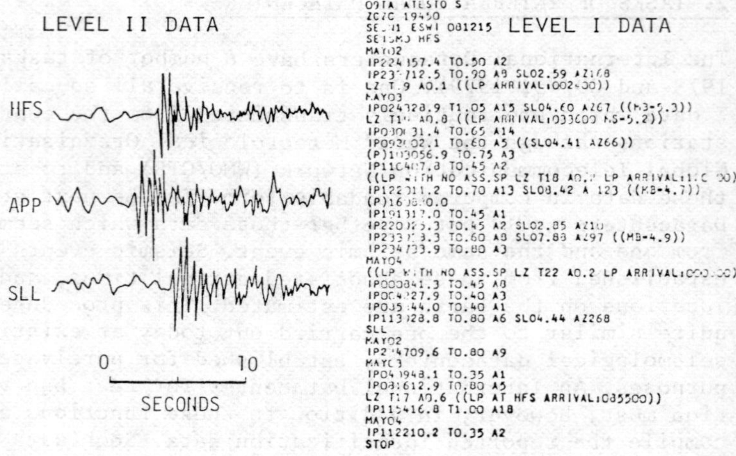

*Figure 1. Examples of Level I data (right) and Level
II data (left) which should be exchanged
within a global monitoring system.*

In addition to these tasks, International Datacenters should
be able to conduct other service functions in connexion with
test ban verification that might be demanded by the countries
parties to a CTBT.

For co-ordinating the efforts of the individual Datacenters
and ensuring that the tasks agreed upon are carried out
properly, the service of an appropriate international body
is needed. The trilaterally negotiating parties have agreed
to set up such a body called a Committee of Experts (CD/130
1980). This Committee will have ongoing responsibility
for facilitating the implementation of the International Monitoring System, for reviewing its operation and considering
improvements to it, and for considering technological developments that have a bearing on its operation. The Committee will
serve as a forum in which treaty parties may exchange technical
information and cooperate in promoting the effectiveness of the
system.

3. DEMONSTRATION FACILITIES

The purpose of the demonstration facilities established at
the Hagfors Observatory of the National Defense Research
Institute, as part of the research work in detection
seismology, was to demonstrate one possible way to carry out
the main tasks of International Seismological Datacenters.
The computational procedures and the data handling routines
developed are only examples of such procedures and routines.
The intention is that these examples will initiate and
facilitate a detailed and thorough technical specification
of the Datacenter procedures which are necessary for the
preparation of the establishment of the International Datacenters. The temporary Datacenter facilities established
include:

- A temporary computer connexion to the World Meteorological Organization global telecommunication network.

- Implementation on large computers at the Stockholm
  University Computer Center of quite extensive programs
  to carry out the main functions foreseen for the
  daily analyses at the International Datacenters.

- Compilation of an experimental data base containing
  information from earthquakes recorded during one week
  at 60 globally distributed seismological stations.

3.1 Connexion to WMO global telecommunication system

A temporary connexion to the WMO global telecommunication
system (WMO/GTS) has been established in co-operation with
the Swedish Meteorological and Hydrological Institute.
Technically, this connexion is a telecommunication link
between a small computer at the demonstration facility in
Stockholm and the computer at the Swedish Meteorological and
Hydrological Institute at Norrköping, some 200 kilometres
from Stockholm. This latter computer is part of the WMO/GTS.
During the demonstration seismological data were transmitted
from the WMO communication center in Tokyo, Paris, Bracknell
(outside London) and Offenbach in the Federal Republic of
Germany and received in Stockholm. The data were transmitted
as fairly comprehensive station bulletins similar to those
foreseen to be transmitted from the stations in the monitoring
systems. Analysed data, in a form of a Datacenter bulletin
foreseen for the verification system, were also transmitted
from the demonstration facility to the above-mentioned WMO
communication centers.

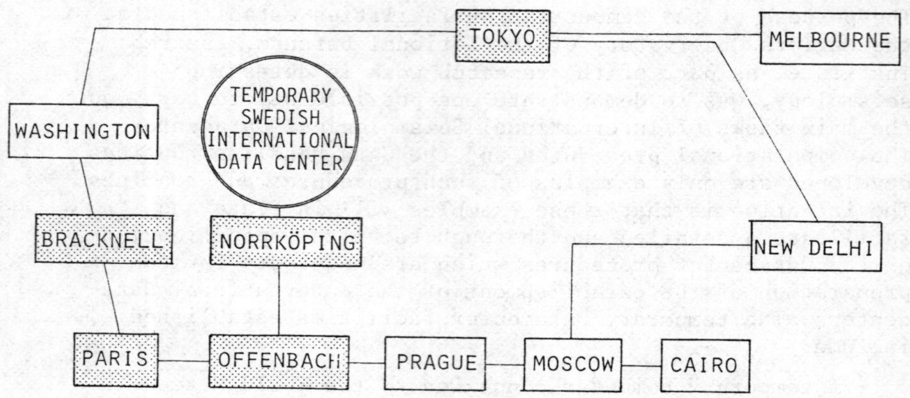

Figure 2. The main trunk of the World Meteorological Global Telecommunication Network (WMO/GTS) and the connexion to the Datacenter Demonstration Facilities in Stockholm. The WMO/GTS communication centers used during the demonstration are shaded.

3.2 Data analysis programs

The programs compiled for the analysis and handling of reported Level I data are designed to carry out a number of the tasks of an International Datacenter. The arrival times reported from individual stations are grouped by the computer in a systematic way to form potential seismic events. This procedure, by which potential seismic events are established from the individual station reports, is the fundamental process for further analysis at the Datacenters. It is, however, a procedure which sometimes can produce unreal events from arrivaltimes that randomly fits together. Once the individual station arrivaltimes are associated with an event, the position of the event on the globe and its depth below the surface of the earth is estimated. The association and location procedures utilize the preliminary event location or, more specifically, the estimated azimuths and apparent velocities reported by array stations.

A special checking procedure has been developed by which the
computer judges whether the established events should be
regarded as real or just the result of accidentally agreeing
reported arrival times. This checking procedure utilizes
dynamic information, e.g. signal or noise amplitudes at
reporting stations and the estimated magnitude of the
possible event. Based on the actual or expected sensitivities
of the individual stations a likelihood judgement is made
whether or not it is likely that the reporting stations, and
no others, should have detected the event, if the event was
real. It is indeed most important to prevent, in this way or
by other methods, the introduction of unreal events that
might create unfounded suspicions.

The reported long period surface wave data, which are
essential for the identification of events, are associated
with the located events by an automatic procedure. This
takes into account not only the expected arrival time of the
long period waves in relation to the observed times, but
also the differences between the expected and observed
arrival directions of the signals at individual stations. A
computer program for the compilation of short period identification parameters, which should be reported in a verification system, has also been developed.

One key question concerning International Datacenters and
also one which has been extensively discussed in the CD
Seismic Expert Group, is whether the processing at International Datacenters should be fully automatic or whether
interference or assistance by seismologists should be
allowed. Fully automatic programs at International Datacenters could provide identical output bulletins which,
however, might include spurious events, even if further
developed checking routines might substantially reduce
the probability of such mistakes. Analysis routines containing judgement by analysis personnel might reduce the
number of spurious events but would, on the other hand,
provide somewhat different outputs from the centers. To
facilitate a thorough discussion of this important point two
analysis programs have been established, one automatic and
one interactive, allowing seismological experts to interfere
with the processing.

3.3 Experimental Data Base

To provide a data base for this demonstration the interval
of one week - 15-21 January 1978 - was selected, and a
global network of 60 seismological stations, corresponding
as closely as possible to the network suggested by the CD
Seismic Expert Group.

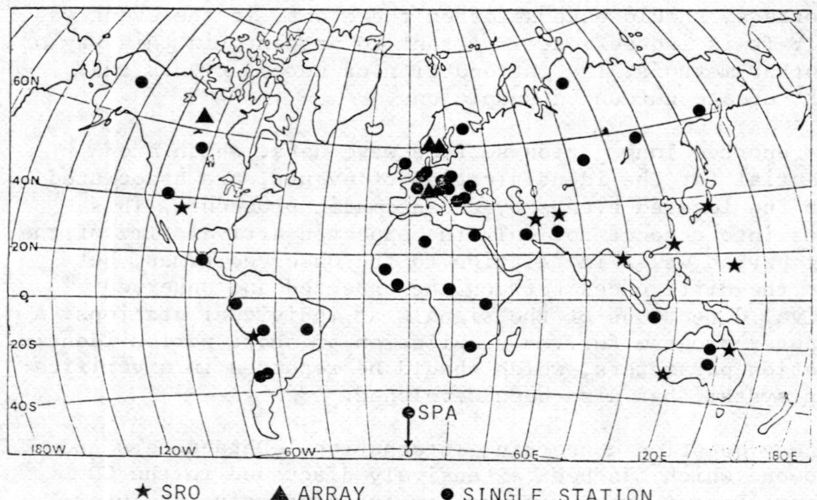

*Figure 3. Seismological stations used in the
experiment.*

For most of the stations the compiled data were those
reported from the individual stations to the United States
Geological Survey and in widely distributed bulletins. These
data contain the information for the definition, location
and magnitude estimation of events but not for the identification of the events as earthquakes or explosions. Data for
events identification were obtained from Grafenberg array
station in the Federal Republic of Germany and from the
Hagfors Observatory in Sweden. Original records were obtained from 11 so-called Seismic Research Observatories
(SRO), established by the United States in co-operation with
the host countries around the world. These records were
analysed at the Hagfors Observatory for long period surface
waves data and short period identification data. The data
thus obtained from the 11 SRO´s correspond closely to those
recommended by the CD Seismic Expert Group.

Complete records, co-called Level II data, were compiled for
the events recorded at the Hagfors Observatory and at the 11
SRO´s. Both the short period and the long period records are
available on computer plotted paper records. The short
period data are also available on the computer system, for
presentation on a graphic display. The compilation of these
Level II data was made rather to facilitate the interpretation of the Level I data than to demonstrate a possible
handling procedure of Level II data at Datacenters.

4. EXPERIENCE GAINED FROM THE DEMONSTRATION FACILITIES

During the establishment and demonstration of the facilities
in Stockholm experience has been gained on several problems
connected with the establishment of International Datacenters and also with the verification system as a whole.
This chapter gives a short summary of the experience gained
so far.

4.1 Level I data

For the Level I data, which are to be routinely reported
from the individual stations, there is a considerable
difference between the data routinely reported today and the
data that should be reported for test ban verification. This
difference is specially pronounced for the long period
surface waves and the short identification data, where today
essentially no data are routinely reported. It is most
important that procedures are developed at individual stations to extract and report those additional data, which are
necessary for test ban verification. The analyses of the SRO
data show that this could be a quite extensive and tedious
work.

Some of the data suggested to be reported by the CD Seismic
Expert Group were not as valuable in the Datacenter analysis
as expected. Such data should be reconsidered or be replaced
by others. One example is the dispersion data for surface
waves, which turned out to be of limited use in the analysis, whereas measurements giving the direction of the
incoming surface waves were quite valuable for the proper
association of the long period waves with events.

The information on the downtimes of the individual stations
and of their detection capability or actual noise values
proved to be of great importance and almost as essential as
the reporting of observed data signals.

4.2 Station network

The 60 station network turned out to be quite efficient in
defining and locating seismic events. The number of defined
events depends of course on the criteria used for the definition
of the events. In all 123 events were defined and located during
the demonstration using the automatic procedure. About 1,600
of in all around 4,000 reported short period arrival times were
associated with these events. About 2,400 signals or 60 per
cent remained unassociated. This is a high proportion but quite
similar to those obtained by other datacenters working with
data from global networks with substatially more stations.
16 of the defined events were obviously "false events" as they
were based on combinations of stations used in defining other
events or phases that probably are associated with such events.
Such "false events" can probably be excluded using a somewhat
more sofisticated automatic processing. 5 more events are probably also "false", giving 102 real events.

ISC reports in its bulletin 331 events for the acutal time period.
Most of these events or 193, are local events being based on
stations at distances less than 10 degrees. For as many as 153
of these local events all stations were within 5 degrees of the
event. The ISC thus reports 138 teleseismic events, having at
least two reporting stations at distances larger than 10 degrees.
88 of these events were common with those defined during the
demonstration. 14 of the events located during the demonstration,
6 of which were reported also by SDAC (see below), were not reported by ISC and 50 teleseismic ISC events were not defined during
the demonstration. 33 or 2/3 of these events were located in the
southern hemisphere. Magnitudes were reported for only 16 of those
50 events and the magnitude distribution of these and the 88
common events are shown in figure 4.

The number of events defined and located during the demonstration
(DEMO) is in table 1 compared also with the number of events in
the US Geological Survey (USGS) monthly bulletin and also with
the number of events given in the weekly event summary produced
by the Seismic Data Analysis Centre (SDAC) in Washington. The
station network used for the SDAC experimental bulletin is not
specified.

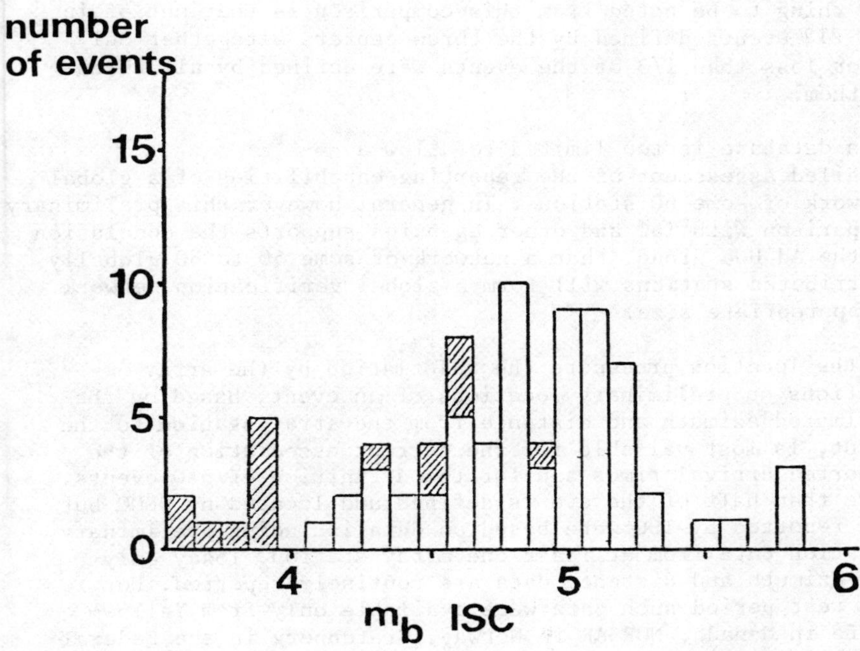

Figure 4. *The dashed parts of the columns mark events not detected by the 60 station network used in the application. The epicenter of the undetected 4.9 event is south of Kermadec Islands. (From Slunga (1980)).*

Table 1. *Comparison of the number of events reported January 15-21, 1978 by the US Geological Survey (USGS), the Seismic Data Analysis Center in Washington (SDAC) and the International Datacenter Demonstration Facilities in Stockholm (DEMO). In all 212 events were defined by the three centers together.*

Center	All event	Event common to all	Event reported only one center	Event common with USGS	Event common with DEMO	Event common with SDAC
USGS	129	57	47	–	70	69
DEMO	123	57	32	70	–	78
SDAC	120	57	30	69	78	–

One thing to be noted from this comparison is that out of in
all 212 events defined by the three centers altogether only
57 or less than 1/3 of the events were defined by all three
of them.

This database is too limited to allow a
detailed assessment of the reporting capabilities of a global
network of some 60 stations. In general however this preliminary
comparison with ISC and other agencies supports the conclusion
of the Ad Hoc group, that a network of some 50 to 60 globally
distributed stations will form a global verification network
of appropriate size.

In the location procedure the information by the array
stations on preliminary locations of an event, based on the
estimated azimuth and distance from the array station to the
event, is most valuable for the correct association of the
reported arrival times and for the definition of new events.
More than half of the events defined and located by DEMO but
not reported by ISC were based on data including preliminary
location data from at least one array station. Today only
few azimuth and distance data are routinely reported. For
the test period such data were available only from Yellowknife in Canada, NORSAR in Norway, Grafenberg in the Federal
Republic of Germany and Hagfors in Sweden. The Datacenters
would substantially improve their ability to associate short
period signals if preliminary location data were reported
from more individual stations in the global monitoring network, which would reduce the number of spurious events and
might also increase the number of correctly defined events.
This would mean that also rather small array stations having
only three recording points separated by some 10-50 kilometres,
would be of great value for a global network.

4.3 Location, checking and magnitude estimation procedures

It was found reasonable to allow for the possibility that a
reported short period arrival time might be associated with more
than one event. A clear difference must also be made between those
station data which define an event and those which are only
associated with an event. It is also most important that full
use is made of the preliminary location data reported from
array stations. A comparison between the locations reported
by USGS and those made during the demonstration showed all
location differences were less than one degree (110 km) and that
for 2/3 of the common events the differences were less than half
a degree. Depth estimates were available for a fairly small
number of common events and for some the differences were larger,
especially for events with few reporting stations.

The result of the automatic checking procedure are still
tentative and further studies must be made. The preliminary
results suggest, however, that the procedure developed so
far, using the reported amplitude and noise information, can
be used for the elimination of spurious events and mis-
associated arrival times. About 40 probably "false events"
were rejected by the dynamic checking procedure. The procedure on the other hand rejected 10 events which were closely
associated with earthquakes reported by ISC. This checking
procedure can be used both for P-wave and surface wave data.

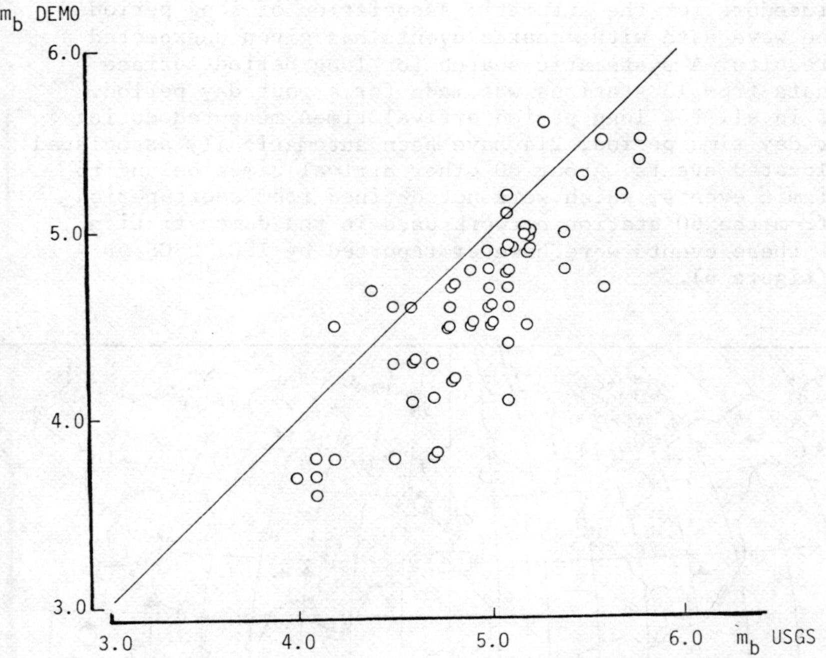

*Figure 5. Comparison of m_b-values estimated by USGS
and DEMO for common events.*

The implemented magnitude estimation procedure takes into
account both the reported signal values and the noise values
at those stations which have not seen the event. For P-wave
data this magnitude is substantially smaller than the
magnitude estimated from reported signals only. A comparison
with USGS magnitudes in figure 4 shows an average difference

of 0.3–0.4 magnitude units. For surface wave data the
difference between the magnitudes estimated in the two ways
is much less. One reason for this might be that the noise
values for long period records were measured every hour,
whereas the short period noise data employed were essentially
those reported in the CD Seismic Expert Group (CCD/558,
1978), which appear to underestimate the actual noise
values.

4.4 Long period data

The procedure for the automatic association of long period
surface wave data with located events has given unexpected
good results. A systematic search for long period surface
wave data from 13 stations was made for a four day period.
Out of in all 344 long period arrival times measured during
a four day time period, 214 have been automatically associated
with located events. About 80 other arrival times belong to
20 seismic events, which were not defined from short period
data from the 60 station network used in the demonstration.
Ten of these events were however reported by ISC, USGS or
SDAC (figure 6).

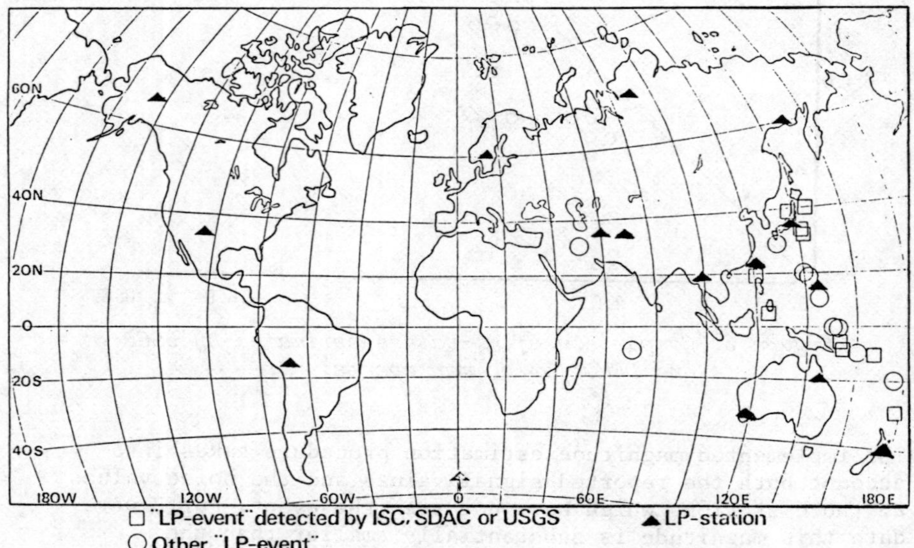

Figure 6. Events located from long period data alone.

If these signals are included among the associated, that
leaves only about 50 or about 15 per cent of the long period
data unassociated.

Long period data are lacking for only 22 of the in all 78
events defined from short period data during the four day
period. The number of events with and without Rayleigh wave
associations are shown versus body wave magnitude in figure
7. Only two events in the northern hemisphere with m_b above
4 had no surface wave data and both these events were deeper
than 100 km. This interesting result shows that routine
analysis and reporting of long period surface waves are most
valuable, as these data can be confidently associated with
short period data. The result also shows that this association
can be carried out by an automatic process.

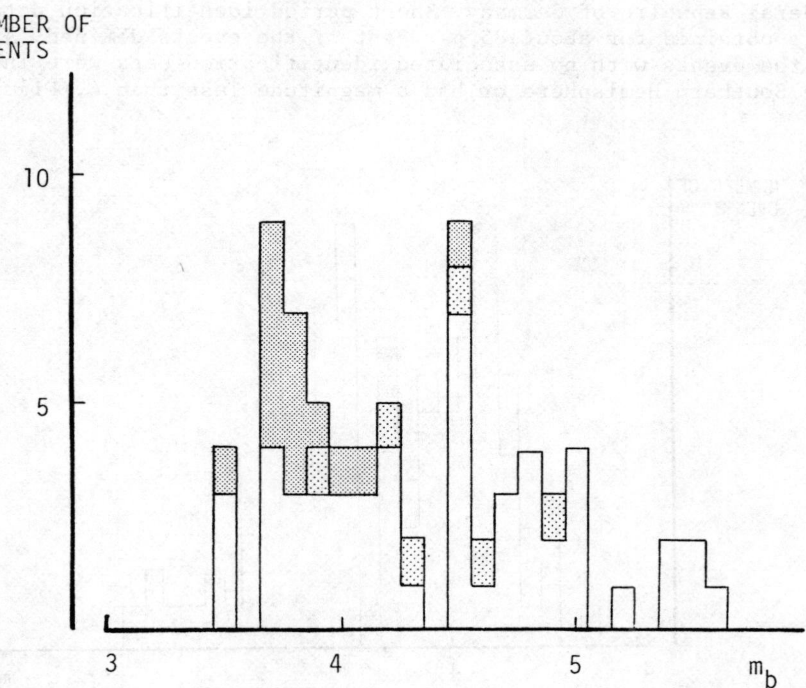

Figure 7. Number of events at various short period magnitudes
for which long period data have or have not been
obtained from the Hagfors Observatory or any of
the SRO's. The events for which no long period
data are obtained are separated in the northern
(▓▓) and southern (▒▒) hemisphere.

4.5 Short period identification data

One of the important new functions of the proposed International Datacenters concerns the compilation and reporting of
what is known as identification data. These data are intended
for the discrimination between underground explosions and earthquakes. International Datacenters should however not make any
assessment as to the nature of the events. The procedures to
compile short period identification data are still tentative
but it has been demonstrated that such data, obtained from a
number of stations, can be compiled without assessing the
nature of the event.

Short period identification data such as complexity and spectral
moment were calculated using data from nine SRO stations, the
Hagfors Observatory in Sweden and the Grafenberg array in the
Federal Republic of Germany. Short period identification data
were obtained for about 85 per cent of the events defined. All
of the events with no associated identification data were in
the Southern Hemisphere or had a magnitude less than 4. Figure

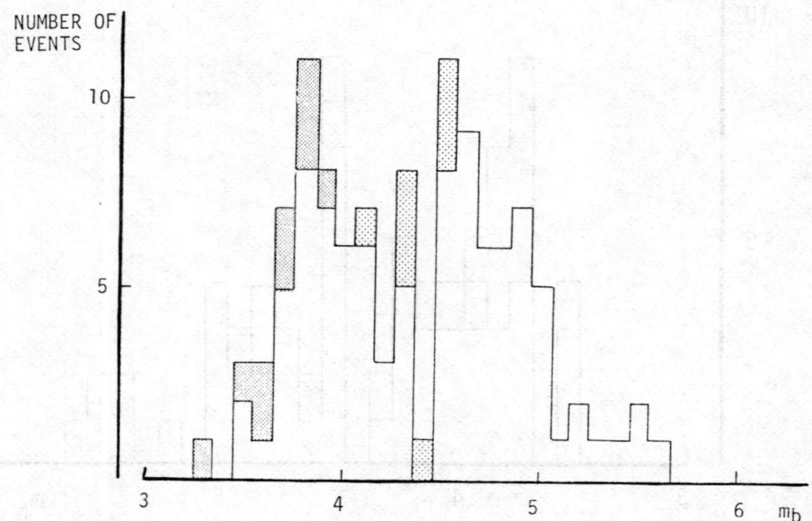

Figure 8. Number of events at various magnitudes for which
short period identification data have or have not
been obtained from the Hagfors Observatory or any
of the SRO's. The events for which no short
period identification data are obtained are
separated in the northern (   ) and southern (   )
hemisphere.

8 shows the distribution of events for which data from at least
one station could be obtained. Further studies should, however,
be carried out on a national level to estimate the identification
capability and the applicability of such compiled data. The main
obstacle here, as for the long period signals, is the present
lack of routinely reported identification data.

4.6 Data handling and exchange

In writing the data handling routines it was found that general
so-called data base systems for data handling, storage and
retrieval are inferior to more specialized routines, developed
for these specific purposes. The computer programs developed
for this demonstration have been implemented on large computers,
IBM 370 and DEC-10. The program could, however, have been
implemented on a specially designated computer of the size
described in the first report of the Seismic Ad Hoc Group.

The connexion to the WMO/GTS was made possible through a
close co-operation with the Swedish Meteorological and
Hydrological Institute. No specific technical problems were
encountered. The handling of seismic data is still unfamiliar
to operators on the WMO/GTS and our experience is that
testing, to familiarize these operators with seismic data,
is needed to get reliable data transmission.

The output bulletin has been prepared in two formats, one
containing only basic event information and one that contains also the information that has been reported from
stations which have been associated with the events (figure
9).

The compilation of complete records of both short and long
period signals, so-called Level II data from the 11 SRO
stations clearly shows the value of having the full records
obtained by the individual stations available, when assessing and interpreting a seismic event. It is therefore felt
that there will be a need for a substantial exchange of such
Level II data in a global verification system and that
efficient routines for the exchange and compilation of such
data should be established.

```
-  3.7 16.8S+- 0.1 173.8W+- 0.1  62KM+-  32 BASED ON  9 STAT.
ANDS
  ASSOC. SP-TIMES 15 NUMBER OF ASSOC. LP-TIMES  8
  5.0 BASED CN   7( I ) STAT   2( II) STAT  2(III) STAT 13( IV) STAT
  4.0 BASED ON   9( I ) STAT   1(III) STAT
  0.46    STD:      0.0   BASED ON   1 VALUES
  0.91    STD:      0.0   BASED ON   1 VALUES
  0.97    STD:      0.0   BASED ON   1 VALUES
VECTORS                    BASED ON   1 STAT.
      INITIAL PHASE               CODA
      0  -9 -24 -36 -43           0 -13 -26 -36 -45
      0   0   0   0   0           0   0   0   0   0


-  3.7 16.9S+- 0.1 173.8W+- 0.1  62KM+-  32 BASED ON  9 STAT.
ANDS
  ASSOC. SP-TIMES 15 NUMBER OF ASSOC. LP-TIMES  8
  5.0 BASED CN   7( I ) STAT   2( II) STAT  2(III) STAT 13( IV) STAT
  4.0 BASED CN   8( I ) STAT   1(III) STAT
  0.46    STD:      0.0   BASED ON   1 VALUES
  0.91    STD:      0.0   BASED CN   1 VALUES
  0.97    STD:      0.0   BASED ON   1 VALUES
VECTORS                    BASED ON   1 STAT.
      INITIAL PHASE               CODA
      0  -9 -24 -36 -43           0 -13 -26 -36 -45
      0   0   0   0   0           0   0   0   0   0

(P   )  73412.0   -0.9*   3.5   34.3 213.8  13.96                       2 0.80
(P   )  74034.6    0.6* 38.0  258.7  91.8    8.39        21.4 0.79 5.1 1 0.30
(P   )  74203.4   -0.6* 49.2  258.0  95.3    7.68        12.0 0.60 5.0 1 0.86
(P   )  74205.7    0.4* 49.3  253.0  92.2    7.67       193.7 1.00 6.0 1 0.01
(P   )  74447.1   -0.3* 73.4  180.0 185.3    5.82        14.3 0.50 5.3 1 0.42
(P   )  74512.0    2.9  77.2  266.7 105.6    5.54        25.0 0.60
(P   )  74526.1    0.9* 80.1   42.8 238.2    5.33                       2 0.67
(P   )  74535.2    0.3* 82.0   50.0 243.2    5.20        24.0 1.47 5.1 1 0.59
(P   )  74544.0   -0.4* 83.8   10.8 205.0    5.07        19.7 0.80 5.3 1 0.43
(P   )  74652.0   -0.5* 98.4   11.3 232.0    4.52         4.0 0.80 5.3 1 0.14
(PKP )  75252.6    0.2 145.0  344.5  23.4    3.88
(PKP )  75256.0    0.0 146.0  353.8   9.3    3.92
(PKP )  75258.4   -0.8 146.8    1.9 357.2    3.95
(PKP )  75259.0   -1.9 147.2  351.1  13.1    3.97
(PKP )  75305.6   -4.7 149.5    3.2 355.5    4.05        13.4 0.70

CMPLX              TMFI              TMFC
 0.46      0.0     0.91      0.0     0.97      0.0

(LZ  )  74750       75  26.4 195.8  25.5             253.0 29. 3.6 1 0.06
RTED:                                45.0
(LZ  )  75305       78  38.0 258.7  91.8             504.0 33. 4.1 1 0.83
RTED:                               100.0
(LZ  )  80520        0  63.6 241.2  93.6             247.0 34. 4.2 1 0.70
RTED:                                85.0
(LZ  )  80910       -5  69.8 320.4 130.7              93.0 21. 4.0 1 0.26
RTED:                               125.0
(LZ  )  80955       69  75.5 302.0 116.5             195.0 33. 4.2 1 0.61
RTED:                                95.0
(LZ  )  81145      -40  82.0  50.0 243.2             215.0 34. 4.3 1 0.43
(LZ  )  82000      112  92.8 288.7 106.7              63.0 32. 3.9 1 0.03
(LZ  )  83825      256 121.5 300.5  89.3              78.0 32. 4.1 1 0.11
RTED:                                90.0
```

Figure 9. *Examples of two output bulletins to be*
distributed from Datacenters, one con-
taining only basic event information
(above) and the other containing also
information reported from individual
stations (below).

5. FUTURE WORK

The temporary facilities established at the Hagfors Observatory for the purpose of the demonstration, will be
maintained and developed. The intention is that these
facilities will contribute to the work still to be done in
specifying the data handling and analysis routines which
should be implemented at International Datacenters. One
important task is to further develop the automatic checking
routines and also to compare the results of such automatic
processing with the results obtained through the interaction
in the processing by seismologists. In this connexion the
influence of seismologist interaction on the final result of
the datacenter analysis should be studied. The capability of
a global network of a limited number of stations to define
and locate events in different part of the world and to
obtain identification parameters should also be further
studied. The facilities could also be used to test, compare
and develop methods and ideas put forward by interested
seismological experts. It is our hope that this work, which
remains to be done to prepare the final implementation of a
global verification system, will be carried out in close
international co-operation.

ACKNOWLEDGEMENTS

We would like to thank the following institutes that provided special data for the experimental database:
Central Seismological Observatory Grafenberg in Federal
Republic of Germany, the NEIS of the US Geological Survey
and Teledyne Geotech in Alexandria, USA.

Discussions with members of the staffs at the NEIS and at
the International Seismological Center (ISC) in the UK are
much appreciated.

We are also thankful to the Stockholm University Computer
Center for its assistance in the work with the computer and
to the Swedish Meteorological and Hydrolocial Institute for
its cooperation in establishing the connection to the WMO/GTS.

REFERENCES

Dahlman, O. and Israelson H., 1977, Monitoring Underground
Nuclear Explosions, Elsevier, Amsterdam, 440 pp.

CCD/558, 1978. Report of the Ad Hoc group of scientific
experts to consider international co-operative measures to
detect and identify seismic events, Disarmament Conference

CD/43, 1979. Second Report of the Ad Hoc group of scientific
experts to consider international co-operative measures to
detect and identify seismic events, Committee on Disarmament
Document.

CD/130, 1980. Letter dated 30 July 1980 from the Permanent
Representatives of the Union of Soviet Socialist Republics,
the United Kingdom of Great Britain and Northern Ireland and
the United States of America transmitting a document entitled
"Tripartite Report to the Committee on Dirsarmament".

APPENDIX

List of publications intended to give a more comprehensive
technical documentation of the various components of the
demonstration facilities and of the preliminary results
obtained. The list also gives the scientists, who has been
responsible for the development of each component. According
to the present plans the reports will be published in the
autumn 1980.

Gunnel Barkeby   International Seismological Datacenter:
                      Input for an experimental database.

Gunnel Barkeby   International Seismological Datacenter:
                      Output of an experimental database.

Gunnel Barkeby   International Seismological Datacenter:
                      Database structure, computer facilities,
                      automatic and interactive analysis.

Eva Elvers   International Seismological Datacenter:
                      Event checking procedures using dynamic
                      information. Magnitude estimation.

Hans Israelson,   International Seismological Datacenter:
Ingvar Jeppsson,   Preparation of an experimental database.
Gunnel Barkeby

Hans Israelson   International Seismological Datacenter:
                      Compilation of identification data.

Ragnar Slunga   International Seismological Datacenter:
                      An algorithm for associating reported arrivals
                      to a global seismic network into groups de-
                      fining seismic events.

DESIGN AND DEVELOPMENT OF A SEISMIC DATA CENTER

Michael A. Chinnery and Alvin G. Gann

Applied Seismology Group, Lincoln Laboratory, M.I.T.,
Cambridge, Massachusetts 02142, U.S.A.

Ann U. Kerr

Defense Advanced Research Projects Agency, Arlington,
Virginia 22209, U.S.A.

The U. S. has embarked on the design and development of a
sophisticated data center for the analysis and management of
seismic data. There are two broad motivations for this project.
First, we need to develop a new generation system for the
management and retrieval of digital seismic data, and to provide
a modern data resource for the research community. Second, we
need a facility that will be able to respond to any U. S.
obligations that might be incurred under future agreements on
international seismic data exchange and global seismic monitoring,
such as those currently under discussion by the Committee on
Disarmament. This paper discusses the functions that the data
center will be required to perform, and its basic computer
architecture.

INTRODUCTION

The need for a modern seismic data center has arisen from
two recent developments. First, as seismology enters the
transition from reliance on analog paper and film data to full
use of high dynamic range digital data, the existing methods
for the storage and distribution of data are inadequate and
must be replaced by more sophisticated systems. Second, international discussions that may lead to agreements on international seismic data exchange and global seismic monitoring
will require the development of systems that will accept both

*E. S. Husebye and S. Mykkeltveit (eds.), Identification of Seismic Sources – Earthquake or Underground
Explosion, 831–842.*

parametric data reports and seismic waveform data, perhaps in
close to real time, and rapidly prepare an event bulletin for
general distribution.

In response to these needs, the U. S. in developing a
state-of-the-art Seismic Data Center (SDC) that will fulfill
the following objectives in support of the seismic research
community.

1.   To provide a state-of-the-art seismic data management
      facility for the storage and retrieval of digital
      seismic data.

2.   To provide a test bed for the evaluation of a variety
      of automatic signal processing techniques for digital
      seismic data.

3.   To develop the most efficient techniques for the
      rapid preparation of a comprehensive event bulletin,
      using both digital waveform data and station seismic
      arrival reports.

4.   To provide a variety of services to the research
      community, including both local and remote access to
      all archived data and event bulletins.

The objectives in support of international exchange of
seismic data are as follows:

1.   To provide a facility that will meet all anticipated
      requirements for international exchange of seismic
      data.

2.   To provide a test bed for evaluating the hardware,
      software and processing techniques that may be
      required under contemplated agreements on inter-
      national exchange of seismic data.

It is not possible, at present, to formulate a detailed
set of the functions that this data center will be required to
perform. New digital stations are still being installed, and
considerable research is needed before we can fully understand
the best ways to use this new type of data. In the international
arena, agreements on global seismic monitoring have yet to be
concluded, and we cannot predict the detailed requirements
that these agreements will place on the center.

In this paper, therefore, we describe in rather general
terms the types of data that the SDC will receive, the types
of products that it must output, and the storage, retrieval

and processing facilities that must be available. We also
briefly describe the basic computer architecture of the proto-
type SDC now being developed at Lincoln Laboratory.

INPUT DATA CHARACTERISTICS

The Seismic Data Center will be required to accept data
from a wide variety of sources (see Figure 1). Three main
categories of data are anticipated. The first consists of
data reports or level I data (1). These are measurements made
by station operators on observed seismic arrivals, and include
a variety of parameters such as arrival time, amplitude and
phase identification. The second consists of digital waveform
data from both single stations and arrays. Some of these
stations may be U. S. participants in some form of International
Exchange of Seismic Data (IESD); for these stations, one task
of the data center will be to extract data reports from the
waveforms, and distribute them to all IESD participants. The
third may consist of actual lists of events, with preliminary
event parameters. Such lists are frequently prepared by array
stations and certain existing earthquake information organizations.

These data types will arrive at the SDC in a variety of
different ways, and with highly variable delays (see Figure
1). One significant task of the SDC will be to ensure that
each of these sets of data is properly entered into the SDC
data base.

The rates at which these data enter the system will place
an important constraint on the sytem design. A modern three
component digital seismic station may generate data at a rate
in excess of 2 kbits/second. If a network of such stations
were to feed continuous data into the Seismic Data Center,
this data source would dominate all others by far, in terms of
both processing and storage requirements. We have selected a
system design input rate of 125 kbits/second, since this would
appear to provide ample capacity for the forseeable future.
We also require that the system be able to handle twice this
data rate during catch-up periods (e.g., following a hardware
failure, and that the system could be adapted to an average
input rate of 250 kbits/second by simply adding more hardware,
without invalidating the existing hardware.

DATA STORAGE AND RETRIEVAL

Requirements on the data base management system concern
its capacity and its speed of retrieval. The design data rate

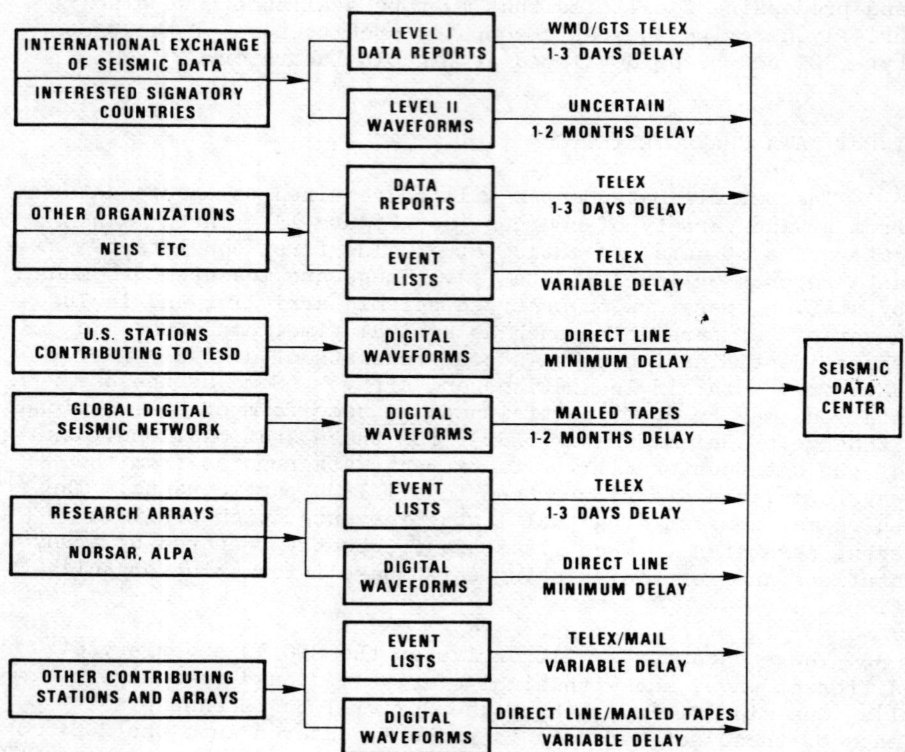

Figure 1. Seismic Data Center Input Data Streams.

of 125 kbits/second corresponds to 10^{10} bits/day. This requirement can be met by a combination of on-line disk storage and off-line tape storage (commercially available disk units can store about 5×10^9 bits, and high density tape units can store about 10^9 bits on a 2400 foot tape). We have selected this tape/disk solution for early SDC development. We also recognize that other systems (such as optical disks) may become available within the next few years. The SDC architecture must have the flexibility that such devices can be incorporated into the system with little problem at a later stage.

The required rate of retrieval of data from on-line disk storage depends somewhat on the type of processing carried out at the SDC. We do not anticipate that there will normally be a requirement to retrieve large quantities (say, all data for an entire day) in one transmission. However, in order not to exclude this possibility, we require that data can be

moved from disk storage to other parts of the SDC at a minimum
of an order of magnitude faster than real time (this corresponds
to a net signalling rate of at least 2 Mbits/second). With
this signalling rate, small quantities of data (say, a few
waveforms), can be retrieved from disk storage in a fraction
of a second.

Retrieval of data from off-line tape storage will inevitably
take longer. We, therefore, set the requirement that, under
normal conditions, all input data will be stored on-line
until analysis of that data is completed. Retrieval from
tape will then be limited to a small number of requests,
primarily from research scientists, and we require simply
that all the waveforms for a given event be on a single tape.
The time needed to respond to a request for such data will
then be the time necessary to select a tape, mount it and
read it into the system. Depending on the total volume of
data input to the center, each tape will contain all available
data for a period ranging from 2 hours to 1 day.

In addition to routine functions, we also require that
the data base have the capacity to store certain special data
sets on-line. These would include reference events, used to
aid an analyst in seismic interpretation, and data sets
accumulated in response to a user request.

ROUTINE OUTPUT PRODUCTS

The SDC will produce a wide variety of routine output
data and provide a range of user services. These are summarized
in Figure 2.

There are two principal routine products. First, the
SDC must produce a comprehensive event bulletin. This will
provide a data base in itself for some users; to others, it
will function as an index to the waveform data stored at the
SDC. International discussions (1) have suggested that this
bulletin should be available with a net delay of 3-5 days
from real time. A bulletin produced this quickly will also
be of substantial use to the research community. However, we
note that some data may not be available at the SDC on this
time scale, and this suggests that it may be necessary to
prepare several bulletins. We require that the SDC be able
to compile an event bulletin within 24 hours of the receipt
of the last piece of input data used in its preparation (this
assumes 7 days a week SDC operation). This constraint will
satisfy all anticipated requirements.

The contents of the event bulletin are less well defined.

Certainly, the bulletin must contain all the event parameters
currently included in such publications as the Earthquake
Data Reports (of U. S. Geological Survey). We anticipate
that current and future research may lead to substantial
improvements in this list of parameters, and parameters
relating for example, to source mechanism may be included.
One immediate addition will be information about the availability
of waveform data for listed seismic arrivals.

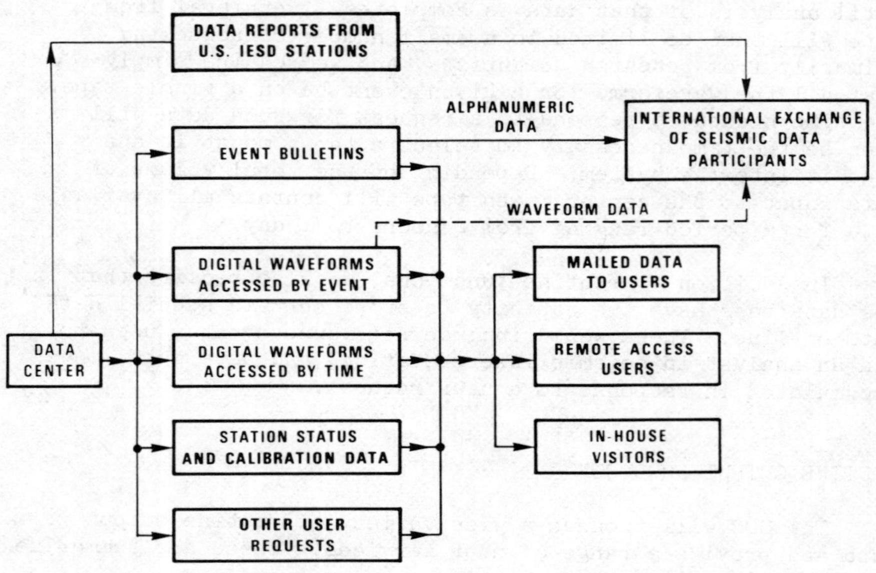

Figure 2.   Data Center Products and User Interfaces.

     The second important routine product will be access to
the waveform data base. We anticipate that user requests for
waveform data will be of two kinds. The user may request
waveform data for a particular event, as listed in the event
bulletin; or he may request the waveform data from a given
station (or stations) for a given time interval. Where
building up a large data set, he may leave a standing order
with the SDC for certain specific event types.

     IESD output products will follow specifications laid out
in international agreements. At present, we anticipate
general distribution of data reports from U. S. stations
which are included in IESD and an event bulletin, both via
the WMO/GTS network. Level II waveform data will be supplied

by whatever methods are specified by the international agreements.

THE USER INTERFACE

We plan to provide a variety of methods whereby the user
can interact with the data center. For those who can visit
the center, in-house computer facilities will be provided for
interaction with the data base, local interactive display and
analysis.

For users who are unable to reach the center several
systems will be provided. Data requests sent by mail or
telephone will be satisfied by mailed tapes or transmission
over a low data rate channel, such as a dialup phone connection.
The latter will be adequate for the transmission of event
bulletins, or small quantities of digital data.

Where larger amounts of data are required, a more
sophisticated form of access would be needed for local recording and computational facilities. We anticipate that the
capability required would range from a simple microcomputer
based terminal to something like a Seismic Analysis Station,
described later in this paper, depending on the level of
support required. Such a terminal would be connected to the
data center by a high data rate communications link.

Certain types of analytical tools will also be made
available to the user. Examples include searching the event
bulletin on one or more of the event parameters, carrying out
filtering spectral analysis, rotation of horizontal components
and other kinds of time series analysis, and analytically
modifying the instrument response of the stored data (e.g.,
to produce a broad-band signal). Within the limitations of
system capacity, users will also be able to apply user-
generated processing methods to the data.

DATA PROCESSING AT THE SDC

In order to translate the input data (Figure 1) into the
output data products (Figure 2), it will be necessary to
carry out a series of seismic processing steps within the
SDC. These processing steps will set some important requirements on the SDC design.

Figure 3 shows, in schematic form, the processing steps
involved in the preparation of a final event bulletin. There
are two basic sets of procedures. The first is carried out
completely automatically on the input data streams; the

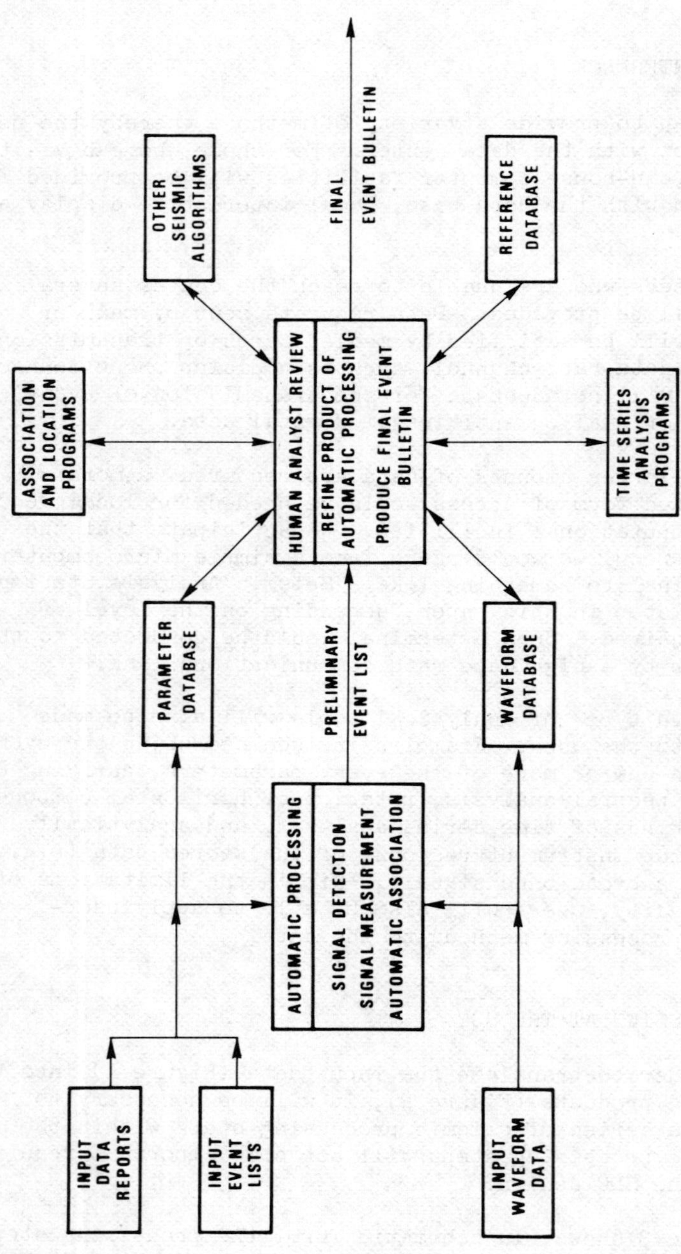

Figure 3.   Data Center Functions and Procedures.

second is carried out by a human analyst, who will refine the results of the automatic processing steps using a variety of software tools and with access to all available data.

The actual processing tasks that will be carried out automatically are not yet clear. Many existing systems include only signal detection in this category; however, there are many indications in the literature that parameters such as arrival time, amplitude, dominant period and signal character can usefully be measured by the application of automatic algorithms. A program of research is being carried out in the U. S. with the objective of evaluating and implementing these concepts. We anticipate that this will substantially reduce the amount of human analyst time necessary in event bulletin preparation.

Based on our present experience, however, there is no way in which the human analyst can be eliminated from the processing procedures. The subjective judgement and experience of the mind are essential to the production of an acceptable event bulletin. This places a requirement on the system for a sophisticated interactive man-machine interface. The analyst must be able to interact directly with event lists, lists of signal characterization parameters, and digital waveforms. He must be supplied with a wide variety of software tools, including time series analysis programs, association and location programs, and programs to aid him in the determination of event parameters such as magnitude and source mechanism.

Such an interactive "seismic analyst station" will include all of the basic features that a researcher will need in order to interact with the system. Our development of a seismic analyst station is, therefore, aimed at devising a very general purpose interactive seismic terminal, which (with minor software variations) will satisfy the needs of the operational analyst, the visiting research scientist and the remote terminal operator.

SYSTEM ARCHITECTURE

The computer architecture which has been selected for the implementation of the Seismic Data Center consists of multiple minicomputers interconnected by a high data rate local computer network. Such an arrangement provides great flexibility, and easy expansibility to meet requirements that may develop in the future.

Within this architecture, it is convenient to subdivide the overall system into a series of subsystems, each of which

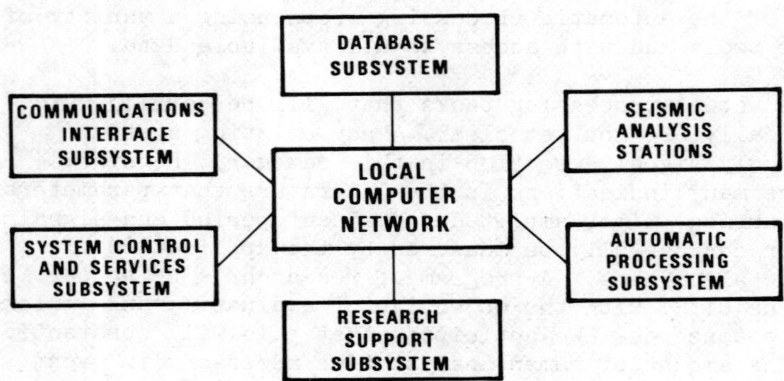

Figure 4. Conceptual Seismic Data Center Configuration.

corresponds to a major functional component of the system
(see Figure 4). In particular, each subsystem will consist
of one or more computers with attached peripheral devices.
The principal features of these subsystems are summarized
below:

1. The Database Subsystem consists of two parts: the
    Parameter Database and the Waveform Database. The
    Parameter Database stores all of the alphanumeric
    data and provides a data management and retrieval
    system for events and arrivals. The Waveform
    Database provides a facility for maintaining and
    accessing the very large amounts of waveform data
    which the system acquires and processes.

2. The Research Support Subsystem will carry out a
    variety of computational tasks in support of research
    users, both local and remote. This support will
    include the selection of data sets for study, the
    analysis of both parametric and waveform data and
    the development of new algorithms and computer
    programs.

3. The System Control and Services Subsystem will
    carry out a variety of tasks, including the prepara-
    tion of data for distribution as requested by
    authorized outside users, development of new software
    for the system, and providing other basic system
    services.

4. The Automatic Processing Subsystem will carry out

automatic (non-interactive) computational tasks
related to detecting the arrival of seismic signals
from the digital waveform data and the automatic
association of arrival information into the preliminary characterization of events.

5. The Communications Interface Subsystem is responsible
for all data input, and certain types of data
output.

6. The Local Computer Network Subsystem interconnects
the computers and provides all intercomputer communication of data and control information. The local
computer network is a key element in the system.
It is required to be extremely reliable since it
carries all intercomputer communications including
all the data and all the command and status traffic
in the system.

7. The Seismic Analysis Stations provide both seismic
analysts and users with the ability to perform
extensive computations on both the waveforms and
the elements of the parameter database. Each of
these is a single-user interactive computer system,
with display devices, local input-output peripherals,
and substantial disk storage.

Further details of these subsystems can be found elsewhere
(2).

CONCLUDING REMARKS

Initiatives arising from international discussions about
seismic data exchange and global seismic monitoring have
stimulated the design and development of a data storage and
retrieval facility that will fully utilize state-of-the-art
computer techniques. The challenge now is to ensure that
this data center will be fully responsive to the data needs
of the seismic research community. We seek the advice and
assistance of the research community, as we attempt to formulate
these requirements.

ACKNOWLEDGEMENTS

This research has been supported by the Defense Advanced
Research Projects Agency.

REFERENCES

1. *Report of the Conference of the Committee on Disarmament of the Ad Hoc Group of Scientific Experts to Consider International Co-operative Measures to Detect and to Identify Seismic Events*, Document CCD/558, Conference of the Committee on Disarmament, Geneva, 8 March 1978.

2. Chinnery, M. A., and Gann, A. G., *Design and Development of a Seismic Data Center*, Technical Note 1980-39, Lincoln Laboratory, M.I.T., 11 August 1980.

EVALUATION OF SEISMIC NETWORK CAPABILITY FOR EVENT LOCATION

William L. Rodi

Systems, Science and Software
P. O. Box 1620
La Jolla, California 92038
U.S.A.

INTRODUCTION

An important element in the evaluation of a seismic network
for nuclear monitoring is measuring the network's capability for
event location. Described here is an algorithm, based on linear
inverse theory, for predicting the location capability of a real
or hypothetical network and determining the importance of individual data observed by the network. The data may include the
arrival times of regional and teleseismic P and S phases, including pP, and backazimuth estimates derived from Lg. The data
importances allow one to rank stations or phases by the information they contribute to the network, and thus are a valuable
guide in the design of improved networks.

METHOD

The network evaluation algorithm is a multiphase extension
of an earlier algorithm developed by M. H. Wirth (1). The new
algorithm assumes a linear inverse formulation of the event
location problem. An optimal estimate of the four-vector of
location parameters (epicentral coordinates, focal depth and
origin time) is defined as the minimum-variance linear estimate
derived simultaneously from all the network data. The 4 by 4
parameter covariance matrix of the estimate can be determined
knowing only the variances of the data errors and the partial
derivatives of the data with respect to the location parameters.
By assuming a probability distribution for the data errors
(Gaussian), the parameter covariance matrix can be expressed as
a four-dimensional confidence region for the location estimate.

843

E. S. Husebye and S. Mykkeltveit (eds.), Identification of Seismic Sources - Earthquake or Underground Explosion, 843–850.

Instead of using the four-dimensional confidence region, the algorithm measures the location capability of a network in terms of separate confidence regions determined for the epicentral coordinates and focal depth. These are obtained from the marginal variances of the separate parameter estimates; therefore, they reflect trade-offs between all four of the location parameters (i.e., none of the parameters is assumed to be known or fixed). The distinction between network capabilities for epicenter and depth determination is important in nuclear monitoring problems.

In practice, the location capability of a network is degraded by the absence of data from undetected phases at some stations. This source of uncertainty in network location estimates is described by detection probabilities assigned to the phases at each station. The detection probabilities depend on the event magnitude as well as its true location. The effect of the detection probabilities P_i on the expected average network performance can be treated approximately in the linear inverse formulation by increasing the variance of each datum by the factor P_i^{-1}.

The algorithm for data importances is based on concepts developed by Minster, et al. (2). For a given event, two importances are calculated for each network datum: its importance for determining the event epicenter and its importance for determining the focal depth. They are defined by the increase in the size of the confidence regions that would result by omitting the datum from the data set and locating the event with the remaining data. Denoting the area of the epicenter confidence ellipse as E and the length of the depth confidence interval as d, the importances of the i'th datum are defined by

$$\text{Epicenter Importance} = \frac{E \text{ (without i'th datum)}}{E \text{ (with i'th datum)}}$$

$$\text{Depth Importance} = \frac{d \text{ (without i'th datum)}}{d \text{ (with i'th datum)}} .$$

The importances take values between one and infinity, and larger values imply more important data.

Importances can be computed directly from the full network data variances and the partial derivative matrix A, or equivalently, from the eigenvalues and eigenvectors of A. It is not necessary to repeat confidence region calculations with each datum dropped in turn. The importances defined here are, in fact, related to the information density matrix S (3), which is obtained from the column eigenvectors of A. The diagonal elements of S measure the first-order sensitivity of the confidence regions to the data variances.

EXAMPLES

Figures 1 through 3 give examples of ranking stations in a network based on their importances for determining the epicenter and depth of an event. In each example, it is assumed that only the first motion P wave arrival time is determined at each station. Thus, the data importances directly reflect the importances of the stations themselves. At each station the probability of detecting the P wave is assumed to be one and the arrival-time standard error to be 1.0 second. Therefore, in these examples, the depth and epicenter importances of a station depend on the station's location relative to the event and the other stations in the network, rather than on data quality.

Figure 1 ranks the stations in each of three hypothetical networks distributed in simple patterns about an event. Each network is shown twice to give the station ranking by epicenter importance and depth importance. The ranks are plotted at the station locations and they are assigned in order of decreasing importance (Station 1 is the most important in the network). Because of the symmetry in the networks, two stations often have equal importances and rank. One should note that the importance values themselves are not shown and the rank defines only an ordering of the stations by importance.

The networks in Figure 1 each have six stations at three epicentral distances from the event (Δ = 20°, 40°, 60°), but they differ in their azimuthal distributions. Network 1 provides the best, and Network 3 the worst, azimuthal coverage. As a result, the epicenter confidence ellipse (not shown) is largest for Network 3 and smallest for Network 1. The networks determine very nearly equal focal depth confidence intervals, however, since depth determination depends primarily on distance coverage, rather than azimuthal coverage.

The importance rankings shown in Figure 1 display clearly the influence of station location on station importance. For epicenter determination, the stations closest to the event and those near any gaps in azimuthal coverage are the most important stations in a network. In Network 3, one can see that when a large gap in azimuthal coverage occurs, proximity to the gap overrides epicentral closeness in controlling the importance of a station. The depth importance rankings show a different effect. The stations closest to the event and farthest from the event are the most important for determining focal depth. Stations at intermediate epicentral distances, compared to the distance range spanned by the network, are least important. Furthermore, station azimuth is not a critical factor for determining focal depth from P arrival times.

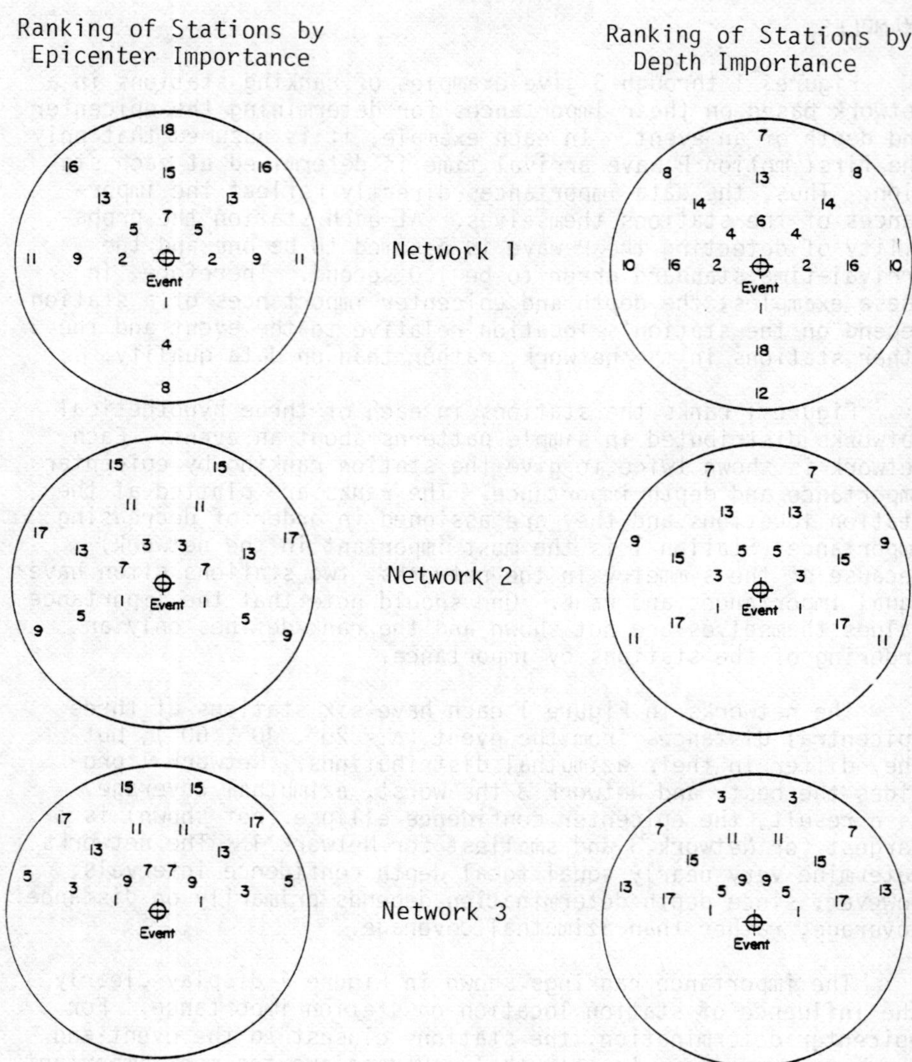

Figure 1. The stations within each of three simple networks are ranked by their importances for determining the epicenter and focal depth of an event based on P wave arrival times. The networks each have 18 stations distributed at three distances from the event ($\Delta = 20°$, $40°$, and $60°$) but the azimuthal distributions of the networks differ. The rank of a station is plotted at the station location (1 for the most important station in the network, 2 for the second most important station, etc.).

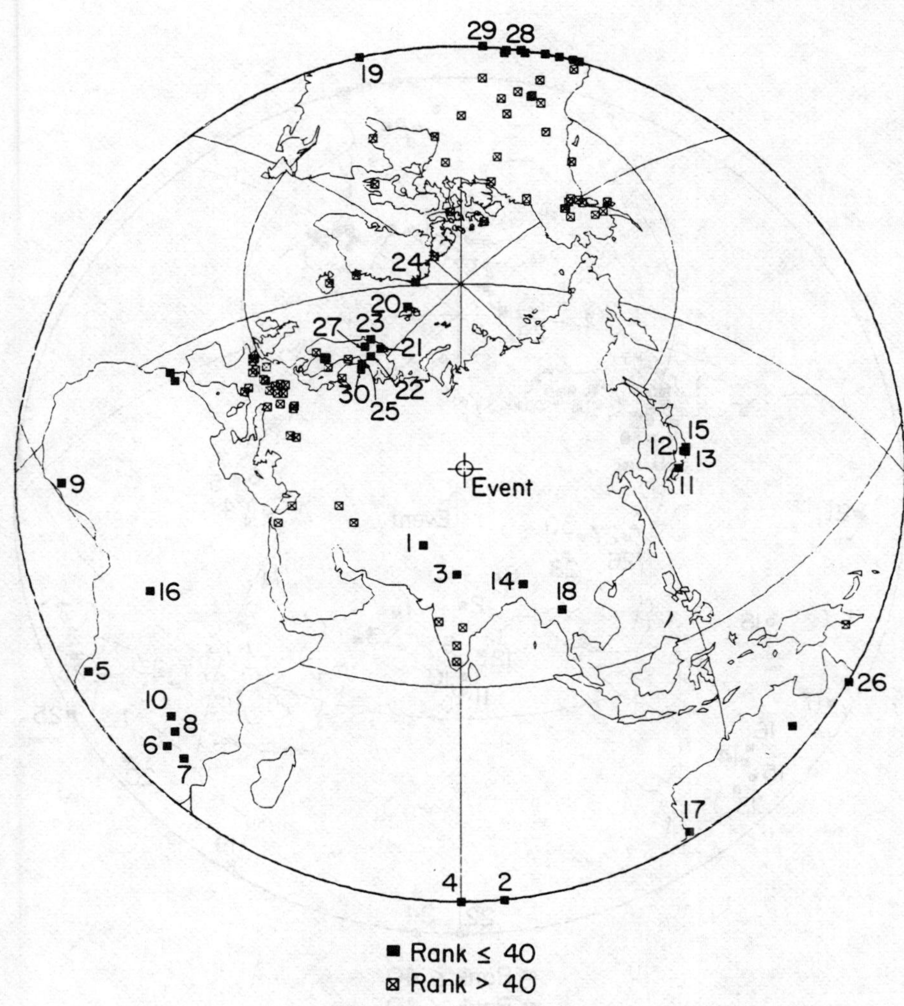

Figure 2. Ranking of 113 worldwide stations by their importance for locating the epicenter of an event (⊕) in Eurasia. The 40 most important stations are plotted as solid squares and the ranks of the 30 most important stations are indicated. Stations at epicentral distances greater than 90° are plotted on the circumference of the plot (Δ = 90°).

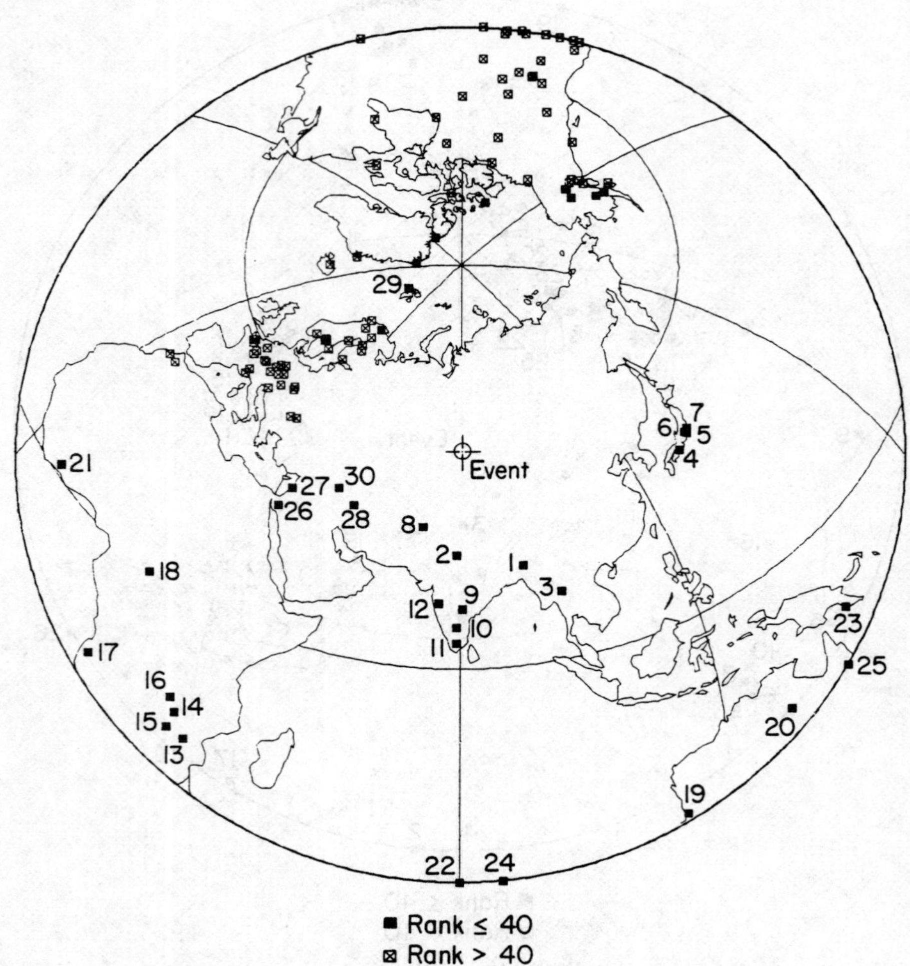

Figure 3. Ranking of 113 worldwide stations by their importance for determining the focal depth of an event in Eurasia. The 40 most important stations are plotted as solid squares and the ranks of the 30 most important stations are indicated. Stations at epicentral distances greater than 90° are plotted on the circumference of the plot (Δ = 90°).

Figures 2 and 3 show the importance rankings of actual worldwide stations for locating an event in Eurasia. The 113 stations in this network were selected as follows. Three events with similar magnitudes ($m_b \sim 5$) and nearly identical locations were selected from the 1969-1970 Bulletins of the International Seismological Centre (ISC). Stations reporting P arrivals from one or more of these events were included in the network for this example. While this criterion defines a representative network of stations operating in 1969-1970, it should be noted that considerably fewer stations typically report a single $m_b = 5$ central Asian event (≤ 65 for the three events chosen). Furthermore, in this example the stations were treated as equal in quality, although many reported only one or two of the events or reported emergent P arrivals. Therefore, this example serves mainly to illustrate the effects of worldwide station geometry, rather than to evaluate the true capability of worldwide stations for locating $m_b = 5$ Eurasian events.

In Figures 2 and 3, the 40 highest ranking stations are shown as solid squares and the rest as open squares with crosses. The ranks of the 30 most important stations are labeled. The rankings display a dependence on station distribution similar to that of the hypothetical networks. The stations ranking highest in epicenter importance (Figure 2) are generally either at the closest epicentral distances (e.g., southern Asia) or in sparsely covered azimuthal ranges (e.g., Japan and South America). Stations at both long and short epicentral distances rank high in importance for determining event depth (Figure 3).

The variation in station density in this example (Figures 2 and 3) complicates the relationship between station importance and station distribution, an effect that was not a factor in the earlier examples. Stations in a cluster tend to have small importances individually, although as a group their joint importance may be quite large. The European and Fennoscandian stations in Figure 2 are the best example of this. They are fairly close to the event and cover an otherwise empty azimuthal range, but due to their number and dense spacing each individual station is relatively unimportant for determining the event epicenter.

CONCLUSION

The examples presented here are rather simple applications of the network evaluation algorithm. Only P wave data were used and detection probabilities and station quality were not taken into account. Moreover, the method at this time does not fully account for biases in arrival-time data caused by path anomalies, which are known to be a significant source of error in event locations. Nonetheless, the examples demonstrate that this

quantitative approach to measuring network and station performance
for event location is a useful tool for the design of improved
networks, and also for acquiring a better understanding of location techniques.

ACKNOWLEDGMENTS

This work was supported by the Advanced Research Projects
Agency (ARPA), Contract No. MDA903-79-C-0670 and performed at
Systems, Science and Software with Boris Shkoller and John M.
Savino.

REFERENCES

(1) Wirth, M. H.: 1970, "Estimation of Network Detection and
     Location Capability," Research Memorandum, Air Force
     Technical Applications Center, August (revised October 1977
     by R. Blandford and H. Husted).

(2) Minster, J. B., Jordan, T. H., Molnar, P. and Haines, E.:
     1974, "Numerical Modelling of Instantaneous Plate Tectonics,"
     Geophys. J. R. astr. Soc., 36, pp. 541-576.

(3) Wiggins, R. A.: 1972, "The General Linear Inverse Problem:
     Implication of Surface Waves and Free Oscillations for Earth
     Structure," Rev. Geophys., 10, pp. 251-285.

APPENDIX

Scientific Directors, Lecturers, Participants

SCIENTIFIC DIRECTORS

Best, W.   AFOSR/NP
                            Bolling Air Force Base
                            DC 20332
                            USA

Husebye, E.S.   NTNF/NORSAR
                            Post Box 51
                            N-2007 Kjeller
                            NORWAY

Rodean, H.C.   Lawrence Livermore National Laboratory
                            Post Box 808
                            Livermore, CA 94550
                            USA

LECTURERS

Aki, K.   Dept. of Earth and Planetary Sciences
                            54-525
                            M.I.T.
                            Cambridge, Mass. 02139
                            USA

Blandford, R.R.   Teledyne-Geotech
                            Post Box 334
                            Alexandria, VA 22313
                            USA

Cara, M.   Institut de physique du globe
                            4 Place Jussieu
                            F-75230 Paris Cedex 5
                            FRANCE

Christoffersson, A.   Dept. of Statistics
                            Uppsala University
                            Post Box 513
                            S-751 20 Uppsala
                            SWEDEN

Dahlman, O.   Hagfors Observatory
                            FOA 202
                            S-10450 Stockholm
                            SWEDEN

Doornbos, D.

NTNF/NORSAR
Post Box 51
N-2007 Kjeller
NORWAY

Douglas, A.

MOD/PE
Blacknest, Brimpton
Nr. Reading RG7 4RS
UNITED KINGDOM

Dziewonski, A.M.

Department of Geological Sciences
Harvard University
Cambridge, Massachusetts 02138
USA

Gubbins, D.

Dept. of Geodesy & Geophysics
University of Cambridge
Madingley Rise, Madingley Road
Cambridge CB3 0EZ
UNITED KINGDOM

Harjes, H.-P.

Ruhr-Univesität Bochum
Institut für Geophysik
Postfach 10 21 48
D-4530 Bochum 1
FEDERAL REPUBLIC OF GERMANY

Harkrider, D.G.

California Institute of Technology
Seismological Laboratory   252-21
Pasadena, California   91125
USA

Kennett, B.L.N.

Dept. of Applied Mathematics and
  Theoretical Physics
Silver Street
Cambridge CB3 9EW
UNITED KINGDOM

Knopoff, L.

University of California, Los Angeles
Institute of Geophysics and Planetary
  Physics
Los Angeles, CA 90024
USA

Madariaga, R.

Institut de physique du globe
4 Place Jussieu
F-75230 Paris Cedex 5
FRANCE

Marzetta, T.   Schlumberger-Doll Research
P.O. Box 307
Ridgefield, CT 06877
USA

Müller, G.   Institut für Meteorlogie und
  Geophysik
Feldbergstrasse 47
D-6000 Frankfurt a.M.
FEDERAL REPUBLIC OF GERMANY

Peterson, J.   U.S. Dept. of Interior
Geological Survey, Branch of Global
  Seismology
Building 10002, Kirtland AFB-East
Albuquerque, New Mexico 87115
USA

Ringdal, F.   NTNF/NORSAR
Post Box 51
N-2007 Kjeller
NORWAY

Rodean, H.C.   Lawrence Livermore Laboratory
Post Box 808
Livermore, California 94550
USA

Tjøstheim, D.   University of Bergen
Mathematics Institute
Allegt. 53-55
N-5014 Bergen U
NORWAY

PARTICIPANTS

Alptekin, O   Black Sea Technical University
Faculty of Earth Sciences
Department of Geophysics
Trabzon
TURKEY

Asudeh, I.   Dept. of Earth Sciences
Bullard Laboratories
Madingley Road, Madingley Rise
Cambridge CB3 0EZ
UNITED KINGDOM

Baer, M.	Institute for Geophysics ETH CH-8093 Zurich SWITZERLAND
Bastians, R.	Vening Meinesz Lab/IVAU P.O. Box 80.021 3508 TA Utrecht THE NETHERLANDS
Berteussen, K.-A.	Dept. of Geophysics University of Oslo Post box 1022, Blindern Oslo 3 NORWAY
Brandes, W.	Erdbebenstation Bensberg Geologisches Institut Zülpicherstr. 49 5 Köln Lindenthal FEDERAL REPUBLIC OF GERMANY
Brüstle, W.	Universitätsinstitut für Meteorologie und Geophysik D-6000 Frankfurt a.M. Feldbergstrasse 47 FEDERAL REPUBLIC OF GERMANY
Buchen, P.	The University of Sydney Department of Applied Mathematics Sydney, N.S.W. 2006 AUSTRALIA
Bulin, G.	Defense Advanced Research Projects Agency 1400 Wilson Blvd Arlington, VA 22209 USA
Burdick, L.	Lamont-Doherty Geological Observatory Columbia University Palisades, NY 10964 USA
Chinnery, M.A.	Massachusetts Institute of Technology Lincoln Laboratory/Applied Seismology Group 42 Carleton Street Cambridge, Mass. 02142 USA

Clarke, T.   D.A.M.T.P.
                              Silver Street
                              Cambridge CB3 9EW
                              UNITED KINGDOM

Cleary, J.   Australian National University
                              Research School of Earth Sciences
                              Canberra, ACT
                              AUSTRALIA

Dainty, A.   Georgia Institute of Technology
                              School of Geophysical Sciences
                              225 North Avenue
                              Atlanta, Georgia   30332
                              USA

Day, S.   Systems, Science and Software
                              P.O. Box 1620
                              La Jolla, California   92037
                              USA

Deschamps, A.   Institut de Physique du globe
                              4 Place Jussieu
                              F-75230 Paris Cedex 05
                              FRANCE

Eisenhauer, T.   AFTAC/TJF
                              Patrick Air Force Base
                              Florida 32925
                              USA

Ekern, J.   GECO
                              Boks 2508 Ullandhaug
                              4001 Stavanger
                              NORWAY

Faber, S.   Universitätsinstitut für
                               Meteorologie und Geophysik
                              Feldbergstrasse 47
                              D-6000 Frankfurt a.M.
                              FEDERAL REPUBLIC OF GERMANY

Ferguson, J.   Southern Methodist University
                              Dallas Seismological Observatory
                              Dallas, Texas 75222
                              USA

Finckh, P.	Institute for Geophysics ETH CH-8093 Zurich SWITZERLAND
Fitch, T.	Massachusetts Institute of Technology Lincoln Lab/Applied Seismology Group 42 Carleton Street Cambridge, Mass. 02142 USA
Green, A.	Energy, Mines and Resources Earth Physics Branch Division of Seismology and Geothermal Studies 1 Observatory Crescent Ottawa Canada K1A 0Y3
Greenfield, R.	The Pennsylvania State University College of Earth and Mineral Sciences University Park, Pennsylvania 16802 USA
Häge, H.	Universitätsinstitut für Meteorologie und Geophysik Feldbergstrasse 47 D-6000 Frankfurt a.M. FEDERAL REPUBLIC OF GERMANY
Handelman, M.	The Aerospace Corporation Building A2 - Room 2043 Post Box 92957 Los Angeles, California 90009 USA
Hart, R.	Sierra Geophysics 150 N. Santa Anita Avenue Arcadia, California 91006 USA
Hauksson, E.	Lamont-Doherty Geological Observatory Columbia University Palisades, NY 10964 USA
Hjelme, J.	Geodetic Institute Seismic Division 22, Gamlehave alle DK-2920 Charlottenlund DANMARK

Hjortenberg, E.   Geodetic Institute
                          Seismic Division
                          22, Gamlehave alle
                          DK-2920 Charlottenlund
                          DANMARK

Illingworth, M.   DAMTP
                          Silver Street
                          Cambridge CB3 9 EW
                          UNITED KINGDOM

Israelson, H.   FOA 202
                          Fack
                          S-104 50 Stockholm
                          SWEDEN

Korhonen, H.   Institute of Seismology
                          University of Helsinki
                          Et. Hesperiankatu 4
                          SF-00100 Helsinki 10
                          FINLAND

Kullinger, B.   Solid Earth Physics
                          Box 556
                          S-751 22 Uppsala
                          SWEDEN

Loyer, M.   L.D.G.
                          B.P. 136
                          F-92124 Montrouge Cedex
                          FRANCE

Malin, P.E.   Department of Geological Sciences
                          University of Southern California
                          Los Angeles, California 90007
                          USA

Massinon, B.   L.D.G.
                          B.P. 136
                          F-92124 Montrouge Cedex
                          FRANCE

da Silva de Matos, D.   Instituto Geofísico da Universidade do
                              Porto
                          Serra do Pilar
                          4400 Vila Nova de Gaia
                          PORTUGAL

Mauk, F.   Teledyne-Geotech
                            3401 Shiloh Road
                            Garland, Texas
                            USA

Mereu, R.F.   University of Western Ontario
                            London, Ontario
                            CANADA   N6A 5B7

Murphy, J.   Systems, Science and Software
                            P.O. Box 1620
                            La Jolla, California   92037
                            USA

Mykkeltveit, S.   NTNF/NORSAR
                            Post Box 51
                            N-2007 Kjeller
                            NORWAY

Nakanishi, K.   Earth Sciences Division, L-205
                            Lawrence Livermore Laboratory
                            P.O. Box 808
                            Livermore, California   94550
                            USA

Necioglu, A.   Dept. of Geological Engineering
                            Middle East Technical University
                            Ankara
                            TURKEY

Necioglu, G.   Dept. of Geological Engineering
                            Middle East Technical University
                            Ankara
                            TURKEY

Nikolaev, A.V.   Institute of Physics of the Earth
                            Bolshaya Grusinskaya 10
                            Moscow D-242
                            USSR

Noponen, I.   Teledyne Geotech
                            Alexandria, Virginia 22314
                            USA

Nordstrand, I.   FOA 202
                            Fack
                            S-10450 Stockholm
                            SWEDEN

Olsen, K.H.   University of California
                      Geophysics Group G-7, Mail Stop 676
                      Los Alamos Scientific Laboratory
                      Los Alamos, New Mexico 87545
                      USA

Özdemir, H.   Maden Fakultesi
                      Jeofizik Kursusu
                      Tesvikiye Istanbul
                      TURKEY

Peseckis, L.   Lamont-Doherty Geological Observatory
                      Columbia University
                      Palisades, New York 10964
                      USA

Pisarenko, V.L.   Institute of Physics of the Earth
                      Bolshaya Grusinskaya 10
                      Moscow D-242
                      USSR

Pomeroy, P.   Roundout Associates
                      P.O. Box 224
                      Stone Ridge, NY 12484
                      USA

Rademacher, H.   Institute for Geophysics
                      University of Cologne
                      Albertus-Magnus Platz
                      D-5000 Köln 1
                      FEDERAL REPUBLIC OF GERMANY

Rodi, W.   Systems, Science and Software
                      Box 1620
                      La Jolla, California 92038
                      USA

Rygg, E.   Seismological Observatory
                      University of Bergen
                      Allegt. 41
                      5014 BERGEN U

Sabina, F.J.   IIMAS-UNAM
                      Apartado Postal 20-726
                      MEXICO 20, D.F.

Schlittenhardt, J.	Universitätsinstitut für Meteorologie und Geophysik Feldbergstrasse 47 D-6000 Frankfurt a.M. FEDERAL REPUBLIC OF GERMANY
Shore, M.J.	VELA Seismological Center 312 Montgomery Street Alexandria, Va. 22314 USA
Strehlau, J.	Institute of Geophysics New University D-2300 Kiel FEDERAL REPUBLIC OF GERMANY
Strelitz, R.	Department of Geological Sciences University of Southern California Los Angeles, California 90007 USA
Stump, B.W.	AFWL/NTESG Kirtland AFB, New Mexico 87117 USA
Thomson, C.	Department of Earth Sciences Bullard Laboratories Madingley Rise, Madingley Road Cambridge CB3 0EZ UNITED KINGDOM
Troitskiy, P.	Institute of Physics of the Earth B. Grusinskaya 10 Moscow D-242 USSR
Turnbull, L.S.	U.S. Arms Control and Disarmament Agency Room 5499, 23rd and E Street Washington, D.C. USA
Unger, R.	Instituto de Geodesia y Topografia Facultad de ciencias exactas y tecnologia Tucuman ARGENTINA

Ursin, B.   SINTEF
                             Division of Petroleum Technology
                             N-7034 Trondheim-NTH
                             NORWAY

Vaage, S.   Seismological Observatory
                             Allegt. 41
                             5014 BERGEN U.

van Gils, J.   Observatoire Royale Belgique
                             3 av. Circulaire
                             B-1180 Bruxelles
                             BELGIUM

Ward, S.N.   Dept. of Geology
                             Hoffman Lab, Harvard University
                             24 Oxford Street
                             Cambridge, Massachusetts 02138
                             USA

Zürn, W.   Geowissenschaftliches Observatorium
                               Schiltach
                             Heubach 206
                             D-7620 Wolfach
                             FEDERAL REPUBLIC OF GERMANY